Quantum Field Theory for the Gifted Amateur

Quantum Field Theory for the Gifted Amateur

Tom Lancaster

Department of Physics, University of Durham

Stephen J. Blundell

Department of Physics, University of Oxford

OXFORD

UNIVERSITY PRESS

OXFORD
UNIVERSITY PRESS

Great Clarendon Street, Oxford, OX2 6DP,
United Kingdom

Oxford University Press is a department of the University of Oxford.
It furthers the University's objective of excellence in research, scholarship,
and education by publishing worldwide. Oxford is a registered trade mark of
Oxford University Press in the UK and in certain other countries

© Tom Lancaster and Stephen J. Blundell 2014

The moral rights of the authors have been asserted

First Edition published in 2014

Reprinted 2014 (five times, once with corrections), 2015 (twice once with corrections), 2016 (with corrections),
2018 (with corrections), 2019, 2020, 2021, 2022

Published in the United States of America by Oxford University Press
198 Madison Avenue, New York, NY 10016, United States of America

British Library Cataloguing in Publication Data

Data available

Library of Congress Control Number: 2013950755

ISBN 978–0–19–969932–2 (hbk.)
ISBN 978–0–19–969933–9 (pbk.)

Printed and bound by
CPI Group (UK) Ltd, Croydon CR0 4YY

Preface

BRICK: Well, they say nature hates a vacuum, Big Daddy.
BIG DADDY: That's what they say, but sometimes I think
that a vacuum is a hell of a lot better than some of the stuff
that nature replaces it with.
Tennessee Williams (1911–1983) *Cat on a Hot Tin Roof*

Quantum field theory is arguably the most far-reaching and beautiful physical theory ever constructed. It describes not only the quantum vacuum, but also the stuff that nature replaces it with. Aspects of quantum field theory are also more stringently tested, as well as verified to greater precision, than any other theory in physics. The subject nevertheless has a reputation for difficulty which is perhaps well-deserved; its practitioners not only manipulate formidable equations but also depict physical processes using a strange diagrammatic language consisting of bubbles, wiggly lines, vertices, and other geometrical structures, each of which has a well defined quantitative significance. Learning this mathematical and geometrical language is an important initiation rite for any aspiring theoretical physicist, and a quantum field theory graduate course is found in most universities, aided by a large number of weighty quantum field theory textbooks. These books are written by professional quantum field theorists and are designed for those who aspire to join them in that profession. Consequently they are frequently thorough, serious minded and demand a high level of mathematical sophistication.

The motivation for our book is the idea that quantum field theory is too important, too beautiful and too engaging to be restricted to the professionals. Experimental physicists, or theoretical physicists in other fields, would benefit greatly from knowing some quantum field theory, both to understand research papers that use these ideas and also to comprehend and appreciate the important insights that quantum field theory has to offer. Quantum field theory has given us such a radically different and revolutionary view of the physical world that we think that more physicists should have the opportunity to engage with it. The problem is that the existing texts require far too much in the way of advanced mathematical facility and provide too little in the way of physical motivation to assist those who want to learn quantum field theory but not to be professional quantum field theorists. The gap between an undergraduate course on quantum mechanics and a graduate level quantum field theory textbook is a wide and deep chasm, and one of the aims of this book is to provide a bridge to cross it. That being said, we are not assuming the readers of this are simple-minded folk who

[1] After all, with the number of chapters we ended up including, we could have called it 'Fifty shades of quantum field theory'.

can be fobbed off with a trite analogy as a substitute for mathematical argument. We aim to introduce all the maths but, by using numerous worked examples and carefully worded motivations, to smooth the path for understanding in a manner we have not found in the existing books.

We have chosen this book's title with great care.[1] Our imagined reader is an amateur, wanting to learn quantum field theory without (at least initially) joining the ranks of professional quantum field theorists; but (s)he is gifted, possessing a curious and adaptable mind and willing to embark on a significant intellectual challenge; (s)he has abundant curiosity about the physical world, a basic grounding in undergraduate physics, and a desire to be told an entertaining and intellectually stimulating story, but will not feel patronized if a few mathematical niceties are spelled out in detail. In fact, we suspect and hope that our book will find wide readership amongst the graduate trainee quantum field theorists who will want to use it in conjunction with one of the traditional texts (for learning most hard subjects, one usually needs at least two books in order to get a more rounded picture).

One feature of our book is the large number of worked examples, which are set in slightly smaller type. They are integral to the story, and flesh out the details of calculations, but for the more casual reader the guts of the argument of each chapter is played out in the main text. To really get to grips with the subject, the many examples should provide transparent demonstrations of the key ideas and understanding can be confirmed by tackling the exercises at the end of each chapter. The chapters are reasonably short, so that the development of ideas is kept at a steady pace and each chapter ends with a summary of the key ideas introduced.

Though the vacuum plays a big part in the story of quantum field theory, we have not been writing in one. In many ways the present volume represents a compilation of some of the best ideas from the literature and, as a result, we are indebted to these other books for providing the raw material for many of our arguments. There is an extensive list of further reading in Appendix A where we acknowledge our sources, but we note here, in particular, the books by Zee and by Peskin and Schroeder and their legendary antecedent: the lectures in quantum field theory by Sidney Coleman. The latter are currently available online as streamed videos and come highly recommended. Also deserving of special mention is the text by Weinberg which is 'a book to which we are all indebted, and from which none of us can escape.'[2]

[2] T.S. Eliot on *Ulysses*.

It is a pleasure to acknowledge the help we have received from various sources in writing this book. Particular mention is due to Sönke Adlung at Oxford University Press who has helped steer this project to completion. No authors could wish for a more supportive editor and we thank him, Jessica White and the OUP team, particularly Mike Nugent, our eagle-eyed copy editor. We are very grateful for the comments and corrections we received from a number of friends and colleagues who kindly gave up their time to read drafts of various chapters: Peter Byrne, Claudio Castelnovo, John Chalker, Martin Galpin, Chris Maxwell, Tom

McLeish, Johannes Möller, Paul Tulip and Rob Williams. They deserve much credit for saving us from various embarrassing errors, but any that remain are due to us; those that we find post-publication will be posted on the book's website:

http://www.dur.ac.uk/physics/qftgabook

For various bits of helpful information, we thank Hideo Aoki, Nikitas Gidopoulos, Paul Goddard and John Singleton. Our thanks are also due to various graduate students at Durham and Oxford who have unwittingly served as guinea pigs as we tried out various ways of presenting this material in graduate lectures. Finally we thank Cally and Katherine for their love and support.

TL & SJB
Durham & Oxford
January 2, 2014

Note added (November 2017): We are grateful to the following for pointing out various errors that have been corrected in this reprinting: Ray Artz, Petra Axolotl, Toby Baldwin, Stafford Baines, Chet Balestra, Kamil Bradler, Christoph Bruder, Jean Pierre Brunel, Yu-Ching Chen, Zu-Cheng Chen, Anatoliy Dovbnya, David Drysdale, Ernie Eld, Benjamin Frigan, Alexandros Gezerlis, Christian Gütschow, Wei-Han Hsiao, Martin Hohenadler, Patrik Iannotti, Ioannis Karafyllidis, Bernard Leikind, Li Junkang, Árpád Lukács, Perry Ma, James Manton, David Morel, Robin Oliver-Jones, Steve Palzewicz, João Penedo, Randy Roberts, Lewis Robinson, Peter Rottengatter, Marius Schaper, Franz Schindler, Sebastian Schwerdhoefer, Jimmy Snyder, Andrew Solomon, Benjamin Strekha, Sean J Swarbrick, Thomas Veness, Ming-Tso Wei, Charles Weiner and Albert Zhou.

Contents

Overture

0

To begin at the beginning
Dylan Thomas (1914–1953)

Beginnings are always troublesome
George Eliot (1819–1880)

0.1 What is quantum field theory?

> Every particle and every wave in the Universe is simply an excitation of a quantum field that is defined over all space and time.

That remarkable assertion is at the heart of **quantum field theory**. It means that any attempt to understand the fundamental physical laws governing elementary particles has to first grapple with the fundamentals of quantum field theory. It also means that any description of complicated interacting systems, such as are encountered in the many-body problem and in condensed matter physics, will involve quantum field theory to properly describe the interactions. It may even mean, though at the time of writing no-one knows if this is true, that a full theory of quantum gravity will be some kind of quantum upgrade of general relativity (which is a classical field theory). In any case, quantum field theory is the best theory currently available to describe the world around us and, in a particular incarnation known as quantum electrodynamics (QED), is the most accurately tested physical theory. For example, the magnetic dipole moment of the electron has been tested to ten significant figures.

The ideas making up quantum field theory have profound consequences. They explain why all electrons are identical (the same argument works for all photons, all quarks, etc.) because each electron is an excitation of the same electron quantum field and therefore it's not surprising that they all have the same properties. Quantum field theory also constrains the symmetry of the representations of the permutation symmetry group of any class of identical particles so that some classes obey Fermi–Dirac statistics and others Bose–Einstein statistics. Interactions in quantum field theory involve products of operators which are found to create and annihilate particles and so interactions correspond to processes in which particles are created or annihilated; hence there

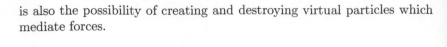

is also the possibility of creating and destroying virtual particles which mediate forces.

0.2   What is a field?

This is all very well, but what is a field? We will think of a **field** as some kind of machine that takes a position in spacetime and outputs an object representing the amplitude of something at that point in spacetime (Fig. 1). The amplitude could be a scalar, a vector, a complex number, a spinor or a tensor. This concept of a field, an unseen entity which pervades space and time, can be traced back to the study of gravity due to Kepler and ultimately Newton, though neither used the term and the idea of action-at-a-distance between two gravitationally attracting bodies seemed successful but nevertheless utterly mysterious. Euler's fluid dynamics got closer to the matter by considering what we would now think of as a velocity field which modelled the movement of fluid at every point in space and hence its capacity to do work on a test particle imagined at some particular location. Faraday, despite (or perhaps because of) an absence of mathematical schooling, grasped intuitively the idea of an electric or magnetic field that permeates all space and time, and although he first considered this a convenient mental picture he began to become increasingly convinced that his lines of force had an independent physical existence. Maxwell codified Faraday's idea and the electromagnetic field, together with all the paraphernalia of field theory, was born.

Thus in classical physics we understand that gravity is a field, electromagnetism is a field, and each can be described by a set of equations which governs their behaviour. The field can oscillate in space and time and thus wave-like excitations of the field can be found (electromagnetic waves are well-known; gravity waves were only detected in 2016). The advent of quantum mechanics removed the distinction between what had been thought of as wave-like objects and particle-like objects. Therefore even matter itself is an excitation of a quantum field and quantum fields become the fundamental objects which describe reality.

0.3   Who is this book for?

Quantum field theory is undoubtedly important, but it is also notoriously difficult. Forbidding-looking integrals and a plethora of funny squiggly Feynman diagrams are enough to strike fear in many a heart and stomach. The situation is not helped by the fact that the many excellent existing books are written by exceedingly clever practitioners who structure their explanations with the aspiring professional in mind. This book is designed to be different. It is written by experimental physicists and aimed at the interested amateur. Quantum field theory is too interesting and too important to be reserved for professional theorists. However, though our imagined reader is not necessarily an aspiring pro-

Fig. 1 A field is some kind of machine that takes a position in spacetime, given by the coordinates x^μ, and outputs an object representing the amplitude of something at that point in spacetime. Here the output is the scalar $\phi(x^\mu)$ but it could be, for example, a vector, a complex number, a spinor or a tensor.

fessional (though we hope quite a few will be) we will assume that (s)he is enthusiastic and has some familiarity with non-relativistic quantum mechanics, special relativity and Fourier transforms at an undergraduate physics level. In the remainder of this chapter we will review a few basic concepts that will serve to establish some conventions of notation.

0.4   Special relativity

Quantum fields are defined over space and time and so we need a proper description of spacetime, and so we will need to use Einstein's special theory of relativity which asserts that the speed c of light is the same in every inertial frame. This theory implies that the coordinates of an event in a frame S and a frame $\bar{S}$ (moving relative to frame S at speed v along the x-axis) are related by the **Lorentz transformation**

$$
\begin{aligned}
\bar{t} &= \gamma\left(t - \frac{vx}{c^2}\right), \\
\bar{x} &= \gamma(x - vt), \\
\bar{y} &= y, \\
\bar{z} &= z,
\end{aligned}
\tag{1}
$$

where $\gamma = (1 - \beta^2)^{-1/2}$ and $\beta = v/c$. Because the speed of light sets the scale for all speeds, we will choose units such that $c = 1$. For similar reasons[1] we will also set $\hbar = 1$.

A good physical theory is said to be **covariant** if it transforms sensibly under coordinate transformations.[2] In particular, we require that quantities should be **Lorentz covariant** if they are to transform appropriately under the elements of the Lorentz group (which include the Lorentz transformations of special relativity, such as eqn 1). This will require us to write our theory in terms of certain well-defined mathematical objects, such as scalars, vectors and tensors.[3]

- **Scalars:** A scalar is a number, and will take the same value in every inertial frame. It is thus said to be **Lorentz invariant**. Examples of scalars include the electric charge and rest mass of a particle.

- **Vectors:** A vector can be thought of as an arrow. In a particular basis it can be described by a set of components. If the basis is rotated, then the components will change, but the length of the arrow will be unchanged (the length of a vector is a scalar). In spacetime, vectors have four components and are called **four-vectors**. A four-vector is an object which has a single time-like component and three space-like components. Three-vectors will be displayed in bold italics, such as $\boldsymbol{x}$ or $\boldsymbol{p}$ for position and momentum respectively. The components of three-vectors are listed with a Roman index taken from the middle of the alphabet: e.g. x^i, with $i = 1, 2, 3$. Four-vectors are made from a time-like part and a space-like part and are displayed in italic script, so position in

[1] In some of the early chapters we will include the factors of $\hbar$ and c so that the reader can make better contact with what they already know, and will give notice when we are going to remove them.

[2] A good counterexample is the two-component 'shopping vector' that contains the price of fish and the price of bread in each component. If you approach the supermarket checkout with the trolley at 45° to the vertical, you will soon discover that the prices of your shopping will not transform appropriately.

[3] We will postpone discussion of tensors until the end of this section.

spacetime is written x where $x = (t, \boldsymbol{x})$. Components for four-vectors will be given a Greek index, so for example x^μ where $\mu = 0, 1, 2, 3$. We say that the zeroth component, x^0, is time-like.

Example 0.1

Some other examples of four-vectors[4] are:

- the energy-momentum four-vector $p = (E, \boldsymbol{p})$,
- the current density four-vector $j = (\rho, \boldsymbol{j})$,
- the vector potential four-vector $A = (V, \boldsymbol{A})$.

The four-dimensional derivative operator ∂_μ is also a combination of a time-like part and a space-like part, and is defined[5] by

$$\partial_\mu \equiv \frac{\partial}{\partial x^\mu} = \left(\frac{\partial}{\partial t}, \boldsymbol{\nabla}\right) = \left(\frac{\partial}{\partial t}, \frac{\partial}{\partial x}, \frac{\partial}{\partial y}, \frac{\partial}{\partial z}\right). \tag{2}$$

Note the lower index written in ∂_μ, contrasting with the upper index on four-vectors like x^μ, which means that the four-dimensional derivative is 'naturally lowered'. This is significant, as we will now describe.

[4] These are written with $c = 1$.

[5] Though strictly ∂_μ refers only to the μth component of the four-vector operator ∂, rather than to the whole thing, we will sometimes write a subscript (or superscript) in expressions like this to indicate whether coordinates are listed with the indices lowered or with them raised.

[6] In full, this equation would be
$$\bar{a}^\mu = \sum_\nu \left(\frac{\partial \bar{x}^\mu}{\partial x^\nu}\right) a^\nu.$$

A general coordinate transformation from one inertial frame to another maps $\{x^\mu\} \to \{\bar{x}^\mu\}$, and the vector a^μ transforms as

$$\bar{a}^\mu = \left(\frac{\partial \bar{x}^\mu}{\partial x^\nu}\right) a^\nu. \tag{3}$$

Here we have used the **Einstein summation convention**, by which twice repeated indices are assumed to be summed.[6] Certain other vectors will transform differently. For example, the gradient vector $\partial_\mu \phi \equiv \partial \phi / \partial x^\mu$ transforms as

$$\frac{\partial \phi}{\partial \bar{x}^\mu} = \left(\frac{\partial x^\nu}{\partial \bar{x}^\mu}\right) \frac{\partial \phi}{\partial x^\nu}. \tag{4}$$

The jargon is that a^μ transforms like a **contravariant vector** and $\partial \phi / \partial x^\mu \equiv \partial_\mu \phi$ transforms like a **covariant vector**,[7] though we will avoid these terms and just note that a^μ has its indices 'upstairs' and $\partial_\mu \phi$ has them 'downstairs' and they will then transform accordingly.

The Lorentz transformation (eqn 1) can be rewritten in matrix form as

$$\begin{pmatrix} \bar{t} \\ \bar{x} \\ \bar{y} \\ \bar{z} \end{pmatrix} = \begin{pmatrix} \gamma & -\beta\gamma & 0 & 0 \\ -\beta\gamma & \gamma & 0 & 0 \\ 0 & 0 & 1 & 0 \\ 0 & 0 & 0 & 1 \end{pmatrix} \begin{pmatrix} t \\ x \\ y \\ z \end{pmatrix}, \tag{5}$$

or for short as

$$\bar{x}^\mu = \Lambda^\mu{}_\nu x^\nu, \tag{6}$$

where $\Lambda^\mu{}_\nu \equiv (\partial \bar{x}^\mu / \partial x^\nu)$ is the Lorentz transformation matrix. In the same way, the energy-momentum four-vector transforms as

$$\bar{p}^\mu = \Lambda^\mu{}_\nu p^\nu. \tag{7}$$

[7] These unfortunate terms are due to the English mathematician J. J. Sylvester (1814–1897). Both types of vectors transform covariantly, in the sense of 'properly', and we wish to retain this sense of the word 'covariant' rather than using it to simply label one type of object that transforms properly. Thus we will usually specify whether the indices on a particular object are 'upstairs' (like a^μ) or 'downstairs' (like $\partial_\mu \phi$) and their transformation properties can then be deduced accordingly.

A downstairs vector a_μ transforms as

$$\bar{a}_\mu = \Lambda_\mu{}^\nu a_\nu, \tag{8}$$

where $\Lambda_\mu{}^\nu \equiv (\partial x^\nu/\partial \bar{x}^\mu)$ is the inverse of the Lorentz transformation matrix $\Lambda^\mu{}_\nu$.[8]

The Lorentz transformation changes components but leaves the length of the four-vector x unchanged. This length is given by the square root[9] of

$$|x|^2 = x \cdot x = (x^0)^2 - (x^1)^2 - (x^2)^2 - (x^3)^2. \tag{10}$$

In general, the **four-vector inner product**[10] is

$$a \cdot b = a^0 b^0 - \boldsymbol{a} \cdot \boldsymbol{b}, \tag{11}$$

which we can write

$$a \cdot b = g_{\mu\nu} a^\mu b^\nu, \tag{12}$$

where the **metric tensor** $g_{\mu\nu}$ is given by

$$g_{\mu\nu} = \begin{pmatrix} 1 & 0 & 0 & 0 \\ 0 & -1 & 0 & 0 \\ 0 & 0 & -1 & 0 \\ 0 & 0 & 0 & -1 \end{pmatrix}. \tag{13}$$

Upstairs and downstairs vectors are related by the metric tensor via

$$a_\mu = g_{\mu\nu} a^\nu, \tag{14}$$

so that we can lower or raise an index by inserting the metric tensor. The form of the metric tensor in eqn 13 allows us to write

$$a^0 = a_0 \quad a^i = -a_i, \tag{15}$$

and hence

$$a \cdot b = g_{\mu\nu} a^\mu b^\nu = a_\mu b^\mu = a^\mu b_\mu. \tag{16}$$

Note also that $a \cdot b = g^{\mu\nu} a_\mu b_\nu$ and $g_{\mu\nu} = g^{\mu\nu}$.

Example 0.2

(i) An example of an inner product is

$$p \cdot p = (E, \boldsymbol{p}) \cdot (E, \boldsymbol{p}) = p_\mu p^\mu = E^2 - \boldsymbol{p}^2 = m^2, \tag{17}$$

where m is the rest mass of the particle.

(ii) The combination $\partial_\mu x^\nu = \frac{\partial x^\nu}{\partial x^\mu} = \delta^\nu_\mu$ and hence the inner product $\partial_\mu x^\mu = 4$ (remember the Einstein summation convention).

(iii) The **d'Alembertian operator** ∂^2 is given by a product of two derivative operators (and is the four-dimensional generalization of the Laplacian operator). It is written as

$$\partial^2 = \partial^\mu \partial_\mu \quad = \quad \frac{\partial^2}{\partial t^2} - \frac{\partial^2}{\partial x^2} - \frac{\partial^2}{\partial y^2} - \frac{\partial^2}{\partial z^2} \tag{18}$$

$$= \quad \frac{\partial^2}{\partial t^2} - \boldsymbol{\nabla}^2. \tag{19}$$

---

[8] Note that in the equation $\bar{x}_\mu = \Lambda_\mu{}^\nu x_\nu$ we have

$$\Lambda_\mu{}^\nu = \begin{pmatrix} \gamma & \beta\gamma & 0 & 0 \\ \beta\gamma & \gamma & 0 & 0 \\ 0 & 0 & 1 & 0 \\ 0 & 0 & 0 & 1 \end{pmatrix}. \tag{9}$$

In fact

$$\Lambda^\mu{}_\nu \Lambda_\mu{}^\rho = \left(\frac{\partial \bar{x}^\mu}{\partial x^\nu}\right)\left(\frac{\partial x^\rho}{\partial \bar{x}^\mu}\right) = \delta^\rho_\nu,$$

where the **Kronecker delta** δ^j_i is defined by

$$\delta^j_i = \begin{cases} 1 & i = j \\ 0 & i \neq j. \end{cases}$$

The δ^j_i symbol is named after Leopold Kronecker (1823–1891).

[9] Of course $|x|^2$ can be negative in special relativity, so it is better to deal with $|x|^2$, the square of the length, rather than its square root $|x|$.

[10] Note that other conventions are possible and some books write $a \cdot b = -a^0 b^0 + \boldsymbol{a} \cdot \boldsymbol{b}$ and define their metric tensor differently. This is an entirely legitimate alternative lifestyle choice, but it's best to stick to one convention in a single book.

Exercise: You can check that

$$g_{\mu\nu} g^{\nu\rho} = \delta^\rho_\mu$$

and also that

$$\Lambda_\mu{}^\nu = g_{\mu\kappa} \Lambda^\kappa{}_\rho g^{\rho\nu}.$$

Named in honour of the French mathematician Jean le Rond d'Alembert (1717–1783). In some texts the d'Alembertian is written as □ and in some as □². Because of this confusion, we will avoid the □ symbol altogether.

To complete this discussion, we can also define a general **tensor** $T^{i\cdots k}_{\ell\cdots n}$ with an arbitrary set of upstairs and downstairs indices. This transforms as

$$\bar{T}^{i'\cdots k'}_{\ell'\cdots n'} = \frac{\partial \bar{x}^{i'}}{\partial x^i} \cdots \frac{\partial \bar{x}^{k'}}{\partial x^k} \frac{\partial x^\ell}{\partial \bar{x}^{\ell'}} \cdots \frac{\partial x^n}{\partial \bar{x}^{n'}} T^{i\cdots k}_{\ell\cdots n}. \tag{20}$$

Example 0.3

(i) The Kronecker delta δ^i_j is a 'mixed tensor of second rank',[11] and one can check that it transforms correctly as follows:

$$\bar{\delta}^i_j = \frac{\partial \bar{x}^i}{\partial x^k} \frac{\partial x^\ell}{\partial \bar{x}^j} \delta^k_\ell = \frac{\partial \bar{x}^i}{\partial x^k} \frac{\partial x^k}{\partial \bar{x}^j} = \delta^i_j. \tag{21}$$

Note that whenever δ_{ij} or δ^{ij} are written, they are not tensors and are simply a shorthand for the scalar 1, in the case when $i=j$, or 0 when $i \neq j$.

(ii) The antisymmetric symbol or **Levi-Civita symbol**[12] $\varepsilon^{ijk\ell}$ is defined in four dimensions by (i) all even permutations $ijk\ell$ of 0123 (such as $ijk\ell = 2301$) have $\varepsilon^{ijk\ell} = 1$; (ii) all odd permutations $ijk\ell$ of 0123 (such as $ijk\ell = 0213$) have $\varepsilon^{ijk\ell} = -1$; (iii) all other terms are zero (e.g. $\varepsilon^{0012} = 0$). The Levi-Civita symbol can be defined in other dimensions.[13] We will not treat this symbol as a tensor, so the version with downstairs indices ε_{ijk} is identical to ε^{ijk}.

0.5 Fourier transforms

We will constantly be needing to swap between representations of an object in spacetime and in the corresponding frequency variables, that is spatial and temporal frequency. Spatial frequency $\boldsymbol{k}$ and temporal frequency ω also form a four-vector $(\omega, \boldsymbol{k})$ and using $E = \hbar\omega$ and $\boldsymbol{p} = \hbar\boldsymbol{k}$ we see that this is the energy-momentum four-vector $(E, \boldsymbol{p})$. (In fact, with our convention $\hbar = 1$ the two objects are identical!) To swap between representations, we define the four-dimensional Fourier transform $\tilde{f}(k)$ of a function $f(x)$ of spacetime x as

$$\tilde{f}(k) = \int d^4x \, e^{ik\cdot x} f(x), \tag{22}$$

where four-dimensional integration is defined by

$$\int d^4x = \int dx^0 dx^1 dx^2 dx^3. \tag{23}$$

The inverse transform is

$$f(x) = \int \frac{d^4k}{(2\pi)^4} e^{-ik\cdot x} \tilde{f}(k), \tag{24}$$

and contains four factors of 2π that are needed for each of the four integrations. Another way of writing eqn 22 is

$$\tilde{f}(\omega, \boldsymbol{k}) = \int d^3x dt \, e^{i(\omega t - \boldsymbol{k}\cdot\boldsymbol{x})} f(t, \boldsymbol{x}). \tag{25}$$

In the spirit of this definition, we will try to formulate our equations so that every factor of dk comes with a (2π), hopefully eliminating one of the major causes of insanity in the subject, the annoying factors[14] of 2π.

[11] It has two indices (hence second rank) and one is upstairs, one is downstairs (hence mixed).

[12] This is named after Italian physicist Tullio Levi-Civita (1873–1941). Useful relationships with the Levi-Civita symbol include results for three-dimensional vectors:

$$(\boldsymbol{b} \times \boldsymbol{c})^i = \varepsilon^{ijk} b^j c^k,$$

$$\boldsymbol{a} \cdot (\boldsymbol{b} \times \boldsymbol{c}) = \varepsilon^{ijk} a^i b^j c^k,$$

and matrix algebra:

$$\det \mathbf{A} = \varepsilon_{i_1 i_2 \cdots i_n} A_{1i_1} A_{2i_2} \cdots A_{ni_n},$$

where $\mathbf{A}$ is a $n \times n$ matrix with components A_{ij}.

[13] The version in two dimensions is rather simple:

$$\varepsilon^{01} = -\varepsilon^{10} = 1,$$
$$\varepsilon^{00} = \varepsilon^{11} = 0.$$

In three dimensions, the nonzero components are:

$$\varepsilon^{012} = \varepsilon^{201} = \varepsilon^{120} = 1,$$
$$\varepsilon^{021} = \varepsilon^{210} = \varepsilon^{102} = -1.$$

Jean Baptiste Joseph Fourier (1768–1830)

[14] Getting these right is actually important: if you have $(2\pi)^4$ on the top of an equation and not the bottom, your answer will be out by a factor of well over two million.

Example 0.4

The **Dirac delta function** $\delta(x)$ is a function localized at the origin and which has integral unity. It is the perfect model of a localized particle. The integral of a d-dimensional **Dirac delta function** $\delta^{(d)}(x)$ is given by

$$\int \mathrm{d}^d x \, \delta^{(d)}(x) = 1. \tag{26}$$

It is defined by

$$\int \mathrm{d}^d x \, f(x)\delta^{(d)}(x) = f(0). \tag{27}$$

Consequently, its Fourier transform is given by

$$\tilde{\delta}^{(d)}(k) = \int \mathrm{d}^d x \, \mathrm{e}^{\mathrm{i}k \cdot x}\delta^{(d)}(x) = 1. \tag{28}$$

Hence, the inverse Fourier transform in four-dimensions is

$$\int \frac{\mathrm{d}^4 k}{(2\pi)^4} \mathrm{e}^{-\mathrm{i}k \cdot x} = \delta^{(4)}(x). \tag{29}$$

0.6   Electromagnetism

In SI units Maxwell's equations in free space can be written:

$$\begin{aligned} \boldsymbol{\nabla} \cdot \boldsymbol{E} &= \tfrac{\rho}{\epsilon_0}, & \boldsymbol{\nabla} \times \boldsymbol{E} &= -\tfrac{\partial \boldsymbol{B}}{\partial t}, \\ \boldsymbol{\nabla} \cdot \boldsymbol{B} &= 0, & \boldsymbol{\nabla} \times \boldsymbol{B} &= \mu_0 \boldsymbol{J} + \tfrac{1}{c^2}\tfrac{\partial \boldsymbol{E}}{\partial t}. \end{aligned} \tag{30}$$

In this book we will choose[15] the **Heaviside–Lorentz**[16] system of units (also known as the 'rationalized Gaussian CGS' system) which can be obtained from SI by setting $\epsilon_0 = \mu_0 = 1$. Thus the electrostatic potential $V(\boldsymbol{x}) = q/4\pi\epsilon_0|\boldsymbol{x}|$ of SI becomes

$$V(\boldsymbol{x}) = \frac{q}{4\pi|\boldsymbol{x}|}, \tag{31}$$

in Heaviside–Lorentz units, and Maxwell's equations can be written

$$\begin{aligned} \boldsymbol{\nabla} \cdot \boldsymbol{E} &= \rho, & \boldsymbol{\nabla} \times \boldsymbol{E} &= -\tfrac{1}{c}\tfrac{\partial \boldsymbol{B}}{\partial t}, \\ \boldsymbol{\nabla} \cdot \boldsymbol{B} &= 0, & \boldsymbol{\nabla} \times \boldsymbol{B} &= \tfrac{1}{c}\left(\boldsymbol{J} + \tfrac{\partial \boldsymbol{E}}{\partial t}\right). \end{aligned} \tag{32}$$

Using our other choice of $c = \hbar = 1$ obviously removes the factors of c from these equations. In addition, the fine structure constant $\alpha = e^2/4\pi\hbar c \approx \tfrac{1}{137}$ simplifies to

$$\alpha = \frac{e^2}{4\pi}. \tag{33}$$

Note that we will give electromagnetic charge q in units of the electron charge e by writing $q = Q|e|$. The charge on the electron corresponds to $Q = -1$.

James Clerk Maxwell (1831–1879)

[15]Although SI units are preferable for many applications in physics, the desire to make our (admittedly often complicated) equations as simple as possible motivates a different choice of units for the discussion of electromagnetism in quantum field theory. Almost all books on quantum field theory use Heaviside–Lorentz units, though the famous textbooks on electrodynamics by Landau and Lifshitz and by Jackson do not.

[16]These units are named after the English electrical engineer O. Heaviside (1850–1925) and the Dutch physicist H. A. Lorentz (1853–1928).

Part I

The Universe as a set of harmonic oscillators

In this introductory part of the book, we trace the development of the picture of the Universe which underpins quantum field theory. This picture is one in which harmonic oscillators are ubiquitous, and such oscillators form a central paradigm of quantum systems. We show that non-interacting harmonic oscillators have eigenfunctions which look and behave like particles and find some elegant methods for describing systems of coupled oscillators.

- In Chapter 1 we provide a formulation of classical mechanics which is suitable for a quantum upgrade. This allows us to talk about *functionals* and *Lagrangians*.

- The *simple harmonic oscillator*, presented in Chapter 2, is well-known from basic quantum physics as an elementary model of an oscillating system. We solve this simple model using *creation and annihilation operators* and show that the solutions have the characteristics of *particles*. Linking masses by springs into a chain allows us to generalize this problem and the solutions are *phonons*.

- The next step is to change our viewpoint and get rid of wave functions entirely and develop the *occupation number representation* which we do in Chapter 3. We show that bosons are described by commuting operators and fermions are described by anticommuting operators.

- Already we have many of the building blocks of a useful theory. In Chapter 4 we consider how to build single-particle operators out of creation and annihilation operators, and this already gives us enough information to discuss the *tight-binding model* of solid state physics and the *Hubbard model*.

<table>
<tr><td style="vertical-align: top; width: 25%;">

1

Lagrangians

</td></tr>
</table>

Lagrangians

In this chapter, we give an introduction to Lagrangians in classical mechanics and explain why the Lagrangian formulation is a sensible way to describe quantum fields.

1.1 Fermat's principle

We begin with an example from the study of optics. Consider the passage of a light ray through a slab of glass as shown in Fig. 1.1. The bending of the light ray near the air/glass interface can be calculated using the known refractive indices of air and glass using the famous Snell's law of refraction, named after Willebrord Snellius, who described it in 1621, but discovered first by the Arabian mathematician Ibn Sahl in 984. In 1662, Pierre de Fermat produced an ingenious method of deriving Snell's law on the basis of his **principle of least time**. This states that the path taken between two points A and B by a ray of light is the path that can be traversed by the light in the least time. Because light travels more slowly in glass than in air, the ray of light crosses the glass at an altered angle so it doesn't have so much path length in the glass. If the light ray were to take a straight line path from A to B this would take longer. This was all very elegant, but it didn't really explain why light would choose to do this; why should light take the path which took the least time? Why is light in such a hurry?

Fermat's principle of least time is cute, and seems like it is telling us something, but at first sight it looks unhelpful. It attempts to replace a simple formula (Snell's law), into which you can plug numbers and calculate trajectories, with a principle characterizing the solution of the problem but for which you need the whole apparatus of the calculus of variations to solve any problems. Fermat's principle is however the key to understanding quantum fields, as we shall see.

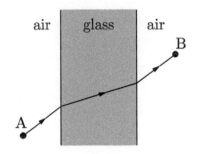

Fig. 1.1 Refraction of a light ray through a slab of glass. The ray finds the path of least travel time from A to B.

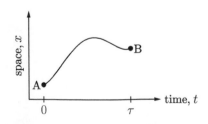

Fig. 1.2 A particle moves from A to B in time τ and its path is described by Newton's laws of motion.

1.2 Newton's laws

A somewhat similar problem is found in the study of dynamics. Consider a particle of mass m subject to a force F which moves in one spatial dimension x from point A to B, as shown in the spacetime diagram in Fig. 1.2. Here time is on the horizontal axis and space is on the vertical axis. The exact path $x(t)$ that the particle takes is given by Newton's laws of motion, i.e.

$$F = m\ddot{x}. \tag{1.1}$$

This equation can be integrated to find $x(t)$. However, when you stop and think about it, this is a very quantum-unfriendly approach. The solution gives you the position of the particle *at every time t* from $t = 0$ to $t = \tau$. Quantum mechanics tells us that you might measure the particle's position at $t = 0$ and find it at A, and you might measure it again at $t = \tau$ and find it at B, but you'd not be able to know precisely what it did in between. Having a method which lets you calculate $x(t)$ from a differential equation is not a good starting point.

Dynamics need to be formulated in a completely different way if the subject is to be generalized to quantum mechanics. This is exactly what Joseph-Louis Lagrange and William Rowan Hamilton did, although they had no idea that what they were doing would make dynamics more quantum-friendly. We will take a slightly different approach to theirs and arrive at the final answer by asking ourselves how kinetic energy T and potential energy V vary during the trajectory of the particle. We know that they must sum to the total energy $E = T + V$ which must be a constant of the motion. But during the trajectory, the balance between kinetic and potential energy might change.

It is simple enough to write down the average kinetic energy $\bar{T}$ during the trajectory, which is given by

$$\bar{T} = \frac{1}{\tau} \int_0^\tau \frac{1}{2} m[\dot{x}(t)]^2 \, \mathrm{d}t. \tag{1.2}$$

The average potential energy $\bar{V}$ during the trajectory is given by

$$\bar{V} = \frac{1}{\tau} \int_0^\tau V[x(t)] \, \mathrm{d}t. \tag{1.3}$$

These two quantities must sum to give the total energy $E = \bar{E} = \bar{T} + \bar{V}$. However, what we want to do is to consider how $\bar{T}$ and $\bar{V}$ *vary* as you alter the trajectory. To do this, we need a little bit of mathematics, which we cover in the next section.

1.3 Functionals

The expressions in eqns 1.2 and 1.3 are functionals of the trajectory $x(t)$. What does this mean?

Let us recall that a **function** is a machine (see Fig. 1.3) which turns a number into another number. For example, the function $f(x) = 3x^2$ will turn the number 1 into the number 3, or the number 3 into the number 27. Give a function a number and it will return another number.

A **functional** is a machine which turns a function into a number. You feed the machine with an entire function, like $f(x) = x^2$ or $f(x) = \sin x$ and it returns a number.

J.-L. Lagrange (1736–1813) was an Italian-born French mathematician and physicist.

W. R. Hamilton (1805–1865) was an Irish mathematician and physicist.

function

NUMBER
3

NUMBER
27

functional

NUMBER
19

FUNCTION
$\frac{e^{-x}x^2}{3}$

Fig. 1.3 A function turns a number into another number. A functional operates on an entire function and produces a number.

Example 1.1

Here are some examples of functionals.

- The functional $F[f]$ operates on the function f as follows:

$$F[f] = \int_0^1 f(x)\,\mathrm{d}x. \tag{1.4}$$

  Hence, given the function $f(x) = x^2$, the functional returns the number

$$F[f] = \int_0^1 x^2\,\mathrm{d}x = \frac{1}{3}. \tag{1.5}$$

- The functional $G[f]$ operates on the function f as follows:

$$G[f] = \int_{-a}^a 5[f(x)]^2\,\mathrm{d}x. \tag{1.6}$$

  Hence, given the function $f(x) = x^2$, the functional returns the number

$$G[f] = \int_{-a}^a 5x^4\,\mathrm{d}x = 2a^5. \tag{1.7}$$

- A function can be thought of as a trivial functional. For example, the functional $F_x[f]$ given by

$$F_x[f] = \int_{-\infty}^{\infty} f(y)\delta(y - x)\mathrm{d}y = f(x), \tag{1.8}$$

  returns the value of the function evaluated at x.

We now want to see how a functional changes as you adjust the function which is fed into it. The key concept here is **functional differentiation**. Recall that a derivative of a function is defined as follows:

$$\frac{\mathrm{d}f}{\mathrm{d}x} = \lim_{\epsilon \to 0} \frac{f(x + \epsilon) - f(x)}{\epsilon}. \tag{1.9}$$

The derivative of the function tells you how the number returned by the function $f(x)$ changes as you slightly change the number x that you feed into the 'machine'. In the same way, we can define a **functional derivative** of a functional $F[f]$ as follows:

$$\frac{\delta F}{\delta f(x)} = \lim_{\epsilon \to 0} \frac{F[f(x') + \epsilon\delta(x' - x)] - F[f(x')]}{\epsilon}. \tag{1.10}$$

The functional derivative tells you how the number returned by the functional $F[f(x)]$ changes as you slightly change the function $f(x)$ that you feed into the machine.

Example 1.2

Here are some examples of calculations of functional derivatives. You can work through these if you want to acquire the skill of functional differentiation, or skip to the next bit of text if you are happy to take the results on trust.

- The functional $I[f] = \int_{-1}^{1} f(x)\,dx$ has a functional derivative given by

$$
\begin{aligned}
\frac{\delta I[f]}{\delta f(x_0)} &= \lim_{\epsilon \to 0} \frac{1}{\epsilon} \left[\int_{-1}^{1} [f(x) + \epsilon\delta(x - x_0)]\,dx - \int_{-1}^{1} f(x)\,dx \right] \\
&= \int_{-1}^{1} \delta(x - x_0)\,dx \\
&= \begin{cases} 1 & -1 \le x_0 \le 1 \\ 0 & \text{otherwise.} \end{cases}
\end{aligned} \tag{1.11}
$$

- The functional $J[f] = \int [f(y)]^p \phi(y)\,dy$ has a functional derivative with respect to $f(x)$ given by

$$
\begin{aligned}
\frac{\delta J[f]}{\delta f(x)} &= \lim_{\epsilon \to 0} \frac{1}{\epsilon} \left[\int [f(y) + \epsilon\delta(y - x)]^p \phi(y)\,dy - \int [f(y)]^p \phi(y)\,dy \right] \\
&= p[f(x)]^{p-1}\phi(x).
\end{aligned} \tag{1.12}
$$

- The functional $H[f] = \int_a^b g[f(x)]\,dx$, where g is a function whose derivative is $g' = dg/df$, has a functional derivative given by

$$
\begin{aligned}
\frac{\delta H[f]}{\delta f(x_0)} &= \lim_{\epsilon \to 0} \frac{1}{\epsilon} \left[\int g[f(x) + \epsilon\delta(x - x_0)]\,dx - \int g[f(x)]\,dx \right] \\
&= \lim_{\epsilon \to 0} \frac{1}{\epsilon} \left[\int (g[f(x)] + \epsilon\delta(x - x_0)g'[f(x)])\,dx - \int g[f(x)]\,dx \right] \\
&= \int \delta(x - x_0)\,g'[f(x)]\,dx \\
&= g'[f(x_0)].
\end{aligned} \tag{1.13}
$$

- Using the result of the previous example, the functional $\bar{V}[x] = \frac{1}{\tau}\int_0^\tau V[x(t)]\,dt$ so that

$$
\frac{\delta \bar{V}[x]}{\delta x(t)} = \frac{1}{\tau} V'[x(t)]. \tag{1.14}
$$

- The functional $J[f]$ is defined by $J[f] = \int g(f')\,dy$ where $f' = df/dy$. Hence

$$
\frac{\delta J[f]}{\delta f(x)} = \lim_{\epsilon \to 0} \frac{1}{\epsilon} \left[\int dy\, g\left(\frac{\partial}{\partial y}[f(y) + \epsilon\delta(y - x)] \right) - \int dy\, g\left(\frac{\partial f}{\partial y} \right) \right], \tag{1.15}
$$

and using

$$
g\left(\frac{\partial}{\partial y}[f(y) + \epsilon\delta(y - x)] \right) = g\left(f' + \epsilon\delta'(y - x) \right) \approx g(f') + \epsilon\delta'(y - x)\frac{dg(f')}{df'}, \tag{1.16}
$$

then the calculation can be reduced to an integral by parts

$$
\frac{\delta J[f]}{\delta f(x)} = \int dy\, \delta'(y-x)\frac{dg(f')}{df'} = \left[\delta(y-x)\frac{dg(f')}{df'} \right] - \int dy\, \delta(y-x)\frac{d}{dy}\left(\frac{dg(f')}{df'} \right). \tag{1.17}
$$

The term in square brackets vanishes assuming x is inside the limits of the integral, and we have simply

$$
\frac{\delta J[f]}{\delta f(x)} = -\frac{d}{dx}\left(\frac{dg(f')}{df'} \right). \tag{1.18}
$$

- An example of the previous result is that for $F[\phi] = \int \left(\frac{\partial \phi}{\partial y} \right)^2 dy$,

$$
\frac{\delta F[\phi]}{\delta \phi(x)} = -2\frac{\partial^2 \phi}{\partial x^2}. \tag{1.19}
$$

- Another example is for $\bar{T} = \frac{1}{\tau}\int_0^\tau \frac{1}{2}m[\dot{x}(t)]^2\,dt$,

$$
\frac{\delta \bar{T}[x]}{\delta x(t)} = -\frac{m\ddot{x}}{\tau}. \tag{1.20}
$$

Equation 1.19 can be easily generalized to three dimensions and leads to a very useful result which is worth memorizing, namely that if

$$
I = \int (\boldsymbol{\nabla}\phi)^2\,d^3x,
$$

then

$$
\frac{\delta I}{\delta \phi} = -2\boldsymbol{\nabla}^2\phi.
$$

1.4   Lagrangians and least action

With these mathematical results under our belt, we are now ready to return to the main narrative. How does the average kinetic energy and the average potential energy vary as we adjust the particle trajectory? We now have the main results from eqns 1.14 and 1.20 which are:

$$\frac{\delta \bar{V}[x]}{\delta x(t)} = \frac{V'[x(t)]}{\tau}, \qquad \frac{\delta \bar{T}[x]}{\delta x(t)} = -\frac{m\ddot{x}}{\tau}. \qquad (1.21)$$

In fact, Newton's laws tell us that for the classical trajectory of the particle we have that $m\ddot{x} = -\mathrm{d}V/\mathrm{d}x$. This means that for variations about the classical trajectory

$$\frac{\delta \bar{V}[x]}{\delta x(t)} = \frac{\delta \bar{T}[x]}{\delta x(t)}. \qquad (1.22)$$

This means that if you slightly deviate from the classical trajectory, then both the average kinetic energy and the average potential energy will increase[1] and *by the same amount*. This equation can be rewritten as

$$\frac{\delta}{\delta x(t)} (\bar{T}[x] - \bar{V}[x]) = 0, \qquad (1.23)$$

i.e. that the *difference* between the average kinetic energy and the average potential energy is stationary about the classical trajectory. This shows that there is something rather special about the difference between kinetic energy and potential energy, and motivates us to define a quantity known as the **Lagrangian** L as

$$\boxed{L = T - V.} \qquad (1.24)$$

The integral of the Lagrangian over time is known as the **action** S

$$S = \int_0^\tau L \, \mathrm{d}t, \qquad (1.25)$$

and so the action has dimensions of energy×time, and hence is measured in Joule-seconds. This is the same units as Planck's constant h, and we will see later in this chapter why it is often appropriate to think of measuring S in units of Planck's constant. Our variational principle (eqn 1.23) connecting variations of average kinetic energy and average potential energy can now be rewritten in a rather appealing way since for this problem $S = \int_0^\tau (T - V) \, \mathrm{d}t = \tau(\bar{T}[x] - \bar{V}[x])$, so that

$$\boxed{\frac{\delta S}{\delta x(t)} = 0,} \qquad (1.26)$$

and this is known as **Hamilton's principle of least action.**[2] It states that the classical trajectory taken by a particle is such that the action is stationary; small adjustments to the path taken by the particle (Fig. 1.4) only increase the action (in the same way that small adjustments to the path taken by a ray of light from the one determined by Snell's law lengthen the time taken by the light ray).

[1]They might both decrease by the same amount if the classical trajectory maximizes (rather than minimizes) the action, see below, but this case is not the one usually encountered.

[2]To be pedantic, the principle only shows the action is stationary. It could be a maximum, or a saddle point, just as easily as a minimum. 'Stationary action' would be better than 'least action'. But we are stuck with the name.

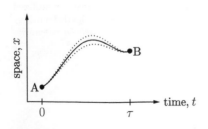

Fig. 1.4 Small adjustments to the path of a particle from its classical trajectory lead to an increase in the action S.

Example 1.3

The Lagrangian L can be written as a function of both position and velocity. Quite generally, one can think of it as depending on a generalized position coordinate $x(t)$ and its derivative $\dot{x}(t)$, called the velocity. Then the variation of S with $x(t)$ is $\delta S/\delta x(t)$ and can be written as

$$
\begin{aligned}
\frac{\delta S}{\delta x(t)} &= \int du \left[\frac{\delta L}{\delta x(u)} \frac{\delta x(u)}{\delta x(t)} + \frac{\delta L}{\delta \dot{x}(u)} \frac{\delta \dot{x}(u)}{\delta x(t)} \right] \\
&= \int du \left[\frac{\delta L}{\delta x(u)} \delta(u-t) + \frac{\delta L}{\delta \dot{x}(u)} \frac{d}{dt} \delta(u-t) \right] \\
&= \frac{\delta L}{\delta x(t)} + \left[\delta(u-t) \frac{\delta L}{\delta \dot{x}(u)} \right]_{t_i}^{t_f} - \int du\, \delta(u-t) \frac{d}{dt} \frac{\delta L}{\delta \dot{x}(u)} \\
&= \frac{\delta L}{\delta x(t)} - \frac{d}{dt} \frac{\delta L}{\delta \dot{x}(t)},
\end{aligned}
\tag{1.27}
$$

and hence the principle of least action (eqn 1.26) yields

$$
\boxed{\frac{\delta L}{\delta x(t)} - \frac{d}{dt} \frac{\delta L}{\delta \dot{x}(t)} = 0,}
\tag{1.28}
$$

which is known as the **Euler–Lagrange equation**.

Leonhard Euler (1707–1783). In the words of Pierre-Simon Laplace (1749–1827): *Read Euler, read Euler, he is the master of us all.*

The Lagrangian L is related to the **Lagrangian density** $\mathcal{L}$ by

$$
L = \int dx\, \mathcal{L},
\tag{1.29}
$$

so that the action S is given by

$$
S = \int dt\, L = \int dt\, dx\, \mathcal{L}.
\tag{1.30}
$$

The following example introduces the idea of a Lagrangian density, a concept we will come back to frequently, but also provides a nice way to derive the classical wave equation.

Example 1.4

Consider waves on a string of mass m and length ℓ. Let us define the mass density $\rho = m/\ell$, tension $\mathcal{T}$ and displacement from the equilibrium $\psi(x,t)$ (see Fig. 1.5). The kinetic energy T can then be written as $T = \frac{1}{2} \int_0^\ell dx\, \rho (\partial \psi/\partial t)^2$ and the potential energy $V = \frac{1}{2} \int_0^\ell dx\, \mathcal{T}(\partial \psi/\partial x)^2$. The action is then

$$
S[\psi(x,t)] = \int dt\,(T-V) = \int dt\, dx\, \mathcal{L}\left(\psi, \frac{\partial \psi}{\partial t}, \frac{\partial \psi}{\partial x} \right),
\tag{1.31}
$$

where

$$
\mathcal{L}\left(\psi, \frac{\partial \psi}{\partial t}, \frac{\partial \psi}{\partial x} \right) = \frac{\rho}{2} \left(\frac{\partial \psi}{\partial t} \right)^2 - \frac{\mathcal{T}}{2} \left(\frac{\partial \psi}{\partial x} \right)^2
\tag{1.32}
$$

is the Lagrangian density. We then have immediately

$$
\begin{aligned}
0 = \frac{\delta S}{\delta \psi} &= \frac{\partial \mathcal{L}}{\partial \psi} - \frac{d}{dx} \frac{\partial \mathcal{L}}{\partial (\partial \psi/\partial x)} - \frac{d}{dt} \frac{\partial \mathcal{L}}{\partial (\partial \psi/\partial t)} \\
&= 0 + \mathcal{T} \frac{\partial^2 \psi}{\partial x^2} - \rho \frac{\partial^2 \psi}{\partial t^2},
\end{aligned}
\tag{1.33}
$$

and so the wave equation $(\partial^2 \psi/\partial x^2) = (1/v^2)(\partial^2 \psi/\partial t^2)$ with $v = \sqrt{\mathcal{T}/\rho}$ emerges almost effortlessly.

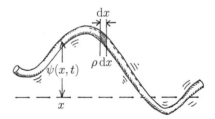

Fig. 1.5 Waves on a string. The displacement from equilibrium is $\psi(x,t)$ and the equation of motion can be derived by considering an element of the string of length dx and mass $\rho\, dx$. The figure shows a short section in the middle of the string which is assumed to be tethered at either end so that $\psi(0,t) = \psi(\ell,t)$.

As a final trick, let us put the Euler–Lagrange equation on a fully relativistic footing, bracing ourselves to use some four-vector notation and the Einstein summation convention (see Section 0.4). If the Lagrangian density $\mathcal{L}$ depends on a function $\phi(x)$ (where x is a point in spacetime) and its derivative[3] $\partial_\mu \phi$, then the action S is given by

$$S = \int \mathrm{d}^4 x \, \mathcal{L}(\phi, \partial_\mu \phi). \tag{1.34}$$

[3]Recall from the argument in Section 0.4 that the index μ in $\partial_\mu \phi$ is naturally lowered; see eqn 0.2.

By analogy with eqn 1.27, the action principle gives[4]

$$\boxed{\frac{\delta S}{\delta \phi} = \frac{\partial \mathcal{L}}{\partial \phi} - \partial_\mu \left(\frac{\partial \mathcal{L}}{\partial(\partial_\mu \phi)} \right) = 0,} \tag{1.35}$$

the four-vector version of the Euler–Lagrange equation.

[4]Remember that we are using the Einstein summation convention, by which twice repeated indices are assumed summed. Equation 1.35 is simply the four-dimensional generalisation of eqns 1.27 and 1.28.

Example 1.5

As a simple example of this, consider the case when $\mathcal{L}$ is given by

$$\mathcal{L} = \frac{1}{2}(\partial_\mu \phi)^2 - \frac{1}{2} m^2 \phi^2, \tag{1.36}$$

where $(\partial_\mu \phi)^2 = (\partial_\mu \phi)(\partial^\mu \phi)$. Simple differentiation then gives

$$\frac{\partial \mathcal{L}}{\partial \phi} = -m^2 \phi, \quad \text{and} \quad \frac{\partial \mathcal{L}}{\partial(\partial_\mu \phi)} = \partial^\mu \phi. \tag{1.37}$$

Use of the action principle (eqn 1.35) gives

$$\frac{\delta S}{\delta \phi} = -m^2 \phi - \partial_\mu \partial^\mu \phi = 0 \quad \text{and hence} \quad (\partial^2 + m^2)\phi = 0. \tag{1.38}$$

1.5   Why does it work?

In this chapter, we have considered two variational principles: Fermat's principle of least time and Hamilton's principle of least action. One describes the path taken by a ray of light, the other describes the path taken by a classical particle. They are very elegant, but why not stick to using Snell's law and Newton's law? And why do they both work?

The answer to both of these questions is quantum mechanics. We will talk about this in more detail later in the book, but the motion of a particle (photon or billiard ball) going from A to B involves *all possible paths*, the sensible classical ones and completely insane ones. You have to sum them all up, but they are each associated with a phase factor, and for most sets of paths the different phases mean that the contributions cancel out. It is only when the phase is stationary that nearby paths all give a non-cancelling contribution. The wave function for a particle has a phase factor[5] given by

$$e^{iS/\hbar}, \tag{1.39}$$

[5]Fermat's principle of least time tells us something about optical path length of a ray, that is the difference between the phase at the beginning and end of a ray. By analogy with Hamilton's principle of least action of a classical mechanical system, one can posit that the action S is given by a constant multiplied by the phase of a wave function. This defines the constant which is given the symbol $\hbar$. Thus we take the phase to be $S/\hbar$.

where $S = \int L\,dt$ is the action, so a stationary phase equates to a stationary action (eqn 1.26). (Running the argument in reverse, the phase factor for a photon of energy E is $e^{-iEt/\hbar}$, and so stationary phase equates to stationary time, which is Fermat's principle.) We will see how this approach leads naturally to Feynman's path-integral approach later in the book (Chapter 23). But for now, notice simply that if the action is stationary then the classical path which minimizes the action is the one that is observed and everything else cancels.

Snell's law and Newton's law are enough to solve classical systems. But neither allow the generalization to quantum systems to be performed with ease. Thus, to formulate quantum field theory (the grand task of this book), we have to start with a Lagrangian picture. Our next step is to gain some insight into what happens in non-classical systems, and so in the next chapter we will turn our attention to an archetypal quantum system: the simple harmonic oscillator.

Chapter summary

- Fermat's principle of least time states that light takes a path which takes the least time.
- The Lagrangian for a classical particle is given by $L = T - V$.
- Classical mechanics can be formulated using Hamilton's principle of least action.

$$\frac{\delta S}{\delta x(t)} = 0, \qquad (1.40)$$

  where the action $S = \int L\,dt$.

- Both Fermat's and Hamilton's principles show how the classical paths taken by a photon or massive particle are ones in which the phase of the corresponding wave function is stationary.

Exercises

(1.1) Use Fermat's principle of least time to derive Snell's law.

(1.2) Consider the functionals $H[f] = \int G(x,y)f(y)\,dy$, $I[f] = \int_{-1}^{1} f(x)\,dx$ and $J[f] = \int \left(\frac{\partial f}{\partial y}\right)^2 dy$ of the function f. Find the functional derivatives $\frac{\delta H[f]}{\delta f(z)}$, $\frac{\delta^2 I[f^3]}{\delta f(x_0)\delta f(x_1)}$ and $\frac{\delta J[f]}{\delta f(x)}$.

(1.3) Consider the functional $G[f] = \int g(y,f)\,dy$. Show that

$$\frac{\delta G[f]}{\delta f(x)} = \frac{\partial g(x,f)}{\partial f}. \qquad (1.41)$$

Now consider the functional $H[f] = \int g(y,f,f')\,dy$ and show that

$$\frac{\delta H[f]}{\delta f(x)} = \frac{\partial g}{\partial f} - \frac{d}{dx}\frac{\partial g}{\partial f'}, \qquad (1.42)$$

where $f' = \partial f/\partial y$. For the functional $J[f] = \int g(y, f, f', f'') \, \mathrm{d}y$ show that

$$\frac{\delta J[f]}{\delta f(x)} = \frac{\partial g}{\partial f} - \frac{\mathrm{d}}{\mathrm{d}x}\frac{\partial g}{\partial f'} + \frac{\mathrm{d}^2}{\mathrm{d}x^2}\frac{\partial g}{\partial f''}, \qquad (1.43)$$

where $f'' = \partial^2 f/\partial y^2$.

(1.4) Show that

$$\frac{\delta \phi(x)}{\delta \phi(y)} = \delta(x - y), \qquad (1.44)$$

and

$$\frac{\delta \dot{\phi}(t)}{\delta \phi(t_0)} = \frac{\mathrm{d}}{\mathrm{d}t}\delta(t - t_0). \qquad (1.45)$$

(1.5) For a three-dimensional elastic medium, the potential energy is

$$V = \frac{\tau}{2}\int \mathrm{d}^3x \, (\boldsymbol{\nabla}\psi)^2, \qquad (1.46)$$

and the kinetic energy is

$$T = \frac{\rho}{2}\int \mathrm{d}^3x \left(\frac{\partial \psi}{\partial t}\right)^2. \qquad (1.47)$$

Use these results, and the functional derivative approach, to show that ψ obeys the wave equation

$$\boldsymbol{\nabla}^2\psi = \frac{1}{v^2}\frac{\partial^2 \psi}{\partial t^2}, \qquad (1.48)$$

where v is the velocity of the wave.

(1.6) Show that if $Z_0[J]$ is given by

$$Z_0[J] = \exp\left(-\frac{1}{2}\int \mathrm{d}^4x \, \mathrm{d}^4y \, J(x)\Delta(x - y)J(y)\right), \qquad (1.49)$$

where $\Delta(x) = \Delta(-x)$ then

$$\frac{\delta Z_0[J]}{\delta J(z_1)} = -\left[\int \mathrm{d}^4y \, \Delta(z_1 - y)J(y)\right] Z_0[J]. \quad (1.50)$$

Simple harmonic oscillators

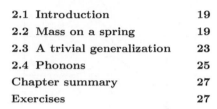

2.1 Introduction

The advent of quantum mechanics convinced people that things that had previously been thought of as particles were in fact waves. For example, it was found that electrons and neutrons were not simply little rigid bits of matter but they obey a wave equation known as the Schrödinger equation. This idea is known as **first quantization**. To summarize:

$$\boxed{\textbf{First quantization: } \text{Particles behave like waves.}} \quad (2.1)$$

However, this is not the end of the story. It was also realized that things that had been previously thought of as waves were in fact particles. For example, electromagnetic waves and lattice waves were not simply periodic undulations of some medium but they could actually behave like particles. These phenomena were given particle-like names: photons and phonons. This idea is known as **second quantization**. To summarize:

$$\boxed{\textbf{Second quantization: } \text{Waves behave like particles.}} \quad (2.2)$$

Quite how these ideas link up is one of main themes in quantum field theory which sees the underlying reality behind both waves and particles as a **quantum field**. However, before we get to this point, it is worth spending some time reviewing second quantization in a bit more detail as it is the less familiar idea. In this chapter, we focus on the most famous example of a wave phenomenon in physics: the oscillations of a mass on a spring.

2.2 Mass on a spring

Consider a mass on a spring (as shown in Fig. 2.1), one of the simplest physical problems there is. Assume we have mass m, spring constant K, the displacement of the mass from equilibrium is given by x and the momentum of the mass is given by $p = m\dot{x}$. The total energy E is the sum of the kinetic energy $p^2/2m$ and the potential energy $\frac{1}{2}Kx^2$.

In quantum mechanics, we replace p by the operator $-\mathrm{i}\hbar\partial/\partial x$, and we then have the Schrödinger equation for a harmonic oscillator

$$\left(-\frac{\hbar^2}{2m}\frac{\partial^2}{\partial x^2} + \frac{1}{2}Kx^2\right)\psi = E\psi. \quad (2.3)$$

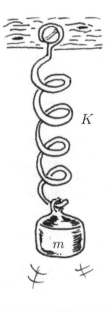

Fig. 2.1 A mass m suspended on a spring, of spring constant K.

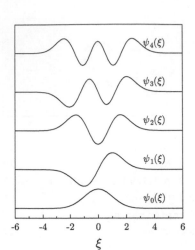

Fig. 2.2 Eigenfunctions of the simple harmonic oscillator.

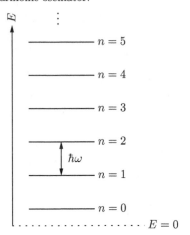

Fig. 2.3 The ladder of energy levels for the simple harmonic oscillator.

The solutions to this equation can be obtained by a somewhat involved series-solution method and are given by

$$\psi_n(\xi) = \frac{1}{\sqrt{2^n n!}} \left(\frac{m\omega}{\pi\hbar}\right)^{1/4} H_n(\xi) e^{-\xi^2/2}, \qquad (2.4)$$

where $H_n(\xi)$ is a Hermite polynomial and $\xi = \sqrt{m\omega/\hbar x}$. As shown in Fig. 2.2, these eigenfunctions look very wave-like. However, they do have a 'particle-like' quality which is apparent from the eigenvalues, which turn out to be

$$E_n = \left(n + \frac{1}{2}\right)\hbar\omega, \qquad (2.5)$$

where $\omega = \sqrt{K/m}$. This is a ladder of energy levels (see Fig. 2.3), though note when $n = 0$ the energy is not zero but $\hbar\omega/2$. This is called the **zero-point energy**. Because the energy levels form a ladder, you must add a quantum of energy $\hbar\omega$ to move up the ladder by one rung. It is as if you are adding a lump, or particle, of energy. Can we make this vague feeling any more concrete? Yes we can. Moreover, we can do it quite elegantly and without dirtying our hands with Hermite polynomials!

To accomplish this, we start with the Hamiltonian for the simple harmonic oscillator written out as follows:

$$\hat{H} = \frac{\hat{p}^2}{2m} + \frac{1}{2}m\omega^2\hat{x}^2, \qquad (2.6)$$

where we have expressed the spring constant as $K = m\omega^2$. One feels the seductive temptation to factorize this Hamiltonian and write it as

$$\frac{1}{2}m\omega^2\left(\hat{x} - \frac{i}{m\omega}\hat{p}\right)\left(\hat{x} + \frac{i}{m\omega}\hat{p}\right), \qquad (2.7)$$

but we hit a problem: in quantum mechanics, the operators $\hat{x}$ and $\hat{p}$ do not commute. Hence, multiplying out the brackets in eqn 2.7 doesn't quite give us back $\hat{H}$ but instead produces

$$\frac{1}{2}m\omega^2\left(\hat{x} - \frac{i}{m\omega}\hat{p}\right)\left(\hat{x} + \frac{i}{m\omega}\hat{p}\right) = \frac{1}{2}m\omega^2\hat{x}^2 + \frac{\hat{p}^2}{2m} + \frac{i\omega}{2}[\hat{x}, \hat{p}], \quad (2.8)$$

where $[\hat{x}, \hat{p}] \equiv \hat{x}\hat{p} - \hat{p}\hat{x} = i\hbar$ is the commutator of $\hat{x}$ and $\hat{p}$. The right-hand side of eqn 2.8 is then just $\hat{H} - \frac{\hbar\omega}{2}$, which is nearly $\hat{H}$ but has the correction $-\frac{\hbar\omega}{2}$ due to the zero-point energy being subtracted. The factorization can therefore be made to work and we realize that the operators $\hat{x} - \frac{i}{m\omega}\hat{p}$ and $\hat{x} + \frac{i}{m\omega}\hat{p}$ are clearly going to be useful. Since $\hat{p}$ and $\hat{x}$ are Hermitian, the operators $\hat{x} - \frac{i}{m\omega}\hat{p}$ and $\hat{x} + \frac{i}{m\omega}\hat{p}$ are adjoints of each other (and therefore are not themselves Hermitian, so cannot correspond to any observable). We will give them the special names $\hat{a}^\dagger$ and $\hat{a}$, although we will include a multiplicative constant $\sqrt{m\omega/2\hbar}$ in our definition so that they have particularly nice properties. Hence we will write

$$\hat{a} = \sqrt{\frac{m\omega}{2\hbar}}\left(\hat{x} + \frac{i}{m\omega}\hat{p}\right), \qquad (2.9)$$

$$\hat{a}^\dagger = \sqrt{\frac{m\omega}{2\hbar}}\left(\hat{x} - \frac{i}{m\omega}\hat{p}\right). \qquad (2.10)$$

Example 2.1

The commutator $[\hat{a}, \hat{a}^\dagger]$ can be evaluated as follows (remembering that $[\hat{x}, \hat{p}] = i\hbar$):

$$
\begin{aligned}
[\hat{a}, \hat{a}^\dagger] &= \frac{m\omega}{2\hbar}\left(-\frac{i}{m\omega}[\hat{x}, \hat{p}] + \frac{i}{m\omega}[\hat{p}, \hat{x}]\right) \\
&= \frac{m\omega}{2\hbar}\left(\frac{\hbar}{m\omega} + \frac{\hbar}{m\omega}\right) \\
&= 1.
\end{aligned}
\tag{2.11}
$$

The definitions of $\hat{a}$ and $\hat{a}^\dagger$ can be inverted to give

$$
\hat{x} = \sqrt{\frac{\hbar}{2m\omega}}(\hat{a} + \hat{a}^\dagger),
\tag{2.12}
$$

$$
\hat{p} = -i\sqrt{\frac{\hbar m\omega}{2}}(\hat{a} - \hat{a}^\dagger).
\tag{2.13}
$$

Putting our new definitions of $\hat{a}$ and $\hat{a}^\dagger$ into our equation for the Hamiltonian yields

$$
\hat{H} = \hbar\omega\left(\hat{a}^\dagger\hat{a} + \frac{1}{2}\right).
\tag{2.14}
$$

The active ingredient in this Hamiltonian is the combination $\hat{a}^\dagger\hat{a}$. If $\hat{a}^\dagger\hat{a}$ has an eigenstate $|n\rangle$ with eigenvalue n, then $\hat{H}$ will also have an eigenstate $|n\rangle$ with eigenvalue $\hbar\omega(n + \frac{1}{2})$, so that we have recovered the eigenvalues of a simple harmonic oscillator in eqn 2.5. However, we need to prove that n takes the values $0, 1, 2, \ldots$. The first step is to show that $n \geq 0$. We can do that by noting that

$$
n = \langle n|\hat{a}^\dagger\hat{a}|n\rangle = |\hat{a}|n\rangle|^2 \geq 0.
\tag{2.15}
$$

Next, we have to show that n takes only integer values, and we will do that below but beforehand let us introduce a bit of notation to save some writing. We define the **number operator** $\hat{n}$ by

$$
\hat{n} = \hat{a}^\dagger\hat{a},
\tag{2.16}
$$

and hence write

$$
\hat{n}|n\rangle = n|n\rangle.
\tag{2.17}
$$

Number of what? The quantity n labels the energy level on the ladder (see Fig. 2.3) that the system has reached, or equivalently the number of quanta (each of energy $\hbar\omega$) that must have been added to the system when it was in its ground state. We can rewrite the Hamiltonian as

$$
\hat{H} = \hbar\omega\left(\hat{n} + \frac{1}{2}\right),
\tag{2.18}
$$

and therefore

$$
\hat{H}|n\rangle = \left(n + \frac{1}{2}\right)\hbar\omega|n\rangle,
\tag{2.19}
$$

so that $|n\rangle$ is also an eigenstate of the Hamiltonian. Thus $|n\rangle$ is a convenient shorthand for the more complicated form of $\psi_n(\xi)$ shown in eqn 2.4. The next examples show that the eigenvalue n indeed takes integer values.

Example 2.2

This example looks at the property of the state defined by $\hat{a}^\dagger|n\rangle$. One of the things we can do is to operate on this with the number operator:

$$\hat{n}\hat{a}^\dagger|n\rangle = \hat{a}^\dagger\hat{a}\hat{a}^\dagger|n\rangle. \tag{2.20}$$

Using the commutator in eqn 2.11 gives $\hat{a}\hat{a}^\dagger = 1 + \hat{a}^\dagger\hat{a}$ and hence

$$\hat{n}\hat{a}^\dagger|n\rangle = (n+1)\hat{a}^\dagger|n\rangle. \tag{2.21}$$

The above example shows that the state $\hat{a}^\dagger|n\rangle$ is an eigenstate of $\hat{H}$ but with an eigenvalue one higher than the state $|n\rangle$. In other words, the operator $\hat{a}^\dagger$ has the effect of adding one quantum of energy. For this reason $\hat{a}^\dagger$ is called a **raising operator**.

Example 2.3

This example looks at the property of the state defined by $\hat{a}|n\rangle$. We can operate on this with the number operator:

$$\hat{n}\hat{a}|n\rangle = \hat{a}^\dagger\hat{a}\hat{a}|n\rangle. \tag{2.22}$$

Using the commutator in eqn 2.11 gives $\hat{a}^\dagger\hat{a} = \hat{a}\hat{a}^\dagger - 1$ and hence

$$\hat{n}\hat{a}|n\rangle = (n-1)\hat{a}|n\rangle. \tag{2.23}$$

The above example shows that the state $\hat{a}|n\rangle$ is an eigenstate of $\hat{H}$ but with an eigenvalue one lower than the state $|n\rangle$. In other words, the operator $\hat{a}$ has the effect of subtracting one quantum of energy. For this reason $\hat{a}$ is called a **lowering operator**.

Example 2.4

Question: Normalize the operators $\hat{a}$ and $\hat{a}^\dagger$.
Solution: We have shown that $\hat{a}|n\rangle = k|n-1\rangle$, where k is a constant. Hence, taking the norm of this state (i.e. premultiplying it by its adjoint) gives

$$|\hat{a}|n\rangle|^2 = \langle n|\hat{a}^\dagger\hat{a}|n\rangle = |k|^2\langle n-1|n-1\rangle = |k|^2, \tag{2.24}$$

where the last equality is because the simple harmonic oscillator states are normalized (so that $\langle n-1|n-1\rangle = 1$). However, we notice that $\hat{a}^\dagger\hat{a} = \hat{n}$ is the number operator and hence

$$\langle n|\hat{a}^\dagger\hat{a}|n\rangle = \langle n|\hat{n}|n\rangle = n. \tag{2.25}$$

Equations 2.24 and 2.25 give $k = \sqrt{n}$. (This assumes k to be real, but any state contains an arbitrary phase factor, so we are free to choose k to be real.)

In the same way, we have shown that $\hat{a}^\dagger|n\rangle = c|n+1\rangle$, where c is a constant. Hence,

$$|\hat{a}^\dagger|n\rangle|^2 = \langle n|\hat{a}\hat{a}^\dagger|n\rangle = |c|^2\langle n+1|n+1\rangle = |c|^2, \tag{2.26}$$

and using $\hat{a}\hat{a}^\dagger = 1 + \hat{a}^\dagger\hat{a} = 1 + \hat{n}$, we have that

$$\langle n|\hat{a}\hat{a}^\dagger|n\rangle = \langle n|1+\hat{n}|n\rangle = n+1, \tag{2.27}$$

and hence $c = \sqrt{n+1}$ (choosing c to be real). In summary, our results are:

$$\hat{a}|n\rangle = \sqrt{n}|n-1\rangle, \tag{2.28}$$
$$\hat{a}^\dagger|n\rangle = \sqrt{n+1}|n+1\rangle. \tag{2.29}$$

If we keep hitting the state $|n\rangle$ with $\hat{a}$, we eventually get to $|0\rangle$, the ground state of the simple harmonic oscillator. At this point, we will annihilate the state completely with a further application of $\hat{a}$ because $\hat{a}|0\rangle = 0$. Notice that

$$\hat{H}|0\rangle = \hbar\omega\left(\hat{n} + \frac{1}{2}\right)|0\rangle = \frac{1}{2}\hbar\omega|0\rangle, \qquad (2.30)$$

so this really is the ground state and we see that the energy is $\frac{1}{2}\hbar\omega$, the zero-point energy.[1]

It is also possible to go the other way, operating on $|0\rangle$ with the raising operator $\hat{a}^\dagger$. This leads to

$$\hat{a}^\dagger|0\rangle = |1\rangle, \qquad (2.31)$$

$$\hat{a}^\dagger|1\rangle = \sqrt{2}|2\rangle \implies |2\rangle = \frac{(\hat{a}^\dagger)^2}{\sqrt{2}}|0\rangle, \qquad (2.32)$$

$$\hat{a}^\dagger|2\rangle = \sqrt{3}|3\rangle \implies |3\rangle = \frac{(\hat{a}^\dagger)^3}{\sqrt{3 \times 2}}|0\rangle, \qquad (2.33)$$

and in general

$$|n\rangle = \frac{(\hat{a}^\dagger)^n}{\sqrt{n!}}|0\rangle, \qquad (2.34)$$

so the state $|n\rangle$ can be obtained by repeated application of the operator $\hat{a}^\dagger$. This leads to a completely different way of thinking about these new operators: we call $\hat{a}^\dagger$ a **creation operator** and $\hat{a}$ an **annihilation operator**. We imagine $\hat{a}^\dagger$ acting to create a quantum of energy $\hbar\omega$ and move the oscillator up one rung of the ladder. Its adjoint, $\hat{a}$, acts to annihilate a quantum of energy $\hbar\omega$ and move the oscillator down one rung of the ladder (see Fig. 2.4). These quanta of energy behave like particles; we are adding and subtracting particles by the application of these operators and we have realized the dream of second quantization. A wave problem has spontaneously produced particles!

2.3 A trivial generalization

The next thing we can do is a completely trivial generalization of what we have done before, but it is worth thinking about because certain complicated problems reduce to it. Consider a set of N uncoupled simple harmonic oscillators. They are uncoupled, so one could be on your desk, another sitting in your bathroom, another one out in the park. They don't talk to each other, influence each other or affect each other in any way. Nevertheless, just for fun, we are going to consider their joint Hamiltonian $\hat{H}$ which is simply the sum of individual Hamiltonians $\hat{H}_k$ where k runs from 1 to N. Hence

$$\hat{H} = \sum_{k=1}^{N} \hat{H}_k, \qquad (2.35)$$

[1]To recap, we can write

$$
\begin{aligned}
|\hat{a}|n\rangle|^2 &= \langle n|\hat{a}^\dagger\hat{a}|n\rangle \\
&= \langle n|\frac{\hat{H}}{\hbar\omega} - \frac{1}{2}|n\rangle \\
&= \frac{E_n}{\hbar\omega} - \frac{1}{2},
\end{aligned}
$$

where $E_n = (n + \frac{1}{2})\hbar\omega$. Thus the condition $\hat{a}|0\rangle = 0$ implies $\frac{E_0}{\hbar\omega} - \frac{1}{2} = 0$, and hence the energy of the ground state is given by $E_0 = \frac{1}{2}\hbar\omega$.

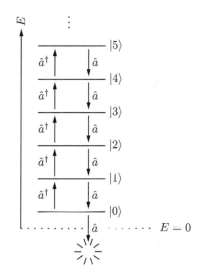

Fig. 2.4 The ladder of energy levels for the simple harmonic oscillator. The operator $\hat{a}^\dagger$ moves the oscillator up one rung of the ladder. Its adjoint, $\hat{a}$, acts to move the oscillator down one rung of the ladder. Note that $\hat{a}|0\rangle = 0$, so dropping off the bottom of the ladder results in zero.

and

$$\hat{H}_k = \frac{\hat{p}_k^2}{2m_k} + \frac{1}{2}m_k\omega_k^2\hat{x}_k^2. \tag{2.36}$$

So far, nothing complicated. We now define the operator $\hat{a}_k^\dagger$ which creates a quantum of energy in the kth oscillator (and leaves all the others unaffected). We also define the operator $\hat{a}_k$ which annihilates a quantum of energy in the kth oscillator (and leaves all the others unaffected). Acting with an operator a_k only affects the number of quanta for the kth oscillator. We can therefore write

$$\hat{a}_k^\dagger|n_1, n_2, ..., n_k, ...\rangle \propto |n_1, n_2, ..., n_k + 1, ...\rangle, \tag{2.37}$$
$$\hat{a}_k|n_1, n_2, ..., n_k, ...\rangle \propto |n_1, n_2, ..., n_k - 1, ...\rangle.$$

The operators will have the commutation rules:

$$[\hat{a}_k, \hat{a}_q] = 0, \tag{2.38}$$
$$\left[\hat{a}_k^\dagger, \hat{a}_q^\dagger\right] = 0, \tag{2.39}$$
$$\left[\hat{a}_k, \hat{a}_q^\dagger\right] = \delta_{kq}. \tag{2.40}$$

In eqn 2.40, the point is that operators acting on different oscillators commute (and therefore do not affect each other). Hence we can follow the results of the previous section and write the Hamiltonian in eqn 2.35 as

$$\hat{H} = \sum_{k=1}^{N} \hbar\omega_k \left(\hat{a}_k^\dagger\hat{a}_k + \frac{1}{2}\right). \tag{2.41}$$

Again we have to define a vacuum state $|0, 0, 0, 0, 0, ...\rangle$ (usually just called $|0\rangle$), such that

$$\hat{a}_k|0\rangle = 0, \tag{2.42}$$

for all k. This is the state in which every one of the oscillators is in its ground state.

A general state of the system, written as $|n_1, n_2, \cdots, n_N\rangle$ is known as the **occupation number representation**. Using the same techniques as in Section 2.2, we can write this general state as

$$|n_1, n_2, \cdots, n_N\rangle = \frac{1}{\sqrt{n_1!n_2!\cdots n_N!}}(\hat{a}_1^\dagger)^{n_1}(\hat{a}_2^\dagger)^{n_2}\cdots(\hat{a}_N^\dagger)^{n_N}|0, 0, \cdots 0\rangle. \tag{2.43}$$

The idea here is that we are operating on the vacuum state with a product of creation operators necessary to put n_1 quanta of energy into oscillator number 1, n_2 quanta of energy into oscillator number 2, etc. We can express this even more succinctly using the following notation

$$|\{n_k\}\rangle = \prod_k \frac{1}{\sqrt{n_k!}}(\hat{a}_k^\dagger)^{n_k}|0\rangle, \tag{2.44}$$

where $|0\rangle$ is the vacuum state, as explained above.

In this model, all of the oscillators have been completely independent. It is now time to tackle a more challenging problem, coupling the oscillators together, and that is covered in the next section.

2.4   Phonons

Consider a linear chain of N atoms (see Fig. 2.5), each of mass m, and connected by springs of unstretched length a and which have spring constant K. The masses are normally at position $R_j = ja$, but can be displaced slightly by an amount x_j. The momentum of the jth mass is p_j. The Hamiltonian for this system is

$$\hat{H} = \sum_j \frac{\hat{p}_j^2}{2m} + \frac{1}{2}K(\hat{x}_{j+1} - \hat{x}_j)^2. \tag{2.45}$$

In contrast to the model in the previous section, we are now dealing with a coupled problem. Each mass is strongly coupled to its neighbour by the springs and there is no way in which we can consider them as independent. However, we will show that the excitations in this system behave exactly as a set of totally independent oscillators. This works because we can Fourier transform the problem, so that even though the masses are coupled in real space, the excitations are uncoupled in reciprocal space. How this is done is covered in the following example.

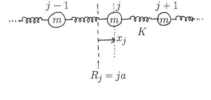

Fig. 2.5 A linear chain of atoms.

Example 2.5

We begin by Fourier transforming both x_j and p_j, by writing

$$x_j = \frac{1}{\sqrt{N}} \sum_k \tilde{x}_k e^{ikja}, \tag{2.46}$$

$$p_j = \frac{1}{\sqrt{N}} \sum_k \tilde{p}_k e^{ikja}, \tag{2.47}$$

and equivalently of course

$$\tilde{x}_k = \frac{1}{\sqrt{N}} \sum_j x_j e^{-ikja}, \tag{2.48}$$

$$\tilde{p}_k = \frac{1}{\sqrt{N}} \sum_j p_j e^{-ikja}. \tag{2.49}$$

We impose periodic boundary conditions[2] forcing $e^{ikja} = e^{ik(j+N)a}$ The wave vector k therefore takes the values $2\pi m/Na$, where m is an integer in the range $-N/2 < m \leq N/2$. Note that

$$\sum_j e^{ikja} = N\delta_{k,0}. \tag{2.50}$$

The commutation relations are

$$[x_j, p_{j'}] = i\hbar\delta_{jj'}, \tag{2.54}$$

$$[\tilde{x}_k, \tilde{p}_{k'}] = \frac{1}{N} \sum_j \sum_{j'} e^{-ikja} e^{-ik'j'a} [x_j, p_{j'}]$$

$$= \frac{i\hbar}{N} \sum_j e^{-i(k+k')ja}$$

$$= i\hbar\delta_{k,-k'}. \tag{2.55}$$

We can now work out terms in the Hamiltonian, so that

$$\sum_j p_j^2 = \sum_j \left(\frac{1}{\sqrt{N}} \sum_k \tilde{p}_k e^{ikja}\right)\left(\frac{1}{\sqrt{N}} \sum_{k'} \tilde{p}_{k'} e^{ik'ja}\right)$$

$$= \frac{1}{N} \sum_j \sum_k \sum_{k'} \tilde{p}_k \tilde{p}_{k'} e^{i(k+k')ja}. \tag{2.56}$$

For the time being, we will drop the 'hats' on $\hat{x}_j$ and $\hat{p}_j$ to save cluttering the algebra. We note that there is another way of solving this problem which involves writing the operators x_j and p_j in terms of creation and annihilation operators of particular modes. It is arguably a more elegant route to the answer and will generalize more easily to additional cases, but as it is a little more abstract we will save it for later in the book.

[2]Periodic boundary conditions are described in detail in the next chapter.

The second quantization trick to evaluate sums of the form

$$\frac{1}{N} \sum_j \sum_{kq} \tilde{p}_k \tilde{p}_q e^{i(k+q)ja} : \tag{2.51}$$

Step 1: Perform the spatial sum, making use of the identity $\sum_j e^{i(k-k')ja} = N\delta_{kk'}$. This gives

$$\sum_{kq} \tilde{p}_k \tilde{p}_q \delta_{k,-q}. \tag{2.52}$$

Step 2: Use the Kronecker delta to do one of the momentum sums. This has the effect of setting $q = -k$, leaving us with a sum over a single index:

$$\sum_k \tilde{p}_k \tilde{p}_{-k}. \tag{2.53}$$

This combination of sums over one spatial and two momentum indices will occur frequently and is easily dealt with using the trick shown in the box. The result is

$$\sum_j p_j^2 = \sum_k \tilde{p}_k \tilde{p}_{-k}. \tag{2.57}$$

The other term in the Hamiltonian may be treated similarly and we have

$$
\begin{aligned}
\sum_j (x_{j+1} - x_j)^2 &= \frac{1}{N} \sum_j \sum_k \sum_{k'} \tilde{x}_k \tilde{x}_{k'} e^{i(k+k')ja} (e^{ika} - 1)(e^{ik'a} - 1) \\
&= \sum_k \tilde{x}_k \tilde{x}_{-k} \left(4 \sin^2 \frac{ka}{2} \right), \tag{2.58}
\end{aligned}
$$

where use has been made of the identity $1 - \cos ka = 2 \sin^2 \frac{ka}{2}$. We can thus express[3] the Hamiltonian as

$$\hat{H} = \sum_k \left[\frac{1}{2m} \hat{p}_k \hat{p}_{-k} + \frac{1}{2} m \omega_k^2 \hat{x}_k \hat{x}_{-k} \right], \tag{2.59}$$

where $\omega_k^2 = (4K/m) \sin^2(ka/2)$.

[3] *Notational point:* For clarity, we will now omit the tilde from $\tilde{x}_k$ and $\tilde{p}_k$, but to remind ourselves that they are operators we will reinstate the 'hats', so they become $\hat{x}_k$ and $\hat{p}_k$.

With our definitions we automatically have that $\hat{p}_k^\dagger = \hat{p}_{-k}$ and $\hat{x}_k^\dagger = \hat{x}_{-k}$ (which follow from the requirement that $\hat{p}_j$ and $\hat{x}_j$ are Hermitian), so we can write down the commutation relation as $[\hat{x}_k, \hat{p}_{k'}^\dagger] = i\hbar \delta_{kk'}$ and we can also write down creation and annihilation operators as follows:

$$\hat{a}_k = \sqrt{\frac{m\omega_k}{2\hbar}} \left(\hat{x}_k + \frac{i}{m\omega_k} \hat{p}_k \right), \tag{2.60}$$

$$\hat{a}_k^\dagger = \sqrt{\frac{m\omega_k}{2\hbar}} \left(\hat{x}_{-k} - \frac{i}{m\omega_k} \hat{p}_{-k} \right), \tag{2.61}$$

which have commutation relations $[\hat{a}_k^\dagger, \hat{a}_{k'}^\dagger] = [\hat{a}_k, \hat{a}_{k'}] = 0$ and $[\hat{a}_k, \hat{a}_{k'}^\dagger] = \delta_{k,k'}$. We can invert eqns 2.60 and 2.61 to yield

$$\hat{x}_k = \sqrt{\frac{\hbar}{2m\omega_k}} \left(\hat{a}_k + \hat{a}_{-k}^\dagger \right), \tag{2.62}$$

$$\hat{p}_k = -i\sqrt{\frac{m\hbar\omega_k}{2}} \left(\hat{a}_k - \hat{a}_{-k}^\dagger \right). \tag{2.63}$$

The Hamiltonian becomes

$$\hat{H} = \sum_k \frac{\hbar\omega_k}{2} \left(\hat{a}_k \hat{a}_k^\dagger + \hat{a}_{-k}^\dagger \hat{a}_{-k} \right), \tag{2.64}$$

and reindexing this gives

$$\hat{H} = \sum_k \frac{\hbar\omega_k}{2} \left(\hat{a}_k \hat{a}_k^\dagger + \hat{a}_k^\dagger \hat{a}_k \right). \tag{2.65}$$

[4] i.e. $\hat{a}_k \hat{a}_k^\dagger - \hat{a}_k^\dagger \hat{a}_k = 1$.

Using the commutator[4] then produces

$$\hat{H} = \sum_k \hbar\omega_k \left(\hat{a}_k^\dagger \hat{a}_k + \frac{1}{2} \right). \tag{2.66}$$

This example has demonstrated that the Hamiltonian for a set of coupled masses on a chain can be expressed in terms of a sum over modes, labelled by wave vector k. Comparing this result (eqn 2.66) with that obtained near the end of the previous section (eqn 2.41), one sees that they are identical. Thus these modes behave as if they are *entirely independent and uncoupled* simple harmonic oscillators. We call these modes **phonons**, and each phonon mode, labelled by k, can be given integer multiples of the quantum of energy $\hbar\omega_k$.

This is the key point. Each phonon mode behaves like a simple harmonic oscillator and so can accept energy in an integer number of lumps, each lump being of size $\hbar\omega_k$. This is because the energy eigenvalues of a simple harmonic oscillator form a ladder of energy levels where the rung-size is fixed. These lumps of energy look like particles, and so we think of phonons themselves as particles. Later on, we play exactly the same trick with the electromagnetic field and show that it behaves like a set of uncoupled simple harmonic oscillators and call the quanta *photons*. This is the heart of second quantization: wave problems have oscillator solutions and hence produce particles! The key insight in this chapter is essentially that the oscillator picture of physical systems leads to a particle picture of those systems. This notion is summarized in Fig. 2.6.

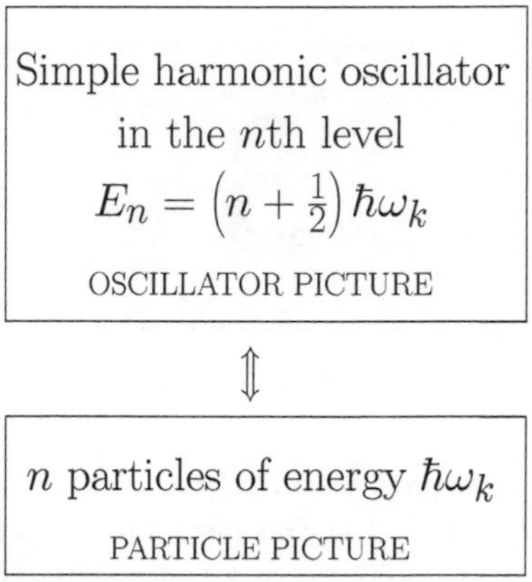

Simple harmonic oscillator in the nth level
$E_n = \left(n + \frac{1}{2}\right)\hbar\omega_k$
OSCILLATOR PICTURE

$\updownarrow$

n particles of energy $\hbar\omega_k$
PARTICLE PICTURE

Fig. 2.6 The equivalence of the oscillator and particle pictures.

Chapter summary

- First quantization shows that particles behave like waves; second quantization shows that waves behave like particles.
- The simple harmonic oscillator has energy eigenvalues given by $E_n = (n + \frac{1}{2})\hbar\omega$. Eigenstates can be written in the occupation number representation.
- Using creation and annihilation operators the Hamiltonian for the simple harmonic oscillator can be written $\hat{H} = \hbar\omega(\hat{a}^\dagger\hat{a} + \frac{1}{2})$.
- The phonon problem can be re-expressed as a sum over noninteracting modes, each one of which behaves like a simple harmonic oscillator.

Exercises

(2.1) For the one-dimensional harmonic oscillator, show that with creation and annihilation operators defined as in eqns 2.9 and 2.10, $[\hat{a}, \hat{a}] = 0$, $[\hat{a}^\dagger, \hat{a}^\dagger] = 0$, $[\hat{a}, \hat{a}^\dagger] = 1$ and $\hat{H} = \hbar\omega(\hat{a}^\dagger\hat{a} + \frac{1}{2})$.

(2.2) For the Hamiltonian $\hat{H} = \frac{\hat{p}^2}{2m} + \frac{1}{2}m\omega^2\hat{x}^2 + \lambda\hat{x}^4$, where λ is small, show by writing the Hamiltonian in terms of creation and annihilation operators and using perturbation theory, that the energy eigenvalues of all the levels are given by

$$E_n = \left(n + \frac{1}{2}\right)\hbar\omega + \frac{3\lambda}{4}\left(\frac{\hbar}{m\omega}\right)^2(2n^2 + 2n + 1).$$
(2.67)

(2.3) Use eqns 2.46 and 2.62 to show that

$$\hat{x}_j = \frac{1}{\sqrt{N}}\left(\frac{\hbar}{m}\right)^{\frac{1}{2}}\sum_k \frac{1}{(2\omega_k)^{1/2}}[\hat{a}_k e^{ikja} + \hat{a}_k^\dagger e^{-ikja}].$$
(2.68)

(2.4) Using $\hat{a}|0\rangle = 0$ and eqns 2.9 and 2.10 together with $\langle x|\hat{p}|\psi\rangle = -i\hbar\frac{d}{dx}\langle x|\psi\rangle$, show that

$$0 = \left(x + \frac{\hbar}{m\omega}\frac{d}{dx}\right)\langle x|0\rangle,$$
(2.69)

and hence

$$\langle x|0\rangle = \left(\frac{m\omega}{\pi\hbar}\right)^{1/4}e^{-m\omega x^2/2\hbar}.$$
(2.70)

3

Occupation number representation

I am not a number. I am a free man.
Patrick McGoohan in *The Prisoner*

In the previous chapter we considered simple harmonic oscillators, showed that such oscillators have particle-like eigenfunctions, and introduced *en passant* the occupation number representation to describe sets of uncoupled oscillators. Now we are going to make further use of this idea to describe systems of many particles more generally. The use of the occupation number representation to describe a system of identical particles is partly a matter of notational administration and partly a radical change in the way we look at the world. We (i) change the way we label our states (this is administration) and (ii) get rid of wave functions altogether (this is the radical part). We replace wave functions with a version of the creation and annihilation operators we used to solve the quantum oscillator problems. This turns out to simplify things enormously. In setting up this new method we will need to ensure that we never violate the sacred laws of indistinguishable particles. These are as follows:

- There are two types of particle in the Universe: Bose[1] particles and Fermi[2] particles.
- If you exchange two identical bosons you get the same state again.
- If you exchange two fermions, you get minus the state you started with.[3]

To make things simple we'll confine our particles to a box for most of this chapter. Although this makes the notation ugly (and non-covariant) it has several advantages (not least that we can normalize the states in a simple way). We recap the physics of particle in a box in the next section.

3.1 A particle in a box

We will now choose units equivalent to setting $\hbar = 1$. Therefore the momentum operator for motion along the x-direction $\hat{p}$ can be written

$$\hat{p} = -\mathrm{i}\frac{\partial}{\partial x}. \tag{3.1}$$

[1]Satyendra Nath Bose (1894–1974) wrote a paper in 1924 which laid the foundations for quantum statistics.

[2]In addition to his many contributions to other branches of physics, Enrico Fermi (1901–1954) made numerous contributions to quantum field theory. His 1932 review of Quantum Electrodynamics (Rev. Mod. Phys. **4**, 87) had enormous influence on the field.

[3]A consequence of this principle is that fermions obey the Pauli exclusion principle: It is impossible to put two in the same quantum state. This affects the statistical distribution among a set of discrete energy levels of these indistinguishable quantum particles, and so is referred to as **quantum statistics**. In Sections 3.1–3.3 we won't worry about quantum statistics, but return to the issue in Section 3.4.

Solutions to the Schrödinger equation for a particle in a box of size L are running waves $\psi(x) = \frac{1}{\sqrt{L}}\mathrm{e}^{\mathrm{i}px}$, and eigenstates of the momentum operator $[\hat{p}\psi(x) = -\mathrm{i}\frac{\mathrm{d}\psi}{\mathrm{d}x} = p\psi(x)]$. We use periodic boundary conditions and so the wave function for our system must have the property that $\psi(x) = \psi(x + L)$. Therefore $\mathrm{e}^{\mathrm{i}px} = \mathrm{e}^{\mathrm{i}p(x+L)}$, and hence $\mathrm{e}^{\mathrm{i}pL} = 1$. In order to satisfy this condition, we need $pL = 2\pi m$, with m an integer. This imposes a quantization condition on the possible momentum states that particles in the box can take. They must satisfy

$$p_m = \frac{2\pi m}{L}. \qquad (3.2)$$

We'll call the possible momentum states things like p_1 and p_2, etc. The energy of a single particle in a state p_m is E_{p_m}, which depends on the momentum state, but it doesn't matter how yet.

To make a simple many-particle system we'll put more and more non-interacting Bose particles in the box. How does this affect the total momentum and energy of the system? Applying the momentum operator to the two-particle state (see Fig. 3.1) yields

$$\hat{p}|p_1 p_2\rangle = (p_1 + p_2)|p_1 p_2\rangle, \qquad (3.3)$$

and applying the Hamiltonian operator gives us

$$\hat{H}|p_1 p_2\rangle = (E_{p_1} + E_{p_2})|p_1 p_2\rangle. \qquad (3.4)$$

Since the particles don't interact, having two particles in a particular energy state (p_3 say) costs an energy $2E_{p_3}$, i.e. double the single-particle energy. The total energy is given by $\sum_m n_{p_m} E_{p_m}$, where n_{p_m} is the total number of particles in the state $|p_m\rangle$.

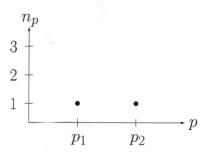

Fig. 3.1 A two-particle state given by $|p_1 p_2\rangle$.

3.2 Changing the notation

As advertised, the first thing we'll do is change the notation. In conventional quantum mechanics we label each identical particle (e.g. particle A, particle B, particle C) and list its momentum. This gives rise to states labelled like $|ABC\rangle = |p_1 p_2 p_1\rangle$ for a three-particle state. We read this as 'particle A in momentum state p_1, particle B in state p_2 and particle C in state p_1'. This is shown in Fig. 3.2 How could we streamline this notation? We haven't yet used the facts that the particles are indistinguishable (making the assignment A, B, C, etc. rather meaningless) and that there are only certain allowable values of momentum (i.e. $2\pi m/L$). We can therefore decide that instead of listing which particle (A, B, C, etc.) is in which state p_m, we could just list the values of the momentum p_m, saying *how many particles occupy each momentum state*. For our example, we'd say 'two particles are in momentum state p_1, one is in momentum state p_2'. We therefore define a state of N particles by listing how many are in each of the momentum states. We will write these as $|n_1 n_2 n_3 \ldots\rangle$. Our example is written $|2100\ldots\rangle$. Specifying a state by

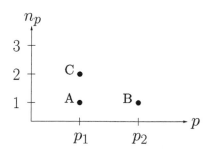

Fig. 3.2 A three-particle state given by $|p_1 p_2 p_1\rangle$ can be written as $|2100\cdots\rangle$ in the occupation number representation.

Particles	old	new
0	$|0\rangle$	$|00\rangle$
1	$|q_1\rangle$	$|10\rangle$
1	$|q_2\rangle$	$|01\rangle$
2	$|q_1 q_1\rangle$	$|20\rangle$
2	$|q_1 q_2\rangle$	$|11\rangle$
2	$|q_2 q_2\rangle$	$|02\rangle$
3	$|q_1 q_1 q_1\rangle$	$|30\rangle$
$\vdots$	$\vdots$	$\vdots$

and so on.

Table of old and new notations.

listing the number of identical particles in each quantum state is another instance of the **occupation number representation**, which we met in the last chapter.

Example 3.1

Let's compare the old and new notation for a system with only two possible states: $p_1 = 2\pi/L$ and $p_2 = 4\pi/L$. Into these states we can put any number of Bose particles that we choose. The table in the margin compares the old and new notations.

What happens when we act on a state in occupation number representation with the Hamiltonian? As we saw above, the answer needs to be

$$\hat{H}|n_1 n_2 n_3 \dots\rangle = \left[\sum_m n_{p_m} E_{p_m}\right] |n_1 n_2 n_3 \dots\rangle, \qquad (3.5)$$

that is, we multiply the number of particles in each state by the energy of that state and sum over all of the states. This is obviously the right thing to do to find the total energy.

Now for the reason why occupation numbers are useful. Equation 3.5 can be made to look quite similar to something we've seen before. Recall that a quantum harmonic oscillator had a total energy given by $E = \left(n + \frac{1}{2}\right)\hbar\omega$ with n being some integer. Dropping the zero-point energy (which just amounts to a new choice in our zero of energy) we can write this as $E = n\hbar\omega$. In words, the oscillator has n quanta in it. Crucially, the energy levels are equally spaced, so having two quanta in the system gives an energy which is twice that for a single quantum in the system. Now imagine N *independent* oscillators labelled by the index k. These might not be identical, so we say that the energy level spacing of the kth oscillator is $\hbar\omega_k$. The total energy of this system is now $E = \sum_{k=1}^{N} n_k \hbar\omega_k$. So the kth oscillator has n_k quanta in it, contributing an energy $n_k \hbar\omega_k$ to the sum. Compare this to our system of identical particles for which the total energy is $E = \sum_{p_m} n_{p_m} E_{p_m}$. In words, this says that the momentum state p_m has n_{p_m} particles in it. Each particle contributes an energy E_{p_m} giving a contribution of $n_{p_m} E_{p_m}$ for the momentum mode p_m. These two systems therefore have the same energy level structure.

We have built an analogy between two completely different systems: harmonic oscillators and identical particles.

SHO		Identical particles	
Quanta in oscillators	$\rightarrow$	Particles in momentum states	(3.6)
kth oscillator	$\rightarrow$	mth momentum mode p_m	
$E = \sum_{k=1}^{N} n_k \hbar\omega_k$	$\rightarrow$	$E = \sum_{m=1}^{N} n_{p_m} E_{p_m}.$	

As we'll see in the next section, the real value to this analogy is that it can be pushed further.[4]

[4]We should be a little careful here; we haven't physically replaced particles with pendulums. We've simply exploited the fact that the simple harmonic oscillator energy levels are equally spaced, so, for example, two quanta in an oscillator have an energy that is twice that of a single quantum in that oscillator.

3.3   Replace state labels with operators

The notation change to occupation number representation is just window dressing. We're now going to make a more dramatic change in an attempt to almost remove state vectors altogether! We'll do this by having operators describing the physics rather than state vectors. We'll have to retain only one special state, the vacuum $|0\rangle$. Following eqn 2.43,[5] we will build up a general state of several harmonic oscillators by acting on the vacuum state $|0\rangle$ and write

$$|n_1 n_2 \cdots\rangle = \prod_k \frac{1}{(n_k!)^{\frac{1}{2}}} (\hat{a}_k^\dagger)^{n_k} |0\rangle. \tag{3.7}$$

[5] For the oscillator problem in Chapter 2 we had the very useful technology of creation and annihilation operators to add and remove quanta from the SHO system, and we use the same idea here.

Example 3.2

We can build up a state with quanta arranged among various oscillators. A state of the system in which there are two quanta in oscillator number 1 and one quantum in oscillator number 2 (written in occupation number representation as $|21000\ldots\rangle$) can be written as

$$|21000\ldots\rangle = \left[\frac{1}{\sqrt{2!}} (\hat{a}_1^\dagger)^2 \right] \left[\frac{1}{\sqrt{1!}} \hat{a}_2^\dagger \right] |0\rangle. \tag{3.8}$$

However, what we are doing here is not just talking about creating quanta in oscillators, but also particles in particular momentum states. We want a creation operator $\hat{a}_{p_m}^\dagger$ to create one particle in a momentum state $|p_m\rangle$. It's hopefully now just a matter of changing $k \rightarrow p_m$ to go from operators describing oscillators to operators describing particles in momentum states and we say that $\hat{a}_{p_m}^\dagger$ creates a particle in momentum state $|p_m\rangle$ rather than a quantum in oscillator k. Before we can do this, we need to check that doing this doesn't violate any of the symmetry principles we stated at the start of this chapter. That is, we need to know how this new formulation deals with Bose and Fermi particles.

3.4   Indistinguishability and symmetry

We now have to make sure that introducing creation and annihilation of states will respect the exchange symmetry principles for identical particles. It will turn out that the basic technology is supplied by the commutation relations of the creation $\hat{a}_{p_m}^\dagger$ and annihilation $\hat{a}_{p_m}$ operators for a particular state. The key point is that it matters which order you use these operators to put particles into states.

Example 3.3

Consider a system with two momentum states p_1 and p_2 described in the occupation number representation as $|n_1 n_2\rangle$ formed from creation operators acting on a vacuum state $|0\rangle$. We'll define

$$\hat{a}_{p_1}^\dagger |0\rangle = |10\rangle, \quad \hat{a}_{p_2}^\dagger |0\rangle = |01\rangle. \tag{3.9}$$

Now let's add another particle in the unoccupied state. This gives us

$$\hat{a}_{p_2}^\dagger \hat{a}_{p_1}^\dagger |0\rangle \propto |11\rangle, \quad \hat{a}_{p_1}^\dagger \hat{a}_{p_2}^\dagger |0\rangle \propto |11\rangle. \tag{3.10}$$

The proportionality sign is because the constant remains to be determined.

Whether you first put one particle in state p_1 and then another in state p_2, or do the same thing in the reverse order, you must end up with the same final state: $|11\rangle$. This means that

$$\hat{a}_{p_1}^\dagger \hat{a}_{p_2}^\dagger = \lambda \hat{a}_{p_2}^\dagger \hat{a}_{p_1}^\dagger, \tag{3.11}$$

[6]In Chapter 29 we examine the possibility of other choices.

where λ is a constant. We assume two possibilities,[6] $\lambda = \pm 1$, and they correspond to wave functions which are either symmetric or antisymmetric under particle exchange. The particles are then called **bosons** or **fermions** and we will find that the two cases correspond to two different types of commutation relation for the creation and annihilation operators. We will examine each case in turn.

Case I: $\lambda = 1$, *bosons:* In this case

$$\hat{a}_{p_2}^\dagger \hat{a}_{p_1}^\dagger = \hat{a}_{p_1}^\dagger \hat{a}_{p_2}^\dagger. \tag{3.12}$$

Rearranging (and generalizing the labels of the momentum states to i and j) we find

$$\left[\hat{a}_i^\dagger, \hat{a}_j^\dagger\right] = \hat{a}_i^\dagger \hat{a}_j^\dagger - \hat{a}_j^\dagger \hat{a}_i^\dagger = 0, \tag{3.13}$$

which means that the creation operators for different particle states commute. We also have $[\hat{a}_i, \hat{a}_j] = 0$ and, importantly, we define

$$\left[\hat{a}_i, \hat{a}_j^\dagger\right] = \delta_{ij}, \tag{3.14}$$

so that we can use our harmonic oscillator analogy. We conclude that these commutation relations are the same as those for simple harmonic oscillators above. We've succeeded in developing a system to describe many particles! The formalism is identical to that we developed for many independent oscillators in the previous section. We can therefore build up a general state of many particles in momentum states using

$$|n_1 n_2 \cdots\rangle = \prod_m \frac{1}{(n_{p_m}!)^{\frac{1}{2}}} (\hat{a}_{p_m}^\dagger)^{n_{p_m}} |0\rangle. \tag{3.15}$$

The particles this formalism describes are Bose particles: we can put any number into quantum states and, as we'll see later, they are symmetric upon exchange of particles.

Finally we note that the commutation of different operators implies that

$$\hat{a}_{p_1}^\dagger \hat{a}_{p_2}^\dagger |0\rangle = \hat{a}_{p_2}^\dagger \hat{a}_{p_1}^\dagger |0\rangle = |1_{p_1} 1_{p_2}\rangle, \tag{3.16}$$

that is, it doesn't matter which order you put particles in the states, you get exactly the same in either case.

In general we can write

$$\hat{a}_i^\dagger|n_1\cdots n_i\cdots\rangle = \sqrt{n_i+1}|n_1\cdots n_i+1\cdots\rangle, \quad (3.17)$$
$$\hat{a}_i|n_1\cdots n_i\cdots\rangle = \sqrt{n_i}|n_1\cdots n_i-1\cdots\rangle. \quad (3.18)$$

Case II: $\lambda = -1$, *fermions:* In this case we'll write the fermion creation operators $\hat{c}_i^\dagger$ to distinguish them from the boson operators (written $\hat{a}_i^\dagger$). Plugging in $\lambda = -1$, we find that

$$\left\{\hat{c}_i^\dagger, \hat{c}_j^\dagger\right\} \equiv \hat{c}_i^\dagger\hat{c}_j^\dagger + \hat{c}_j^\dagger\hat{c}_i^\dagger = 0, \quad (3.19)$$

where the curly brackets indicate an **anticommutator**. Fermion operators anticommute: when you change their order you pick up a minus sign since $\hat{c}_i^\dagger\hat{c}_j^\dagger = -\hat{c}_j^\dagger\hat{c}_i^\dagger$. In particular, setting $i=j$ we find

$$\hat{c}_i^\dagger\hat{c}_i^\dagger + \hat{c}_i^\dagger\hat{c}_i^\dagger = 0 \quad \text{and hence} \quad \hat{c}_i^\dagger\hat{c}_i^\dagger = 0. \quad (3.20)$$

In other words, trying to shoehorn two fermions into the same state annihilates it completely. This is the **Pauli exclusion principle**[7] which means that each state can either be unoccupied or contain a single fermion. We also have

$$\{\hat{c}_i, \hat{c}_j\} = 0. \quad (3.21)$$

Finally we define an anticommutator

$$\left\{\hat{c}_i, \hat{c}_j^\dagger\right\} = \delta_{ij}, \quad (3.22)$$

which enables us to use the simple harmonic oscillator analogy here too. Again we can describe putting particles in momentum states in the same way as putting quanta in independent oscillators. The only difference is that we use anticommutators rather than commutators.

Finally we note that since $\hat{c}_i^\dagger\hat{c}_j^\dagger|0\rangle = -\hat{c}_j^\dagger\hat{c}_i^\dagger|0\rangle$, it really does matter which order you put the particles into the states.

[7] The Pauli exclusion principle is named after Wolfgang Pauli (1900–1958), the Austrian theoretical physicist who was one of the pioneers of quantum mechanics, and also of quantum field theory and quantum electrodynamics, which were his main concerns after 1927. He referred to fellow-physicist Wolfgang Paul (1913–1993) as his 'real part'. Pauli was merciless with his criticism of other physicists and their work, famous for his put-downs, one of the most well-known of which was that a particular scientific paper was 'not even wrong'.

Example 3.4

Let's check that the number operator still works for fermions. The number operator is $\hat{n}_i = \hat{c}_i^\dagger\hat{c}_i$. We'll see the action of $\hat{n}_1$ on a state $|11\rangle \equiv \hat{c}_1^\dagger\hat{c}_2^\dagger|0\rangle$. We expect that this is an eigenstate with eigenvalue 1 if all goes to plan because there is one particle in the $i=1$ state. We'll now proceed to check this:

$$\begin{aligned}\hat{n}_1|11\rangle &= \hat{c}_1^\dagger\hat{c}_1|11\rangle \\ &= \hat{c}_1^\dagger\hat{c}_1(\hat{c}_1^\dagger\hat{c}_2^\dagger|0\rangle) \\ &= \hat{c}_1^\dagger(1 - \hat{c}_1^\dagger\hat{c}_1)\hat{c}_2^\dagger|0\rangle \quad \text{(using the anticommutation relation)} \\ &= \hat{c}_1^\dagger\hat{c}_2^\dagger|0\rangle \equiv |11\rangle, \quad (3.23)\end{aligned}$$

i.e. $|11\rangle$ is an eigenstate of $\hat{n}_1$ with eigenvalue 1 as we expected.

For fermions, in order to get the signs right, the normalization of creation and annihilation operators are defined as follows

$$\hat{c}_i^\dagger |n_1 \cdots n_i \cdots \rangle = (-1)^{\Sigma_i} \sqrt{1 - n_i} |n_1 \cdots n_i + 1 \cdots \rangle, \quad (3.24)$$
$$\hat{c}_i |n_1 \cdots n_i \cdots \rangle = (-1)^{\Sigma_i} \sqrt{n_i} |n_1 \cdots n_i - 1 \cdots \rangle, \quad (3.25)$$

The square roots are redundant in eqns 3.24 and 3.25 since $n_i = 0$ or 1, so they can be omitted if desired.

where

$$(-1)^{\Sigma_i} = (-1)^{(n_1 + n_2 + \cdots + n_{i-1})}. \quad (3.26)$$

This gives us a factor of -1 for every particle standing to the left of the state labelled n_i in the state vector.

Example 3.5

Let's check to see we've got it right when we come to exchange the particles. Remember that for bosons we get the same state upon exchange and for fermions we pick up a minus sign. To exchange particles we need to imagine a sequence of operations which swap particles around. An example would be

$$|110\rangle \qquad \text{a state with particles in states 1 and 2} \qquad (3.27)$$
$$\rightarrow |101\rangle \qquad \text{move particle from state 2 to 3}$$
$$\rightarrow |011\rangle \qquad \text{move particle from state 1 to 2}$$
$$\rightarrow |110\rangle \qquad \text{move particle from state 3 to 1.}$$

To move a particle we annihilate it from its original state and then create it in its new state. Moving a particle from state 2 to 3 is achieved using $\hat{a}_3^\dagger \hat{a}_2 |110\rangle$ The full sequence of events can be described with the string of operators

$$\hat{a}_1^\dagger \hat{a}_3 \hat{a}_2^\dagger \hat{a}_1 \hat{a}_3^\dagger \hat{a}_2 |110\rangle = \pm |110\rangle, \quad (3.28)$$

where we want the $+$ sign for bosons and the $-$ sign for fermions.

Let's check that the commutation relations give the correct results for swapping Bose particles. We can swap operators that refer to different states (since $[\hat{a}_i, \hat{a}_j] = 0$) to get

$$\hat{a}_1^\dagger \hat{a}_3 \hat{a}_2^\dagger \hat{a}_1 \hat{a}_3^\dagger \hat{a}_2 |110\rangle = \hat{a}_3 \hat{a}_3^\dagger \hat{a}_1^\dagger \hat{a}_1 \hat{a}_2^\dagger \hat{a}_2 |110\rangle. \quad (3.29)$$

But $\hat{a}_i^\dagger \hat{a}_i$ is just the number operator $\hat{n}_i$, counting the number of states in a mode i.

$$\hat{a}_3 \hat{a}_3^\dagger \hat{n}_1 \hat{n}_2 |110\rangle = (1) \times (1) \times \hat{a}_3 \hat{a}_3^\dagger |110\rangle, \quad (3.30)$$

we need to turn $\hat{a}_3 \hat{a}_3^\dagger$ into a number operator. Swap the order with the commutators to get

$$(1 + \hat{a}_3^\dagger \hat{a}_3)|110\rangle = (1 + \hat{n}_3)|110\rangle, \quad (3.31)$$

and notice that $\hat{n}_3 |110\rangle = 0$ to obtain $(1 + \hat{a}_3^\dagger \hat{a}_3)|110\rangle = |110\rangle$, as expected.

Let's now check that the commutation relations give the correct results for swapping Fermi particles. Now when we swap any two fermion operators we pick up a minus sign. Carrying out the same steps as before

$$\hat{c}_1^\dagger \hat{c}_3 \hat{c}_2^\dagger \hat{c}_1 \hat{c}_3^\dagger \hat{c}_2 |110\rangle = -\hat{c}_3 \hat{c}_3^\dagger \hat{c}_1^\dagger \hat{c}_1 \hat{c}_2^\dagger \hat{c}_2 |110\rangle \qquad (3.32)$$
$$= -(1 - \hat{c}_3^\dagger \hat{c}_3)|110\rangle$$
$$= -|110\rangle.$$

Exchanging the particles gives us the required minus sign.

3.5   The continuum limit

So far our particles have been confined in a box, but we will frequently wish to work in the continuum limit of the formalism. As the box size increases, the momentum states become more closely spaced until, in the limit, the momentum becomes a continuous variable. The Kronecker δ_{ij} functions we've been using become Dirac $\delta^{(3)}(\boldsymbol{p})$ functions and the sums become integrals. We have, for the continuous case, that

$$[\hat{a}_{\boldsymbol{p}}, \hat{a}_{\boldsymbol{q}}^\dagger] = \delta^{(3)}(\boldsymbol{p} - \boldsymbol{q}), \tag{3.33}$$

and for energies

$$\hat{H} = \int \mathrm{d}^3 p \, E_{\boldsymbol{p}} \hat{a}_{\boldsymbol{p}}^\dagger \hat{a}_{\boldsymbol{p}}. \tag{3.34}$$

Now that we have a working theory where the commutation properties of the operators contain all of the information about the states, let's check that it works and gets the correct rules for inner products and for position space wave functions.

Example 3.6

For single-particle states, we have

$$\langle \boldsymbol{p} | \boldsymbol{p}' \rangle = \langle 0 | \hat{a}_{\boldsymbol{p}} \hat{a}_{\boldsymbol{p}'}^\dagger | 0 \rangle. \tag{3.35}$$

We use the commutation relations to get

$$\langle \boldsymbol{p} | \boldsymbol{p}' \rangle \;=\; \langle 0 | \left[\delta^{(3)}(\boldsymbol{p} - \boldsymbol{p}') \pm \hat{a}_{\boldsymbol{p}'}^\dagger \hat{a}_{\boldsymbol{p}} \right] | 0 \rangle \tag{3.36}$$
$$=\; \delta^{(3)}(\boldsymbol{p} - \boldsymbol{p}'),$$

which is the right answer. We can check this by working out a position space wave function. We'll start by making a change of basis[8] $|\boldsymbol{x}\rangle = \int \mathrm{d}^3 q \, \phi_{\boldsymbol{q}}^*(\boldsymbol{x}) |\boldsymbol{q}\rangle$, which involves expanding a position state in terms of momentum states. Using this expansion, we obtain

$$\langle \boldsymbol{x} | \boldsymbol{p} \rangle = \int \mathrm{d}^3 q \, \phi_{\boldsymbol{q}}(\boldsymbol{x}) \langle \boldsymbol{q} | \boldsymbol{p} \rangle = \phi_{\boldsymbol{p}}(\boldsymbol{x}). \tag{3.37}$$

This is clearly okay, but rather trivial. More interesting is the case of a two-particle state

$$\langle \boldsymbol{p}' \boldsymbol{q}' | \boldsymbol{q} \boldsymbol{p} \rangle = \langle 0 | \hat{a}_{\boldsymbol{p}'} \hat{a}_{\boldsymbol{q}'} \hat{a}_{\boldsymbol{q}}^\dagger \hat{a}_{\boldsymbol{p}}^\dagger | 0 \rangle. \tag{3.38}$$

Commuting, we obtain

$$\langle \boldsymbol{p}' \boldsymbol{q}' | \boldsymbol{q} \boldsymbol{p} \rangle = \delta^{(3)}(\boldsymbol{p}' - \boldsymbol{p}) \delta^{(3)}(\boldsymbol{q}' - \boldsymbol{q}) \pm \delta^{(3)}(\boldsymbol{p}' - \boldsymbol{q}) \delta^{(3)}(\boldsymbol{q}' - \boldsymbol{p}), \tag{3.39}$$

where we take the plus sign for bosons and minus sign for fermions. Again we can work out the position space wave function by making a change of basis $|\boldsymbol{x}\boldsymbol{y}\rangle = \frac{1}{\sqrt{2!}} \int \mathrm{d}^3 p' \mathrm{d}^3 q' \, \phi_{\boldsymbol{p}'}^*(\boldsymbol{x}) \phi_{\boldsymbol{q}'}^*(\boldsymbol{y}) |\boldsymbol{p}' \boldsymbol{q}'\rangle$ (where the factor of $\frac{1}{\sqrt{2!}}$ is needed to prevent the double counting that results from the unrestricted sums). This gives us

$$\frac{1}{\sqrt{2!}} \int \mathrm{d}^3 p' \mathrm{d}^3 q' \, \phi_{\boldsymbol{p}'}(\boldsymbol{x}) \phi_{\boldsymbol{q}'}(\boldsymbol{y}) \langle \boldsymbol{p}' \boldsymbol{q}' | \boldsymbol{p} \boldsymbol{q} \rangle = \frac{1}{\sqrt{2}} \left[\phi_{\boldsymbol{p}}(\boldsymbol{x}) \phi_{\boldsymbol{q}}(\boldsymbol{y}) \pm \phi_{\boldsymbol{q}}(\boldsymbol{x}) \phi_{\boldsymbol{p}}(\boldsymbol{y}) \right], \tag{3.40}$$

which is the well-known expression for two-particle states.

---

[8]The state $|\boldsymbol{x}\rangle$ can be written in a new basis by inserting the resolution of the identity in the form

$$1 = \int \mathrm{d}^3 q \, |\boldsymbol{q}\rangle \langle \boldsymbol{q}|,$$

so that

$$|\boldsymbol{x}\rangle = \int \mathrm{d}^3 q \, |\boldsymbol{q}\rangle \langle \boldsymbol{q} | \boldsymbol{x} \rangle,$$

and writing $\langle \boldsymbol{x} | \boldsymbol{q} \rangle = \phi_{\boldsymbol{q}}(\boldsymbol{x})$ [and hence $\langle \boldsymbol{q} | \boldsymbol{x} \rangle = \phi_{\boldsymbol{q}}^*(\boldsymbol{x})$] yields

$$|\boldsymbol{x}\rangle = \int \mathrm{d}^3 q \, \phi_{\boldsymbol{q}}^*(\boldsymbol{x}) |\boldsymbol{q}\rangle.$$

It works. We have successfully reformulated quantum mechanics. The effort we've spent here writing everything in terms of creation and annihilation operators will soon pay off. The machinery they represent is so useful that we'll try to build the quantum fields we're searching for out of these objects. As you might suspect, this search will be successful.

Chapter summary

- The occupation number representation describes states by listing the number of identical particles in each quantum state.
- We focus on the vacuum state $|0\rangle$ and then construct many-particle states by acting on $|0\rangle$ with creation operators.
- To obey the symmetries of many-particle mechanics, bosons are described by commuting operators and fermions are described by anticommuting operators.

Exercises

(3.1) For boson operators satisfying

$$\left[\hat{a}_{\boldsymbol{p}}, \hat{a}_{\boldsymbol{q}}^{\dagger}\right] = \delta_{\boldsymbol{pq}}, \qquad (3.41)$$

show that

$$\frac{1}{\mathcal{V}} \sum_{\boldsymbol{pq}} e^{i(\boldsymbol{p}\cdot\boldsymbol{x}-\boldsymbol{q}\cdot\boldsymbol{y})} \left[\hat{a}_{\boldsymbol{p}}, \hat{a}_{\boldsymbol{q}}^{\dagger}\right] = \delta^{(3)}(\boldsymbol{x}-\boldsymbol{y}), \quad (3.42)$$

where $\mathcal{V}$ is the volume of space over which the system is defined. Repeat this for fermion commutation operators.

(3.2) Show that for the simple harmonic oscillator:

   (a) $[\hat{a}, (\hat{a}^{\dagger})^{n}] = n(\hat{a}^{\dagger})^{n-1}$,

   (b) $\langle 0|\hat{a}^{n}(\hat{a}^{\dagger})^{m}|0\rangle = n!\delta_{nm}$,

   (c) $\langle m|\hat{a}^{\dagger}|n\rangle = \sqrt{n+1}\,\delta_{m,n+1}$,

   (d) $\langle m|\hat{a}|n\rangle = \sqrt{n}\,\delta_{m,n-1}$.

(3.3) The three-dimensional harmonic oscillator is described by the Hamiltonian

$$\hat{H} = \frac{1}{2m}(\hat{p}_1^2 + \hat{p}_2^2 + \hat{p}_3^2) + \frac{1}{2}m\omega^2(\hat{x}_1^2 + \hat{x}_2^2 + \hat{x}_3^2). \quad (3.43)$$

Define creation operators $\hat{a}_1^{\dagger}$, $\hat{a}_2^{\dagger}$ and $\hat{a}_3^{\dagger}$ so that $[\hat{a}_i, \hat{a}_j^{\dagger}] = \delta_{ij}$ and show that $\hat{H} = \hbar\omega \sum_{i=1}^{3}(\frac{1}{2} + \hat{a}_i^{\dagger}\hat{a}_i)$. Angular momentum can be defined as

$$\hat{L}^i = -i\hbar\epsilon^{ijk}\hat{a}_j^{\dagger}\hat{a}_k, \qquad (3.44)$$

so that for example $\hat{L}^3 = -i\hbar(\hat{a}_1^{\dagger}\hat{a}_2 - \hat{a}_2^{\dagger}\hat{a}_1)$. Define new creation operators

$$
\begin{aligned}
\hat{b}_1^{\dagger} &= -\frac{1}{\sqrt{2}}(\hat{a}_1^{\dagger} + i\hat{a}_2^{\dagger}), \\
\hat{b}_0^{\dagger} &= \hat{a}_3^{\dagger}, \\
\hat{b}_{-1}^{\dagger} &= \frac{1}{\sqrt{2}}(\hat{a}_1^{\dagger} - i\hat{a}_2^{\dagger}), \quad (3.45)
\end{aligned}
$$

and show that $[\hat{b}_i, \hat{b}_j^{\dagger}] = \delta_{ij}$, the Hamiltonian is $\hat{H} = \hbar\omega \sum_{m=-1}^{1}(\frac{1}{2} + \hat{b}_m^{\dagger}\hat{b}_m)$ and $\hat{L}^3 = \hbar \sum_{m=-1}^{1} m\hat{b}_m^{\dagger}\hat{b}_m$.

(3.4) Since for fermions we have that $\hat{c}_1^{\dagger}\hat{c}_2^{\dagger}|0\rangle = -\hat{c}_2^{\dagger}\hat{c}_1^{\dagger}|0\rangle$, then the state $|11\rangle$ depends on which fermion is added first. If we put the first fermion into the state $\psi_i(r_1)$ and the second fermion into the state $\psi_j(r_2)$, a representation for $|11\rangle$ is

$$\Psi(r_1, r_2) = \frac{1}{\sqrt{2}} \left[\psi_1(r_1)\psi_2(r_2) - \psi_2(r_1)\psi_1(r_2)\right]. \quad (3.46)$$

Show that a generalization of this result for N fermions is $\Psi(r_1 \cdots r_N)$, which can be written out as

$$\frac{1}{\sqrt{N!}} \begin{vmatrix} \psi_1(r_1) & \psi_2(r_1) & \cdots & \psi_N(r_1) \\ \psi_1(r_2) & \psi_2(r_2) & \cdots & \psi_N(r_2) \\ \vdots & \vdots & & \vdots \\ \psi_1(r_N) & \psi_2(r_N) & \cdots & \psi_N(r_N) \end{vmatrix}. \quad (3.47)$$

This expression is known as a **Slater determinant**.

Making second quantization work

In this chapter we will bridge the gap between first and second quantization. This will result in new operators that act on states given in the occupation number representation. For this discussion we'll limit ourselves to non-relativistic particles in a box.[1] By working in the non-relativistic limit, we are taking a shortcut and thereby avoiding some important complications. A fuller, relativistic treatment will be presented in Chapter 11, but the approach taken in this chapter gives a route to some immediately useful results and in particular to the behaviour of electrons in a solid.

4.1 Field operators

In the discussion so far we have introduced creation and annihilation operators $\hat{a}_{\boldsymbol{p}}^{\dagger}$ and $\hat{a}_{\boldsymbol{p}}$ which create and annihilate particles into particular momentum states. So if we have the vacuum state $|0\rangle$ and we apply $\hat{a}_{\boldsymbol{p}}^{\dagger}$ to it, then there is a big puff of smoke and a flash of light and the resulting state $\hat{a}_{\boldsymbol{p}}^{\dagger}|0\rangle$ contains a single particle sitting in the momentum state $\boldsymbol{p}$. Being a momentum eigenstate it is of course extended in space (but completely localized in momentum space). Now, by making appropriate linear combinations of operators, specifically using Fourier sums, we can construct operators, called **field operators**, that create and annihilate particles, but this time they don't create/annihilate particles in particular momentum states but instead they create/annihilate particles localized at particular *spatial locations*. Thus the operator $\hat{\psi}^{\dagger}(\boldsymbol{x})$ defined by

$$\hat{\psi}^{\dagger}(\boldsymbol{x}) = \frac{1}{\sqrt{\mathcal{V}}} \sum_{\boldsymbol{p}} \hat{a}_{\boldsymbol{p}}^{\dagger} e^{-i\boldsymbol{p}\cdot\boldsymbol{x}}, \qquad (4.7)$$

creates a particle at position $\boldsymbol{x}$, while $\hat{a}_{\boldsymbol{p}}^{\dagger}$ creates a particle in a state with three-momentum $\boldsymbol{p}$. Similarly, the operator $\hat{\psi}(\boldsymbol{x})$ defined by

$$\hat{\psi}(\boldsymbol{x}) = \frac{1}{\sqrt{\mathcal{V}}} \sum_{\boldsymbol{p}} \hat{a}_{\boldsymbol{p}} e^{i\boldsymbol{p}\cdot\boldsymbol{x}}, \qquad (4.8)$$

annihilates a particle at position $\boldsymbol{x}$, while $\hat{a}_{\boldsymbol{p}}$ annihilates a particle in a state with three-momentum $\boldsymbol{p}$. In summary:

[1] We retain the 'particle-in-a-box' normalization of Chapter 3. States are defined in a box of volume $\mathcal{V}$ (using periodic boundary conditions). For later convenience, we list some useful results using this normalization:

A state $|\alpha\rangle$ can be described in position coordinates $\psi_\alpha(\boldsymbol{x}) = \langle \boldsymbol{x}|\alpha\rangle$ or momentum coordinates $\tilde{\psi}_\alpha(\boldsymbol{p}) = \langle \boldsymbol{p}|\alpha\rangle$, and since

$$\langle \boldsymbol{p}|\alpha\rangle = \int \mathrm{d}^3 x \, \langle \boldsymbol{p}|\boldsymbol{x}\rangle\langle \boldsymbol{x}|\alpha\rangle, \qquad (4.1)$$

then the Fourier transform

$$\tilde{\psi}_\alpha(\boldsymbol{p}) = \frac{1}{\sqrt{\mathcal{V}}} \int \mathrm{d}^3 x \, e^{-i\boldsymbol{p}\cdot\boldsymbol{x}} \psi_\alpha(\boldsymbol{x}), \qquad (4.2)$$

implies that

$$\langle \boldsymbol{p}|\boldsymbol{x}\rangle = \frac{1}{\sqrt{\mathcal{V}}} e^{-i\boldsymbol{p}\cdot\boldsymbol{x}}. \qquad (4.3)$$

The inverse transform is

$$\psi_\alpha(\boldsymbol{x}) = \frac{1}{\sqrt{\mathcal{V}}} \sum_{\boldsymbol{p}} e^{i\boldsymbol{p}\cdot\boldsymbol{x}} \tilde{\psi}_\alpha(\boldsymbol{p}), \qquad (4.4)$$

since $\boldsymbol{p}$ takes discrete values. We also have

$$\int \mathrm{d}^3 x \, e^{-i\boldsymbol{p}\cdot\boldsymbol{x}} = \mathcal{V}\delta_{\boldsymbol{p},0} \qquad (4.5)$$

and

$$\frac{1}{\mathcal{V}} \sum_{\boldsymbol{p}} e^{i\boldsymbol{p}\cdot\boldsymbol{x}} = \delta^{(3)}(\boldsymbol{x}). \qquad (4.6)$$

> The field operators $\hat{\psi}^\dagger(\boldsymbol{x})$ and $\hat{\psi}(\boldsymbol{x})$ respectively create or destroy a single particle at a point $\boldsymbol{x}$.

The following examples explore some properties of field operators.

Example 4.1

We examine the state $|\Psi\rangle = \hat{\psi}^\dagger(\boldsymbol{x})|0\rangle$ and check that it does indeed correspond to a single particle at a particular location. Note that

$$|\Psi\rangle = \hat{\psi}^\dagger(\boldsymbol{x})|0\rangle = \frac{1}{\sqrt{\mathcal{V}}}\sum_{\boldsymbol{p}} e^{-i\boldsymbol{p}\cdot\boldsymbol{x}}\hat{a}_{\boldsymbol{p}}^\dagger|0\rangle. \tag{4.9}$$

To calculate the total number of particles in this state we can use the number operator $\hat{n}_{\boldsymbol{p}} = \hat{a}_{\boldsymbol{p}}^\dagger\hat{a}_{\boldsymbol{p}}$ (which measures the number of particles in state $\boldsymbol{p}$) and then sum over all momentum states. Consider

$$\sum_{\boldsymbol{q}} \hat{n}_{\boldsymbol{q}}|\Psi\rangle = \frac{1}{\sqrt{\mathcal{V}}}\sum_{\boldsymbol{q}\boldsymbol{p}}\hat{a}_{\boldsymbol{q}}^\dagger\hat{a}_{\boldsymbol{q}}\hat{a}_{\boldsymbol{p}}^\dagger|0\rangle e^{-i\boldsymbol{p}\cdot\boldsymbol{x}}, \tag{4.10}$$

and again using $\hat{a}_{\boldsymbol{q}}\hat{a}_{\boldsymbol{p}}^\dagger|0\rangle = \delta_{\boldsymbol{q}\boldsymbol{p}}|0\rangle$ we deduce that

$$\sum_{\boldsymbol{q}} \hat{n}_{\boldsymbol{q}}|\Psi\rangle = |\Psi\rangle, \tag{4.11}$$

and the state Ψ is then an eigenstate of the number operator with eigenvalue 1.

To find out exactly where the particle is created, first define $|\boldsymbol{p}\rangle = \hat{a}_{\boldsymbol{p}}^\dagger|0\rangle$ and then we can look at the projection of $|\Psi\rangle$ on another position eigenstate $|\boldsymbol{y}\rangle$:

$$\begin{aligned}
\langle\boldsymbol{y}|\Psi\rangle = \langle\boldsymbol{y}|\hat{\psi}^\dagger(\boldsymbol{x})|0\rangle &= \frac{1}{\sqrt{\mathcal{V}}}\sum_{\boldsymbol{p}} e^{-i\boldsymbol{p}\cdot\boldsymbol{x}}\langle\boldsymbol{y}|\boldsymbol{p}\rangle \\
&= \frac{1}{\mathcal{V}}\sum_{\boldsymbol{p}} e^{-i\boldsymbol{p}\cdot(\boldsymbol{x}-\boldsymbol{y})} \\
&= \delta^{(3)}(\boldsymbol{x}-\boldsymbol{y}),
\end{aligned} \tag{4.12}$$

showing that the single particle created by $\hat{\psi}^\dagger(\boldsymbol{x})$ may only be found at $\boldsymbol{y} = \boldsymbol{x}$.

We use the result

$$\langle\boldsymbol{y}|\boldsymbol{p}\rangle = \frac{1}{\sqrt{\mathcal{V}}}e^{i\boldsymbol{p}\cdot\boldsymbol{y}},$$

which follows from eqn 4.3 that states that the projection of a momentum eigenstate in position space is a plane wave. The last equality in eqn 4.12 follows directly from eqn 4.6.

Example 4.2

The field operators satisfy commutation relations. For example

$$\left[\hat{\psi}(\boldsymbol{x}),\hat{\psi}^\dagger(\boldsymbol{y})\right] = \frac{1}{\mathcal{V}}\sum_{\boldsymbol{p}\boldsymbol{q}} e^{i(\boldsymbol{p}\cdot\boldsymbol{x}-\boldsymbol{q}\cdot\boldsymbol{y})}[\hat{a}_{\boldsymbol{p}},\hat{a}_{\boldsymbol{q}}^\dagger], \tag{4.13}$$

and using $[\hat{a}_{\boldsymbol{p}},\hat{a}_{\boldsymbol{q}}^\dagger] = \delta_{\boldsymbol{p}\boldsymbol{q}}$ we have that

$$\left[\hat{\psi}(\boldsymbol{x}),\hat{\psi}^\dagger(\boldsymbol{y})\right] = \frac{1}{\mathcal{V}}\sum_{\boldsymbol{p}\boldsymbol{q}} e^{i(\boldsymbol{p}\cdot\boldsymbol{x}-\boldsymbol{q}\cdot\boldsymbol{y})}\delta_{\boldsymbol{p}\boldsymbol{q}} = \frac{1}{\mathcal{V}}\sum_{\boldsymbol{p}} e^{i\boldsymbol{p}\cdot(\boldsymbol{x}-\boldsymbol{y})} = \delta^{(3)}(\boldsymbol{x}-\boldsymbol{y}), \tag{4.14}$$

where the last equality follows from eqn 4.6. By the same method you can show

$$\left[\hat{\psi}^\dagger(\boldsymbol{x}),\hat{\psi}^\dagger(\boldsymbol{y})\right] = \left[\hat{\psi}(\boldsymbol{x}),\hat{\psi}(\boldsymbol{y})\right] = 0. \tag{4.15}$$

These results are for boson operators, but we can derive analogous results for fermion operators:

$$\begin{aligned}
\left\{\hat{\psi}(\boldsymbol{x}),\hat{\psi}^\dagger(\boldsymbol{y})\right\} &= \delta^{(3)}(\boldsymbol{x}-\boldsymbol{y}), \\
\left\{\hat{\psi}^\dagger(\boldsymbol{x}),\hat{\psi}^\dagger(\boldsymbol{y})\right\} &= \left\{\hat{\psi}(\boldsymbol{x}),\hat{\psi}(\boldsymbol{y})\right\} = 0. \tag{4.16}
\end{aligned}$$

4.2 How to second quantize an operator

How can we upgrade the operators that we had in first-quantized quantum mechanics into second-quantized operators? Let us first remind ourselves what an operator $\hat{A}$ does. It maps a quantum state $|\psi\rangle$ (which lives in a **Hilbert space**[2] and describes the quantum mechanical state of a single particle) into another state $\hat{A}|\psi\rangle$. We will focus in this section on $\hat{A}$ being a **single-particle operator**, an example of which would be $\hat{p}_1 = -\mathrm{i}\hbar\nabla = -\mathrm{i}\hbar\frac{\partial}{\partial x_1}$, which operates on a *single* coordinate x_1 that can only describe the position of one particle. Thus we will ignore for now objects like the Coulomb potential operator $\hat{V}(x_1, x_2) = \frac{q}{4\pi|x_1-x_2|}$ which depends on the coordinates of *two* particles. An operator that acts on n particles is called an n-**particle operator**.

The operator $\hat{A}$ can be described quite generally using a variety of coordinate systems, as can be seen if you insert two resolutions of the identity, $1 = \sum_\alpha |\alpha\rangle\langle\alpha|$ and $1 = \sum_\beta |\beta\rangle\langle\beta|$ into the right-hand side of the trivial identity $\hat{A} = \hat{A}$:

$$\hat{A} = \sum_{\alpha\beta} |\alpha\rangle\langle\alpha|\hat{A}|\beta\rangle\langle\beta| = \sum_{\alpha\beta} \mathcal{A}_{\alpha\beta}|\alpha\rangle\langle\beta|, \qquad (4.17)$$

where the matrix element $\mathcal{A}_{\alpha\beta} = \langle\alpha|\hat{A}|\beta\rangle$. This is all well and good, but quantum field theory is acted out on a larger stage than Hilbert space. We need a space that includes not only single-particle states, but the possibility of many-particle states. This is known as **Fock space**.[3]

Remarkably, the second-quantized many-body upgrade $\hat{A}$ of the single-particle operator $\hat{A}$ to Fock space is a very compact expression:

$$\boxed{\hat{A} = \sum_{\alpha\beta} \mathcal{A}_{\alpha\beta}\,\hat{a}_\alpha^\dagger\hat{a}_\beta.} \qquad (4.18)$$

This equation has the same single-particle matrix element $\mathcal{A}_{\alpha\beta} = \langle\alpha|\hat{A}|\beta\rangle$ that we had in eqn 4.17, even though it is now valid for the many-particle case. The interpretation of eqn 4.18 is beautifully simple. The operator $\hat{A}$ is a sum over all processes in which you use $\hat{a}_\beta$ to remove a single particle in state $|\beta\rangle$, multiply it by the matrix element $\mathcal{A}_{\alpha\beta}$, and then use $\hat{a}_\alpha^\dagger$ to place that particle into a final state $|\alpha\rangle$. This operator $\hat{A}$ thus describes all possible single-particle processes that can occur and operates on a many-particle state (see Fig. 4.1). The proof of eqn 4.18 is rather technical and is covered in the following example, which can be skipped if you are happy to take the result on trust.

Example 4.3

To prove eqn 4.18, we need to introduce some formalism. Let us first write down an expression for an N-particle state

$$|\psi_1, \ldots, \psi_N\rangle = \frac{1}{\sqrt{N!}} \sum_P \xi^P \prod_{i=1}^N |\psi_{P(i)}\rangle. \qquad (4.19)$$

[2] A Hilbert space is a complex vector space endowed with an inner product, whose elements look like $\sum_n a_n|\phi_n\rangle$ where a_n are complex numbers. It can thus describe particular superpositions of states such as, just for an example, $(|\phi_0\rangle + \mathrm{i}|\phi_1\rangle)/\sqrt{2}$ in which the system might equally be found in state $|\phi_0\rangle$ or $|\phi_1\rangle$, but the quantum amplitudes for being in each state are $\pi/2$ out of phase. This is a superposition of possibilities, but it only describes a single particle.

[3] Fock space $\mathcal{F}$ includes N-particle states for all possible values of N. Mathematically, one can write this as

$$\mathcal{F} = \bigoplus_{N=0}^\infty \mathcal{F}_N = \mathcal{F}_0 \oplus \mathcal{F}_1 \oplus \cdots,$$

where $\mathcal{F}_0 = \{|0\rangle\}$ is a set containing only the vacuum state and $\mathcal{F}_1$ is the single-particle Hilbert space. The subspaces $\mathcal{F}_{N\geq2}$, which describe N-particle states with $N \geq 2$, have to be symmetrized for bosons and antisymmetrized for fermions (so, for example, $\mathcal{F}_2$ is not simply $\mathcal{F}_1 \otimes \mathcal{F}_1$, but only the symmetric or antisymmetric part of $\mathcal{F}_1 \otimes \mathcal{F}_1$). Creation operators move an element of $\mathcal{F}_N$ into an element of $\mathcal{F}_{N+1}$, while annihilation operators perform the reverse manoeuvre. An element of $\mathcal{F}$ can describe a superposition of states with different particle numbers.

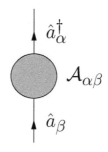

Fig. 4.1 A process represented by the operator in eqn 4.18.

Q						
5	5	0	1	2	3	4
4	4	5	0	1	2	3
3	3	4	5	0	1	2
2	2	3	4	5	0	1
1	1	2	3	4	5	0
0	0	1	2	3	4	5
	0	1	2	3	4	5
				P		

Fig. 4.2 For $N = 3$, there are $N! = 6$ permutations of $\{1, 2, 3\}$, which one could label by the integers $0, 1, \ldots, 5$. A sum over permutations P and over Q is equivalent to $N!$ copies of a sum over permutations $P' = P + Q$.

[4]The right-hand side of eqn 4.21 can be written $\det(\langle \chi_i | \psi_j \rangle)$ for fermions or $\mathrm{per}(\langle \chi_i | \psi_j \rangle)$ for bosons. The **determinant** of a matrix A_{ij} is

$$\det(A) = \sum_P (-1)^P \prod_{i=1}^N A_{iP(i)}.$$

The **permanent** of a matrix A_{ij} is

$$\mathrm{per}(A) = \sum_P \prod_{i=1}^N A_{iP(i)}.$$

Fig. 4.3 The action of a creation operator $\hat{a}_\phi^\dagger$ on a state $|\psi_1, \ldots, \psi_N\rangle$.

This is a sum over all $N!$ permutations of single-particle states $|\psi_i\rangle$. There are $N!$ permutations of the sequence of integers $i = 1, \ldots, N$ and $P(i)$ labels a permutation of this sequence. The factor $\xi = 1$ for bosons and $\xi = -1$ for fermions. For fermions, the notation ξ^P gives a plus or minus sign for an even or odd permutation respectively (and for bosons, of course, the factor can be ignored as it is always $+1$). The inner product of two such N-particles states is

$$\langle \chi_1, \ldots, \chi_N | \psi_1, \ldots, \psi_N \rangle = \frac{1}{N!} \sum_P \sum_Q \xi^{P+Q} \prod_{i=1}^N \langle \chi_{Q(i)} | \psi_{P(i)} \rangle. \quad (4.20)$$

We can rewrite this using $P' = P + Q$ which spans all the permutations $N!$ times, see Fig. 4.2, and hence

$$\langle \chi_1, \ldots, \chi_N | \psi_1, \ldots, \psi_N \rangle = \sum_{P'} \xi^{P'} \prod_{i=1}^N \langle \chi_i | \psi_{P'(i)} \rangle. \quad (4.21)$$

If desired, this rather complicated expression can be written rather compactly as a determinant (for fermions, or a 'permanent' for bosons).[4]

Let's now try using a creation operator $\hat{a}_\phi^\dagger$ to act on one of these N-particle states:

$$\hat{a}_\phi^\dagger | \psi_1, \ldots, \psi_N \rangle = | \phi, \psi_1, \ldots, \psi_N \rangle. \quad (4.22)$$

Thus the cuckoo lays its $|\phi\rangle$ egg in the $|\psi_1, \ldots, \psi_N\rangle$ nest, see Fig. 4.3. Annihilation is somewhat harder to formulate, mainly because you have to cater for the possibility that the state you are annihilating is some admixture of states already present. To proceed, it is helpful to realize that $\hat{a}_\phi$ is the Hermitian conjugate of $\hat{a}_\phi^\dagger$. Thus

$$
\begin{aligned}
\langle \chi_1, \ldots, \chi_{N-1} | \hat{a}_\phi | \psi_1, \ldots, \psi_N \rangle &= \langle \psi_1, \ldots, \psi_N | \hat{a}_\phi^\dagger | \chi_1, \ldots, \chi_{N-1} \rangle^* \quad (4.23) \\
&= \langle \psi_1, \ldots, \psi_N | \phi, \chi_1, \ldots, \chi_{N-1} \rangle^* \\
&= \begin{vmatrix} \langle \psi_1 | \phi \rangle & \langle \psi_1 | \chi_1 \rangle & \cdots & \langle \psi_1 | \chi_{N-1} \rangle \\ \vdots & \vdots & & \vdots \\ \langle \psi_N | \phi \rangle & \langle \psi_N | \chi_1 \rangle & \cdots & \langle \psi_N | \chi_{N-1} \rangle \end{vmatrix}_\xi^*.
\end{aligned}
$$

This last line has been obtained by recognizing that the expression to evaluate is a determinant (fermions) or permanent (bosons). It can be expanded as a series of minors:

$$
\langle \chi_1, \ldots, \chi_{N-1} | \hat{a}_\phi | \psi_1, \ldots, \psi_N \rangle
$$
$$
= \sum_{k=1}^N \xi^{k-1} \langle \psi_k | \phi \rangle^* \langle \psi_1, \ldots (\text{no } \psi_k) \ldots \psi_N | \chi_1, \ldots, \chi_{N-1} \rangle^* \quad (4.24)
$$
$$
= \sum_{k=1}^N \xi^{k-1} \langle \phi | \psi_k \rangle \langle \chi_1, \ldots, \chi_{N-1} | \psi_1, \ldots (\text{no } \psi_k) \ldots \psi_N \rangle.
$$

This is true for any premultiplying $\langle \chi_1, \ldots, \chi_{N-1} |$, and so

$$\hat{a}_\phi | \psi_1, \ldots, \psi_N \rangle = \sum_{k=1}^N \xi^{k-1} \langle \phi | \psi_k \rangle | \psi_1, \ldots (\text{no } \psi_k) \ldots \psi_N \rangle. \quad (4.25)$$

This means that a combination of annihilation of $|\beta\rangle$ and creation of $|\alpha\rangle$ gives

$$\hat{a}_\alpha^\dagger \hat{a}_\beta | \psi_1, \ldots, \psi_N \rangle = \sum_{k=1}^N \xi^{k-1} \langle \beta | \psi_k \rangle | \alpha, \psi_1, \ldots (\text{no } \psi_k) \ldots \psi_N \rangle, \quad (4.26)$$

but now we can simply translate the inserted state $|\alpha\rangle$ into the position formerly held by the annihilated state $|\psi_k\rangle$ and thereby remove the annoying ξ^{k-1} factor, since

$$\xi^{k-1} | \alpha, \psi_1, \ldots (\text{no } \psi_k) \ldots \psi_N \rangle = | \psi_1, \ldots \psi_{k-1}, \alpha, \psi_{k+1}, \ldots \psi_N \rangle \quad (4.27)$$

and hence

$$\hat{a}_\alpha^\dagger \hat{a}_\beta | \psi_1, \ldots, \psi_N \rangle = \sum_{k=1}^N \langle \beta | \psi_k \rangle | \psi_1, \ldots \psi_{k-1}, \alpha, \psi_{k+1}, \ldots \psi_N \rangle. \quad (4.28)$$

After all this effort, we are now ready to tackle the real problem. Given a state $|\psi_1, \ldots, \psi_N\rangle$ and a single-particle operator $\hat{A} = \sum_{\alpha\beta} A_{\alpha\beta} |\alpha\rangle\langle\beta|$, the obvious second-quantized upgrade is to let $\hat{A}$ act separately on each individual component state inside $|\psi_1, \ldots, \psi_N\rangle$ and then sum the result. For example, if $\hat{A}$ is the single-particle momentum operator, we would want $\hat{A}$ to be the operator for the total momentum, the sum of all the momenta of individual particles within the many-body state.

Let's make it very simple to start with and set $\hat{A} = |\alpha\rangle\langle\beta|$. In this case, for a single-particle state $|\psi_j\rangle$ we would have $\hat{A}|\psi_j\rangle = \langle\beta|\psi_j\rangle|\alpha\rangle$ and hence

$$\hat{A}|\psi_1 \ldots \psi_N\rangle = \sum_{k=1}^{N} \langle\beta|\psi_k\rangle|\psi_1, \ldots \psi_{k-1}, \alpha, \psi_{k+1}, \ldots \psi_N\rangle. \qquad (4.29)$$

But this is simply $\hat{a}_\alpha^\dagger \hat{a}_\beta |\psi_1, \ldots, \psi_N\rangle$ as we found in eqn 4.28. Hence $\hat{A} = \hat{a}_\alpha^\dagger \hat{a}_\beta$. In the general case of $\hat{A} = \sum_{\alpha\beta} A_{\alpha\beta}|\alpha\rangle\langle\beta|$, we easily then arrive at

$$\hat{A} = \sum_{\alpha\beta} A_{\alpha\beta}\, \hat{a}_\alpha^\dagger \hat{a}_\beta. \qquad (4.30)$$

After the hard work of the previous example, which justified eqn 4.18, we now have the power at our fingertips to do some splendidly elegant things.

Example 4.4

(i) The resolution of the identity $\hat{1} = \sum_\alpha |\alpha\rangle\langle\alpha|$ is a special case of $\hat{A} = \sum_{\alpha\beta} A_{\alpha\beta}|\alpha\rangle\langle\beta|$ with $A_{\alpha\beta} = \delta_{\alpha\beta}$. Its second-quantized upgrade is immediately

$$\hat{n} = \sum_\alpha \hat{a}_\alpha^\dagger \hat{a}_\alpha, \qquad (4.31)$$

the number operator, that counts one for every particle in the state on which it operates.

(ii) The usual momentum operator can be written $\hat{A} = \hat{p} = \sum_{\boldsymbol{p}} \boldsymbol{p}|\boldsymbol{p}\rangle\langle\boldsymbol{p}|$ and this upgrades immediately to

$$\hat{p} = \sum_{\boldsymbol{p}} \boldsymbol{p}\, \hat{a}_{\boldsymbol{p}}^\dagger \hat{a}_{\boldsymbol{p}} = \sum_{\boldsymbol{p}} \boldsymbol{p}\, \hat{n}_{\boldsymbol{p}}. \qquad (4.32)$$

In much the same way, a function of the momentum operator $\hat{A} = f(\boldsymbol{p})$ becomes

$$\hat{A} = \sum_{\boldsymbol{p}} f(\boldsymbol{p})\, \hat{a}_{\boldsymbol{p}}^\dagger \hat{a}_{\boldsymbol{p}} = \sum_{\boldsymbol{p}} f(\boldsymbol{p})\, \hat{n}_{\boldsymbol{p}}. \qquad (4.33)$$

One particular example of this is the free-particle Hamiltonian which one can write down immediately in second-quantized form as

$$\hat{H} = \sum_{\boldsymbol{p}} \frac{p^2}{2m}\, \hat{n}_{\boldsymbol{p}}. \qquad (4.34)$$

Equation 4.34 can be thought about as the simple diagram shown in Fig. 4.4. Since the occupation number states (with which we want to work) are eigenstates of $\hat{n}_{\boldsymbol{p}}$ then so is $\hat{H}$. We have found that the operator $\hat{H}$ is diagonal in the basis in which we're working. In fact, to diagonalize any Hamiltonian (which means to find the energies of the eigenstates) we simply express it in terms of number operators. Having the Hamiltonian in this form makes perfect sense. We're saying that the total energy in the system is given by the sum of the energies $p^2/2m$ of all of the particles.

(iii) What happens if our operator is a function of $\hat{\boldsymbol{x}}$, rather than $\hat{\boldsymbol{p}}$? Well, eqn 4.18 is valid even if our states $|\alpha\rangle$ and $|\beta\rangle$ are position states. Our creation and annihilation operators are simply field operators, and the sum over momentum values becomes an integral over space. We then can write down our second-quantized operator $\hat{V}$ as

$$\hat{V} \quad = \quad \int \mathrm{d}^3x\, \hat{\psi}^\dagger(\boldsymbol{x})V(\boldsymbol{x})\,\hat{\psi}(\boldsymbol{x}). \qquad (4.35)$$

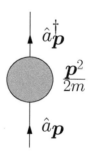

Fig. 4.4 A process represented by the Hamiltonian $\hat{H} = \sum_{\boldsymbol{p}} \frac{p^2}{2m}\hat{n}_{\boldsymbol{p}}$. Summing over all momentum states involves repeating this counting process which can be imagined as counting sheep crossing through a gate (annihilate a sheep on the left-hand side of the gate, count its energy in the sum, create the sheep again on the right-hand side of the gate, now move on to the next sheep $\ldots$).

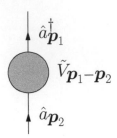

Fig. 4.5 A process represented by eqn 4.36. The sequence of events is: annihilate a particle with momentum $\boldsymbol{p}_2$, count the potential energy $\tilde{V}_{\boldsymbol{p}_1-\boldsymbol{p}_2}$, create a particle with momentum $\boldsymbol{p}_1$.

If we want, we can express this back in momentum space using eqns 4.7 and 4.8, and hence

$$\hat{V} = \frac{1}{\mathcal{V}} \int \mathrm{d}^3 x \sum_{\boldsymbol{p}_1 \boldsymbol{p}_2} \hat{a}^\dagger_{\boldsymbol{p}_1} \mathrm{e}^{-\mathrm{i}\boldsymbol{p}_1 \cdot \boldsymbol{x}} V(\hat{\boldsymbol{x}}) \hat{a}_{\boldsymbol{p}_2} \mathrm{e}^{\mathrm{i}\boldsymbol{p}_2 \cdot \boldsymbol{x}} = \sum_{\boldsymbol{p}_1 \boldsymbol{p}_2} \tilde{V}_{\boldsymbol{p}_1-\boldsymbol{p}_2} \hat{a}^\dagger_{\boldsymbol{p}_1} \hat{a}_{\boldsymbol{p}_2}, \qquad (4.36)$$

where $\tilde{V}_{\boldsymbol{p}} = \frac{1}{\mathcal{V}} \int \mathrm{d}^3 x \, V(\boldsymbol{x}) \mathrm{e}^{-\mathrm{i}\boldsymbol{p} \cdot \boldsymbol{x}}$ is the Fourier transform of $V(\boldsymbol{x})$. The operator $\hat{V}$ isn't diagonal, since the $\hat{a}$ operators create and annihilate particles with different momenta. We can think of this term as representing a process of an incoming particle with momentum $\boldsymbol{p}_2$ which interacts with a potential field (represented by $\tilde{V}_{\boldsymbol{p}_1-\boldsymbol{p}_2}$) and then leaves again with momentum $\boldsymbol{p}_1$ (see Fig. 4.5).

Example 4.5

Let's have a look at the influence of the potential on a simple system described by a Hamiltonian

$$\hat{H} = E_0 \sum_{\boldsymbol{p}} \hat{d}^\dagger_{\boldsymbol{p}} \hat{d}_{\boldsymbol{p}} - \frac{V}{2} \sum_{\boldsymbol{p}_1 \boldsymbol{p}_2} \hat{d}^\dagger_{\boldsymbol{p}_1} \hat{d}_{\boldsymbol{p}_2}. \qquad (4.37)$$

We'll further constrain the system to have three energy levels, so states are expressed using a basis $|n_{\boldsymbol{p}_1} n_{\boldsymbol{p}_2} n_{\boldsymbol{p}_3}\rangle$. For simplicity, let's first set $V = 0$. In this case, if we put a particle into the system then the three possible eigenstates of the system are $|100\rangle$, $|010\rangle$, $|001\rangle$. Now we put $V \neq 0$, and so

$$\frac{V}{2} \sum_{\boldsymbol{p}_1 \boldsymbol{p}_2} \hat{d}^\dagger_{\boldsymbol{p}_1} \hat{d}_{\boldsymbol{p}_2} |100\rangle = \frac{V}{2} \left(|100\rangle + |010\rangle + |001\rangle \right), \qquad (4.38)$$

and similarly for the other states (try them).

At this point it's easiest to slip into matrix notation, for which the entire Hamiltonian takes the form

$$H = \left[E_0 \begin{pmatrix} 1 & 0 & 0 \\ 0 & 1 & 0 \\ 0 & 0 & 1 \end{pmatrix} - \frac{V}{2} \begin{pmatrix} 1 & 1 & 1 \\ 1 & 1 & 1 \\ 1 & 1 & 1 \end{pmatrix} \right]. \qquad (4.39)$$

The eigenvalues for this equation are $\varepsilon = E_0, E_0, E_0 - \frac{3V}{2}$. The ground state of the system (called $|\Omega\rangle$) therefore has an energy $E_0 - \frac{3V}{2}$. The corresponding eigenstate is

$$|\Omega\rangle = \frac{1}{\sqrt{3}} \left(|100\rangle + |010\rangle + |001\rangle \right). \qquad (4.40)$$

We conclude that the ground state is a special 1:1:1 superposition of the three states we started with.

Example 4.6

What is the equivalent of the probability density of the wave function in second-quantized language? The answer is that the number density of particles at a point $\boldsymbol{x}$ is described by the **number density operator** $\hat{\rho}(\boldsymbol{x})$ given by

$$\begin{aligned} \hat{\rho}(\boldsymbol{x}) &= \hat{\psi}^\dagger(\boldsymbol{x}) \hat{\psi}(\boldsymbol{x}) \\ &= \frac{1}{\mathcal{V}} \sum_{\boldsymbol{p}_1 \boldsymbol{p}_2} \left[\mathrm{e}^{-\mathrm{i}(\boldsymbol{p}_1-\boldsymbol{p}_2) \cdot \boldsymbol{x}} \right] \hat{a}^\dagger_{\boldsymbol{p}_1} \hat{a}_{\boldsymbol{p}_2}. \end{aligned} \qquad (4.41)$$

This operator enables us to write the potential energy operator for a single particle in an external potential as

$$\hat{V} = \int \mathrm{d}^3 x \, V(\boldsymbol{x}) \hat{\rho}(\boldsymbol{x}). \qquad (4.42)$$

4.3   The kinetic energy and the tight-binding Hamiltonian

One application of this formalism is to the tight-binding Hamiltonian which is used in condensed matter physics. The basic idea behind this model is very straightforward: electrons don't like being confined. We know this from the well-known problem of a one-dimensional infinite square well with width L. There the kinetic part of the energy is given by $E_n = \frac{1}{2m}\left(\frac{n\pi}{L}\right)^2$. The smaller we make L, the larger the energy. Conversely, kinetic energy can be saved by allowing the electron to move in a larger volume. In the tight-binding model, one considers a lattice of fixed atoms with electrons moving between them (as shown in Fig. 4.6). These electrons can lower their kinetic energies by hopping from lattice site to lattice site. To deal with the discrete lattice in the model, we need to work in a basis where the fermion creation operator $\hat{c}_i^\dagger$ creates a particle at a particular lattice site labelled by i. The kinetic energy saving for a particle to hop between points j and i is called t_{ij}. Clearly t_{ij} will have some fundamental dependence on the overlap of atomic wave functions. The Hamiltonian $\hat{H}$ contains a sum over all processes in which an electron hops between sites, and so is a sum over pairs of sites:

$$\hat{H} = \sum_{ij}(-t_{ij})\hat{c}_i^\dagger\hat{c}_j. \tag{4.43}$$

Each term in this sum consists of processes in which we annihilate a particle at site j and create it again at site i (thus modelling a hopping process), thus saving energy t_{ij}. For now, we'll consider the simplest possible case of $t_{ij} = t$ for nearest neighbours and $t = 0$ otherwise. Thus we rewrite $\hat{H}$ as

$$\hat{H} = -t\sum_{i\tau}\hat{c}_i^\dagger\hat{c}_{i+\tau}, \tag{4.44}$$

where the sum over τ counts over nearest neighbours.

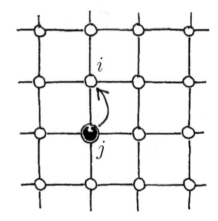

Fig. 4.6 The operator $\hat{c}_i^\dagger\hat{c}_j$ causing an electron to hop from site j to i. Such a process saves kinetic energy t_{ij}.

Example 4.7

The tight-binding Hamiltonian is not diagonal since we have a bilinear combination of operators that look like $\hat{c}_i^\dagger\hat{c}_j$. We need to diagonalize it and that can be achieved by making a change of basis (again, essentially using a Fourier transform). Thus

$$\hat{c}_i = \frac{1}{\sqrt{\mathcal{V}}}\sum_{k}\mathrm{e}^{\mathrm{i}\boldsymbol{k}\cdot\boldsymbol{r}_i}\hat{c}_{\boldsymbol{k}} \quad \text{and} \quad \hat{c}_i^\dagger = \frac{1}{\sqrt{\mathcal{V}}}\sum_{q}\mathrm{e}^{-\mathrm{i}\boldsymbol{q}\cdot\boldsymbol{r}_i}\hat{c}_{\boldsymbol{q}}^\dagger. \tag{4.45}$$

Substituting this into the Hamiltonian yields

$$\hat{H} = -\frac{t}{\mathcal{V}}\sum_{i\tau}\sum_{\boldsymbol{kq}}\mathrm{e}^{-\mathrm{i}(\boldsymbol{q}-\boldsymbol{k})\cdot\boldsymbol{r}_i+\mathrm{i}\boldsymbol{k}\cdot\boldsymbol{r}_\tau}\hat{c}_{\boldsymbol{q}}^\dagger\hat{c}_{\boldsymbol{k}}. \tag{4.46}$$

We play our usual trick and exploit the fact that $\frac{1}{\mathcal{V}}\sum_i\mathrm{e}^{-\mathrm{i}(\boldsymbol{q}-\boldsymbol{k})\cdot\boldsymbol{r}_i} = \delta_{\boldsymbol{qk}}$, which removes the sum over i. Lastly we do the sum over $\boldsymbol{q}$. After jumping through these familiar mathematical hoops we end up with

$$\hat{H} = -t\sum_{\boldsymbol{k}}\sum_{\tau}\mathrm{e}^{\mathrm{i}\boldsymbol{k}\cdot\boldsymbol{r}_\tau}\hat{c}_{\boldsymbol{k}}^\dagger\hat{c}_{\boldsymbol{k}}, \tag{4.47}$$

which is diagonal in momentum.

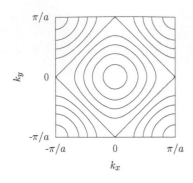

Fig. 4.7 The tight-binding model for a two-dimensional square lattice. Contours showing lines of constant energy $E_{\mathbf{k}}$ based on eqn 4.50 are plotted as a function of k_x and k_y. For $t > 0$, the lowest $E_{\mathbf{k}}$ occurs at $(k_x, k_y) = (0,0)$, the centre of this plot. For $t < 0$, the lowest $E_{\mathbf{k}}$ occurs at the corners, $(k_x, k_y) = (\pm\frac{\pi}{a}, \pm\frac{\pi}{a})$.

We can tidy up this result by writing the Hamiltonian as

$$\hat{H} = \sum_{\mathbf{k}} E_{\mathbf{k}} \hat{c}_{\mathbf{k}}^{\dagger} \hat{c}_{\mathbf{k}}, \tag{4.48}$$

where the dispersion $E_{\mathbf{k}}$ is given by

$$E_{\mathbf{k}} = -\sum_{\tau} t \, \mathrm{e}^{\mathrm{i}\mathbf{k}\cdot\mathbf{r}_{\tau}}. \tag{4.49}$$

Example 4.8

For a two-dimensional square lattice, the sum over τ counts over vectors $(a, 0)$, $(-a, 0)$, $(0, a)$ and $(0, -a)$. This leads to a dispersion relation (see also Fig. 4.7)

$$E_{\mathbf{k}} = -2t(\cos k_x a + \cos k_y a). \tag{4.50}$$

4.4 Two particles

We have been focussing on single-particle operators. What happens for the more interesting case of two particles interacting with each other? From the discussion above we can guess (quite correctly) that a second-quantized two-particle operator $\hat{A}$ is given by

$$\hat{A} = \sum_{\alpha\beta\gamma\delta} \mathcal{A}_{\alpha\beta\gamma\delta} \, \hat{a}_{\alpha}^{\dagger} \hat{a}_{\beta}^{\dagger} \hat{a}_{\gamma} \hat{a}_{\delta}, \tag{4.51}$$

where $\mathcal{A}_{\alpha\beta\gamma\delta} = \langle \alpha, \beta | \hat{\mathcal{A}} | \gamma, \delta \rangle$ is the matrix element for the Hilbert-space version. This expression embodies the idea that two-particle operators involve a sum over all processes involving the annihilation of two particles in particular states, multiplication by the matrix element, and creation of two particles in two new states. We will frequently encounter two-particle operators $\hat{V}$ written as a function of spatial coordinates, and the corresponding expression involving field operators is

$$\hat{V} = \frac{1}{2} \int \mathrm{d}^3 x \, \mathrm{d}^3 y \, \hat{\psi}^{\dagger}(\mathbf{x}) \hat{\psi}^{\dagger}(\mathbf{y}) V(\mathbf{x}, \mathbf{y}) \hat{\psi}(\mathbf{y}) \hat{\psi}(\mathbf{x}). \tag{4.52}$$

The factor of $\frac{1}{2}$ ensures that we don't double count the interactions. Notice the way we have arranged the operators: we put the creation operators to the left and the annihilation operators to the right, a convention which is known as **normal ordering**. Why is this important? Normal ordering makes the evaluation of matrix elements much easier and in particular it makes sure that our operator, $\hat{V}$, has zero vacuum expectation value:

$$\langle 0 | \hat{V} | 0 \rangle = 0. \tag{4.53}$$

This avoids us to having to blush over embarrassing infinite self-energy terms that can otherwise creep into our answers. With creation operators to the left and annihilation operators to the right we make sure that $\langle 0|\hat{V}|0\rangle$ is fixed at zero.

An additional convention we have adopted in eqn 4.52 is that reading the coordinates from right-to-left gives us x, y, y, x. (This can be remembered as the dance step: left-in, right-in, right-out, left-out.) For bosons the order would not matter, but fermions anticommute and so for example interchanging the first two creation operators would introduce a minus sign.

It is useful to see what would happen if we hadn't introduced normal ordering. For example, an alternative guess for the form of the two-particle interaction might have been

$$\hat{V}_{\text{wrong}} = \frac{1}{2}\int d^3x\, d^3y\, V(\boldsymbol{x},\boldsymbol{y})\hat{\rho}(\boldsymbol{y})\hat{\rho}(\boldsymbol{x}). \qquad (4.54)$$

This form is wrong and, because we've left out the all-important normal ordering, will lead to an extra self-energy term, as shown in the following example.

Example 4.9

We will consider the case of fermions and will need the anticommutation relations $\left\{\hat{\psi}(\boldsymbol{x}),\hat{\psi}^\dagger(\boldsymbol{y})\right\} = \delta^{(3)}(\boldsymbol{x}-\boldsymbol{y})$ and $\left\{\hat{\psi}(\boldsymbol{x}),\hat{\psi}(\boldsymbol{y})\right\} = 0$. Now consider the operator $\hat{\rho}(\boldsymbol{x})\hat{\rho}(\boldsymbol{y})$, which is given by

$$\hat{\rho}(\boldsymbol{x})\hat{\rho}(\boldsymbol{y}) = \hat{\psi}^\dagger(\boldsymbol{x})\hat{\psi}(\boldsymbol{x})\hat{\psi}^\dagger(\boldsymbol{y})\hat{\psi}(\boldsymbol{y}). \qquad (4.55)$$

Notice that this is certainly not normal ordered, but we can make it so as follows. Swap the second and third operators. This involves using an anticommutator. We obtain

$$\begin{aligned}
\hat{\rho}(\boldsymbol{x})\hat{\rho}(\boldsymbol{y}) &= -\hat{\psi}^\dagger(\boldsymbol{x})\hat{\psi}^\dagger(\boldsymbol{y})\hat{\psi}(\boldsymbol{x})\hat{\psi}(\boldsymbol{y}) + \delta^{(3)}(\boldsymbol{x}-\boldsymbol{y})\hat{\psi}^\dagger(\boldsymbol{x})\hat{\psi}(\boldsymbol{y}) \\
&= \hat{\psi}^\dagger(\boldsymbol{x})\hat{\psi}^\dagger(\boldsymbol{y})\hat{\psi}(\boldsymbol{y})\hat{\psi}(\boldsymbol{x}) + \delta^{(3)}(\boldsymbol{x}-\boldsymbol{y})\hat{\psi}^\dagger(\boldsymbol{x})\hat{\psi}(\boldsymbol{y}),
\end{aligned} \qquad (4.56)$$

where, in the last line, we've made another swap (this time of the two right-most operators in the first term, involving a second sign change). We are left with a normal ordered term plus a term involving a delta function. Comparing this to the correct expression for the two-particle interaction, we have

$$\hat{V}_{\text{wrong}} = \hat{V} + \frac{1}{2}\int d^3x\, V(\boldsymbol{x},\boldsymbol{x})\hat{\psi}^\dagger(\boldsymbol{x})\hat{\psi}(\boldsymbol{x}). \qquad (4.57)$$

The 'wrong' form of the potential energy includes an unwanted self-energy term. Normal ordering ensures we dispose of this abomination.

As before, we continue the procedure of second quantization by substituting the mode expansion into our equation

$$\begin{aligned}
\hat{V} &= \frac{1}{2}\int d^3x\, d^3y\, \hat{\psi}^\dagger(\boldsymbol{x})\hat{\psi}^\dagger(\boldsymbol{y})V(\boldsymbol{x},\boldsymbol{y})\hat{\psi}(\boldsymbol{y})\hat{\psi}(\boldsymbol{x}) \\
&= \frac{1}{2\mathcal{V}^2}\int d^3x\, d^3y\, \sum_{\boldsymbol{p}_1\boldsymbol{p}_2\boldsymbol{p}_3\boldsymbol{p}_4} e^{i(-\boldsymbol{p}_1\cdot\boldsymbol{x}-\boldsymbol{p}_2\cdot\boldsymbol{y}+\boldsymbol{p}_3\cdot\boldsymbol{y}+\boldsymbol{p}_4\cdot\boldsymbol{x})}\hat{a}^\dagger_{\boldsymbol{p}_1}\hat{a}^\dagger_{\boldsymbol{p}_2}V(\boldsymbol{x}-\boldsymbol{y})\hat{a}_{\boldsymbol{p}_3}\hat{a}_{\boldsymbol{p}_4}.
\end{aligned} \qquad (4.58)$$

At this point we restrict the potential to the form $V(\boldsymbol{x},\boldsymbol{y}) = V(\boldsymbol{x}-\boldsymbol{y})$, so that it only depends on the relative separation of the particles. This will guarantee that momentum is conserved in the interaction.

Example 4.10

To deal with this, we define $z = x - y$ and eliminate x to obtain

$$\hat{V} = \frac{1}{2\mathcal{V}^2} \sum_{p_1 p_2 p_3 p_4} \hat{a}^\dagger_{p_1} \hat{a}^\dagger_{p_2} \hat{a}_{p_3} \hat{a}_{p_4} \int \mathrm{d}^3 z\, V(z) \mathrm{e}^{\mathrm{i}(p_4 - p_1)\cdot z} \int \mathrm{d}^3 y\, \mathrm{e}^{\mathrm{i}(-p_1 - p_2 + p_3 + p_4)\cdot y}.$$

(4.59)

The last integral gives us a Kronecker delta $\mathcal{V}\delta_{p_1 + p_2 - p_3, p_4}$, which can be used to eat up one of the momentum sums (the p_4 one here) and sets $p_4 = p_1 + p_2 - p_3$. Now we have

$$\frac{1}{2\mathcal{V}} \sum_{p_1 p_2 p_3} \hat{a}^\dagger_{p_1} \hat{a}^\dagger_{p_2} \hat{a}_{p_3} \hat{a}_{p_1 + p_2 - p_3} \int \mathrm{d}^3 z\, V(z) \mathrm{e}^{-\mathrm{i}(p_3 - p_2)\cdot z}.$$

(4.60)

The last integral is the Fourier transform of the potential

$$\frac{1}{\mathcal{V}} \int \mathrm{d}^3 z\, V(z) \mathrm{e}^{-\mathrm{i}(p_3 - p_2)\cdot z} = \tilde{V}_{p_3 - p_2}.$$

(4.61)

Now for some interpretation: set $q = p_3 - p_2$ and eliminate p_3, yielding

$$\hat{V} = \frac{1}{2} \sum_{p_1 p_2 q} \tilde{V}_q \hat{a}^\dagger_{p_1} \hat{a}^\dagger_{p_2} \hat{a}_{p_2 + q} \hat{a}_{p_1 - q}.$$

(4.62)

Lastly we reindex the sum subtracting q from p_2 and adding it to p_1.

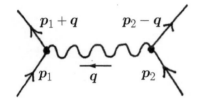

Fig. 4.8 A Feynman diagram for the process described in the text. Interpreting this diagram, one can think of time increasing in the vertical direction.

The result of these index acrobatics is a two-particle interaction potential that looks like

$$\hat{V} = \frac{1}{2} \sum_{p_1 p_2 q} \tilde{V}_q \hat{a}^\dagger_{p_1 + q} \hat{a}^\dagger_{p_2 - q} \hat{a}_{p_2} \hat{a}_{p_1}.$$

(4.63)

This can be interpreted as a scattering problem in momentum space. This scattering is illustrated in a cartoon known as a *Feynman diagram*. The diagram in Fig. 4.8 is translated into a story as follows: A particle comes in with momentum p_2. It sends out a force-carrying particle with momentum q, reducing its final momentum to $p_2 - q$. The force-carrying particle is absorbed by a second particle, which originally had momentum p_1 and ends up with a final momentum of $p_1 + q$. Notice we conserve momentum at the vertices of the diagrams.

Our aim is going to be to turn all of our combinations of operators into number operators. This makes evaluating the energy a simple matter. The first thing to notice about eqn 4.63 is that it is not diagonal. In general, problems involving the potential energy (as in eqn 4.63) cannot be solved exactly. Despite this, the two-particle potential leads to a remarkably rich seam of physics.[5]

4.5 The Hubbard model

As this chapter's final example of a second-quantized Hamiltonian, we turn to the **Hubbard model**.[6] It is difficult to overemphasize the importance of this model in solid state physics, due to the fact that it captures the essential physics of the competition between kinetic energy (favouring electrons to become delocalized) and potential energy (favouring electrons to become localized). Though relatively easy to state, the

[5]As we will see, not only does this tell us about seemingly pedestrian matters concerning the energies of electrons in metals, but it also describes interactions that lead to superfluidity, superconductivity, magnetism, charge density waves and pretty much every other phenomenon you've ever encountered in condensed matter physics. Of course, the fact that eqn 4.63 cannot generally be diagonalized should give us pause. We need a programme to turn the two-body potential diagonal. As we will see in Part X, we do this in two stages: firstly we massage our Hamiltonian until it's made up of bilinear combinations of operators. We may then make transformations until we have number operators.

[6]The Hubbard model is named after John Hubbard (1931–1980).

Hubbard model is surprisingly complicated to solve and exact results are only available in certain scenarios. The Hubbard model is based on a tight-binding Hamiltonian (to model the kinetic energy) plus a two-body potential energy term (to model electron–electron interactions):

$$\hat{H} = \sum_{ij}(-t_{ij})\hat{c}_i^\dagger \hat{c}_j + \frac{1}{2}\sum_{ijkl}\hat{c}_i^\dagger \hat{c}_j^\dagger V_{ijkl}\hat{c}_k \hat{c}_l. \qquad (4.64)$$

Recognizing that electrons carry a spin σ which can point up ($|\uparrow\rangle$) or down ($|\downarrow\rangle$) this becomes

$$\hat{H} = \sum_{ij\sigma}(-t_{ij})\hat{c}_{i\sigma}^\dagger \hat{c}_{j\sigma} + \frac{1}{2}\sum_{ijkl\sigma\sigma'}\hat{c}_{i\sigma}^\dagger \hat{c}_{j\sigma'}^\dagger V_{ijkl}\hat{c}_{k\sigma'}\hat{c}_{l\sigma}. \qquad (4.65)$$

We assume in this model that no spin-flip processes can occur; thus spins never flip from up-to-down or vice versa on hopping. However, the interaction between electrons is based upon the Coulomb interaction between their charges and so electrons with opposite spin will interact with each other just as much as electrons with the same spins. Now for Hubbard's model: *the Coulomb interaction is assumed to be significant only between two electrons that are on the same site.* These electrons interact via a constant potential energy $U = V_{iiii}$. However, the Pauli exclusion principle ensures that should two spins hop into the same site they must have different spins, and this is where spin *does* make a difference. The Hamiltonian for the Hubbard model is then

$$\hat{H} = \sum_{ij\sigma}(-t_{ij})\hat{c}_{i\sigma}^\dagger \hat{c}_{j\sigma} + U\sum_{i}\hat{n}_{i\uparrow}\hat{n}_{i\downarrow}, \qquad (4.66)$$

which looks simple, but the eigenstates tend to be complex and highly correlated. It is often only necessary to consider hopping elements for nearest neighbours.

Example 4.11

Consider the Hubbard model with nearest-neighbour hopping only, and to make things as simple as possible, consider a lattice of only two sites. If, in addition, we only have one, spin-up electron, then that single electron can either be on the first site (a configuration we will denote as $|\uparrow, 0\rangle$, meaning 'spin-up electron on site 1, nothing on site 2') or on the second site (a configuration we will denote as $|0,\uparrow\rangle$). Hence a general state can be written as the superposition $|\psi\rangle = a|\uparrow, 0\rangle + b|0,\uparrow\rangle$, where a and b are complex numbers. In this case, there's no possibility of potential terms (because there is only one electron) and in this basis the Hubbard Hamiltonian can be written

$$\hat{H} = \begin{pmatrix} 0 & -t \\ -t & 0 \end{pmatrix}. \qquad (4.67)$$

Solving gives us a ground state $|\psi\rangle = \frac{1}{\sqrt{2}}(|\uparrow, 0\rangle + |0,\uparrow\rangle)$ with energy $E = -t$ and an excited state $|\psi\rangle = \frac{1}{\sqrt{2}}(|\uparrow, 0\rangle - |0,\uparrow\rangle)$ with energy $E = t$.

Now we'll have two electrons in the system. If they have the same spin the answer's simple and $E = 0$ (try it). If the electrons have different spins we write a general state as $|\psi\rangle = a|\uparrow\downarrow, 0\rangle + b|\uparrow,\downarrow\rangle + c|\downarrow,\uparrow\rangle + d|0,\uparrow\downarrow\rangle$ and in this basis the Hubbard Hamiltonian is

$$\hat{H} = \begin{pmatrix} U & -t & t & 0 \\ -t & 0 & 0 & -t \\ t & 0 & 0 & t \\ 0 & -t & t & U \end{pmatrix}. \qquad (4.68)$$

This can be diagonalized to give a ground state energy of $E = \frac{U}{2} - \frac{1}{2}\left(U^2 + 16t^2\right)^{\frac{1}{2}}$ which corresponds to a wave function $|\psi\rangle = N\left(|\uparrow\downarrow, 0\rangle + W|\uparrow,\downarrow\rangle - W|\downarrow,\uparrow\rangle + |0,\uparrow\downarrow\rangle\right)$ with N some normalization constant and $W = \frac{U}{4t} + \frac{1}{4t}\left(U^2 + 16t^2\right)^{\frac{1}{2}}$.

In this chapter we have considered models on discrete lattices which are relevant to condensed matter physics. In the following chapter, we will consider how to generalize this approach to the continuum. However, readers who are eager to see some uses of the machinery developed in this chapter should now be able to tackle the first parts of the chapters in Part X, which deal with some applications from condensed matter physics.

Chapter summary

- Field operators $\hat{\psi}^\dagger$ and $\hat{\psi}$ are the analogues of creation and annihilation operators in position space.
- Second quantization of models can be carried out by writing down operators as products of creation and annihilation operators that encode the physical particle processes. We write these products in normal-ordered form (meaning the field creation operators are written to the left and the field annihilation operators to the right).
- The vacuum expectation of normal-ordered products is always zero.
- Examples of this technique have been given for the tight-binding model and the Hubbard model.

Exercises

(4.1) We can define a generalized commutator $[\hat{A}, \hat{B}]_\zeta$ as
$$[\hat{A}, \hat{B}]_\zeta = \hat{A}\hat{B} - \zeta\hat{B}\hat{A}, \qquad (4.69)$$
so that $\zeta = 1$ yields $[\hat{A}, \hat{B}]_\zeta = [\hat{A}, \hat{B}]$ for bosons and $\zeta = -1$ yields $[\hat{A}, \hat{B}]_\zeta = \{\hat{A}, \hat{B}\}$ for fermions. The generalized commutation relations can then be written
$$[\hat{\psi}(\boldsymbol{x}), \hat{\psi}^\dagger(\boldsymbol{y})]_\zeta = \delta^{(3)}(\boldsymbol{x} - \boldsymbol{y}), \qquad (4.70)$$
and
$$[\hat{\psi}(\boldsymbol{x}), \hat{\psi}(\boldsymbol{y})]_\zeta = 0. \qquad (4.71)$$
Repeat the argument given in Example 4.9 using the generalized commutator and show that $\hat{V}_{\text{wrong}}$ yields the same result for both bosons and fermions.

(4.2) One can define a **single-particle density matrix** as
$$\hat{\rho}_1(\boldsymbol{x} - \boldsymbol{y}) = \langle\hat{\psi}^\dagger(\boldsymbol{x})\hat{\psi}(\boldsymbol{y})\rangle. \qquad (4.72)$$
By substituting the expansion of $\hat{\psi}^\dagger(\boldsymbol{x})$ and $\hat{\psi}(\boldsymbol{y})$ in terms of $\hat{a}_{\boldsymbol{q}}^\dagger$ and $\hat{a}_{\boldsymbol{q}}$ into this expression, show that
$$\hat{\rho}_1(\boldsymbol{x} - \boldsymbol{y}) = \frac{1}{\mathcal{V}}\sum_{\boldsymbol{p}\boldsymbol{q}} e^{-i(\boldsymbol{q}\cdot\boldsymbol{x} - \boldsymbol{p}\cdot\boldsymbol{y})}\langle\hat{a}_{\boldsymbol{q}}^\dagger\hat{a}_{\boldsymbol{p}}\rangle. \qquad (4.73)$$

(4.3) Evaluate the eigenvalues and eigenvectors of the Hubbard Hamiltonian given in eqn 4.68 in Example 4.11.
(a) What are the energy eigenvalues in the limit $t = 0$?
(b) How do these energy levels change as $t \neq 0$ in the limit $t/U \ll 1$?

Part II

Writing down Lagrangians

An important step in writing down a physical model is to construct the appropriate Lagrangian. This part of the book is concerned with how to do that and is structured as follows:

- In Chapter 5 we describe how the arguments in the previous part which worked on discrete systems can be generalized to the *continuum limit*. After reviewing Hamilton's formulation of classical mechanics and Poisson brackets, we shift our attention from the Lagrangian L to the *Lagrangian density* $\mathcal{L}$. We use the electromagnetic field as a first example of this approach.

- We have our first stab at constructing a *relativistic* quantum wave equation, the *Klein–Gordon equation* in Chapter 6. This turns out to have some unsavoury characteristics that means that it is not the right equation to describe electrons, but it is nevertheless illuminating and illustrates some of the issues we are going to come across later.

- In Chapter 7 we present a set of example Lagrangians and show how they work. After reading this chapter the reader should have a good working knowledge of the simplest examples of quantum field theories.

5 Continuous systems

In the last few chapters we have seen examples of how various systems can be reduced to a set of simple harmonic oscillators, each one describing a different normal mode of the system. Each normal mode may be thought of as a momentum state for non-interacting, identical particles, with the number of particles corresponding to the number of quantized excitations of that mode. We have defined operators that have created or annihilated these particles. In the previous chapter, we particularly focussed on discrete models (tight-binding, Hubbard, etc.) in which particles moved around on a lattice. In this chapter we want to extend this notion to the **continuum limit**, that is when the discretization of the lattice can be ignored. This will lead to the concept of a classical field. Before we scale those dizzy heights, there is some unfinished business from classical mechanics to deal with, thereby introducing some useful formalism.

5.1 Lagrangians and Hamiltonians

In this section, we return to the Lagrangian formulation of classical mechanics and introduce the Hamiltonian and the Poisson bracket. The Hamiltonian will give us an expression for the energy which appears as a conserved quantity if the Lagrangian doesn't change with time. The Poisson bracket is also telling us about how quantities are conserved. Let's begin by considering the Lagrangian $L(q_i, \dot{q}_i)$ for a system described by coordinates q_i. The rate of change of L is given by

$$\frac{\mathrm{d}L}{\mathrm{d}t} = \frac{\partial L}{\partial q_i}\dot{q}_i + \frac{\partial L}{\partial \dot{q}_i}\ddot{q}_i, \tag{5.1}$$

where the summation convention is assumed. Using the Euler–Lagrange equation we can turn this into

$$\frac{\mathrm{d}L}{\mathrm{d}t} = \frac{\mathrm{d}}{\mathrm{d}t}\left(\frac{\partial L}{\partial \dot{q}_i}\right)\dot{q}_i + \frac{\partial L}{\partial \dot{q}_i}\ddot{q}_i = \frac{\mathrm{d}}{\mathrm{d}t}\left(\frac{\partial L}{\partial \dot{q}_i}\dot{q}_i\right). \tag{5.2}$$

We define the **canonical momentum** by

$$p_i = \frac{\partial L}{\partial \dot{q}_i}, \tag{5.3}$$

and then eqn 5.2 can be written

$$\frac{\mathrm{d}}{\mathrm{d}t}(p_i\dot{q}_i - L) = 0. \tag{5.4}$$

This is clearly a conservation law, and we write it as

$$\frac{\mathrm{d}H}{\mathrm{d}t} = 0, \tag{5.5}$$

thereby defining the **Hamiltonian** H by

$$H = p_i \dot{q}_i - L. \tag{5.6}$$

We will see later that the Hamiltonian corresponds to the energy of the system. Since H is conserved it is instructive to look at variations in H. Thus

$$\delta H = p_i \delta \dot{q}_i + \delta p_i \dot{q}_i - \frac{\partial L}{\partial q_i} \delta q_i - \frac{\partial L}{\partial \dot{q}_i} \delta \dot{q}_i. \tag{5.7}$$

The first and fourth terms in this expression cancel because of eqn 5.3, and so

$$\delta H = \delta p_i \dot{q}_i - \frac{\partial L}{\partial q_i} \delta q_i. \tag{5.8}$$

Since H is a function of q_i and p_i, we expect that

$$\delta H = \frac{\partial H}{\partial q_i} \delta q_i + \frac{\partial H}{\partial p_i} \delta p_i, \tag{5.9}$$

and so matching terms in eqns 5.8 and 5.9 produces $\dot{q}_i = \partial H / \partial p_i$ and $\partial H / \partial q_i = -\partial L / \partial q_i = -\dot{p}_i$, and these two equalities are known as **Hamilton's equations**:

Note that because $p_i = (\partial L / \partial \dot{q}_i)$ then $\dot{p}_i = \mathrm{d}/\mathrm{d}t(\partial L / \partial \dot{q}_i) = \partial L / \partial q_i$, where the last equality follows from the Euler–Lagrange equation.

$$\frac{\partial H}{\partial p_i} = \dot{q}_i, \qquad \frac{\partial H}{\partial q_i} = -\dot{p}_i. \tag{5.10}$$

Hamilton's equations represent another way to find the equations of motion of a system.

We now turn to the other piece of nineteenth century classical mechanics which we want to review. We define the **Poisson bracket** $\{A, B\}_{\mathrm{PB}}$ by

Siméon Denis Poisson (1781–1840)

$$\{A, B\}_{\mathrm{PB}} = \frac{\partial A}{\partial q_i} \frac{\partial B}{\partial p_i} - \frac{\partial A}{\partial p_i} \frac{\partial B}{\partial q_i}. \tag{5.11}$$

Now consider any function F which depends on the coordinates q_i and p_i. The rate of change of F is given by

$$\frac{\mathrm{d}F}{\mathrm{d}t} = \frac{\partial F}{\partial t} + \frac{\partial F}{\partial q_i} \dot{q}_i + \frac{\partial F}{\partial p_i} \dot{p}_i = \frac{\partial F}{\partial t} + \{F, H\}_{\mathrm{PB}}, \tag{5.12}$$

where the last equality is achieved using Hamilton's equations. Thus if the field is not itself a function of time, we have

$$\frac{\mathrm{d}F}{\mathrm{d}t} = \{F, H\}_{\mathrm{PB}}, \tag{5.13}$$

so that if $\{F, H\}_{\mathrm{PB}} = 0$ then F is a constant of the motion. We have therefore found a link between a conservation law (F is a conserved quantity) with the Poisson bracket of the Hamiltonian and F being zero. This is highly evocative of the commutator in quantum mechanics, but note that this result is entirely classical!

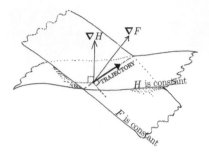

Fig. 5.1 The trajectory of the system moves on the intersection of surfaces of constant H and F. This is because $(\dot{q}, \dot{p})$ is perpendicular to both $\nabla F \equiv \left(\frac{\partial F}{\partial q}, \frac{\partial F}{\partial p}\right)$ (eqn 5.14) and $\nabla H \equiv \left(\frac{\partial H}{\partial q}, \frac{\partial H}{\partial p}\right)$ (eqn 5.15).

Example 5.1

There is an interesting graphical interpretation of this result of a conserved F (see Fig. 5.1). If $F = F(q, p)$, then the vector $(\dot{q}, \dot{p}) = (\partial H/\partial p, -\partial H/\partial q)$ is tangent to the surface $F(q, p) = $ constant, because

$$\nabla F \cdot (\dot{q}, \dot{p}) = \left(\frac{\partial F}{\partial q}, \frac{\partial F}{\partial p}\right) \cdot (\dot{q}, \dot{p}) = \frac{\partial F}{\partial q_i}\frac{\partial H}{\partial p_i} - \frac{\partial F}{\partial p_i}\frac{\partial H}{\partial q_i} = \{F, H\}_{\text{PB}} = 0, \quad (5.14)$$

and $(\dot{q}, \dot{p})$ is also tangent to the surface $H(q, p) = $ constant, because

$$\nabla H \cdot (\dot{q}, \dot{p}) = \left(\frac{\partial H}{\partial q}, \frac{\partial H}{\partial p}\right) \cdot (\dot{q}, \dot{p}) = -\dot{p} \cdot \dot{q} + \dot{q} \cdot \dot{p} = 0, \quad (5.15)$$

and so the trajectory of the system moves on the intersection of these two surfaces.

In quantum mechanics, the rate of change of the expectation value of an operator $\hat{F}$ is given by

$$\frac{d\langle \hat{F}\rangle}{dt} = \frac{1}{i\hbar}\langle [\hat{F}, \hat{H}]\rangle, \quad (5.16)$$

and since this looks very much like the classical version

$$\frac{dF}{dt} = \{F, H\}_{\text{PB}}, \quad (5.17)$$

we are led to visualize the classical to quantum crossover as involving the replacement

$$\{F, H\}_{\text{PB}} \to \frac{1}{i\hbar}\langle [\hat{F}, \hat{H}]\rangle, \quad (5.18)$$

and in fact for any Poisson bracket

$$\{A, B\}_{\text{PB}} \to \frac{1}{i\hbar}\langle [\hat{A}, \hat{B}]\rangle. \quad (5.19)$$

Example 5.2

The Poisson bracket between position and momentum coordinates is

$$\begin{aligned}\{q_j, p_k\}_{\text{PB}} &= \frac{\partial q_j}{\partial q_i}\frac{\partial p_k}{\partial p_i} - \frac{\partial p_k}{\partial q_i}\frac{\partial q_j}{\partial p_i} \\ &= \delta_{ij}\delta_{ik} - 0 \times 0 \\ &= \delta_{jk}.\end{aligned} \quad (5.20)$$

Using eqn 5.19 we have $[\hat{q}_j, \hat{p}_k] = i\hbar\delta_{jk}$, familiar from quantum mechanics.

5.2 A charged particle in an electromagnetic field

To practise some of the ideas in the previous section, we will consider a very simple example: a free particle of mass m. Shortly we will give it a charge q and turn on an electromagnetic field, but for now we will keep all electromagnetic fields switched off. We want to include relativity, and so therefore the action S must be a Lorentz-invariant scalar. We can write $S = \int_{\tau_1}^{\tau_2} L\gamma\,\mathrm{d}\tau$ where τ is the proper time, and so γL must be a Lorentz invariant. We can thus write $L = \text{constant}/\gamma$ and we will choose

$$L = -\frac{mc^2}{\gamma}, \qquad (5.21)$$

where the constant mc^2 is chosen so that L has the correct low-velocity limit.[1] The action S can now be written

$$S = \int_{\tau_1}^{\tau_2} -mc^2\mathrm{d}\tau = -mc\int_a^b \mathrm{d}s, \qquad (5.22)$$

where the interval[2] $\mathrm{d}s = \sqrt{c^2\mathrm{d}t^2 - \mathrm{d}x^2 - \mathrm{d}y^2 - \mathrm{d}z^2}$. By the principle of least action $\delta S = 0$ and so $\delta\int_a^b \mathrm{d}s = 0$. The integral $\int_a^b \mathrm{d}s$ takes its maximum value (see Fig. 5.2) along a straight world-line (see Exercise 5.5) and so this implies reassuringly that free particles move along straight lines.

[1]See Exercise 5.4.

[2]The interval $\mathrm{d}s$ is an invariant, and so takes the same form in all inertial frames. In the rest frame of the particle $\mathrm{d}s = c\,\mathrm{d}\tau$ and so this is true in all inertial frames.

Example 5.3

Find expressions for the momentum and energy of a free particle.
Answer:

$$\boldsymbol{p} = \frac{\partial L}{\partial \boldsymbol{v}} = \frac{\partial}{\partial \boldsymbol{v}}\left(-mc^2(1 - v^2/c^2)^{1/2}\right) = \gamma m\boldsymbol{v}, \qquad (5.23)$$

$$E = H = \boldsymbol{p}\cdot\boldsymbol{v} - L = \gamma m v^2 + \frac{mc^2}{\gamma} = \gamma mc^2\left[\frac{v^2}{c^2} + \left(1 - \frac{v^2}{c^2}\right)\right] = \gamma mc^2. \qquad (5.24)$$

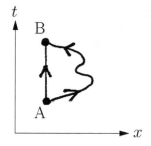

Fig. 5.2 The straight-line path from A to B has a *larger* $\int_a^b \mathrm{d}s$ than the wiggly one (because $c\,\mathrm{d}t > \sqrt{c^2\,\mathrm{d}t^2 - \mathrm{d}x^2}$), even though the wiggly one looks longer.

In special relativity, we can assemble the energy and momentum into a momentum four-vector $p^\mu = \left(\frac{E}{c}, \boldsymbol{p}\right)$ [or equivalently $p_\mu = \left(\frac{E}{c}, -\boldsymbol{p}\right)$]. To make our free particle more interesting, let's give it a charge q and couple it to an electromagnetic field. The electromagnetic field can be described by a four-vector field $A^\mu(x) = \left(\frac{V(x)}{c}, \boldsymbol{A}(x)\right)$ where $V(x)$ is the scalar potential and $\boldsymbol{A}(x)$ is the magnetic vector potential. The action becomes[3]

$$S = \int -mc\,\mathrm{d}s - qA_\mu\,\mathrm{d}x^\mu = \int_{t_1}^{t_2}\left(\frac{-mc^2}{\gamma} + q\boldsymbol{A}\cdot\boldsymbol{v} - qV\right)\mathrm{d}t. \qquad (5.25)$$

The Lagrangian is the integrand of this equation, and so the canonical momentum in this case becomes

$$\boldsymbol{p} = \frac{\partial L}{\partial \boldsymbol{v}} = \gamma m\boldsymbol{v} + q\boldsymbol{A}. \qquad (5.26)$$

Now setting $c = 1$, the familiar $\boldsymbol{E}$ and $\boldsymbol{B}$ fields can be obtained from the four-vector potential $A^\mu = (V, \boldsymbol{A})$ using the equations

$$\boldsymbol{B} = \boldsymbol{\nabla}\times\boldsymbol{A} \quad \text{and} \quad \boldsymbol{E} = -\frac{\partial \boldsymbol{A}}{\partial t} - \boldsymbol{\nabla}V, \qquad (5.27)$$

[3]The potential energy of the electromagnetic field interaction is linear in the field A^μ and takes the Lorentz-invariant form $-qA_\mu\mathrm{d}x^\mu$. Note that $A_\mu = \left(\frac{V}{c}, -\boldsymbol{A}\right)$ and $\mathrm{d}x^\mu = (c\,\mathrm{d}t, \mathrm{d}\boldsymbol{x}) = (c, \boldsymbol{v})\mathrm{d}t$ and so $-qA_\mu\mathrm{d}x^\mu = -q(V - \boldsymbol{A}\cdot\boldsymbol{v})\mathrm{d}t$.

Note that:

$$\partial^\mu = \left(\frac{\partial}{\partial t}, -\boldsymbol{\nabla}\right),$$

$$\partial_\mu = \left(\frac{\partial}{\partial t}, \boldsymbol{\nabla}\right),$$

$$A^\mu = (V, \boldsymbol{A}),$$

$$A_\mu = (V, -\boldsymbol{A}),$$

Also, keep in mind that

$$\boldsymbol{E} = (E^1, E^2, E^3),$$

and

$$\boldsymbol{B} = (B^1, B^2, B^3),$$

are ordinary three-vectors.

The equation is $F_{\mu\nu}F^{\mu\nu} = 2\left(\boldsymbol{B}^2 - (\boldsymbol{E}/c)^2\right)$ if you include the factors of c.

[4]It's useful to note that the components of the electric field may be extracted from $F^{\mu\nu}$ using

$$E^i = -F^{0i} = F^{i0}, \quad (5.33)$$

and the magnetic field is given by

$$B^i = -\frac{1}{2}\varepsilon^{ijk}F^{jk}, \quad (5.34)$$

where ε^{ijk} is the antisymmetric symbol.

but to put this in a relativistically covariant form we define the second-rank antisymmetric tensor $F_{\mu\nu}$ by

$$F_{\mu\nu} = \partial_\mu A_\nu - \partial_\nu A_\mu. \quad (5.28)$$

This is known as the **electromagnetic field tensor** and looks like a four-dimensional version of a curl. In components, this yields

$$F_{\mu\nu} = \begin{pmatrix} 0 & E^1 & E^2 & E^3 \\ -E^1 & 0 & -B^3 & B^2 \\ -E^2 & B^3 & 0 & -B^1 \\ -E^3 & -B^2 & B^1 & 0 \end{pmatrix}. \quad (5.29)$$

Similarly, the tensor

$$F^{\mu\nu} = \begin{pmatrix} 0 & -E^1 & -E^2 & -E^3 \\ E^1 & 0 & -B^3 & B^2 \\ E^2 & B^3 & 0 & -B^1 \\ E^3 & -B^2 & B^1 & 0 \end{pmatrix}. \quad (5.30)$$

Searching for a Lagrangian for the electromagnetic field, we again look for a Lorentz scalar. One very suitable choice is

$$F_{\mu\nu}F^{\mu\nu} = 2(\boldsymbol{B}^2 - \boldsymbol{E}^2). \quad (5.31)$$

The Lagrangian may then be written

$$L = -\frac{1}{4}\int \mathrm{d}^3x\, F_{\mu\nu}F^{\mu\nu}, \quad (5.32)$$

where the choice of the factor $-\frac{1}{4}$ will be explained later in the chapter.[4]

Finally we note a particularly important feature of electromagnetism is that electric charges are *locally* conserved. This fact is enshrined in the **continuity equation** relating the charge density $\rho(x)$ and current density $\boldsymbol{J}(x)$, which is written

$$\frac{\partial\rho}{\partial t} + \boldsymbol{\nabla}\cdot\boldsymbol{J} = \partial_\mu J^\mu = 0, \quad (5.35)$$

where the electromagnetic current four-vector $J^\mu = (\rho, \boldsymbol{J})$. The continuity equation prevents the destruction of charge followed by its creation at another arbitrarily distant point. Such a process would obey *global* conservation of charge, but would allow information to travel at an arbitrary velocity, which we know from special relativity is forbidden.

5.3   Classical fields

We will now take the continuum limit of a simple model from classical mechanics and see how this leads to the concept of a classical field. We consider the case of a classical system of balls connected by springs (whose quantum mechanics was examined in Chapter 2). In the continuum limit we are going to make the distance ℓ between the balls shrink to zero and simultaneously let the mass m of each ball shrink to zero, while keeping the mass per unit length $\rho = m/\ell$ along the chain constant. The displacement $q_j(t)$ of the jth mass at time t becomes a continuous variable $\phi(x, t)$ which depends on both spatial position x and time t. At each point x we have an independent degree of freedom $\phi(x, t)$, so we say that the field has a 'continuous infinity' of degrees of freedom. The function $\phi(x, t)$ is one example of a classical field, but the concept can be much more general than this. For example, rather than describing just a simple displacement, our field can describe more complicated objects such as electromagnetic fields. In each case though, the field depends on position in spacetime. We can now make a working definition of the idea of a classical field:[5]

> A classical **field** is a machine that takes a position in spacetime and outputs an object representing the amplitude of the field at that point. The output might be a scalar (in which case we refer to a **scalar field**), a complex number (a **complex scalar field**), a vector (in which case we refer to a **vector field**), a tensor or something more complicated.

[5]We make the fussy assignment of a *classical* field here to distinguish it from a *quantum* field. As we will see shortly the latter inputs a position in spacetime and outputs an *operator* rather than an amplitude.

Fields are *locally* defined objects. The field 'machine' that we have described is like a computer subroutine which returns the value of ϕ at the position x and time t that we've inputted. It is local in the sense that it doesn't tell you about the value of ϕ somewhere else.

Example 5.4

The temperature $T(\boldsymbol{x}, t)$ is an example of a scalar field. The $\boldsymbol{E}(\boldsymbol{x}, t)$ and $\boldsymbol{B}(\boldsymbol{x}, t)$ fields of electromagnetism are familiar examples of vector fields, which could be time-dependent. The four-vector potential $A^\mu(x)$ is an example of a four-vector field (and here we have written it as a function of spacetime coordinates x^ν). The object defined in eqn 5.28, $F_{\mu\nu}(x)$, is an example of a second-rank tensor field.

5.4 Lagrangian and Hamiltonian density

The next task is to use this language of classical fields to formulate Lagrangians and Hamiltonians. We will do this first for a simple example.

Example 5.5

In the discrete case considered in Section 2.4 we had a Hamiltonian given by

$$H = \sum_j \frac{p_j^2}{2m} + \frac{1}{2}K(q_{j+1} - q_j)^2 \tag{5.36}$$

and a Lagrangian given by

$$L = \sum_j \frac{p_j^2}{2m} - \frac{1}{2}K(q_{j+1} - q_j)^2. \tag{5.37}$$

In the continuum limit, these need to be replaced as the number of masses goes to infinity while $\ell \to 0$. The sums become integrals, so we make the substitution

$$\sum_j \to \frac{1}{\ell} \int dx. \tag{5.38}$$

Using the substitution in eqn 5.38 the kinetic energy term in H and L becomes

$$\sum_j \frac{1}{2}m \left(\frac{\partial q_j}{\partial t}\right)^2 \to \frac{1}{\ell} \int dx \frac{1}{2}m \left(\frac{\partial \phi(x,t)}{\partial t}\right)^2 = \int dx \frac{1}{2}\rho \left(\frac{\partial \phi}{\partial t}\right)^2, \tag{5.39}$$

where we've used the string density $\rho = m/\ell$. For the potential energy term we also make a rule for the differences in coordinates: these become spatial derivatives and hence

$$\frac{(q_{j+1} - q_j)}{\ell} \to \frac{\partial \phi(x,t)}{\partial x}. \tag{5.40}$$

This means that the potential energy term becomes

$$\sum_j \frac{1}{2}K(q_{j+1} - q_j)^2 \to \sum_j \frac{1}{2}K\ell^2 \left(\frac{\partial \phi(x,t)}{\partial x}\right)^2 = \int dx \frac{1}{2}\mathcal{T} \left(\frac{\partial \phi}{\partial x}\right)^2 \tag{5.41}$$

where we've replaced the spring constant K and the lattice spacing ℓ with the tension in the string $\mathcal{T} = K\ell$. Putting this all together (and generalizing to a three-dimensional system) gives a Hamiltonian

$$H = \int d^3x \left[\frac{1}{2}\rho \left(\frac{\partial \phi}{\partial t}\right)^2 + \frac{1}{2}\mathcal{T} (\boldsymbol{\nabla}\phi)^2\right], \tag{5.42}$$

and a Lagrangian

$$L = \int d^3x \left[\frac{1}{2}\rho \left(\frac{\partial \phi}{\partial t}\right)^2 - \frac{1}{2}\mathcal{T} (\boldsymbol{\nabla}\phi)^2\right]. \tag{5.43}$$

Our expressions for H and L are integrals over a position coordinate and so we can define a Hamiltonian density $\mathcal{H}$ by

$$H = \int d^3x \, \mathcal{H}, \tag{5.44}$$

and similarly a Lagrangian density $\mathcal{L}$ as

$$L = \int d^3x \, \mathcal{L}. \tag{5.45}$$

In general, both the Hamiltonian density $\mathcal{H}$ and the Lagrangian density $\mathcal{L}$ are functions of ϕ, $\dot{\phi}$ and ϕ'. We can define a **conjugate momentum** $\pi(x)$ in terms of the functional derivative[6]

$$\pi(x) = \frac{\delta L}{\delta \dot{\phi}} = \frac{\partial \mathcal{L}}{\partial \dot{\phi}}, \tag{5.46}$$

and then this implies that $\mathcal{H}$ and $\mathcal{L}$ can be related by[7]

$$\mathcal{H} = \pi\dot{\phi} - \mathcal{L}. \tag{5.47}$$

Now recall from eqn 1.34 that the action S is related to $\mathcal{L}$ by the four-dimensional integral $S = \int d^4x \, \mathcal{L}(\phi, \partial_\mu\phi)$, and the action principle, given by $\delta S = 0$, implies that (eqn 1.35)

$$\frac{\partial \mathcal{L}}{\partial \phi} - \partial_\mu \left(\frac{\partial \mathcal{L}}{\partial(\partial_\mu\phi)}\right) = 0, \tag{5.48}$$

the four-vector version of the Euler–Lagrange equation.

[6]Just like $p_i = \dfrac{\partial L}{\partial \dot{q}_i}$. For a careful justification of the last equality see Weinberg, Chapter 7.

[7]In Example 5.5 the particular expressions are:

$$\mathcal{L} = \frac{1}{2}\rho \left(\frac{\partial \phi}{\partial t}\right)^2 - \frac{1}{2}\mathcal{T} (\boldsymbol{\nabla}\phi)^2,$$

$$\mathcal{H} = \frac{1}{2}\rho \left(\frac{\partial \phi}{\partial t}\right)^2 + \frac{1}{2}\mathcal{T} (\boldsymbol{\nabla}\phi)^2,$$

and

$$\pi = \rho\frac{\partial \phi}{\partial t}.$$

You can check that eqn 5.47 holds.

Example 5.6

For the electromagnetic field, the Lagrangian density is written

$$\mathcal{L} = -\frac{1}{4}F_{\mu\nu}F^{\mu\nu} = \frac{1}{2}(\boldsymbol{E}^2 - \boldsymbol{B}^2). \tag{5.49}$$

In the absence[8] of an electric potential ($V = 0$), we have $\mathcal{L} = \frac{1}{2}(\boldsymbol{E}^2 - \boldsymbol{B}^2) = \frac{1}{2}(\dot{\boldsymbol{A}}^2 - \boldsymbol{B}^2)$. The conjugate momentum is $\pi^i = \partial\mathcal{L}/\partial(\partial_0 A_i)$, and so $\boldsymbol{\pi} = -\dot{\boldsymbol{A}} = \boldsymbol{E}$, so that

$$\mathcal{H} = \pi^i \dot{A}_i - \mathcal{L} = \frac{1}{2}(\boldsymbol{E}^2 + \boldsymbol{B}^2). \tag{5.50}$$

The Euler–Lagrange equation (eqn 1.35) gives

$$\frac{\partial\mathcal{L}}{\partial A_\mu} - \partial_\lambda\left(\frac{\partial\mathcal{L}}{\partial(\partial_\lambda A_\mu)}\right) = 0. \tag{5.51}$$

The first term in this expression is zero since $\mathcal{L} = -\frac{1}{4}F_{\mu\nu}F^{\mu\nu}$ contains only derivatives of A_μ. The second term can be rewritten as

$$\partial_\lambda F^{\lambda\mu} = 0, \tag{5.52}$$

which is a compact way of writing two of Maxwell's equations in free space, namely $\boldsymbol{\nabla}\cdot\boldsymbol{E} = 0$ and $\boldsymbol{\nabla}\times\boldsymbol{B} = \dot{\boldsymbol{E}}$. The current density four-vector $J^\mu = (\rho, \boldsymbol{J})$ can also be included, and couples linearly to the electromagnetic potential, and so the Lagrangian becomes

$$\mathcal{L} = -\frac{1}{4}F_{\mu\nu}F^{\mu\nu} - J^\mu A_\mu. \tag{5.53}$$

In this case

$$\frac{\partial\mathcal{L}}{\partial A_\mu} = -J^\mu, \tag{5.54}$$

and so

$$\partial_\lambda F^{\lambda\mu} = J^\mu, \tag{5.55}$$

which generates the other two Maxwell equations: $\boldsymbol{\nabla}\cdot\boldsymbol{E} = \rho$ and $\boldsymbol{\nabla}\times\boldsymbol{B} = \boldsymbol{J} + \dot{\boldsymbol{E}}$.

In SI units $\mathcal{L} = -\frac{1}{4\mu_0}F_{\mu\nu}F^{\mu\nu}$ and so $\mathcal{L} = \frac{1}{2}\epsilon_0\boldsymbol{E}^2 - \frac{1}{2\mu_0}\boldsymbol{B}^2$.

[8]Recall eqn 5.27 for the relationship between fields and potentials.

In SI units $\mathcal{H} = \frac{1}{2\mu_0}\boldsymbol{B}^2 + \frac{1}{2}\epsilon_0\boldsymbol{E}^2$.

In SI units $\partial_\lambda F^{\lambda\mu} = \mu_0 J^\mu$.

Solving the Euler–Lagrange equation will produce all the allowable *modes of oscillation* and hence the allowed wave vectors, the particular values k_n, that may occur without dissipation. By the principle of superposition the most general wave is one made up of a weighted sum of all of the possible modes

$$\phi(x, t) = \sum_{\boldsymbol{k}_n} a_{\boldsymbol{k}_n} e^{-i(\omega t - \boldsymbol{k}_n \cdot \boldsymbol{x})}. \tag{5.56}$$

Thus we have seen how many classical continuous systems may be expressed in a Lagrangian formalism so that the Euler–Lagrange equation allows us to derive the normal modes. Our experience with the simple harmonic oscillator in Chapter 2 motivates the idea that these normal modes are simply harmonic oscillators themselves and so can be quantized, leading to particle-like solutions. For the electromagnetic field, those quanta are of course photons.[9] Now with this classical field theory under our belt, we are ready to return to the quantum world, and in the next chapter we will make our first stab at relativistic quantum mechanics.

[9]We will put this on a firmer footing in Chapter 14 and show in detail how photons emerge from quantizing the classical electromagnetic field.

Chapter summary

- A field takes a position in spacetime and outputs an object representing the amplitude (be it scalar, vector or complex) of the field at that point.
- The Lagrangian and Hamiltonian density are related by $\mathcal{H} = \pi\dot{\phi} - \mathcal{L}$ where $\pi = \partial\mathcal{L}/\partial\dot{\phi}$ is the generalized momentum.
- The Lagrangian for the electromagnetic field is $\mathcal{L} = -\frac{1}{4}F_{\mu\nu}F^{\mu\nu}$, where $F_{\mu\nu} = \partial_\mu A_\nu - \partial_\nu A_\mu$ is the electromagnetic field tensor.

Exercises

(5.1) If the Lagrangian does depend explicitly on time, then

$$\frac{dL}{dt} = \frac{\partial L}{\partial t} + \frac{\partial L}{\partial x_i}\dot{x}_i + \frac{\partial L}{\partial \dot{x}_i}\ddot{x}_i. \tag{5.57}$$

In this case show that

$$\frac{\partial L}{\partial t} = -\frac{dH}{dt}. \tag{5.58}$$

(5.2) Show that Poisson brackets anticommute

$$\{A, B\}_{\text{PB}} = -\{B, A\}_{\text{PB}}, \tag{5.59}$$

and also satisfy the Jacobi identity

$$\{\{A, B\}_{\text{PB}}, C\}_{\text{PB}} + \{\{C, A\}_{\text{PB}}, B\}_{\text{PB}} + \{\{B, C\}_{\text{PB}}, A\}_{\text{PB}} = 0, \tag{5.60}$$

and show that quantum mechanical commutators also have the same properties.

(5.3) Show that the commutator of two Hermitian operators $\hat{A}$ and $\hat{B}$ is anti-Hermitian, i.e. that

$$[\hat{A}, \hat{B}]^\dagger = -[\hat{A}, \hat{B}]. \tag{5.61}$$

The factor of i in many commutator expressions (e.g. $[\hat{x}, \hat{p}] = i\hbar$, $[\hat{L}_x, \hat{L}_y] = i\hbar\hat{L}_z$, and $[\hat{A}, \hat{B}] = \frac{1}{i\hbar}\{A, B\}_{\text{PB}}$) makes sure that this property is obeyed.

(5.4) The Lagrangian for a free particle is $L = -mc^2/\gamma$. Find an expression for L, p and H when $v \ll c$.

(5.5) Show that $\int_a^b ds$ takes its maximum value when the integration is done over a straight world-line between a and b [NB $ds^2 = c^2dt^2 - dx^2 - dy^2 - dz^2$].

(5.6) Use the Lagrangian $L = \frac{-mc^2}{\gamma} + q\mathbf{A}\cdot\mathbf{v} - qV$ for a free particle of charge q and mass m in an electromagnetic field to derive the Lorentz force, i.e. show

$$\frac{d}{dt}(\gamma m\mathbf{v}) = q(\mathbf{E} + \mathbf{v} \times \mathbf{B}). \tag{5.62}$$

[*Hint:* $\nabla(\mathbf{a}\cdot\mathbf{b}) = (\mathbf{a}\cdot\nabla)\mathbf{b} + (\mathbf{b}\cdot\nabla)\mathbf{a} + \mathbf{b}\times\text{curl }\mathbf{a} + \mathbf{a}\times\text{curl }\mathbf{b}$.]

(5.7) Show that for the Lagrangian

$$L = \frac{-mc^2}{\gamma} + q\mathbf{A}\cdot\mathbf{v} - qV, \tag{5.63}$$

when $v \ll c$ the momentum becomes $\mathbf{p} = m\mathbf{v} + q\mathbf{A}$ and the energy becomes

$$E = mc^2 + \frac{1}{2m}(\mathbf{p} - q\mathbf{A})^2 + qV. \tag{5.64}$$

(5.8) Another Lorentz scalar that can be obtained from $F_{\mu\nu}$ is

$$\varepsilon^{\alpha\beta\gamma\delta}F_{\alpha\beta}F_{\gamma\delta}. \tag{5.65}$$

Show that this leads to $\mathbf{E}\cdot\mathbf{B}$ being an invariant.

(5.9) Show that $\partial_\mu F^{\mu\nu} = J^\nu$ reduces to two of Maxwell's equations, and that the other two can be generated from the equation

$$\partial_\lambda F_{\mu\nu} + \partial_\mu F_{\nu\lambda} + \partial_\nu F_{\lambda\mu} = 0. \tag{5.66}$$

(5.10) Show that $\partial_\beta\partial_\alpha F^{\alpha\beta} = 0$ and hence that $\partial_\mu J^\mu = 0$. Interpret this as a continuity equation.

A first stab at relativistic quantum mechanics

Bohr: What are you working on?
Dirac: I'm trying to get a relativistic theory of the electron.
Bohr: But Klein has already solved that problem.
(Conversation at the 1927 Solvay Conference.)

We now take a break from the formalism of classical mechanics and turn back to quantum mechanics. In this chapter, we will attempt to find a quantum mechanical equation of motion that is properly relativistic.[1] This will lead to an equation known as the Klein–Gordon equation, though in fact it was derived first by Schrödinger, who rejected it because it had what seemed like a fatal flaw (which we will explore later); he then went on to a second attempt at the problem and produced what is known as the Schrödinger equation, which works brilliantly but is non-relativistic.

[1] It's a first stab because the right answer is due to Dirac, and we will deal with that later. This chapter is about the attempt introduced by Klein and also by Gordon, which is why in the opening quote Bohr didn't think Dirac was working on something worthwhile.

6.1 The Klein–Gordon equation

We first review the logical chain of ideas which leads to the Schrödinger equation. In the non-relativistic quantum mechanics of a free particle, we start with the dispersion relationship, which links the energy and momentum of the particle using the simple expression for kinetic energy $E = \frac{p^2}{2m}$. We turn the variables E and p into operators ($E \to \hat{E}, p \to \hat{p}$), and then we make the substitutions[2] $\hat{E} = i\hbar\frac{\partial}{\partial t}$ and $\hat{p} = -i\hbar\boldsymbol{\nabla}$. This then leads to the Schrödinger equation

[2] Just to make some of the equations a bit more familiar, we will briefly reintroduce all the factors of $\hbar$ and c.

$$i\hbar\frac{\partial\phi(\boldsymbol{x},t)}{\partial t} = -\frac{\hbar^2}{2m}\boldsymbol{\nabla}^2\phi(\boldsymbol{x},t), \tag{6.1}$$

where the object that the operators act on is $\phi(\boldsymbol{x},t)$, the wave function.

The free-particle solutions to the Schrödinger equation are plane waves of the form $\phi(\boldsymbol{x},t) = Ne^{-i(\omega t - \boldsymbol{k}\cdot\boldsymbol{x})}$, where N is a normalization constant. We can write this solution in four-vector form $\phi(x) = Ne^{-ip\cdot x}$. We'll call a wave with this sign combination in the exponential an **incoming wave**. Operating on it with momentum and energy operators yields

$$\begin{aligned}\hat{p}\phi(\boldsymbol{x},t) &= -i\hbar\boldsymbol{\nabla}(Ne^{-i(\omega t - \boldsymbol{k}\cdot\boldsymbol{x})}) = \hbar\boldsymbol{k}\phi(\boldsymbol{x},t), \\ \hat{E}\phi(\boldsymbol{x},t) &= i\hbar\frac{\partial}{\partial t}(Ne^{-i(\omega t - \boldsymbol{k}\cdot\boldsymbol{x})}) = \hbar\omega\phi(\boldsymbol{x},t).\end{aligned} \tag{6.2}$$

That is, an incoming wave has a positive momentum and energy.

Now, to get a relativistic wave equation we'll try the same trick. For a relativistic particle we have the dispersion relationship

$$E = (\boldsymbol{p}^2 c^2 + m^2 c^4)^{\frac{1}{2}}, \tag{6.3}$$

and by making the same operator substitutions as we had before ($E \to \hat{E} = i\hbar\frac{\partial}{\partial t}, \boldsymbol{p} \to \hat{\boldsymbol{p}} = -i\hbar\boldsymbol{\nabla}$) we obtain

$$i\hbar\frac{\partial\phi}{\partial t} = (-\hbar^2 c^2 \boldsymbol{\nabla}^2 + m^2 c^4)^{\frac{1}{2}}\phi. \tag{6.4}$$

This has two big problems. Number one: it doesn't look covariant (the spatial and temporal derivatives appear in different forms). Number two: how do we cope with the square root sign which seems to imply that we take the square root of a differential operator?

Instead of tackling these problems we take a side-step. To avoid the square root we'll simply square the dispersion relation and start again. The squared form of the dispersion is $E^2 = \boldsymbol{p}^2 c^2 + m^2 c^4$. Making the same operator substitutions as before, we obtain

$$-\hbar^2 \frac{\partial^2 \phi(\boldsymbol{x},t)}{\partial t^2} = (-\hbar^2 c^2 \boldsymbol{\nabla}^2 + m^2 c^4)\phi(\boldsymbol{x},t). \tag{6.5}$$

Oskar Klein (1894–1977)
Walter Gordon (1893–1939)

This equation of motion for a wave function is called the **Klein–Gordon equation**. To make things as easy to read as possible we'll revert to using natural units where $\hbar = c = 1$ from now on, and this gives us the following form of the Klein–Gordon equation:

$$-\frac{\partial^2 \phi(\boldsymbol{x},t)}{\partial t^2} = (-\boldsymbol{\nabla}^2 + m^2)\phi(\boldsymbol{x},t). \tag{6.6}$$

The Klein–Gordon equation fits in nicely with our relativistically covariant language, since we can write

$$(\partial^2 + m^2)\phi(x) = 0, \tag{6.7}$$

[3] Note that $\partial^2 = \partial_\mu \partial^\mu = \frac{\partial^2}{\partial t^2} - \boldsymbol{\nabla}^2$.

thereby amalgamating the space and time derivatives.[3]

Example 6.1

[4] In natural units $\boldsymbol{k} = \boldsymbol{p}$ and $\omega = E$ and since we're mostly interested in the energies and momenta of particles we'll use $\boldsymbol{p}$ and E.

Let's solve the Klein–Gordon equation for a free particle. We'll try a solution[4] $\phi(\boldsymbol{x},t) = Ne^{-iEt+i\boldsymbol{p}\cdot\boldsymbol{x}} = Ne^{-ip\cdot x}$. Upon substituting we find that $\phi(x)$ is indeed a solution, yielding the expected dispersion

$$E^2 = \boldsymbol{p}^2 + m^2. \tag{6.8}$$

This looks fine, until we realize that we have to take the square root to get the energy which gives $E = \pm(\boldsymbol{p}^2 + m^2)^{\frac{1}{2}}$. We might be tempted to just throw away the negative solutions as unphysical, but in fact the appearance of negative energy solutions is going to be telling us something.

6.2 Probability currents and densities

One of the reasons that Schrödinger wasn't happy with the Klein–Gordon equation after he'd derived it was that something rather nasty happens when you think about the flow of probability density. The probability of a particle being located somewhere depends on $\phi^*(x)\phi(x)$ and so if this quantity is time-dependent then particles must be sloshing around. The probability density ρ and probability current density[5] j obey a continuity equation

$$\frac{\mathrm{d}\rho}{\mathrm{d}t} + \boldsymbol{\nabla} \cdot \boldsymbol{j} = 0, \tag{6.9}$$

which is more easily written in four-vector notation as

$$\partial_\mu j^\mu = 0. \tag{6.10}$$

If, as is usual in non-relativistic quantum mechanics,[6] we take the spatial part to be

$$\boldsymbol{j}(x) = -\mathrm{i}\left[\phi^*(x)\boldsymbol{\nabla}\phi(x) - \phi(x)\boldsymbol{\nabla}\phi^*(x)\right], \tag{6.11}$$

then, for eqn 6.10 to work,[7] we require the probability density to look like[8]

$$\rho(x) = \mathrm{i}\left[\phi^*(x)\frac{\partial\phi(x)}{\partial t} - \frac{\partial\phi^*(x)}{\partial t}\phi(x)\right]. \tag{6.12}$$

The resulting covariant probability current for the Klein–Gordon equation is then given by

$$j^\mu(x) = \mathrm{i}\left\{\phi^*(x)\partial^\mu\phi(x) - \left[\partial^\mu\phi^*(x)\right]\phi(x)\right\}, \tag{6.13}$$

which, as the notation suggests, is a four-vector. Substituting in our wave function $\phi(x) = N\mathrm{e}^{-\mathrm{i}p\cdot x}$ gives the time-like component of the probability current as

$$j^0 = \rho = 2|N|^2 E. \tag{6.14}$$

Since E can be positive or negative we can't interpret ρ as a probability density, since there's no such thing as negative probability! This seems to ring the death knell for the Klein–Gordon equation as a half-decent single-particle quantum mechanical equation.

6.3 Feynman's interpretation of the negative energy states

Fortunately, all is not lost. Richard Feynman[9] and, independently, Ernst Stueckelberg[10] came up with an interpretation of the negative energy states as particles moving backward in time. We call these states **antiparticles**. Consider the classical equation of motion for a charged particle in an electromagnetic field which looks like

$$m\frac{d^2x^\mu}{d\tau^2} = qF^\mu{}_\nu\frac{dx^\nu}{d\tau}, \tag{6.15}$$

[5]Here j is the current density of particles, whereas in Example 5.6 we used J as the current density of charge. If both appear together, as they will do in Section 6.3 below, we will write the former as j and the latter as J_{em}.

[6]In non-relativistic quantum mechanics, j is defined with some additional constants:
$$j = -(\mathrm{i}\hbar/2m)(\psi^*\boldsymbol{\nabla}\psi - \psi\boldsymbol{\nabla}\psi^*).$$
We dispense with the $\hbar/2m$ constant here.

[7]It will work, and you can prove it as follows. Take the Klein–Gordon equation (eqn 6.5) and premultiply it by ϕ^*. Then take the complex conjugate of eqn 6.5 and premultiply by ϕ. Subtracting these two results will give an equation of the form of eqn 6.9 with j and ρ as given.

[8]Note that this is *very* different from non-relativistic quantum mechanics where $\rho = |\psi|^2$.

[9]Richard Feynman (1918–1988) invented much of the physics in this book. The bongo-playing theoretical physicist once defined science as 'the belief in the ignorance of experts'. He was never satisfied by an explanation until he had worked it out for himself, and it is perhaps fitting that his quirky shorthand for doing quantum field theory calculations, the little cartoons we now call Feynman diagrams, decorate the pages of every book on quantum field theory, including this one. (He also noted that bongo players didn't feel the need to point out that he was a theoretical physicist.)

[10]Brilliant, eccentric and difficult, the Swiss mathematician and physicist Ernst Stueckelberg (1905–1984) may also claim credit for the invention of a significant amount of the physics in this book, although the opaqueness of many of his arguments and his insistence on inventing novel notation have prevented this from being widely recognized. His achievements include: this interpretation of antiparticles; the invention of spacetime ('Feynman') diagrams; the explanation of ('Yukawa') force-carrying particles; the discovery of the renormalization group and the conservation of baryon number. See Schweber, Chapter 10 for more details.

where τ is proper time, q is charge and $F^{\mu\nu}$ is the electromagnetic field tensor. Changing the sign of the proper time τ has the same effect as changing the sign of the charge. Therefore, a particle travelling backward in time looks the same as an oppositely charged antiparticle moving forward in time. One strategy to eliminate negative energy states is therefore to reverse the charge and momentum of all negative energy solutions, turning them into antiparticles. How does this eliminate the negative sign of the energy? Since we write the phase of the wave function as $Et - \boldsymbol{p} \cdot \boldsymbol{x}$, then making the substitution $t \to -t$ means that we have to change $-E \to E$ for the first term to make sense. Of course reversing the time (like playing a film backwards) reverses all momenta so we also need to swap $\boldsymbol{p} \to -\boldsymbol{p}$ for consistency.

We can go further by invoking some quantum mechanics. Let's examine the electromagnetic current density for a Klein–Gordon particle which we will define as $J_{em}^{\mu} = qj^{\mu}$ where q is the charge of the particle. The current four-vector J_{em}^{μ} for a positively charged, incoming particle[11] with positive energy is given by

$$\begin{aligned} J_{em}^{\mu} &= (+q)2|N|^2 p^{\mu} & (6.16) \\ &= (+q)2|N|^2 (E, \boldsymbol{p}). & (6.17) \end{aligned}$$

In contrast, a negative energy particle with positive charge has $J_{em}^{\mu} = (+q)2|N|^2(-E, \boldsymbol{p})$. Since we've decided that negative energies are disgusting because they mess up the probability, we want to turn the energy positive. We do this by taking the minus sign out to the front of the equation to yield

$$J_{em}^{\mu} = (-q)2|N|^2 (E, -\boldsymbol{p}). \qquad (6.18)$$

Notice that we have to swap the sign of charge and the three-momentum $\boldsymbol{p}$, changing an incoming particle into an outgoing antiparticle, in agreement with our argument above.

We now have a rule to turn unphysical negative energies into physical positive energies. Simply treat the particle as one with positive energy, but change the particle into an antiparticle by reversing the charge and swapping the sign of the three-momentum.

Is this really permissible for thinking about interacting quantum particles, which we know from quantum mechanics can be created and destroyed and emitted and absorbed? It is, as we'll see from Fig. 6.1. In a typical process we might have an incoming particle absorbed by a system, donating its energy, momentum and charge to the system [Fig. 6.1(i)]. Alternatively a particle could be emitted, depleting the system of energy, momentum and charge [Fig. 6.1(ii)]. The key point is that absorption of a particle in a negative energy state [Fig. 6.1(iii)] is equivalent to the emission of an antiparticle in a positive energy state [Fig. 6.1(iv)], at least as far as the system is concerned.

To summarize then, although the dispersion relation $E = \pm\sqrt{\boldsymbol{p}^2 + m^2}$ admits positive and negative solutions, only positive energies are physical. We therefore make all energies positive, but we can't just ignore the formerly negative energy states. The formerly negative energy states are

[11]Incoming implies inclusion of a factor $e^{-ip\cdot x}$.

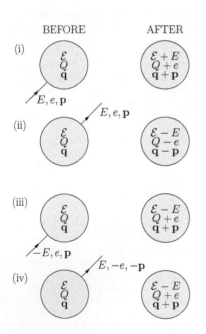

Fig. 6.1 (i) An incoming particle (energy E, charge e ($\equiv |e|$), momentum $\boldsymbol{p}$) is absorbed by a system (energy $\mathcal{E}$, charge Q, momentum $\boldsymbol{q}$). Also shown are (ii) emission of particle, (iii) absorption of a particle in a negative energy state and (iv) emission of an antiparticle in a positive energy state.

interpreted as positive energy *antiparticles* with momenta in the opposite direction to the corresponding particle. A general solution to the Klein–Gordon equation with a particular positive energy now corresponds to a superposition of two states[12]

$$\phi(x) = \left[\begin{array}{c} \text{Incoming positive} \\ \text{energy particle} \\ \propto e^{-i(Et - \boldsymbol{p} \cdot \boldsymbol{x})} \end{array} \right] + \left[\begin{array}{c} \text{Outgoing positive} \\ \text{energy antiparticle} \\ \propto e^{+i(Et - \boldsymbol{p} \cdot \boldsymbol{x})} \end{array} \right]. \quad (6.19)$$

This general solution is therefore not an inherently single-particle description, because we are forced to consider particles and antiparticles.

6.4   No conclusions

Although the Klein–Gordon equation has now been made semi-respectable with Feynman's interpretation it still has an uncomfortable feeling about it. What is implied by the new interpretation? Does it correspond to anything sensible in real life? We'll leave these questions for a moment. The Klein–Gordon equation will return later and we will find that it is, in fact, a perfectly respectable equation but in a slightly different context, namely that of describing the dynamics of fields whose excitations are spinless particles. Our next step is to systematize our procedure for writing down theories.

Chapter summary

- The Klein–Gordon equation is $(\partial^2 + m^2)\phi = 0$.
- It suffers from having negative energy states and negative probabilities.
- We can think of these negative energy states as positive energy antiparticles

[12]The assignments 'incoming' and 'outgoing' are made to link up with the construction of Feynman diagrams in Chapter 19.

Exercises

(6.1) This problem looks forward to ideas which we will consider in more detail in the next chapter. Show that the Lagrangian density $\mathcal{L}$ given by

$$\mathcal{L} = \frac{1}{2} \left(\partial_\mu \phi \right)^2 - \frac{1}{2} m^2 \phi^2, \quad (6.20)$$

yields the Klein–Gordon equation when using the Euler–Lagrange equation. Also derive expressions for the momentum $\pi = \partial \mathcal{L}/\partial \dot{\phi}$ and the Hamiltonian density $\mathcal{H} = \pi \dot{\phi} - \mathcal{L}$ for this Lagrangian and interpret the result.

7

Examples of Lagrangians, or how to write down a theory

[1] Lagrangian density is a bit of a mouthful, so we are using the universal shorthand of referring to Lagrangian densities simply as 'Lagrangians.' Pedants beware.

[2] Symmetry will turn out to be essential to our understanding of field theory. However, we postpone its discussion until Chapter 10.

[3] The assignment of masslessness will be explained shortly.

I regard as quite useless the reading of large treatises of pure analysis: too large a number of methods pass at once before the eyes. It is in the works of applications that one must study them; one judges their ability there and one apprises the manner of making use of them.
Joseph-Louis Lagrange (1736–1813)

Theoretical physics is all about writing down models to describe the behaviour of particular systems in the Universe. The insight that we have gained from the previous chapters is that the main starting point for a field theory is writing down an appropriate Lagrangian.[1] This is such an important step in the process that we are devoting an entire chapter to it, partly to reiterate some of the ideas introduced earlier and partly to demystify the act of plucking a Lagrangian out of thin air as practised by many professional theorists.

Why do we choose a particular Lagrangian $\mathcal{L}$? Usually because it's the simplest model that contains all the key physical ideas and has the correct symmetry for the problem.[2] Its artful construction allows it to spit out the right equation of motion once fed into the waiting jaws of the Euler–Lagrange equation, given by

$$\frac{\partial \mathcal{L}}{\partial \phi} - \partial_\mu \left(\frac{\partial \mathcal{L}}{\partial(\partial_\mu \phi)} \right) = 0. \tag{7.1}$$

Then, the wavelike solutions of this equation of motion can be interpreted as particles with the appropriate dispersion relation, i.e. the relation between energy and momentum. This process of upgrading classical modes into quantum particles will be described in detail in the next part of the book. Here we will just assume that it can be done somehow.

In this chapter, we will examine a succession of simple examples of choices of Lagrangian, for massless and massive scalar fields, a scalar field coupled to an external source, a couple of scalar fields and a complex field (deferring vector and spinor fields to later in the book).

7.1 A massless scalar field

We start with the simplest possible Lagrangian, one for a massless[3] scalar field[4] $\phi(x)$. The scalar field assigns a scalar amplitude to each position x in spacetime. It can change as a function of spacetime coordinate and its gradient is $\partial_\mu \phi \equiv \partial \phi(x)/\partial x^\mu$. The Lagrangian will depend only on the rate of change of ϕ in time $\partial_0 \phi$ and in space $\boldsymbol{\nabla}\phi$. We have to make a choice of $\mathcal{L}$ which is relativistically covariant and so we choose

$$\mathcal{L} = \frac{1}{2}\partial^\mu \phi \, \partial_\mu \phi = \frac{1}{2}\left(\partial_\mu \phi\right)^2. \qquad (7.2)$$

This can be expanded as $\mathcal{L} = \frac{1}{2}(\partial_0 \phi)^2 - \frac{1}{2}\boldsymbol{\nabla}\phi \cdot \boldsymbol{\nabla}\phi$, which has the look of[5] $\mathcal{L} = (\text{kinetic energy}) - (\text{potential energy})$. The factor of $\frac{1}{2}$ is included so that later results come out nicely, as we shall see.

Example 7.1

Using the Lagrangian in eqn 7.2, we have

$$\frac{\partial \mathcal{L}}{\partial \phi} = 0, \quad \frac{\partial \mathcal{L}}{\partial(\partial_\mu \phi)} = \partial^\mu \phi, \qquad (7.3)$$

and hence plugging into the Euler–Lagrange equation (eqn 7.1) we have

$$\partial_\mu \partial^\mu \phi = 0. \qquad (7.4)$$

This is the wave equation $\partial^2 \phi = 0$, or

$$\frac{\partial^2 \phi}{\partial t^2} - \boldsymbol{\nabla}^2 \phi = 0, \qquad (7.5)$$

and has wave-like solutions

$$\phi(x, t) = \sum_{\boldsymbol{p}} a_{\boldsymbol{p}} e^{-i(E_{\boldsymbol{p}}t - \boldsymbol{p} \cdot \boldsymbol{x})}, \qquad (7.6)$$

with dispersion relation $E_{\boldsymbol{p}} = c|\boldsymbol{p}|$ [though in our units $c = 1$], see Fig. 7.1.

Note that $E_{\boldsymbol{p}} = 0$ at $|\boldsymbol{p}| = 0$ and so the dispersion relation is *gapless*. If we take this equation to describe the energies of the particles that will (later) represent quantum excitations of this system then, since this is the $m = 0$ version of the relativistic dispersion $E_{\boldsymbol{p}} = \sqrt{\boldsymbol{p}^2 + m^2}$, we call the particles massless and the field a massless scalar field. The wave solutions of this linear equation obey the principle of superposition. We say that the particles which correspond to these waves are free or non-interacting. If we send these particles towards each other they will simply pass on through each other without scattering.

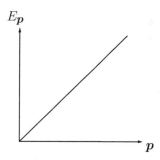

Fig. 7.1 The dispersion relation $E_{\boldsymbol{p}} = |\boldsymbol{p}|$.

7.2   A massive scalar field

The next step is to try and include mass in the problem. To do this, we make $\mathcal{L}$ depend not only on $\partial_\mu \phi$ (how the scalar field *changes* through spacetime) but also on ϕ (the magnitude of the scalar field). This can be done by introducing a potential energy term $U(\phi)$ (and since Lagrangians are kinetic energy *minus* potential energy, this will come with a minus sign). The potential energy term expresses the cost of having a field there at all, rather than just vacuum. By choosing $U(\phi) \propto \phi^2$ we will have a quadratic potential energy with a minimum at $\phi = 0$. Thus we write down

$$\mathcal{L} = \frac{1}{2}\left(\partial_\mu\phi\right)^2 - \frac{1}{2}m^2\phi^2,\qquad(7.7)$$

where the $\frac{1}{2}m^2$ factor is chosen to make the next bit come out right. The parameter m has yet to be shown to be anything physical, but will turn out (of course) to be the particle mass.

Example 7.2
Using the Lagrangian in eqn 7.7, we have

$$\frac{\partial\mathcal{L}}{\partial\phi} = -m^2\phi, \quad \frac{\partial\mathcal{L}}{\partial(\partial_\mu\phi)} = \partial^\mu\phi,\qquad(7.8)$$

and hence plugging into the Euler–Lagrange equation (eqn 7.1) we have

$$(\partial_\mu\partial^\mu + m^2)\phi = 0.\qquad(7.9)$$

The equation of motion for this field theory is the Klein–Gordon equation! The solution of these equations is again $\phi(\boldsymbol{x},t) = ae^{-\mathrm{i}(E_{\boldsymbol{p}}t - \boldsymbol{p}\cdot\boldsymbol{x})}$, with dispersion $E_{\boldsymbol{p}}^2 = \boldsymbol{p}^2 + m^2$ (see Fig. 7.2).

Obviously if $m = 0$ we revert to the case of Section 7.1. With $m \neq 0$ we have a gap in the dispersion relation (at $\boldsymbol{p} = 0$, $E_{\boldsymbol{p}} = \pm m$) corresponding to the particle's mass. Again, the equations of motion are linear and therefore the particles described by this theory don't interact.

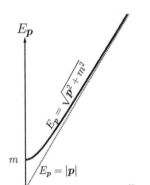

Fig. 7.2 The dispersion relation $E_{\boldsymbol{p}} = \sqrt{\boldsymbol{p}^2 + m^2}$ for a massive particle.

7.3   An external source

We now want to introduce interactions. The simplest way to do this is to have the scalar field interact with an external potential. This potential is described by a function known as a **source current** $J(x)$ which interacts with the field, giving a potential energy term $-J(x)\phi(x)$. The resulting Lagrangian is written

$$\mathcal{L} = \frac{1}{2}[\partial_\mu\phi(x)]^2 - \frac{1}{2}m^2[\phi(x)]^2 + J(x)\phi(x).\qquad(7.10)$$

The equations of motion become[6]

$$(\partial_\mu\partial^\mu + m^2)\phi(x) = J(x).\qquad(7.11)$$

[6] A shorthand version of the same equation is
$$(\partial^2 + m^2)\phi(x) = J(x).$$

This is now an inhomogeneous partial differential equation. The use of sources (or sorcery) will later turn up as a method of probing quantum fields following a technique invented by Julian Schwinger.

7.4 The ϕ^4 theory

How do we get particles to interact with other particles (or, equivalently, fields to interact with other fields)? Here's the simplest recipe. The ϕ^4 Lagrangian is like the scalar case, but with an extra potential term $U(\phi)$ proportional to ϕ^4. This term makes the theory interacting. Unfortunately it also makes it unsolvable. As we'll see, we have to use a sort of perturbation theory to make predictions from this theory. The Lagrangian is[7]

$$\mathcal{L} = \frac{1}{2}\partial^\mu\phi\partial_\mu\phi - \frac{1}{2}m^2\phi^2 - \frac{1}{4!}\lambda\phi^4, \qquad (7.12)$$

which leads to the (not-very-memorable) equation of motion

$$(\partial^2 + m^2)\phi = -\frac{\lambda}{3!}\phi^3. \qquad (7.13)$$

The constant of proportionality λ is known as the interaction strength. Clearly, when $\lambda = 0$ we return to the massive scalar field Lagrangian and the fields don't interact.

7.5 Two scalar fields

Here's another way to make particles interact. Why not have two different types of particle in the Universe, described by two fields $\phi_1(x)$ and $\phi_2(x)$? In our simple theory they have the same mass and will interact with themselves and each other via a potential energy $U(\phi_1, \phi_2) = g\left(\phi_1^2 + \phi_2^2\right)^2$ where g is the interaction strength here. Notice that multiplying out the bracket gives self-interacting ϕ^4 terms and, crucially for us, a cross-term $2\phi_1^2\phi_2^2$ which forces the two types of field to interact. The resulting Lagrangian density is

$$\mathcal{L} = \frac{1}{2}(\partial_\mu\phi_1)^2 - \frac{1}{2}m^2\phi_1^2 + \frac{1}{2}(\partial_\mu\phi_2)^2 - \frac{1}{2}m^2\phi_2^2 - g\left(\phi_1^2 + \phi_2^2\right)^2. \qquad (7.14)$$

We notice that we get the same Lagrangian as in eqn 7.14 even if we change our definition of the fields. (We say that the Lagrangian has a symmetry.) If we transform the two fields using the mapping $\phi_1 \to \phi_1'$ and $\phi_2 \to \phi_2'$ where

$$\begin{pmatrix} \phi_1' \\ \phi_2' \end{pmatrix} = \begin{pmatrix} \cos\theta & -\sin\theta \\ \sin\theta & \cos\theta \end{pmatrix} \begin{pmatrix} \phi_1 \\ \phi_2 \end{pmatrix}, \qquad (7.15)$$

then the Lagrangian is unchanged. This looks a lot like we've rotated the fields in ϕ_1-ϕ_2 space. We say that the particles have an **internal degree of freedom**. It's as if they carry round a dial labelled by θ and by turning this dial (i.e. changing θ) a ϕ_1 particle may be turned into a superposition of ϕ_1 and ϕ_2 particles. The invariance of the physics (via the Lagrangian) with respect to rotations by θ in the ϕ_1-ϕ_2 plane expresses an $SO(2)$ symmetry.[8]

[7]The numerical coefficient of the ϕ^4 term is often chosen with convenience in mind. We will use several in the course of the book, but the coefficient $\frac{\lambda}{4!}$ will be most common.

[8]$SO(2)$ is the two-dimensional special orthogonal group, corresponding to the group of 2×2 orthogonal matrices with unit determinant. This is a continuous symmetry and we will later show that continuous symmetries lead to conserved quantities. More of this in Chapters 13 and 15.

(a)
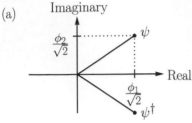

(b)
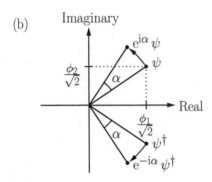

Fig. 7.3 The complex scalar field on an Argand diagram.

[9]$U(1)$ is the one-dimensional group of unitary transformations. Mathematically we note that $SO(2)$ is isomorphic to $U(1)$ (see Fig. 7.4), so nothing has really changed at a fundamental level.

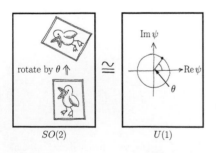

Fig. 7.4 The group $SO(2)$, describing rotations in two-dimensions, is isomorphic to $U(1)$, the one-dimensional group of unitary transformations described by the phase of the complex number $\psi = |\psi|e^{i\theta}$.

7.6 The complex scalar field

We can also make a transformation that simplifies the two-scalar-fields Lagrangian considered in the previous section. We define two new fields ψ and $\psi^\dagger$ given by

$$\psi = \frac{1}{\sqrt{2}}[\phi_1 + i\phi_2],$$
$$\psi^\dagger = \frac{1}{\sqrt{2}}[\phi_1 - i\phi_2]. \tag{7.16}$$

This transformation can be pictured by an Argand diagram with ϕ_1 along the real axis and ϕ_2 along the imaginary axis [Fig. 7.3(a)]. Using the new fields we obtain a new Lagrangian

$$\mathcal{L} = \partial^\mu\psi^\dagger\partial_\mu\psi - m^2\psi^\dagger\psi - g(\psi^\dagger\psi)^2. \tag{7.17}$$

This is known as a complex scalar field theory. Although we originally made it up from two sorts of field, we can imagine that it describes one sort of complex-valued field ψ. This new field is constructed from a bit of ϕ_1 and a bit of ϕ_2 and retains two degrees of freedom. The new Lagrangian (eqn 7.17) is invariant with respect to rotations in the complex plane [Fig. 7.3(b)]

$$\psi \to \psi e^{i\alpha}, \quad \psi^\dagger \to e^{-i\alpha}\psi^\dagger, \tag{7.18}$$

which expresses a $U(1)$ symmetry.[9]

Example 7.3

The $\psi^\dagger\psi\phi$ theory: Finally, as a rather more complicated example, we consider a theory with three types of particle. We add together Lagrangians for the complex scalar field theory (with mass m) and for the scalar field theory (with mass μ) and also arrange for the ϕ and ψ particles to interact via a potential $g\psi^\dagger\psi\phi$. The Lagrangian is

$$\begin{aligned}\mathcal{L} &= \partial^\mu\psi^\dagger\partial_\mu\psi - m^2\psi^\dagger\psi \\ &\quad + \frac{1}{2}(\partial_\mu\phi)^2 - \frac{1}{2}\mu^2\phi^2 - g\psi^\dagger\psi\phi.\end{aligned} \tag{7.19}$$

As we will see a little later, this theory looks a lot like quantum electrodynamics!

The fields in this chapter have all been classical and although we've claimed that particles are the excitations of these fields we have yet to see how they arise from quantum mechanical versions of the fields. This is the subject of the next part of the book.

Chapter summary

- Lagrangians are deduced for a massless scalar field, massive scalar field, the inclusion of a source, two scalar fields and a complex field.
- For example, the Lagrangian for a massive scalar field coupled to a source is

$$\mathcal{L} = \frac{1}{2}[\partial_\mu \phi(x)]^2 - \frac{1}{2}m^2[\phi(x)]^2 + J(x)\phi(x).$$

- The Euler–Lagrange equation can be used to deduce an equation of motion.
- For example, the equation of motion for a massive scalar field coupled to a source comes out to be

$$(\partial^2 + m^2)\phi(x) = J(x).$$

- The complex scalar field $\psi = \frac{1}{\sqrt{2}}(\phi_1 + i\phi_2)$ possesses a $U(1)$ symmetry.

Exercises

(7.1) For the Lagrangian $\mathcal{L}$ given by

$$\mathcal{L} = \frac{1}{2}\partial^\mu \phi \partial_\mu \phi - \frac{1}{2}m^2\phi^2 - \sum_{n=1}^{\infty} \lambda_n \phi^{2n+2}, \quad (7.20)$$

show that the equation of motion is given by

$$(\partial^2 + m^2)\phi + \sum_{n=1}^{\infty} \lambda_n(2n+2)\phi^{2n+1} = 0. \quad (7.21)$$

(7.2) Consider a massive scalar field $\phi(x)$ coupled to a source $J(x)$, described by the Lagrangian of eqn 7.10. Show that the equations of motion are those of eqn 7.11.

(7.3) Show that the equations of motion following from the Lagrangian in eqn 7.14 are the coupled equations

$$\partial_\mu \partial^\mu \phi_1 + m^2\phi_1 + 4g\phi_1(\phi_1^2 + \phi_2^2) = 0, (7.22)$$
$$\partial_\mu \partial^\mu \phi_2 + m^2\phi_2 + 4g\phi_2(\phi_1^2 + \phi_2^2) = 0. (7.23)$$

(7.4) Show that eqn 7.7 is equivalent to

$$\mathcal{L} = \frac{1}{2}\dot{\phi}^2 - \frac{1}{2}(\boldsymbol{\nabla}\phi)^2 - \frac{1}{2}m^2\phi^2. \quad (7.24)$$

Show that

$$\pi = \frac{\partial \mathcal{L}}{\partial \dot{\phi}} = \dot{\phi} \quad (7.25)$$

and

$$\mathcal{H} = \frac{1}{2}\dot{\phi}^2 + \frac{1}{2}(\boldsymbol{\nabla}\phi)^2 + \frac{1}{2}m^2\phi^2. \quad (7.26)$$

Define the quantity

$$\Pi^\mu = \frac{\partial \mathcal{L}}{\partial(\partial_\mu \phi)}, \quad (7.27)$$

and show that $\Pi^\mu = \partial^\mu \phi$ and $\Pi^0 = \pi = \dot{\phi}$.

Part III

The need for quantum fields

In this part we introduce quantum fields, how to construct quantum field theories, and look at the transformations that can be applied to them. Understanding the underlying symmetries of a quantum field turns out to be crucial for exploring the properties of quantum field theory. This part is structured as follows:

- To study quantum fields, we must first describe how to include time dependence, which is covered in Chapter 8. In doing this, we inadvertently stumble on a reason why single-particle quantum mechanics is insufficient for fully describing the world in which we live.

- In Chapter 9 we discuss how to describe the response of quantum mechanical states to transformations: translations, rotations and Lorentz boosts.

- This sets up the treatment of Noether's theorem in Chapter 10, demonstrating the fundamental relationship between symmetry and conservation laws.

- We are then ready in Chapter 11 to plunge into the deep waters of constructing quantum field theories out of classical field theories, a manufacturing process termed *canonical quantization*. We give an extended example of this process for complex scalar field theory in Chapter 12

- These ideas are extended to multicomponent fields in Chapter 13 and we work this through for the case of massive electromagnetism, thereby studying a spin-1 vector field.

- Transformations so far have been global. When examining local transformations we find ourselves introducing *gauge theory* and this is covered in Chapter 14.

- In Chapter 15 we cover discrete symmetries: C, P and T. We look at the connection between group theory and the symmetries we have been studying in Part III.

8

The passage of time

[1]Single-particle quantum mechanics, as the name suggests, describes the quantum behaviour of a single particle confined in some potential. As we will describe, even in such a starting situation it is possible to have to worry about the appearance of particle–antiparticle pairs, and thus the assumption that 'you are just dealing with a single particle' is not as straightforward as you might think.

[2]This is of course named after Austrian physicist Erwin Schrödinger (1887–1961).

> *The innocent and the beautiful*
> *Have no enemy but time*
> W. B. Yeats (1865–1939)

> *He that will not apply new remedies must expect new evils;*
> *for time is the great innovator*
> Francis Bacon (1561–1626)

There is more than one way of putting time into quantum mechanics. In this chapter we'll discuss Heisenberg's version of quantum theory in which all the time dependence is located in the operators, while the wave functions stay fixed and unchanging. This contrasts with the more familiar version that Schrödinger formulated in which the wave functions carry the time dependence, the formulation that is taught in most introductory courses. Far from being a dry exercise in formalism, this will allow us to kill single-particle quantum mechanics[1] stone-dead and guide us in a direction towards quantum fields.

8.1 Schrödinger's picture and the time-evolution operator

The most familiar way of doing quantum mechanics is to think about time-dependent wave functions. The manner in which the wave function changes with time is given by the Schrödinger[2] equation

$$i\frac{\partial \psi(\boldsymbol{x},t)}{\partial t} = \hat{H}\psi(\boldsymbol{x},t), \tag{8.1}$$

which is an equation of motion for the wave function. Dynamic variables like momentum and position are accessed with operators $\hat{\boldsymbol{p}} = -i\nabla$ and $\hat{\boldsymbol{x}} = \boldsymbol{x}$ respectively, which act on the wave function. This way of doing quantum mechanics is known as the **Schrödinger picture** and is completely different from classical mechanics, where the dynamic variables depend on time [e.g. $\boldsymbol{p}(t), \boldsymbol{x}(t)$, etc.] and are themselves described by equations of motion. The root of this difference is that time itself is a funny thing in quantum mechanics. There is no operator that tells us when some event occurred; instead time is a parameter that in the Schrödinger picture is located purely in our wave function $\psi(\boldsymbol{x},t)$.

Although no operator tells us the time, there is an operator $\hat{U}(t_2, t_1)$ that can evolve a particle forward through time, from time t_1 to t_2. Thus we can write

$$\psi(t_2) = \hat{U}(t_2, t_1)\psi(t_1), \tag{8.2}$$

and we will call $\hat{U}(t_2, t_1)$ a **time-evolution operator**. Notice that the time-evolution operator has two slots in which times can be inserted: the start-time t_1 and the stop-time t_2.

The time-evolution operator has five very useful properties:

(1) $\hat{U}(t_1, t_1) = 1$.

   If the two times are the same, then nothing can happen and $\hat{U}$ is just the identity operator.

(2) $\hat{U}(t_3, t_2)\hat{U}(t_2, t_1) = \hat{U}(t_3, t_1)$.

   This is the composition law and shows that we can build up a time translation by multiplying a set of smaller translations that do the same evolution, but in steps [see Fig. 8.1(a)].

(3) $\mathrm{i}\dfrac{\mathrm{d}}{\mathrm{d}t_2}\hat{U}(t_2, t_1) = \hat{H}\hat{U}(t_2, t_1)$.

   The proof of this equation is simple. Differentiating eqn 8.2 with respect to t_2 gives

$$\frac{\mathrm{d}\psi(t_2)}{\mathrm{d}t_2} = \frac{\mathrm{d}\hat{U}(t_2, t_1)}{\mathrm{d}t_2}\psi(t_1), \tag{8.3}$$

   but eqn 8.1 gives

$$\mathrm{i}\frac{\mathrm{d}\psi(t_2)}{\mathrm{d}t_2} = \hat{H}\psi(t_2) = \hat{H}\hat{U}(t_2, t_1)\psi(t_1), \tag{8.4}$$

   and the result follows. This shows that the time-evolution operator itself obeys the Schrödinger equation.

(4) $\hat{U}(t_1, t_2) = \hat{U}^{-1}(t_2, t_1)$,

   so by taking the inverse of a time-evolution operator, one can turn back time [see Fig. 8.1(b)].

(5) $\hat{U}^\dagger(t_2, t_1)\hat{U}(t_2, t_1) = 1$,

   i.e. the time-evolution operator is unitary. The proof is short. When $t_2 = t_1$, the result is trivial. If we vary t_2 away from t_1, then $\hat{U}^\dagger(t_2, t_1)\hat{U}(t_2, t_1)$ could change, but since[3]

$$\begin{aligned}
\frac{\mathrm{d}}{\mathrm{d}t_2}\left[\hat{U}^\dagger(t_2, t_1)\hat{U}(t_2, t_1)\right] &= \frac{\mathrm{d}\hat{U}^\dagger}{\mathrm{d}t_2}U + U^\dagger\frac{\mathrm{d}\hat{U}}{\mathrm{d}t_2} \\
&= -\frac{\hat{U}^\dagger\hat{H}\hat{U}}{\mathrm{i}} + \frac{\hat{U}^\dagger\hat{H}\hat{U}}{\mathrm{i}} = 0, \tag{8.5}
\end{aligned}$$

   it won't. QED. One consequence of this result is that $\hat{U}^\dagger(t_2, t_1) = \hat{U}^{-1}(t_2, t_1)$, that is, the adjoint is the inverse.

Property number 3 allows us to write an explicit expression for $\hat{U}(t_2, t_1)$, since integrating this equation gives[4]

$$\hat{U}(t_2, t_1) = \mathrm{e}^{-\mathrm{i}\hat{H}(t_2 - t_1)}. \tag{8.7}$$

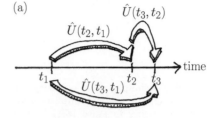

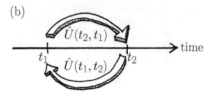

Fig. 8.1 (a) The composition law: $\hat{U}(t_3, t_2)\hat{U}(t_2, t_1) = \hat{U}(t_3, t_1)$. (b) The inverse of a time-evolution operator is equivalent to reversing the direction of time: $\hat{U}(t_1, t_2) = \hat{U}^{-1}(t_2, t_1)$.

[3]Note that in deriving eqn 8.5 we are using property 3 of the time evolution operator in the form

$$\frac{\mathrm{d}\hat{U}}{\mathrm{d}t} = \frac{\hat{H}\hat{U}}{\mathrm{i}}, \qquad \frac{\mathrm{d}\hat{U}^\dagger}{\mathrm{d}t} = -\frac{\hat{U}^\dagger\hat{H}}{\mathrm{i}}.$$

[4]This exponential expression for an operator is actually shorthand for an expansion:

$$\mathrm{e}^{\hat{A}} = 1 + \hat{A} + \frac{1}{2!}\hat{A}\hat{A} + \frac{1}{3!}\hat{A}\hat{A}\hat{A} + \dots \tag{8.6}$$

We will meet many such exponential expressions.

8.2   The Heisenberg picture

The Schrödinger picture puts all the time-dependence into the wave functions, but is that the best way to look at things? Consider the expectation value of an operator

$$\langle \hat{O}(t) \rangle = \langle \psi(t) | \hat{O} | \psi(t) \rangle. \tag{8.8}$$

This tells us the average value that we'll get if we make a measurement of the quantity corresponding to the operator $\hat{O}$ at a time t. However we construct our quantum mechanics, we must end up with this equation for expectation values to guarantee that we get the same prediction for the result of any measurement.[5]

We can also formulate things as follows by only worrying about the wave function at $t = 0$. After all, if we know $\psi(0)$, we can always calculate $\psi(t)$ using

$$\psi(t) = \hat{U}(t,0)\psi(0) = e^{-i\hat{H}t}\psi(0). \tag{8.9}$$

Applying this to the expectation value in eqn 8.8, we can write

$$\langle \psi(t) | \hat{O} | \psi(t) \rangle = \langle \psi(0) | \hat{U}^{\dagger}(t,0) \, \hat{O} \, \hat{U}(t,0) | \psi(0) \rangle. \tag{8.10}$$

We can interpret this equation in two ways. In the Schrödinger picture we think of time-independent operators $\hat{O}_S \equiv \hat{O}$ and time-dependent wave function states $|\psi_S(t)\rangle \equiv \hat{U}(t,0)|\psi(0)\rangle$, so that

$$\langle \psi(0) | \hat{U}^{\dagger}(t,0) \left[\hat{O} \right] \hat{U}(t,0) | \psi(0) \rangle = \langle \psi_S(t) | \hat{O}_S | \psi_S(t) \rangle, \tag{8.11}$$

where the 'S' subscripts stand for 'Schrödinger picture'. To work out how the states evolve with time, we use the Schrödinger equation, eqn 8.1. However, we can also move the square brackets in eqn 8.11, so that it reads

$$\langle \psi(0) | \left[\hat{U}^{\dagger}(t,0) \, \hat{O} \, \hat{U}(t,0) \right] | \psi(0) \rangle = \langle \psi_H | \hat{O}_H(t) | \psi_H \rangle, \tag{8.12}$$

where the 'H' stands for the **Heisenberg picture**. We have made the wave functions $\psi_H \equiv \psi(0)$ time-independent (essentially by freezing the wave functions at their $t = 0$ values) and all the time dependence is now formally located in the operators $\hat{O}_H(t)$ which are given by

$$\hat{O}_H(t) \equiv \hat{U}^{\dagger}(t,0) \, \hat{O}_S \, \hat{U}(t,0), \tag{8.13}$$

a situation which is analogous to the dynamical variables in classical mechanics. The time dependence of $\hat{O}_H(t)$ is simply given by differentiating eqn 8.13, so that[6]

$$\frac{d\hat{O}_H(t)}{dt} = \frac{d\hat{U}^{\dagger}}{dt}\hat{O}_S\hat{U} + \hat{U}^{\dagger}\hat{O}_S\frac{d\hat{U}}{dt} = \frac{1}{i}(-\hat{U}^{\dagger}\hat{H}\hat{O}_S\hat{U} + \hat{U}^{\dagger}\hat{O}_S\hat{H}\hat{U}), \tag{8.14}$$

and hence we arrive[7] at the result

$$\frac{d\hat{O}_H(t)}{dt} = \frac{1}{i\hbar}[\hat{O}_H(t), \hat{H}], \tag{8.15}$$

which is known as the **Heisenberg equation of motion**.

[5]Another example of a quantity we want to agree on is the amplitude for a system being in state $|\phi(t_2)\rangle$ at a time t_2, having been prepared in a state $|\psi(t_1)\rangle$ at an earlier time t_1, which is given by $\langle \phi(t_2)|\psi(t_1)\rangle$. (Strictly we should only insist on the probability being the same, so we could allow a difference in phase.)

Werner Heisenberg (1901–1976)

[6]Note that we are once again using property 3 of the time evolution operator in the form

$$\frac{d\hat{U}}{dt} = \frac{\hat{H}\hat{U}}{i}, \qquad \frac{d\hat{U}^{\dagger}}{dt} = -\frac{\hat{U}^{\dagger}\hat{H}}{i}.$$

[7]We do this using eqn 8.13 and the fact that $\hat{U}$ and $\hat{H}$ commute.

8.3   The death of single-particle quantum mechanics

The time-evolution operator gives us enough ammunition to kill off single-particle quantum mechanics forever. We ask the question: 'what is the amplitude for a particle to travel outside its forward light-cone?' (see Fig. 8.2), or in other words: 'what is the amplitude for a particle at the origin in spacetime ($\boldsymbol{x} = 0$, $t = 0$) to travel to position $\boldsymbol{x}$ at time t which would require it to travel faster than light?' Faster than light means $|\boldsymbol{x}| > t$ and the interval $(\boldsymbol{x} - 0, t - 0)$ is space-like.[8] If the amplitude is nonzero then there will be a nonzero probability for a particle to be found outside its forward light-cone. This is unacceptable and would spell the death of quantum theory as we've known it so far.

To answer the question we are going to evaluate the expression

$$\mathcal{A} = \langle \boldsymbol{x} | e^{-i\hat{H}t} | \boldsymbol{x} = 0 \rangle, \tag{8.16}$$

for the space-like interval. We'll work with a basis of momentum states $\hat{H}|\boldsymbol{p}\rangle = E_{\boldsymbol{p}}|\boldsymbol{p}\rangle$ where the particle has a relativistic dispersion $E_{\boldsymbol{p}} = \sqrt{\boldsymbol{p}^2 + m^2}$. We'll also need the transformation between position and momentum bases:[9] $\langle \boldsymbol{x} | \boldsymbol{p} \rangle = \frac{1}{(2\pi)^{3/2}} e^{i\boldsymbol{p}\cdot\boldsymbol{x}}$.

We start by inserting a resolution of the identity[10]

$$\begin{aligned}
\langle \boldsymbol{x} | e^{-i\hat{H}t} | \boldsymbol{x} = 0 \rangle &= \int d^3 p \, \langle \boldsymbol{x} | e^{-i\hat{H}t} | \boldsymbol{p} \rangle \langle \boldsymbol{p} | \boldsymbol{x} = 0 \rangle \\
&= \int d^3 p \, \langle \boldsymbol{x} | \boldsymbol{p} \rangle e^{-iE_{\boldsymbol{p}}t} \frac{1}{(2\pi)^{3/2}} e^{-i\boldsymbol{p}\cdot 0} \\
&= \int d^3 p \, \frac{1}{(2\pi)^3} e^{i\boldsymbol{p}\cdot\boldsymbol{x}} e^{-iE_{\boldsymbol{p}}t}. \tag{8.17}
\end{aligned}$$

This has been fairly straightforward so far. Now we need to do the integral.

Example 8.1

The technique for doing an integral like this is very useful in quantum field theory, so we'll go through it in detail. We start by converting to spherical polars

$$\begin{aligned}
\mathcal{A} &= \int_0^{2\pi} d\phi \int_0^\infty \frac{d|\boldsymbol{p}|}{(2\pi)^3} |\boldsymbol{p}|^2 \int_{-1}^1 d(\cos\theta) \, e^{i|\boldsymbol{p}||\boldsymbol{x}|\cos\theta} e^{-iE_{\boldsymbol{p}}t} \\
&= \frac{1}{(2\pi)^2} \frac{1}{i|\boldsymbol{x}|} \int_0^\infty d|\boldsymbol{p}| \, |\boldsymbol{p}| \left(e^{i|\boldsymbol{p}||\boldsymbol{x}|} - e^{-i|\boldsymbol{p}||\boldsymbol{x}|} \right) e^{-iE_{\boldsymbol{p}}t} \\
&= \frac{-i}{(2\pi)^2 |\boldsymbol{x}|} \int_{-\infty}^\infty d|\boldsymbol{p}| \, |\boldsymbol{p}| e^{i|\boldsymbol{p}||\boldsymbol{x}|} e^{-it\sqrt{|\boldsymbol{p}|^2 + m^2}}. \tag{8.18}
\end{aligned}$$

To do this integral we need to resort to complex analysis.[11] The integration path is shown in Fig. 8.3(a). There are cuts on the imaginary axis starting at $\pm im$ and heading off to $\pm\infty$. We can deform the contour into a large semicircle in the upper half-plane that goes down the upper cut on the left side and up the cut on the right [Fig. 8.3(b)].

[8]This means that $|\boldsymbol{x}| > ct$ in unnatural units. Our convention is that if $x = (t, \boldsymbol{x})$ then $x^2 = t^2 - (x^1)^2 - (x^2)^2 - (x^3)^2$ and so the interval is space-like if $x^2 < 0$.

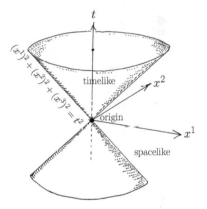

Fig. 8.2 A particle shouldn't be observed outside its forward light-cone. If it starts at the origin, it should only be able to make time-like journeys.

[9]This assumes a particular normalization. For our current purpose, where all we will do is to show a particular quantity is nonzero, the value of the normalization constant is irrelevant.

[10]i.e. since we have a complete set of states then $1 = \int d^3 p \, |\boldsymbol{p}\rangle\langle\boldsymbol{p}|$.

[11]See Appendix B for a primer.

(a)

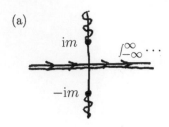

(b)

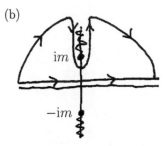

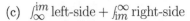

(c) $\int_\infty^{im}$ left-side $+ \int_{im}^\infty$ right-side

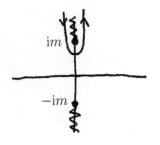

Fig. 8.3 The contour integration.

[12]The Compton wavelength λ of a particle is given by $\lambda = h/mc$, i.e. h/m in units where $c = 1$.

[13]The same sentiment was also expressed by Hippocrates and by Shakespeare (in Hamlet).

[14]The Andromeda galaxy is some 2.5 million light-years from Earth.

The term $e^{i|\boldsymbol{p}||\boldsymbol{x}|}$ decreases exponentially as we head off to large imaginary values of $|\boldsymbol{p}|$ because we have $|\boldsymbol{x}| > 0$. The term $e^{-it\sqrt{|\boldsymbol{p}|^2+m^2}}$ also decreases exponentially as you head up the left side of the cut but increases as you go up on the right side. (This is because $\mathrm{Im}\left(\sqrt{|\boldsymbol{p}|^2+m^2}\right) < 0$ on the left side and $\mathrm{Im}\left(\sqrt{|\boldsymbol{p}|^2+m^2}\right) > 0$ on the right side.) The increase on the right side isn't a problem since it's multiplied by $e^{i|\boldsymbol{p}||\boldsymbol{x}|}$ and for our condition that $|\boldsymbol{x}| > t$ the product $e^{-i|\boldsymbol{p}||\boldsymbol{x}|}e^{-it\sqrt{|\boldsymbol{p}|^2+m^2}}$ decreases. Jordan's lemma therefore tells us that as we make the half-circle large it doesn't contribute to the integral, therefore, and we only have to worry about the part going up and down the cut. This allows us to deform the integral further [Fig. 8.3(c)]. We now substitute $|\boldsymbol{p}| = iz$ and the integral becomes

$$
\begin{aligned}
\mathcal{A} &= \frac{-i}{(2\pi)^2|\boldsymbol{x}|} \int_m^\infty d(iz)\, ize^{-z|\boldsymbol{x}|} \left(e^{t\sqrt{z^2-m^2}} - e^{-t\sqrt{z^2-m^2}} \right) \\
&= \frac{i}{2\pi^2|\boldsymbol{x}|} e^{-m|\boldsymbol{x}|} \int_m^\infty dz\, ze^{-(z-m)|\boldsymbol{x}|} \sinh\left(t\sqrt{z^2-m^2} \right).
\end{aligned} \tag{8.19}
$$

The integrand is positive, definite and the integral is nonzero. This means that $|\mathcal{A}|^2$ yields a nonzero number, heralding doom for the single-particle picture.

We've discovered that the probability of finding a particle outside its forward light-cone is proportional to $e^{-m|\boldsymbol{x}|}$. Although this is small for large $|\boldsymbol{x}|$ it's certainly unacceptable that it exists at all. We are unable to reconcile single-particle quantum mechanics and special relativity. Single-particle quantum mechanics is dead!

There is another way of understanding why single-particle quantum mechanics fails when faced with special relativity. Imagine squeezing a particle of mass m into a box with side smaller than the particle's Compton wavelength.[12] The uncertainty in position $\Delta x \ll \lambda$ so $\Delta p \gg h/\lambda = m$ and so the energy of the particle is now much greater than m. This is enough, according to special relativity, to lead to pair production (i.e. the creation of particle–antiparticle pairs). Thus, on average, the box has more than one particle in it. Single-particle quantum mechanics can never describe this situation, by definition!

8.4   Old quantum theory is dead; long live fields!

A desperate disease requires a dangerous remedy
Guy Fawkes (1570–1606)[13]

What are we to do instead? We resort to fields. In classical mechanics we assume that an observer has access to all possible measurements of a system. Special relativity teaches us that no one person can measure everything. If information would have to travel faster than c for a measurement to be made it is impossible to make such a measurement. In fact, you make a measurement at a single point in spacetime where you, yourself, are localized. You can't then make a measurement at a point that has a space-like separation from you. More simply, an observer can't make a measurements here on Earth and then make a measurement of an event in Andromeda[14] that occurred one second later. The Andromeda information would have to travel to the Earthbound observer faster than c, which is forbidden by special relativity.

Have we seen this limitation elsewhere in physics? Absolutely. In electromagnetism we calculate and measure the value of an electric or magnetic vector at a spacetime point x. We call these $\boldsymbol{E}(x)$ and $\boldsymbol{B}(x)$ respectively and they are examples of *fields*. They are machines that take spacetime coordinates $x^{\mu} = (t, \boldsymbol{x})$ and output the value of the electric or magnetic vector at that point. The field machines are defined *locally*, and so they only tell us about the electromagnetic vectors at the point x. We know that we can't design an experiment that measures $\boldsymbol{E}(x)$ for $x = $ (now, **here**) and for $x = $ (now, **Andromeda**).[15]

We need to fix up this idea of a locally defined field to make it quantum mechanical. In quantum mechanics we have operators that correspond to observables. It seems, therefore, that we will need **operator-valued fields** if we're to reconcile quantum mechanics and special relativity. We know two things about these field operators already: (i) crucially, they must not allow information to travel faster than c; (ii) they must be Hermitian if they are to correspond to observables.[16] We can state the former property quantum mechanically. If we say that the operator $\hat{O}_1$ corresponds to an observable measured at spacetime point x and $\hat{O}_2$ corresponds to an observable measured at y, then if x and y have a space-like separation $[(x - y)^2 < 0]$ then the operators must commute:

$$[\hat{O}(x), \hat{O}(y)] = 0, \tag{8.20}$$

that is, the result of one measurement won't affect the result of the other. How could they if the separation between the experiments is space-like?

To make quantum fields we need to find operators that are defined at positions in spacetime. We'll call these $\hat{\phi}(x)$. We hope that we can build everything from these locally defined fields.[17] In the next few chapters we will continue our search for these objects.

[15] Of course, if we solve some equations of motion for the fields we may be able to work out what's happening elsewhere.

[16] It is not essential that a general field should necessarily correspond to an observable. Indeed we will meet many examples where this is not the case. For electric and magnetic fields we do seek Hermitian fields.

[17] This turns out to be the case!

Chapter summary

- The time-evolution operator is given by $\hat{U}(t_2, t_1) = \mathrm{e}^{-\mathrm{i}\hat{H}(t_2 - t_1)}$, and in the Schrödinger picture acts to time-evolve states. The operators $\hat{O}_{\mathrm{S}}$ are time-independent.

- In the Heisenberg picture states don't evolve in time, but the operators do: $\hat{O}_{\mathrm{H}}(t) = \hat{U}^{\dagger}(t, 0) \, \hat{O}_{\mathrm{S}} \, \hat{U}(t, 0)$. Operators obey the Heisenberg equation of motion $\frac{\mathrm{d}\hat{O}_{\mathrm{H}}(t)}{\mathrm{d}t} = \frac{1}{\mathrm{i}\hbar}[\hat{O}_{\mathrm{H}}(t), \hat{H}]$.

- Squeezing a particle of mass m into a region smaller that its Compton wavelength h/m can lead to pair production, invalidating single-particle quantum mechanics.

- To fix this, we need operator-valued fields.

Exercises

(8.1) Show that the form of the time evolution operator $\hat{U}(t_2, t_1) = \exp[-i\hat{H}(t_2 - t_1)]$ (as given in eqn 8.7) exhibits properties 1–5 in Section 8.1.

(8.2) For the Hamiltonian

$$\hat{H} = \sum_k E_k \hat{a}_k^\dagger \hat{a}_k, \qquad (8.21)$$

use the Heisenberg equation of motion to show that the time dependence of the operator $\hat{a}_k^\dagger$ is given by

$$\hat{a}_k^\dagger(t) = \hat{a}_k^\dagger(0)\, e^{iE_k t/\hbar}, \qquad (8.22)$$

and find a similar expression for $\hat{a}_k(t)$.

(8.3) For the Hamiltonian in the previous problem find an expression for the time-dependence of the operator $\hat{X} = X_{\ell m} \hat{a}_\ell^\dagger \hat{a}_m$.

(8.4) A spin-$\frac{1}{2}$ particle is in a magnetic field aligned along the y-direction. The Hamiltonian can be written as $\hat{H} = \omega \hat{S}_y$ where ω is a constant (proportional to the magnetic field). Use the Heisenberg equation of motion to show that

$$\frac{\mathrm{d}\hat{S}_{\mathrm{H}}^z}{\mathrm{d}t} = -\omega \hat{S}_{\mathrm{H}}^x, \qquad (8.23)$$

$$\frac{\mathrm{d}\hat{S}_{\mathrm{H}}^x}{\mathrm{d}t} = \omega \hat{S}_{\mathrm{H}}^z, \qquad (8.24)$$

and give a physical interpretation of this result.

Quantum mechanical transformations

When Gregor Samsa awoke one morning from uneasy dreams he found himself transformed in his bed into a giant insect.
Frank Kafka (1883–1924), *The Metamorphosis*

Quantum mechanical states transform (though usually less dramatically than the transformation that affected Gregor Samsa) when you translate or rotate coordinates. In this chapter we are going to look at how quantum states, operators and quantum fields will transform when subjected to translations, rotations or Lorentz boosts. Where do we start? One thing we certainly know how to do is transform coordinates in spacetime. For example, we can translate a position three-vector $\boldsymbol{x}$ by a three-vector $\boldsymbol{a}$, which is achieved simply by adding $\boldsymbol{a}$ to the vector. To put things formally, we act on $\boldsymbol{x}$ with a translation operator $\mathbf{T}(\boldsymbol{a})$ and transform it into $\boldsymbol{x}'$ given by

$$\boldsymbol{x}' = \mathbf{T}(\boldsymbol{a})\boldsymbol{x} = \boldsymbol{x} + \boldsymbol{a}. \tag{9.1}$$

We also know how to rotate a three-vector through an angle θ about the z-axis, using

$$\boldsymbol{x}' = \mathbf{R}(\theta)\boldsymbol{x} = \begin{pmatrix} \cos\theta & -\sin\theta & 0 \\ \sin\theta & \cos\theta & 0 \\ 0 & 0 & 1 \end{pmatrix} \begin{pmatrix} x^1 \\ x^2 \\ x^3 \end{pmatrix}. \tag{9.2}$$

We need to find operators that can transform quantum states, which live in Hilbert space (rather than the Euclidean space of three-vectors or Minkowski space of four-vectors). Here we investigate what form these operators take.

Remark about notation:
We will be careful to distinguish between, on one hand, operators $\mathbf{T}(\boldsymbol{a})$ and $\mathbf{R}(\theta)$ which act on single vectors (like $\boldsymbol{x}$) and, on the other hand, quantum mechanical transformation operators like $\hat{U}(\boldsymbol{a})$ and $\hat{U}(\theta)$ which act on quantum states (like $|\boldsymbol{x}\rangle$) and also (as we shall see) on operators (like $\hat{\boldsymbol{x}}$) and on quantum fields (like $\hat{\phi}(x)$). These quantum mechanical transformation operators will turn out to be unitary, so we shall use the symbol $\hat{U}$ for all of them.

9.1 Translations in spacetime

Let's start with looking at the properties of single-particle quantum systems when we translate them in space. Suppose we're examining a particle localized at a coordinate and we want to translate it to another point in space. We need a quantum mechanical operator $\hat{U}$ that takes a state localized at $\boldsymbol{x}$ and transforms it to a position $\boldsymbol{x}+\boldsymbol{a}$. The translation operator is written as

$$\hat{U}(\boldsymbol{a})|\boldsymbol{x}\rangle = |\boldsymbol{x} + \boldsymbol{a}\rangle. \tag{9.3}$$

[1]The transformation in eqn 9.2 is a right-handed, *active* rotation or a left-handed, *passive* rotation.

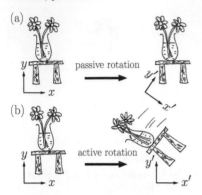

(a) ... (b) ...

Fig. 9.1 (a) A passive rotation acts on the coordinates. (b) An active rotation acts on the object.

[2]A unitary matrix is defined as one for which $U^\dagger = U^{-1}$. Note that some transformations in quantum mechanics must be performed by an *antiunitary* operator as discussed in Chapter 15.

[3]A **group** is a set G together with an operation $\bullet$ which connects two elements of the set (let's say $a, b \in G$) and produces another element of the set (written $a \bullet b$). This has to satisfy four conditions:

(1) $\forall a, b \in G$, $a \bullet b \in G$ (closure). The symbol $\forall$ means 'for all'. The first property written in full is 'For all members of the group a and b, the combination $a \bullet b$ is also a member of the group'.

(2) $\forall a, b, c \in G$, $a \bullet (b \bullet c) = (a \bullet b) \bullet c$ (associativity).

(3) $\exists e \in G$ such that $\forall a \in G$, $a \bullet e = e \bullet a = a$ (identity).

(4) $\forall a \in G$, $\exists a^{-1} \in G$ such that $a \bullet a^{-1} = a^{-1} \bullet a = e$ (inverse).

Note that none of the conditions was that $a \bullet b = b \bullet a$. If that condition (commutativity) holds, then one has a special type of group called an **abelian group**, but many groups (including ones which are important in quantum field theory) are non-abelian.

[4]Named after the Norwegian mathematician Sophus Lie (1842–1899). Note that Lie is pronounced 'lee'.

We will treat this as an *active transformation* where we move a particle to a new position (instead of changing the coordinates under the particle, which is known as a *passive transformation*)[1] (see Fig. 9.1).

Next we examine the properties we want in a translation operator. Translating the particle shouldn't change the probability density. Thus

$$\langle \psi(\boldsymbol{x})|\psi(\boldsymbol{x})\rangle = \langle \psi(\boldsymbol{x}+\boldsymbol{a})|\psi(\boldsymbol{x}+\boldsymbol{a})\rangle = \langle \psi(\boldsymbol{x})|\hat{U}^\dagger(\boldsymbol{a})\hat{U}(\boldsymbol{a})|\psi(\boldsymbol{x})\rangle, \quad (9.4)$$

which means that we may write

$$\hat{U}^\dagger(\boldsymbol{a})\hat{U}(\boldsymbol{a}) = 1, \quad (9.5)$$

that is, the translation operator is **unitary**.[2]

We know that a translation through a distance $\boldsymbol{a}$ followed by one through a distance $\boldsymbol{b}$ can be represented by a single translation through a distance $\boldsymbol{a}+\boldsymbol{b}$. We want our operators to have this property too, so we add to our wish list

$$\hat{U}(\boldsymbol{a})\hat{U}(\boldsymbol{b}) = \hat{U}(\boldsymbol{a}+\boldsymbol{b}). \quad (9.6)$$

We should also have the trivial property that the transformation can do nothing to the particle, that is $\hat{U}(0) = 1$.

To recap then, we have identified three rules for our translation operator:

- $\hat{U}(\boldsymbol{a})\hat{U}^\dagger(\boldsymbol{a}) = 1$ ($\hat{U}$ is unitary),
- $\hat{U}(\boldsymbol{a})\hat{U}(\boldsymbol{b}) = \hat{U}(\boldsymbol{a}+\boldsymbol{b})$ (composition rule),
- $\hat{U}(0) = 1$ (a zero translation does nothing).

Notice that these rules are telling us that our transformations should form a *group*.[3] It is a rather special group because each element depends on a vector $\boldsymbol{a}$, and thus the group is continuous, differentiable and has an infinite number of elements. Such a group is called a **Lie group**.[4]

Example 9.1

Let's check that $\hat{U}(\boldsymbol{a})$ has the required conditions of a group. A group has four properties: the existence of an identity [$\hat{U}(0) = 1$], closure (multiplying any two elements in the group makes another element of the group), associativity [$\hat{U}(\boldsymbol{a}+\boldsymbol{b})\hat{U}(\boldsymbol{c}) = \hat{U}(\boldsymbol{a})\hat{U}(\boldsymbol{b}+\boldsymbol{c})$] and the existence of an inverse for each element of the group. We also know that the inverse of a translation $\hat{T}^{-1}(\boldsymbol{a})$, is just a translation through $-\boldsymbol{a}$. Putting this together with the unitarity of the operator, we can write that $\hat{U}^{-1}(\boldsymbol{a}) = \hat{U}^\dagger(\boldsymbol{a}) = \hat{U}(-\boldsymbol{a})$.

Let's now try out the new operator. It's supposed to move a state though $\boldsymbol{a}$ and we can use the position operator $\hat{\boldsymbol{x}}$ to check where it is:

$$\hat{U}(\boldsymbol{a})|\boldsymbol{x}\rangle = |\boldsymbol{x}+\boldsymbol{a}\rangle$$
$$\hat{\boldsymbol{x}}\hat{U}(\boldsymbol{a})|\boldsymbol{x}\rangle = \hat{\boldsymbol{x}}|\boldsymbol{x}+\boldsymbol{a}\rangle = (\boldsymbol{x}+\boldsymbol{a})|\boldsymbol{x}+\boldsymbol{a}\rangle$$
$$\hat{U}^\dagger(\boldsymbol{a})\hat{\boldsymbol{x}}\hat{U}(\boldsymbol{a})|\boldsymbol{x}\rangle = (\boldsymbol{x}+\boldsymbol{a})\hat{U}^\dagger(\boldsymbol{a})|\boldsymbol{x}+\boldsymbol{a}\rangle = (\boldsymbol{x}+\boldsymbol{a})|\boldsymbol{x}\rangle, \quad (9.7)$$

or comparing the effects of operators on the state $|\boldsymbol{x}\rangle$, we see that

$$\hat{U}^\dagger(\boldsymbol{a})\hat{\boldsymbol{x}}\hat{U}(\boldsymbol{a}) = (\hat{\boldsymbol{x}}+\boldsymbol{a}). \quad (9.8)$$

That is, the space translation operator can be thought of as a transformation of an operator like $\hat{\boldsymbol{x}}$ instead of a transformation of a wave function like $|\boldsymbol{x}\rangle$.

Example 9.2

Often translating will have no effect on a property that we determine using an operator $\hat{O}$. In that case the property probed by that operator is an **invariant**.[5] In that case we would have

$$\langle\psi(\boldsymbol{x})|\hat{O}|\psi(\boldsymbol{x})\rangle = \langle\psi(\boldsymbol{x})|\hat{U}^{-1}(\boldsymbol{a})\hat{O}\hat{U}(\boldsymbol{a})|\psi(\boldsymbol{x})\rangle. \tag{9.9}$$

Pulling out the operators we see that the condition for an invariant is

$$\hat{U}^{-1}(\boldsymbol{a})\hat{O}\hat{U}(\boldsymbol{a}) = \hat{O}. \tag{9.10}$$

Act on the left with $\hat{U}$ and thus invariance implies that $\hat{O}\hat{U} = \hat{U}\hat{O}$, or $[\hat{O},\hat{U}] = 0$. Recall from eqn 8.15 that operators that commute with the Hamiltonian operator $\hat{H}$ represent constants of the motion, i.e. quantities that are conserved.

[5]See Chapter 10.

Can we come up with an explicit expression for a translation operator that acts on quantum states? We will do this for the special case of a position wave function $\psi(x) = \langle x|p\rangle$ (the result will turn out to also allow us to transform general quantum mechanical states). We increment the point at which we evaluate $\psi(x)$ by an infinitesimal amount δa in the x-direction, for which we may write

$$\psi(x+\delta a) = \psi(x) + \frac{d\psi(x)}{dx}\delta a + \dots. \tag{9.11}$$

Remembering that the momentum operator $\hat{p} = -\mathrm{i}\frac{d}{dx}$, we have

$$\psi(x+\delta a) = (1+\mathrm{i}\hat{p}\delta a)\,\psi(x). \tag{9.12}$$

We say that the momentum operator $\hat{p}$ is the **generator** for the space translation. To make a translation through a distance a we can translate through δa a large number N times, such that

$$\begin{aligned}\psi(x+a) &= \lim_{N\to\infty}(1+\mathrm{i}\hat{p}\delta a)^N\,\psi(x)\\ &= \mathrm{e}^{\mathrm{i}\hat{p}a}\psi(x).\end{aligned} \tag{9.13}$$

This gives us a space evolution operator. However, this is not quite the $\hat{U}(\boldsymbol{a})$ that we're trying to find! We want to *translate* the entire quantum state through a distance a not study how it evolves over a distance a! The translation operator is, therefore, $\hat{U}(\boldsymbol{a}) = \mathrm{e}^{-\mathrm{i}\hat{\boldsymbol{p}}\cdot\boldsymbol{a}}$.

Example 9.3

Acting on a momentum state with our operator we obtain

$$\begin{aligned}\hat{U}(\boldsymbol{a})|\boldsymbol{q}\rangle &= \mathrm{e}^{-\mathrm{i}\hat{\boldsymbol{p}}\cdot\boldsymbol{a}}|\boldsymbol{q}\rangle\\ &= \mathrm{e}^{-\mathrm{i}\boldsymbol{q}\cdot\boldsymbol{a}}|\boldsymbol{q}\rangle.\end{aligned} \tag{9.14}$$

Projecting along the coordinate direction gives us a translated wave function

$$\langle\boldsymbol{x}|\hat{U}(\boldsymbol{a})|\boldsymbol{q}\rangle = \langle\boldsymbol{x}|\boldsymbol{q}\rangle\mathrm{e}^{-\mathrm{i}\boldsymbol{q}\cdot\boldsymbol{a}} = \frac{1}{\sqrt{\mathcal{V}}}\mathrm{e}^{\mathrm{i}\boldsymbol{q}\cdot(\boldsymbol{x}-\boldsymbol{a})}. \tag{9.15}$$

Previously we had a time-evolution operator $\hat{U}(t_2,t_1) = \mathrm{e}^{-\mathrm{i}\hat{H}(t_2-t_1)}$. We can see how this relates to the argument in the previous section. Let's evolve a system through a time δt_a. We obtain

$$\psi(t + \delta t_a) = \psi(t) + \frac{d\psi(t)}{dt}\delta t_a + \dots \tag{9.16}$$

Now we need to remember that $\hat{H} = \mathrm{i}\frac{\mathrm{d}}{\mathrm{d}t}$, which gives us

$$\psi(t + \delta t_a) = \left(1 - \mathrm{i}\hat{H}\delta t_a\right)\psi(t). \tag{9.17}$$

Remembering the change of sign required to turn this from a time-evolution operator[6] into a time translation operator we easily obtain

$$\hat{U}(t_a) = \mathrm{e}^{\mathrm{i}\hat{H}t_a}. \tag{9.18}$$

[6]It might be helpful to view time evolution as *passive*: one studies how ψ changes as coordinate t increases. Time translation is then *active*: one moves a quantum state through time.

We can put the space and time translations together and define a space-time translation operator $\hat{U}(a) = \mathrm{e}^{\mathrm{i}\hat{p}\cdot a} = \mathrm{e}^{\mathrm{i}\hat{H}t_a - \mathrm{i}\hat{\boldsymbol{p}}\cdot\boldsymbol{a}}$, where we choose the definition of the four-momentum operator to be $\hat{p} = (\hat{H}, \hat{\boldsymbol{p}})$. We will return to the consequences of this exponential form for the expression for $\hat{U}(a)$ later in the chapter.

9.2   Rotations

Next we ask how to rotate objects. A rotation matrix $\mathbf{R}(\boldsymbol{\theta})$ acts on a vector quantity, such as the momentum of a particle as follows: $\boldsymbol{p}' = \mathbf{R}(\boldsymbol{\theta})\boldsymbol{p}$. We specify the rotation as $\mathbf{R}(\boldsymbol{\theta})$, where the direction of the vector $\boldsymbol{\theta}$ is the axis of rotation and its magnitude is the angle. For rotations of quantum states, we propose an operator

$$|\boldsymbol{p}'\rangle = \hat{U}(\boldsymbol{\theta})|\boldsymbol{p}\rangle = |\mathbf{R}(\boldsymbol{\theta})\boldsymbol{p}\rangle. \tag{9.19}$$

Remark about notation:
Spacetime translations, rotations, Lorentz boosts and combinations of the above are all elements of the Poincaré group. They can all be represented by unitary operators. To keep ourselves from an overload of brackets and indices, we will simplify the notation by denoting all these unitary operators by $\hat{U}$ and let the operator's argument indicate whether we are dealing with a translation $\hat{U}(\boldsymbol{a})$, rotation $\hat{U}(\boldsymbol{\theta})$ or Lorentz boost $\hat{U}(\boldsymbol{\phi})$. This should not cause confusion as the context will make it clear which type of transformation we are dealing with.

As before, we require that $\hat{U}(\boldsymbol{\theta})$ has: (i) unitarity, (ii) a composition rule, and (iii) an identity element. What we're looking for then is that

- $\hat{U}^\dagger(\boldsymbol{\theta})\hat{U}(\boldsymbol{\theta}) = 1$,
- $\hat{U}(\boldsymbol{\theta}_1)\hat{U}(\boldsymbol{\theta}_2) = \hat{U}(\boldsymbol{\theta}_{12})$ where $\mathbf{R}(\boldsymbol{\theta}_{12}) = \mathbf{R}(\boldsymbol{\theta}_1)\mathbf{R}(\boldsymbol{\theta}_2)$,
- $\hat{U}(0) = 1$,

and, just as before, our operators form a Lie group, called the rotation group.

Example 9.4
We can prove that the operator defined this way is unitary

$$\begin{aligned}
\hat{U}(\boldsymbol{\theta})\hat{U}^\dagger(\boldsymbol{\theta}) &= \hat{U}(\boldsymbol{\theta})\left(\int \mathrm{d}^3 p\, |\boldsymbol{p}\rangle\langle\boldsymbol{p}|\right)\hat{U}^\dagger(\boldsymbol{\theta}) \\
&= \int \mathrm{d}^3 p\, |\mathbf{R}(\boldsymbol{\theta})\boldsymbol{p}\rangle\langle\mathbf{R}(\boldsymbol{\theta})\boldsymbol{p}| \tag{9.20}
\end{aligned}$$

and by defining $\boldsymbol{p}' = \mathbf{R}(\boldsymbol{\theta})\boldsymbol{p}$, so that $\mathrm{d}^3 p' = \mathrm{d}^3 p$, we have

$$\int \mathrm{d}^3 p\, |\mathbf{R}(\boldsymbol{\theta})\boldsymbol{p}\rangle\langle\mathbf{R}(\boldsymbol{\theta})\boldsymbol{p}| = \int \mathrm{d}^3 p'\, |\boldsymbol{p}'\rangle\langle\boldsymbol{p}'| = 1, \qquad (9.21)$$

as required. Notice that our proof rests on the fact that $\mathrm{d}^3 p' = \mathrm{d}^3 p$, which is true since $\det \mathbf{R}(\boldsymbol{\theta}) = 1$ for a proper rotation and so the Jacobian is unity.

[7]The proof for this goes through much like the version in Example 9.1 and is left as an exercise.

For the translation case we had

$$\hat{U}^\dagger(\boldsymbol{a})\, \hat{\boldsymbol{x}}\, \hat{U}(\boldsymbol{a}) = \hat{\boldsymbol{x}} + \boldsymbol{a}, \qquad (9.22)$$

and we find[7] an analogous expression for the rotations, i.e.

$$\hat{U}^\dagger(\boldsymbol{\theta})\, \hat{\boldsymbol{p}}\, \hat{U}(\boldsymbol{\theta}) = \mathbf{R}(\boldsymbol{\theta})\hat{\boldsymbol{p}}. \qquad (9.23)$$

Thus the momentum operator is transformed in just the same way as one would rotate a momentum vector.

Finally, we may find an explicit expression for the rotation operator that acts on wave functions. Let's examine a rotation of a wave function about the z-axis and expand

$$\psi(\theta^z + \delta\theta^z) = \psi(\theta^z) + \frac{\mathrm{d}\psi(\theta^z)}{\mathrm{d}\theta^z}\delta\theta^z + \dots \qquad (9.24)$$

Here we recall that $\hat{J}^z = -\mathrm{i}\frac{\mathrm{d}}{\mathrm{d}\theta^z}$, from which we see that rotations are generated by the angular momentum operator. We obtain

$$\psi(\theta + \delta\theta^z) = \left(1 + \mathrm{i}\hat{J}^z\delta\theta^z\right)\psi(\theta^z), \qquad (9.25)$$

and finally, repeating $N \to \infty$ times and changing the sign gives $\hat{U}(\theta^z) = \mathrm{e}^{-\mathrm{i}\hat{J}^z\theta^z}$, and for a rotation about an arbitrary axis we have

$$\hat{U}(\boldsymbol{\theta}) = \mathrm{e}^{-\mathrm{i}\boldsymbol{\hat{J}}\cdot\boldsymbol{\theta}}. \qquad (9.26)$$

9.3   Representations of transformations

We have seen how to transform spacetime coordinates by translations and rotations. But spacetime is an empty coordinate system without filling it with quantum fields. So we now have to ask how the quantum fields themselves transform. A quantum field $\hat{\phi}(x)$ takes a position in spacetime and returns an operator whose eigenvalues can be a scalar, a vector (the $W^\pm$ and Z^0 particles are described by vector fields), a spinor (the object that describes a spin-$\frac{1}{2}$ particle such as an electron), or a tensor. A spatial rotation simply rotates the coordinates for a scalar field [Fig. 9.2(a)] but for a vector field simply rotating the coordinates is not enough [Fig. 9.2(b)], you also have to rotate the vectors [Fig. 9.2(c)].

For this reason we need to think a bit more deeply about how transformations are represented, and for concreteness let us keep focussing

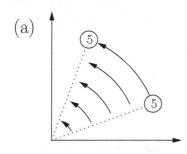

(a)

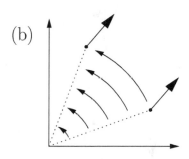

(b)

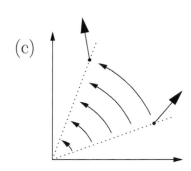

(c)

Fig. 9.2 (a) A spatial rotation simply rotates the coordinates of the scalar fields but does not alter the value of the scalar. (b) Simply rotating the coordinates of a vector field is not a correct transformation since (c) you have to rotate the vectors at each point in space as well.

on rotations. Any rotation $\mathbf{R}(\boldsymbol{\theta})$ can be represented by a square matrix $D(\boldsymbol{\theta})$. This takes a very similar form to $\hat{U}(\boldsymbol{\theta}) = e^{-i\boldsymbol{\hat{J}}\cdot\boldsymbol{\theta}}$, namely

$$D(\boldsymbol{\theta}) = e^{-i\boldsymbol{J}\cdot\boldsymbol{\theta}}. \tag{9.27}$$

[8]A **representation of a group** D is obtained by mapping each element g_i of the group G into a continuous linear operator that acts on some vector space. This mapping preserves the group product, so that if $g_1 \bullet g_2 = g_3$, then $D(g_1)D(g_2) = D(g_3)$.

In this equation, $\boldsymbol{J}$ is a square matrix, a representation[8] of the operator $\boldsymbol{\hat{J}}$. Equation 9.27 can be rewritten as

$$J^i = -\frac{1}{i}\frac{\partial D(\theta^i)}{\partial \theta^i}\bigg|_{\theta^i=0}. \tag{9.28}$$

Example 9.5

(a) Consider rotations about the z-axis. A trivial representation is $D(\theta^z) = 1$. This means that

$$J^z = -\frac{1}{i}\frac{\partial D(\theta^z)}{\partial \theta^z}\bigg|_{\theta^z=0} = 0. \tag{9.29}$$

[9]This makes perfect sense because a scalar can have no angular momentum.

By extension $J^x = J^y = 0$. This representation is appropriate for a scalar field.[9]

(b) It turns out that the representation of a rotation about the z-axis for a spin-$\frac{1}{2}$ particle is

$$D(\theta^z) = \begin{pmatrix} e^{-i\theta^z/2} & 0 \\ 0 & e^{i\theta^z/2} \end{pmatrix}, \tag{9.30}$$

and hence

$$J^z = -\frac{1}{i}\frac{\partial D(\theta^z)}{\partial \theta^z}\bigg|_{\theta^z=0} = \frac{1}{2}\begin{pmatrix} 1 & 0 \\ 0 & -1 \end{pmatrix}, \tag{9.31}$$

The representation for a rotation of a field with angular momentum quantum number j is frequently given the symbol $D^{(j)}(\boldsymbol{\theta})$. Thus in this example, part (a) describes $D^{(0)}(\boldsymbol{\theta}) = 1$, part (b) describes $D^{(1/2)}(\boldsymbol{\theta})$, and part (c) describes a representation related to $D^{(1)}(\boldsymbol{\theta})$ which uses the Cartesian axes $\boldsymbol{x}$, $\boldsymbol{y}$ and $\boldsymbol{z}$ as the basis of the spatial components, rather than the three azimuthal quantum numbers. The general formulae for the matrix elements of $D^{(j)}(\boldsymbol{\theta})$ are

$$\begin{aligned}[D^{(j)}(\boldsymbol{\theta})]_{m,m'} &= \langle jm'|\hat{U}(\boldsymbol{\theta})|jm\rangle \\ &= \langle jm'|e^{-i\boldsymbol{\hat{J}}\cdot\boldsymbol{\theta}}|jm\rangle.\end{aligned}$$

We can then use the standard relations for angular momentum operators:

$$\hat{J}^z|jm\rangle = m|jm\rangle,$$
$$\hat{J}^{\pm}|jm\rangle = \sqrt{(j\mp m)(j+1\pm m)}|jm{\pm}1\rangle,$$
$$\hat{J}^{\pm} = \hat{J}^x \pm i\hat{J}^y.$$

which we recognise as the operator for the z-component of angular momentum for a spin-$\frac{1}{2}$ particle. This representation of the rotation group will be useful for the spinor fields introduced in Chapter 37.

(c) For rotations about the x-axis we have

$$\mathbf{R}(\theta^x) = \begin{pmatrix} 1 & 0 & 0 & 0 \\ 0 & 1 & 0 & 0 \\ 0 & 0 & \cos\theta^x & -\sin\theta^x \\ 0 & 0 & \sin\theta^x & \cos\theta^x \end{pmatrix}, \tag{9.32}$$

so we obtain

$$J^x = -\frac{1}{i}\frac{\partial\mathbf{R}(\theta^x)}{\partial\theta^x}\bigg|_{\theta^x=0} = i\begin{pmatrix} 0 & 0 & 0 & 0 \\ 0 & 0 & 0 & 0 \\ 0 & 0 & 0 & -1 \\ 0 & 0 & 1 & 0 \end{pmatrix}. \tag{9.33}$$

Repeating for the y- and z-axes yields

$$J^y = i\begin{pmatrix} 0 & 0 & 0 & 0 \\ 0 & 0 & 0 & 1 \\ 0 & 0 & 0 & 0 \\ 0 & -1 & 0 & 0 \end{pmatrix} \quad J^z = i\begin{pmatrix} 0 & 0 & 0 & 0 \\ 0 & 0 & -1 & 0 \\ 0 & 1 & 0 & 0 \\ 0 & 0 & 0 & 0 \end{pmatrix}. \tag{9.34}$$

---

[10]This means that there are elements of the Lie group which are arbitrarily close to the identity and so an infinitesimal group element can be written

$$g(\boldsymbol{\alpha}) = 1 + i\alpha^i T^i + O(\alpha^2),$$

where T^i are the generators of the group. The Lie algebra is expressed by the commutator

$$[T^i, T^j] = if^{ijk}T^k,$$

where f^{ijk} are called **structure constants**. For rotations $T^i = J^i$ and $f^{ijk} = \varepsilon^{ijk}$.

The important point about these representations is that they all share the same underlying algebraic structure as the rotation operator. This algebra is called a **Lie algebra**, and can turn up whenever you have a continuous group.[10] The Lie algebra is encoded within the commutation relations of the generators. In other words, whether a generator of rotations rotates abstract quantum states in Hilbert space (like $\hat{J}^z = -i\frac{d}{d\theta^z}$), real space vectors (as in the previous example) or complex spinors (as in Chapter 37) each set of generators should have the same commutation relations: $[J^i, J^j] = i\varepsilon^{ijk}J^k$.

9.4 Transformations of quantum fields

We have seen that a transformation associated with a unitary operator $\hat{U}$ will turn a locally-defined operator $\hat{O}$ (such as the position operator) into $\hat{U}^\dagger \hat{O} \hat{U}$ (see eqn 9.8). Now we need to determine how such transformations affect quantum fields. We'll start by examining a scalar field: an operator-valued field whose matrix elements are scalars.

Example 9.6

If we translate both a state and an operator by the same vector distance $\boldsymbol{a}$ then nothing should change. In equations

$$\langle \boldsymbol{y} | \hat{\phi}(\boldsymbol{x}) | \boldsymbol{y} \rangle = \langle \boldsymbol{y} + \boldsymbol{a} | \hat{\phi}(\boldsymbol{x} + \boldsymbol{a}) | \boldsymbol{y} + \boldsymbol{a} \rangle. \tag{9.35}$$

Since $| \boldsymbol{y} + \boldsymbol{a} \rangle = \hat{U}(\boldsymbol{a}) | \boldsymbol{y} \rangle$ we must have

$$\hat{U}(\boldsymbol{a}) \hat{\phi}(\boldsymbol{x}) \hat{U}^\dagger(\boldsymbol{a}) = \hat{\phi}(\boldsymbol{x} + \boldsymbol{a}). \tag{9.36}$$

Notice that this looks like $\hat{U}$ and $\hat{U}^\dagger$ have been bolted on the wrong way round when compared to the expression

$$\hat{U}^\dagger(\boldsymbol{a}) \hat{\boldsymbol{x}} \hat{U}(\boldsymbol{a}) = (\hat{\boldsymbol{x}} + \boldsymbol{a}), \tag{9.37}$$

but these two expressions are talking about different things. The position operator $\hat{\boldsymbol{x}}$ tells us where a particle is localized, the field operator $\hat{\phi}(\boldsymbol{x})$ acts on a localized particle at $\boldsymbol{x}$. There's no reason that they should transform in the same way.

The previous example demonstrates that there are two ways of thinking about translations operators, and we shouldn't confuse them. They can be pictured:

- As acting on states: $\hat{U}(\boldsymbol{a}) | \boldsymbol{x} \rangle = | \boldsymbol{x} + \boldsymbol{a} \rangle$ moves a locally defined state from being localized at $\boldsymbol{x}$ to being localized at $\boldsymbol{x} + \boldsymbol{a}$.

- As acting on locally defined operators:

$$\hat{U}^\dagger(\boldsymbol{a}) \hat{\phi}(\boldsymbol{x}) \hat{U}(\boldsymbol{a}) = \hat{\phi}(\boldsymbol{x} - \boldsymbol{a}). \tag{9.38}$$

These are not the same as we can see from Fig. 9.3. We can imagine a field operator $\hat{\phi}(\boldsymbol{x})$ as a sniper's rifle shooting a state at a position $\boldsymbol{x}$ [Fig. 9.3(a)]. If we translate the *state* from $|\boldsymbol{x}\rangle$ to $|\boldsymbol{x}+\boldsymbol{a}\rangle$, $\hat{\phi}(\boldsymbol{x})$ misses the state because we've moved the state [Fig. 9.3(b)]. If, on the other hand, we translate the operator from $\hat{\phi}(\boldsymbol{x})$ to $\hat{\phi}(\boldsymbol{x}-\boldsymbol{a})$, we miss again, but this time because we've changed the position we're aiming at [Fig. 9.3(c)]. Since we're interested in the fields, changing the operator will be the most useful procedure and so the equation to focus on is eqn 9.38.

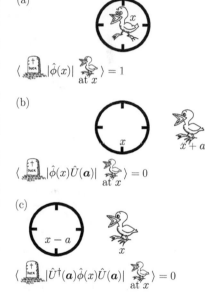

(a)

$$\langle \text{🪦} | \hat{\phi}(\boldsymbol{x}) | \underset{\text{at } x}{🦆} \rangle = 1$$

(b)

$$\langle \text{🪦} | \hat{\phi}(\boldsymbol{x}) \hat{U}(\boldsymbol{a}) | \underset{\text{at } x}{🦆} \rangle = 0$$

(c)

$$\langle \text{🪦} | \hat{U}^\dagger(\boldsymbol{a}) \hat{\phi}(\boldsymbol{x}) \hat{U}(\boldsymbol{a}) | \underset{\text{at } x}{🦆} \rangle = 0$$

Fig. 9.3 The field operator $\hat{\phi}(\boldsymbol{x})$ acting on a localized state is pictured as a sniper's bullet shooting a duck located at position $\boldsymbol{x}$. (a) Dead on target, the state of a dead duck has perfect overlap with the state produced by acting $\hat{\phi}(\boldsymbol{x})$ on the live duck at $\boldsymbol{x}$. (b) Translating the state results in a miss, as does (c) translating the operator.

Example 9.7

The creation and annihilation operators are also actually quantum field operators, albeit in momentum space. That is, the operator $\hat{a}_{\boldsymbol{m}}^\dagger$ takes a momentum $\boldsymbol{m}$ and outputs an operator that creates a particle at that position in momentum space.

We'll examine what we get if we transform according to $\hat{U}^\dagger(\boldsymbol{a})\hat{a}^\dagger_m\hat{U}(\boldsymbol{a})$, which is a translation in spacetime by a vector $\boldsymbol{a}$. We can try out this translated operator on a momentum eigenstate $|\boldsymbol{q}\rangle$ (assumed different to $\boldsymbol{m}$):

$$
\begin{aligned}
\hat{U}^\dagger(\boldsymbol{a})\hat{a}^\dagger_m\hat{U}(\boldsymbol{a})|\boldsymbol{q}\rangle &= \mathrm{e}^{\mathrm{i}\hat{p}\cdot a}\hat{a}^\dagger_m\mathrm{e}^{-\mathrm{i}\hat{p}\cdot a}|\boldsymbol{q}\rangle \\
&= \mathrm{e}^{\mathrm{i}\hat{p}\cdot a}\hat{a}^\dagger_m|\boldsymbol{q}\rangle\mathrm{e}^{-\mathrm{i}q\cdot a} \\
&= \mathrm{e}^{\mathrm{i}\hat{p}\cdot a}|\boldsymbol{m},\boldsymbol{q}\rangle\mathrm{e}^{-\mathrm{i}q\cdot a} \\
&= |\boldsymbol{m},\boldsymbol{q}\rangle\mathrm{e}^{\mathrm{i}(m+q)\cdot a}\mathrm{e}^{-\mathrm{i}q\cdot a} \\
&= |\boldsymbol{m},\boldsymbol{q}\rangle\mathrm{e}^{\mathrm{i}m\cdot a}.
\end{aligned}
\tag{9.39}
$$

So the transformed operator $\hat{a}^\dagger_m$ still creates a state with momentum $\boldsymbol{m}$, but the result of the transformation is an additional phase $\mathrm{e}^{\mathrm{i}m\cdot a}$. We conclude that

$$
\hat{U}^\dagger(\boldsymbol{a})\hat{a}^\dagger_m\hat{U}(\boldsymbol{a}) = \mathrm{e}^{\mathrm{i}m\cdot a}\hat{a}^\dagger_m.
\tag{9.40}
$$

This discussion has been for scalar fields, but we must remember that fields come in many forms, including vector fields as discussed in the following example.

Example 9.8

A classical scalar field $\phi(\boldsymbol{x})$ (like the distribution of temperature across a metal block) has a large maximum $\phi(\boldsymbol{d}) = \phi_{\max}$ at some point in space removed from the origin by a vector $\boldsymbol{d}$. Rotating the distribution with an operator $\mathbf{R}(\boldsymbol{\theta})$ moves the maximum to a position $\mathbf{R}(\boldsymbol{\theta})\boldsymbol{d}$. This is embodied in the operator equation $\hat{U}^\dagger(\boldsymbol{\theta})\hat{\phi}(\boldsymbol{x})\hat{U}(\boldsymbol{\theta}) = \hat{\phi}(\mathbf{R}^{-1}(\boldsymbol{\theta})\boldsymbol{x})$, which demonstrates that we need to enter the position $\boldsymbol{x} = \mathbf{R}(\boldsymbol{\theta})\boldsymbol{d}$ in the transformed field to find the maximum [see Fig. 9.4(a) and (b)].

For a classical vector field $\boldsymbol{V}$, such as the velocity distribution in a liquid, let's imagine that we have a maximum velocity $\boldsymbol{V}(\boldsymbol{d}) = \boldsymbol{V}_{\max}$. Being a vector, this has a direction. A rotation $\mathbf{R}(\boldsymbol{\theta})$ now moves the position of the maximum *and also its direction*. The analogous mathematical description is now $\hat{U}^\dagger(\boldsymbol{\theta})\hat{\boldsymbol{V}}(\boldsymbol{x})\hat{U}(\boldsymbol{\theta}) = \mathbf{R}(\boldsymbol{\theta})\hat{\boldsymbol{V}}(\mathbf{R}^{-1}(\boldsymbol{\theta})\boldsymbol{x})$, demonstrating that the transformed field has a maximum at $\mathbf{R}(\boldsymbol{\theta})\boldsymbol{d}$ and that this vector is rotated to $\mathbf{R}(\boldsymbol{\theta})\boldsymbol{V}_{\max}$ [see Fig. 9.4(c) and (d)].

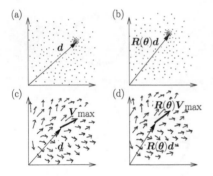

Fig. 9.4 (a) A classical scalar field has a maximum at some point in space removed from the origin by a vector $\boldsymbol{d}$. (b) Rotating the distribution with an operator $\mathbf{R}(\boldsymbol{\theta})$ moves the maximum to a position $\mathbf{R}(\boldsymbol{\theta})\boldsymbol{d}$. (c) A classical vector field also has a maximum value at $\boldsymbol{d}$. (d) The transformed field has a maximum at $\mathbf{R}(\boldsymbol{\theta})\boldsymbol{d}$ and this vector is rotated to $\mathbf{R}(\boldsymbol{\theta})\boldsymbol{V}_{\max}$.

It is therefore important to note that if we transform a field in some way, we generally change the point at which the field is evaluated and also its polarization. We can write a more general rotation of a general field operator $\hat{\boldsymbol{\Phi}}(x)$ as

$$
\hat{U}^\dagger(\boldsymbol{\theta})\hat{\boldsymbol{\Phi}}(x)\hat{U}(\boldsymbol{\theta}) = D(\boldsymbol{\theta})\hat{\boldsymbol{\Phi}}(\mathbf{R}^{-1}(\boldsymbol{\theta})x),
\tag{9.41}
$$

where $D(\boldsymbol{\theta})$ is the appropriate representation of the rotation operator, as discussed in Section 9.3 (e.g. for scalar fields $D(\boldsymbol{\theta}) = 1$).

9.5   Lorentz transformations

A similar philosophy carries over to the case of Lorentz transformations of states and quantum fields.[11] Consider a boost of a four-vector in the

---

[11] As we are now explicitly considering spacetime, we will now switch from labelling our spatial axes x, y and z and go back to using 1, 2 and 3, with 0 for the time axis.

x-direction whose Lorentz transformation is given by $x'^\mu = \Lambda(\beta^1)^\mu{}_\nu x^\nu$, where

$$\Lambda(\beta^1) = \begin{pmatrix} \gamma^1 & \beta^1\gamma^1 & 0 & 0 \\ \beta^1\gamma^1 & \gamma^1 & 0 & 0 \\ 0 & 0 & 1 & 0 \\ 0 & 0 & 0 & 1 \end{pmatrix}. \tag{9.42}$$

This transformation connects two inertial frames moving with relative speed $v = c\beta^1$ along x. Using the substitutions $\gamma^i = \cosh\phi^i$, $\gamma^i\beta^i = \sinh\phi^i$ and $\tanh\phi^i = \beta^i$, where ϕ^i is called the **rapidity**, this matrix becomes

$$\Lambda(\phi^1) = \begin{pmatrix} \cosh\phi^1 & \sinh\phi^1 & 0 & 0 \\ \sinh\phi^1 & \cosh\phi^1 & 0 & 0 \\ 0 & 0 & 1 & 0 \\ 0 & 0 & 0 & 1 \end{pmatrix}. \tag{9.43}$$

We write an operator that acts in Hilbert space:[12] $\hat{U}(\boldsymbol{\phi})|\boldsymbol{p}\rangle = |\boldsymbol{\Lambda}(\boldsymbol{\phi})\boldsymbol{p}\rangle$. We also want a generalized matrix form like we had with the rotation $e^{-i\boldsymbol{J}\cdot\boldsymbol{\theta}}$. From our experience with rotations we suppose that we can write a generalized Lorentz transformation matrix as[13]

$$D(\boldsymbol{\phi}) = e^{i\boldsymbol{K}\cdot\boldsymbol{\phi}}, \tag{9.44}$$

and the generators of the Lorentz transformations are given by

$$K^i = \frac{1}{i}\frac{\partial D(\phi^i)}{\partial\phi^i}\bigg|_{\phi^i=0}. \tag{9.45}$$

As shown in the exercises, we can find the explicit forms for these generators for the case of vectors by taking $D(\phi^i)$ to be the $\boldsymbol{\Lambda}(\phi^i)$ matrices such as that in eqn 9.43 for a boost along x. We therefore require our quantum fields to Lorentz transform according to

$$\hat{U}^\dagger(\boldsymbol{\phi})\hat{\boldsymbol{\Phi}}(x)\hat{U}(\boldsymbol{\phi}) = D(\boldsymbol{\phi})\hat{\boldsymbol{\Phi}}(\boldsymbol{\Lambda}^{-1}(\boldsymbol{\phi})x). \tag{9.46}$$

Finally, we recall that the fundamentally important feature of the generators of a transformation is their commutation relations. Upon working out a set of generators for the Lorentz transformations, as you're invited to in the exercises, it comes as a shock to find

$$[K^1, K^2] = K^1K^2 - K^2K^1 = -iJ^3. \tag{9.47}$$

In other words, the difference between (i) boosting along x and then boosting along y and (ii) boosting along y and then boosting along x is a rotation about z. Mathematically speaking, the fact that $[K^1, K^2] = -iJ^3$ implies that the Lie algebra of the Lorentz transformations isn't *closed* and their generators do not form a group. Let's see what happens if we examine some other commutators. Pressing on, we find

$$\begin{aligned} \left[J^1, K^1\right] &= 0, & (9.48) \\ \left[J^1, K^2\right] &= iK^3, & (9.49) \end{aligned}$$

[12]We will not pursue this operator further here. Unlike linear and angular momentum the boost three-vector is not conserved, so we do not use its eigenvalues to label states.

[13]Note the plus sign in the exponent compared to the rotation case.

along with cyclic permutations, that is to say $\left[J^i, K^j\right] = i\varepsilon^{ijk}K^k$. This implies that the boosts and rotations *taken together* form a closed Lie algebra and it is this larger group that is called the **Lorentz group**. We may write a general Lorentz transformation combining both boosts and rotations as

$$D(\boldsymbol{\theta}, \boldsymbol{\phi}) = e^{-i(\boldsymbol{J}\cdot\boldsymbol{\theta} - \boldsymbol{K}\cdot\boldsymbol{\phi})}, \qquad (9.50)$$

where, as always, the generators $\boldsymbol{J}$ and $\boldsymbol{K}$ are those appropriate for the object being transformed. If you also include the spacetime translations in the mix, you end up with an even larger group called the **Poincaré group**.

Chapter summary

- A transformation is associated with a unitary operator $\hat{U}$ which acts on states via $|x'\rangle = \hat{U}|x\rangle$.
- A general spacetime translation has $\hat{U}(a) = e^{i\hat{p}\cdot a}$.
- A rotation has $\hat{U}(\boldsymbol{\theta}) = e^{-i\hat{\boldsymbol{J}}\cdot\boldsymbol{\theta}}$.
- A Lorentz transformation has $\hat{U}(\boldsymbol{\phi}) = e^{i\hat{\boldsymbol{K}}\cdot\boldsymbol{\phi}}$.
- Transformations can be represented by a matrix $D(a, \boldsymbol{\theta}, \boldsymbol{\phi})$ appropriate to the type of field. For scalar fields $D = 1$.
- The quantum fields that we seek will have to have the following transformation properties:

$$\hat{U}^\dagger(a)\hat{\boldsymbol{\Phi}}(x)\hat{U}(a) = D(a)\,\hat{\boldsymbol{\Phi}}(x - a),$$
$$\hat{U}^\dagger(\boldsymbol{\theta})\hat{\boldsymbol{\Phi}}(x)\hat{U}(\boldsymbol{\theta}) = D(\boldsymbol{\theta})\,\hat{\boldsymbol{\Phi}}(\boldsymbol{R}^{-1}(\boldsymbol{\theta})x),$$
$$\hat{U}^\dagger(\boldsymbol{\phi})\hat{\boldsymbol{\Phi}}(x)\hat{U}(\boldsymbol{\phi}) = D(\boldsymbol{\phi})\,\hat{\boldsymbol{\Phi}}(\boldsymbol{\Lambda}^{-1}(\boldsymbol{\phi})x),$$

i.e. they obey the translation, rotation and Lorentz transformation properties of a local operator.

Exercises

(9.1) Deduce that the generators of the translation operator are given by

$$\hat{p} = -\frac{1}{i}\frac{\partial \hat{U}(a)}{\partial a}\bigg|_{a=0}. \qquad (9.51)$$

(9.2) Show that explicit forms of the generators of the Lorentz group for four-vectors are

$$K^1 = \frac{1}{i}\frac{\partial D(\phi^1)}{\partial \phi^1}\bigg|_{\phi^1=0} = -i\begin{pmatrix} 0 & 1 & 0 & 0 \\ 1 & 0 & 0 & 0 \\ 0 & 0 & 0 & 0 \\ 0 & 0 & 0 & 0 \end{pmatrix}, \qquad (9.52)$$

and similarly

$$K^2 = -i \begin{pmatrix} 0 & 0 & 1 & 0 \\ 0 & 0 & 0 & 0 \\ 1 & 0 & 0 & 0 \\ 0 & 0 & 0 & 0 \end{pmatrix}, \qquad (9.53)$$

and

$$K^3 = -i \begin{pmatrix} 0 & 0 & 0 & 1 \\ 0 & 0 & 0 & 0 \\ 0 & 0 & 0 & 0 \\ 1 & 0 & 0 & 0 \end{pmatrix}. \qquad (9.54)$$

(9.3) Show that an infinitesimal boost by v^j along the x^j-axis is given by the Lorentz transformation

$$\Lambda^\mu{}_\nu = \begin{pmatrix} 1 & v^1 & v^2 & v^3 \\ v^1 & 1 & 0 & 0 \\ v^2 & 0 & 1 & 0 \\ v^3 & 0 & 0 & 1 \end{pmatrix}. \qquad (9.55)$$

Show also that an infinitesimal rotation by θ^j about x^j is given by

$$\Lambda^\mu{}_\nu = \begin{pmatrix} 1 & 0 & 0 & 0 \\ 0 & 1 & \theta^3 & -\theta^2 \\ 0 & -\theta^3 & 1 & \theta^1 \\ 0 & \theta^2 & -\theta^1 & 1 \end{pmatrix}. \qquad (9.56)$$

Hence show that a general infinitesimal Lorentz transformation can be written $x'^\mu = \Lambda^\mu{}_\nu x^\nu$ where $\Lambda = 1 + \omega$ where

$$\omega^\mu{}_\nu = \begin{pmatrix} 0 & v^1 & v^2 & v^3 \\ v^1 & 0 & \theta^3 & -\theta^2 \\ v^2 & -\theta^3 & 0 & \theta^1 \\ v^3 & \theta^2 & -\theta^1 & 0 \end{pmatrix}. \qquad (9.57)$$

Show that $\omega^{\mu\nu} = \omega^\mu{}_\lambda g^{\lambda\nu}$ and $\omega_{\mu\nu} = g_{\mu\lambda}\omega^\lambda{}_\nu$ are antisymmetric. An antisymmetric 4×4 matrix has six independent parameters. This makes sense as we are encoding three rotations and three boosts.

Show that the relationships between θ^i, v^i and ω^{ij} are

$$\theta^i = -\frac{1}{2}\varepsilon^{ijk}\omega^{jk} \text{ and } v^i = \omega^{0i}. \qquad (9.58)$$

(9.4) A transformation of the Poincaré group combines translations (by vector a^μ), rotations and Lorentz boosts and can be written $x'^\mu = a^\mu + \Lambda^\mu{}_\nu x^\nu$. An infinitesimal transformation of the Poincaré group is thus $x'^\mu = x^\mu + a^\mu + \omega^\mu{}_\nu x^\nu$, where $\omega^\mu{}_\nu$ is given in the previous problem. Therefore a function $f(x)$ transforms to

$$\begin{aligned} f(x') &= f(x + a + \omega x) & (9.59) \\ &= f(x) + a^\mu \partial_\mu f(x) + \omega^\mu{}_\nu x^\nu \partial_\mu f(x). \end{aligned}$$

Use the antisymmetry of $\omega_{\mu\nu}$ to write

$$f(x') = [1 + a^\mu\partial_\mu - \frac{1}{2}\omega_{\mu\nu}(x^\mu\partial^\nu - x^\nu\partial^\mu)]f(x), \qquad (9.60)$$

and hence

$$f(x') = [1 - ia^\mu p_\mu + \frac{i}{2}\omega_{\mu\nu}M^{\mu\nu}]f(x), \qquad (9.61)$$

where $p_\mu = i\partial_\mu$ and $M^{\mu\nu} = i(x^\mu\partial^\nu - x^\nu\partial^\mu)$ are generators of the Poincaré group. Note that $M^{\mu\nu}$ is antisymmetric and is related to the generators $\boldsymbol{J}$ and $\boldsymbol{K}$ by

$$J^i = \frac{1}{2}\varepsilon^{ijk}M^{jk}, \qquad (9.62)$$

so that $\boldsymbol{J} = (M^{23}, M^{31}, M^{12})$, and

$$K^i = M^{0i}, \qquad (9.63)$$

so that $\boldsymbol{K} = (M^{01}, M^{02}, M^{03})$. Also show that

$$\Lambda = \exp(-\frac{i}{2}\omega_{\mu\nu}M^{\mu\nu}), \qquad (9.64)$$

with $\Lambda = e^{-i(\boldsymbol{J}\cdot\boldsymbol{\theta} - \boldsymbol{K}\cdot\boldsymbol{\phi})}$.

10

Symmetry

[1]Emmy Noether (1882–1935). Her name should be pronounced to rhyme with Goethe, Alberta and Norris McWhirter.

[2]Consideration of local symmetries, where a transformation changes from point to point, will be postponed to a later chapter.

The Universe is built on a plan the profound symmetry of which is somehow present in the inner structure of our intellect
Paul Valéry (1871–1945)

In the previous chapter we saw that translations were achieved with an operator that contains the momentum operator, and rotations were achieved using the angular momentum operator. A coincidence? Actually not, and in fact we can go further. If a system possesses some kind of invariance (so that for example it is invariant to translations) then a particular quantity will be conserved. This idea is bound up with a result called Noether's theorem, named after the great mathematician Emmy Noether.[1] This chapter will explore this theorem in detail.

A point to note from the outset. In this chapter we will be using the word symmetry to mean **global symmetry**, i.e. a symmetry possessed by the entire system under consideration. Thus rotational symmetry refers to a symmetry associated with rotation of the coordinates applying to every point in our system.[2]

10.1 Invariance and conservation

Let's get the terminology straight. A quantity is **invariant** when it takes the same value when subjected to some particular transformation, and is then said to exhibit a particular symmetry. A translationally invariant system looks the same however you translate it. A rotationally invariant system appears identical however it is rotated around a particular axis. Electric charge is said to be Lorentz invariant because it takes the same value when viewed in different inertial frames of reference. A system is said to possess a particular symmetry if a property is invariant under a particular transformation, so for example a cylinder possesses rotational symmetry about its axis. A quantity is said to be **conserved** when it takes the same value before and after a particular event. Thus in a collision between particles, the four-momentum is conserved *within* a frame of reference but it is not invariant *between* frames of reference.

Having stressed that invariance and conservation are entirely different concepts, it is now time to point out that they are in fact connected! For example:

- Conservation of *linear momentum* is related to invariance under *spatial translations*.

- Conservation of *angular momentum* is related to invariance under *rotations*.

- Conservation of *energy* is related to invariance under *time translations*.

The general idea is clear: invariances lead to conservation laws.

Example 10.1

(i) The Euler–Lagrange equation (eqn 1.28) can be written as

$$\frac{\partial L}{\partial x^\mu} = \frac{\mathrm{d}}{\mathrm{d}t}\left(\frac{\partial L}{\partial \dot{x}^\mu}\right),\tag{10.1}$$

and using the canonical momentum $p_\mu = \partial L/\partial \dot{x}^\mu$ (eqn 5.3), we can write eqn 10.1 as

$$\frac{\partial L}{\partial x^\mu} = \dot{p}^\mu.\tag{10.2}$$

This has the immediate consequence that if L does not depend on x^μ, then $\partial L/\partial x^\mu = 0$ and hence p^μ is a constant. In other words if the Lagrangian is invariant under a particular component of spacetime translation, the corresponding component of four-momentum is a conserved quantity.

(ii) Another example of this follows from eqn 5.58 of Exercise 5.1 which showed that

$$\frac{\partial L}{\partial t} = -\frac{\mathrm{d}H}{\mathrm{d}t}.\tag{10.3}$$

Thus if L does not depend explicitly on time, then the Hamiltonian H (i.e. the energy) is a conserved quantity.

Under a symmetry transformation, various quantities will change.[3] One efficient way to describe this is to say that a field $\phi(x^\mu)$ will change under a symmetry transformation by an amount parameterized by a quantity λ. For example, for a translation along a spacetime vector a^μ we could write the transformation

$$\phi(x^\mu) \to \phi(x^\mu + \lambda a^\mu).\tag{10.4}$$

The larger λ, the larger the degree the transformation is applied. It will be useful to consider infinitesimal transformations, and for this purpose we adopt the notation

$$D\phi = \left.\frac{\partial \phi}{\partial \lambda}\right|_{\lambda=0}.\tag{10.5}$$

An infinitesimal change in ϕ, induced by an infinitesimal $\delta\lambda$, can be written as

$$\delta\phi = D\phi\,\delta\lambda.\tag{10.6}$$

[3] In this chapter, we are going to assume that the transformations are *continuous*. Thus we are ignoring discrete symmetries which will be considered in Chapter 15.

Example 10.2

In the example of our spacetime translation $\phi(x^\mu) \to \phi(x^\mu + \lambda a^\mu)$, we have

$$D\phi = a^\mu \partial_\mu \phi.\tag{10.7}$$

This result arises as follows: by writing $y^\mu = x^\mu + \lambda a^\mu$, we have

$$\frac{\partial \phi}{\partial \lambda} = \frac{\partial \phi}{\partial y^\mu} \frac{\partial y^\mu}{\partial \lambda}, \tag{10.8}$$

and $\frac{\partial y^\mu}{\partial \lambda} = a^\mu$. Taking the limit of $\lambda \to 0$ gives the required result.

10.2   Noether's theorem

We now turn to Noether's theorem, which shows that where we have a continuous symmetry we also have a conservation law. To identify this conservation law we will look for a divergenceless current (i.e. obeying $\partial_\mu J^\mu = 0$). Such a current is locally conserved (see eqn. 5.35 if in doubt) and its time-like component will give rise to a conserved charge.

Let the field $\phi(x)$ change to $\phi(x) + \delta\phi(x)$, where the variation $\delta\phi(x)$ vanishes on the boundary of the region in spacetime that we're considering.[4] The change in the Lagrangian density $\mathcal{L}$ is then given by

$$\delta\mathcal{L} = \frac{\partial \mathcal{L}}{\partial \phi}\delta\phi + \frac{\partial \mathcal{L}}{\partial(\partial_\mu \phi)}\delta(\partial_\mu \phi). \tag{10.9}$$

We will slightly simplify this equation using the substitution

$$\Pi^\mu(x) = \frac{\partial \mathcal{L}}{\partial(\partial_\mu \phi)}. \tag{10.10}$$

This new object $\Pi^\mu(x)$ is a generalization of what, in eqn 5.46, we called the conjugate momentum $\pi(x) = \delta\mathcal{L}/\delta\dot\phi$. Our new object $\Pi^\mu(x)$, the **momentum density**, is the four-vector generalization of $\pi(x)$. In fact $\pi(x)$ is the time-like component of $\Pi^\mu(x)$, i.e.

$$\Pi^0(x) = \pi(x). \tag{10.11}$$

This substitution allows us to write eqn 10.9 as

$$\delta\mathcal{L} = \frac{\partial \mathcal{L}}{\partial \phi}\delta\phi + \Pi^\mu \delta(\partial_\mu \phi). \tag{10.12}$$

Using $\delta(\partial_\mu \phi) = \partial_\mu(\delta\phi)$, and the simple differentiation of a product

$$\partial_\mu(\Pi^\mu \delta\phi) = \Pi^\mu \partial_\mu(\delta\phi) + (\partial_\mu \Pi^\mu)\delta\phi, \tag{10.13}$$

we have

$$\delta\mathcal{L} = \left(\frac{\partial \mathcal{L}}{\partial \phi} - \partial_\mu \Pi^\mu\right)\delta\phi + \partial_\mu(\Pi^\mu \delta\phi). \tag{10.14}$$

The action S should not change under the transformation and so

$$\delta S = \int \mathrm{d}^4 x \left(\frac{\partial \mathcal{L}}{\partial \phi} - \partial_\mu \Pi^\mu\right)\delta\phi = 0, \tag{10.15}$$

[4]This is a continuous transformation. Noether's theorem applies to these continuous transformations but not to the discrete transformations examined in Chapter 15.

where the second term in eqn 10.14 vanishes when integrated over a large enough volume (using the divergence theorem), and so

$$\frac{\partial \mathcal{L}}{\partial \phi} = \partial_\mu \Pi^\mu, \tag{10.16}$$

which is the Euler–Lagrange equation. Now let's play this argument the other way round and state that ϕ obeys the equation of motion. In this case eqn 10.14 becomes

$$\delta \mathcal{L} = \partial_\mu (\Pi^\mu \delta \phi), \tag{10.17}$$

and eqn 10.6 then allows us to write

$$\delta \mathcal{L} = \partial_\mu (\Pi^\mu D\phi)\delta\lambda. \tag{10.18}$$

Because we are applying a symmetry transformation, the action $S = \int \mathrm{d}^4 x\, \mathcal{L}$ is unchanged as we have said, and so we don't expect $\mathcal{L}$ to change, i.e. $\delta\mathcal{L} = 0$. However, this condition is too restrictive and so in some circumstances we could be persuaded to allow $\mathcal{L}$ to change by the four-divergence of some function $W^\mu(x)$, so that

$$\delta \mathcal{L} = (\partial_\mu W^\mu)\delta\lambda. \tag{10.19}$$

This is because

$$\delta S = \int \mathrm{d}^4 x\, \delta \mathcal{L} = \delta\lambda \int \mathrm{d}^4 x\, \partial_\mu W^\mu = \delta\lambda \int \mathrm{d}\mathcal{A}\, n_\mu W^\mu, \tag{10.20}$$

where $\mathcal{A}$ is the surface of the region of spacetime and n_μ is the outward normal (see Fig. 10.1). Since the fields are held constant on the surface in order to have a well-defined variational principle, the change in action is at most a constant, and so $S' = \int \mathrm{d}^4 x\, (\delta\mathcal{L} + \partial_\mu W^\mu \delta\lambda)$ will be stationary under the same conditions that $S = \int \mathrm{d}^4 x\, \delta\mathcal{L}$ will be stationary. Putting eqns 10.18 and 10.19 together leads to

$$\partial_\mu (\Pi^\mu D\phi - W^\mu) = 0, \tag{10.21}$$

or more simply $\partial_\mu J_N^\mu = 0$ where

$$\boxed{J_N^\mu(x) = \Pi^\mu(x)D\phi(x) - W^\mu(x),} \tag{10.22}$$

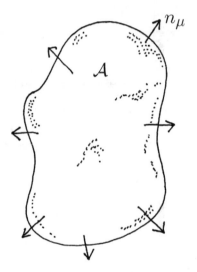

Fig. 10.1 A region of spacetime with surface $\mathcal{A}$. The outward normal is n_μ.

is a locally conserved **Noether current**. We have deduced a form of **Noether's theorem** that states the following:

If a continuous symmetry transformation $\phi \to \phi + D\phi$ only changes $\mathcal{L}$ by the addition of a four-divergence (i.e. $D\mathcal{L} = \partial_\mu W^\mu$) for arbitrary ϕ, then this implies the existence of a current $J_N^\mu = \Pi^\mu D\phi - W^\mu(x)$. If ϕ obeys the equations of motion then the current is conserved, i.e. $\partial_\mu J_N^\mu = 0$.

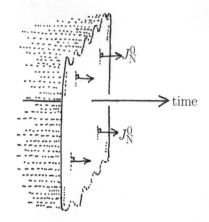

Conserved currents are important because they give rise to **conserved charges** $Q_N = \int J_N^\mu \, \mathrm{d}\mathcal{A}_\mu$ (also called **Noether charges**). A helpful way to apply this idea is to perform the integration over a surface at constant time (see Fig. 10.2). In four dimensions, this 'surface' is a three-volume. Thus

$$Q_N = \int \mathrm{d}^3x \, J_N^0, \qquad (10.23)$$

where J_N^0 is the time-like component normal to the surface. Now since $\partial_\mu J_N^\mu = 0$, then

$$0 = \int \mathrm{d}^3x \, (\partial_\mu J_N^\mu) = \int \mathrm{d}^3x \, (\partial_0 J_N^0 + \partial_k J_N^k). \qquad (10.24)$$

The second term in the integral is $\int \mathrm{d}^3x \, \partial_k J_N^k = \int \mathrm{d}\mathcal{A}_k \, J_N^k$ by the divergence theorem, and this will vanish if the volume is big enough. Thus $\int \mathrm{d}^3x \, \partial_0 J_N^0 = 0$ and since $\partial_0 = \mathrm{d}/\mathrm{d}t$, this implies $\mathrm{d}Q_N/\mathrm{d}t = 0$, i.e. a conserved charge.

Fig. 10.2 The integration is performed over a surface in spacetime at constant time. The component of the Noether current which is perpendicular to this surface is J_N^0, the time-like part.

Everything we have done so far is valid for classical field theories. We might wonder if this discussion can be carried over to the quantum mechanical fields that are the main subject of this book. Fortunately for us Noether's theorem can be carried over to the quantum realm[5] with one addition: the so-called 'normal ordering interpretation' which we introduce in the next chapter. Noether's theorem then provides us with a conserved charge operator $\hat{Q}_N$, whose eigenvalues are (observable) conserved charges. In fact, there is also a quantum mechanical corollary to Noether's theorem which says that if we know the conserved charge operator $\hat{Q}_N$ for a quantum mechanical field $\hat{\phi}$ then we can use it to generate the symmetry transformation $D\hat{\phi}$ via[6]

$$\left[\hat{Q}_N, \hat{\phi}\right] = -\mathrm{i}D\hat{\phi}. \qquad (10.25)$$

This is illustrated in Fig. 10.3. For now we return to classical field theory and continue to investigate the benefits we gain from Noether's theorem.

Using Noether's theorem to find conserved charges

I: Find $D\phi = \left.\frac{\partial \phi}{\partial \lambda}\right|_{\lambda \to 0}$.

II: Find $\Pi^\mu(x) = \frac{\partial \mathcal{L}}{\partial(\partial_\mu \phi)}$.

III: Find $\partial_\mu W^\mu = D\mathcal{L}$.

IV: Write $J_N^\mu = D\phi \Pi^\mu - W^\mu$.

V: Find $Q_N = \int \mathrm{d}^3x J_N^0$.

[5]This assumes that all symmetries of the classical theory will turn out to be symmetries of the quantum theory too. This is true for many important symmetries. It turns out not to be the case for a small number of cases, which are known as *anomalies* and are discussed in more advanced books such as Zee or Peskin and Schroeder.

[6]This result is easily seen by recognizing that conserved charge may be written $Q_N = \int \mathrm{d}^3x \, \Pi^0 D\phi$ and recognizing that Π^0 is the conjugate momentum to ϕ.

10.3 Spacetime translation

The field theories that we will examine in this book will be symmetric with respect to spacetime translations. That is to say that the physical consequences of the theories will be the same whether an experiment takes place at a point $x^\mu = (1066\text{AD}, \textbf{Hastings})$ or $y^\mu = (1863\text{AD}, \textbf{Gettysburg})$. Here we examine the consequences of this using Noether's theorem.

If we subject the coordinates to a variation, so that $x'^\mu = x^\mu + \lambda a^\mu$, then we have that $D\phi = a^\mu \partial_\mu \phi$ and also that

$$D\mathcal{L} = a^\mu \partial_\mu \mathcal{L} = \partial_\mu(a^\mu \mathcal{L}). \qquad (10.26)$$

We recognize therefore that $D\mathcal{L} = \partial_\mu W^\mu$ where $W^\mu = a^\mu \mathcal{L}$ and so we can write down the conserved current

$$
\begin{aligned}
J_N^\mu &= \Pi^\mu D\phi - W^\mu \\
&= \Pi^\mu a^\nu \partial_\nu \phi - a^\mu \mathcal{L} \\
&= a^\nu [\Pi^\mu \partial_\nu \phi - \delta_\nu^\mu \mathcal{L}] \\
&= a_\nu T^{\mu\nu},
\end{aligned}
\tag{10.27}
$$

where $T^{\mu\nu} = \Pi^\mu \partial^\nu \phi - g^{\mu\nu}\mathcal{L}$ is a quantity that is known as the **energy-momentum tensor**. The conserved charges that arise from this current can be written

$$
P^\alpha = \int \mathrm{d}^3 x\, T^{0\alpha}.
\tag{10.28}
$$

Thus the time-like component of this conserved charge is

$$
P^0 = \int \mathrm{d}^3 x\, T^{00} = \int \mathrm{d}^3 x\, [\pi(x)\dot\phi(x) - \mathcal{L}(x)] = \int \mathrm{d}^3 x\, \mathcal{H},
\tag{10.29}
$$

which we recognize as the energy. The space-like components give us

$$
P^k = \int \mathrm{d}^3 x\, T^{0k} = \int \mathrm{d}^3 x\, \pi(x)\partial^k \phi(x),
\tag{10.30}
$$

which we recognize as the momentum. (Note that $g^{00} = 1$ and $g^{0k} = 0$.)

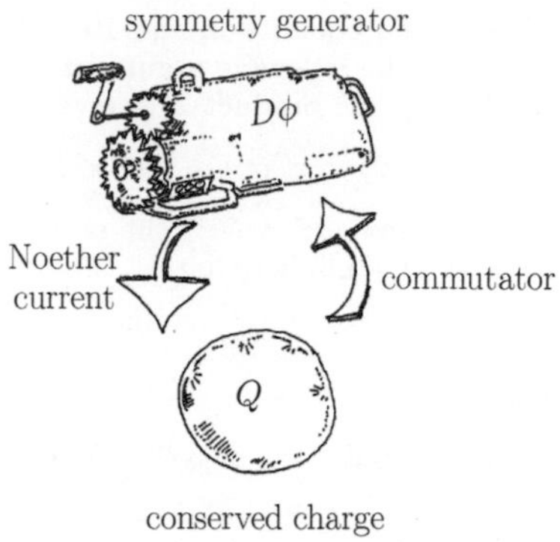

symmetry generator

$D\phi$

Noether current

commutator

Q

conserved charge

Fig. 10.3 Noether's theorem for the field ϕ.

Example 10.3

The energy-momentum tensor $T^{\mu\nu}$ is not uniquely defined. If one adds a term $\partial_\lambda X^{\lambda\mu\nu}$ to $T^{\mu\nu}$, where $X^{\lambda\mu\nu} = -X^{\mu\lambda\nu}$, then show that the new tensor is still divergenceless.
Solution: $\partial_\mu(T^{\mu\nu} + \partial_\lambda X^{\lambda\mu\nu}) = \partial_\mu T^{\mu\nu} + \partial_\mu \partial_\lambda X^{\lambda\mu\nu} = 0 + 0 = 0$. This works because (i) $\partial_\mu T^{\mu\nu} = 0$ because $T^{\mu\nu}$ is a conserved current and (ii) $\partial_\mu \partial_\lambda X^{\lambda\mu\nu} = 0$ follows from the antisymmetry of $X^{\lambda\mu\nu}$ with respect to swapping the first two indices and the fact that you are summing over both indices μ and λ.

The energy-momentum tensor $T^{\mu\nu}$, as we have constructed it, is not symmetric, and that turns out to cause problems when it is used in general relativity,[7] but we can use the trick in the previous example to redefine $T^{\mu\nu} \to T^{\mu\nu} + \partial_\lambda X^{\lambda\mu\nu}$ and force it to be symmetric. This symmetric nature of $T^{\mu\nu}$ is also important when we come to look at Lorentz transformations.

Example 10.4

We will extract some information from $T^{\mu\nu}$. For an infinitesimal Lorentz transformation[8] we can write

$$
\delta x^\mu = \omega^{\mu\nu} x_\nu\, \delta\lambda,
\tag{10.31}
$$

where $\omega^{\mu\nu}$ is an antisymmetric tensor. Hence the change in field ϕ is given by

$$
\delta\phi = \delta x^\mu \partial_\mu \phi = \omega^{\mu\nu} x_\nu \partial_\mu \phi \delta\lambda,
\tag{10.32}
$$

[7] Einstein's field equations are

$$
R_{\mu\nu} - \frac{1}{2}g_{\mu\nu}R = -\frac{8\pi G}{c^4}T_{\mu\nu},
$$

where $R_{\mu\nu}$ is the Ricci tensor, expressing the curvature of spacetime, and R is the Ricci scalar. For further details, see any good book on general relativity, e.g. Penrose (2004).

[8] See Exercise 9.3.

and so

$$D\phi = \omega^{\mu\nu} x_\nu \partial_\mu \phi = \omega_{\mu\nu} x^\nu \partial^\mu \phi \qquad (10.33)$$

so that the effect on the Lagrangian density $\mathcal{L}$ is[9]

$$D\mathcal{L} = \omega_{\rho\sigma} x^\sigma \partial^\rho \mathcal{L} = \partial^\rho \omega_{\rho\sigma} x^\sigma \mathcal{L} = \partial_\mu [g^{\mu\rho} \omega_{\rho\sigma} x^\sigma \mathcal{L}] = \partial_\mu W^\mu, \qquad (10.34)$$

where $W^\mu = g^{\mu\rho} \omega_{\rho\sigma} x^\sigma \mathcal{L}$. Thus we can go straight to our conserved current

$$
\begin{aligned}
J_N^\mu &= \Pi^\mu D\phi - W^\mu \\
&= \omega_{\rho\sigma} [\Pi^\mu x^\sigma \partial^\rho \phi - g^{\mu\rho} x^\sigma \mathcal{L}] \\
&= \omega_{\rho\sigma} x^\sigma T^{\rho\mu} \\
&= \omega_{\rho\sigma} (\tilde{J}^\mu)^{\rho\sigma}, \qquad (10.35)
\end{aligned}
$$

where $(\tilde{J}^\mu)^{\rho\sigma} = x^\sigma T^{\rho\mu}$. There are six parameters in $\omega^{\rho\sigma}$ (remember that it is antisymmetric) and so there are six conserved currents. We will write

$$(J^\mu)^{\rho\sigma} = x^\rho T^{\mu\sigma} - x^\sigma T^{\mu\rho}, \qquad (10.36)$$

where the currents have been antisymmetrized[10] compared with eqn 10.35. These are conserved currents and so $\partial_\mu (J^\mu)^{\rho\sigma} = 0$, although this only works because $T^{\mu\nu}$ is symmetric. The conserved charges[11] are

$$Q^{\lambda\rho} = \int \mathrm{d}^3 x \, (J^0)^{\lambda\rho}, \qquad (10.37)$$

from which we can extract the angular momentum components

$$Q^{ij} = \int \mathrm{d}^3 x \, (x^i T^{j0} - x^j T^{i0}) \qquad (10.38)$$

and the boosts

$$Q^{0i} = \int \mathrm{d}^3 x \, (x^0 T^{i0} - x^i T^{00}). \qquad (10.39)$$

Note that since $\mathrm{d}Q^{0i}/\mathrm{d}t = 0$ and $x^0 = t$, then

$$0 = \int \mathrm{d}^3 x \, T^{i0} + t \int \mathrm{d}^3 x \, \dot{T}^{i0} - \frac{\mathrm{d}}{\mathrm{d}t} \int \mathrm{d}^3 x \, x^i T^{00} = P^i + t\dot{P}^i - \frac{\mathrm{d}}{\mathrm{d}t} \int \mathrm{d}^3 x \, x^i T^{00}, \quad (10.40)$$

and hence $\frac{\mathrm{d}}{\mathrm{d}t} \int \mathrm{d}^3 x \, x^i T^{00}$ is a constant, showing that the centre of mass of the system moves in uniform motion.

[9]The second equality in eqn 10.34 follows from the following argument. Differentiating the product $x^\sigma \mathcal{L}$ gives

$$\partial^\rho (x^\sigma \mathcal{L}) = x^\sigma \partial^\rho \mathcal{L} + \mathcal{L} \partial^\rho x^\sigma,$$

and because $\partial^\rho x^\sigma = \delta^{\rho\sigma}$, the antisymmetry of $\omega_{\rho\sigma}$ implies that

$$\omega_{\rho\sigma} \partial^\rho x^\sigma = \omega_{\rho\sigma} \delta^{\rho\sigma} = 0.$$

Hence

$$
\begin{aligned}
\omega_{\rho\sigma} x^\sigma \partial^\rho \mathcal{L} &= \omega_{\rho\sigma} \partial^\rho (x^\sigma \mathcal{L}) \\
&= \partial^\rho (\omega_{\rho\sigma} x^\sigma \mathcal{L}).
\end{aligned}
$$

[10]We have turned $(\tilde{J}^\mu)^{\rho\sigma}$ into $(J^\mu)^{\rho\sigma}$ using $(J^\mu)^{\rho\sigma} = (\tilde{J}^\mu)^{\rho\sigma} - (\tilde{J}^\mu)^{\sigma\rho}$.

[11]These are Noether conserved charges, but we drop subscript N's here to avoid clutter.

10.4   Other symmetries

We have seen how translations, rotations and boosts can transform spacetime coordinates and also Lagrangians. However, the only effect on the fields has been due to the fact that the fields are functions of the spacetime coordinates. A field, you will remember, is a machine that when fed a coordinate x^μ returns a number $\phi(x^\mu)$. What we have been considering is a scalar field, but of course when the field is more complicated (for example, a vector or a spinor) then we have to worry about how the symmetry transformation affects the field itself, in addition to the effect on the spacetime coordinates of which the field is a function. That will lead to additional layers of complexity which we will save for later. One of the most exciting consequences of this is that if we have a complex scalar field we will be able to show that the $U(1)$ symmetry we noticed in Section 7.6 leads to the conservation of particle number. We will save that gem for Chapter 12, but before that, in the following chapter, we will outline a procedure for quantizing our field theories.

Chapter summary

- Noether's theorem states that if a continuous symmetry transformation $\phi \to \phi + D\phi$ only changes $\mathcal{L}$ by the addition of a four-divergence (i.e. $D\mathcal{L} = \partial_\mu W^\mu$) for arbitrary ϕ, then this implies the existence of a current $J_N^\mu = \Pi^\mu D\phi - W^\mu(x)$ and if ϕ obeys the equations of motion then the current is conserved, i.e. $\partial_\mu J_N^\mu = 0$.

- In short, Noether's theorem states that continuous symmetries lead to conserved currents.

Exercises

(10.1) Show that $[\phi(x), P^\alpha] = i\partial^\alpha \phi(x)$, where P^α is the conserved charges from spacetime translation (eqn 10.28).

(10.2) Consider a system characterized by N fields $\phi_1, \ldots, \phi_N$. The Lagrangian density is then $\mathcal{L}(\phi_1, \ldots, \phi_N; \partial_\mu \phi_1, \ldots, \partial_\mu \phi_N; x^\mu)$. Show that the Noether current is

$$J^\mu = \sum_a \Pi_a^\mu D\phi^a - W^\mu(x), \qquad (10.41)$$

where $D\mathcal{L} = \partial_\mu W^\mu$.

(10.3) For the Lagrangian

$$\mathcal{L} = \frac{1}{2}(\partial_\mu \phi)^2 - \frac{1}{2}m^2 \phi^2, \qquad (10.42)$$

evaluate $T^{\mu\nu}$ and show that T^{00} agrees with what you would expect from the Hamiltonian for this Lagrangian. Show that $\partial_\mu T^{\mu\nu} = 0$. Derive expressions for $P^0 = \int \mathrm{d}^3 x\, T^{00}$ and $P^i = \int \mathrm{d}^3 x\, T^{0i}$.

(10.4) For the Lagrangian

$$\mathcal{L} = -\frac{1}{4}F_{\mu\nu}F^{\mu\nu} = \frac{1}{2}(\boldsymbol{E}^2 - \boldsymbol{B}^2), \qquad (10.43)$$

show that

$$\Pi^{\sigma\rho} \equiv \frac{\partial \mathcal{L}}{\partial(\partial_\sigma A_\rho)} = -F^{\sigma\rho}. \qquad (10.44)$$

Hence show that the energy-momentum tensor

$$T^\mu_\nu = \Pi^{\mu\sigma}\partial_\nu A_\sigma - \delta^\mu_\nu \mathcal{L}, \qquad (10.45)$$

can be written as

$$T^{\mu\nu} = -F^{\mu\sigma}\partial^\nu A_\sigma + \frac{1}{4}g^{\mu\nu}F^{\alpha\beta}F_{\alpha\beta}. \qquad (10.46)$$

This tensor is not symmetric but we can symmetrize it by adding $\partial_\lambda X^{\lambda\mu\nu}$ where $X^{\lambda\mu\nu} = F^{\mu\lambda}A^\nu$. Show that $X^{\lambda\mu\nu} = -X^{\mu\lambda\nu}$. Show further that the symmetrized energy-momentum tensor $\tilde{T}^{\mu\nu} = T^{\mu\nu} + \partial_\lambda X^{\lambda\mu\nu}$ can be written

$$\tilde{T}^{\mu\nu} = F^{\mu\sigma}F_\sigma^{\ \nu} + \frac{1}{4}g^{\mu\nu}F^{\alpha\beta}F_{\alpha\beta}. \qquad (10.47)$$

Hence show that $\tilde{T}^{00} = \frac{1}{2}(\boldsymbol{E}^2 + \boldsymbol{B}^2)$, the energy density in the electromagnetic field, and $\tilde{T}^{i0} = (\boldsymbol{E} \times \boldsymbol{B})^i$, which is the Poynting vector and describes the energy flow.

11 Canonical quantization of fields

[1]Field operators were introduced in Chapter 4.

You've got to have a system
Harry Hill (1964–)

Quantum field theory allows us to consider a Universe in which there exist different, yet indistinguishable, copies of elementary particles. These particles can be created and destroyed by interactions amongst themselves or with external entities. Quantum field theory allows us to describe such phenomena because particles themselves are simply excitations of quantum fields.

To see how particles emerge from fields we need to develop a way to *quantize* the classical fields we have looked at previously. The good news is that there is a machine available that takes a classical field theory and, after we turn the handle, it spits out a quantum field theory: that is, a theory where quantities are described in terms of the number of quantum particles in the system. The name of the machine is **canonical quantization**. In developing canonical quantization we'll see that particles are added or removed from a system using **field operators**[1] and these are formed from the creation and annihilation operators we found so useful in previous chapters.

11.1 The canonical quantization machine

Canonical quantization is the turn-the-handle method of obtaining a quantum field theory from a classical field theory. The method runs like this:

- **Step I**: Write down a classical Lagrangian density in terms of fields. This is the creative part because there are lots of possible Lagrangians. After this step, everything else is automatic.

- **Step II**: Calculate the momentum density and work out the Hamiltonian density in terms of fields.

- **Step III**: Now treat the fields and momentum density as operators. Impose commutation relations on them to make them quantum mechanical.

- **Step IV**: Expand the fields in terms of creation/annihilation operators. This will allow us to use occupation numbers and stay sane.

- **Step V**: That's it. Congratulations, you are now the proud owner of a working quantum field theory, provided you remember the *normal ordering* interpretation.

We'll illustrate the method with one of the simplest field theories: the theory of the massive scalar field.

Step I: We write down a Lagrangian density for our theory. For massive scalar field theory this was given in eqn 7.7, which we rewrite as

$$\mathcal{L} = \frac{1}{2} \left[\partial_\mu \phi(x)\right]^2 - \frac{1}{2} m^2 \left[\phi(x)\right]^2. \tag{11.1}$$

[Recall that the equation of motion for this theory is the Klein–Gordon equation $(\partial^2 + m^2)\phi = 0$ leading to a dispersion $E_{\boldsymbol{p}}^2 = \boldsymbol{p}^2 + m^2$.]

Step II: Find the momentum density (eqn 10.10) $\Pi^\mu(x)$ given by

$$\Pi^\mu(x) = \frac{\partial \mathcal{L}}{\partial(\partial_\mu \phi(x))}. \tag{11.2}$$

For our Lagrangian in eqn 11.1 this gives $\Pi^\mu(x) = \partial^\mu \phi(x)$. The time-like component[2] of this tensor is $\Pi^0(x) = \pi(x) = \partial^0 \phi(x)$. This allows us to define the Hamiltonian density in terms of the momentum density

$$\mathcal{H} = \Pi^0(x)\partial_0 \phi(x) - \mathcal{L}, \tag{11.3}$$

and using our Lagrangian[3] leads to a Hamiltonian density

$$\begin{aligned}
\mathcal{H} &= \partial^0 \phi(x)\partial_0 \phi(x) - \mathcal{L} \\
&= \frac{1}{2}\left[\partial_0 \phi(x)\right]^2 + \frac{1}{2}\left[\boldsymbol{\nabla}\phi(x)\right]^2 + \frac{1}{2}m^2\left[\phi(x)\right]^2.
\end{aligned} \tag{11.5}$$

The last line in eqn 11.5 tells us that the energy has contributions from (i) a kinetic energy term reflecting changes in the configuration in time, (ii) a 'shear term' giving an energy cost for spatial changes in the field and (iii) a 'mass' term reflecting the potential energy cost of there being a field in space at all. Taken together this has a reassuring look of $E = $ (kinetic energy) $+$ (potential energy), which is what we expect from a Hamiltonian.

Step III: We turn fields into field operators. That is to say, we make them **operator-valued fields**: one may insert a point in spacetime into such an object and obtain an operator. We therefore take $\phi(x) \to \hat{\phi}(x)$ and $\Pi^0(x) \to \hat{\Pi}^0(x)$. To make these field operators quantum mechanical we need to impose commutation relations between them. In single-particle quantum mechanics we have $[\hat{x}, \hat{p}] = i\hbar$. By analogy, we quantize the field theory by defining the **equal-time commutator** for the field operators

$$[\hat{\phi}(t, \boldsymbol{x}), \hat{\Pi}^0(t, \boldsymbol{y})] = i\delta^{(3)}(\boldsymbol{x} - \boldsymbol{y}). \tag{11.6}$$

As the name suggests, this applies at equal times only and otherwise the fields commute. We also have that $[\hat{\phi}(x), \hat{\phi}(y)] = [\hat{\Pi}^0(x), \hat{\Pi}^0(y)] = 0$ (and likewise for the daggered versions). Expressed in terms of these fields, the Hamiltonian density $\mathcal{H}$ is now an operator $\hat{\mathcal{H}}$ which acts on

---

[2]The metric $(+, -, -, -)$, which allows us to say $A^0 = A_0$, lets us swap up and down indices for the zeroth component of any object without incurring the penalty of a minus sign.

[3]which may be helpfully rewritten as

$$\mathcal{L} = \frac{1}{2}\left\{ \left(\frac{\partial \phi}{\partial t}\right)^2 - (\boldsymbol{\nabla}\phi)^2 - m^2\phi^2 \right\}. \tag{11.4}$$

state vectors. This is all well and good, except that we don't know how operators like $\hat{\phi}(x)$ act on occupation number states like $|n_1 n_2 n_3 \ldots\rangle$. What we do know is how creation and annihilation operators act on these vectors. If only we could build fields out of these operators!

Step IV: How do we get any further? The machinery of creation and annihilation operators is so attractive that we'd like to define everything in terms of them. In particular, we found a neat analogy between particles in momentum eigenstates and quanta in oscillators. What we do, therefore, is to expand the field operators in terms of the creation and annihilation operators. We've already seen this in the coupled oscillator problem (eqn 2.68) where we obtained a time-independent position operator of the form

$$\hat{x}_j = \left(\frac{\hbar}{m}\right)^{\frac{1}{2}} \sum_k \frac{1}{(2\omega_k N)^{\frac{1}{2}}} (\hat{a}_k \mathrm{e}^{\mathrm{i}jka} + \hat{a}_k^\dagger \mathrm{e}^{-\mathrm{i}jka}). \tag{11.7}$$

By analogy we write down a time-independent field operator for the continuous case which will look like

We are rewriting our momenta $\boldsymbol{k} \to \boldsymbol{p}$ and our energies $\omega_{\boldsymbol{k}} \to E_{\boldsymbol{p}}$, so the factor $(2\omega_{\boldsymbol{k}})^{1/2} \to (2E_{\boldsymbol{p}})^{1/2}$. The normalization will be discussed below.

$$\hat{\phi}(\boldsymbol{x}) = \int \frac{\mathrm{d}^3 p}{(2\pi)^{\frac{3}{2}}} \frac{1}{(2E_{\boldsymbol{p}})^{\frac{1}{2}}} \left(\hat{a}_{\boldsymbol{p}} \mathrm{e}^{\mathrm{i}\boldsymbol{p}\cdot\boldsymbol{x}} + \hat{a}_{\boldsymbol{p}}^\dagger \mathrm{e}^{-\mathrm{i}\boldsymbol{p}\cdot\boldsymbol{x}}\right), \tag{11.8}$$

with $E_{\boldsymbol{p}} = +(\boldsymbol{p}^2 + m^2)^{\frac{1}{2}}$ (that is, we consider positive roots only, a feature discussed below) and where, as before, our creation and annihilation operators have a commutation relation $[\hat{a}_{\boldsymbol{p}}, \hat{a}_{\boldsymbol{q}}^\dagger] = \delta^{(3)}(\boldsymbol{p} - \boldsymbol{q})$.

The field operators that we've written down are intended to work in the Heisenberg picture. To obtain their time dependence, we hit them with time-evolution operators. We use the Heisenberg prescription for making an operator time dependent:

$$\hat{\phi}(x) = \hat{\phi}(t, \boldsymbol{x}) = \hat{U}^\dagger(t, 0)\hat{\phi}(\boldsymbol{x})\hat{U}(t, 0) = \mathrm{e}^{\mathrm{i}\hat{H}t}\hat{\phi}(\boldsymbol{x})\mathrm{e}^{-\mathrm{i}\hat{H}t}. \tag{11.9}$$

The only part that the $\hat{U}(t, 0) = \mathrm{e}^{-\mathrm{i}\hat{H}t}$ operators affect are the creation and annihilation operators, and we have

$$\hat{U}^\dagger(t, 0)\hat{a}_{\boldsymbol{p}}\hat{U}(t, 0) = \mathrm{e}^{-\mathrm{i}E_{\boldsymbol{p}}t}\hat{a}_{\boldsymbol{p}}, \tag{11.10}$$

which shows that the $\hat{a}_{\boldsymbol{p}}$ picks up $\mathrm{e}^{-\mathrm{i}E_{\boldsymbol{p}}t}$. Similarly, the $\hat{a}_{\boldsymbol{p}}^\dagger$ part will pick up $\mathrm{e}^{\mathrm{i}E_{\boldsymbol{p}}t}$.

Example 11.1

To see this unfolding we'll consider a simple example of one component of the field acting on an example state. The component we'll consider is $\hat{a}_{\boldsymbol{q}}\mathrm{e}^{\mathrm{i}\boldsymbol{q}\cdot\boldsymbol{x}}$ and the example state is $|n_{\boldsymbol{p}} n_{\boldsymbol{q}} n_{\boldsymbol{r}}\rangle$. Taking it one step at a time:

$$\mathrm{e}^{\mathrm{i}\hat{H}t}\hat{a}_{\boldsymbol{q}}\mathrm{e}^{\mathrm{i}\boldsymbol{q}\cdot\boldsymbol{x}}\mathrm{e}^{-\mathrm{i}\hat{H}t}|n_{\boldsymbol{p}} n_{\boldsymbol{q}} n_{\boldsymbol{r}}\rangle$$

$$= \mathrm{e}^{\mathrm{i}\hat{H}t}\hat{a}_{\boldsymbol{q}}|n_{\boldsymbol{p}} n_{\boldsymbol{q}} n_{\boldsymbol{r}}\rangle\mathrm{e}^{\mathrm{i}\boldsymbol{q}\cdot\boldsymbol{x}}\mathrm{e}^{-\mathrm{i}(n_{\boldsymbol{p}}E_{\boldsymbol{p}} + n_{\boldsymbol{q}}E_{\boldsymbol{q}} + n_{\boldsymbol{r}}E_{\boldsymbol{r}})t}$$

$$= \sqrt{n_{\boldsymbol{q}}}\mathrm{e}^{\mathrm{i}\hat{H}t}|n_{\boldsymbol{p}}(n_{\boldsymbol{q}} - 1)n_{\boldsymbol{r}}\rangle\mathrm{e}^{\mathrm{i}\boldsymbol{q}\cdot\boldsymbol{x}}\mathrm{e}^{-\mathrm{i}(n_{\boldsymbol{p}}E_{\boldsymbol{p}} + n_{\boldsymbol{q}}E_{\boldsymbol{q}} + n_{\boldsymbol{r}}E_{\boldsymbol{r}})t}$$

$$= \sqrt{n_{\boldsymbol{q}}}|n_{\boldsymbol{p}}(n_{\boldsymbol{q}} - 1)n_{\boldsymbol{r}}\rangle\mathrm{e}^{\mathrm{i}(n_{\boldsymbol{p}}E_{\boldsymbol{p}} + (n_{\boldsymbol{q}}-1)E_{\boldsymbol{q}} + n_{\boldsymbol{r}}E_{\boldsymbol{r}})t}\mathrm{e}^{\mathrm{i}\boldsymbol{q}\cdot\boldsymbol{x}}\mathrm{e}^{-\mathrm{i}(n_{\boldsymbol{p}}E_{\boldsymbol{p}} + n_{\boldsymbol{q}}E_{\boldsymbol{q}} + n_{\boldsymbol{r}}E_{\boldsymbol{r}})t}$$

$$= \sqrt{n_{\boldsymbol{q}}}|n_{\boldsymbol{p}}(n_{\boldsymbol{q}} - 1)n_{\boldsymbol{r}}\rangle\mathrm{e}^{-\mathrm{i}E_{\boldsymbol{q}}t}\mathrm{e}^{\mathrm{i}\boldsymbol{q}\cdot\boldsymbol{x}}. \tag{11.11}$$

Part of what we're left with, namely $\sqrt{n_q}|n_p(n_q - 1)n_r\rangle$, is exactly the same as if we'd just acted on the original state with $\hat{a}_q$. The effect of dynamicizing this operator has just been to multiply by a factor $e^{-iE_q t}$, so we conclude that the operator we seek is $\hat{a}_q e^{-i(E_q t - q \cdot x)} = \hat{a}_q e^{-iq \cdot x}$.

In summary, what we call the **mode expansion** of the scalar field is given by[4]

$$\hat{\phi}(x) = \int \frac{\mathrm{d}^3 p}{(2\pi)^{\frac{3}{2}}} \frac{1}{(2E_p)^{\frac{1}{2}}} \left(\hat{a}_p e^{-ip \cdot x} + \hat{a}_p^\dagger e^{ip \cdot x}\right), \qquad (11.12)$$

$$\text{with } E_p = +(p^2 + m^2)^{\frac{1}{2}}.$$

[4]This looks a little different from what we had in Chapter 4 when we introduced field operators. We will explain the reason for the difference in Section 11.5.

Note that by expanding out the position field we'll get the momentum expansion for free, since for our scalar field example $\Pi^\mu(x) = \partial^\mu \phi(x)$. Also note that because the field in our classical Lagrangian is a real quantity, the field operator $\hat{\phi}(x)$ should be Hermitian. By inspection $\hat{\phi}^\dagger(x) = \hat{\phi}(x)$, so this is indeed the case.

11.2 Normalizing factors

Before proceeding, we will justify the normalization factors in eqn 11.12. In evaluating integrals over momentum states we have the problem that $\mathrm{d}^3 p$ is not a Lorentz-invariant quantity. We can use $\mathrm{d}^4 p$ where $p = (p^0, \boldsymbol{p})$ is the four-momentum, but for a particle of mass m then only values of the four-momentum which satisfy[5] $p^2 = m^2$ need to be considered. This is known as the **mass shell condition** (see Fig. 11.1). Consequently we can write our integration measure

$$\mathrm{d}^4 p \, \delta(p^2 - m^2) \, \theta(p^0). \qquad (11.13)$$

We have included a Heaviside step function $\theta(p^0)$ to select only positive mass particles.

This section can be skipped if you are happy to take eqn 11.12 on trust. The purpose here is simply to justify the factors $(2\pi)^{3/2}(2E_p)^{1/2}$ which otherwise seem to appear by magic.

[5]The condition $p^2 = m^2$ means that $(p^0)^2 - \boldsymbol{p}^2 = E_p^2 - \boldsymbol{p}^2 = m^2$.

Example 11.2

Show that $\delta(p^2 - m^2)\,\theta(p_0) = \frac{1}{2E_p}\delta(p_0 - E_p)\theta(p_0)$.
We use the identity

$$\delta[f(x)] = \sum_{\{y|f(y)=0\}} \frac{1}{|f'(y)|} \delta(x - y), \qquad (11.14)$$

where the notation tells us that the sum is evaluated for those values of x that make $f(x) = 0$. We take $x = p^0$ and $f(p^0) = p^2 - m^2 = (p^0)^2 - \boldsymbol{p}^2 - m^2$. This gives us that $|f'(p^0)| = 2|p^0|$ and we use the fact that the zeros of $f(p^0)$ occur for $p^0 = \pm(\boldsymbol{p}^2 + m^2)^{\frac{1}{2}} = \pm E_p$ to write

$$\delta(p^2 - m^2)\,\theta(p_0) = \frac{1}{2E_p}[\delta(p_0 - E_p)\theta(p_0) + \delta(p_0 + E_p)\theta(p_0)], \qquad (11.15)$$

and so the result follows (since the second term in eqn 11.15 is zero).

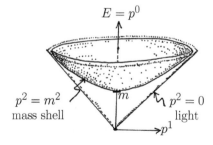

Fig. 11.1 The mass shell $p^2 = m^2$ is a hyperboloid in four-momentum space. Also shown is the equivalent surface for light, $p^2 = 0$.

Thus we will write our Lorentz-invariant measure as

$$\frac{\mathrm{d}^3 p}{(2\pi)^3 2E_{\boldsymbol{p}}}, \tag{11.16}$$

where the additional factor $\delta(p_0 - E_{\boldsymbol{p}})\theta(p_0)$ is there in every calculation and so we suppress it, and we have included the factor $1/(2\pi)^3$ because the mode expansion is essentially an inverse Fourier transform (we have one factor of $1/(2\pi)$ for every component of three-momentum). We are now in a position to write down integrals, for example:

$$1 = \int \frac{\mathrm{d}^3 p}{(2\pi)^3 2E_{\boldsymbol{p}}} |p\rangle\langle p|. \tag{11.17}$$

This requires us to have Lorentz-covariant four-momentum states $|p\rangle$. We previously normalized momentum states according to

$$\langle \boldsymbol{p}|\boldsymbol{q}\rangle = \delta^{(3)}(\boldsymbol{p} - \boldsymbol{q}), \tag{11.18}$$

so our new four-momentum states $|p\rangle$ will need to be related to the three-momentum states $|\boldsymbol{p}\rangle$ by

$$|p\rangle = (2\pi)^{3/2}(2E_{\boldsymbol{p}})^{1/2}|\boldsymbol{p}\rangle, \tag{11.19}$$

and then their normalization can be written

$$\langle p|q\rangle = (2\pi)^3 2E_{\boldsymbol{p}} \delta^{(3)}(\boldsymbol{p} - \boldsymbol{q}). \tag{11.20}$$

Similarly, to make creation operators $\hat{a}^\dagger$ appropriately normalized so that they create Lorentz-covariant states, we must define them by

$$\hat{\alpha}_{\boldsymbol{p}}^\dagger = (2\pi)^{3/2}(2E_{\boldsymbol{p}})^{1/2}\hat{a}_{\boldsymbol{p}}^\dagger, \tag{11.21}$$

so that $\hat{\alpha}_{\boldsymbol{p}}^\dagger|0\rangle = |p\rangle$. In this case our mode expansion would take the form of a simple inverse Fourier transform using our Lorentz-invariant measure

$$\hat{\phi}(x) = \int \frac{\mathrm{d}^3 p}{(2\pi)^3} \frac{1}{(2E_{\boldsymbol{p}})} \left(\hat{\alpha}_{\boldsymbol{p}} \mathrm{e}^{-\mathrm{i}p\cdot x} + \hat{\alpha}_{\boldsymbol{p}}^\dagger \mathrm{e}^{\mathrm{i}p\cdot x} \right), \tag{11.22}$$

or writing in terms of $\hat{a}_{\boldsymbol{p}}^\dagger$ and $\hat{a}_{\boldsymbol{p}}$ rather than $\hat{\alpha}_{\boldsymbol{p}}^\dagger$ and $\hat{\alpha}_{\boldsymbol{p}}$:

$$\hat{\phi}(x) = \int \frac{\mathrm{d}^3 p}{(2\pi)^{\frac{3}{2}}} \frac{1}{(2E_{\boldsymbol{p}})^{\frac{1}{2}}} \left(\hat{a}_{\boldsymbol{p}} \mathrm{e}^{-\mathrm{i}p\cdot x} + \hat{a}_{\boldsymbol{p}}^\dagger \mathrm{e}^{\mathrm{i}p\cdot x} \right), \tag{11.23}$$

which is identical to eqn 11.12.

11.3   What becomes of the Hamiltonian?

We can now substitute our expansion of the field operator $\hat{\phi}(x)$ into the Hamiltonian to complete our programme of canonical quantization. This will provide us with an expression for the energy operator in terms of the creation and annihilation operators. The Hamiltonian is given by the volume integral of the Hamiltonian density,

$$\hat{H} = \int \mathrm{d}^3x \, \frac{1}{2} \left\{ \left[\partial_0 \hat{\phi}(x) \right]^2 + [\boldsymbol{\nabla}\hat{\phi}(x)]^2 + m^2 [\hat{\phi}(x)]^2 \right\}. \qquad (11.24)$$

All we need do is substitute in the mode expansion and use the commutation relations to simplify. We'll start by computing the momentum density $\hat{\Pi}_\mu(x) = \partial_\mu \hat{\phi}(x)$ which is given by

$$\hat{\Pi}_\mu(x) = \partial_\mu \hat{\phi}(x) = \int \frac{\mathrm{d}^3p}{(2\pi)^{\frac{3}{2}}(2E_{\boldsymbol{p}})^{\frac{1}{2}}} (-\mathrm{i}p_\mu) \left(\hat{a}_{\boldsymbol{p}} \mathrm{e}^{-\mathrm{i}p\cdot x} - \hat{a}_{\boldsymbol{p}}^\dagger \mathrm{e}^{\mathrm{i}p\cdot x} \right). \qquad (11.25)$$

To obtain an expression for $\partial_0 \hat{\phi}(x)$, simply take the time-like component of $\Pi_\mu(x)$:

$$\partial_0 \hat{\phi}(x) = \int \frac{\mathrm{d}^3p}{(2\pi)^{\frac{3}{2}}(2E_{\boldsymbol{p}})^{\frac{1}{2}}} (-\mathrm{i}E_{\boldsymbol{p}}) \left(\hat{a}_{\boldsymbol{p}} \mathrm{e}^{-\mathrm{i}p\cdot x} - \hat{a}_{\boldsymbol{p}}^\dagger \mathrm{e}^{\mathrm{i}p\cdot x} \right), \qquad (11.26)$$

while the space-like components provide us with $\boldsymbol{\nabla}\hat{\phi}(x)$:

$$\boldsymbol{\nabla}\hat{\phi}(x) = \int \frac{\mathrm{d}^3p}{(2\pi)^{\frac{3}{2}}(2E_{\boldsymbol{p}})^{\frac{1}{2}}} (\mathrm{i}\boldsymbol{p}) \left(\hat{a}_{\boldsymbol{p}} \mathrm{e}^{-\mathrm{i}p\cdot x} - \hat{a}_{\boldsymbol{p}}^\dagger \mathrm{e}^{\mathrm{i}p\cdot x} \right). \qquad (11.27)$$

We now have all of the ingredients to calculate the Hamiltonian.

Example 11.3

Substitution of the mode expansion leads to a multiple integral of the form

$$
\begin{aligned}
\hat{H} &= \frac{1}{2} \int \frac{\mathrm{d}^3x \, \mathrm{d}^3p \, \mathrm{d}^3q}{(2\pi)^3 (2E_{\boldsymbol{p}})^{\frac{1}{2}}(2E_{\boldsymbol{q}})^{\frac{1}{2}}} \left[(-E_{\boldsymbol{p}}E_{\boldsymbol{q}} - \boldsymbol{p}\cdot\boldsymbol{q})[\hat{a}_{\boldsymbol{p}}\mathrm{e}^{-\mathrm{i}p\cdot x} - \hat{a}_{\boldsymbol{p}}^\dagger \mathrm{e}^{\mathrm{i}p\cdot x}] \right. \\
&\quad \left. \times [\hat{a}_{\boldsymbol{q}}\mathrm{e}^{-\mathrm{i}q\cdot x} - \hat{a}_{\boldsymbol{q}}^\dagger \mathrm{e}^{\mathrm{i}q\cdot x}] + m^2 [\hat{a}_{\boldsymbol{p}}\mathrm{e}^{-\mathrm{i}p\cdot x} + \hat{a}_{\boldsymbol{p}}^\dagger \mathrm{e}^{\mathrm{i}p\cdot x}][\hat{a}_{\boldsymbol{q}}\mathrm{e}^{-\mathrm{i}q\cdot x} + \hat{a}_{\boldsymbol{q}}^\dagger \mathrm{e}^{\mathrm{i}q\cdot x}] \right]. \; (11.28)
\end{aligned}
$$

We use our favourite trick again: first, we integrate over $\boldsymbol{x}$ and use $\int \mathrm{d}^3x \, \mathrm{e}^{\mathrm{i}\boldsymbol{p}\cdot\boldsymbol{x}} = (2\pi)^3 \delta^{(3)}(\boldsymbol{p})$ and obtain

$$
\begin{aligned}
\hat{H} &= \frac{1}{2} \int \frac{\mathrm{d}^3p \, \mathrm{d}^3q}{(2E_{\boldsymbol{p}})^{\frac{1}{2}}(2E_{\boldsymbol{q}})^{\frac{1}{2}}} \qquad\qquad\qquad\qquad\qquad\qquad\qquad (11.29) \\
&\quad \times \left[\delta^{(3)}(\boldsymbol{p}-\boldsymbol{q})(E_{\boldsymbol{p}}E_{\boldsymbol{q}} + \boldsymbol{p}\cdot\boldsymbol{q} + m^2)[\hat{a}_{\boldsymbol{p}}^\dagger \hat{a}_{\boldsymbol{q}} \mathrm{e}^{\mathrm{i}(E_{\boldsymbol{p}}-E_{\boldsymbol{q}})t} + \hat{a}_{\boldsymbol{p}}\hat{a}_{\boldsymbol{q}}^\dagger \mathrm{e}^{-\mathrm{i}(E_{\boldsymbol{p}}-E_{\boldsymbol{q}})t}] \right. \\
&\quad \left. + \delta^{(3)}(\boldsymbol{p}+\boldsymbol{q})(-E_{\boldsymbol{p}}E_{\boldsymbol{q}} - \boldsymbol{p}\cdot\boldsymbol{q} + m^2)[\hat{a}_{\boldsymbol{p}}^\dagger \hat{a}_{\boldsymbol{q}}^\dagger \mathrm{e}^{\mathrm{i}(E_{\boldsymbol{p}}+E_{\boldsymbol{q}})t} + \hat{a}_{\boldsymbol{p}}\hat{a}_{\boldsymbol{q}} \mathrm{e}^{-\mathrm{i}(E_{\boldsymbol{p}}+E_{\boldsymbol{q}})t}] \right].
\end{aligned}
$$

Then we do the integral over $\boldsymbol{q}$ to mop up the delta functions:

$$
\begin{aligned}
\hat{H} &= \frac{1}{2} \int \mathrm{d}^3p \, \frac{1}{2E_{\boldsymbol{p}}} \left[(E_{\boldsymbol{p}}^2 + \boldsymbol{p}^2 + m^2)(\hat{a}_{\boldsymbol{p}}^\dagger \hat{a}_{\boldsymbol{p}} + \hat{a}_{\boldsymbol{p}}\hat{a}_{\boldsymbol{p}}^\dagger) \right. \\
&\quad \left. + (-E_{\boldsymbol{p}}^2 + \boldsymbol{p}^2 + m^2)(\hat{a}_{\boldsymbol{p}}^\dagger \hat{a}_{-\boldsymbol{p}}^\dagger \mathrm{e}^{2\mathrm{i}E_{\boldsymbol{p}}t} + \hat{a}_{\boldsymbol{p}}\hat{a}_{-\boldsymbol{p}}\mathrm{e}^{-2\mathrm{i}E_{\boldsymbol{p}}t}) \right], \qquad (11.30)
\end{aligned}
$$

and since $E_{\boldsymbol{p}}^2 = \boldsymbol{p}^2 + m^2$ this quickly simplifies to

$$\hat{H} = \frac{1}{2} \int \mathrm{d}^3p \, E_{\boldsymbol{p}} (\hat{a}_{\boldsymbol{p}}\hat{a}_{\boldsymbol{p}}^\dagger + \hat{a}_{\boldsymbol{p}}^\dagger \hat{a}_{\boldsymbol{p}}). \qquad (11.31)$$

Using the commutation relations on the result in eqn 11.31 we obtain

$$E = \int \mathrm{d}^3 p \, E_{\boldsymbol{p}} \left(\hat{a}_{\boldsymbol{p}}^\dagger \hat{a}_{\boldsymbol{p}} + \frac{1}{2} \delta^{(3)}(0) \right). \qquad (11.32)$$

The last term should fill us with dread. The term $\frac{1}{2} \int \mathrm{d}^3 p \, \delta^3(0)$ will give us an infinite contribution to the energy! However we should keep in mind that by evaluating the total energy E we're asking the theory to tell us about something we can't measure. This is a nonsensical question and we have paid the price by getting a nonsensical answer. What we can measure is *differences* in energy between two configurations. This is fine in our formalism, since upon taking the differences between two energies the $\frac{1}{2} \delta^{(3)}(0)$ factors will cancel obligingly. However, it's still rather unsatisfactory to have infinite terms hanging around in all of our equations. We need to tame this infinity.

11.4   Normal ordering

To get around the infinity encountered at the end of the last section we define the act of **normal ordering**. Given a set of free fields, we define the normal ordered product as

$$N \left[\hat{A}\hat{B}\hat{C}^\dagger \dots \hat{X}^\dagger \hat{Y} \hat{Z} \right] = \left(\begin{array}{c} \text{Operators rearranged with all} \\ \text{creation operators on the left} \end{array} \right). \quad (11.33)$$

The N is not an operator in the sense that is doesn't act on a state to provide some new information. Instead, normal ordering is an interpretation that we use to eliminate the meaningless infinities that occur in field theories. So the rule goes that 'if you want to tell someone about a string of operators in quantum field theory, you have to normal order them first'. If you don't, you're talking nonsense.[6]

Rearranging operators is fine for Bose fields, but swapping the order of Fermi fields results in the expression picking up a minus sign. We therefore need to multiply by a factor $(-1)^P$, where P is the number of swaps needed to normally order a product of operators.

[6]The origin of the infinity in eqn 11.32 is due to an ordering ambiguity in the classical Lagrangian. Consider a general classical theory formed from complex values fields $\psi(x)$. The Lagrangian describing these fields will contain bilinear terms like $\psi^\dagger \psi$, which could equally well be written in a classical theory as $\psi \psi^\dagger$. When we quantize and insert the mode expansion the order that was chosen suddenly becomes crucial. Normal ordering removes the ambiguity to give a meaningful quantum theory.

Example 11.4

Some examples of normal ordering a string of Bose operators follow. We first consider a single mode operator

$$N \left[\hat{a}\hat{a}^\dagger \right] = \hat{a}^\dagger \hat{a}, \qquad N \left[\hat{a}^\dagger \hat{a} \right] = \hat{a}^\dagger \hat{a}, \qquad (11.34)$$

and

$$N \left[\hat{a}^\dagger \hat{a} \hat{a} \hat{a}^\dagger \hat{a}^\dagger \right] = \hat{a}^\dagger \hat{a}^\dagger \hat{a}^\dagger \hat{a} \hat{a}. \qquad (11.35)$$

Next, consider the case of many modes

$$N \left[\hat{a}_{\boldsymbol{p}} \hat{a}_{\boldsymbol{q}}^\dagger \hat{a}_{\boldsymbol{r}} \right] = \hat{a}_{\boldsymbol{q}}^\dagger \hat{a}_{\boldsymbol{p}} \hat{a}_{\boldsymbol{r}}. \qquad (11.36)$$

The order of $\hat{a}_{\boldsymbol{p}}$ and $\hat{a}_{\boldsymbol{r}}$ doesn't matter since they commute.

For Fermi fields, $N[\hat{c}_{\boldsymbol{p}} \hat{c}_{\boldsymbol{q}}^\dagger \hat{c}_{\boldsymbol{r}}] = -\hat{c}_{\boldsymbol{q}}^\dagger \hat{c}_{\boldsymbol{p}} \hat{c}_{\boldsymbol{r}}$, the sign change occurring because we have performed a single permutation (and hence an odd number of them).

Step V: We can finally complete the programme of canonical quantization. With the normal ordering interpretation the old expression

$$\hat{H} = \frac{1}{2} \int \mathrm{d}^3 p \, E_{\boldsymbol{p}} (\hat{a}_{\boldsymbol{p}} \hat{a}_{\boldsymbol{p}}^\dagger + \hat{a}_{\boldsymbol{p}}^\dagger \hat{a}_{\boldsymbol{p}}), \tag{11.37}$$

becomes

$$\begin{aligned} N[\hat{H}] &= \frac{1}{2} \int \mathrm{d}^3 p \, E_{\boldsymbol{p}} N[\hat{a}_{\boldsymbol{p}} \hat{a}_{\boldsymbol{p}}^\dagger + \hat{a}_{\boldsymbol{p}}^\dagger \hat{a}_{\boldsymbol{p}}] \\ &= \frac{1}{2} \int \mathrm{d}^3 p \, E_{\boldsymbol{p}} 2 \hat{a}_{\boldsymbol{p}}^\dagger \hat{a}_{\boldsymbol{p}} \\ &= \int \mathrm{d}^3 p \, E_{\boldsymbol{p}} \hat{n}_{\boldsymbol{p}}, \end{aligned} \tag{11.38}$$

where $\hat{n}_{\boldsymbol{p}} = \hat{a}_{\boldsymbol{p}}^\dagger \hat{a}_{\boldsymbol{p}}$ is the number operator. Acting on a state it tells you how many excitations there are in that state with momentum $\boldsymbol{p}$.

We now have a Hamiltonian operator that makes sense. It turns out that the Hamiltonian for the scalar field theory is exactly that which we obtained for independent particles in Chapter 3. This isn't so surprising since we started with a Lagrangian describing waves that didn't interact. What we've seen though, is that the excited states of the wave equation can be thought of as particles possessing quantized momenta. These particles could be called scalar phions.[7] They are Bose particles with spin[8] $S = 0$.

Example 11.5

We have claimed that the vacuum energy is unobservable, so may be ignored. However *changes* in vacuum energy terms are physically significant and lead to measurable effects. Such a change results if you adjust the boundary conditions for your field and this is the basis of the **Casimir effect**.[9] This is a small, attractive force between two closely spaced metal plates, which results from the vacuum energy of the electromagnetic field. The essential physics can be understood using a toy model[10] involving a *massless* scalar field in one dimension.

Consider two metal plates I and II separated by a distance L. We put a third plate (III) in between them, a distance x from plate I (see Fig. 11.2). We will derive the force on plate III resulting from the field on either side of it. The presence of the plates forces the field to be quantized according to $k_n = n\pi/x$ or $n\pi/(L-x)$. The dispersion is $E_n = k_n$ and so the total zero-point energy is given by

$$E = \sum_{n=1}^{\infty} \left[\frac{1}{2} \left(\frac{n\pi}{x} \right) + \frac{1}{2} \left(\frac{n\pi}{L-x} \right) \right] = f(x) + f(L-x), \tag{11.39}$$

that is, $\frac{1}{2} \hbar \omega_n$ per mode.[11] These sums both diverge, just as we expect since we are evaluating the infinite vacuum energy.

However, real plates can't reflect radiation of arbitrarily high frequency: the highest energy modes leak out. To take account of this we cut off these high-energy modes thus:

$$\frac{n\pi}{2x} \to \frac{n\pi}{2x} e^{-n\pi a/x}, \tag{11.40}$$

which removes those modes with wavelength significantly smaller than the cut-off a. The value of this cut-off is arbitrary, so we hope that it won't feature in any measurable quantity.

[7]This is because they are excitations of the scalar field ϕ (Greek letter 'phi').

[8]We shall see that spin information is encoded by multiplying the creation and annihilation operators by objects that tell us about the particle's spin polarization. These objects are vectors for $S = 1$ particles and spinors for a $S = 1/2$ particles. For $S = 0$ they do not feature.

[9]The effect was predicted by the Dutch physicist Hendrik Casimir (1909–2000).

[10]This approach follows Zee. A more detailed treatment may be found in Itzykson and Zuber, Chapter 3.

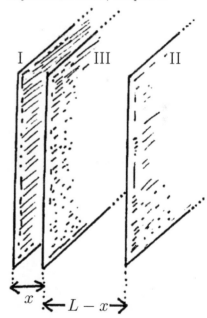

Fig. 11.2 Two metal plates (I and II) separated by a distance L, with a third plate (III) inserted in between.

[11]Remember, for photons $\omega_n = ck_n$, and the zero-point energy is $\frac{1}{2}\hbar\omega_n$, so in units in which $\hbar = c = 1$ this becomes $\frac{1}{2}k_n$, and so $E = \sum_n \frac{1}{2}k_n$. Equation 11.39 contains two such sums.

The sums may now be evaluated:

$$
\begin{aligned}
f(x) &= \sum_n \frac{n\pi}{2x} \mathrm{e}^{-n\pi a/x} \\
&= -\frac{1}{2}\frac{\partial}{\partial a}\sum_n \mathrm{e}^{-n\pi a/x} \\
&= -\frac{1}{2}\frac{\partial}{\partial a}\frac{1}{1 - \mathrm{e}^{-\pi a/x}} \\
&= \frac{\pi}{2x}\frac{\mathrm{e}^{\pi a/x}}{(\mathrm{e}^{\pi a/x} - 1)^2} \approx \frac{x}{2\pi a^2} - \frac{\pi}{24x} + O(a^2).
\end{aligned}
\tag{11.41}
$$

The total energy between plates I and II is

$$
E = f(x) + f(L - x) = \frac{L}{2\pi a^2} - \frac{\pi}{24}\left(\frac{1}{x} + \frac{1}{L-x}\right) + O(a^2).
\tag{11.42}
$$

If x is small compared to L, then we find a force

$$
F = -\frac{\partial E}{\partial x} = -\frac{\pi}{24x^2},
\tag{11.43}
$$

which is independent of a, as we hoped. Thus, there is an attractive force between the closely spaced plates I and III. This is the Casimir force. We can understand this force intuitively by realizing that as the two plates are pulled together we lose the high-energy modes. This reduces the energy between the plates and leads to an attractive force. A more quantum-field-theory friendly interpretation is that the effect results from quantum fluctuations in the vacuum, in which particles are spontaneously created and annihilated. These processes give rise to the vacuum Feynman diagrams described in Chapter 19.

11.5   The meaning of the mode expansion

In the next chapter we'll turn the crank on our canonical quantization machine for the second simplest field theory that we can imagine: that of the complex scalar field. Before doing that, let's have a closer look at the mode expansion for this theory. Our first guess for the field operator might have been the simple Fourier transform

$$
\hat{\phi}^-(x) = \int \frac{\mathrm{d}^3 p}{(2\pi)^{\frac{3}{2}}}\frac{1}{(2E_{\boldsymbol{p}})^{\frac{1}{2}}}\hat{a}_{\boldsymbol{p}}\mathrm{e}^{-\mathrm{i}p\cdot x},
\tag{11.44}
$$

which looks like the one we had in Chapter 4, where the $\mathrm{e}^{-\mathrm{i}p\cdot x}$ tells us that the particles are incoming. Unfortunately, this expansion won't work for the relativistic theory we're considering. The problem is the existence of negative energy solutions in the relativistic equations of motion: each momentum state gave rise to two energies $E_{\boldsymbol{p}} = \pm(\boldsymbol{p}^2 + m^2)$ and we can't just leave out half of the solutions.[12] Looking back at our discussion of the Klein–Gordon equation we saw the resolution of this problem, and we'll employ the same solution here.

What we need is a mode expansion that includes these negative energy modes

$$
\hat{\phi}(x) = \sum_{\boldsymbol{p}}\left(\begin{array}{c}\text{positive } E_{\boldsymbol{p}} \\ \text{mode}\end{array}\right) + \sum_{\boldsymbol{p}}\left(\begin{array}{c}\text{negative } E_{\boldsymbol{p}} \\ \text{mode}\end{array}\right).
\tag{11.45}
$$

[12]Why not? The result would be a field for which $[\hat{\phi}(x), \hat{\phi}(y)] \neq 0$ for space-like $|x - y|$. See Weinberg, Chapter 5 for the details.

The expansion is carried out in terms of incoming plane waves $e^{-ip\cdot x}$. Now recall Feynman's interpretation of the negative frequency modes in which negative energies are assumed to be meaningless, all energies are set to be positive and the signs of three-momenta for such modes are flipped. In this picture, the formerly negative energy states are interpreted as *outgoing* antiparticles. The positive energy modes continue to represent incoming particles. Incoming here means, not only a factor $e^{-ip\cdot x}$, but also that the particle is annihilated: it comes into the system and is absorbed by it. Conversely, outgoing means a factor $e^{ip\cdot x}$ and that the particle is created. We therefore interpret the mode expansion as

$$\phi(x) = \sum_{\boldsymbol{p}} \left(\begin{array}{c} \text{incoming positive } E_{\boldsymbol{p}} \\ \text{particle annihilated} \end{array} \right) + \sum_{\boldsymbol{p}} \left(\begin{array}{c} \text{outgoing positive } E_{\boldsymbol{p}} \\ \text{antiparticle created} \end{array} \right).$$

(11.46)

The resulting expansion of a field annihilation operator is then

$$\hat{\phi}(x) = \int \frac{\mathrm{d}^3 p}{(2\pi)^{\frac{3}{2}}} \frac{1}{(2E_{\boldsymbol{p}})^{\frac{1}{2}}} \left(\hat{a}_{\boldsymbol{p}} e^{-ip\cdot x} + \hat{b}_{\boldsymbol{p}}^{\dagger} e^{ip\cdot x} \right),$$

(11.47)

where the particle and antiparticle energies are given by $E_{\boldsymbol{p}} = +(\boldsymbol{p}^2 + m^2)^{\frac{1}{2}}$. The rules are that $\hat{a}_{\boldsymbol{p}}$ annihilates particles, $\hat{a}_{\boldsymbol{p}}^{\dagger}$ creates particles and the operators $\hat{b}_{\boldsymbol{p}}$ and $\hat{b}_{\boldsymbol{p}}^{\dagger}$ respectively destroy and create antiparticles. (For our scalar field theory the particles and antiparticles are the same: we say that 'each particle is its own antiparticle'. This means that $\hat{a}_{\boldsymbol{p}}$ can be thought of as either an operator that annihilates a particle or one that annihilates an antiparticle.)[13]

[13] As we will see, in other theories we will need separate operators to perform these two distinct roles, but in scalar field theory there is no difference. Thus in this case $\hat{a}_{\boldsymbol{p}} = \hat{b}_{\boldsymbol{p}}$, because in a scalar field theory a particle is its own antiparticle: particles and antiparticles are the same. This will not necessarily be the case for other field theories, an example of which will be described in the next chapter.

Example 11.6

Let's see what happens if we hit the vacuum state with the field annihilation operator from eqn 11.23. Since $\hat{a}_{\boldsymbol{p}}^{\dagger}|0\rangle = |\boldsymbol{p}\rangle$ we have

$$\hat{\phi}(x)|0\rangle = \int \frac{\mathrm{d}^3 p}{(2\pi)^{\frac{3}{2}} (2E_{\boldsymbol{p}})^{\frac{1}{2}}} e^{ip\cdot x} |\boldsymbol{p}\rangle.$$

(11.48)

We have therefore created an outgoing superposition of momentum states. Pick out one of these states and make an amplitude by folding in a relativistically normalized state $\langle q| = (2\pi)^{\frac{3}{2}} (2E_{\boldsymbol{q}})^{\frac{1}{2}} \langle \boldsymbol{q}|$:

$$(2\pi)^{\frac{3}{2}} (2E_{\boldsymbol{p}})^{\frac{1}{2}} \langle \boldsymbol{q}|\hat{\phi}(x)|0\rangle = \int \mathrm{d}^3 p\, e^{ip\cdot x} \langle \boldsymbol{q}|\boldsymbol{p}\rangle = \int \mathrm{d}^3 p\, e^{ip\cdot x} \delta^{(3)}(\boldsymbol{q} - \boldsymbol{p}) = e^{iq\cdot x}.$$

(11.49)

In the language of second quantization, $e^{iq\cdot x}$ tells us how much amplitude there is in the qth momentum mode if we create a scalar particle at spacetime point x. Note that $e^{ip\cdot x} \equiv e^{i(E_{\boldsymbol{p}}t - \boldsymbol{p}\cdot\boldsymbol{x})}$ and so

$$\int \mathrm{d}^3 p\, e^{ip\cdot x} \delta^{(3)}(\boldsymbol{q} - \boldsymbol{p}) = e^{i(E_{\boldsymbol{q}}t - \boldsymbol{q}\cdot\boldsymbol{x})} = e^{iq\cdot x},$$

(11.50)

demonstrating once again that our integral over three-momentum coordinates nevertheless results in a single particle in a four-momentum state.

[14]or, for the complex valued fields, as bilinear in the field and its adjoint.

Canonical quantization for a scalar field:

I Write down a Lagrangian density $\mathcal{L}$.

II Evaluate the momentum density $\Pi^\mu(x) = \partial\mathcal{L}/\partial(\partial_\mu\phi)$ and the Hamiltonian density

$$\mathcal{H} = \Pi^0\partial_0\phi - \mathcal{L}.$$

III Turn fields into operators and enforce the commutation relation $[\hat{\phi}(x), \hat{\Pi}^0(y)] = i\delta^{(3)}(\boldsymbol{x} - \boldsymbol{y})$ at equal times.

IV Express the field in terms of a mode expansion of the form

$$\hat{\phi}(x) = \int \frac{\mathrm{d}^3p}{(2\pi)^{\frac{3}{2}}} \frac{1}{(2E_{\boldsymbol{p}})^{\frac{1}{2}}}$$
$$\times \left(\hat{a}_{\boldsymbol{p}}\mathrm{e}^{-\mathrm{i}p\cdot x} + \hat{a}_{\boldsymbol{p}}^\dagger\mathrm{e}^{\mathrm{i}p\cdot x}\right),$$

where $E_{\boldsymbol{p}} = +(\boldsymbol{p}^2 + m^2)^{\frac{1}{2}}$.

V Evaluate $\hat{H}$ and normal order the result, leading to an expression like:

$$N[\hat{H}] = \int \mathrm{d}^3p\, E_{\boldsymbol{p}}\hat{a}_{\boldsymbol{p}}^\dagger\hat{a}_{\boldsymbol{p}}.$$

It is important to note that canonical quantization will not succeed in diagonalizing all field theories. Roughly speaking it works only for those Lagrangians which can be written as quadratic in a field and its derivatives.[14] (We will revisit the mathematics of this point in Chapter 23.) The result of canonical quantization is a system described by single particles in momentum states which don't interact with each other. For this reason Lagrangians that may be canonically quantized are called **non-interacting theories**. In contrast, those theories which cannot be diagonalized with canonical quantization are called **interacting theories**; these will be described in terms of single particles in momentum states which do interact with each other. Interacting theories are the subject of much of the rest of this book. For now we will continue to look at non-interacting theories and in the next chapter we examine some more uses of the canonical quantization routine that we have built.

Chapter summary

- Canonical quantization is an automated method for turning a classical field theory into a quantum field theory.
- Field operators are formed from mode expansions with amplitudes made of creation and annihilation operators.
- Normal ordering is needed to make sense of quantum field theories.
- The results of each step of the canonical quantization procedure for the scalar field theory are shown in the box in the margin.

Exercises

(11.1) One of the criteria we had for a successful theory of a scalar field was that the commutator for space-like separations would be zero. Let's see if our scalar field has this feature. Show that

$$[\hat{\phi}(x), \hat{\phi}(y)] = \int \frac{\mathrm{d}^3p}{(2\pi)^3} \frac{1}{2E_{\boldsymbol{p}}} \left(\mathrm{e}^{-\mathrm{i}p\cdot(x-y)} - \mathrm{e}^{-\mathrm{i}p\cdot(y-x)}\right).$$
(11.51)

For space-like separation we are able to swap $(y-x)$ in the second term to $(x-y)$. This gives us zero, as required.

(11.2) Show that, at equal times $x^0 = y^0$,

$$[\hat{\phi}(x), \hat{\Pi}^0(y)] = \frac{\mathrm{i}}{2} \int \frac{\mathrm{d}^3p}{(2\pi)^3} \left(\mathrm{e}^{\mathrm{i}\boldsymbol{p}\cdot(\boldsymbol{x}-\boldsymbol{y})} + \mathrm{e}^{-\mathrm{i}\boldsymbol{p}\cdot(\boldsymbol{x}-\boldsymbol{y})}\right).$$
(11.52)

In this expression there's nothing stopping us swapping the sign of $\boldsymbol{p}$ in the second term, and show that this leads to

$$[\hat{\phi}(x), \hat{\Pi}^0(y)] = \mathrm{i}\delta^{(3)}(\boldsymbol{x} - \boldsymbol{y}).$$
(11.53)

Examples of canonical quantization

Examples are like cans of over-strength lager. One is both too many and never enough.
Anonymous

In this chapter we will apply the methods from the previous chapter to the complex scalar field. This theory illustrates several ideas that are very important in quantum field theory. We will also find the Noether current and finally, we will take the non-relativistic limit of the theory. This last point is particularly important for thinking about applications to condensed matter, where much of the physics is firmly in the non-relativistic domain.

12.1 Complex scalar field theory

The Lagrangian for the complex scalar field[1] is an example of a theory with more than one component. Here we have two: $\psi(x)$ and $\psi^\dagger(x)$. We'll now canonically quantize this Lagrangian using the procedure presented in Chapter 11.

Step I: We start by writing down the Lagrangian we want to quantize. The Lagrangian for complex scalar field theory is

$$\mathcal{L} = \partial^\mu \psi^\dagger(x) \partial_\mu \psi(x) - m^2 \psi^\dagger(x)\psi(x). \qquad (12.1)$$

Step II: Identify the momentum density and write down the Hamiltonian. Each component[2] σ of the field (i.e. $\sigma = \psi$ and $\psi^\dagger$) has a different momentum density. The momentum of the ψ-field and of the $\psi^\dagger$-field are given, as usual by the time-like (zeroth) component of the momentum density Π_σ^μ:

$$\Pi_{\sigma=\psi}^0 = \frac{\partial \mathcal{L}}{\partial(\partial_0 \psi)} = \partial^0 \psi^\dagger, \quad \Pi_{\sigma=\psi^\dagger}^0 = \frac{\partial \mathcal{L}}{\partial(\partial_0 \psi^\dagger)} = \partial^0 \psi. \qquad (12.2)$$

Now that we have the momenta we can write the Hamiltonian. For the case of more than one field component we can write down

$$\begin{aligned}
\mathcal{H} &= \sum_\sigma \Pi_\sigma^0(x) \partial_0 \sigma(x) - \mathcal{L} \\
&= \partial_0 \psi^\dagger(x) \partial_0 \psi(x) + \boldsymbol{\nabla}\psi^\dagger(x) \cdot \boldsymbol{\nabla}\psi(x) + m^2 \psi^\dagger(x)\psi(x). \quad (12.3)
\end{aligned}$$

[1] Recall where it came from. We considered the addition of two scalar fields

$$\begin{aligned}
\mathcal{L} &= \tfrac{1}{2}[\partial_\mu \phi_1(x)]^2 - \tfrac{1}{2}m^2[\phi_1(x)]^2 \\
&\quad + \tfrac{1}{2}[\partial_\mu \phi_2(x)]^2 - \tfrac{1}{2}m^2[\phi_2(x)]^2.
\end{aligned}$$

We noticed that we can make a transformation that simplified this Lagrangian. We defined

$$\begin{aligned}
\psi &= \tfrac{1}{\sqrt{2}}[\phi_1(x) + i\phi_2(x)], \\
\psi^\dagger &= \tfrac{1}{\sqrt{2}}[\phi_1(x) - i\phi_2(x)],
\end{aligned}$$

and we obtained the Lagrangian in eqn 12.1. This Lagrangian looks rather like that of the massive scalar field, but without the factors of $1/2$.

[2] Note carefully that σ isn't a tensor component, it simply labels the component of the field: ψ or $\psi^\dagger$. This implies that $\Pi^{\mu\sigma} = \Pi_\sigma^\mu$.

Step III: Make the fields into quantum mechanical operators. We promote the fields to operators and impose equal-time commutation relations thus

$$\left[\hat{\psi}(t, \boldsymbol{x}), \hat{\Pi}^0_\psi(t, \boldsymbol{y})\right] = \left[\hat{\psi}^\dagger(t, \boldsymbol{x}), \hat{\Pi}^0_{\psi^\dagger}(t, \boldsymbol{y})\right] = \mathrm{i}\delta^{(3)}(\boldsymbol{x} - \boldsymbol{y}), \qquad (12.4)$$

and all other commutators vanish.

Step IV: Expand the fields in terms of modes. For the case of the real scalar field, we found that the field operator[3] $\hat{\phi}(x) = \int_{\boldsymbol{p}}(\hat{a}_{\boldsymbol{p}}\mathrm{e}^{-\mathrm{i}p\cdot x} + \hat{a}^\dagger_{\boldsymbol{p}}\mathrm{e}^{\mathrm{i}p\cdot x})$, so that it is Hermitian [i.e. self-adjoint: $\hat{\phi}^\dagger(x) = \hat{\phi}(x)$]. For our complex scalar field there is no reason why the field operator should be Hermitian. The appropriate mode expansion of the complex scalar field operators take the more general form

$$\hat{\psi}(x) = \int \frac{\mathrm{d}^3 p}{(2\pi)^{\frac{3}{2}}} \frac{1}{(2E_{\boldsymbol{p}})^{\frac{1}{2}}} \left(\hat{a}_{\boldsymbol{p}}\mathrm{e}^{-\mathrm{i}p\cdot x} + \hat{b}^\dagger_{\boldsymbol{p}}\mathrm{e}^{\mathrm{i}p\cdot x}\right),$$

$$\hat{\psi}^\dagger(x) = \int \frac{\mathrm{d}^3 p}{(2\pi)^{\frac{3}{2}}} \frac{1}{(2E_{\boldsymbol{p}})^{\frac{1}{2}}} \left(\hat{a}^\dagger_{\boldsymbol{p}}\mathrm{e}^{\mathrm{i}p\cdot x} + \hat{b}_{\boldsymbol{p}}\mathrm{e}^{-\mathrm{i}p\cdot x}\right), \qquad (12.5)$$

where $E_{\boldsymbol{p}} = +(\boldsymbol{p}^2 + m^2)^{\frac{1}{2}}$. Here the operators $\hat{a}_{\boldsymbol{p}}$ and $\hat{b}_{\boldsymbol{p}}$ annihilate two different types of particle. They satisfy the commutation relations $\left[\hat{a}_{\boldsymbol{p}}, \hat{a}^\dagger_{\boldsymbol{q}}\right] = \left[\hat{b}_{\boldsymbol{p}}, \hat{b}^\dagger_{\boldsymbol{q}}\right] = \delta^{(3)}(\boldsymbol{p} - \boldsymbol{q})$ and all others combinations vanish. Substituting these into the Hamiltonian we find that, after some algebra and then (**Step V**) normal ordering, that we obtain:

$$N\left[\hat{H}\right] = \int \mathrm{d}^3 p\, E_{\boldsymbol{p}} \left(\hat{a}^\dagger_{\boldsymbol{p}}\hat{a}_{\boldsymbol{p}} + \hat{b}^\dagger_{\boldsymbol{p}}\hat{b}_{\boldsymbol{p}}\right)$$

$$= \int \mathrm{d}^3 p\, E_{\boldsymbol{p}} \left(\hat{n}^{(a)}_{\boldsymbol{p}} + \hat{n}^{(b)}_{\boldsymbol{p}}\right), \qquad (12.6)$$

where $\hat{n}^{(a)}_{\boldsymbol{p}}$ counts the number of a-particles with momentum $\boldsymbol{p}$ and, similarly, $\hat{n}^{(b)}_{\boldsymbol{p}}$ counts the number of b-particles. We notice that the a and b particles have the same energy $E_{\boldsymbol{p}}$, and so we interpret a and b as particles and antiparticles respectively. This is justified in the next section. Thus $\hat{\psi}(x)$ involves a sum over all momenta of operators that annihilate particles ($\hat{a}_{\boldsymbol{p}}$) and create antiparticles ($\hat{b}^\dagger_{\boldsymbol{p}}$). We therefore understand these new operators as follows:

- $\hat{a}^\dagger_{\boldsymbol{p}}$ is the creation operator for a particle with momentum $\boldsymbol{p}$,

- $\hat{b}^\dagger_{\boldsymbol{p}}$ is the creation operator for an antiparticle with momentum $\boldsymbol{p}$.

We've quantized the theory and we've found that the excitations of the field are scalar particles and antiparticles.

It is worthwhile justifying how we have been able to treat ψ and $\psi^\dagger$ as if they are independent fields. We began with

$$\psi(x) = \frac{1}{\sqrt{2}}[\phi_1(x) + \mathrm{i}\phi_2(x)] \quad \text{and} \quad \psi^\dagger(x) = \frac{1}{\sqrt{2}}[\phi_1(x) - \mathrm{i}\phi_2(x)], \quad (12.7)$$

[3]Here we abbreviate $\int_{\boldsymbol{p}} \equiv \int \frac{\mathrm{d}^3 p}{(2\pi)^{\frac{3}{2}}} \frac{1}{(2E_{\boldsymbol{p}})^{\frac{1}{2}}}$.

and this implies that

$$\phi_1 = \frac{1}{\sqrt{2}}[\psi + \psi^\dagger] \quad \text{and} \quad \phi_2 = \frac{-i}{\sqrt{2}}[\psi - \psi^\dagger]. \qquad (12.8)$$

Thus all we have done in moving from a description based on ϕ_1 and ϕ_2 to one based on ψ and $\psi^\dagger$ is to perform a change of basis. Allowing the action S to vary with respect to ψ and $\psi^\dagger$ gives as much freedom as if we were varying ϕ_1 and ϕ_2. For example, we can adjust ϕ_1 alone using $\delta\psi = \delta\psi^\dagger$ and we can adjust ϕ_2 alone using $\delta\psi = -\delta\psi^\dagger$.

12.2   Noether's current for complex scalar field theory

The complex scalar field has an *internal* $U(1)$ symmetry.[4] This means that the global transformations of the fields

$$\psi \to e^{i\alpha}\psi, \quad \psi^\dagger \to e^{-i\alpha}\psi^\dagger \qquad (12.9)$$

have no effect on the Lagrangian. Remember that Noether's theorem says that every symmetry begets a conserved current. This is our first opportunity to see what is conserved for an internal symmetry. In fact, we will see that the conserved Noether charge is particle number.

To get to Noether's current for the internal $U(1)$ symmetry, it's easiest to write out the transformation for an infinitesimal change in the phase α:

$$\begin{array}{ll} \psi \to \psi + i\psi\,\delta\alpha, & D\psi = +i\psi, \\ \psi^\dagger \to \psi^\dagger - i\psi^\dagger\delta\alpha, & D\psi^\dagger = -i\psi^\dagger. \end{array} \qquad (12.10)$$

Note that $D\mathcal{L} = 0$ and since $D\mathcal{L} = \partial_\mu W^\mu$ this implies that we can take[5] $W^\mu = 0$. The Noether current is therefore given by $J_N^\mu = \sum_\sigma \Pi_\sigma^\mu D\sigma$, (because $W^\mu = 0$) where the index σ again labels the component of the field: in this case whether it's $\sigma = \psi$ or $\psi^\dagger$. As a result we have

$$\begin{aligned} J_N^\mu &= \sum_\sigma \Pi_\sigma^\mu D\sigma = \Pi_\psi^\mu D\psi + \Pi_{\psi^\dagger}^\mu D\psi^\dagger \\ &= i\left[\left(\partial^\mu\psi^\dagger\right)\psi - (\partial^\mu\psi)\psi^\dagger\right]. \end{aligned} \qquad (12.11)$$

We may make this a Noether current operator $\hat{J}_N^\mu$ by upgrading all of our fields into field operators. This creates an ordering ambiguity in writing the fields and their derivatives. We will remove this ambiguity by normal ordering. Substituting the mode expansion for the fields allows us to find the conserved charge operator $\hat{Q}_N$.

[4]Internal symmetries are discussed in more detail in the next chapter.

[5]The property $W^\mu = 0$ turns out to be true of all internal symmetries.

Example 12.1

The charge operator is

$$\hat{Q}_N = \int \mathrm{d}^3x\,\hat{J}_N^0 = i\int \mathrm{d}^3x\,\left[\left(\partial^0\hat{\psi}^\dagger\right)\hat{\psi} - (\partial^0\hat{\psi})\hat{\psi}^\dagger\right]. \qquad (12.12)$$

Of course, there's the ordering ambiguity here, but we'll press on and insert the mode expansion to yield

$$\hat{Q}_{\rm N} = \frac{1}{2} \int {\rm d}^3 p \left(-\hat{a}_{\boldsymbol{p}}^\dagger \hat{a}_{\boldsymbol{p}} + \hat{b}_{\boldsymbol{p}} \hat{b}_{\boldsymbol{p}}^\dagger - \hat{a}_{\boldsymbol{p}} \hat{a}_{\boldsymbol{p}}^\dagger + \hat{b}_{\boldsymbol{p}}^\dagger \hat{b}_{\boldsymbol{p}} \right). \tag{12.13}$$

Finally, we normal order to give a conserved charge of

$$N \left[\hat{Q}_{\rm N} \right] = \int {\rm d}^3 p \left(\hat{b}_{\boldsymbol{p}}^\dagger \hat{b}_{\boldsymbol{p}} - \hat{a}_{\boldsymbol{p}}^\dagger \hat{a}_{\boldsymbol{p}} \right). \tag{12.14}$$

> **Noether's theorem: $U(1)$ internal symmetry**
>
> $$D\psi = {\rm i}\psi \qquad D\psi^\dagger = -{\rm i}\psi^\dagger$$
> $$\Pi_\psi^\mu = \partial^\mu \psi^\dagger \qquad \Pi_{\psi^\dagger}^\mu = \partial^\mu \psi$$
> $$D\mathcal{L} = 0 \qquad W^\mu = 0$$
> $$J_{\rm N}^\mu = {\rm i} \left[(\partial^\mu \psi^\dagger)\psi - (\partial^\mu \psi)\psi^\dagger \right]$$
> $$\hat{Q}_{\rm Nc} = -N \left[\hat{Q}_{\rm N} \right]$$
> $$= \int {\rm d}^3 p \, (\hat{n}_{\boldsymbol{p}}^{(a)} - \hat{n}_{\boldsymbol{p}}^{(b)}).$$

The conserved charge is given by the difference in the number of antiparticles $n_{\boldsymbol{p}}^{(b)}$ and the number of particles $n_{\boldsymbol{p}}^{(a)}$. This reason for this is quite simple. Particles, which are excitations in the field, carry Noether charge of one sign. To have a conserved charge we need other sorts of particle excitation carrying negative charge to cancel out the contribution from the particles. We are left with the conclusion that the reason for the existence of antiparticles is that charge wouldn't be conserved if they didn't exist.

Obviously, if a current $J_{\rm N}^\mu$ is conserved, then so is $-J_{\rm N}^\mu$ and so our assignment of which excitations carry positive Noether charge and which are negative is quite arbitrary. It is only when we attach some observable property, such as electric charge q, to each that we decide which is which. By convention, the **number current** $\hat{J}_{\rm Nc}^\mu$ is defined to be positive for particles (created by $\hat{a}_{\boldsymbol{p}}^\dagger$ operators) and negative for antiparticles (created by $\hat{b}_{\boldsymbol{p}}^\dagger$ operators) and so we will define[6] $\hat{J}_{\rm Nc}^\mu = -N \left[\hat{J}_{\rm N}^\mu \right]$, $\hat{Q}_{\rm Nc} = \int {\rm d}^3 x \, \hat{J}_{\rm Nc}^0$ and therefore $\hat{Q}_{\rm Nc} = -N[\hat{Q}_{\rm N}]$. Hence using eqn 12.14 and recalling that $\hat{n}_{\boldsymbol{p}}^{(a)} = \hat{a}_{\boldsymbol{p}}^\dagger \hat{a}_{\boldsymbol{p}}$ and $\hat{n}_{\boldsymbol{p}}^{(b)} = \hat{b}_{\boldsymbol{p}}^\dagger \hat{b}_{\boldsymbol{p}}$, we can conclude that

$$\hat{Q}_{\rm Nc} = \int {\rm d}^3 p \left(\hat{n}_{\boldsymbol{p}}^{(a)} - \hat{n}_{\boldsymbol{p}}^{(b)} \right). \tag{12.15}$$

[6]With a subscript c for conventional, just like conventional current direction in circuit theory!

12.3 Complex scalar field theory in the non-relativistic limit

In the final part of this chapter, we will road test our new complex scalar field theory and see how it works out in the non-relativistic limit and whether or not it will regenerate the familiar results of non-relativistic quantum mechanics.

In the non-relativistic domain the excitation energies of particles are small compared to the particle mass contribution mc^2, where here we'll temporarily reinstate factors of c and $\hbar$. This means that the mass term $E_{\boldsymbol{p}=0} = mc^2$ provides by far the largest fraction of the energy of an excitation, that is $E = mc^2 + \varepsilon$, where ε is small. To make a theory non-relativistic, the strategy is to replace the relativistic field ϕ with

$$\phi(\boldsymbol{x}, t) \to \Psi(\boldsymbol{x}, t) {\rm e}^{-{\rm i}mc^2 t/\hbar}, \tag{12.16}$$

allowing us to factor out the enormous rest energy.

Example 12.2

To get an idea of how this works, we'll take the non-relativistic limit of the Klein–Gordon equation. We start (in natural units) with

$$(\partial^2 + m^2)\Psi(\boldsymbol{x}, t)e^{-imt} = 0, \tag{12.17}$$

and moving to unnatural units we have

$$\left(\hbar^2 \frac{\partial^2}{\partial t^2} - \hbar^2 c^2 \boldsymbol{\nabla}^2 + m^2 c^4\right)\Psi(\boldsymbol{x}, t)e^{-imc^2 t/\hbar} = 0. \tag{12.18}$$

The first term on the left yields

$$\hbar^2 \frac{\partial^2}{\partial t^2}\Psi(\boldsymbol{x}, t)e^{-imc^2 t/\hbar} = \hbar^2 \left(\frac{\partial^2 \Psi}{\partial t^2} - \frac{2imc^2}{\hbar}\frac{\partial \Psi}{\partial t} - \frac{m^2 c^4}{\hbar^2}\Psi\right)e^{-imc^2 t/\hbar}, \tag{12.19}$$

and substitution gives us

$$\hbar^2 \frac{\partial^2 \Psi}{\partial t^2} - 2imc^2\hbar\frac{\partial \Psi}{\partial t} - \hbar^2 c^2 \boldsymbol{\nabla}^2 \Psi = 0. \tag{12.20}$$

Lastly we notice that the first term, with its lack of factors of c^2, will be much smaller than the other two. We can therefore drop the first term and we end up with

$$i\hbar \frac{\partial}{\partial t}\Psi(\boldsymbol{x}, t) = -\frac{\hbar^2}{2m}\boldsymbol{\nabla}^2 \Psi(\boldsymbol{x}, t). \tag{12.21}$$

The result is, of course, that we recover the Schrödinger equation for a free particle.

To take the non-relativistic limit of the complex scalar field Lagrangian[7] we substitute in $\psi = \frac{1}{\sqrt{2m}}e^{-imt}\Psi$ (with an extra normalization factor $1/\sqrt{2m}$ added with malice aforethought).

[7]We'll use the full interacting form $\mathcal{L} = \partial^\mu \psi^\dagger(x)\partial_\mu \psi(x) - m^2 \psi^\dagger(x)\psi(x) - \lambda[\psi^\dagger(x)\psi(x)]^2$ and return to our convention that $\hbar = c = 1$.

Example 12.3

As in the last example, the guts of taking the limit may be found in the time derivatives. We find

$$\partial_0 \psi^\dagger \partial_0 \psi = \frac{1}{2m}\left[\partial_0 \Psi^\dagger \partial_0 \Psi + im\left(\Psi^\dagger \partial_0 \Psi - (\partial_0 \Psi^\dagger)\Psi\right) + m^2 \Psi^\dagger \Psi\right]. \tag{12.22}$$

The first term going as $1/m$ is negligible in comparison with the others. The third term cancels against the mass term in the original Lagrangian. The dynamic part of the theory is therefore contained in the second term. We can see that a time derivative of Ψ (which we may expand in plane wave modes as something like $\sum a_{\boldsymbol{p}} e^{-ip \cdot x}$) will bring down a factor $-iE_{\boldsymbol{p}}$. Since $\Psi^\dagger$ is obtained from Ψ through taking an adjoint, the time derivative of $\Psi^\dagger$ will therefore involve bringing down a factor $iE_{\boldsymbol{p}}$. The result is that $(\Psi^\dagger \partial_0 \Psi - \Psi \partial_0 \Psi^\dagger)$ may be replaced[8] by $2\Psi^\dagger \partial_0 \Psi$.

[8]We could have replaced this with $-2\Psi(\partial_0 \Psi^\dagger)$ if we had chosen modes varying as $e^{+ip \cdot x}$, but we choose to favour matter over antimatter with our choice.

After these manipulations, we end up with

$$\mathcal{L} = i\Psi^\dagger(x)\partial_0 \Psi(x) - \frac{1}{2m}\boldsymbol{\nabla}\Psi^\dagger(x) \cdot \boldsymbol{\nabla}\Psi(x) - \frac{g}{2}[\Psi^\dagger(x)\Psi(x)]^2, \tag{12.23}$$

with $g = \lambda/2m^2$. This looks a lot less neat and covariant than the relativistic form, because of the asymmetry between the matter and antimatter fields that resulted from the choice we made in simplifying the time derivative term. This is acceptable because life is a lot less symmetrical in the non-relativistic world, but we will see the consequences later. Anyway, we will now continue and feed this Lagrangian into the canonical quantization machine.

In this example we begin to see some of the limitations of canonical quantization. For a general potential $V(x)$ the procedure does not return a diagonalized Hamiltonian (although it will work for a constant potential $V(x) = V_0$). We will see in Chapter 16 how more general potentials may be dealt with using perturbation theory.

Example 12.4

Let's attempt to quantize the non-relativistic Lagrangian, without the self-interaction ($g = 0$), but in the presence of an external potential $V(x)$ which enters as $V\Psi^\dagger\Psi$. There's nothing for it but to press on with the five point programme.

Step I: The Lagrangian density with the external potential is given by

$$\mathcal{L} = i\Psi^\dagger(x)\partial_0\Psi(x) - \frac{1}{2m}\boldsymbol{\nabla}\Psi^\dagger(x)\cdot\boldsymbol{\nabla}\Psi(x) - V(x)\Psi^\dagger(x)\Psi(x). \tag{12.24}$$

Feeding this through the Euler–Lagrange equations yields the Schrödinger equation and, for $V(x) = 0$, a dispersion $E_{\boldsymbol{p}} = \frac{\boldsymbol{p}^2}{2m}$ (Exercise 12.5). This only has positive energy solutions so we might expect that the negative frequency part of the mode expansion won't be needed here. This will turn out to be the case.

Step II: We calculate the momentum densities thus:

$$\Pi_\Psi^0 = \frac{\partial\mathcal{L}}{\partial(\partial_0\Psi)} = i\Psi^\dagger, \quad \Pi_{\Psi^\dagger}^0 = \frac{\partial\mathcal{L}}{\partial(\partial_0\Psi^\dagger)} = 0. \tag{12.25}$$

Notice that there's no momentum density conjugate to the field $\Psi^\dagger$, which is the consequence of our choice in Example 12.3. The momentum allows us to calculate the Hamiltonian density:

$$\begin{aligned}\mathcal{H} &= \Pi_\Psi^0\partial_0\Psi - \mathcal{L} \\ &= \frac{1}{2m}\boldsymbol{\nabla}\Psi^\dagger(x)\cdot\boldsymbol{\nabla}\Psi(x) + V(x)\Psi^\dagger(x)\Psi(x).\end{aligned} \tag{12.26}$$

Lo and behold, we have obtained a Schrödinger-like equation for the Hamiltonian density.

Step III: The equal-time commutation relation between the position and momentum operators is

$$\begin{aligned}\left[\hat{\Psi}(t,\boldsymbol{x}), \hat{\Pi}_\Psi^0(t,\boldsymbol{y})\right] &= i\delta^{(3)}(\boldsymbol{x} - \boldsymbol{y}), \\ \left[\hat{\Psi}(t,\boldsymbol{x}), \hat{\Psi}^\dagger(t,\boldsymbol{y})\right] &= \delta^{(3)}(\boldsymbol{x} - \boldsymbol{y}),\end{aligned} \tag{12.27}$$

where the second equality follows from inserting our expression $\Pi_\Psi^0 = i\Psi^\dagger$.

Step IV: A mode expansion with positive and negative frequency parts won't obey the commutation relation above. The mode expansion which does the trick is simply

$$\hat{\Psi}(x) = \int\frac{\mathrm{d}^3p}{(2\pi)^{\frac{3}{2}}}\hat{a}_{\boldsymbol{p}}e^{-ip\cdot x}, \tag{12.28}$$

with $E_{\boldsymbol{p}} = \frac{\boldsymbol{p}^2}{2m}$.

Step V: Substituting the mode expansion gives us a Hamiltonian operator

$$\hat{H} = \int\mathrm{d}^3p\left(\frac{\boldsymbol{p}^2}{2m}\hat{a}_{\boldsymbol{p}}^\dagger\hat{a}_{\boldsymbol{p}}\right) + \int\frac{\mathrm{d}^3x\,\mathrm{d}^3p\,\mathrm{d}^3q}{(2\pi)^3}\left(V(t,\boldsymbol{x})e^{i(E_{\boldsymbol{p}} - E_{\boldsymbol{q}})t}e^{-i(\boldsymbol{p}-\boldsymbol{q})\cdot\boldsymbol{x}}\hat{a}_{\boldsymbol{p}}^\dagger\hat{a}_{\boldsymbol{q}}\right). \tag{12.29}$$

The time-dependent part will guarantee conservation of energy. It forces the potential to have a time dependence of the form $V(t,\boldsymbol{x}) = e^{-i(E_{\boldsymbol{p}} - E_{\boldsymbol{q}})t}V(\boldsymbol{x})$ if the momentum of the particle is going to change. With this constraint we obtain a Hamiltonian

$$\hat{H} = \int\mathrm{d}^3p\left(\frac{\boldsymbol{p}^2}{2m}\hat{a}_{\boldsymbol{p}}^\dagger\hat{a}_{\boldsymbol{p}}\right) + \int\mathrm{d}^3p\,\mathrm{d}^3q\left(\tilde{V}(\boldsymbol{p} - \boldsymbol{q})\hat{a}_{\boldsymbol{p}}^\dagger\hat{a}_{\boldsymbol{q}}\right), \tag{12.30}$$

where $\tilde{V}(\boldsymbol{p} - \boldsymbol{q}) = \int\mathrm{d}^3x\frac{1}{(2\pi)^3}V(\boldsymbol{x})e^{-i(\boldsymbol{p}-\boldsymbol{q})\cdot\boldsymbol{x}}$ and the potential must impart an energy $E_{\boldsymbol{p}} - E_{\boldsymbol{q}}$. Notice that this is equivalent to the Hamiltonian we argued for in Chapter 4.

We can look at the non-relativistic complex scalar field in another way by considering a different form for the field. We first note that the non-relativistic form of the Lagrangian is still invariant with respect to global $U(1)$ transformations. Motivated by this, we write

$$\Psi(x) = \sqrt{\rho(x)}e^{i\theta(x)}, \tag{12.31}$$

so that $\Psi(x)$ is written in terms of amplitude and phase. The $U(1)$ transformation is now enacted by $\theta \to \theta + \alpha$ (see the box for the conserved charges). Whereas before we had two fields, $\phi_1(x)$ and $\phi_2(x)$, now we have two fields, $\rho(x)$ and $\theta(x)$. Substituting in the Lagrangian in eqn 12.23, we obtain (**Step I**) the Lagrangian in terms of ρ and θ fields

$$\mathcal{L} = \frac{i}{2}\partial_0\rho - \rho\partial_0\theta - \frac{1}{2m}\left[\frac{1}{4\rho}(\boldsymbol{\nabla}\rho)^2 + \rho(\boldsymbol{\nabla}\theta)^2\right] - \frac{g}{2}\rho^2. \tag{12.32}$$

Now we turn off interactions (by setting $g = 0$) and feed the resulting Lagrangian into the canonical quantization machine and make selective use of its features. First we find the momenta (**Step II**):

$$\Pi^0_\rho(x) = \frac{\partial\mathcal{L}}{\partial(\partial_0\rho(x))} = \frac{i}{2},$$

$$\Pi^0_\theta(x) = \frac{\partial\mathcal{L}}{\partial(\partial_0\theta(x))} = -\rho(x), \tag{12.33}$$

from which we see that the Π^0_θ is most interesting since it's not a constant. Let's make this quantum mechanical (**Step III**) by imposing commutation relations

$$\left[\hat{\theta}(\boldsymbol{x},t),\hat{\Pi}^0_\theta(\boldsymbol{y},t)\right] = -\left[\hat{\theta}(\boldsymbol{x},t),\hat{\rho}(\boldsymbol{y},t)\right] = i\delta^{(3)}(\boldsymbol{x}-\boldsymbol{y}) \tag{12.34}$$

The box shows that the time-like component of the conserved current is $\rho(x)$. We therefore define the total number of particles as $\hat{N}(t) = \int \mathrm{d}^3x\,\hat{\rho}(\boldsymbol{x},t)$ and by integrating[9] the commutation relation equation we obtain

$$\left[\hat{N}(t),\hat{\theta}(\boldsymbol{x},t)\right] = i, \tag{12.35}$$

which is an important result.[10] It tells us that in the sort of coherent condensed matter system that can be described by the non-relativistic complex scalar field theory the operator for the number of excitations is conjugate to the operator for the phase angle of the field. We will return to this important idea when we come to discuss superfluids and superconductors.

Noether's theorem: $U(1)$ **internal symmetry**

$$D\theta = 1$$
$$\Pi^0_\theta = -\rho \qquad \Pi^i_\theta = \frac{\rho}{m}\partial^i\theta$$
$$D\mathcal{L} = 0 \qquad W^\mu = 0$$
$$J^0_N = -\rho(x) \qquad \boldsymbol{J}_N = -\frac{\rho}{m}\boldsymbol{\nabla}\theta$$
$$Q_{Nc} = \int \mathrm{d}^3x\,\rho(x) \qquad \boldsymbol{J}_{Nc} = \frac{\rho}{m}\boldsymbol{\nabla}\theta$$

[9] This calculation simply uses

$$\int \mathrm{d}^3y\,[\hat{\rho}(\boldsymbol{y},t),\hat{\theta}(\boldsymbol{x},t)] = [\hat{N}(t),\hat{\theta}(\boldsymbol{x},t)]$$

and

$$\int \mathrm{d}^3y\,\delta^{(3)}(\boldsymbol{x}-\boldsymbol{y}) = 1.$$

[10] Equation 12.35 is called the number-phase uncertainty relation.

Chapter summary

- Quantization of the complex scalar field leads to two types of excitation: particles and antiparticles. The conserved charge is the number of particles minus the number of antiparticles.
- The non-relativistic limit of the theory may be taken. From this we can derive a number-phase uncertainty relation.

Exercises

(12.1) Fill in the missing steps of the algebra that led to eqn 12.6.

(12.2) Evaluate (a) $[\hat{\psi}(x), \hat{\psi}^\dagger(y)]$ and (b) $[\hat{\Psi}(x), \hat{\Psi}^\dagger(y)]$ using the appropriate mode expansions in the text.

(12.3) Consider the Lagrangian for two scalar fields from Section 7.5.
(a) Evaluate $[\hat{Q}_N, \hat{\phi}_1]$ and (b) $[\hat{Q}_N, \hat{\phi}_2]$, where $\hat{Q}_N$ is the Noether charge.
Hint: You could do this by brute force and evaluate $\hat{Q}_N$ and then find the commutator. A preferable method is just to use: $[\hat{Q}_N, \hat{\phi}_i] = -iD\hat{\phi}_i$ from Chapter 10.
(b) Use these results to show that $[\hat{Q}_N, \hat{\psi}] = \hat{\psi}$.

(12.4) Consider the theory described by the Lagrangian in eqn 12.32. Use Noether's theorem in the form $[\hat{Q}_N, \hat{\phi}] = -iD\hat{\phi}$ to provide an alternative derivation of eqn 12.35.

(12.5) Apply the Euler–Lagrange equations to eqn 12.24 and show that it yields $E_{\boldsymbol{p}} = \boldsymbol{p}^2/2m$ when $V = 0$.

(12.6) Find the Noether current for the Lagrangian in eqn 12.24. Check that it is what you expect for non-relativistic quantum mechanics.

(12.7) Consider the complex scalar field again. The internal transformation operator may be written $\hat{U}(\alpha) = e^{i\hat{Q}_{Nc}\alpha}$, where $\hat{Q}_{Nc}$ is the conserved number charge operator. Show that $\hat{U}^\dagger(\alpha)\hat{\psi}(x)\hat{U}(\alpha) = e^{i\alpha}\hat{\psi}(x)$.

Fields with many components and massive electromagnetism

13

In the previous chapter we described a theory with two field components. Here we expand the discussion to theories with more components and we will see that the interesting thing about these components is their symmetries. We'll examine two cases, the first of which has an internal symmetry and a second where the theory is described in terms of four-vectors.

13.1 Internal symmetries

When we find families of closely-related particles in Nature, there is an inevitable lurking suspicion that these particles are really a single physical entity but that there is some internal dial which can be rotated to turn one member of the family into another. The invariance of the Lagrangian with respect to turns on this internal dial is described as an **internal symmetry**, quite unrelated to the symmetry of spacetime in which the particles live, move and have their being.[1]

An early example of the idea of an internal symmetry was **isospin**. The neutron and proton have almost the same mass ($m_p \approx m_n$) and this led Heisenberg to wonder if they were the same particle in two different states with some kind of internal symmetry that he termed isospin. The idea is that neutrons and protons both have isospin $I = \frac{1}{2}$ but that, as for conventional spin,[2] there are two possible eigenvalues of the $\hat{I}_z$ operator: $I_z = 1/2$ (the proton) and $I_z = -1/2$ (the neutron). By analogy with spin angular momentum, we may arrange these into a two-component object $\begin{pmatrix} p \\ n \end{pmatrix}$, known as an isospin doublet. We can rotate these isospin doublets with the same matrices[3] that rotate spins-1/2. Before we get too carried away, we should remember that, strictly, $m_p \neq m_n$ so that isospin is only an approximate description.[4]

The idea of an internal symmetry nevertheless has wide applicability. We will explore this idea using the example of three scalar particles. We imagine arranging them into a vector (t, d, h) which transforms internally according to the three-dimensional rotation group $SO(3)$. Rotating the internal dial by 90° about the h-axis turns a t particle, denoted $(1, 0, 0)$, into a d particle $(0, 1, 0)$ (see Fig. 13.1). This also implies that super-

[1]We saw the consequence of a $U(1)$ internal symmetry in the complex scalar field in the last chapter.

[2]Note that isospin has nothing to do with the actual spin angular momentum of the particles.

[3]As explored in Chapter 37, these rotations form a representation of the group $SU(2)$.

[4]Of course protons and neutrons are actually each composed of quarks.

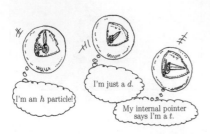

Fig. 13.1 In our toy model, particles can be imagined as having an internal dial. Whether the dial points along $(1, 0, 0)$, $(0, 1, 0)$ or $(0, 0, 1)$ determines which kind of particle it is. The position of the dial is not related to the directions in spacetime through which particles move, but expresses a direction in abstract space describing the internal symmetry.

positions of particles, obtained by some arbitrary rotation of the dial, have just as much validity as the original particles do. Since particles are excitations in fields, we should be able to examine isospin in field theory. In the field case we arrange the fields that generate these particles into a vector such as (ϕ_1, ϕ_2, ϕ_3). Subjecting this column vector to internal rotations by turning the internal dial, one can carry out the equivalent act of turning a ϕ_1 field into a ϕ_2 field by rotating about the ϕ_3-axis. If the Lagrangian describing the theory of these particles is invariant with respect to these rotations, then Noether's theorem gives us a conservation law describing the charges of the fields.

To simplify our notation, let's write the three scalar fields as a column vector $\mathbf{\Phi}(x)$ as follows

$$\mathbf{\Phi}(x) = \begin{pmatrix} \phi_1(x) \\ \phi_2(x) \\ \phi_3(x) \end{pmatrix}. \tag{13.1}$$

The free Lagrangian (**step I** of the Canonical quantization process) for this theory is given by

$$\mathcal{L} = \frac{1}{2}(\partial^\mu \mathbf{\Phi}) \cdot (\partial_\mu \mathbf{\Phi}) - \frac{m^2}{2}\mathbf{\Phi} \cdot \mathbf{\Phi}, \tag{13.2}$$

which is short for

$$\mathcal{L} = \frac{1}{2}\left[(\partial_\mu \phi_1)^2 - m^2\phi_1^2 + (\partial_\mu \phi_2)^2 - m^2\phi_2^2 + (\partial_\mu \phi_3)^2 - m^2\phi_3^2\right], \tag{13.3}$$

that is, the sum of the Lagrangians for the individual fields. Note carefully that the object $\mathbf{\Phi}(x)$ is not a vector field that lives in Minkowski space like the vectors x^μ and p^μ do. This point is examined in the following example.

Example 13.1

Here, unlike elsewhere, the dot products aren't short for a sum $g_{\mu\nu}A^\mu A^\nu$. The vector $\mathbf{\Phi}$ doesn't live in Minkowski space at all, so has no dealings with the tensor $g_{\mu\nu}$. Instead, the dot products here are short for

$$\mathbf{\Phi} \cdot \mathbf{\Phi} = \phi_1\phi_1 + \phi_2\phi_2 + \phi_3\phi_3, \tag{13.4}$$

and

$$\partial^\mu \mathbf{\Phi} \cdot \partial_\mu \mathbf{\Phi} = \partial^\mu\phi_1\partial_\mu\phi_1 + \partial^\mu\phi_2\partial_\mu\phi_2 + \partial^\mu\phi_3\partial_\mu\phi_3. \tag{13.5}$$

This is to say that the α in Φ_α isn't a tensor index and there is consequently no difference between Φ^α and Φ_α.

Our example has an $SO(3)$ internal symmetry: we can transform the vector $\mathbf{\Phi}$ using a three-dimensional rotation matrix and the Lagrangian won't change one iota. That is to say that there is no effect on the Lagrangian on transforming to a field $\mathbf{\Phi}'$ via a transformation such as a rotation about the ϕ_3-axis:

$$\begin{pmatrix} \phi_1' \\ \phi_2' \\ \phi_3' \end{pmatrix} = \begin{pmatrix} \cos\theta & -\sin\theta & 0 \\ \sin\theta & \cos\theta & 0 \\ 0 & 0 & 1 \end{pmatrix} \begin{pmatrix} \phi_1 \\ \phi_2 \\ \phi_3 \end{pmatrix}. \tag{13.6}$$

The canonical quantization for this theory proceeds as before, but we now implement our new notation. This leads to a Hamiltonian (**step II**)

$$\hat{\mathcal{H}} = \sum_\alpha \left[\frac{1}{2}(\partial_0 \hat{\phi}_\alpha)^2 + \frac{1}{2}(\boldsymbol{\nabla}\hat{\phi}_\alpha)^2 + \frac{1}{2}m^2 \hat{\phi}_\alpha^2 \right]. \tag{13.7}$$

Note that $\Pi_\alpha^\mu = \frac{\partial \mathcal{L}}{\partial(\partial_\mu \Phi^\alpha)}$.

The equal-time commutation relations (**step III**) become

$$\left[\hat{\Phi}_\alpha(t, \boldsymbol{x}), \hat{\Pi}_\beta^0(t, \boldsymbol{y}) \right] = \mathrm{i}\delta^{(3)}(\boldsymbol{x} - \boldsymbol{y})\delta_{\alpha\beta}, \tag{13.8}$$

including a Kronecker delta because each component of $\boldsymbol{\Phi}$ has a nonzero commutator with the same component of its own momentum field only. The mode expansion (**step IV**) is then given by

$$\boldsymbol{\Phi}(x) = \int \frac{\mathrm{d}^3 p}{(2\pi)^{\frac{3}{2}}} \frac{1}{(2E_{\boldsymbol{p}})^{\frac{1}{2}}} \begin{pmatrix} \hat{a}_{\boldsymbol{p}1}\mathrm{e}^{-\mathrm{i}p\cdot x} + \hat{a}_{\boldsymbol{p}1}^\dagger \mathrm{e}^{\mathrm{i}p\cdot x} \\ \hat{a}_{\boldsymbol{p}2}\mathrm{e}^{-\mathrm{i}p\cdot x} + \hat{a}_{\boldsymbol{p}2}^\dagger \mathrm{e}^{\mathrm{i}p\cdot x} \\ \hat{a}_{\boldsymbol{p}3}\mathrm{e}^{-\mathrm{i}p\cdot x} + \hat{a}_{\boldsymbol{p}3}^\dagger \mathrm{e}^{\mathrm{i}p\cdot x} \end{pmatrix}, \tag{13.9}$$

where $\hat{a}_{\boldsymbol{p}\alpha}$ are the annihilation operators for the α field and we have the commutators $[\hat{a}_{\boldsymbol{p}\alpha}, \hat{a}_{\boldsymbol{q}\beta}^\dagger] = \delta^{(3)}(\boldsymbol{p} - \boldsymbol{q})\delta_{\alpha\beta}$ and all others vanish.

Example 13.2

We could also abbreviate the notation slightly by writing this as

$$\boldsymbol{\Phi}(x) = \int \frac{\mathrm{d}^3 p}{(2\pi)^{\frac{3}{2}}} \frac{1}{(2E_{\boldsymbol{p}})^{\frac{1}{2}}} \sum_{\alpha=1}^{3} \boldsymbol{h}_\alpha \left(\hat{a}_{\boldsymbol{p}\alpha}\mathrm{e}^{-\mathrm{i}p\cdot x} + \hat{a}_{\boldsymbol{p}\alpha}^\dagger \mathrm{e}^{\mathrm{i}p\cdot x} \right), \tag{13.10}$$

where

$$\boldsymbol{h}_1 = \begin{pmatrix} 1 \\ 0 \\ 0 \end{pmatrix}, \quad \boldsymbol{h}_2 = \begin{pmatrix} 0 \\ 1 \\ 0 \end{pmatrix}, \quad \boldsymbol{h}_3 = \begin{pmatrix} 0 \\ 0 \\ 1 \end{pmatrix}, \tag{13.11}$$

are internal vectors that tell us about the *polarization* of the field in its internal space.

Substituting into the Hamiltonian for this theory then leads[5] to the result that

$$\hat{H} = \int \mathrm{d}^3 p \, E_{\boldsymbol{p}} \sum_{\alpha=1}^{3} \hat{a}_{\boldsymbol{p}\alpha}^\dagger \hat{a}_{\boldsymbol{p}\alpha}, \tag{13.12}$$

and we observe that the energy of the particles may be found by summing over all momenta and polarizations.

[5]One needs to substitute the mode expansion and normal ordering **steps IV** and **V**.

Example 13.3

Noether's theorem reminds us that a symmetry gives us a conserved quantity, so we examine the internal $SO(3)$ symmetry of our fields. The Lagrangian is symmetric with respect to the transformation $\boldsymbol{\Phi} \rightarrow \boldsymbol{\Phi} - \boldsymbol{\theta} \times \boldsymbol{\Phi}$. For example (and assuming sums over repeated indices), rotations about the ϕ_3-axis give $\Phi^a \rightarrow \Phi^a - \varepsilon^{a3c}\theta^3 \Phi^c$ and we have

$$D^3 \phi^1 = \phi^2, \quad D^3 \phi^2 = -\phi^1, \quad D^3 \phi^3 = 0, \tag{13.13}$$

Noether's theorem: $SO(3)$ **internal symmetry**

$$D^b \Phi^a = -\varepsilon^{abc}\Phi^c \qquad \Pi^{a\mu} = \partial^\mu \Phi^a$$
$$D\mathcal{L} = 0 \qquad W^\mu = 0$$
$$J_{\mathrm{Nc}}^{a\mu} = \varepsilon^{abc}\Phi^b(\partial^\mu \Phi^c)$$
$$\boldsymbol{Q}_{\mathrm{Nc}} = \int \mathrm{d}^3 x \, (\boldsymbol{\Phi} \times \partial_0 \boldsymbol{\Phi})$$
$$N\left[\hat{Q}_{\mathrm{Nc}}^a \right] = -\mathrm{i} \int \mathrm{d}^3 p \, \varepsilon^{abc}\hat{a}_{b\boldsymbol{p}}^\dagger \hat{a}_{c\boldsymbol{p}}.$$

where we call the symmetry from rotation around the b-axis $D^b \Phi^a$. Using the fact that, as usual, $W^\mu = 0$ for this internal transformation (since $D\mathcal{L} = 0$), this yields a Noether current $J_\mathrm{N}^{3\mu}$ for rotation around the internal 3-axis:

$$J_\mathrm{N}^{3\mu} = \Pi^{a\mu} D^3 \Phi^a = (\partial^\mu \phi^1)\phi^2 - (\partial^\mu \phi^2)\phi^1. \tag{13.14}$$

[6] We dispense with the normal ordering symbol from this point. It should be assumed in all calculations of energies and conserved charges.

Swapping signs over, the conserved charge is given in terms of (normally ordered[6]) creation and annihilation operators by

$$\hat{Q}_{\mathrm{Nc}}^3 = -\mathrm{i} \int \mathrm{d}^3 p \left(\hat{a}_{1\boldsymbol{p}}^\dagger \hat{a}_{2\boldsymbol{p}} - \hat{a}_{2\boldsymbol{p}}^\dagger \hat{a}_{1\boldsymbol{p}} \right). \tag{13.15}$$

Rotations about the 1- and 2-axes will lead to similar conserved charges in terms of the other field operators. In general the conserved charge will be given by

$$\boldsymbol{Q}_{\mathrm{Nc}} = \int \mathrm{d}^3 x \, (\boldsymbol{\Phi} \times \partial_0 \boldsymbol{\Phi}) \quad \text{and} \quad \hat{Q}_{\mathrm{Nc}}^a = -\mathrm{i} \int \mathrm{d}^3 p \, \varepsilon^{abc} \hat{a}_{b\boldsymbol{p}}^\dagger \hat{a}_{c\boldsymbol{p}}. \tag{13.16}$$

This conserved charge is isospin. As shown in Exercise 13.1, isospin has the structure of angular momentum: in this case an isospin-1. Remember that, despite the similarity between the angular momentum in Exercise 3.3 (which arises since the symmetry is with respect to spatial rotations) and the conserved charge in this example, the symmetry here is *internal* and so this angular momentum-like charge is internal to the field.

13.2 Massive electromagnetism

[7] The theory described in this section is rather like quantum electrodynamics but, it turns out, without the interesting complication of gauge invariance, which we leave until the next chapter.

[8] This is in contrast to our example of the internal $SO(3)$ field $\boldsymbol{\Phi}(x)$.

We now turn to the interesting case of the massive vector field. This theory describes an intriguing world[7] in which photons have mass! The theory is described in terms of $A^\mu(x)$ which is called a 'vector field' because, when we input a spacetime point x, the field outputs a four-vector A^μ corresponding to that point. The key observation about A^μ is that it lives in Minkowski space and so *transforms* in the same way as a four-vector does under the Lorentz transformation.[8] We'll see that the particles created by the quantum version of this field have three possible polarizations, which will lead to the excitations carrying a spin of $S = 1$. The Lagrangian for this theory (**Step I**) is given by

$$\mathcal{L} = -\frac{1}{4} F_{\mu\nu} F^{\mu\nu} + \frac{1}{2} m^2 A_\mu A^\mu, \tag{13.17}$$

where $F_{\mu\nu} = \partial_\mu A_\nu - \partial_\nu A_\mu$. This is just the Lagrangian for the electromagnetic field but with a Klein–Gordon-esque mass term on the right-hand side. Note that the sign of the second term that contains the factor m^2 is positive. In the Klein–Gordon equation the m^2 term was negative, but the reason for our sign choice here will become apparent in the following example.

[9] Equation 13.18 can of course be written out purely in terms of A^μ as

$$\partial_\mu \left(\partial^\mu A^\nu - \partial^\nu A^\mu \right) + m^2 A^\nu = 0.$$

Example 13.4

We can find the equations of motion for the field in this theory by plugging the Lagrangian (eqn 13.17) into the Euler–Lagrange equation. The result is that[9]

$$\partial_\mu F^{\mu\nu} + m^2 A^\nu = 0. \tag{13.18}$$

This is known as the **Proca equation**. Taking the divergence gives us zero from the first term (since $\partial_\mu \partial_\nu F^{\mu\nu} = 0$) and we obtain from the second term $m^2 \partial_\nu A^\nu = 0$. Since $m^2 \neq 0$, we have the additional condition that the field obeys $\partial_\mu A^\mu = 0$. This allows us to write the Proca equations in another form

$$(\partial^2 + m^2) A^\nu = 0, \tag{13.19}$$

which looks a lot like the Klein–Gordon equation[10] for each of the four components of the field $A^\nu(x)$. The dispersion relation is therefore $E_{\boldsymbol{p}}^2 = \boldsymbol{p}^2 + m^2$ for each component.

Alexandru Proca (1897–1955) developed the theory of massive spin-1 particles using field theory.

[10]This shows that we did indeed make a sensible sign choice in eqn 13.17.

We now continue the procedure of quantizing this theory.[11] Finding the momentum density $\Pi_\sigma^0(x)$ for each field component σ is the first part of **Step II**, which is easily done. Extending the definition of $\Pi_\sigma^\mu(x)$ from the last chapter to include the vector index on A^ν, we have:

$$\Pi^{\mu\nu}(x) \;=\; \frac{\partial \mathcal{L}}{\partial(\partial_\mu A_\nu)} = \partial^\nu A^\mu - \partial^\mu A^\nu. \tag{13.20}$$

Note that $\Pi^{\mu\nu}$ is now a second-rank tensor whose two indices can be raised or lowered using the metric tensor $g_{\mu\nu}$. We then have momentum densities

$$\Pi^{00} = 0, \quad \Pi^{0i} = \partial^i A^0 - \partial^0 A^i = E^i, \tag{13.21}$$

where $\boldsymbol{E}$ is the 'electric field' of the theory (a useful name for notational simplicity, but that's all since this is not electromagnetism!). Notice that $\Pi^{00} = 0$ means that A^0 has no dynamics of its own.[12] We will eliminate it in the next step (and justify this in step IV). We find the Hamiltonian density to be[13]

$$\mathcal{H} \;=\; \frac{1}{2}\left(\boldsymbol{E}^2 + \boldsymbol{B}^2 + m^2 \boldsymbol{A}^2 + \frac{1}{m^2}(\boldsymbol{\nabla}\cdot\boldsymbol{E})^2\right). \tag{13.22}$$

Notice the positive contribution from the mass term, which justifies the choice of sign in the Lagrangian in eqn 13.17.

Next we need some equal-time commutation relations (**step III**) to make all of this quantum mechanical. Clearly we want a way to deal with the different components of the vector field which, when we substitute in a sensible mode expansion, will yield up a total energy which is the sum of the positive energies of all of the particles. Since we have eliminated A^0 from the Hamiltonian in eqn 13.22 we only need relations between the space-like components (i and j). We choose

$$\left[\hat{A}^i(t,\boldsymbol{x}), \hat{E}^j(t,\boldsymbol{y})\right] = -\mathrm{i}\delta^{(3)}(\boldsymbol{x}-\boldsymbol{y})g^{ij}, \tag{13.23}$$

which has the form we had for previous cases (namely $[\phi, \partial_0\phi] = \mathrm{i}\delta$ since $g^{ij} = -\delta_{ij}$) for each spatial component.

We may now expand the field A^μ (**step IV**) in terms of plane waves. Since these will look something like $A \sim a_{\boldsymbol{p}}\mathrm{e}^{-\mathrm{i}p\cdot x}$, the condition $\partial_\mu A^\mu = 0$ implies that $p_\mu A^\mu = 0$. This tells us that the four-vector A^μ is *orthogonal* to p_μ, just as we have in three dimensions if $\boldsymbol{c}\cdot\boldsymbol{d} = 0$ for vectors $\boldsymbol{c}$ and $\boldsymbol{d}$. Moreover, the condition $p_\mu A^\mu = 0$ introduces a constraint on the components of A^μ: they are no longer independent and

[11]The rest of this chapter works carefully through the nuts and bolts of getting from the Lagrangian in eqn 13.17 to the final result of the Hamiltonian in eqn 13.31, which is our familiar second-quantized harmonic-oscillator type Hamiltonian.

[12]This lack of dynamics might cause us to worry that the quantization routine won't work on this field component. In fact, we will eliminate A^0 and so this lack of a momentum won't turn out to be a problem. This is quite unlike the case of QED where the problem of having $\Pi^{00} = 0$ is a knotty one. See Aitchison and Hey, Section 7.3 for a discussion.

[13] In four steps:
I. Write

$$\mathcal{H} \;=\; -\boldsymbol{E}\cdot\dot{\boldsymbol{A}} - \frac{1}{2}(\boldsymbol{E}^2 - \boldsymbol{B}^2)$$
$$-\frac{m^2}{2}\left[(A^0)^2 - \boldsymbol{A}^2\right].$$

II. Use eqn 13.21 along with $\boldsymbol{E}\cdot\boldsymbol{\nabla}A^0 = \boldsymbol{\nabla}\cdot(\boldsymbol{E}A^0) - A^0(\boldsymbol{\nabla}\cdot\boldsymbol{E})$.
III. Then use the equations of motion (eqn 13.18) to eliminate A^0 via

$$A^0 = -\frac{1}{m^2}\boldsymbol{\nabla}\cdot\boldsymbol{E}.$$

IV. The Hamiltonian includes a term $\boldsymbol{\nabla}\cdot(\boldsymbol{E}A^0)$, but this is a total derivative. We constrain all of our fields to vanish at infinity, so we may drop the total derivative from the Hamiltonian density.

[14]The polarization of the vector particle tells us about its spin angular momentum. Vector particles have $S = 1$
and the three degrees of freedom correspond to $S_z = 1, 0$ and -1. This is
examined further in the next chapter.

we may express one of the components of A^μ in terms of the others. The
theory therefore describes a field (and particles) with three independent
polarization degrees of freedom. The polarization degrees of freedom
are taken to be A^i with $i = 1, 2, 3$. The component A^0 is completely
determined via $\boldsymbol{p} \cdot \boldsymbol{A} = p^0 A^0$, which shows that we were justified in
eliminating it from eqn 13.22.

Now for the mode expansion. As in the case of the $SO(3)$ fields which
we wrote as $\boldsymbol{\Phi}$, we'll need separate creation and annihilation operators
for each polarization. Unlike that case, however, we need to multiply
each by a **polarization vector**[14] $\epsilon^\mu_\lambda(p)$ which lives in Minkowski space
and whose components depend on the value of the momentum of the
particle that we're considering.[15]

[15]Unlike μ, the label λ on the polarization vector is not a tensor index, it's
part of the name that tells us the polarization vector to which we're referring.
We could equally have called the vectors Tom, Dick and Harry. However,
as we have done here, it's often useful
to use λ as a shorthand to also tell us
along which direction the polarization
vector points in the rest frame of the
particle.

Since there are only three degrees of freedom there will be three polarizations $\lambda = 1, 2, 3$. We choose that there is no $\lambda = 0$ polarization
[that is, we will force the field to be orthogonal to the 0-direction in
the rest frame of a particle, so we won't require a time-like polarization
basis vector $\epsilon^\mu_{\lambda=0}(m, 0) = (1, 0, 0, 0)$ (see below)]. The resulting mode
expansion is written

$$
\hat{A}^\mu(x) = \int \frac{\mathrm{d}^3 p}{(2\pi)^{\frac{3}{2}}} \frac{1}{(2E_{\boldsymbol{p}})^{\frac{1}{2}}} \left[\begin{pmatrix} \epsilon^0_1(p) \\ \epsilon^1_1(p) \\ \epsilon^2_1(p) \\ \epsilon^3_1(p) \end{pmatrix} \hat{a}_{\boldsymbol{p}1} \mathrm{e}^{-ip \cdot x} + \begin{pmatrix} \epsilon^{0*}_1(p) \\ \epsilon^{1*}_1(p) \\ \epsilon^{2*}_1(p) \\ \epsilon^{3*}_1(p) \end{pmatrix} \hat{a}^\dagger_{\boldsymbol{p}1} \mathrm{e}^{ip \cdot x} \right.
$$

$$
+ \begin{pmatrix} \epsilon^0_2(p) \\ \epsilon^1_2(p) \\ \epsilon^2_2(p) \\ \epsilon^3_2(p) \end{pmatrix} \hat{a}_{\boldsymbol{p}2} \mathrm{e}^{-ip \cdot x} + \begin{pmatrix} \epsilon^{0*}_2(p) \\ \epsilon^{1*}_2(p) \\ \epsilon^{2*}_2(p) \\ \epsilon^{3*}_2(p) \end{pmatrix} \hat{a}^\dagger_{\boldsymbol{p}2} \mathrm{e}^{ip \cdot x}
$$

$$
+ \left. \begin{pmatrix} \epsilon^0_3(p) \\ \epsilon^1_3(p) \\ \epsilon^2_3(p) \\ \epsilon^3_3(p) \end{pmatrix} \hat{a}_{\boldsymbol{p}3} \mathrm{e}^{-ip \cdot x} + \begin{pmatrix} \epsilon^{0*}_3(p) \\ \epsilon^{1*}_3(p) \\ \epsilon^{2*}_3(p) \\ \epsilon^{3*}_3(p) \end{pmatrix} \hat{a}^\dagger_{\boldsymbol{p}3} \mathrm{e}^{ip \cdot x} \right], \quad (13.24)
$$

[16]The commutation relations between
the creation and annihilation operators
are

$$
\left[\hat{a}_{\boldsymbol{p}\lambda}, \hat{a}^\dagger_{\boldsymbol{q}\lambda'} \right] = \delta^{(3)}(\boldsymbol{p}-\boldsymbol{q})\delta_{\lambda\lambda'}. \quad (13.25)
$$

or, more compactly[16]

$$
\hat{A}^\mu(x) = \int \frac{\mathrm{d}^3 p}{(2\pi)^{\frac{3}{2}}} \frac{1}{(2E_{\boldsymbol{p}})^{\frac{1}{2}}} \sum_{\lambda=1}^{3} \left(\epsilon^\mu_\lambda(p)\hat{a}_{\boldsymbol{p}\lambda} \mathrm{e}^{-ip \cdot x} + \epsilon^{\mu*}_\lambda(p)\hat{a}^\dagger_{\boldsymbol{p}\lambda} \mathrm{e}^{ip \cdot x} \right).
$$

$$(13.26)$$

Since $p_\mu A^\mu = 0$, we'll require $p_\mu \epsilon^\mu_\lambda(p) = 0$, which shows how the polarization vectors depend on the momentum.

Example 13.5

The next job is to work out what we want the polarization vectors to be. Being vectors
and living in Minkowski space, these will have to transform like vectors under the
Lorentz transformations. It will be a good idea to have these vectors orthonormal so
we want

$$
\epsilon^*_\lambda(p) \cdot \epsilon_{\lambda'}(p) = g_{\mu\nu}\epsilon^{\mu*}_\lambda \epsilon^\nu_{\lambda'} = -\delta_{\lambda\lambda'}, \quad (13.27)
$$

where the minus sign comes from our metric.[17] Consider a particle in its rest frame, where it has momentum $p^\mu = (m, 0, 0, 0)$. As stated above, we need $p^\mu \epsilon_{\lambda\mu}(p) = 0$, for all λ. That is, we want the polarization vectors in this frame to be normal to this p^μ. One possible choice is to use linear polarization vectors, given by

$$\epsilon_1(m,0) = \begin{pmatrix} 0 \\ 1 \\ 0 \\ 0 \end{pmatrix}, \ \epsilon_2(m,0) = \begin{pmatrix} 0 \\ 0 \\ 1 \\ 0 \end{pmatrix}, \ \epsilon_3(m,0) = \begin{pmatrix} 0 \\ 0 \\ 0 \\ 1 \end{pmatrix}. \tag{13.28}$$

We can now work out the value of $\epsilon_\lambda(p)$ in an arbitrary frame of reference by boosting with Lorentz transformation $\Lambda(p)$. For example, a particle moving with momentum $p_z = |\boldsymbol{p}|$ along the z-direction has $p^\mu = (E_{\boldsymbol{p}}, 0, 0, |\boldsymbol{p}|)$ which is achieved with a boost matrix

$$\Lambda^\mu{}_\nu(p) = \frac{1}{m} \begin{pmatrix} E_{\boldsymbol{p}} & 0 & 0 & |\boldsymbol{p}| \\ 0 & m & 0 & 0 \\ 0 & 0 & m & 0 \\ |\boldsymbol{p}| & 0 & 0 & E_{\boldsymbol{p}} \end{pmatrix}, \tag{13.29}$$

yielding the polarization vectors

$$\epsilon_1(E_{\boldsymbol{p}}, 0, 0, |\boldsymbol{p}|) = \begin{pmatrix} 0 \\ 1 \\ 0 \\ 0 \end{pmatrix}, \ \epsilon_2(E_{\boldsymbol{p}}, 0, 0, |\boldsymbol{p}|) = \begin{pmatrix} 0 \\ 0 \\ 1 \\ 0 \end{pmatrix}, \ \epsilon_3(E_{\boldsymbol{p}}, 0, 0, |\boldsymbol{p}|) = \begin{pmatrix} |\boldsymbol{p}|/m \\ 0 \\ 0 \\ E_{\boldsymbol{p}}/m \end{pmatrix}. \tag{13.30}$$

Finally, using the mode expansion,[18] the Hamiltonian after normal ordering (**step V**) is diagonalized in the form

$$\hat{H} = \int \mathrm{d}^3 p \, E_{\boldsymbol{p}} \sum_{\lambda=1}^{3} \hat{a}^\dagger_{\boldsymbol{p}\lambda} \hat{a}_{\boldsymbol{p}\lambda}, \tag{13.31}$$

meaning that the total energy is the energy of all particles in all polarizations as we might have expected. Although the complication of having three field components is irritating, the algebra involves little more than the tedious job of keeping track of some new indices and the polarization vectors.

13.3   Polarizations and projections

Clearly, the added complexity of the vector field arises because of the different polarizations which are possible. In this section we'll develop a tensor toolkit to deal with this feature.[19]

Let's start with something very simple. To project a three-vector $\boldsymbol{X}$ along the direction of a vector $\boldsymbol{p}$ (see Fig. 13.2), then all we have to do is to take the scalar product between $\boldsymbol{X}$ and $\hat{\boldsymbol{p}} = \boldsymbol{p}/|\boldsymbol{p}|$ (to find the length of the component), and then multiply the result by $\hat{\boldsymbol{p}}$ to point it in the correct direction. The projected vector is then $\boldsymbol{X}_{\mathrm{L}}$ where

$$\boldsymbol{X}_{\mathrm{L}} = \frac{(\boldsymbol{p} \cdot \boldsymbol{X})\boldsymbol{p}}{|\boldsymbol{p}|^2}, \tag{13.32}$$

where the 'L' superscript stands for 'longitudinal'. The transverse part of $\boldsymbol{X}$ is then $\boldsymbol{X}_{\mathrm{T}} = \boldsymbol{X} - \boldsymbol{X}_{\mathrm{L}}$. We can write these equations in component form as[20]

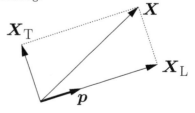

Fig. 13.2 Projecting a three-vector $\boldsymbol{X}$ along the direction of a vector $\boldsymbol{p}$.

[20]Assume summation over repeated indices.

$$X_{\mathrm{L}}^i = P_{\mathrm{L}}^{ij} X^j \quad \text{and} \quad X_{\mathrm{T}}^i = P_{\mathrm{T}}^{ij} X^j, \tag{13.33}$$

where the longitudinal and transverse **projection matrices** P_{L}^{ij} and P_{T}^{ij} are given by

$$P_{\mathrm{L}}^{ij} = \frac{p^i p^j}{|\boldsymbol{p}|^2}, \tag{13.34}$$

and

$$P_{\mathrm{T}}^{ij} = \delta_{ij} - \frac{p^i p^j}{|\boldsymbol{p}|^2}. \tag{13.35}$$

The four-dimensional upgrades of these are immediate: we define projection tensors (returning to tensor notation)

$$P_{\mathrm{L}}^{\mu\nu} = \frac{p^\mu p^\nu}{p^2} \quad \text{and} \quad P_{\mathrm{T}}^{\mu\nu} = g^{\mu\nu} - \frac{p^\mu p^\nu}{p^2}. \tag{13.36}$$

For our massive spin-1 modes p^μ represents the energy-momentum four-vector, so $p^2 = m^2$. If we work in a basis of linear polarizations, the transverse projection tensor can be related to the polarization vectors via the following equality[21]

$$\sum_{\lambda=1}^{3} \epsilon_{\lambda\mu}(p)\epsilon_{\lambda\nu}(p) = -P_{\mu\nu}^{\mathrm{T}}. \tag{13.38}$$

This relationship is proved in the following example:

---

[21] If we work in a basis of circular polarizations a slightly more complicated identity is needed.

$$\frac{1}{2}\sum_\lambda \left[\epsilon_{\mu\lambda}(p)\epsilon_{\nu\lambda}^*(p) + \epsilon_{\mu\lambda}^*(-p)\epsilon_{\nu\lambda}(-p)\right]$$
$$= -P_{\mu\nu}^{\mathrm{T}}. \tag{13.37}$$

See Maggiore, Chapter 4.

Example 13.6

Method 1: In the rest frame $p^\mu = (m, 0, 0, 0)$ and the polarization vectors are given by eqn 13.28 and hence we have

$$\sum_{\lambda=1}^{3} \epsilon_{\lambda\mu}(m,0)\epsilon_{\lambda\nu}(m,0) = \begin{pmatrix} 0 & 0 & 0 & 0 \\ 0 & 1 & 0 & 0 \\ 0 & 0 & 1 & 0 \\ 0 & 0 & 0 & 1 \end{pmatrix}. \tag{13.39}$$

One can then boost this second-rank tensor using the matrix of eqn 13.29 to give

$$\sum_{\lambda=1}^{3} \epsilon_{\lambda\mu}(p)\epsilon_{\lambda\nu}(p) = \Lambda^\rho{}_\mu \Lambda^\sigma{}_\nu \sum_{\lambda=1}^{3} \epsilon_{\lambda\rho}(m,0)\epsilon_{\lambda\sigma}(m,0) \tag{13.40}$$

$$= \frac{1}{m^2}\begin{pmatrix} |\boldsymbol{p}|^2 & 0 & 0 & -E_{\boldsymbol{p}}|\boldsymbol{p}| \\ 0 & m^2 & 0 & 0 \\ 0 & 0 & m^2 & 0 \\ -E_{\boldsymbol{p}}|\boldsymbol{p}| & 0 & 0 & E_{\boldsymbol{p}}^2 \end{pmatrix} = -g_{\mu\nu} + \frac{p_\mu p_\nu}{m^2}.$$

Method 2: The quantity $\sum_{\lambda=1}^{3} \epsilon_{\lambda\mu}\epsilon_{\lambda\nu}$ is a tensor and so it must be built up from other tensors. Since the only other tensors we can build from the variables in the problem are $g_{\mu\nu}$ and $p_\mu p_\nu$ we'll assume it has to be a linear combination of these: $A g_{\mu\nu} + B p_\mu p_\nu$. For a particle at rest we need $A + Bm^2 = 0$ in order that $\sum_{\lambda=1}^{3} \epsilon_{\lambda 0}\epsilon_{\lambda 0} = 0$, which fixes $B = -A/m^2$. Next, for a particle at rest we need the diagonal space-like components to equal 1, giving us $A = -1$.

Chapter summary

- Putting more components into our theory introduces the notion of the polarization of our field. The Hamiltonian for a simple non-interacting Lagrangian again yields harmonic oscillator solutions, but this time the energy requires a sum over all momenta and over polarizations.

- We have applied this to a Lagrangian for massive electromagnetism,

$$\mathcal{L} = -\frac{1}{4}F_{\mu\nu}F^{\mu\nu} + \frac{1}{2}m^2 A_\mu A^\mu,$$

and have shown that the polarization vectors $\epsilon^\mu(p)$ are related to the transverse projection tensor.

Exercises

(13.1) (a) Show that the conserved charge in eqn 13.16 may be written

$$\hat{Q}_{\text{Nc}} = \int \mathrm{d}^3 p \, \hat{A}_p^\dagger \boldsymbol{J} \hat{A}_p, \qquad (13.41)$$

where $\hat{A}_p = (\hat{a}_{1p}, \hat{a}_{2p}, \hat{a}_{3p})$ and $\boldsymbol{J} = (J_x, J_y, J_z)$ are the spin-1 angular momentum matrices from Chapter 9.

(b) Use the transformations from Exercise 3.3 to find the form of the angular momentum matrices appropriate to express the charge as $\hat{Q}_{\text{Nc}} = \int \mathrm{d}^3 p \, \hat{B}_p^\dagger \boldsymbol{J} \hat{B}_p$ where $\hat{B}_p = (\hat{b}_{1p}, \hat{b}_{0p}, \hat{b}_{-1p})$.

(13.2) (a) Confirm that eqn 13.29 is the appropriate matrix to boost a particle along the z-direction.

(b) Show that the boosted vectors in eqn 13.30 are still correctly normalized.

(c) Consider the circular polarization vectors $\epsilon_{\lambda=\text{R}}^{\mu*} = -\frac{1}{\sqrt{2}}(0,1,\mathrm{i},0)$, $\epsilon_{\lambda=\text{L}}^{\mu*} = \frac{1}{\sqrt{2}}(0,1,-\mathrm{i},0)$, $\epsilon_{\lambda=3}^{\mu*} = (0,0,0,1)$. Show that these are correctly normalized according to $g_{\mu\nu}\epsilon_\lambda^{\mu*}\epsilon_{\lambda'}^\nu = -\delta_{\lambda\lambda'}$.

(13.3) Show that P_{L} and P_{T} are indeed projection operators by showing that $P^2 = P$.

(13.4) The Lagrangian for electromagnetism in vacuo is $\mathcal{L} = -\frac{1}{4}F^{\mu\nu}F_{\mu\nu}$. Show that this can be rewritten as

$$\mathcal{L} = -\frac{1}{2}(\partial_\mu A_\nu \partial^\mu A^\nu - \partial_\mu A_\nu \partial^\nu A^\mu), \qquad (13.42)$$

and hence show that using the transverse projection operator, it may be expressed as

$$\mathcal{L} = \frac{1}{2}A^\mu P_{\mu\nu}^{\text{T}} \partial^2 A^\nu. \qquad (13.43)$$

This shows that $\mathcal{L}$ only includes the transverse components of the field, squaring with the idea of electromagnetic waves only representing vibrations transverse to the direction of propagation.

14 Gauge fields and gauge theory

[1]'Gauge' is an awful bit of terminology with which we are unfortunately stuck. Einstein's general relativity showed that spacetime geometry has a dynamical role and Hermann Weyl wondered if the scale of length could itself be dynamical, varying through spacetime. In this picture, one could make a choice of gauge which would be a choice of scale-length: metal wire comes in different thickness or gauges, so the term seemed entirely appropriate. He later adapted his scale argument to one involving phase, as outlined here, but the name 'gauge' stuck.

Nobody ever reads a paper in which someone has done an experiment involving photons with the footnote that says 'this experiment was done in Coulomb gauge'.
Sidney Coleman (1937–2007)

14.1 What is a gauge field?

We continue our discussion of invariances possessed by physical systems and arrive at gauge invariance. The idea here is that our field theory may admit different configurations of the fields which yield identical observables. Our physical description thus contains some inherent vagueness and so we can make a choice about which particular formulation, out of the many possible ones, to adopt in a given situation (this is called 'a choice of gauge'). A transformation from one description to another is called a **gauge transformation**, and the underlying invariance is called a **gauge invariance**.[1] Note that gauge invariance is not a symmetry. Particles do not carry around a knob called 'gauge' that allow us to change them into other particles. Gauge invariance is merely a statement of our inability to find a unique description of a system.

Example 14.1

(i) In electrical theory we have to choose a zero for our potential V (i.e. choose which potential we call 'ground'). This is because we only measure potential *differences* (e.g. with a voltmeter) and so the choice of zero potential is an arbitrary one.

(ii) In quantum mechanics we have to make a choice of what we mean by the zero of phase of a wave function. The choice is arbitrary because we only measure phase *differences* (e.g. in an interference experiment). There is nothing stopping one transforming all wave functions according to $\psi(x) \rightarrow \psi(x)e^{i\alpha}$ and the physics is unchanged.

(iii) In electromagnetism, the magnetic vector potential $\boldsymbol{A}$ can be transformed according to $\boldsymbol{A} \rightarrow \boldsymbol{A} + \boldsymbol{\nabla}\chi$, where $\chi(\boldsymbol{x})$ is a function of position, and $\boldsymbol{B} = \boldsymbol{\nabla} \times \boldsymbol{A}$ will be unaffected (because $\boldsymbol{\nabla} \times \boldsymbol{\nabla}\chi = 0$). Thus one has to choose the form of $\boldsymbol{A}$.

The preceding example raises some important issues.

- The potential V and $\boldsymbol{A}$, and the quantum mechanical phase, all possess an intrinsic vagueness in description, and we have to make specific choices in each case. We might wonder if the choices are connected. It will turn out that they are.

- We thought about our choice of zero for V and phase as being something we make once for our particular description of a particular problem, but which is valid for the whole Universe. If zero volts is set differently in New York and in San Francisco, then problems will occur if you run a conducting wire between their respective electrical grounds. However, $\boldsymbol{A}(\boldsymbol{x})$ is chosen with a function $\chi(\boldsymbol{x})$ which varies from place to place. Thus we will need to distinguish between global (affecting everywhere in the Universe) and local (affecting only a particular point) gauge transformations.

With these points in mind, it is time to start exploring this subtle invariance, and by doing so we will shed new light on the origin of electromagnetism.

We start with another look at the complex scalar field theory

$$\mathcal{L} = (\partial^\mu \psi)^\dagger (\partial_\mu \psi) - m^2 \psi^\dagger \psi. \tag{14.1}$$

Recall from Section 7.6 that this has a $U(1)$ symmetry. We can make the replacement

$$\psi(x) \to \psi(x) \mathrm{e}^{i\alpha}, \tag{14.2}$$

and the Lagrangian (and by extension the equations of motion) aren't changed. The most important point to note about eqn 14.2 is that it's a **global transformation**: it changes the field $\psi(x)$ by the same amount at every spacetime point. We say that the theory, given in eqn 14.1, is invariant with respect to *global* $U(1)$ transformations.

It turns out that an incredibly rich seam of physics is revealed if one asks what happens if the value of α given in eqn 14.2 depends on spacetime. That is, we investigate the transformation

$$\psi(x) \to \psi(x) \mathrm{e}^{i\alpha(x)}, \tag{14.3}$$

where the function $\alpha(x)$ allows us to transform the field by a different amount at every point in spacetime. This is a very significant change. Your first reaction to this might be to say that surely no theory could be invariant with respect to a different change in phase at every point in spacetime! Let's see what actually happens.

Example 14.2

The good news is that the mass term $m^2 \psi^\dagger \psi$ doesn't change when we transform locally. The bad news is that the term involving the derivatives does. This is because the derivative now acts on $\alpha(x)$. In fact, we have

$$
\begin{aligned}
\partial_\mu \psi(x) \quad &\to \quad \partial_\mu \psi(x) \mathrm{e}^{i\alpha(x)} \\
&= \quad \mathrm{e}^{i\alpha(x)} \partial_\mu \psi(x) + \psi(x) \mathrm{e}^{i\alpha(x)} i\partial_\mu \alpha(x) \\
&= \quad \mathrm{e}^{i\alpha(x)} \left[\partial_\mu + i\partial_\mu \alpha(x) \right] \psi(x).
\end{aligned}
\tag{14.4}
$$

Similarly

$$\partial^\mu \psi^\dagger(x) \rightarrow e^{-i\alpha(x)}\left[\partial^\mu - i\partial^\mu\alpha(x)\right]\psi^\dagger(x). \tag{14.5}$$

The first term in the Lagrangian becomes

$$(\partial^\mu\psi^\dagger)(\partial_\mu\psi) - i(\partial^\mu\alpha)\psi^\dagger(\partial_\mu\psi) + i(\partial^\mu\psi^\dagger)(\partial_\mu\alpha)\psi + (\partial^\mu\alpha)(\partial_\mu\alpha)\psi^\dagger\psi, \tag{14.6}$$

which is certainly not what we started with.

As expected, having α depend on x has robbed the theory of its $U(1)$ symmetry. We say that it is not invariant with respect to *local $U(1)$* transformations. Undaunted, we're going to see if we can restore this local symmetry. We'll do this by adding a new field $A^\mu(x)$, designed to roll around spacetime cancelling out terms that stop the theory from being invariant. The way this happens will depend on the manner in which the new field transforms. The way to proceed is to introduce the field $A^\mu(x)$ via a new object D_μ given by

$$D_\mu = \partial_\mu + iqA_\mu(x). \tag{14.7}$$

This new object D_μ is called the **covariant derivative**. This will fix up the local $U(1)$ symmetry if we insist that the new field A_μ transforms according to $A_\mu \rightarrow A_\mu - \frac{1}{q}\partial_\mu\alpha(x)$. The parameter q is known as the coupling strength and will later tell us how strongly the A_μ field interacts with other fields.[2]

[2]For the case of QED examined in Chapter 39 we will be able to identify q as the electromagnetic charge. For now we treat it as a parameter.

Example 14.3

If $\psi(x) \rightarrow \psi(x)e^{i\alpha(x)}$, then $\partial_\mu\psi \rightarrow (\partial_\mu\psi)e^{i\alpha} + i(\partial_\mu\alpha)\psi e^{i\alpha}$ and so

$$\begin{aligned} D_\mu\psi = (\partial_\mu + iqA_\mu)\psi &\rightarrow (\partial_\mu\psi)e^{i\alpha} + i(\partial_\mu\alpha)\psi e^{i\alpha} + iqA_\mu\psi e^{i\alpha} - i(\partial_\mu\alpha)\psi e^{i\alpha} \\ &= D_\mu(\psi e^{i\alpha}). \end{aligned} \tag{14.8}$$

This property makes the whole Lagrangian invariant if we replace ordinary derivatives by covariant ones:

$$\mathcal{L} = (D^\mu\psi)^\dagger(D_\mu\psi) - m^2\psi^\dagger\psi, \tag{14.9}$$

since now with $D_\mu\psi \rightarrow D_\mu\psi e^{i\alpha}$, the first term is invariant.

Thus we conclude that we need our theory to be invariant with respect to *two* sets of transformations, which must be implemented together:

$$\psi(x) \rightarrow \psi(x)e^{i\alpha(x)}, \tag{14.10}$$

$$A_\mu(x) \rightarrow A_\mu(x) - \frac{1}{q}\partial_\mu\alpha(x). \tag{14.11}$$

A theory which has a field $A^\mu(x)$ introduced to produce an invariance with respect to local transformations is known as a **gauge theory**. The field $A^\mu(x)$ is known as a **gauge field**.

We might expect the field $A^\mu(x)$ could make a contribution to the Lagrangian itself. It only exists in our description because we've invented it to satisfy our demand for a locally invariant theory, but if such ambitions have any groundings in reality then the field A^μ should have dynamics of its own! As we will see in the next section, it gives us electromagnetism.

14.2 Electromagnetism is the simplest gauge theory

What are the possible forms of the part of the Lagrangian describing the field $A^\mu(x)$? It must be a theory for which the Lagrangian is invariant under transformations of the form $A_\mu(x) \to A_\mu(x) - \frac{1}{q}\partial_\mu\alpha(x)$. Electromagnetism provides an example of such a theory. It is, after all, a theory described in terms of a vector field $A^\mu(x) = (V(x), \boldsymbol{A}(x))$ which is used to form a Lagrangian

$$\mathcal{L} = -\frac{1}{4}(\partial_\mu A_\nu - \partial_\nu A_\mu)(\partial^\mu A^\nu - \partial^\nu A^\mu) - J^\mu_{\text{em}} A_\mu, \qquad (14.12)$$

from which the equations of motion follow as

$$\partial^2 A^\nu - \partial^\nu(\partial_\mu A^\mu) = J^\nu_{\text{em}}, \qquad (14.13)$$

which are known as the first two Maxwell equations.[3]

One of the most important observations about this formulation of electromagnetism is that neither the Lagrangian nor the equations of motion are changed if one makes the swap

$$A_\mu(x) \to A_\mu(x) - \partial_\mu\chi(x), \qquad (14.14)$$

which is short for the changes

$$
\begin{aligned}
V &\to V - \partial_0\chi, \\
\boldsymbol{A} &\to \boldsymbol{A} + \boldsymbol{\nabla}\chi.
\end{aligned}
\qquad (14.15)
$$

This observation, which is known as gauge invariance, means that if A_μ does the job of describing the electromagnetic fields in some situation then so does $A_\mu - \partial_\mu\chi$. This also means that electromagnetism is a gauge theory since we can call $\chi(x)$ by the name $\alpha(x)/q$ and the condition in eqn 14.11 will clearly be satisfied.

Example 14.4

The theory can also be written in terms of the tensor $F_{\mu\nu} = \partial_\mu A_\nu - \partial_\nu A_\mu$ with a Lagrangian written as

$$\mathcal{L} = -\frac{1}{4}F_{\mu\nu}F^{\mu\nu}. \qquad (14.16)$$

Gauge invariance amounts to the statement that we can change the field according to $A_\mu \to A_\mu - \partial_\mu\chi(x)$ and $F^{\mu\nu}$ is unchanged.

It may be helpful to think of the freedom of the choice of gauge as a choice of language. We can speak French, German or Venusian, but we are able to communicate the same underlying message no matter what language we speak. If we're sensible, we can choose $\chi(x)$ (that is, the language we're speaking) in such a way as to make whatever we're

[3]Recall that we wrote these in Chapter 5 as $\partial_\lambda F^{\lambda\nu} = J^\nu_{\text{em}}$ (where $F^{\lambda\nu} = \partial^\lambda A^\nu - \partial^\nu A^\lambda$), which is clearly the same. More interestingly, it is worth considering the role of the Maxwell equations in this context. In the next section we will quantize the theory in terms of electromagnetic particles called photons, which are excitations of the quantum field $\hat{A}(x)$. The Maxwell equations are the equations of motion for $\hat{A}(x)$ and the closest we have to a Schrödinger equation for the photon. As it is not possible to construct a probability density from the field $A(x)$, the photons that the quantum field theory of electromagnetism describes do not have an interpretation in terms of a probability amplitude of their spatial localization. See Berestetskii, Lifshitz and Pitaevskii (Introduction and Chapter 1) for further details.

[4] Actually it's more usually (and incorrectly) known as the Lorentz condition due to its misattribution to Hendrik Lorentz (1853–1928) rather than to the less famous Ludvig Lorenz (1829–1891) who used it first. See J. D. Jackson and L. B. Okun, Rev. Mod. Phys. **73**, 663 (2001) for details of the history.

Common gauges for electromagnetism

Lorenz gauge	$\partial_\mu A^\mu = 0$
Coulomb gauge	$\boldsymbol{\nabla} \cdot \boldsymbol{A} = 0$
Axial gauge	$A^3 = 0$
Weyl gauge	$A^0 = 0$

describing as simple as possible. There are several common choices, the most famous of which are listed in the box in the margin.

Let's try making a choice of $\chi(x)$. We'll try choosing $\chi(x)$ in such a way that

$$\partial_\mu A^\mu(x) = 0, \qquad (14.17)$$

which is known as the Lorenz condition or as **Lorenz gauge**.[4]

Example 14.5

To do this we write $A_\mu \to A'_\mu = A_\mu - \partial_\mu \chi$. We want

$$\partial^\mu A'_\mu = \partial^\mu A_\mu - \partial^\mu \partial_\mu \chi = 0, \qquad (14.18)$$

which is achieved by setting $\partial^2 \chi = \partial^\mu A_\mu$. This results in the desired condition $\partial_\mu A'^\mu = 0$.

Lorenz gauge is useful because, in the absence of a current J^μ_{em}, it results in a massless Klein–Gordon equation for each component of the electromagnetic field. That is, with $\partial_\mu A'^\mu = 0$, eqn 14.13 becomes

$$\partial^2 A'^\mu = 0, \qquad (14.19)$$

whose solutions are plane waves of the form $A^\mu = \epsilon^\mu(p)\mathrm{e}^{-\mathrm{i}p \cdot x}$ with $E_{\boldsymbol{p}} = |\boldsymbol{p}|$. The Lorenz condition therefore makes electromagnetism resemble the case of vector field theory. Recall that there the Lorenz condition $\partial_\mu A^\mu$ wasn't a choice, it was mandated by the mass term $m^2 A^\mu A_\mu$ in the Lagrangian of that theory. In any case, the result is the same: it reduces the number of independent components of A'^μ from four to three.

However, this still doesn't make A'^μ unique! This is because we can make a further shift $A'_\mu \to A''_\mu = A'_\mu - \partial_\mu \xi$ as long as $\partial^2 \xi = 0$ (so that both A'_μ and A''_μ satisfy the Lorenz condition). To make A''^μ unique, we will choose

$$\partial_0 \xi = A'_0, \qquad (14.20)$$

which implies $A''_0 = 0$. With this further choice, the Lorenz condition then reduces to $\boldsymbol{\nabla} \cdot \boldsymbol{A}'' = 0$. This is known as **Coulomb gauge** and further reduces the number of independent field components by one. This makes it clear that although the field A^μ has four components, the physics allows only *two* independent components. This can lead to trouble when the field is quantized.

Example 14.6

[5] Remember that the polarization vectors introduced here will carry the information about the spin state of the photon.

The equations of motion in Lorenz gauge read $\partial^2 A^\mu = 0$, which, with $A^0 = 0$, have plane wave solutions[5] $\boldsymbol{A} = \boldsymbol{\epsilon}\mathrm{e}^{-\mathrm{i}p \cdot x}$. The equation encoding the Coulomb gauge condition, $\boldsymbol{\nabla} \cdot \boldsymbol{A} = 0$, leads to

$$\boldsymbol{p} \cdot \boldsymbol{A} = \boldsymbol{p} \cdot \boldsymbol{\epsilon} = 0, \qquad (14.21)$$

which tells us that the direction of propagation of the wave is perpendicular to the polarization. For a wave propagating along z with momentum $q^\mu = (|\boldsymbol{q}|, 0, 0, |\boldsymbol{q}|)$ we could have

$$\epsilon_1(q) = \begin{pmatrix} 1 \\ 0 \\ 0 \end{pmatrix}, \quad \epsilon_2(q) = \begin{pmatrix} 0 \\ 1 \\ 0 \end{pmatrix}, \tag{14.22}$$

corresponding to linear polarization. We could choose the waves to have circular polarization instead, in which case we could have

$$\epsilon_R^*(q) = -\tfrac{1}{\sqrt{2}} \begin{pmatrix} 1 \\ i \\ 0 \end{pmatrix}, \quad \epsilon_L^*(q) = \tfrac{1}{\sqrt{2}} \begin{pmatrix} 1 \\ -i \\ 0 \end{pmatrix}. \tag{14.23}$$

In order to observe the effects of electromagnetism, the electromagnetic field must couple to a matter field. We might wonder how to write down this coupling. It turns out that the most simple form may be achieved with a simple recipe known as the **minimal coupling prescription**. This simply involves swapping derivatives ∂_μ of the matter field for covariant derivatives D_μ and is illustrated in the example below.

Example 14.7

Consider complex scalar field theory in the presence of an electromagnetic field. The Lagrangian is written as the sum of the Lagrangians for the two separate theories:

$$\mathcal{L} = (\partial^\mu \psi)^\dagger (\partial_\mu \psi) - m^2 \psi^\dagger \psi - \frac{1}{4} F_{\mu\nu} F^{\mu\nu}. \tag{14.24}$$

To enact the minimal coupling prescription we upgrade derivatives of the matter field ψ to covariant derivatives:

$$\begin{aligned} \mathcal{L} &= (D^\mu \psi)^\dagger (D_\mu \psi) - m^2 \psi^\dagger \psi - \frac{1}{4} F_{\mu\nu} F^{\mu\nu} \\ &= (\partial^\mu \psi^\dagger - iq A^\mu \psi^\dagger)(\partial_\mu \psi + iq A_\mu \psi) - m^2 \psi^\dagger \psi - \frac{1}{4} F_{\mu\nu} F^{\mu\nu} \\ &= \partial^\mu \psi^\dagger \partial_\mu \psi - m^2 \psi^\dagger \psi - \frac{1}{4} F_{\mu\nu} F^{\mu\nu} \\ &\quad + \left(-iq A^\mu \psi^\dagger (\partial_\mu \psi) + iq (\partial^\mu \psi^\dagger) A_\mu \psi + q^2 \psi^\dagger \psi A^\mu A_\mu \right). \end{aligned} \tag{14.25}$$

The final line shows the coupling between the A_μ field and the ψ and $\psi^\dagger$ fields. The strength of the coupling is set by the coupling strength q, which is also known as electromagnetic charge.

The notion that a gauge field, introduced to guarantee a local symmetry, will dictate the form of the coupling, or *interactions*, in a theory is known as the **gauge principle** and is a philosophy to which we will return later.[6]

[6]See Chapter 46.

14.3 Canonical quantization of the electromagnetic field

The previous section reminded us that even though A^μ has four components, only two are needed for a physical description (which makes good

[7]The gauge field A^μ should not, therefore, be confused with the vector field discussed in the last chapter. The vector field was massive and this property robs it of the possibility of gauge invariance. It also led to its having three degrees of freedom. The massless nature of the gauge field leads to its gauge invariance and its only having two free components.

[8]This second method may be found in the book by Aitchison and Hey and the one by Maggiore.

[9]You should find

$$\mathcal{H} = \frac{1}{2}(\boldsymbol{E}^2 + \boldsymbol{B}^2) + \boldsymbol{E} \cdot \boldsymbol{\nabla} A^0.$$

The last term is removed using the same method as employed in the previous chapter, noting that $\boldsymbol{\nabla} \cdot \boldsymbol{E} = 0$ and that all fields should vanish at infinity.

physical sense because light is a transverse wave and therefore has two components).[7]

Having two redundant components of A^μ flapping around, reflecting the freedom to choose which language that we speak, is something we might worry about when we quantize the theory. There are now two ways to go. We can fix the gauge from the outset in order to reduce the number of degrees of freedom from four to two, making a choice of which language we're going to speak throughout, or we can continue and see how far we get before the redundancy comes back to bite us.[8] We choose the first option in the following example, and working in the Coulomb gauge we will canonically quantize electromagnetism simply by following the usual five steps.

Example 14.8

Step I: The Lagrangian is still

$$\mathcal{L} = -\frac{1}{4}F_{\mu\nu}F^{\mu\nu} = -\frac{1}{4}(\partial_\mu A_\nu - \partial_\nu A_\mu)(\partial^\mu A^\nu - \partial^\nu A^\mu), \qquad (14.26)$$

although we'll need to remember that our choice of gauge dictates $A^0 = 0$ and $\boldsymbol{\nabla} \cdot \boldsymbol{A} = 0$. Actually, this will be implemented at step III.
Step II: We find that the $\Pi^{\mu\nu}$ tensor has components

$$\Pi^{\mu\nu} = \frac{\partial \mathcal{L}}{\partial(\partial_\mu A_\nu)} = -(\partial^\mu A^\nu - \partial^\nu A^\mu), \qquad (14.27)$$

just as in the case of massive vector field theory. Again, the momentum density of the zeroth component $\Pi^{00} = 0$, but we don't care since we have decided to eliminate $A^0 = 0$. The momentum component conjugate to the ith component of the field is $\Pi^{0i} = E^i$, that is, the electric vector field $\boldsymbol{E}(x)$. The Hamiltonian[9] is then

$$\mathcal{H} = \frac{1}{2}(\boldsymbol{E}^2 + \boldsymbol{B}^2). \qquad (14.28)$$

Step III: Next we need commutation relations. Our first guess might be that $\left[\hat{A}^\mu(\boldsymbol{x}), \hat{\Pi}^{0\nu}(\boldsymbol{y})\right] = -\mathrm{i}\delta^{(3)}(\boldsymbol{x}-\boldsymbol{y})g^{\mu\nu}$, similar to the massive vector field. This won't work here though. Rewriting yields $\left[\hat{A}^i(\boldsymbol{x}), \hat{E}^j(\boldsymbol{y})\right] = -\mathrm{i}\delta^{(3)}(\boldsymbol{x}-\boldsymbol{y})g^{ij}$, which looks fine until you take the divergence (with respect to $\boldsymbol{x}$) of this equation and get

$$\left[\partial_i \hat{A}^i(\boldsymbol{x}), \hat{E}^j(\boldsymbol{y})\right] = \mathrm{i}\partial^j \delta^{(3)}(\boldsymbol{x}-\boldsymbol{y}). \qquad (14.29)$$

But this should be zero! (This is because $\partial_i \hat{A}^i = \boldsymbol{\nabla} \cdot \hat{\boldsymbol{A}} = 0$.) The problem is that we haven't yet ensured that only transverse components of $\hat{\boldsymbol{A}}$ are permitted to enter the solution. We need a way of projecting out the unnecessary ones to implement the gauge condition. The answer to this problem turns out to necessitate the use of our projection tensor P^{T} (see Chapter 13). The commutation relation we need is given by

$$\begin{aligned}\left[\hat{A}^i(\boldsymbol{x}), \hat{E}^j(\boldsymbol{y})\right] &= \mathrm{i}\int \frac{\mathrm{d}^3 p}{(2\pi)^3} \mathrm{e}^{\mathrm{i}\boldsymbol{p}\cdot(\boldsymbol{x}-\boldsymbol{y})}\left(\delta^{ij} - \frac{p^i p^j}{p^2}\right) \qquad (14.30)\\ &= \mathrm{i}\delta^{(3)}_{\mathrm{tr}}(\boldsymbol{x}-\boldsymbol{y}),\end{aligned}$$

which gives zero on the right when you take the divergence (exercise). The symbol $\delta^{(3)}_{\mathrm{tr}}$ is known as a *transverse delta function* due to the inclusion of the transverse projection tensor P^{T}. This guarantees that our creation and annihilation operators satisfy the required relation $[\hat{a}_{\boldsymbol{p}\lambda}, \hat{a}^\dagger_{\boldsymbol{q}\lambda'}] = \delta^{(3)}(\boldsymbol{p}-\boldsymbol{q})\delta_{\lambda\lambda'}$.

Step IV: Next the mode expansion, which only contains those two polarizations that are transverse to the direction of propagation:[10]

$$\hat{A}^\mu(x) = \int \frac{\mathrm{d}^3 p}{(2\pi)^{\frac{3}{2}}} \frac{1}{(2E_{\boldsymbol{p}})^{\frac{1}{2}}} \sum_{\lambda=1}^{2} \left(\epsilon^\mu_\lambda(p) \hat{a}_{\boldsymbol{p}\lambda} \mathrm{e}^{-\mathrm{i}p\cdot x} + \epsilon^{\mu*}_\lambda(p) \hat{a}^\dagger_{\boldsymbol{p}\lambda} \mathrm{e}^{\mathrm{i}p\cdot x} \right). \tag{14.31}$$

Inserting this and (**Step V**) normally ordering, we obtain the expected final result:

$$\hat{H} = \int \mathrm{d}^3 p \sum_{\lambda=1}^{2} E_{\boldsymbol{p}} \hat{a}^\dagger_{\boldsymbol{p}\lambda} \hat{a}_{\boldsymbol{p}\lambda}, \tag{14.32}$$

with $E_{\boldsymbol{p}} = |\boldsymbol{p}|$. We conclude that the excitations of the electromagnetic field are **photons**, which may be observed with two transverse polarizations. Apart from the slight complication of making the commutation relations compatible with the choice of gauge and the hassle of the indices, there's nothing hard about the quantization of electromagnetism in Coulomb gauge.

[10]Note that the two polarization vectors obey $p_\mu \epsilon^\mu_\lambda(p) = 0$ and are normalized according to $g_{\mu\nu} \epsilon^{\mu*}_\lambda(p) \epsilon^\nu_{\lambda'}(p) = -\delta_{\lambda\lambda'} = g_{\lambda\lambda'}$.

The last example showed us that quantized particles of the electromagnetic field are photons. These particles have spin $S = 1$ and come in two types: $\hat{a}^\dagger_{\boldsymbol{p}1}|0\rangle$ and $\hat{a}^\dagger_{\boldsymbol{p}2}|0\rangle$, corresponding to the two transverse polarizations of the electromagnetic field. Consider a photon propagating along the z-direction with momentum $q^\mu = (|\boldsymbol{q}|, 0, 0, |\boldsymbol{q}|)$. If we work in a basis of circularly polarized vectors, we may write $\epsilon^*_{\lambda=\mathrm{R}}(q) = -\frac{1}{\sqrt{2}}(0, 1, \mathrm{i}, 0)$ (corresponding to $S^z = 1$) and $\epsilon^*_{\lambda=\mathrm{L}}(q) = \frac{1}{\sqrt{2}}(0, 1, -\mathrm{i}, 0)$ (corresponding to $S^z = -1$). There are no photons with $S^z = 0$ as this would correspond to an unphysical longitudinal polarization $\epsilon^*_{\lambda=3}(p) = (0, 0, 0, 1)$. We examine this in Exercise 14.2 and in the following example.

Example 14.9

We calculated the rotation matrices $\hat{\boldsymbol{J}}$ for the vector field in Chapter 9. These also function as angular momentum operators for photon states. We will create a photon travelling along the z-direction, so that $q^\mu = (|\boldsymbol{q}|, 0, 0, |\boldsymbol{q}|)$. We write

$$|\gamma_\lambda\rangle = \epsilon^*_\lambda(q) \hat{a}^\dagger_{\boldsymbol{q}\lambda}|0\rangle. \tag{14.33}$$

Let's calculate the helicity[11] of photons with $\lambda = \mathrm{R}$. The helicity operator is $\hat{h} = \hat{\boldsymbol{J}} \cdot \boldsymbol{q}/|\boldsymbol{q}|$, leading to an operator $\hat{h} = \hat{J}^z$. Operating with the angular momentum matrix $\hat{J}^z$ suitable for $S = 1$ spins, we have

$$\hat{J}^z|\gamma_{\lambda=\mathrm{R}}\rangle = -\frac{1}{\sqrt{2}} \begin{pmatrix} 0 & 0 & 0 & 0 \\ 0 & 0 & -\mathrm{i} & 0 \\ 0 & \mathrm{i} & 0 & 0 \\ 0 & 0 & 0 & 0 \end{pmatrix} \begin{pmatrix} 0 \\ 1 \\ \mathrm{i} \\ 0 \end{pmatrix} \hat{a}^\dagger_{\boldsymbol{q}\mathrm{R}}|0\rangle = +|\gamma_{\lambda=\mathrm{R}}\rangle. \tag{14.34}$$

We see that the right-circularly polarized photon is in an $S^z = 1$ eigenstate and also the $h = +1$ (positive) eigenstate of helicity. It's easy to show that the left-circularly polarized photon is in the $S^z = -1$ state and has negative helicity. You may confirm that the disallowed polarization $\epsilon^*_{\lambda=3}(q) = (0, 0, 0, 1)$ would correspond to a longitudinally polarized photon. Note that this polarization is allowed for the vector particle from Chapter 13 as vector fields have three degrees of freedom

[11]Helicity, the projection of the spin along the momentum direction, is explained in detail in Chapter 36. Since the photon does not have a rest frame the helicity is the most useful quantity to use in talking about its spin.

Chapter summary

- A theory which has a field $A_\mu(x)$ introduced to produce an invariance with respect to local transformations is known as a *gauge theory*. The field $A_\mu(x)$ is known as a *gauge field*. Theories must be invariant with respect to the transformations:

$$\psi(x) \rightarrow \psi(x)e^{i\alpha(x)},$$
$$A_\mu(x) \rightarrow A_\mu(x) - \frac{1}{q}\partial_\mu\alpha(x).$$

$$(14.35)$$

- Applied to the Lagrangian for electromagnetism, $\mathcal{L} = -\frac{1}{4}F_{\mu\nu}F^{\mu\nu}$, we have canonically quantized electromagnetism.

Exercises

(14.1) Fill in the algebra leading to eqn 14.32.

(14.2) *A demonstration that the photon has spin-1, with only two spin polarizations.*

A photon γ propagates with momentum $q^\mu = (|\boldsymbol{q}|, 0, 0, |\boldsymbol{q}|)$. Working with a basis where the two transverse photon polarizations are $\epsilon^\mu_{\lambda=1}(q) = (0, 1, 0, 0)$ and $\epsilon^\mu_{\lambda=2}(q) = (0, 0, 1, 0)$, it may be shown, using Noether's theorem, that the operator $\hat{S}^z$, whose eigenvalue is the z-component spin angular momentum of the photon, obeys the commutation relation

$$\left[\hat{S}^z, \hat{a}^\dagger_{\boldsymbol{q}\lambda}\right] = i\epsilon^{\mu=1*}_\lambda(q)\hat{a}^\dagger_{\boldsymbol{q}\lambda=2} - i\epsilon^{\mu=2*}_\lambda(q)\hat{a}^\dagger_{\boldsymbol{q}\lambda=1}.$$
$$(14.36)$$

(i) Define creation operators for the circular polarizations via

$$\hat{b}^\dagger_{\boldsymbol{q}R} = -\frac{1}{\sqrt{2}}\left(\hat{a}^\dagger_{\boldsymbol{q}1} + i\hat{a}^\dagger_{\boldsymbol{q}2}\right),$$
$$\hat{b}^\dagger_{\boldsymbol{q}L} = \frac{1}{\sqrt{2}}\left(\hat{a}^\dagger_{\boldsymbol{q}1} - i\hat{a}^\dagger_{\boldsymbol{q}2}\right). \quad (14.37)$$

Show that

$$\left[\hat{S}^z, \hat{b}^\dagger_{\boldsymbol{q}R}\right] = \hat{b}^\dagger_{\boldsymbol{q}R},$$
$$\left[\hat{S}^z, \hat{b}^\dagger_{\boldsymbol{q}L}\right] = -\hat{b}^\dagger_{\boldsymbol{q}L}. \quad (14.38)$$

(ii) Consider the operation of $\hat{S}^z$ on a state $|\gamma_{\boldsymbol{q}\lambda}\rangle = \hat{b}^\dagger_{\boldsymbol{q}\lambda}|0\rangle$ containing a single photon propagating along z:

$$\hat{S}^z|\gamma_\lambda\rangle = \hat{S}^z\hat{b}^\dagger_{\boldsymbol{q}\lambda}|0\rangle, \quad \lambda = R, L. \quad (14.39)$$

Use the results of (i) to argue that the projection of the photon spin along its direction of propagation must be $S^z = \pm 1$.

See Bjorken and Drell Chapter 14 for the full version of this argument.

Discrete transformations

15

In Chapter 9 we have explored how quantum fields behave under various spacetime translations and rotations. Such transformations are continuous and are thus represented by continuous groups (Lie groups). However, discrete transformations are also possible, represented this time by finite groups, and this chapter examines some of the most commonly encountered and important discrete transformations.

15.1 Charge conjugation

A particularly interesting discrete transformation is one which changes all particles into their antiparticles (see Fig. 15.1). This is accomplished by an operator C: this operator flips not only the sign of a particle's charge, but also its lepton number, its hypercharge, and all the 'charge-like' numbers which characterize a particular particle. We can write

$$C|p\rangle = |\bar{p}\rangle, \tag{15.1}$$

where $\bar{p}$ is the antiparticle of a particle p. Let's say that p has charge[1] q. If this charge can be measured by an operator $\hat{Q}$ then $\hat{Q}|p\rangle = q|p\rangle$, whereas $\hat{Q}|\bar{p}\rangle = -q|\bar{p}\rangle$. Thus $C\hat{Q}|p\rangle = qC|p\rangle = q|\bar{p}\rangle$, but $\hat{Q}C|p\rangle = \hat{Q}|\bar{p}\rangle = -q|\bar{p}\rangle$. This implies that $\hat{Q}C = -C\hat{Q}$, or equivalently

$$C^{-1}\hat{Q}C = -\hat{Q}. \tag{15.2}$$

For the operator C to exchange particles and antiparticles we need

$$C^{-1}\hat{a}_{\boldsymbol{p}}C = \hat{b}_{\boldsymbol{p}} \quad \text{and} \quad C^{-1}\hat{b}_{\boldsymbol{p}}^{\dagger}C = \hat{a}_{\boldsymbol{p}}^{\dagger}. \tag{15.3}$$

Since a complex scalar field $\hat{\psi}(x)$ can be written $\hat{\psi}(x) = \int_{\boldsymbol{p}}(\hat{a}_{\boldsymbol{p}}e^{-ip\cdot x} + \hat{b}_{\boldsymbol{p}}^{\dagger}e^{ip\cdot x})$, we must have[2] $C^{-1}\hat{\psi}C = \hat{\psi}^{\dagger}$.

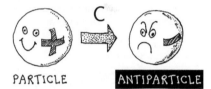

Fig. 15.1 The charge conjugation operator C turns a particle into its antiparticle, reversing its charge and all other 'charge-like' numbers.

[1] By 'charge' we could mean any property like charge, or lepton number, or baryon number, etc. These are generically called **quantum charges**.

[2] Note that

$$\hat{\psi}^{\dagger} = \int_{\boldsymbol{p}}(\hat{a}_{\boldsymbol{p}}^{\dagger}e^{ip\cdot x} + \hat{b}_{\boldsymbol{p}}e^{-ip\cdot x}).$$

Example 15.1

Since $C^2 = I$, the eigenvalues of C can only be ± 1. Most particles are not eigenstates of C, since if they were then $C|p\rangle = |\bar{p}\rangle = \pm|p\rangle$, but this would mean that $|\bar{p}\rangle$ is the same state as $|p\rangle$ and so the particle is its own antiparticle. This is true in some cases, specifically for particles which have no quantum charges.[3] In the case of the photon (γ), it is an eigenstate of C with eigenvalue -1, since if you change all particles to their antiparticles, the electromagnetic field reverses ($A^\mu \to -A^\mu$). It is also the case with the neutral pion π^0 but here the eigenvalue is $+1$. This explains why the reaction $\pi^0 \to \gamma + \gamma$ is allowed [the eigenvalues of C go from $+1 \to (-1) \times (-1)$], but you cannot have $\pi^0 \to \gamma + \gamma + \gamma$.

[3] In other words for particles which have zero electrical charge, zero lepton number, zero baryon number, etc.

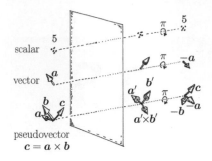

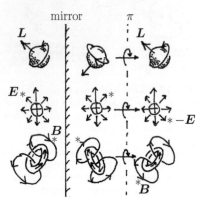

Fig. 15.2 A parity operation consists of reflection in a mirror followed by a π-rotation about an axis perpendicular to the mirror. The mirror reverses a single spatial coordinate (the one perpendicular to the plane of the mirror), the rotation reverses the other two. A scalar is unaffected by this transformation, whereas an ordinary (polar) vector changes sign. A pseudovector (axial vector), such as a vector product of two ordinary vectors, is unchanged.

Fig. 15.3 Electric field E behaves like an ordinary (polar) vector, while magnetic field B and angular momentum L (derived from vector products of polar vectors) are pseudovectors (axial vectors). The asterisk indicates a particular vector in the electric field from a point charge, and in the magnetic field from a current loop, so that you can focus on how a specified vector transforms.

The group multiplication table for the parity operator P and the identity operator I:

	I	P
I	I	P
P	P	I

15.2   Parity

When you look in a mirror, things look the same but there are subtle differences. Five screws in your hand are reflected into five screws, so numbers of things (scalars) don't change. But look carefully and you'll see that the threads in the mirror screws turn in the opposite sense to those in the real world. A mirror image has the spatial coordinate perpendicular to the surface of the mirror reversed, with the other two spatial coordinates left the same. This effect cannot be achieved by a rotation. If you follow a mirror operation by a 180° rotation about an axis perpendicular to the mirror, you will have reversed all three coordinates and this transformation is known as **inversion**. This transformation can be performed using the parity operator P which acts to invert spatial coordinate, i.e. mapping $x \to -x$. This means that the position operator $\hat{x}$ will anticommute with the parity operator:

$$\hat{x}P = -P\hat{x}, \tag{15.4}$$

or equivalently

$$P^{-1}\hat{x}P = -\hat{x}. \tag{15.5}$$

The effect on the momentum coordinates is also $p \to -p$, and hence

$$P^{-1}\hat{p}P = -\hat{p}. \tag{15.6}$$

However, the commutation relation $[\hat{x}, \hat{p}_x] = i$ will be preserved. Since P is a Hermitian operator, and since also it is its own inverse ($P^2 = I$, where I is the identity) then P is also unitary. The parity operation has no effect on scalars but reverses vectors (see Fig. 15.2, and for physical examples see Fig. 15.3). In fact, there are a class of scalars and vectors for which this is not true: a **pseudovector** (sometimes called an **axial vector**) is formed by a cross product between two ordinary vectors (which are sometimes called **polar vectors**) and a **pseudoscalar** is formed via a scalar triple product; these behave oppositely. In summary:

P(scalar) = scalar	P(**vector**) = −**vector**
P(pseudoscalar) = −pseudoscalar	P(**pseudovector**) = **pseudovector**.

Since $P^2 = I$ then the group formed by the elements $\{I, P\}$ under multiplication is isomorphic to $\mathbb{Z}_2$, the cyclic group of order 2, pretty much the simplest non-trivial group you can imagine. As for C, this means the eigenvalues are ±1. Scalars and pseudovectors have parity eigenvalue +1, while vectors and pseudoscalars have parity eigenvalue −1. For example, the photon is an excitation in a massless vector field and therefore has intrinsic parity −1. The pion is described by a pseudoscalar field and also has parity −1. We will later find (from using the Dirac equation and considering what are called spinor fields) that the parity of a fermion is opposite to that of its antiparticle.

In general, the problem we face is that our quantum fields can be scalar-valued fields or vector-valued fields or even more complicated objects. Thus a parity transformation will affect the nature of the field itself, as well as acting on the coordinates in which the field is defined.

Example 15.2

For now, let's just keep things simple and focus on a scalar field. If we operate on a scalar field then $\phi(t, \boldsymbol{x}) \to \phi(t, -\boldsymbol{x})$. Let's look at what it does to the creation and annihilation operators. We have

$$\mathsf{P}^{-1}\hat{\phi}(t, \boldsymbol{x})\mathsf{P} = \hat{\phi}(t, -\boldsymbol{x}) = \int_{\boldsymbol{p}} \hat{a}_{\boldsymbol{p}} e^{-i(Et+\boldsymbol{p}\cdot\boldsymbol{x})} + \hat{a}_{\boldsymbol{p}}^{\dagger} e^{i(Et+\boldsymbol{p}\cdot\boldsymbol{x})}, \qquad (15.7)$$

and this will work if

$$\mathsf{P}^{-1}\hat{a}_{\boldsymbol{p}}\mathsf{P} = \hat{a}_{-\boldsymbol{p}}, \quad \mathsf{P}^{-1}\hat{a}_{\boldsymbol{p}}^{\dagger}\mathsf{P} = \hat{a}_{-\boldsymbol{p}}^{\dagger}, \qquad (15.8)$$

which means that in the case of scalar fields the parity operator simply reverses the momentum on the creation and annihilation operators.

15.3   Time reversal

'If I could turn back time...' With the time-reversal operator T you can! It maps a scalar field $\phi(t, \boldsymbol{x}) \to \phi(-t, \boldsymbol{x})$ and so while it leaves the position operator well alone:

$$\mathsf{T}^{-1}\hat{\boldsymbol{x}}\mathsf{T} = \hat{\boldsymbol{x}}, \qquad (15.9)$$

it reverses momentum coordinates:

$$\mathsf{T}^{-1}\hat{\boldsymbol{p}}\mathsf{T} = -\hat{\boldsymbol{p}}. \qquad (15.10)$$

This means that the commutation relation $[\hat{x}, \hat{p}_x] = i$ will only be preserved if $\mathsf{T}^{-1}i\mathsf{T} = -i$, and hence T must be antiunitary.[4] The archetypal antiunitary operator is K, the complex conjugation operator and in fact you can make a general antiunitary operator by forming the product of K and U, where U is a unitary operator. Accordingly we write

$$\mathsf{T} = \mathsf{UK}, \qquad (15.11)$$

which means that[5] $\mathsf{U} = \mathsf{TK}$.

Example 15.3

(i) For spinless particles you can take U to be the identity and $\mathsf{T} = \mathsf{K}$. This is not so surprising since the complex conjugate of the Schrödinger equation

$$\hat{H}\psi = i\frac{\partial\psi}{\partial t} \qquad (15.12)$$

is

$$\hat{H}\psi^* = -i\frac{\partial\psi^*}{\partial t} = i\frac{\partial\psi^*}{\partial(-t)}, \qquad (15.13)$$

[4]Many symmetry operators S have the property of being **unitary**

$$\langle \mathsf{S}\phi | \mathsf{S}\psi \rangle = \langle \phi | \psi \rangle$$

and **linear**

$$\mathsf{S}(a|\phi\rangle + b|\psi\rangle) = a\mathsf{S}|\phi\rangle + b\mathsf{S}|\psi\rangle.$$

The adjoint of S is written $\mathsf{S}^{\dagger}$ and defined by

$$\langle \phi | \mathsf{S}^{\dagger}\psi \rangle = \langle \mathsf{S}\phi | \psi \rangle.$$

However, it is also possible for a symmetry operator to be **antiunitary**

$$\langle \mathsf{S}\phi | \mathsf{S}\psi \rangle = \langle \phi | \psi \rangle^* = \langle \psi | \phi \rangle$$

and **antilinear**

$$\mathsf{S}(a|\phi\rangle + b|\psi\rangle) = a^*\mathsf{S}|\phi\rangle + b^*\mathsf{S}|\psi\rangle.$$

In this case, we have to define the adjoint $\mathsf{S}^{\dagger}$ by

$$\langle \phi | \mathsf{S}^{\dagger}\psi \rangle = \langle \mathsf{S}\phi | \psi \rangle^* = \langle \psi | \mathsf{S}\phi \rangle.$$

Using these expressions, it is straightforward to show that $\langle \phi | \mathsf{S}^{\dagger}\mathsf{S}|\psi \rangle = \langle \phi | \psi \rangle$ for S being either unitary or antiunitary, and hence $\mathsf{S}^{\dagger} = \mathsf{S}^{-1}$ in both cases. However, note that for an antiunitary operator

$$\mathsf{S}c = c^*\mathsf{S},$$

for any complex number c (because of the antilinear property). A special case is if $c = i$ and then

$$\mathsf{S}i = -i\mathsf{S},$$

so that i anticommutes with S. (For the case $\mathsf{S} = \mathsf{T}$, this leads to $\mathsf{T}^{-1}i\mathsf{T} = -i$.) This looks nonsensical at first sight: how can a simple number like i anticommute with anything? However, it is just a special case of the antilinear property. In the special case that the time-reversal operator $\mathsf{T} = \mathsf{K}$, where K is the complex conjugation operator, this condition is easily demonstrated since

$$\mathsf{T}(i\psi) = -i\mathsf{T}\psi,$$

where ψ is any function.

[5]By postmultiplying both sides of eqn 15.11 by K.

and so the combined operation of time reversal *and* complex conjugation leaves the form of the Schrödinger equation invariant. Thus indeed it looks like K and T are the same thing.

(ii) Adding spin complicates things because angular momentum is reversed when time reverses (it corresponds in some general sense to things going round in a particular sense as a function of time). Thus

$$\mathsf{T}^{-1}\hat{\boldsymbol{S}}\mathsf{T} = -\hat{\boldsymbol{S}}, \tag{15.14}$$

where $\hat{\boldsymbol{S}}$ is the spin operator. The operation of complex conjugation is a bit more complicated since[6]

$$\mathsf{K}^{-1}\hat{S}_x\mathsf{K} = \hat{S}_x, \quad \mathsf{K}^{-1}\hat{S}_y\mathsf{K} = -\hat{S}_y, \quad \mathsf{K}^{-1}\hat{S}_z\mathsf{K} = \hat{S}_z. \tag{15.15}$$

[6]Remember the form of the Pauli spin matrices in which only σ_y has imaginary components, while σ_x and σ_z are real.

An appropriate form for U is $\mathsf{U} = \exp(-i\pi\hat{S}_y)$ so that

$$\mathsf{T} = \exp(-i\pi\hat{S}_y)\mathsf{K}. \tag{15.16}$$

[7]The proof of this goes as follows: $\mathsf{T}^2 = \mathsf{UKUK} = \mathsf{UU}^*$ but since U is unitary $\mathsf{U}^\dagger = (\mathsf{U}^*)^\mathrm{T} = \mathsf{U}^{-1}$ and so $\mathsf{T}^2 = \mathsf{U}(\mathsf{U}^\mathrm{T})^{-1} = \mathsf{X}$ where X is a diagonal matrix containing only phase factors. Thus $\mathsf{U} = \mathsf{XU}^\mathrm{T}$ and thus $\mathsf{U}^\mathrm{T} = \mathsf{UX}^\mathrm{T} = \mathsf{UX}$ and thus $\mathsf{U} = \mathsf{XUX}$ and hence the diagonal elements of X are all $+1$ or all -1. Thus $\mathsf{T}^2 = \pm 1$.

This operator has the property[7] $\mathsf{T}^2 = \pm 1$. For a system containing many particles with spin, the time-reversal operator can be written

$$\mathsf{T} = \prod_i \exp(-i\pi\hat{S}_{iy})\mathsf{K}, \tag{15.17}$$

where the product is evaluated over all the particles. Let's put some flesh on these bones. For a single electron, the spin is one-half and so $\exp(-i\pi\hat{S}_y)$ will have eigenvalues $\pm i$ and so $\mathsf{T}^2 = -1$. This will also be the case for an odd number of electrons but if the number of electrons is even then $\mathsf{T}^2 = 1$.

These expressions have some interesting consequences for the case in which $\mathsf{T}^2 = -1$ (for an odd number of electrons). For a Hamiltonian $\mathcal{H}$ that is invariant under time-reversal (so that $\mathcal{H}$ commutes with T), both a state $|\psi\rangle$ and its time-reversed version $\mathsf{T}|\psi\rangle$ will have the same energy. Are they the same state? If they were, then $\mathsf{T}|\psi\rangle = \alpha|\psi\rangle$ where α is a complex number. But in that case[8] $\mathsf{T}^2|\psi\rangle = \mathsf{T}\alpha|\psi\rangle = \alpha^*\mathsf{T}|\psi\rangle = |\alpha|^2|\psi\rangle$ but since we have assumed $\mathsf{T}^2 = -1$ then we would deduce $|\alpha|^2 = -1$ and that's not possible since $|\alpha|^2 > 0$. We have arrived at a contradiction of our initial assumption, so we can deduce that $|\psi\rangle$ and $\mathsf{T}|\psi\rangle$ are linearly-independent states and are known as a **Kramers doublet**. We have deduced **Kramers' theorem** which states that the energy levels of a time-reversal invariant system with an odd number of electrons are n-fold degenerate where n is even. Essentially the energy levels come in pairs of Kramers doublets, and you can only split these pairs by introducing a perturbation that breaks time-reversal, such as a magnetic field.

[8]We are using the fact that T is antiunitary and this implies $\mathsf{T}\alpha|\psi\rangle = \alpha^*\mathsf{T}|\psi\rangle$.

Hendrik Kramers (1894–1952).

Example 15.4

Reversing time t using T *and* reversing spatial coordinate x using P produces a reversal of spacetime x. On a scalar field this can be written

$$(\mathsf{PT})^{-1}\hat{\phi}(x)(\mathsf{PT}) = \hat{\phi}(-x). \tag{15.18}$$

This operation leaves the creation and annihilation operators unchanged:

$$(\mathsf{PT})^{-1}\hat{a}_{\boldsymbol{p}}(\mathsf{PT}) = \hat{a}_{\boldsymbol{p}}, \quad (\mathsf{PT})^{-1}\hat{a}_{\boldsymbol{p}}^\dagger(\mathsf{PT}) = \hat{a}_{\boldsymbol{p}}^\dagger, \tag{15.19}$$

and this occurs because a parity transformation flips all the momenta and the time reversal flips them all back. In the mode expansion of the scalar field the only effect is to flip the sign of i in the exponentials. It thus acts as an operator to perform a complex conjugation of the scalar field.

Each of C, P and T are conserved in many-particle physics processes, but not all. Famously, P is 'violated' in the weak interaction. However, a quantum field theory that satisfies a fairly minimal set of assumptions (ones in which $\mathcal{L}$ is Lorentz invariant, local, Hermitian and normal ordered) possesses the symmetry CPT. In other words, if you reverse spacetime *and* replace particles by antiparticles, the theory should be invariant. The proof of this **CPT theorem** is (in brief outline only) based on showing that $(CPT)^{-1}\mathcal{L}(x)(CPT) = \mathcal{L}(-x)$, and hence deducing that CPT commutes with the Hamiltonian and is hence a symmetry.[9]

A consequence of the CPT theorem is that the mass and lifetime of any particle is identical to that of its antiparticle. If experiment ever provides conclusive evidence of a deviation from this then CPT symmetry will have been shown not to hold in some particular case, casting doubt on Lorentz invariance. So far, CPT symmetry has survived the test.

[9]The first part can be deduced using the fact that $\mathcal{L}$ is a Lorentz scalar and hence any term in it contains terms in which all tensor indices are contracted, so the terms contain an even number of tensor indices and under CPT these produce an even number of minus signs. More details can be found in the books by Itzykson and Zuber, Srednicki and Weinberg.

15.4 Combinations of discrete and continuous transformations

We have previously examined the group of three-dimensional rotations. These can be represented by orthogonal 3×3 matrices (let's call them $\mathbf{R}$) but with the condition that the determinant of the matrix is $\det \mathbf{R} = +1$. The group which describes this is known as $SO(3)$, the **special orthogonal group**, the 'special' being the determinant being $+1$. These rotations are called **proper rotations**.

The parity transformation can be represented by a matrix

$$\begin{pmatrix} -1 & 0 & 0 \\ 0 & -1 & 0 \\ 0 & 0 & -1 \end{pmatrix}, \tag{15.20}$$

[or, for short, $\mathrm{diag}(-1,-1,-1)$] because it maps $(x,y,z) \to (-x,-y,-z)$ and this matrix clearly has determinant -1. By combining this single transformation with our group $SO(3)$, we get a much larger group $O(3)$, the group of all orthogonal 3×3 matrices.[10] This includes the proper rotations of $SO(3)$ (with $\det \mathbf{R} = +1$) as well as **improper rotations** in which $\det \mathbf{R} = -1$. Improper rotations are ordinary, proper rotations followed by a parity operation (or equivalently by a reflection).

[10]Note that orthogonality implies $R^{\mathrm{T}}R = I$. Hence taking the determinant, $\det \mathbf{R} \times \det \mathbf{R}^{\mathrm{T}} = 1$. Now $\det \mathbf{R} = \det \mathbf{R}^{\mathrm{T}}$ and hence $(\det \mathbf{R})^2 = 1$ or $\det \mathbf{R} = \pm 1$, showing that $O(3)$ contains both matrices with $\det \mathbf{R} = +1$ and matrices with $\det \mathbf{R} = -1$.

Fig. 15.4 The group $O(3)$ consists of two disjoint sets, which can be pictured as two disconnected islands. These sets are distinguished by their elements either having determinant $+1$ [this island is $SO(3)$] and -1. Only the first island is independently a group because the identify transformation has determinant $+1$ and lives exclusively on the left island.

Example 15.5

(i) Reflection by a mirror in the x-y plane is accomplished by $\mathrm{diag}(1,1,-1)$. This has determinant -1 (an improper rotation).

(ii) Rotation about the z-axis by π is accomplished by $\mathrm{diag}(-1,-1,1)$. This has determinant $+1$ (proper rotation).

(iii) The parity operation, $\mathrm{diag}(-1,-1,-1)$, is accomplished by the product of (i) and (ii). It is therefore an improper rotation.

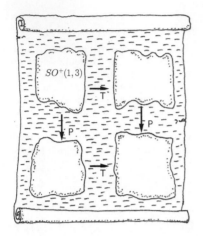

Fig. 15.5 The Lorentz group $O(1,3)$ consists of four separated components. The identity element resides in the component $SO^+(1,3)$, the proper, orthochronous Lorentz subgroup. The other 'islands' can be reached using the operators P, T or the combination PT.

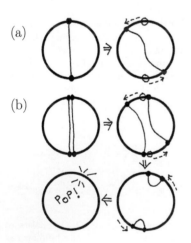

Fig. 15.6 Closed paths in the space $SO(3)$, represented as a ball in which antipodal points on the surface are identified. (a) A path from the north to south pole cannot be continuously deformed to a point, while (b) a double path can.

The group $SO(3)$ is a **connected group**, meaning that you can imagine continuously wandering through all the elements of the group; the group is not the union of two or more disjoint sets. The same cannot be said of $O(3)$. It consists of two disjoint sets, one set with determinant $+1$ [which is $SO(3)$] and the other with determinant -1 and is a **disconnected group** (see Fig. 15.4).

We find something similar when we consider the **Lorentz group**, the group containing all rotations, reflections and Lorentz boosts. This is often given the symbol $O(1,3)$, emphasizing that it has time-like (one-dimensional) and space-like (three-dimensional) parts. This consists of four topologically separated components because as well as P we have to consider T. In a four-dimensional representation we have

$$\mathsf{P} = \mathrm{diag}(1,-1,-1,-1), \qquad \mathsf{T} = \mathrm{diag}(-1,1,1,1). \qquad (15.21)$$

The subgroup of the Lorentz group that doesn't reverse spatial or temporal coordinates is the **proper, orthochronous Lorentz subgroup** $SO^+(1,3)$. This connected group is one of the four components of the Lorentz group. The other three components can be accessed by taking elements of $SO^+(1,3)$ and (i) including P, or (ii) including T, or (iii) including both P and T (see Fig. 15.5).

While wading in these deep topological waters, it is worth making some important remarks about simple rotations which will have important consequences later. Let's consider proper rotations which we have been representing using the group $SO(3)$. Let's think about the topology of this. A rotation is characterized by a rotation axis and an angle. Thus we could imagine a sphere with radius π and then any point inside the sphere could represent a rotation: the direction from the origin would determine the rotation axis and the distance from the origin would determine the rotation angle. However, a rotation of π about a particular axis is equivalent to a rotation of $-\pi$ about the same axis. Thus the topology of $SO(3)$ is equivalent to that of a ball in which antipodal points on the surface are identified.

This topological space is clearly connected but is not **simply connected**, which means that a closed path through the space cannot be continuously shrunk to a point. To show this consider a path running from the north pole to the south pole of the sphere [Fig. 15.6(a)]. You can't deform this continuously to a point. If you move the path near the north pole to the right, the point near the south pole moves to the left. You can't get the two to join up. However, if you run the path twice from the north pole to the south pole, you can do it [Fig. 15.6(b)]. This shows that a 4π rotation is continuously deformable to a point, while a 2π rotation is not. This is the basis of the famous Dirac scissors parlour-game trick, an amusing demonstration in which you thread string around the eyes of a pair of scissors and a nearby chair. Two full rotations of the scissors leave the string apparently highly tangled but with a few deft movements which do not involve further rotations you can untangle the string. The string serves to make the trick more complicated than it need be, and a more transparent version can be carried out using a scarf or a belt, with one end fixed by, e.g. placing it under a book, as shown in Fig. 15.7(a). The free end is rotated by two full turns (i.e. by 720 degrees) in the same sense and looks very twisted, as shown in Fig. 15.7(b). It can be untwisted without rotating the free end in the opposite sense, simply by passing the free end around the middle of the belt and pulling taut, see Fig. 15.7(c)-(f). Try it!

To show this point up more clearly in group theory, we could also use another group to represent rotations: $SU(2)$. This is the **special unitary group** of 2×2 matrices, where again the 'special' means the determinant is one. This can be used to rotate spinors.

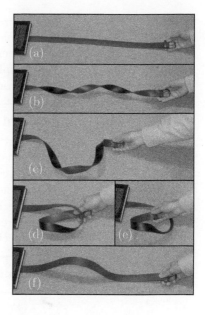

Fig. 15.7 Dirac's scissor trick is more easily demonstrated with a belt with one end free and the other end held fixed by placing it under a book. [Figure from S. J. Blundell, *Magnetism: A Very Short Introduction*, OUP (2012).]

Example 15.6

To rotate by an angle θ about an axis defined by $\hat{n}$ we can use the rotation matrix which we've previously written as $\mathbf{R}(\theta)$. Here we will write $\boldsymbol{\theta} = \theta\hat{n}$, and we will employ a useful identity which is

$$\boldsymbol{\sigma} \cdot \boldsymbol{n} = \begin{pmatrix} n_z & n_x - \mathrm{i}n_y \\ n_x + \mathrm{i}n_y & -n_z \end{pmatrix}, \tag{15.22}$$

where $\boldsymbol{\sigma} = (\sigma_x, \sigma_y, \sigma_z)$ are the Pauli spin matrices. Our rotation matrix, which we will now write as $\mathbf{R}(\hat{n}, \theta)$, is given by

$$\mathbf{R}(\hat{n}, \theta) = \exp\left(-\mathrm{i}\frac{\theta}{2}\boldsymbol{\sigma} \cdot \boldsymbol{n}\right) = I\cos\frac{\theta}{2} - \mathrm{i}\sin\frac{\theta}{2}\boldsymbol{\sigma} \cdot \boldsymbol{n}, \tag{15.23}$$

where I is the identity matrix. These matrices have an interesting feature, which is that while as you would expect

$$\mathbf{R}(\hat{n}, 0) = I, \tag{15.24}$$

you also have

$$\mathbf{R}(\hat{n}, 2\pi) = -I, \tag{15.25}$$

and you need a full 4π rotation to recover

$$\mathbf{R}(\hat{n}, 4\pi) = I. \tag{15.26}$$

The previous example shows that $SU(2)$ is actually a **double cover** of $SO(3)$, meaning that for a particular rotation there are two representations in $SU(2)$ for every one representation in $SO(3)$. For example, the identity (no rotation at all) is just diag$(1,1,1)$ in $SO(3)$, but is represented by both diag$(1,1)$ and diag$(-1,-1)$ in $SU(2)$.

A spinor can be written as a two-component entity $\begin{pmatrix} a \\ b \end{pmatrix}$ where a and b are complex numbers and $|a|^2 + |b|^2 = 1$. Writing $a = x_0 + \mathrm{i}x_1$ and $b = x_2 + \mathrm{i}x_3$ where x_i are real we see that this condition is equivalent to $x_0^2 + x_1^2 + x_2^2 + x_3^2 = 1$ and so $SU(2)$ is isomorphic to S^3, that is the 3-sphere.[11] Thus $SU(2)$ is simply connected, in contrast to $SO(3)$. The formal relationship between the two groups is $SO(3) \cong SU(2)/\mathbb{Z}_2$, meaning that $SO(3)$ is a quotient group.

These arguments can be generalized to the connected component of the Lorentz group: $SO(1,3) \cong SL(2, \mathbb{C})/\mathbb{Z}_2$, where $SL(2, \mathbb{C})$ is the group of 2×2 complex matrices with unit determinant. This gives you some flavour of the way in which the mathematical structure of the spaces describing the spacetime transformations can be described using group theory.

[11]It may be helpful to remember:

- S^1 is a circle, a one-dimensional space that we normally draw on a two-dimensional piece of paper. It can be described with the equation $x^2 + y^2 = 1$.

- S^2 is a 2-sphere, or more commonly just a sphere (meaning the surface of a ball), a two-dimensional space that can be embedded in three-dimensions. It can be described with the equation $x^2 + y^2 + z^2 = 1$.

- S^3 is a 3-sphere, a three-dimensional space that can be embedded in four-dimensions. It can be described with the equation $x^2 + y^2 + z^2 + w^2 = 1$.

For more details on topological notation, see Chapter 29.

circle	S^1	$(x^0)^2 + (x^1)^2 = 1$ $\qquad [S^1 \cong U(1)]$
sphere	S^2	$(x^0)^2 + (x^1)^2 + (x^2)^2 = 1$
n-sphere	S^n	$(x^0)^2 + (x^1)^2 + \cdots + (x^n)^2 = 1$
n-torus	T^n	$S^1 \times S^1 \times \cdots \times S^1 \ (n \geq 2)$

real $n \times n$ matrices

general linear group	$GL(n, \mathbb{R})$	$n \times n$ real matrices
special linear group	$SL(n, \mathbb{R})$	$GL(n, \mathbb{R})$ with $\det M = 1$
orthogonal group	$O(n)$	$GL(n, \mathbb{R})$ with $MM^{\mathrm{T}} = M^{\mathrm{T}}M = I$
special orthogonal group	$SO(n)$	$O(n)$ with $\det M = 1$
Lorentz group	$O(1,3)$	$GL(n, \mathbb{R})$ with $MgM^{\mathrm{T}} = g$
special Lorentz group	$SO(1,3)$	$O(1,3)$ with $\det M = 1$

complex $n \times n$ matrices

general linear group	$GL(n, \mathbb{C})$	$n \times n$ complex matrices
special linear group	$SL(n, \mathbb{C})$	$GL(n, \mathbb{C})$ with $\det M = 1$
unitary group	$U(n)$	$GL(n, \mathbb{C})$ with $MM^{\dagger} = M^{\dagger}M = I$
special unitary group	$SU(n)$	$U(n)$ with $\det M = 1$

Table 15.1 Some useful mathematical objects.

Chapter summary

- Discrete symmetries include charge conjugation C (which changes all particles into their antiparticles and vice versa), parity P which inverts spatial coordinates and time-reversal T which reverses time.
- The combined symmetry CPT holds for a Lorentz invariant, local, Hermitian quantum field theory.
- For reference, some useful mathematical objects are tabulated in Table 15.1.

Exercises

(15.1) Why is the reaction $\pi^0 \to \gamma + \gamma + \gamma$ not allowed?

(15.2) Classify the following as scalars, pseudoscalars, vectors (polar vectors) or pseudovectors (axial vectors): (a) magnetic flux; (b) angular momentum; (c) charge; (d) the scalar product of a vector and a pseudovector; (e) the scalar product of two vectors; (f) the scalar product of two pseudovectors.

(15.3) Find representations for the spinor rotation matrices (a) $\mathbf{R}(\hat{\boldsymbol{x}}, \theta)$, (b) $\mathbf{R}(\hat{\boldsymbol{y}}, \theta)$ and (c) $\mathbf{R}(\hat{\boldsymbol{z}}, \theta)$.

Part IV

Propagators and perturbations

Quantum field theory is particularly challenging when one has to deal with interacting systems. One method to calculate results uses perturbation theory and this can be evaluated using various techniques involving Green's functions and Feynman diagrams. This part introduces some of these techniques.

- In Chapter 16 we introduce *Green's functions* and show how they are related to a propagator, the amplitude that a particle in some spacetime position will be later found at another spacetime position.

- We connect Green's functions to propagators in a quantum field picture in Chapter 17 and introduce the *Feynman propagator* which involves time-ordering of operators. This is applied to Yukawa's model of virtual particle exchange.

- Interactions are added to the picture in Chapter 18 and in particular the idea of the *S-matrix*. We work this out in the *interaction representation*, another way of writing down states and operators which turns out to be very useful for these problems. These ideas are applied to a perturbation expansion and use is made of *Wick's theorem* which provides a method for simplifying products of operators.

- In Chapter 19 we introduce *Feynman diagrams* which provide a wonderful pictorial way of visualizing terms in the perturbation expansion. Rather than evaluating umpteen integrals we can summarize many complex calculations in diagrammatic form and evaluate the results using *Feynman rules* for a particular situation.

- These ideas are applied to scattering theory in Chapter 20 for a simple $\psi^\dagger \psi \phi$ model which serves as a toy model of QED and the *scattering cross-section* is defined and then evaluated.

16 Propagators and Green's functions

[1] As Feynman notes, quantum amplitudes are like Hebrew in that you read them right-to-left.

[2] These are named after George Green of Sneinton, Nottinghamshire (1793–1841) who was perhaps the greatest self-taught mathematical physicist of modern times. His achievements are all the more remarkable since, until 1829, his day job consisted of running a windmill. While still a miller he published his most celebrated work 'An Essay on the Application of Mathematical Analysis to the Theories of Electricity and Magnetism' (1828). The essay included the first use of what are now called Green's functions.

> *When will the world know that peace and propagation are the two most delightful things in it?*
> Horace Walpole, 4th Earl of Oxford (1717–1797)

One way to do quantum mechanics is to calculate a wave function and operate on it with quantum operators. Another way is to directly consider amplitudes for a given process, such as 'the amplitude that my particle starts at point y at a time t_y and ends up at point x at time t_x', which we write down[1] as $\langle x(t_x)|y(t_y)\rangle$. This amplitude is known as a **propagator**. We've spent time in previous chapters removing the wave function from its role as the most useful entity in quantum physics and we'll see that propagators represent an alternative to wave functions and enable us to extract all of the same information. In fact, we'll see that wave functions are actually special cases of propagators and contain less information! For our purposes, propagators represent the most economical way to calculate all of the properties of quantum fields in an interacting system of many particles. In addition propagators for single particles have a neat mathematical property: they are the **Green's functions**[2] of the equation of motion for a particle.

The plan for this chapter is as follows: (i) we'll explain what's meant by a Green's function; (ii) we'll demonstrate that propagator amplitudes are the Green's' functions of quantum mechanics: they tell us the amplitudes for a particle to start at a spacetime point (y, t_y) and then be detected at a point (x, t_x); (iii) finally we'll show that quantum mechanical perturbation theory is conveniently expressed in terms of propagators and that these propagators are most conveniently expressed in terms of cartoons.

16.1 What is a Green's function?

Here's a very general looking differential equation formed from a linear differential operator $\hat{L}$:

$$\hat{L}\, x(t) = f(t). \tag{16.1}$$

The Green's function $G(t, u)$ of the linear operator $\hat{L}$ is defined by the equation

$$\hat{L}\, G(t, u) = \delta(t - u). \tag{16.2}$$

Notice that the Green's function $G(t, u)$ is a function of two variables, t and u.

Example 16.1

Let's consider a familiar example from mechanics. We'll take an oscillator of mass m and spring constant K evolving under the influence of a time-dependent force $f(t)$. The equation of motion for this is

$$m\frac{\mathrm{d}^2}{\mathrm{d}t^2}x(t) + Kx(t) = f(t), \qquad (16.3)$$

so here the linear operator is $\hat{L} = m\frac{\mathrm{d}^2}{\mathrm{d}t^2} + K$. We can build up a force function $f(t)$ by adding together lots of delta functions.

$$f(t) = \int_0^\infty \mathrm{d}u\, f(u)\delta(t - u), \qquad (16.4)$$

where u is a dummy variable which allows us to break $f(t)$ into deltas. This is, of course, merely an example of superposition and is allowed because our differential equation is linear.

Instead of hitting the problem head-on, we sidestep and solve eqn 16.3 for just one of the delta functions. We'll call this solution the Green's function $G(t, u)$:

$$\left[m\frac{\mathrm{d}^2}{\mathrm{d}t^2} + K\right]G(t, u) = \delta(t - u). \qquad (16.5)$$

The complete solution to eqn 16.3 is then given by the integral of the Green's function, weighted by the force function $f(u)$:

$$x(t) = \int_0^\infty \mathrm{d}u\, G(t, u)f(u). \qquad (16.6)$$

The solution is therefore simply a sum of the responses of the system in x to the stimuli $f(u)$.

Why is the solution to eqn 16.3 given by eqn 16.6? Start by acting on eqn 16.6 with the operator $\hat{L}$ (which passes through the integral sign on the right-hand side),

$$
\begin{aligned}
\hat{L}x(t) &= \int \mathrm{d}u\, \hat{L}\, G(t, u)f(u) \\
&= \int \mathrm{d}u\, \delta(t - u)f(u) \text{ by eqn 16.2} \\
&= f(t) \text{ by eqn 16.4,} \qquad (16.7)
\end{aligned}
$$

which is where we started! This implies we can solve an inhomogeneous differential equation by finding the Green's function $G(t, u)$ and then integrating over $f(u)$ to get the solution $x(t)$.

It's now clear why the Green's function $G(t, u)$ needs two arguments: one is the variable we're interested in (here it's the time t), the other u is the variable that describes the position of the delta function $\delta(t - u)$ that we use to define the Green's function via $\hat{L}\, G(t, u) = \delta(t - u)$. The inhomogeneous part of the equation [here $f(t)$] is built up from a set of delta functions weighted by an amplitude function $f(u)$, that is a function of the dummy variable u.

Now we know what use they are, we can find the Green's functions for some commonly encountered $\hat{L}$ operators.[3]

[3]Remember that $\delta(x) = \delta(-x)$ so we can swap the order of the arguments in the delta function for convenience.

Example 16.2

Examples of finding some Green's functions.

- The Green's function for $\hat{L} = \boldsymbol{\nabla}^2$ can be read off using an argument from electromagnetism. We know Poisson's equation (reverting to SI units): $\boldsymbol{\nabla}^2 V(\boldsymbol{x}) = -\frac{\rho(\boldsymbol{x})}{\epsilon_0}$, which for a point charge of magnitude unity, located at $\boldsymbol{u}$, reads

$$\epsilon_0 \boldsymbol{\nabla}^2 V(\boldsymbol{x}) = -\delta^{(3)}(\boldsymbol{x} - \boldsymbol{u}). \qquad (16.8)$$

The potential that solves this equation is known from electromagnetism to be $V(\boldsymbol{x}) = \frac{1}{4\pi\epsilon_0 |\boldsymbol{x}-\boldsymbol{u}|}$, and so

$$\boldsymbol{\nabla}^2 \left[\frac{1}{4\pi|\boldsymbol{x} - \boldsymbol{u}|} \right] = -\delta^{(3)}(\boldsymbol{x} - \boldsymbol{u}), \qquad (16.9)$$

or

$$G(\boldsymbol{x}, \boldsymbol{u}) = -\frac{1}{4\pi|\boldsymbol{x} - \boldsymbol{u}|}. \qquad (16.10)$$

- The Green's function for $\hat{L} = (\boldsymbol{\nabla}^2 + \boldsymbol{k}^2)$ is defined by

$$\left(\boldsymbol{\nabla}^2 + \boldsymbol{k}^2\right) G_{\boldsymbol{k}}(\boldsymbol{x}, \boldsymbol{u}) = \delta^{(3)}(\boldsymbol{x} - \boldsymbol{u}). \qquad (16.11)$$

The solution which will interest us is

$$G_{\boldsymbol{k}}(\boldsymbol{x} - \boldsymbol{u}) = -\frac{e^{i|\boldsymbol{k}||\boldsymbol{x}-\boldsymbol{u}|}}{4\pi|\boldsymbol{x} - \boldsymbol{u}|}, \qquad (16.12)$$

which describes the amplitude of an outgoing, spherical wave.

---

[4] (1+1)-dimensional spacetime means one spatial dimension and one time dimension. We normally think about (3+1)-dimensional spacetime with three spatial dimensions and one time dimension.

[5] We don't need to integrate over all time because the multiplication $G(x, t_x, y, t_y)\phi(y)$ does that part automatically, as in the case of the time-evolution operator which, as we'll shortly see, is closely related to the Green's function.

[6] With this in mind we call G^+ the **time-retarded Green's function**, defined as

$$G^+ = \begin{cases} G & t_x > t_y \\ 0 & t_x < t_y, \end{cases} \qquad (16.14)$$

or more simply

$$G^+ = \theta(t_x - t_y)G. \qquad (16.15)$$

We can also define the **time-advanced Green's function** by

$$G^- = \begin{cases} 0 & t_x > t_y \\ G & t_x < t_y, \end{cases} \qquad (16.16)$$

or more simply

$$G^- = \theta(t_y - t_x)G. \qquad (16.17)$$

16.2 Propagators in quantum mechanics

We know that the equation that governs the change of a wave function $\phi(x, t)$ is the Schrödinger equation $\hat{H}\phi(x, t) = i\frac{\partial \phi(x,t)}{\partial t}$, where $\hat{H}$ is an operator function of x and we'll only consider $(1 + 1)$-dimensional spacetime for now.[4] Why might the Green's function of the Schrödinger equation be useful, and what interpretation might such a function have?

The real beauty of a Green's function is the property that we had in eqn 16.6, namely that

$$\phi(x, t_x) = \int \mathrm{d}y\, G^+(x, t_x, y, t_y)\phi(y, t_y). \qquad (16.13)$$

(The + sign notation will be explained shortly.) Here the Green's function takes a wave function at some time and place and evolves it to another time and place. To find out how it does this we need to integrate over space (but not time).[5]

In fact, we say that the Green's function *propagates* the particle from the spacetime point (y, t_y) to (x, t_x), which explains why we call G^+ a propagator. Here the plus-sign superscript in G^+ means that we constrain $t_x > t_y$ and we define $G^+ = 0$ for $t_x < t_y$, which prevents particles going back in time.[6]

Looking at eqn 16.13 we can interpret $\phi(y, t_y)$ as the amplitude to find a particle at (y, t_y) and $\phi(x, t_x)$ as the amplitude to find a particle at

(x, t_x). It follows that the propagator $G^+(x, t_x, y, t_y)$ is the probability amplitude that a particle in state $|y\rangle$ at time t_y, ends up in a state $|x\rangle$ at time t_x. The interpretation means that the Green's function may be written

$$G^+(x, t_x, y, t_y) = \theta(t_x - t_y)\langle x(t_x)|y(t_y)\rangle. \qquad (16.18)$$

Note that, using this language, our old friend the wave function $\phi(x, t_x) = \langle x|\phi(t)\rangle$ is simply the amplitude that a particle is found at (x, t_x) irrespective of where it started. The propagator therefore contains more information, since it cares where the particle started.

Example 16.3

From the basic definition of the propagator $G^+(x, t_x, y, t_y) = \theta(t_x - t_y)\langle x(t_x)|y(t_y)\rangle$ we can invoke the time-evolution operator so that the time dependence is taken away from the states

$$
\begin{aligned}
G^+(x, t_x, y, t_y) &= \theta(t_x - t_y)\langle x|\hat{U}(t_x - t_y)|y\rangle \\
&= \theta(t_x - t_y)\langle x|e^{-i\hat{H}(t_x - t_y)}|y\rangle. \qquad (16.19)
\end{aligned}
$$

We can also expand the amplitudes in terms of the eigenstates of the operator $\hat{H}$, which we call $|n\rangle$, which have eigenvalues E_n,

$$
\begin{aligned}
G^+(x, t_x, y, t_y) &= \theta(t_x - t_y)\langle x|e^{-i\hat{H}(t_x - t_y)}|y\rangle \\
&= \theta(t_x - t_y)\sum_n \langle x|e^{-i\hat{H}(t_x - t_y)}|n\rangle\langle n|y\rangle. \qquad (16.20)
\end{aligned}
$$

Remembering that $\langle x|n\rangle$ is just a fancy way of writing a wave function $\phi_n(x)$, we see that, in general, the propagator may be written in terms of the eigenfunctions of $\hat{H}$ as

$$G^+(x, t_x, y, t_y) = \theta(t_x - t_y)\sum_n \phi_n(x)\phi_n^*(y)e^{-iE_n(t_x - t_y)}, \qquad (16.21)$$

allowing us to relate wave functions and propagators.

So far we've assumed the propagator $G^+(x, t_x, y, t_y)$ is some sort of Green's function of the Schrödinger equation: now we'll confirm this. We define the Green's function for the Schrödinger equation by considering $G^+(x, t_x, y, t_y)$ as a function of x and t_x only and say

$$\left[\hat{H}_x - i\frac{\partial}{\partial t_x}\right] G^+(x, t_x, y, t_y) = -i\delta(x - y)\delta(t_x - t_y) = -i\delta^{(2)}(x - y), \qquad (16.22)$$

where $\hat{H}_x$ only touches x-coordinates and not y-coordinates. Here we hold y and t_y fixed as dummy variables. Finally, we want to confirm that the amplitude $\langle x(t_x)|y(t_y)\rangle$ is truly the Green's function of the Schrödinger equation.

Example 16.4

Here's how it goes. The Green's function is given by

$$G^+(x, t_x, y, t_y) = \theta(t_x - t_y)\langle x(t_x)|y(t_y)\rangle = \theta(t_x - t_y)\sum_n \phi_n(x)\phi_n^*(y)e^{-iE_n(t_x - t_y)}, \qquad (16.23)$$

where we've explicitly included the θ function to ensure that the particle can't propagate backward in time. Substituting the Green's function into eqn 16.22, we find that we want to work out

$$\left[\hat{H}_x - i\frac{\partial}{\partial t_x}\right] \theta(t_x - t_y) \sum_n \phi_n(x)\phi_n^*(y) e^{-iE_n(t_x - t_y)}. \qquad (16.24)$$

Take this in two stages. **Stage I**: We use the fact that $\frac{\partial}{\partial t_x}\theta(t_x - t_y) = \delta(t_x - t_y)$, to find that the time derivative acting on the retarded Green's function gives us

$$
\begin{aligned}
i\frac{\partial}{\partial t_x}G^+ &= i\delta(t_x - t_y) \sum_n \phi_n(x)\phi_n^*(y) e^{-iE_n(t_x - t_y)} \\
&+ \theta(t_x - t_y) \sum_n E_n \phi_n(x)\phi_n^*(y) e^{-iE_n(t_x - t_y)}. \qquad (16.25)
\end{aligned}
$$

Stage II: Now considering $\hat{H}_x$ acting on the Green's function, we note that $\hat{H}_x\phi_n(x) = E_n\phi_n(x)$ is the only thing that is affected by the $\hat{H}_x$ operator. We have

$$\hat{H}_xG^+ = \hat{H}_x\theta(t_x - t_y)\langle x(t_x)|y(t_y)\rangle = \theta(t_x - t_y) \sum_n E_n\phi_n(x)\phi_n^*(y) e^{-iE_n(t_x - t_y)}. \qquad (16.26)$$

Putting all of this together, we obtain

$$
\begin{aligned}
\left[\hat{H}_x - i\frac{\partial}{\partial t_x}\right] G^+(x, t_x, y, t_y) &= -i\delta(t_x - t_y) \sum_n \phi_n(x)\phi_n^*(y) e^{-iE_n(t_x - t_y)} \\
&= -i\delta(t_x - t_y)\delta(x - y), \qquad (16.27)
\end{aligned}
$$

which is what we set out to prove.

Now let's try to find a propagator for free particles. Our first stop will be the free-particle propagator in position space and the time domain, which we call $G^+(x, t_x, y, t_y)$.

Example 16.5

A non-relativistic, free particle has a Hamiltonian $\hat{H} = \hat{p}^2/2m$, with eigenfunctions $\phi(x) = \frac{1}{\sqrt{L}}e^{ipx}$ and eigenvalues $E_p = \frac{p^2}{2m}$. Treating p as a continuous variable, we start with eqn 16.21, and transforming the sum into an integral[7] we have

$$
\begin{aligned}
G^+(x, t_x, y, t_y) &= \theta(t_x - t_y) L \int \frac{dp}{2\pi} \phi_p(x)\phi_p^*(y) e^{-iE_p(t_x - t_y)} \\
&= \theta(t_x - t_y) \int \frac{dp}{2\pi} e^{ip(x-y)} e^{-i\frac{p^2}{2m}(t_x - t_y)}. \qquad (16.28)
\end{aligned}
$$

This integral can be done: it's a Gaussian integral, of which, much more later (see Chapter 23). For now, we use the result $\int_{-\infty}^{\infty} dx\, e^{-\frac{ax^2}{2}+bx} = \sqrt{\frac{2\pi}{a}}e^{\frac{b^2}{2a}}$ with $a = \frac{i(t_x - t_y)}{m}$ and $b = i(x - y)$. This gives

$$G^+(x, t_x, y, t_y) = \theta(t_x - t_y)\sqrt{\frac{m}{2\pi i(t_x - t_y)}}\, e^{\frac{im(x-y)^2}{2(t_x - t_y)}}. \qquad (16.29)$$

There are more examples of finding propagators in the exercises at the end of this chapter.

[7]We use the replacement

$$\sum_n \to L \int \frac{dp}{2\pi}.$$

Note for later that we will often make the replacement in three dimensions

$$\sum_p \to \mathcal{V} \int \frac{d^3p}{(2\pi)^3},$$

where $\mathcal{V}$ is the volume.

16.3 Turning it around: quantum mechanics from the propagator and a first look at perturbation theory

So far we've used things we know about quantum mechanics to derive some properties of a single-particle propagator. In fact, we want to turn this around. If we start from the propagator, what can we learn about the particle? This is most easily seen by considering yet another form of the Green's function, again as a function of space but this time in the frequency/energy domain. This one looks like

$$G^+(x, y, E) = \sum_n \frac{i\phi_n(x)\phi_n^*(y)}{E - E_n}. \qquad (16.30)$$

(You're invited to prove this relation in Exercises 16.2.)

We can notice two things about this equation when we consider it as a function of the complex[8] variable E: (i) the singularities (or *poles*) on the real axis occur when $E = E_n$, that is, when the parameter E equals the energies of the eigenstates $\phi_n(x)$; and (ii) the residues at the poles are (i times) the wave functions. We see that by being able to write down the Green's function of a system we have access to the energies of the system and its wave functions.[9] Propagators, with their 'from here to there' definition, also have the appealing property that they can be drawn in a cartoon form showing a particle travelling from y to x. This isn't quite as trivial as it sounds: Bohr was doubtful whether the trajectory of a quantum particle could be thought about due to position–momentum uncertainty, but it can. The quantum propagator, which gives this trajectory some meaning, is illustrated in Fig. 16.1.

This doesn't explain what all the fuss is about. Why are Green's functions so useful? One place to look is perturbation theory. In most interesting cases we can't solve a quantum mechanical problem exactly. What we do is write the problem by splitting up the Hamiltonian into two parts: $H = H_0 + V$, where H_0 is the solvable part and V is the perturbing potential. Perturbation theory involves the tedious task of evaluating changes to eigenfunctions and eigenvalues in a series of increasingly complicated corrections.

Let's look again at our Green's functions and see if they lend any insight to the perturbation problem. For simplicity, we'll write the original Green's function equation in a *symbolic*, matrix-like form as $(H - E)G = -1$.[10] (Remember throughout that G just describes a particle propagating from y to x.) The symbolic Green's function equation is solved by writing $G = \frac{1}{E-H}$, which bears a resemblance to the form in eqn 16.31.

Green's functions allow us to interpret a perturbation problem in terms of a propagating particle. We think of the solvable part of the problem in terms of a particle propagating from point to point. We think of the perturbation V as a **scattering process** that interrupts the propagation. To visualize this we'll write a symbolic expression for

[8]Readers unfamiliar with complex analysis can refer to Appendix B.

[9]Actually, to ensure causality it's necessary to replace E_n with $E_n - i\epsilon$ so we have

$$G^+(x, y, E) = \lim_{\epsilon \to 0^+} \sum_n \frac{i\phi_n(x)\phi_n^*(y)}{E - E_n + i\epsilon}.$$
$$(16.31)$$

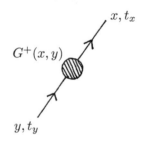

Fig. 16.1 A cartoon representation of the propagator. It will turn out to be the basic unit of the Feynman diagram.

[10]Objects like G should be treated with caution: they're not well defined mathematical objects, just symbols combined in the shape of the original equations that we're using to suggest relationships. Needless to say these arguments are repeated rigorously in the literature.

the Green's function in the presence of a perturbation as $G = \frac{1}{E-H_0-V}$, where we've used $H = H_0 + V$. In perturbation problems G (with no subscript) is called the **full propagator**. Just considering the solvable part of the Hamiltonian H_0 we can write $G_0 = \frac{1}{E-H_0}$, that is, we can find the Green's function for a solvable problem. The '0' subscript means that we're talking about a particle freely propagating, with no scattering and so, in general, G_0 is known as the **free propagator**. If we now consider the perturbed problem and try to expand the full propagator G as a function of the free propagator G_0 we can say

There's a matrix identity that says

$$\frac{1}{A+B} = \frac{1}{A} - \frac{1}{A}B\frac{1}{A} + \frac{1}{A}B\frac{1}{A}B\frac{1}{A} - \dots$$

$$G = \frac{1}{E-H_0-V} = \frac{1}{E-H_0} + \frac{1}{E-H_0}V\frac{1}{E-H_0}$$
$$+ \frac{1}{E-H_0}V\frac{1}{E-H_0}V\frac{1}{E-H_0} + \dots \quad (16.32)$$

The point here is that, in general, we can't calculate $1/(E-H_0-V)$. However, we can calculate each of the terms on the right-hand side of this expression and we might hope that G may be well approximated by just the first few of these.

[11]In the rest of this section we will write the functions as G, rather than G^+, since the expressions in eqns 16.33 and 16.34 apply generally to G, G^+ and G^-.

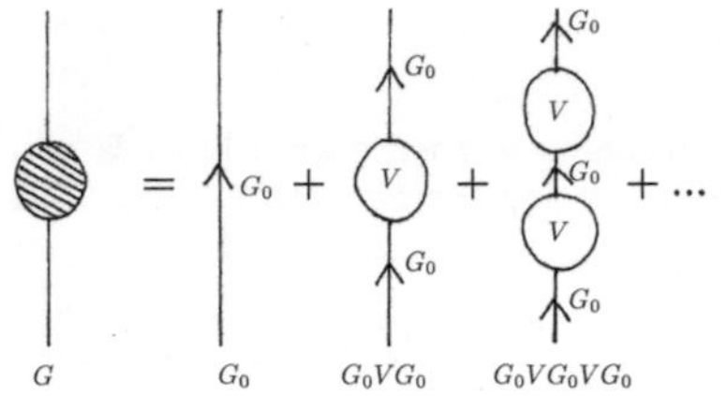

Switching back from symbolic relations to real functions[11] shows that we've succeeded in writing the full Green's function as an expansion in free propagators and potentials. The form of the expansion is

$$G = G_0 + G_0 V G_0 + G_0 V G_0 V G_0 + \dots \quad (16.33)$$

This is a memorable equation and easily turned into pictures as shown in Fig. 16.2. The perturbation problem, previously involving an infinite series of unmemorable terms, has been turned into a vivid picture of particle scattering. The amplitude for a particle to go from y to x is G. This is a superposition of the amplitude for moving freely from y to x (which is G_0); added to the amplitude for making the trip with a single scattering event at some point along the way $G_0 V G_0$; added to the amplitude for making the trip with two scatters $G_0 V G_0 V G_0$ and so on. Thus the algebra of Green's functions is highly amenable to visualization of the physical processes.

Fig. 16.2 The scattering processes that lead to the full propagator in perturbation theory.

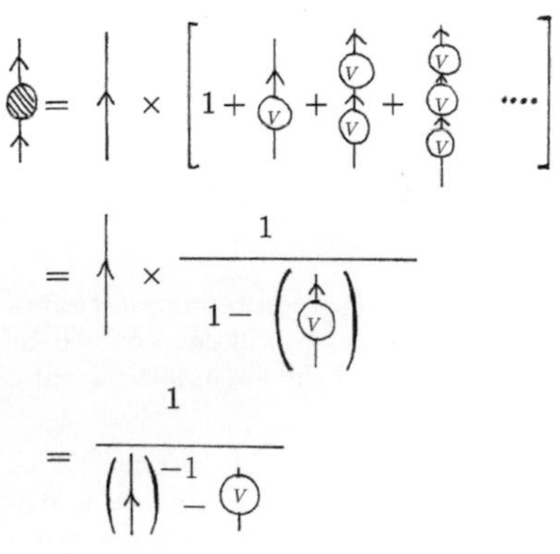

Fig. 16.3 The derivation of Dyson's equation in graphical form.

Example 16.6

The perturbation expansion in eqn 16.33 is just a geometric series which we can rewrite as

$$G = G_0(1 + VG_0 + VG_0VG_0 + \dots)$$
$$= \frac{G_0}{1-VG_0} = \frac{1}{G_0^{-1}-V}. \quad (16.34)$$

This is known as **Dyson's equation**. This equation is also derived from the algebraic manipulation of the cartoons, as shown in Fig. 16.3. This concept will turn out to be very useful. Notice that it is non-perturbative: we don't consider just the first few of the terms in a perturbation series, we sum all of them to infinity.

16.4 The many faces of the propagator

So far we have encountered the free-particle propagator in the space and time domain $G_0^+(x, t_x, y, t_y)$ and in the space and energy domain $G_0^+(x, y, E)$. There are other useful forms that we can derive.

Example 16.7

The physical interpretation of G^+ allows us to work out another expression for the Green's function, this time in the momentum space and in the time domain, so we write $G_0^+(p, t_x, q, t_y)$. Remember the interpretation: we want the amplitude that a particle that starts off in a state $|q\rangle$ will, after a time $t_x - t_y$, end up in a state $|p\rangle$. Thus[12]

$$
\begin{aligned}
G_0^+(p, t_x, q, t_y) &= \theta(t_x - t_y)\langle p|\hat{U}(t_x - t_y)|q\rangle \\
&= \theta(t_x - t_y)\langle p|q\rangle e^{-iE_q(t_x - t_y)} \\
&= \theta(t_x - t_y)\delta(p - q)e^{-iE_q(t_x - t_y)}. \quad (16.35)
\end{aligned}
$$

We see that the free particle cannot change its momentum state, so having both p and q is redundant, so we can write the same equation in the following shorthand:

$$
G_0^+(p, t_x, t_y) = \theta(t_x - t_y)e^{-iE_p(t_x - t_y)}. \quad (16.36)
$$

Notice that we often miss out the delta functions if we decrease the number of arguments we list in G^+. Strictly we should write all arguments and delta functions.

---

[12]Note that, for simplicity in these examples, we're using the non-relativistic normalization so that $\langle p|q\rangle = \delta(p - q)$.

For dealing with perturbation theory problems, propagators are often most useful to know in the momentum and energy domain.

Example 16.8

To get to $G_0^+(p, E)$ we'll take a Fourier transform of $G_0^+(p, t, 0)$

$$
\begin{aligned}
G_0^+(p, E) &= \int dt\, e^{iEt} G_0^+(p, t, 0) \\
&= \int dt\, e^{iEt} \theta(t) e^{-i(E_p - i\epsilon)t} \\
&= \left. \frac{-i e^{i(E - E_p + i\epsilon)t}}{E - E_p + i\epsilon} \right|_0^{\infty} \\
&= \frac{i}{E - E_p + i\epsilon}. \quad (16.37)
\end{aligned}
$$

Remember again, that if we are being strict we should write this as

$$
G_0^+(p, E, q, E') = \frac{i}{E - E_p + i\epsilon}\delta(p - q), \quad (16.38)
$$

We make the replacement $E_p \to E_p - i\epsilon$ to ensure the convergence of the integral at $t = \infty$. It also effectively ensures causality is obeyed. See Appendix B and Exercises 16.2 and 16.4 for the details.

telling us the amplitude for a particle with energy E' and momentum q to enter and a particle with energy E and momentum p to leave. The delta function conserves energy and momentum.

As a helpful summary, we assemble a table of all of the flavours of Green's function that we've introduced so far.

$$G_0^+(x, t_x, y, t_y) \;=\; \theta(t_x - t_y) \sum_n \phi_n(x)\phi_n^*(y) \mathrm{e}^{-\mathrm{i}E_n(t_x - t_y)}$$

$$G_0^+(x, y, E) \;=\; \sum_n \frac{\mathrm{i}\phi_n(x)\phi_n^*(y)}{E - E_n + \mathrm{i}\epsilon}$$

$$G_0^+(p, t_x, t_y) \;=\; \theta(t_x - t_y)\mathrm{e}^{-\mathrm{i}E_p(t_x - t_y)}$$

$$G_0^+(p, E) \;=\; \frac{\mathrm{i}}{E - E_p + \mathrm{i}\epsilon}$$

Chapter summary

- Green's functions are used as propagators which express the amplitude that a particle goes from spacetime point y to spacetime point x. Various forms of the Green's function are shown in the table above.
- Dyson's equation for the full propagator G is the perturbation expansion: $G = G_0 + G_0 V G_0 + \cdots = (G_0^{-1} - V)^{-1}$ where G_0 is the free propagator and V describes a scattering process.

Exercises

(16.1) (a) Solve the Schrödinger equation to find the wave functions $\phi_n(x)$ for a particle in a one-dimensional square well defined by $V(x) = 0$ for $0 \leq x \leq a$ and $V(x) = \infty$ for $x < 0$ and $x > a$.
(b) Show that the retarded Green's function for this particle is given by

$$G^+(n, t_2, t_1) = \theta(t_2 - t_1)\mathrm{e}^{-\mathrm{i}\left(\frac{n^2\pi^2}{2ma^2}\right)(t_2 - t_1)}. \quad (16.39)$$

(c) Find $G^+(n, \omega)$ for the particle.

(16.2) (a) Using a damping factor $\mathrm{e}^{-\epsilon t}$, show that $G_0^+(x, y, E) = \sum_n \frac{\mathrm{i}\phi_n(x)\phi_n^*(y)}{E - E_n + \mathrm{i}\epsilon}$.
(b) The θ-function may be written as an integral

$$\theta(t) = \mathrm{i}\int_{-\infty}^{\infty} \frac{\mathrm{d}z}{2\pi} \frac{\mathrm{e}^{-\mathrm{i}zt}}{z + \mathrm{i}\epsilon}. \quad (16.40)$$

Use this to derive $G_0^+(p, E) = \frac{\mathrm{i}}{E - E_p + \mathrm{i}\epsilon}$, without recourse to a damping factor.

(16.3) Consider the one-dimensional simple harmonic oscillator with a forcing function $f(t) = \tilde{F}(\omega)\mathrm{e}^{-\mathrm{i}\omega(t-u)}$ described by an equation of motion:

$$m\frac{\partial^2}{\partial t^2}A(t-u) + m\omega_0^2 A(t-u) = \tilde{F}(\omega)\mathrm{e}^{-\mathrm{i}\omega(t-u)}. \quad (16.41)$$

(a) Show that the solution to this equation is

$$A(t-u) = -\frac{\tilde{F}(\omega)}{m}\frac{\mathrm{e}^{-\mathrm{i}\omega(t-u)}}{\omega^2 - \omega_0^2} + B(t), \quad (16.42)$$

where $B(t)$ is the solution to the homogeneous equation of motion.
(b) Use this result to show that the general form of

the Green's function $G(t, u)$ for a simple harmonic oscillator is given by

$$G(t, u) = -\frac{1}{m} \int \frac{d\omega}{2\pi} \frac{e^{-i\omega(t-u)}}{\omega^2 - \omega_0^2} + B(t). \quad (16.43)$$

(c) If $G(t, u)$ is subject to the boundary conditions that at $t = 0$ we have $G = 0$ and $\dot{G} = 0$, show that

$$G^+(t, u) = \frac{1}{m\omega_0} \sin\omega_0(t - u), \quad (16.44)$$

for $0 < u < t$. *Hint: This can be done with a Laplace transform of the differential equation for the Green's function, or by using complex analysis to do the integral in the previous problem subject to the boundary conditions.*

(d) Show that the trajectory of a particle in a simple harmonic potential subject to a force $f(t) = F_0 \sin\omega_0 t$ is given by $\frac{F_0}{2m\omega_0^2}(\sin\omega_0 t - \omega_0 t \cos\omega_0 t)$.

(16.4) (a) By taking the Fourier transform of the equation

$$(\nabla^2 + k^2)G_k(x) = \delta^{(3)}(x), \quad (16.45)$$

show that the momentum Green's function implied is

$$\tilde{G}_k(q) = \frac{1}{k^2 - q^2}. \quad (16.46)$$

Note that $\tilde{G}_k(q)$ is undefined when $q^2 = k^2$. This reflects the fact that there is an ambiguity in $\tilde{G}_k(q)$ in that we don't know if the wave is incoming, outgoing or a standing wave. This is sorted out with

an $i\epsilon$ factor as shown in the next part.

(b) Take the Fourier transform of $G_k^+(x) = -\frac{e^{i|k||x|}}{4\pi|x|}$ to show that, for outgoing waves,

$$\tilde{G}_k^+(q) = \frac{1}{k^2 - q^2 + i\epsilon}. \quad (16.47)$$

Hint: To ensure outgoing waves, you could include a damping factor $e^{-\epsilon|x|}$ to damp out waves for $|x| \to \infty$.

(c) (For aficionados of complex analysis.) Take an inverse Fourier transform of $\tilde{G}_k^+(q) = \frac{1}{k^2-q^2+i\epsilon}$ to show that it corresponds to the outgoing wave solution $G_k^+(x) = -\frac{e^{i|k||x|}}{4\pi|x|}$.

Hint: On doing the angular part of the integral you should obtain:

$$G_k^+(x) = \frac{-1}{4\pi^2} \int_0^\infty \frac{|q|\,d|q|}{q^2 - k^2 - i\epsilon} \left(\frac{e^{i|q||x|} - e^{-i|q||x|}}{i|x|}\right). \quad (16.48)$$

Extend the lower limit of the integration to $|q| = -\infty$ and identify the positions of the poles in the complex $|q|$ plane. For an outgoing wave, the contour over which the $e^{i|q||x|}$ part is integrated must be completed in the upper half-plane; the contour over which the $e^{-i|q||x|}$ part is integrated must be completed in the lower half-plane.

(d) What do you expect for an incoming wave solution?

17

Propagators and fields

Interacting and non-interacting theories

Non-interacting	Interacting
$\hat{H}_0$	$\hat{H} = \hat{H}_0 + \hat{H}'$
$\|0\rangle$	$\|\Omega\rangle$
$G_0(x, y)$	$G(x, y)$
No scattering	Scattering

[1]See Chapter 11.

*An error does not become truth by reason or multiplied prop-
agation, nor does truth become error because nobody sees it.*
Mahatma Gandhi (1869–1948)

We describe the Universe by combining fields to form a Lagrangian density $\mathcal{L}[\phi(x)]$. Our canonical quantization process often allows us to quantize these fields leading to a Universe pictured as a vacuum acted on by
field operators like $\hat{a}_{\boldsymbol{p}}^{\dagger}$. The excitations of the vacuum that the field
operators produce are particles and antiparticles.

Our next task is to come up with a scheme to keep track of these
particles. Having done away with the wave functions we need some
objects that contain the information about excitations of the system.
These objects are propagator amplitudes denoted by the letter G. These
propagators are analogous to the Green's function amplitudes discussed
in the previous chapter. The propagators describe the fate of particles
in a field theory.

An important distinction that we will make in this chapter is between
non-interacting and interacting theories. Non-interacting theories are
those which can be diagonalized using canonical quantization.[1] Canoni-
cal quantization allows us to describe the system in terms of a vacuum $|0\rangle$
and non-interacting particles in momentum states $|\boldsymbol{p}\rangle$ created with oper-
ators like $\hat{a}_{\boldsymbol{p}}^{\dagger}$. The Hamiltonian for a non-interacting theory is called $\hat{H}_0$
which may be written as $\hat{H}_0 = \sum_{\boldsymbol{p}} E_{\boldsymbol{p}} \hat{a}_{\boldsymbol{p}}^{\dagger} \hat{a}_{\boldsymbol{p}}$, and we have $\hat{H}_0|\boldsymbol{p}\rangle = E_{\boldsymbol{p}}|\boldsymbol{p}\rangle$,
where $E_{\boldsymbol{p}}$ gives the energies of the non-interacting particles. In contrast,
interacting theories cannot be diagonalized with canonical quantization.
Put another way, these theories contain extra terms describing interactions between particles that we cannot transform away with canonical
quantization. We call the ground states of interacting theories $|\Omega\rangle$ and
the Hamiltonian $\hat{H}$. If we act on $|\Omega\rangle$ with the operator $\hat{a}_{\boldsymbol{p}}^{\dagger}$ we won't
necessarily get a state $|\boldsymbol{p}\rangle$. Instead we might produce a superposition of
many particles (whose momenta sum to $\boldsymbol{p}$). Since most interacting field
theories cannot be exactly solved to give, for example, the energies of the
excitations or particles of the theory, we need to develop a *perturbation*
process to make approximate calculations. This can be done by analogy
with the perturbation process we suggested in the last chapter, which
was based around a series written out in terms of propagators.

17.1   The field propagator in outline

The definition of the field propagator involves a simple thought experiment. We start with our interacting system in its ground state, which we denote by $|\Omega\rangle$. The thought experiment works as follows: we introduce an extra particle of our choice. We will use this extra particle to probe the system. The new particle is introduced (created) at a spacetime point $(y^0, \boldsymbol{y})$. It *interacts* with the system, possibly causing excitations in the fields and all manner of complicated things. Then we remove (annihilate) the particle at spacetime point $(x^0, \boldsymbol{x})$ and ask if the system has remained in the interacting ground state $|\Omega\rangle$. We're interested in the amplitude $G^+(x, y)$ for the experiment, given by

$$G^+(x,y) = \left\langle \Omega \left| \left(\begin{array}{c} \text{Particle annihilated} \\ \text{at } (x^0, \boldsymbol{x}) \end{array} \right) \left(\begin{array}{c} \text{Particle created} \\ \text{at } (y^0, \boldsymbol{y}) \end{array} \right) \right| \Omega \right\rangle.$$
(17.1)

That is, the probability amplitude that the system is still in its ground state after we create a particle at y and later annihilate it at x. The amplitude for this process will depend on the complicated interaction of the probe particle with the system and, it will turn out, will tell us a great deal about the system itself. By analogy with the results of the previous chapter, the amplitude $G^+(x, y)$ is called the Green's function or propagator.[2] So much for definitions. The mathematical object that tells this story of the propagator outlined above is

$$G^+(x,y) = \theta(x^0 - y^0)\langle \Omega | \hat{\phi}(x)\hat{\phi}^\dagger(y) | \Omega \rangle,$$
(17.2)

where the $+$ sign on the left tells us that the particle is created at y before being annihilated at x and the θ-function on the right guarantees this.

[2] We call the propagator for an interacting theory $G(x, y)$ and its ground state $|\Omega\rangle$. For a free theory (with no interactions) we call the propagator $G_0(x, y)$ [or sometimes we will use the equivalent notation $\Delta(x, y)$] and the ground state $|0\rangle$.

Example 17.1

To see that this function does the job we can break down the action of the operators on the states into stages. Remembering that a Heisenberg operator has a time dependence defined by $\hat{\phi}(t, \boldsymbol{x}) = e^{+i\hat{H}t}\hat{\phi}(\boldsymbol{x})e^{-i\hat{H}t}$, and substitution of this gives an expression for the propagator

$$G(x,y) = \langle \Omega | e^{i\hat{H}x^0} \hat{\phi}(\boldsymbol{x}) e^{-i\hat{H}(x^0 - y^0)} \hat{\phi}^\dagger(\boldsymbol{y}) e^{-i\hat{H}y^0} | \Omega \rangle.$$
(17.3)

Taking this one stage at a time (and temporarily ignoring the possibility of creating and annihilating antiparticles), we see that:

- $e^{-i\hat{H}y^0}|\Omega\rangle$ is the state $|\Omega\rangle$ evolved to a time y^0.
- $\hat{\phi}^\dagger(\boldsymbol{y})e^{-i\hat{H}y^0}|\Omega\rangle$ is that state with a particle added at time y^0 at a position $\boldsymbol{y}$.
- $e^{-i\hat{H}(x^0 - y^0)}\hat{\phi}^\dagger(\boldsymbol{y})e^{-i\hat{H}y^0}|\Omega\rangle$ is that state time evolved to time x^0.
- Now hitting this state with $e^{i\hat{H}x^0}\hat{\phi}(\boldsymbol{x})$ removes the particle at time x^0.
- We terminate this string on the left with a $\langle\Omega|$ to find out how much of the original state $|\Omega\rangle$ is left in the final state.

The propagator is, therefore, the amplitude that we put a particle in the system at position $\boldsymbol{y}$ at time y^0 and get it out at position $\boldsymbol{x}$ time x^0.

17.2   The Feynman propagator

We don't always want to deal with the particle propagator G^+ (which is constrained with the θ-function so that $x^0 > y^0$) when thinking about fields. The reason is that although G^+ accounts for the movement of particles it misses out the essential information about antiparticles.

Example 17.2

Consider, for example, the field creation operator for complex scalar field theory $\hat\psi^\dagger(x) = \int \frac{d^3p}{(2\pi)^{\frac{3}{2}}} \frac{1}{(2E_p)^{\frac{1}{2}}} \left(\hat a_{\boldsymbol p}^\dagger e^{ip\cdot x} + \hat b_{\boldsymbol p} e^{-ip\cdot x} \right)$, which creates a particle and annihilates an antiparticle. If this acts on the free vacuum $\hat\psi^\dagger(y)|0\rangle$ (no interactions, so nothing unpredictable can occur) then we create a particle at y, but the antiparticle part annihilates the vacuum $\hat b_{\boldsymbol p}|0\rangle = 0$. Similarly, only the particle part contributes to $\langle 0|\hat\psi(x)$, so the propagator $\langle 0|\hat\psi(x)\hat\psi^\dagger(y)|0\rangle$ describes a particle propagating from y to x, but tells us nothing about antiparticles.

Notation for propagators	
In general	$G(x,y)$
Free propagator	$G_0(x,y)$
Free scalar fields	$\Delta(x,y)$
Photon fields	$D(x,y)$
Fermion fields	$G(x,y)$ or $iS(x,y)$

Gian-Carlo Wick (1909–1992)

[3]This form applies only to the bosonic (i.e. commuting) scalar fields we're considering in this chapter. For an anti-commuting Fermi field $\hat\psi$, we pick up a minus sign every time we swap the order of the operators, so we have

$T\hat\psi(x^0)\hat\psi(y^0)$
$= \begin{cases} \hat\psi(x^0)\hat\psi(y^0) & x^0 > y^0 \\ -\hat\psi(y^0)\hat\psi(x^0) & x^0 < y^0. \end{cases}$

Richard Feynman struggled over how to include the antiparticle contribution for some time before deciding that the most useful form of the propagator was one that summed both particle and antiparticle parts. This form of the propagator makes up the guts of most quantum field calculations and is called the **Feynman propagator**. How do we get the propagator Feynman used which contains both particle and antiparticle parts? We need to introduce a new piece of machinery: the **Wick time-ordering symbol** T. This isn't an operator, but is much more like the normal ordering symbol N in that it is just an instruction on what to do to a string in order that it makes sense. The time-ordering symbol is defined for scalar fields[3] as

$$T\hat\phi(x^0)\hat\phi(y^0) = \begin{cases} \hat\phi(x^0)\hat\phi(y^0) & x^0 > y^0 \\ \hat\phi(y^0)\hat\phi(x^0) & x^0 < y^0, \end{cases} \tag{17.4}$$

so that the scalar fields are always arranged earliest on the right, latest on the left. The Feynman propagator is then defined as

$$\begin{aligned} G(x,y) &= \langle\Omega|T\hat\phi(x)\hat\phi^\dagger(y)|\Omega\rangle \\ &= \theta(x^0-y^0)\langle\Omega|\hat\phi(x)\hat\phi^\dagger(y)|\Omega\rangle + \theta(y^0-x^0)\langle\Omega|\hat\phi^\dagger(y)\hat\phi(x)|\Omega\rangle, \end{aligned} \tag{17.5}$$

where $|\Omega\rangle$ is the interacting ground state of the system. The propagator is therefore made up of two parts. The first part applies for x^0 later than y^0: it creates a particle at y and propagates it to x where it is destroyed. The second part applies when y^0 is later than x^0: it creates an antiparticle at x and propagates the system to point y. Both processes are included in the total propagator.

If, as in the example above, the system doesn't contain any interactions then particles just move around passing through each other. In this case we call the ground state $|0\rangle$ and we have the **free propagator** sometimes called $G_0(x,y)$ or, for scalar fields, more usually just $\Delta(x,y)$ where

$$\Delta(x,y) \;=\; \langle 0|T\hat{\phi}(x)\hat{\phi}^\dagger(y)|0\rangle. \qquad (17.6)$$

The free-particle propagator is depicted in Fig. 17.1. The free propagator forms an essential part of the structure of all perturbation calculations. This is because we think of interactions as events that take place at particular spacetime points (the V-blobs in Fig. 17.2) and we imagine that the particles propagate freely between interactions with the free propagator.

We'll now derive an expression for the free Feynman propagator $\Delta(x,y)$.

Fig. 17.1 The free-particle propagator $\Delta(x,y)$ is a line without an interaction blob since a free particle doesn't interact with any other particles on its way from y to x.

Example 17.3

Let's get an expression for the free propagator. If we use the general expression for the annihilation field acting on the free vacuum $\phi^\dagger(y)|0\rangle$, we have

$$\hat{\phi}^\dagger(y)|0\rangle = \int \frac{\mathrm{d}^3 p}{(2\pi)^{\frac{3}{2}}(2E_{\boldsymbol p})^{\frac{1}{2}}} \left(\hat{a}_{\boldsymbol p}^\dagger |0\rangle \mathrm{e}^{\mathrm{i}p\cdot y} + \hat{b}_{\boldsymbol p}|0\rangle \mathrm{e}^{-\mathrm{i}p\cdot y} \right). \qquad (17.7)$$

Only the particle creation part contributes (since $\hat{b}_{\boldsymbol p}|0\rangle = 0$) and gives

$$\hat{\phi}^\dagger(y)|0\rangle = \int \frac{\mathrm{d}^3 p}{(2\pi)^{\frac{3}{2}}(2E_{\boldsymbol p})^{\frac{1}{2}}} |{\boldsymbol p}\rangle \mathrm{e}^{\mathrm{i}p\cdot y}. \qquad (17.8)$$

How do we get the other half? Take a complex conjugate of eqn 17.8 (swapping $y \to x$ and $p \to q$) to get

$$\langle 0|\hat{\phi}(x) = \int \frac{\mathrm{d}^3 q}{(2\pi)^{\frac{3}{2}}(2E_{\boldsymbol q})^{\frac{1}{2}}} \langle {\boldsymbol q}| \mathrm{e}^{-\mathrm{i}q\cdot x}. \qquad (17.9)$$

Sandwiching together eqn 17.8 and eqn 17.9 we obtain

$$\begin{aligned}
\langle 0|\hat{\phi}(x)\hat{\phi}^\dagger(y)|0\rangle &= \int \frac{\mathrm{d}^3 p\,\mathrm{d}^3 q}{(2\pi)^3 (2E_{\boldsymbol p} 2E_{\boldsymbol q})^{\frac{1}{2}}} \mathrm{e}^{-\mathrm{i}q\cdot x + \mathrm{i}p\cdot y} \delta^{(3)}({\boldsymbol q}-{\boldsymbol p}) \\
&= \int \frac{\mathrm{d}^3 p}{(2\pi)^3 (2E_{\boldsymbol p})} \mathrm{e}^{-\mathrm{i}p\cdot(x-y)}, \qquad (17.10)
\end{aligned}$$

which corresponds to a particle being created at $(y^0, {\boldsymbol y})$ and propagating to $(x^0, {\boldsymbol x})$ where it is annihilated at the later time.

We also need to consider the reverse order for the advanced part of the propagator:

$$\hat{\phi}(x)|0\rangle = \int \frac{\mathrm{d}^3 p}{(2\pi)^{\frac{3}{2}}(2E_{\boldsymbol p})^{\frac{1}{2}}} \left(\hat{a}_{\boldsymbol p}|0\rangle \mathrm{e}^{-\mathrm{i}p\cdot x} + \hat{b}_{\boldsymbol p}^\dagger|0\rangle \mathrm{e}^{\mathrm{i}p\cdot x} \right). \qquad (17.11)$$

Here, only the antiparticle creation part contributes (since $\hat{a}_{\boldsymbol p}|0\rangle = 0$), to give

$$\hat{\phi}(x)|0\rangle = \int \frac{\mathrm{d}^3 p}{(2\pi)^{\frac{3}{2}}(2E_{\boldsymbol p})^{\frac{1}{2}}} |{\boldsymbol p}\rangle \mathrm{e}^{\mathrm{i}p\cdot x}. \qquad (17.12)$$

Again we take the complex conjugate (and change $x \to y$ and $p \to q$):

$$\langle 0|\hat{\phi}^\dagger(y) = \int \frac{\mathrm{d}^3 q}{(2\pi)^{\frac{3}{2}}(2E_{\boldsymbol q})^{\frac{1}{2}}} \langle {\boldsymbol q}| \mathrm{e}^{-\mathrm{i}q\cdot y}, \qquad (17.13)$$

yielding, finally

$$\langle 0|\hat{\phi}^\dagger(y)\hat{\phi}(x)|0\rangle = \int \frac{\mathrm{d}^3 p}{(2\pi)^3 (2E_{\boldsymbol p})} \mathrm{e}^{\mathrm{i}p\cdot(x-y)}. \qquad (17.14)$$

This corresponds to an antiparticle being created at point $(x^0, {\boldsymbol x})$ and propagating to $(y^0, {\boldsymbol y})$, where it is destroyed at the later time. Putting the two halves (eqns 17.10 and 17.14) together we have our final answer for the free propagator $\Delta(x,y)$.

Fig. 17.2 A third-order term in the perturbation expansion of G in terms of $\Delta(x,y)$. The amplitude for the process is proportional to $\Delta(x-z_3)V(z_3)\Delta(z_3-z_2)V(z_2)\Delta(z_2-z_1)V(z_1)\Delta(z_1-y)$.

We have the final result that the Feynman propagator is given by

$$\Delta(x, y) = \langle 0|T\hat{\phi}(x)\hat{\phi}^{\dagger}(y)|0\rangle \tag{17.15}$$

$$= \int \frac{d^3p}{(2\pi)^3(2E_{\boldsymbol{p}})}[\theta(x^0 - y^0)e^{-ip\cdot(x-y)} + \theta(y^0 - x^0)e^{ip\cdot(x-y)}].$$

Notice the way that this came together. We added together a particle part from the retarded half of the propagator and an antiparticle from the advanced part. This is illustrated in Fig. 17.3.

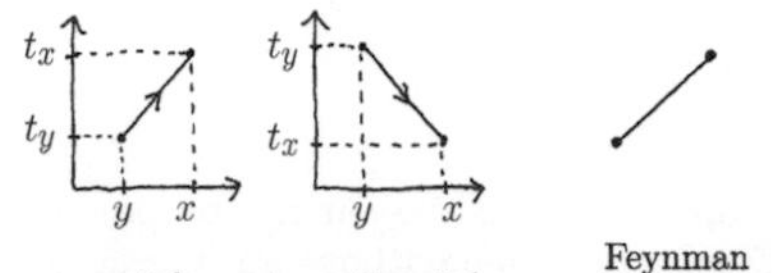

We've been calling these propagators Green's functions, but are they really Green's functions in the sense of Chapter 16? The answer is no for the interacting case. However, the free-propagator functions are the Green's functions of the equations of motion. So for the scalar field we have

$$(\partial^2 + m^2)\Delta(x - y) = -i\delta^{(4)}(x - y). \tag{17.16}$$

This is examined in Exercise 17.2. In general, propagators are defined in such a way that they are true Green's functions in the absence of interactions. That is, free propagators $\Delta(x, y)$ are true Green's functions.

Fig. 17.3 The Feynman propagator is made up of a sum of particle and antiparticle parts.

17.3   Finding the free propagator for scalar field theory

Our expression for the free Feynman propagator for scalar field theory in eqn 17.15 is not very useful in its current form. We would like an expression which doesn't include the rather non-covariant looking Heaviside step functions. A simplification is achieved by using complex analysis, as shown in the next example.

Example 17.4

How does the computation unfold? One of the tools we'll need is the expansion of $\theta(x^0 - y^0)$ in the complex plane.[4] Consider the first term in eqn 17.15, call it $[\Delta(x, y)]_{(1)}$, and substitute this expression for the θ function:

$$[\Delta(x, y)]_{(1)} \equiv \theta(x^0 - y^0)\langle 0|\hat{\phi}(x)\hat{\phi}^{\dagger}(y)|0\rangle$$

$$= \theta(x^0 - y^0)\int \frac{d^3p}{(2\pi)^3(2E_{\boldsymbol{p}})}e^{-iE_{\boldsymbol{p}}(x^0-y^0)+i\boldsymbol{p}\cdot(\boldsymbol{x}-\boldsymbol{y})}$$

$$= i\int_{-\infty}^{\infty} \frac{dz d^3p}{(2\pi)^4(2E_{\boldsymbol{p}})} \frac{e^{-i(E_{\boldsymbol{p}}+z)(x^0-y^0)+i\boldsymbol{p}\cdot(\boldsymbol{x}-\boldsymbol{y})}}{z + i\epsilon}. \tag{17.17}$$

Next we make the substitution $z' = z + E_{\boldsymbol{p}}$, leading to the integral

$$[\Delta(x, y)]_{(1)} = i\int_{-\infty}^{\infty} \frac{dz' d^3p}{(2\pi)^4(2E_{\boldsymbol{p}})} \frac{e^{-iz'(x^0-y^0)+i\boldsymbol{p}\cdot(\boldsymbol{x}-\boldsymbol{y})}}{z' - E_{\boldsymbol{p}} + i\epsilon}. \tag{17.18}$$

Then we change the definition of our four-momentum in the integral. We treat z' as the new p^0, redefining $p = (z', \boldsymbol{p}) = (p^0, \boldsymbol{p})$ leading to

$$[\Delta(x, y)]_{(1)} = i\int_{-\infty}^{\infty} \frac{d^4p}{(2\pi)^4(2E_{\boldsymbol{p}})} \frac{e^{-ip\cdot(x-y)}}{p^0 - E_{\boldsymbol{p}} + i\epsilon}. \tag{17.19}$$

[4]This is given by

$$\theta(x^0 - y^0) = i\int_{-\infty}^{\infty} \frac{dz}{2\pi} \frac{e^{-iz(x^0-y^0)}}{z + i\epsilon}.$$

With the redefinition, $p^0 = E$ doesn't equal $(\boldsymbol{p}^2 + m^2)^{\frac{1}{2}}$ any more. It's now just an integration variable that we happen to have grouped with the three-momentum to clean up the integral. Note that we still have that the energy $E_{\boldsymbol{p}} = (\boldsymbol{p}^2 + m^2)^{\frac{1}{2}}$.

We can carry the same process out on the second term:

$$
\begin{aligned}
[\Delta(x,y)]_{(2)} &\equiv \theta(y^0 - x^0)\langle 0|\hat{\phi}^\dagger(y)\hat{\phi}(x)|0\rangle \\
&= \theta(y^0 - x^0)\int \frac{\mathrm{d}^3 p}{(2\pi)^3(2E_{\boldsymbol{p}})}\mathrm{e}^{-\mathrm{i}E_{\boldsymbol{p}}(y^0 - x^0) + \mathrm{i}\boldsymbol{p}\cdot(\boldsymbol{y}-\boldsymbol{x})} \\
&= \mathrm{i}\int_{-\infty}^{\infty}\frac{\mathrm{d}z\mathrm{d}^3 p}{(2\pi)^4(2E_{\boldsymbol{p}})}\frac{\mathrm{e}^{-\mathrm{i}(E_{\boldsymbol{p}}+z)(y^0 - x^0)+\mathrm{i}\boldsymbol{p}\cdot(\boldsymbol{y}-\boldsymbol{x})}}{z + \mathrm{i}\epsilon}. \quad (17.20)
\end{aligned}
$$

The substitution is now $z' = z + E_{\boldsymbol{p}}$, giving

$$
[\Delta(x,y)]_{(2)} = \mathrm{i}\int_{-\infty}^{\infty}\frac{\mathrm{d}z'\mathrm{d}^3 p}{(2\pi)^4(2E_{\boldsymbol{p}})}\frac{\mathrm{e}^{-\mathrm{i}z'(y^0 - x^0)+\mathrm{i}\boldsymbol{p}\cdot(\boldsymbol{y}-\boldsymbol{x})}}{z' - E_{\boldsymbol{p}} + \mathrm{i}\epsilon}. \quad (17.21)
$$

Lastly, make the substitution $p \to -p$, which is to say $(z',\boldsymbol{p}) \to (p^0,\boldsymbol{p}) \to (-p^0,-\boldsymbol{p})$, we get

$$
\begin{aligned}
[\Delta(x,y)]_{(2)} &= -\mathrm{i}\int_{-\infty}^{\infty}\frac{\mathrm{d}^4 p}{(2\pi)^4(2E_{\boldsymbol{p}})}\frac{\mathrm{e}^{-\mathrm{i}p^0(x^0 - y^0)+\mathrm{i}\boldsymbol{p}\cdot(\boldsymbol{x}-\boldsymbol{y})}}{p^0 + E_{\boldsymbol{p}} - \mathrm{i}\epsilon} \\
&= -\mathrm{i}\int_{-\infty}^{\infty}\frac{\mathrm{d}^4 p}{(2\pi)^4(2E_{\boldsymbol{p}})}\frac{\mathrm{e}^{-\mathrm{i}p\cdot(x-y)}}{p^0 + E_{\boldsymbol{p}} - \mathrm{i}\epsilon}. \quad (17.22)
\end{aligned}
$$

The full propagator $\Delta(x,y) = [\Delta(x,y)]_{(1)} + [\Delta(x,y)]_{(2)}$ is, therefore, the sum of eqns 17.19 and 17.22:

$$
\begin{aligned}
\Delta(x,y) &= \mathrm{i}\int \frac{\mathrm{d}^4 p}{(2\pi)^4(2E_{\boldsymbol{p}})}\mathrm{e}^{-\mathrm{i}p\cdot(x-y)}\left(\frac{1}{p^0 - E_{\boldsymbol{p}} + \mathrm{i}\epsilon} - \frac{1}{p^0 + E_{\boldsymbol{p}} - \mathrm{i}\epsilon}\right) \\
&\quad - \int \frac{\mathrm{d}^4 p}{(2\pi)^4}\mathrm{e}^{-\mathrm{i}p\cdot(x-y)}\frac{\mathrm{i}}{(p^0)^2 - E_{\boldsymbol{p}}^2 + \mathrm{i}\epsilon}. \quad (17.23)
\end{aligned}
$$

Note further that using $E_{\boldsymbol{p}}^2 = \boldsymbol{p}^2 + m^2$ the term in the denominator of the integrand is $(p^0)^2 - E_{\boldsymbol{p}}^2 + \mathrm{i}\epsilon = p^2 - m^2 + \mathrm{i}\epsilon$.

The final result of this example yields the key result that the free Feynman propagator[5] $\Delta(x,y)$ is given by

$$
\Delta(x,y) = \langle 0|T\hat{\phi}(x)\hat{\phi}^\dagger(y)|0\rangle = \int \frac{\mathrm{d}^4 p}{(2\pi)^4}\mathrm{e}^{-\mathrm{i}p\cdot(x-y)}\frac{\mathrm{i}}{p^2 - m^2 + \mathrm{i}\epsilon}. \quad (17.24)
$$

We can immediately extract $\tilde{\Delta}(p)$, the Fourier component[6] of the Feynman propagator corresponding to a particle with momentum p:

$$
\tilde{\Delta}(p) = \frac{\mathrm{i}}{p^2 - m^2 + \mathrm{i}\epsilon}. \quad (17.25)
$$

This turns out to be a very useful equation.

17.4 Yukawa's force-carrying particles

One of the most interesting things about particles is that they interact with each other. Hideki Yukawa's great insight was that this interac-

[5] Because $\Delta(x,y)$ is a function of $x - y$, it is sometimes written $\Delta(x - y)$.

[6] The connection between the quantities $\Delta(x,y)$ and $\tilde{\Delta}(p)$ is a simple Fourier transform:

$$
\Delta(x,y) = \int \frac{\mathrm{d}^4 p}{(2\pi)^4}\mathrm{e}^{-\mathrm{i}p\cdot(x-y)}\tilde{\Delta}(p).
$$

Hideki Yukawa (1907–1981)

tion process itself involved particles. These force-carrying particles are quite different to their more familiar cousins with whom we're already acquainted. They are, however, still described by propagators and provide an immediate application of our field theory propagators. Yukawa's idea centres around one key notion:

> Particles interact by exchanging virtual, force-carrying particles.

Virtual particles are defined as particles existing 'off mass-shell'. The mass-shell is the four-dimensional surface obeying the equation $p^2 = E_{\boldsymbol{p}}^2 - \boldsymbol{p}^2 = m^2$ [see Fig. 11.1]. This, of course, is the usual dispersion for a relativistic particle. For an off mass-shell particle we have $p^2 \neq m^2$. How can this be possible? The argument is that quantum mechanics allows us to violate this classical dispersion, as long as we don't do it for too long! By invoking energy-time uncertainty $\Delta E \Delta t \sim \hbar$, we can say that particles of energy E are allowed to exist off the mass-shell as long as they last for a short time $\Delta t \lesssim \hbar/E$. Virtual particles, therefore, must have a finite range since (i) they can't live forever and (ii) they travel at finite velocity. Yukawa guessed that the potential $U(\boldsymbol{r})$ mediated by the virtual particle would have the form

$$U(\boldsymbol{r}) \propto -\frac{\mathrm{e}^{-|\boldsymbol{r}|/a}}{4\pi|\boldsymbol{r}|}, \tag{17.26}$$

where a has the dimensions of length.

[7]In our units, the length $a = 1/m$. (In SI units, $a = \hbar/mc$.) For a pion (π^0) with mass 135 MeV/c^2, $a \approx 1.5$ fm, roughly the size of a nucleus.

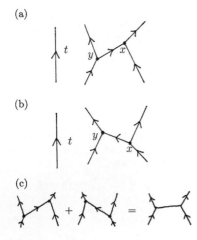

Fig. 17.4 (a) A virtual particle propagates from y to x mediating a force. (b) The same process occurs with a virtual particle propagating from x to y. (c) The processes are summed to make the Feynman propagator.

Example 17.5

Looking back, we see that eqn 17.26 bears a resemblance to $G_{\boldsymbol{k}}^+(\boldsymbol{r}) = -\frac{\mathrm{e}^{\mathrm{i}|\boldsymbol{k}||\boldsymbol{r}|}}{4\pi|\boldsymbol{r}|}$, which is the Green's function for the operator $\hat{L} = (\boldsymbol{\nabla}^2 + \boldsymbol{k}^2)$. From this, we see that if we take $\mathrm{i}|\boldsymbol{k}| = -m$, then we have that the Yukawa potential $U(\boldsymbol{r}) \propto -\frac{\mathrm{e}^{-m|\boldsymbol{r}|}}{4\pi|\boldsymbol{r}|}$ is the Green's function for the equation

$$\left(\boldsymbol{\nabla}^2 - m^2\right) U(\boldsymbol{r}) = \delta^{(3)}(\boldsymbol{r}), \tag{17.27}$$

which is a time-independent version of the Klein–Gordon equation. What we've shown here is that Yukawa's potential is actually the field propagator describing the propagation of virtual scalar field particles. The price that's paid for being off the mass shell is that the virtual particles can travel a distance fixed[7] by their mass m.

The Yukawa particle exchange process is illustrated in Fig. 17.4. We can imagine spacetime processes such as that shown in Fig. 17.4(a) where at time y^0 particle A emits a virtual particle Q, with mass m_Q from position $\boldsymbol{y}$ that at a later time x^0 collides with particle B at position $\boldsymbol{x}$. The energetics of this process may be written

$$E_A = E_A' + E_Q, \tag{17.28}$$

that is, particle A loses energy $E_Q = (\boldsymbol{p}_Q^2 + m_Q^2)^{1/2}$. Of course in this process particle B *gains* energy E_Q.

We may be in serious trouble if Yukawa exchange involves a net transfer of energy from particle A. It makes the interaction rather one-sided for one thing. To maintain the symmetry of the interaction, we must also imagine the additional process shown in Fig. 17.4(b) where particle B (at position $\boldsymbol{x}$) emits an identical particle Q at time x^0, which at later time y^0 collides with particle A at $\boldsymbol{y}$. Both processes are necessary in order to conserve energy so we must have an identical virtual particle heading back to A, which gives back the same amount of energy.

What does the Feynman propagator have to say about Yukawa's notion of force-carrying virtual particles? Actually, the exchange of virtual particles idea is exactly what the Feynman propagator is describing. Recall that the Feynman propagator, describing a particle propagating from spacetime point y to spacetime point x, looks like

$$\Delta(x,y) = \theta(x^0-y^0)\langle 0|\hat\phi(x)\hat\phi^\dagger(y)|0\rangle + \theta(y^0-x^0)\langle 0|\hat\phi(y)^\dagger\hat\phi(x)|0\rangle. \quad (17.29)$$

The first term describes a particle travelling from y to x for $x^0 > y^0$, while the second describes an antiparticle travelling from x to y for $y^0 > x^0$. These situations are what we showed in Fig. 17.4. and we see that both processes are included in the Feynman propagator. If we now turn to the Feynman propagator in momentum space we have

Remember that particles and antiparticles are identical in scalar field theory.

$$\tilde\Delta(p) = \frac{i}{p^2-m^2+i\epsilon} = \frac{i}{(p^0)^2-\boldsymbol{p}^2-m^2+i\epsilon}. \quad (17.30)$$

Notice that the denominator is always off mass shell, that is to say that the expression is only well behaved if $p^0 \neq \sqrt{\boldsymbol{p}^2+m^2}$. The message is that the Green's function describes the propagation of virtual particles.

When we come to calculating Feynman diagrams in a few chapters' time, we'll see that the amplitude $\mathcal{A}$ for two particles to scatter off each other as shown in Fig. 17.5 contains a term $\propto i/(p^2-m^2+i\epsilon)$. We can check this immediately. If we want to calculate the scattering amplitude for two particles in non-relativistic quantum mechanics we can use the Born approximation, which tells us that the amplitude is proportional to the Fourier transform of the scattering potential:

$$\mathcal{A} \propto \tilde U(\boldsymbol{p}) = \int \mathrm{d}^3 r\, U(\boldsymbol{r}) \mathrm{e}^{-i\boldsymbol{p}\cdot\boldsymbol{r}}, \quad (17.31)$$

where $\boldsymbol{p}$ is a scattering vector. If we use Yukawa's suggestion for the interaction potential $U(\boldsymbol{r}) \propto -\frac{\mathrm{e}^{-m|\boldsymbol{r}|}}{4\pi|\boldsymbol{r}|}$ corresponding to an interaction with a range of a, then doing the integral[8] yields

$$\mathcal{A} \propto \frac{1}{-|\boldsymbol{p}|^2 - m^2}, \quad (17.32)$$

which looks rather like the propagator $\tilde\Delta(p)$ with $p^0 = 0$. Our result for the propagator $\tilde\Delta(p)$ is the four-dimensional, relativistic generalization of this result. Indeed, as we saw previously, the propagator $\frac{1}{-|\boldsymbol{p}|^2-m^2}$ is the Green's function for the time-independent Klein–Gordon equation $(-\boldsymbol{\nabla}^2+m^2)\hat\phi(\boldsymbol{x})=0$, while $\tilde\Delta(p)$ is the Green's function for the four-dimensional Klein–Gordon equation $(\partial^2+m^2)\hat\phi(x)=0$.

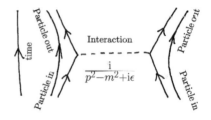

Fig. 17.5 A Feynman diagram for two particles scattering. The amplitude for the process includes a factor $i/(p^2-m^2+i\epsilon)$, which describes the propagation of a virtual, force-carrying particle which can be thought of as mediating a Yukawa force.

[8]The integral can be evaluated using

$$\int \mathrm{d}^3 r \frac{\mathrm{e}^{-i\boldsymbol{p}\cdot\boldsymbol{r}}\mathrm{e}^{-m|\boldsymbol{r}|}}{|\boldsymbol{r}|} = \int_0^\infty \mathrm{d}\boldsymbol{r}\,|\boldsymbol{r}|^2$$
$$\times \int_0^{2\pi}\mathrm{d}\phi\int_0^\pi \mathrm{d}\theta\sin\theta\frac{\mathrm{e}^{-i|\boldsymbol{p}||\boldsymbol{r}|\cos\theta-m|\boldsymbol{r}|}}{|\boldsymbol{r}|}$$
$$= \frac{4\pi}{\boldsymbol{p}^2+m^2}.$$

17.5 Anatomy of the propagator

Let's take a closer look at the anatomy of the free propagator[9] $\Delta(x,y) = \langle 0|T\hat{\phi}(x)\hat{\phi}^\dagger(y)|0\rangle$ for scalar field theory. It arose from the amplitude that a single field excitation is created at spacetime point y and is found at spacetime point x. It contains the sum of amplitudes corresponding to the possibility that this excitation might be a particle or an antiparticle. In position space we have that

$$\Delta(x,y) = \langle 0|T\hat{\phi}(x)\hat{\phi}^\dagger(y)|0\rangle = \int \frac{\mathrm{d}^4 p}{(2\pi)^4} \mathrm{e}^{-\mathrm{i}p\cdot(x-y)} \frac{\mathrm{i}}{p^2 - m^2 + \mathrm{i}\epsilon}. \quad (17.33)$$

It is important to notice that we have a singularity (or pole[10]) in the function when the particle is on mass shell, that is, when $p^2 = E_{\boldsymbol{p}}^2 - \boldsymbol{p}^2 = m^2$.

Now let's return to the full propagator for a general interacting theory $G(x,y) = \langle\Omega|T\hat{\phi}(x)\hat{\phi}^\dagger(y)|\Omega\rangle$. (Remember that we call the interacting ground state $|\Omega\rangle$.) The full name for $G(x,y)$ is the 'single-particle propagator' or the 'two-point Green's function'. We can draw a diagram illustrating $G(x,y)$ as shown in Fig. 17.6(a). The two stumps correspond to the incoming or outgoing particles. The hatched blob in the middle depicts *interactions*: the adventures (involving other particles) that our test particle might have between its creation and annihilation. The name 'two-point' comes from the fact that a single particle is created and a single particle is annihilated, that is, the sum of particles entering or leaving the system, i.e. (number in)+(number out) equals 2. We often emphasise this by writing the propagator as $G^{(2)}(x,y)$ with a superscript '(2)'. We can go further and define the n-point propagator, where n equals the sum of incoming and outgoing lines. These will be denoted $G^{(n)}$. For example we might want to study the four-point propagator [Fig. 17.6(b)]

$$G^{(4)}(x_1, x_2, x_3, x_4) = \langle\Omega|T\hat{\phi}(x_1)\hat{\phi}(x_2)\hat{\phi}^\dagger(x_3)\hat{\phi}^\dagger(x_4)|\Omega\rangle, \quad (17.34)$$

which is useful when we're considering two particles coming in and two particles leaving as might occur in a two-particle scattering experiment.

The aim of much quantum field theory is to find the full interacting Green's functions $G^{(n)}$. One reason is that $G^{(n)}$ can be used to calculate scattering amplitudes using an equation known as the LSZ reduction formula.[11] Another reason is that the form of the interacting Green's function tells us a lot about the system. For example, the two-point propagator, $G^{(2)}(x,y) = \langle\Omega|T\hat{\phi}(x)\hat{\phi}^\dagger(y)|\Omega\rangle$, that tells us about the fate of single particles, is especially useful.

To see this we can consider what happens if we start from a free scalar theory and slowly turn on the interactions to form an interacting theory. Before we turn on the interactions we have the free propagator $\tilde{\Delta}(p) = \frac{\mathrm{i}}{p^2 - m^2 + \mathrm{i}\epsilon}$. The mass of the particle is given by the position of the pole. The residue at the pole (which is i) tells us that the operator

[9]Remember that for a real scalar field $\hat{\phi}(x) = \hat{\phi}^\dagger(x)$ and so we may write the scalar field propagator without the daggers: $\Delta(x,y) = \langle\Omega|T\hat{\phi}(x)\hat{\phi}(y)|\Omega\rangle$. For simplicity's sake, we will do this later in the book.

[10]Another way of making sense of the iϵ is that it is meant to ensure that the integral never hits this pole. It is an instruction to avoid it by adding an infinitesimal imaginary part to the denominator. This is discussed in Appendix B.

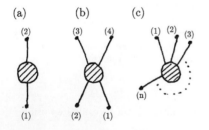

(a) (b) (c)

Fig. 17.6 (a) The two-point propagator $G^{(2)}$. The blob represents the interactions that the particle undergoes with other particles in the system. (b) A four-point propagator $G^{(4)}$. (c) An n-point propagator $G^{(n)}$.

[11]See Peskin and Schroeder for a discussion of the LSZ reduction formula.

$\hat{\phi}^\dagger(y)$ creates precisely one quantum from the vacuum $|0\rangle$ (and later we destroy it). When we turn on the interactions it turns out[12] that the full propagator resembles the free version. The general form of the two-point propagator in momentum space is

[12]The details of the full propagator are discussed in Chapter 33.

$$\tilde{G}^{(2)}(p) = \frac{iZ_{\boldsymbol{p}}}{p^2 - m^2 - \tilde{\Sigma}(p) + i\Gamma_{\boldsymbol{p}}} + \left(\begin{array}{c} \text{Multiparticle} \\ \text{parts} \end{array} \right), \qquad (17.35)$$

where

- $Z_{\boldsymbol{p}}$ tells us how likely it is for a particle with momentum $\boldsymbol{p}$ to exist stably without being destroyed by the interactions,
- $\tilde{\Sigma}(p)$ tells us about how the particle of momentum p interacts with the vacuum,
- $(2\Gamma_{\boldsymbol{p}})^{-1}$ is the particle's lifetime,
- The full propagator also contains a contribution from short-lived multiparticle states which can be scattered out of the vacuum.

Chapter summary

- The Feynman propagator for single-particle excitations in the field is given by $G(x,y) = \langle \Omega | T\hat{\phi}(x)\hat{\phi}^\dagger(y) | \Omega \rangle$.
- For the scalar field theory the momentum space propagator for single particles is $\tilde{\Delta}(p) = \frac{i}{p^2 - m^2 + i\epsilon}$.

Exercises

(17.1) Show that the retarded field propagator for a free particle in momentum space and the time domain: $G_0^+(\boldsymbol{p}, t_x, \boldsymbol{q}, t_y) = \theta(t_x - t_y)\langle 0 | \hat{a}_{\boldsymbol{p}}(t_x) \hat{a}_{\boldsymbol{q}}^\dagger(t_y) | 0 \rangle$ is given by $\theta(t_x - t_y) e^{-i(E_{\boldsymbol{p}} t_x - E_{\boldsymbol{q}} t_y)} \delta^{(3)}(\boldsymbol{p} - \boldsymbol{q})$.

(17.2) Demonstrate that the free scalar propagator is the Green's function of the Klein–Gordon equation. That is, in $(1+1)$ dimensions, show

$$\left(\frac{\partial^2}{\partial(x^0)^2} - \frac{\partial^2}{\partial x^2} + m^2 \right) \langle 0 | T\hat{\phi}(x^0, x)\hat{\phi}^\dagger(y^0, y) | 0 \rangle$$
$$= -i\delta^{(2)}(x - y). \qquad (17.36)$$

(17.3) Consider a scalar field theory defined by the Lagrangian

$$\mathcal{L} = \frac{1}{2}[\partial_\mu \phi(x)]^2 - \frac{m^2}{2}[\phi(x)]^2. \qquad (17.37)$$

By considering the Fourier transform of the field, show that the action may be written

$$S = \frac{1}{2} \int \frac{\mathrm{d}^4 p}{(2\pi)^4} \tilde{\phi}(-p) \left(p^2 - m^2 \right) \tilde{\phi}(p). \qquad (17.38)$$

This provides us with an alternative method for identifying the free propagator $\tilde{G}_0(p)$ as (i/2 times) the inverse of the quadratic term in the momentum-space action.

(17.4) Show that the Feynman propagator for the quantum simple harmonic oscillator with spring constant $m\omega_0^2$ is given by

$$\tilde{G}(\omega) = \left(\frac{1}{m} \right) \frac{i}{\omega^2 - \omega_0^2 + i\epsilon}. \qquad (17.39)$$

(17.5) (a) Consider a one-dimensional system with Lagrangian

$$\mathcal{L} = \frac{1}{2}\left(\frac{\partial\phi(x)}{\partial x}\right)^2 + \frac{m^2}{2}[\phi(x)]^2. \qquad (17.40)$$

The choice of sign makes this a *Euclidean* theory (described in Chapter 25). Discretize this theory (that is, put it on a lattice) by defining $\phi_j = \frac{1}{\sqrt{Na}}\sum_p \tilde{\phi}_p e^{ipja}$, where j labels the lattice site, a is the lattice spacing and N is the number of lattice points. Using the method in Exercise 17.3 show that the action may be written

$$S = \frac{1}{2}\sum_p \tilde{\phi}_{-p}\left(\frac{2}{a^2} - \frac{2}{a^2}\cos pa + m^2\right)\tilde{\phi}_p, \qquad (17.41)$$

and read off the propagator for this theory.

(b) The Lagrangian for a one-dimensional elastic string in (1+1)-dimensional Minkowski space is written

$$\mathcal{L} = \frac{1}{2}\left[(\partial_0\phi)^2 - (\partial_1\phi)^2\right]. \qquad (17.42)$$

Discretize the theory by defining $\phi_j(t) = \frac{1}{\sqrt{Na}}\sum_p \int \frac{d\omega}{2\pi}\tilde{\phi}_p(\omega)e^{-i\omega t}e^{ipja}$. Show that the propagator for the (phonon) excitations that this theory describes is given by

$$\tilde{G}(\omega, p) = \frac{i}{\omega^2 - \omega_0^2(1 - \cos pa)}, \qquad (17.43)$$

where $\omega_0^2 = 2/a^2$.

The S-matrix

18

The Matrix is a system, Neo. That system is our enemy. But when you're inside, you look around, what do you see? Businessmen, teachers, lawyers, carpenters. The very minds of the people we are trying to save. But until we do, these people are still a part of that system, and that makes them our enemy.
Morpheus (Laurence Fishburne) in *The Matrix*

Canonical quantization works well for a small number of theories, where it results in a diagonalized Hamiltonian describing the excitations of the quantum field in terms of non-interacting, or 'free' particles. In fact, the theories that can be solved exactly using the methods of quantum field theory generally describe free particles with no prospect for interactions. A world of free particles would be a very boring place since particle interactions are at the heart of the workings of Nature. Most interesting theories involve particle interactions and cannot be solved exactly. We therefore need to resort to some sort of approximate method, like the perturbation theory of quantum particle mechanics.

We will explore a perturbation theory for quantum fields in the following chapters. The Lagrangian describing the field is often the addition of two parts: a solvable part (frequently describing non-interacting particles only and therefore called the free part) and a part that makes the problem unsolvable (frequently describing interactions and called the interacting part). As before, the free part is able to be solved via canonical quantization and results in non-interacting particles. The interaction part is more interesting because of one key fact:

Interactions involve the creation or destruction of particles.

One example of an interaction process is particle scattering. Particles are fired at each other. At the start of the experiment they are far from each other so don't interact: they are free. When they're smashed together they interact but only for a short time. In this chapter we'll describe a method of dealing with the interactions in a scattering process. This will be time well spent. It turns out that the machinery developed here will be central to creating a more general perturbation theory that can deal with other interactions.

18.1   The S-matrix: a hero for our times

John Wheeler (1911–2008) made numerous contributions to theoretical physics and is also famous for coining the terms 'black hole' and 'wormhole' and 'it from bit'.

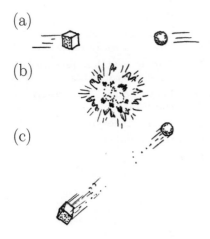

(a)

(b)

(c)

Fig. 18.1 A prototypical scattering experiment. (a) The particles begin well separated. (b) They interact in a complicated way. (c) They recede from each other and end up well separated.

One of the basic building blocks of quantum field theory is the scattering or S-**matrix**. The idea was one of John Wheeler's and it's a great one. The S-matrix is rather like the time-evolution operator, but for an important special case: it describes a *S*cattering experiment.

In a prototypical scattering experiment we start off with well separated particles [Fig. 18.1(a)]. We fire the particles at each other and they interact in some complicated way governed by the Hamiltonian of the real world $\hat{H}$ [Fig. 18.1(b)]. The particles then recede from each other and end up well separated [Fig. 18.1(c)]. The first thing to say about this is that the complicated interactions of the real world make this process impossible to analyse. Working in the Heisenberg picture, we can't even write down convincing expressions for the time-dependent operators. All appears to be lost! What do we do?

Wheeler says that we start by imagining a world without complicated interactions. That is, we split $\hat{H}$ into two parts $\hat{H} = \hat{H}_0 + \hat{H}'$, where $\hat{H}_0$ describes a simple world of non-interacting particles described by some set of state vectors. Remember that we're working in the Heisenberg picture here: these state vectors don't change at all. For two particles one simple-world state might be a momentum state like $|\psi\rangle = |p_2 p_1\rangle_{\text{simpleworld}}$. Now we look at the real world and ask 'Is there a real-world state that looks like the simple-world state $|p_2 p_1\rangle_{\text{simpleworld}}$?' There is such a state at the start of the scattering experiment (at a time $t \to -\infty$) when the particles are very far apart. We'll call this real-world state $|p_2 p_1\rangle_{\text{realworld}}^{\text{in}}$ or an 'in' state, for short. Similarly, we pick out another simple-world state, e.g. $|\phi\rangle = |q_2 q_1\rangle_{\text{simpleworld}}$ and ask if there's a real-world state that looks like this one. There is, but this time at the end of the experiment ($t \to \infty$) when the particles are well separated after their interaction. We'll call this one $|q_2 q_1\rangle_{\text{realworld}}^{\text{out}}$. Notice that the simple-world states only describe the real world in the limits $t \to \pm\infty$.

For the real-world scattering process in which we're interested the amplitude $\mathcal{A}$ for starting with $|p_2 p_1\rangle_{\text{realworld}}^{\text{in}}$ and ending up with $|q_2 q_1\rangle_{\text{realworld}}^{\text{out}}$ is given by

$$\mathcal{A} = {}^{\text{out}}_{\text{realworld}} \langle q_1 q_2 | p_2 p_1 \rangle_{\text{realworld}}^{\text{in}}. \tag{18.1}$$

We must recreate this amplitude using the simple-world states which, after all, are the only ones we can work with. How can this be achieved? Wheeler's answer is the S-matrix. We define

$$\mathcal{A} = {}^{\text{out}}_{\text{realworld}} \langle q_1 q_2 | p_2 p_1 \rangle_{\text{realworld}}^{\text{in}} = {}_{\text{simpleworld}} \langle q_1 q_2 | \hat{S} | p_2 p_1 \rangle_{\text{simpleworld}}. \tag{18.2}$$

So the S-matrix (that is, the matrix elements of the $\hat{S}$-operator) contains the amplitudes for starting with a particular 'in' state and ending up with a particular 'out' state.

Let's take this further and calculate a real amplitude. We now need two things:

- A way of getting a suitable $\hat{H}_0$ to describe some useful simple-world states which resemble the 'in' and 'out' states.

- A way of calculating an expression for the $\hat{S}$-operator. We can then use the eigenstates of the simple Hamiltonian to work out an amplitude.

In the next section we'll examine the machinery needed to get $\hat{H}_0$, which involves yet another new formulation of quantum mechanics. This one is called the **interaction representation**.

18.2   Some new machinery: the interaction representation

So far in our quantum mechanics we've either had a formulation with time-dependent states and time-independent operators (the Schrödinger picture) or one with time-independent states and time-dependent operators (the Heisenberg picture). A third way is available, called the interaction representation, where both the states and the operators have some time dependence. It turns out that this combination is the one we need to obtain $\hat{H}_0$, the simple-world Hamiltonian.

As stated above, to get $\hat{H}_0$ we split up the Hamiltonian into two parts $\hat{H} = \hat{H}_0 + \hat{H}'$. We call these a free part $\hat{H}_0$ and an interaction part $\hat{H}'$. The free part will generally be one that is time-independent and can be easily solved. We then say that operators in the interaction picture $\hat{O}_\text{I}$ evolve in time via the free part $\hat{H}_0$ of the Hamiltonian

$$\hat{O}_\text{I}(t) = \mathrm{e}^{\mathrm{i}\hat{H}_0 t}\hat{O}\mathrm{e}^{-\mathrm{i}\hat{H}_0 t}. \tag{18.3}$$

This is just like the Heisenberg version of quantum mechanics, except that we're only using the free part of the Hamiltonian. We therefore have a Heisenberg-like equation of motion for the operators

$$\mathrm{i}\frac{\mathrm{d}\hat{O}_\text{I}}{\mathrm{d}t} = [\hat{O}_\text{I}(t), \hat{H}_0], \tag{18.4}$$

but again we emphasize that it just involves $\hat{H}_0$.

So far we've not included $\hat{H}'$. Its inclusion results in the wave function having some time dependence, which it wouldn't in the non-interacting case. To see what happens, we compare a matrix element from the Schrödinger picture to one in the interaction picture

$$\langle\phi(t)|\hat{O}|\psi(t)\rangle = \langle\phi_\text{I}(t)|\mathrm{e}^{\mathrm{i}\hat{H}_0 t}\hat{O}\mathrm{e}^{-\mathrm{i}\hat{H}_0 t}|\psi_\text{I}(t)\rangle, \tag{18.5}$$

where the interaction picture states are labelled with a subscript I. We can see that, for the matrix elements to be the same as in the Schrödinger picture, we'll need to define

$$|\psi_\text{I}(t)\rangle = \mathrm{e}^{\mathrm{i}\hat{H}_0 t}|\psi(t)\rangle. \tag{18.6}$$

We can now get to the equation of motion for the interaction picture wave function by differentiating eqn 18.6 with respect to time. We find

$$
\begin{aligned}
i\frac{d}{dt}|\psi_I(t)\rangle &= e^{i\hat{H}_0 t}\left(-\hat{H}_0 + i\frac{d}{dt}\right)|\psi(t)\rangle \\
&= e^{i\hat{H}_0 t}\left(-\hat{H}_0 + \hat{H}\right)|\psi(t)\rangle, \quad (18.7)
\end{aligned}
$$

and recalling that $\hat{H} = \hat{H}_0 + \hat{H}'$ and eqn 18.6 we have

$$
i\frac{d}{dt}|\psi_I(t)\rangle = e^{i\hat{H}_0 t}\hat{H}'e^{-i\hat{H}_0 t}|\psi_I(t)\rangle, \quad (18.8)
$$

which we can rewrite in a recognizable form as

$$
i\frac{d}{dt}|\psi_I(t)\rangle = \hat{H}_I(t)|\psi_I(t)\rangle, \quad (18.9)
$$

where $\hat{H}_I(t) = e^{i\hat{H}_0 t}\hat{H}'e^{-i\hat{H}_0 t}$. This then completely defines the interaction picture.

To recap the description of the interaction picture:

- Both the operators and states can evolve in time.
- The operators evolve by the free part of the Hamiltonian: $\hat{\phi}_I(t) = e^{i\hat{H}_0 t}\hat{\phi}\,e^{-i\hat{H}_0 t}$.
- The states evolve according to the interaction part of the Hamiltonian: $i\frac{\partial}{\partial t}|\psi_I(t)\rangle = \hat{H}_I(t)|\psi_I(t)\rangle$ where $\hat{H}_I(t) = e^{i\hat{H}_0 t}\hat{H}'e^{-i\hat{H}_0 t}$.

Note that all of our different representations coincide at $t = 0$.

18.3   The interaction picture applied to scattering

Why is the interaction picture useful for the scattering problem? The point is that the interaction part of the Hamiltonian is zero at the start and end of the problem. In fact, we imagine that the interaction part of the Hamiltonian $\hat{H}'$ is turned on and off slowly and smoothly as shown in Fig. 18.2. When $\hat{H}_I = 0$ we just have the Heisenberg picture for the free part of the Hamiltonian $\hat{H}_0$. This is good because we choose $\hat{H}_0$ so that we can solve it using our canonical quantization machine and we certainly then know its eigenstates.

What becomes of the states? We can identify the simple-world states as those of the interaction picture at the start and end:

$$
|\phi\rangle_{\text{simpleworld}} = |\phi_I(\pm\infty)\rangle. \quad (18.10)
$$

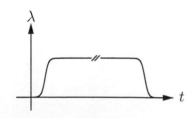

Fig. 18.2 Turning on an interaction Hamiltonian $\hat{H}'$. We multiply it by $\lambda(t)$ which has this profile.

They are eigenstates of $\hat{H}_0$, which is vital since it allows us to build them up from the state $|0\rangle$ which is the vacuum of $\hat{H}_0$ (called variously the non-interacting, free or bare vacuum) using our beloved creation and annihilation operators. During the interaction process we have that

$\hat{H}_\mathrm{I}$ is nonzero so the states evolve in some mysterious and complicated way. However, at the end of the experiment $\hat{H}_\mathrm{I}$ is zero again and the states freeze. It's a bit like musical chairs in this respect. The states are initially sitting down, they rearrange themselves during the music and then freeze on the chairs when the music stops.

We've thought about the states, but what about the operators? Simple: the operators time-evolve according to $\hat{H}_0$ at all times. We can therefore use the freely evolving field operators like $\hat{\phi}(x)$ that we've used previously. These also enjoy an expansion in terms of creation and annihilation operators.

Finally we need to figure out how to work out the $\hat{S}$-operator. We are assisted by the useful fact that all of our quantum mechanical pictures are defined so that they coincide at $t = 0$, so that we're free to write

$$\text{simpleworld}\langle\phi|\hat{S}|\psi\rangle_\text{simpleworld} =^\text{out}_\text{realworld}\langle\phi|\psi\rangle^\text{in}_\text{realworld} = \langle\phi_\mathrm{I}(0)|\psi_\mathrm{I}(0)\rangle.$$
(18.11)

Then, if we knew the time-evolution operator in the interaction picture $\hat{U}_\mathrm{I}(t_2, t_1)$, we could say that

$$\begin{aligned}\text{simpleworld}\langle\phi|\hat{S}|\psi\rangle_\text{simpleworld} &= \langle\phi_\mathrm{I}(\infty)|\hat{U}_\mathrm{I}(\infty,0)\hat{U}_\mathrm{I}(0,-\infty)|\psi_\mathrm{I}(-\infty)\rangle\\ &= \langle\phi_\mathrm{I}(\infty)|\hat{U}_\mathrm{I}(\infty,-\infty)|\psi_\mathrm{I}(-\infty)\rangle \quad (18.12)\\ &= \text{simpleworld}\langle\phi|\hat{U}_\mathrm{I}(\infty,-\infty)|\psi\rangle_\text{simpleworld}.\end{aligned}$$

We get the important result that the $\hat{S}$-operator is the time-evolution operator for the interaction-picture $\hat{U}_\mathrm{I}(t,-t)$ as $t \to \infty$.

18.4 Perturbation expansion of the S-matrix

As in the Schrödinger picture, we have an equation of motion for the interaction picture time-evolution operator

$$i\frac{\mathrm{d}}{\mathrm{d}t_2}\hat{U}_\mathrm{I}(t_2, t_1) = \hat{H}_\mathrm{I}(t_2)\hat{U}_\mathrm{I}(t_2, t_1),$$
(18.13)

where $\hat{U}_I(t,t) = 1$. We might be tempted to treat $\hat{H}_\mathrm{I}(t)$ as a number, and if that were valid we could then integrate and write $\hat{U}_\mathrm{I}(t_2, t_1) = e^{-i\int_{t_1}^{t_2}\mathrm{d}t\,\hat{H}_\mathrm{I}(t)}$, but this is wrong. The interaction Hamiltonian $\hat{H}_\mathrm{I}(t)$ doesn't necessarily commute with itself when evaluated at different times. That is, in general, $[\hat{H}_\mathrm{I}(t_2), \hat{H}_\mathrm{I}(t_1)] \neq 0$. To circumvent this problem, we'll define a new version of the exponential which will make this equation right. The solution turns out to use the **time-ordered product**

$$T[\hat{A}_1(t_1)\hat{A}_2(t_2)\dots\hat{A}_n(t_n)],$$
(18.14)

defined as the string arranged so that the later operators are on the left.[1] As discussed in the previous chapter, the Wick time ordering symbol symbol $T[\,]$ is a little like the normal ordering procedure $N[\,]$ in that it is

[1] Just remember *'later on the left'*. Note that the anticommuting nature of Fermi fields causes us to pick up an additional factor $(-1)^P$ on time ordering, where P is the number of swaps required to rearrange the operators. From here until Chapter 35 we will focus our discussion on (bosonic) scalar fields. Fermi fields, whose behaviour is based on the same general principles we will examine, are discussed specifically in Part IX.

not an operator, it's an instruction on what to do to a string of operators. The reason that using $T[\,]$ helps us here is that everything within a time-ordered product commutes, so we may now take a derivative. We can therefore write an expression for $\hat{U}_{\mathrm{I}}(t_2, t_1)$ that incorporates correctly time-ordered operators as follows:

$$\hat{U}_{\mathrm{I}}(t_2, t_1) = T\left[\mathrm{e}^{-\mathrm{i}\int_{t_1}^{t_2}\mathrm{d}t\,\hat{H}_{\mathrm{I}}(t)}\right]. \tag{18.15}$$

This expression is known as **Dyson's expansion**.

Freeman Dyson (1923–)

Example 18.1

We'll pause to justify eqn 18.15. Using the fact that everything within a time-ordered product commutes we can treat $\hat{U}_{\mathrm{I}}(t_2, t_1)$ as a function of t_2 and take a derivative with respect to this time to obtain

$$\mathrm{i}\frac{\mathrm{d}}{\mathrm{d}t_2}T\left[\mathrm{e}^{-\mathrm{i}\int_{t_1}^{t_2}\mathrm{d}t\,\hat{H}_{\mathrm{I}}(t)}\right] = T\left[\hat{H}_{\mathrm{I}}(t_2)\mathrm{e}^{-\mathrm{i}\int_{t_1}^{t_2}\mathrm{d}t\,\hat{H}_{\mathrm{I}}(t)}\right]. \tag{18.16}$$

Next we notice that t_2 is the latest time in the problem (since $\hat{U}$ evolves the system from t_1 to t_2) and the time ordering therefore puts the operator $\hat{H}_{\mathrm{I}}(t_2)$ left-most. We can then pull $\hat{H}_{\mathrm{I}}(t_2)$ out of the time-ordered product to obtain

$$\mathrm{i}\frac{\mathrm{d}}{\mathrm{d}t_2}T\left[\mathrm{e}^{-\mathrm{i}\int_{t_1}^{t_2}\mathrm{d}t\,\hat{H}_{\mathrm{I}}(t)}\right] = \hat{H}_{\mathrm{I}}(t_2)T\left[\mathrm{e}^{-\mathrm{i}\int_{t_1}^{t_2}\mathrm{d}t\,\hat{H}_{\mathrm{I}}(t)}\right]. \tag{18.17}$$

Comparing this with eqn 18.13 we see that the time-evolution operator in the interaction picture is given by

$$\hat{U}_{\mathrm{I}}(t_2, t_1) = T\left[\mathrm{e}^{-\mathrm{i}\int_{t_1}^{t_2}\mathrm{d}t\,\hat{H}_{\mathrm{I}}(t)}\right]. \tag{18.18}$$

We also know that the $\hat{S}$-operator is the limit $\hat{S} = \hat{U}_{\mathrm{I}}(t_2 \to \infty, t_1 \to -\infty)$. Using this, we are left with Dyson's expansion of the $\hat{S}$-operator

$$\hat{S} = T\left[\mathrm{e}^{-\mathrm{i}\int_{-\infty}^{\infty}\mathrm{d}^4x\,\hat{\mathcal{H}}_{\mathrm{I}}(x)}\right], \tag{18.19}$$

where we've replaced $\int \mathrm{d}t\,\hat{H}_{\mathrm{I}}(t)$ with $\int \mathrm{d}^4x\,\hat{\mathcal{H}}_{\mathrm{I}}(x)$, an integral of the Hamiltonian density.

The exponential form of Dyson's expansion in eqn 18.19 is beautifully memorable, but not very useful. Usually the integral in the exponent cannot be done exactly so we have to expand out the exponential in Dyson's expansion thus:

$$\hat{S} = T\left[1 - \mathrm{i}\int \mathrm{d}^4z\,\hat{\mathcal{H}}_{\mathrm{I}}(z) + \frac{(-\mathrm{i})^2}{2!}\int \mathrm{d}^4y\,\mathrm{d}^4w\,\hat{\mathcal{H}}_{\mathrm{I}}(y)\hat{\mathcal{H}}_{\mathrm{I}}(w) + \ldots\right]. \tag{18.20}$$

Provided that the interaction Hamiltonian $\hat{\mathcal{H}}_{\mathrm{I}}(x)$ is small compared to the full Hamiltonian density, then this provides the basis for a perturbation expansion of the S-matrix. We'll deal with this expansion in detail in the next chapter. Notice that in eqn 18.20 we integrate over different spacetime coordinates each time the interaction Hamiltonian appears.

18.5   Wick's theorem

To make progress in digesting long strings of operators that appear in Dyson expansions, we are faced with solving the following conundrum:

- We will frequently have to evaluate a term like

$$\langle 0|T[\hat{A}\hat{B}\hat{C}\ldots\hat{Z}]|0\rangle, \tag{18.21}$$

   a **vacuum expectation value** (or **VEV** for short) of a time-ordered string of operators. This is hard.

- On the other hand, what is very easy to evaluate is

$$\langle 0|N[\hat{A}\hat{B}\hat{C}\ldots\hat{Z}]|0\rangle, \tag{18.22}$$

   a VEV of a normal-ordered string of operators. This is trivial because normal ordering places annihilation operators on the right and creation operators on the left, and so this VEV is identically zero. So if we could find a way of relating $N[\text{stuff}]$ to $T[\text{stuff}]$ then the problem would be solved.

The next thing to notice is that the mode expansion of a field operator $\hat{\phi}$ contains two parts, an annihilation part and a creation part (see eqn 11.8) and so can be written $\hat{\phi} = \hat{\phi}^- + \hat{\phi}^+$. Thus we have[2] $\hat{\phi}^-|0\rangle = 0$ and $\langle 0|\hat{\phi}^+ = 0$. Let's now take the simplest non-trivial case of a string of operators and consider two field operators $\hat{A}$ and $\hat{B}$. Their product is given by

$$\hat{A}\hat{B} = (\hat{A}^+ + \hat{A}^-)(\hat{B}^+ + \hat{B}^-) = \hat{A}^+\hat{B}^+ + \hat{A}^-\hat{B}^- + \hat{A}^+\hat{B}^- + \hat{A}^-\hat{B}^+. \tag{18.23}$$

If we normal-order this product, we get

$$N[\hat{A}\hat{B}] = \hat{A}^+\hat{B}^+ + \hat{A}^-\hat{B}^- + \hat{A}^+\hat{B}^- + \hat{B}^+\hat{A}^-, \tag{18.24}$$

where we have only needed to swap operators in the final term. Thus $N[\hat{A}\hat{B}]$ only differs from $\hat{A}\hat{B}$ by

$$\hat{A}\hat{B} - N[\hat{A}\hat{B}] = \hat{A}^-\hat{B}^+ - \hat{B}^+\hat{A}^- = [\hat{A}^-, \hat{B}^+], \tag{18.25}$$

i.e. by a simple commutator of operators (which we know from previous experience[3] will just be a complex number, or *c*-number for short). Now the fields $\hat{A}$ and $\hat{B}$ are functions of spacetime coordinates, so we could also write down the time-ordered product

$$T[\hat{A}(x)\hat{B}(y)] = \begin{cases} \hat{A}(x)\hat{B}(y) & x^0 > y^0 \\ \hat{B}(y)\hat{A}(x) & x^0 < y^0, \end{cases} \tag{18.26}$$

and therefore

$$T[\hat{A}(x)\hat{B}(y)] - N[\hat{A}(x)\hat{B}(y)] = \begin{cases} [\hat{A}^-(x), \hat{B}^+(y)] & x^0 > y^0 \\ [\hat{B}^-(y), \hat{A}^+(x)] & x^0 < y^0. \end{cases} \tag{18.27}$$

[2] Our definition assigns the superscript $+$ to the creation part and $-$ to the annihilation part. Some books, e.g. Peskin and Schroeder, choose an opposite convention.

[3] Commutators turn out to be quantities such as $i\hbar$.

Since the VEV of a normal ordered product is zero, we can immediately write down

$$\langle 0|T[\hat{A}(x)\hat{B}(y)]|0\rangle = \begin{cases} \langle 0|[\hat{A}^-(x),\hat{B}^+(y)]|0\rangle & x^0 > y^0 \\ \langle 0|[\hat{B}^-(y),\hat{A}^+(x)]|0\rangle & x^0 < y^0. \end{cases} \qquad (18.28)$$

If we choose $\hat{A} = \hat{B} = \phi$, then the quantity on the left is simply a Feynman propagator.

Subtracting out a normal-ordered string of operators from the time-ordered version is clearly a useful thing to do and so we formalize this and give it a special name: a **contraction**, defined as

$$\overline{\hat{A}\hat{B}} = T[\hat{A}\hat{B}] - N[\hat{A}\hat{B}]. \qquad (18.29)$$

The contraction $\overline{\hat{A}\hat{B}}$ is simply a commutator (see eqn 18.27) because the only difference between normal ordering and time ordering is that we have shuffled various creation and annihilation operators, thereby accumulating various c-numbers. Thus the contraction is a c-number, and it takes the value

$$\overline{\hat{A}\hat{B}} = \overline{\hat{A}\hat{B}}\langle 0|0\rangle = \langle 0|\overline{\hat{A}\hat{B}}|0\rangle = \langle 0|T[\hat{A}\hat{B}]|0\rangle. \qquad (18.30)$$

Moreover, since it is only a c-number, normal ordering has no effect on it and we can write

$$T[\hat{A}\hat{B}] = N[\hat{A}\hat{B}] + \overline{\hat{A}\hat{B}} = N[\hat{A}\hat{B} + \overline{\hat{A}\hat{B}}]. \qquad (18.31)$$

[4]The proof is by induction, and can be found on page 90 of Peskin and Schroeder.

This result can be generalized[4] to longer strings of operators to yield **Wick's theorem**, which can be stated as follows:

$$T[\hat{A}\hat{B}\hat{C}\ldots\hat{Z}] = N\left[\hat{A}\hat{B}\hat{C}\ldots\hat{Z} + \begin{array}{c}\text{all possible contractions of} \\ \hat{A}\hat{B}\hat{C}\ldots\hat{Z}\end{array}\right]. \qquad (18.32)$$

It's worth noting that Wick's theorem applies for free fields only, such as those in the interaction picture. This will be important in Chapter 22.

Example 18.2

Wick's theorem can be illustrated for the case of four operators

$$\begin{aligned} T[\hat{A}\hat{B}\hat{C}\hat{D}] &= N\left[\hat{A}\hat{B}\hat{C}\hat{D}\right] + N\left[\overline{\hat{A}\hat{B}}\hat{C}\hat{D}\right] + N\left[\hat{A}\overline{\hat{B}\hat{C}}\hat{D}\right] + N\left[\hat{A}\hat{B}\overline{\hat{C}\hat{D}}\right] \\ &+ N\left[\hat{A}\hat{B}\overline{\hat{C}}\hat{D}\right] + N\left[\hat{A}\hat{B}\hat{C}\overline{\hat{D}}\right] + N\left[\hat{A}\hat{B}\hat{C}\hat{D}\right] \\ &+ N\left[\hat{A}\hat{B}\hat{C}\hat{D}\right] + N\left[\hat{A}\hat{B}\hat{C}\hat{D}\right] + N\left[\hat{A}\hat{B}\hat{C}\hat{D}\right]. \end{aligned} \qquad (18.33)$$

This is a sum of normal ordered terms including zero, one or two contractions.

Let us look at a particular term involving a single contraction in a Wick expansion of $T[\hat{A}\hat{B}\hat{C}\ldots\hat{Z}]$, say for example $N[\hat{A}\hat{B}\hat{C}\hat{D}\hat{E}\hat{F}\hat{G}\ldots]$. Since the contracted part is a c-number, we can factor it out of the normal ordered product[5] and write

$$N[\hat{A}\hat{B}\hat{C}\hat{D}\hat{E}\hat{F}\hat{G}\ldots] = \hat{C}\hat{F} \times N[\hat{A}\hat{B}\hat{D}\hat{E}\hat{G}\ldots]. \qquad (18.34)$$

However, when we evaluate the VEV of this term it will vanish because $\langle 0|N[\text{anything}]|0\rangle = 0$. Therefore, the only terms which survive when we work out the VEV of $T[\hat{A}\hat{B}\hat{C}\ldots\hat{Z}]$ are the ones involving contractions of *all* operators.[6] Thus

$$
\begin{aligned}
\langle 0|T[\hat{A}\hat{B}\hat{C}\ldots\hat{Y}\hat{Z}]|0\rangle &= \langle 0|T[\hat{A}\hat{B}\hat{C}\hat{D}\hat{E}\hat{F}\ldots\hat{Y}\hat{Z}]|0\rangle \qquad (18.35)\\
&+ \langle 0|T[\hat{A}\hat{B}\hat{C}\hat{D}\hat{E}\hat{F}\ldots\hat{Y}\hat{Z}]|0\rangle\\
&+ \begin{pmatrix} \text{all other combinations involving} \\ \text{contractions of } \textit{every} \text{ pair of operators} \end{pmatrix},
\end{aligned}
$$

and using eqn 18.30 this reduces to

$$
\begin{aligned}
&\langle 0|T[\hat{A}\hat{B}\hat{C}\ldots\hat{Y}\hat{Z}]|0\rangle \\
&= \langle 0|T[\hat{A}\hat{B}]|0\rangle\langle 0|T[\hat{C}\hat{D}]|0\rangle\langle 0|T[\hat{E}\hat{F}]|0\rangle \ldots \langle 0|T[\hat{Y}\hat{Z}]|0\rangle \\
&+ \langle 0|T[\hat{A}\hat{C}]|0\rangle\langle 0|T[\hat{B}\hat{D}]|0\rangle\langle 0|T[\hat{E}\hat{F}]|0\rangle \ldots \langle 0|T[\hat{Y}\hat{Z}]|0\rangle \qquad (18.36)\\
&+ \ldots
\end{aligned}
$$

that is, the VEV of a time-ordered string of operators is given by the sum of products of all possible combinations of VEVs of time ordered pairs.

Example 18.3

For the case of four bosonic operators.

$$
\begin{aligned}
\langle 0|T\left[\hat{A}\hat{B}\hat{C}\hat{D}\right]|0\rangle &= \langle 0|T\left[\hat{A}\hat{B}\hat{C}\hat{D}\right]|0\rangle + \langle 0|T\left[\hat{A}\hat{B}\hat{C}\hat{D}\right]|0\rangle + \langle 0|T\left[\hat{A}\hat{B}\hat{C}\hat{D}\right]|0\rangle \\
&= \langle 0|T[\hat{A}\hat{B}]|0\rangle\langle 0|T[\hat{C}\hat{D}]|0\rangle + \langle 0|T[\hat{A}\hat{C}]|0\rangle\langle 0|T[\hat{B}\hat{D}]|0\rangle \\
&+ \langle 0|T[\hat{A}\hat{D}]|0\rangle\langle 0|T[\hat{B}\hat{C}]|0\rangle. \qquad (18.37)
\end{aligned}
$$

In particular

$$
\begin{aligned}
\langle 0|T\hat{\phi}(x_1)\hat{\phi}(x_2)\hat{\phi}(x_3)\hat{\phi}(x_4)|0\rangle &= \Delta(x_1-x_2)\Delta(x_3-x_4) + \Delta(x_1-x_3)\Delta(x_2-x_4) \\
&+ \Delta(x_1-x_4)\Delta(x_2-x_3), \qquad (18.38)
\end{aligned}
$$

where we have used the Feynman propagator $\Delta(x_1-x_2) = \langle 0|T\hat{\phi}(x_1)\hat{\phi}(x_2)|0\rangle$.

Wick's theorem is an essential tool for expanding $\hat{S}$ in perturbation theory which we discuss in the next chapter.

[5]There could be a sign change in this expression if the operators are fermionic, so in that case you have to pay attention to any sign changes accumulated with swapping the order of operators.

[6]By extension, the VEV of an odd-numbered string of operators will be zero, since you can only contract pairs of operators and so there will always be one operator left over.

Chapter summary

- Real-world scattering processes can be described by an S-matrix element via $\underset{\text{realworld}}{\overset{\text{out}}{}}\langle\phi|\psi\rangle^{\text{in}}_{\text{realworld}} =_{\text{simpleworld}} \langle\phi|\hat{S}|\psi\rangle_{\text{simpleworld}}$.
- The S-matrix works in the interaction picture where operators time-evolve according to the free part of the Hamiltonian $\hat{H}_0$ and states time-evolve according to the interaction part $\hat{H}_I(t)$.
- The $\hat{S}$-operator is given by the Dyson equation $\hat{S} = T[e^{-i\int d^4x\hat{\mathcal{H}}_I(x)}]$.
- Wick's theorem allows us to grind down long time-ordered strings of operators. A vacuum expectation value (VEV) of a time-ordered string is a sum of products of VEVs of time-ordered pairs.

Exercises

(18.1) Although the interaction picture is most useful to us as a step towards a powerful version of perturbation theory, it is also useful in itself for some problems, such as this famous example from magnetic resonance. Consider a Hamiltonian describing a spin-1/2 particle in a large, static magnetic field B_0 which is subject to a perturbing, perpendicular, oscillating field B_1:

$$\hat{H} = \gamma B_0 \hat{S}_z + \gamma B_1(\hat{S}_x \cos\gamma B_0 t + \hat{S}_y \sin\gamma B_0 t).$$
(18.39)

Here γ is the particle's gyromagnetic ratio. Notice that the frequency γB_0 of the oscillating field with amplitude B_1 is chosen so that it matches the energy level separation caused by the static field B_0.
(a) Write the problem in the interaction representation by splitting the Hamiltonian into free and interacting parts.
(b) Use the identities

$$\begin{aligned}
\hat{S}_{\pm} &= \hat{S}_x \pm i\hat{S}_y, \\
\hat{S}_+ e^{i\omega t} &= e^{i\omega \hat{S}_z t}\hat{S}_+ e^{-i\omega \hat{S}_z t}, \\
\hat{S}_- e^{-i\omega t} &= e^{i\omega \hat{S}_z t}\hat{S}_- e^{-i\omega \hat{S}_z t},
\end{aligned}$$
(18.40)

to simplify the interaction Hamiltonian $\hat{H}_I(t)$. You should find that you're able to remove the time dependence completely.

(c) Find the interaction picture evolution operator $\hat{U}_I(t_2, t_1)$.
(d) What is the probability that a particle initially prepared in a state $|\uparrow\rangle$ at $t = 0$ will still be in that state at time t. What is the probability that it will be found in the $|\downarrow\rangle$ state?
(e) Find the expectation value of the $\hat{S}_z$ operator.

(18.2) Show that $|\psi_I(t = \infty)\rangle = \sum_\phi \langle\phi|\hat{S}|\psi\rangle|\phi\rangle$, where the states on the right-hand side are the simple-world states defined in the chapter.

(18.3) Use Wick's theorem to express the string of Bose operators $\hat{a}_p \hat{a}_q^\dagger \hat{a}_k$ in terms of normal ordered fields and contractions.

(18.4) (a) Normal order the string of Bose operators $\hat{b}\hat{g}\hat{b}\hat{b}^\dagger\hat{b}^\dagger$ using the usual Bose commutation relations.
(b) Show that Wick's theorem gives the same answer.

(18.5) Use Wick's theorem to simplify

$$\langle 0|\hat{c}_{p_1-q}^\dagger \hat{c}_{p_2+q}^\dagger \hat{c}_{p_2} \hat{c}_{p_1}|0\rangle,$$

where the operators $\hat{c}_p^\dagger$ and $\hat{c}_p$ create and annihilate fermions respectively and the ground state $|0\rangle$ contains a non-zero number of fermions (see Chapter 43 for an example of such a system).

Expanding the S-matrix: Feynman diagrams

19

They were funny-looking pictures. And I did think consciously: Wouldn't it be funny if this turns out to be useful and the Physical Review would be all full of these funny looking pictures. It would be very amusing.
Richard Feynman (1918–1988)

Like the silicon chips of more recent years, the Feynman diagram was bringing computation to the masses.
Julian Schwinger (1918–1994)

One of Richard Feynman's greatest achievements was his invention of (what are now called) **Feynman diagrams**. These are cartoons that represent terms in the perturbation expansion of the S-matrix. We will see that Feynman diagrams exist in several related forms, but the most straightforward of these are simply spacetime diagrams describing the trajectory of particles. We imagine time running up the page and represent particles by lines with an arrow going in the direction of time [Fig. 19.1(a)]; antiparticles are represented by lines with arrows going in the opposite direction. (This is consistent with Feynman's interpretation of negative energy states as positive energy antiparticles travelling backwards in time.) We can draw particle and antiparticle pairs being created at some time [Fig. 19.1(b)] or destroyed [Fig. 19.1(c)] at some other time. We can also tell stories: the line in Fig. 19.1(d) may look like it describes a single particle looping backward in time, but Feynman has another interpretation. Just as it's unclear to a bomber pilot flying over a curving road whether (s)he is above a single road or a number of them, we're asked to fly over the diagram experiencing everything in time order. From the point of view of the bombardier, we start with a particle for $t < t_2$, then at $t = t_2$ a new particle–antiparticle pair is created. The new particle and antiparticle coexist with the original particle in the interval $t_2 < t < t_3$. At $t = t_3$ the newly created antiparticle collides with our original particle and annihilates. The newly created particle continues out of the diagram for $t > t_3$ [see Fig. 19.1(e)]. Since the particles are identical, no-one can tell the difference.

In this chapter we introduce the perturbation expansion of the S-matrix and its representation in terms of Feynman diagrams. Recall from the previous chapter that an expression for the $\hat{S}$-operator is given

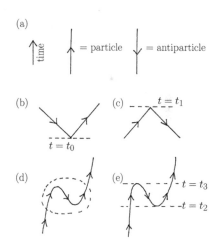

Fig. 19.1 The Feynman diagram. (a) A particle is represented by a line with an arrow going in the direction of time (conventionally up the page). An antiparticle is represented by a line with an arrow going the opposite way. (b) A particle and antiparticle created at time $t = t_0$. (c) A particle and antiparticle are annihilated at time $t = t_1$. (d) Trajectory of a single particle can be interpreted as follows: (e) A particle–antiparticle pair are created at t_2, the antiparticle of which annihilates with a third particle at $t = t_3$.

by Dyson's expansion

$$\hat{S} = T\left[e^{-i\int d^4 z\,\hat{\mathcal{H}}_I(z)}\right]. \tag{19.1}$$

To use this we must expand it as a power series in the operator $\hat{\mathcal{H}}_I(z)$. We'll now consider some of the forms of $\hat{\mathcal{H}}_I(z)$ that are encountered in quantum field theory and see how the expansion of Dyson's equation proceeds. The interaction Hamiltonians will generally involve the products of free-field operators localized at some point z. We'll see later that if we interpret the field operators as describing particles then the interaction[1] can be thought of as a collision of the particles at spacetime point z.

19.1 Meet some interactions

There are a number of models of interactions to try. For each interaction we will draw a diagram in spacetime illustrating the interaction processes. This will eventually lead to the famous diagrams invented by Richard Feynman. At present we won't make any interpretations of the diagrams in terms of particles: this will only be possible a little later when the field operators are made to act on states.

[1]Notice also that we integrate over all z in eqn 19.1 thus taking into account the possibility of the interaction taking place at every point in spacetime.

[2]This has a very similar structure to the interaction used in QED. There the electron corresponds to the excitations in the psi-field and the photon to excitations in the phi-field.

(a)

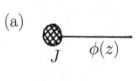

(b)

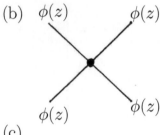

(c)

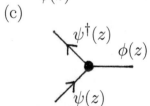

(d)

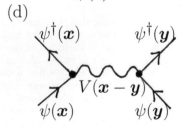

Fig. 19.2 Spacetime diagrams representing some commonly encountered interactions.

Example 19.1

(i) The simplest interaction Hamiltonian $\mathcal{H}_I(z)$ involves a scalar field $\phi(z)$ interacting with an external source field $J(z)$ at a spacetime point z:

$$\mathcal{H}_I(z) = J(z)\phi(z). \tag{19.2}$$

We draw the source $J(z)$ as a blob at a point z. The field is drawn as a stumpy line attached to the source blob [see Fig. 19.2(a)].

(ii) The simplest self-interaction of fields is ϕ^4 (say 'phi-fourth' or 'phi-four') theory described by a Hamiltonian

$$\mathcal{H}_I(z) = \frac{\lambda}{4!}\phi(z)^4. \tag{19.3}$$

This interaction causes four scalar ϕ-fields to meet at a spacetime point z [see Fig. 19.2(b)]. The diagram shows four stumpy lines meeting at z.

(iii) Another simple interaction is

$$\mathcal{H}_I(z) = g\psi^\dagger(z)\psi(z)\phi(z), \tag{19.4}$$

where $\phi(z)$ is, again, a scalar field and $\psi(z)$ is a complex scalar field. This interaction[2] causes an adjoint psi-field [represented by $\psi^\dagger(z)$], a psi-field [$\psi(z)$] and a scalar phi-field [$\phi(z)$] to meet at z. To keep track of fields and their adjoints we use the convention that a field ψ is shown by a stumpy line with an arrow pointing towards the interaction vertex [see Fig. 19.2(c)]. The adjoint field $\psi^\dagger$ is shown with an arrow pointing away from the interaction.

(iv) Another very useful interaction is non-relativistic and describes the Coulomb interaction

$$\mathcal{H}_I(x-y) = \frac{1}{2}\psi^\dagger(\boldsymbol{x})\psi^\dagger(\boldsymbol{y})V(\boldsymbol{x}-\boldsymbol{y})\delta(x^0-y^0)\psi(\boldsymbol{y})\psi(\boldsymbol{x}). \tag{19.5}$$

This one is delocalized. It says that a psi-field and an adjoint psi-field meet at x. These interact, via a potential $V(\boldsymbol{x}-\boldsymbol{y})$, with a psi-field and adjoint psi-field at y. The δ-function ensures that the interaction is instantaneous (that is, occurs at the same time). [This is shown in Fig. 19.2(d).]

19.2   The example of ϕ^4 theory

We'll now carry out the full procedure of calculating an S-matrix element for the simple case of ϕ^4 theory. The Lagrangian density describing the interacting theory is given by

$$\mathcal{L} = \frac{1}{2}[\partial_\mu \phi(x)]^2 - \frac{m^2}{2}\phi(x)^2 - \frac{\lambda}{4!}\phi(x)^4. \qquad (19.6)$$

The free part is given by the first two terms $\mathcal{L}_0 = \frac{1}{2}(\partial_\mu \phi)^2 - \frac{m^2}{2}\phi^2$, which give rise to a free Hamiltonian operator upon canonical quantization

$$\hat{\mathcal{H}}_0 = \frac{1}{2}\left[\left(\frac{\partial\hat{\phi}}{\partial t}\right)^2 + \left(\boldsymbol{\nabla}\hat{\phi}\right)^2 + m^2\hat{\phi}^2\right]. \qquad (19.7)$$

When we use the $\hat{S}$-operator we work in the interaction picture, so the free Hamiltonian time evolves the field operators $\hat{\phi}(x)$. The states evolve via the interaction Hamiltonian which we read off from the interacting part of the Lagrangian $\mathcal{L}_I = -\frac{\lambda}{4!}\phi(x)^4$ as $\hat{\mathcal{H}}_I = \frac{\lambda}{4!}\hat{\phi}(x)^4$. Next we need a programme to expand the S-matrix using this interaction Hamiltonian. It's a glorious five-step plan.

Step I

Decide what S-matrix element to calculate and write it as a vacuum expectation value (VEV) of the free vacuum $|0\rangle$. We do this so we can use the simple form of Wick's theorem later.

As an example we'll take our 'in' state to be a single particle in a momentum state p and the 'out' state will be a single particle in a momentum state q. We'll be calculating the amplitude

$$
\begin{aligned}
\mathcal{A} &= {}^{\text{out}}\langle q|p\rangle^{\text{in}} = \langle q|\hat{S}|p\rangle \\
&= (2\pi)^3 (2E_q)^{\frac{1}{2}}(2E_p)^{\frac{1}{2}}\langle 0|\hat{a}_q \hat{S}\hat{a}_p^\dagger|0\rangle, \qquad (19.8)
\end{aligned}
$$

where we recall that the relativistic normalization of our states means that $|p\rangle = (2\pi)^{\frac{3}{2}}(2E_p)^{\frac{1}{2}}\hat{a}_p^\dagger|0\rangle$.

Step II

We expand the $\hat{S}$-operator using Dyson's expansion

$$
\begin{aligned}
\hat{S} &= T\left[\exp\left(-\mathrm{i}\int \mathrm{d}^4 z\, \hat{\mathcal{H}}_I(z)\right)\right] \\
&= T\left[1 - \mathrm{i}\int \mathrm{d}^4 z\, \hat{\mathcal{H}}_I(z) + \frac{(-\mathrm{i})^2}{2}\int \mathrm{d}^4 y \mathrm{d}^4 w\, \hat{\mathcal{H}}_I(y)\hat{\mathcal{H}}_I(w) + \dots\right] \qquad (19.9) \\
&= T\left[1 - \frac{\mathrm{i}\lambda}{4!}\int \mathrm{d}^4 z\, \hat{\phi}(z)^4 + \frac{(-\mathrm{i})^2}{2!}\left(\frac{\lambda}{4!}\right)^2 \int \mathrm{d}^4 y \mathrm{d}^4 w\, \hat{\phi}(y)^4 \hat{\phi}(w)^4 + \dots\right].
\end{aligned}
$$

Step III

Plug the resulting expression for the $\hat{S}$-operator into the expression for the S-matrix element that we're trying to calculate

$$\mathcal{A} = \langle q|\hat{S}|p\rangle$$

$$= (2\pi)^3 (2E_q)^{\frac{1}{2}} (2E_p)^{\frac{1}{2}} T\left[\langle 0|\hat{a}_q \hat{a}_p^\dagger|0\rangle + \left(\frac{-i\lambda}{4!}\right) \int d^4z \, \langle 0|\hat{a}_q \hat{\phi}(z)^4 \hat{a}_p^\dagger|0\rangle\right.$$

$$\left. + \frac{(-i)^2}{2!} \left(\frac{\lambda}{4!}\right)^2 \int d^4y d^4w \, \langle 0|\hat{a}_q \hat{\phi}(y)^4 \hat{\phi}(w)^4 \hat{a}_p^\dagger|0\rangle + \dots\right].$$

$$(19.10)$$

Since the amplitude is a sum of terms ordered by powers of λ, it makes sense to label the terms $\mathcal{A} = \mathcal{A}^{(0)} + \mathcal{A}^{(1)} + \mathcal{A}^{(2)} \dots$, where $\mathcal{A}^{(n)}$ is the term proportional to λ^n.

Step IV

Use Wick's theorem to grind down the terms. That is, for each term we write the string of operators bookended by the vacuum states as a sum of all of the contractions of pairs.

Let's first consider $\mathcal{A}^{(1)}$, the first-order part of the expansion (the part proportional to λ). Using Wick's theorem on the string $\langle 0|\hat{a}_q \hat{\phi}(z)^4 \hat{a}_p^\dagger|0\rangle \equiv \langle 0|\hat{a}_q \hat{\phi}(z)\hat{\phi}(z)\hat{\phi}(z)\hat{\phi}(z)\hat{a}_p^\dagger|0\rangle$, will yield up two sorts of term. The first contains contractions of the $\hat{a}$-operators and the $\hat{\phi}$-operators separately, like[3]

$$\langle 0|\hat{a}_q \hat{\phi}(z)\hat{\phi}(z)\hat{\phi}(z)\hat{\phi}(z)\hat{a}_p^\dagger|0\rangle \qquad (19.11)$$

$$= \langle 0|\hat{a}_q \hat{a}_p^\dagger|0\rangle \langle 0|T\hat{\phi}(z)\hat{\phi}(z)|0\rangle \langle 0|T\hat{\phi}(z)\hat{\phi}(z)|0\rangle.$$

There are two other combinations of contractions of the $\hat{\phi}$-fields that give this term, making three in total.

The second type of term contracts $\hat{a}$-operators with $\hat{\phi}$ operators. There are twelve ways of doing this. One example is

$$\langle 0|\hat{a}_q \hat{\phi}(z)\hat{\phi}(z)\hat{\phi}(z)\hat{\phi}(z)\hat{a}_p^\dagger|0\rangle \qquad (19.12)$$

$$= \langle 0|\hat{a}_q \hat{\phi}(z)|0\rangle \langle 0|T\hat{\phi}(z)\hat{\phi}(z)|0\rangle \langle 0|\hat{\phi}(z)\hat{a}_p^\dagger|0\rangle.$$

The result of this set of manipulations is that $\mathcal{A}^{(1)}$, the first-order term in the expansion of the S-matrix element for $\langle q|\hat{S}|p\rangle$, is given by

$$\frac{-i\lambda}{4!}(2\pi)^3 (4E_q E_p)^{\frac{1}{2}} \int d^4z \left[3\langle 0|\hat{a}_q \hat{a}_p^\dagger|0\rangle \langle 0|\hat{\phi}(z)\hat{\phi}(z)|0\rangle \langle 0|\hat{\phi}(z)\hat{\phi}(z)|0\rangle\right.$$

$$\left. + 12\langle 0|\hat{a}_q \hat{\phi}(z)|0\rangle \langle 0|T\hat{\phi}(z)\hat{\phi}(z)|0\rangle \langle 0|\hat{\phi}(z)\hat{a}_p^\dagger|0\rangle\right]. \qquad (19.13)$$

Here are some rules for making sense of the contractions:

- Contractions between two fields make a free propagator[4] for that field, that is, $\hat{\phi}(y)\hat{\phi}(z) = \langle 0|T\hat{\phi}(y)\hat{\phi}(z)|0\rangle = \Delta(y-z)$.

---

[3]Note that contractions involving $\hat{a}$-operators don't need T symbols since here the $\hat{a}_p^\dagger$-operators are understood to create particles at $t = -\infty$ only, while the $\hat{a}_p$-operators destroy particles at $t = \infty$ only, which fixes the time ordering unambiguously.

[4]Remember that for a scalar field $\phi(x) = \phi^\dagger(x)$ and

$$\Delta(y-z) = \int \frac{d^4q}{(2\pi)^4} \frac{ie^{-iq\cdot(y-z)}}{q^2 - m^2 + i\epsilon},$$

where q is a dummy momentum, over which we integrate.

- The contraction between a field and the creation operator from the initial particle state $\hat{\phi}(z)\hat{a}_{\boldsymbol{p}}^\dagger$ corresponds to a factor $\dfrac{1}{(2\pi)^{\frac{3}{2}}}\dfrac{1}{(2E_{\boldsymbol{p}})^{\frac{1}{2}}}\mathrm{e}^{-\mathrm{i}p\cdot z}$.

Example 19.2

To see this, we use the expansion on the field $\hat{\phi}(z)$:

$$
\begin{aligned}
\langle 0|\hat{\phi}(z)\hat{a}_{\boldsymbol{p}}^\dagger|0\rangle &= \int \frac{\mathrm{d}^3 q}{(2\pi)^{\frac{3}{2}}}\frac{1}{(2E_{\boldsymbol{q}})^{\frac{1}{2}}}\langle 0|\left(\hat{a}_{\boldsymbol{q}}\mathrm{e}^{-\mathrm{i}q\cdot z}+\hat{a}_{\boldsymbol{q}}^\dagger \mathrm{e}^{\mathrm{i}q\cdot z}\right)\hat{a}_{\boldsymbol{p}}^\dagger|0\rangle \\
&= \int \frac{\mathrm{d}^3 q}{(2\pi)^{\frac{3}{2}}}\frac{1}{(2E_{\boldsymbol{q}})^{\frac{1}{2}}}\langle 0|\left(\hat{a}_{\boldsymbol{q}}\mathrm{e}^{-\mathrm{i}q\cdot z}+\hat{a}_{\boldsymbol{q}}^\dagger \mathrm{e}^{\mathrm{i}q\cdot z}\right)|\boldsymbol{p}\rangle \\
&= \int \frac{\mathrm{d}^3 q}{(2\pi)^{\frac{3}{2}}}\frac{1}{(2E_{\boldsymbol{q}})^{\frac{1}{2}}}\mathrm{e}^{-\mathrm{i}q\cdot z}\delta^{(3)}(\boldsymbol{q}-\boldsymbol{p}) \\
&= \frac{1}{(2\pi)^{\frac{3}{2}}}\frac{1}{(2E_{\boldsymbol{p}})^{\frac{1}{2}}}\mathrm{e}^{-\mathrm{i}p\cdot z}. \qquad (19.14)
\end{aligned}
$$

Notice that, following our conventions, the factor $\mathrm{e}^{-\mathrm{i}p\cdot z}$ corresponds to an incoming particle. Notice also that the factors $\dfrac{1}{(2\pi)^{\frac{3}{2}}}\dfrac{1}{(2E_{\boldsymbol{p}})^{\frac{1}{2}}}$ from the contraction cancel against the factors from the relativistic normalization of the states $|p\rangle = (2\pi)^{\frac{3}{2}}(2E_{\boldsymbol{p}})^{\frac{1}{2}}|\boldsymbol{p}\rangle$, with the result that the contraction $\hat{\phi}(x)|p\rangle = \mathrm{e}^{-\mathrm{i}p\cdot x}|p\rangle$.

- Similarly, the contraction between a final state annihilation operator and a field $\hat{a}_{\boldsymbol{q}}\hat{\phi}(z)$ corresponds to a factor $\dfrac{1}{(2\pi)^{\frac{3}{2}}}\dfrac{1}{(2E_{\boldsymbol{q}})^{\frac{1}{2}}}\mathrm{e}^{\mathrm{i}q\cdot z}$. This is the outgoing particle.

- Contractions between initial and final particles $\hat{a}_{\boldsymbol{q}}\hat{a}_{\boldsymbol{p}}^\dagger = \langle 0|\hat{a}_{\boldsymbol{q}}\hat{a}_{\boldsymbol{p}}^\dagger|0\rangle$ yield a delta function $\delta^{(3)}(\boldsymbol{q}-\boldsymbol{p})$.

Example 19.3

As an example of the use of the rules, consider the term:

$$
12\langle 0|\hat{a}_{\boldsymbol{q}}\hat{\phi}(z)\hat{\phi}(z)\hat{\phi}(z)\hat{a}_{\boldsymbol{p}}^\dagger|0\rangle.
$$

Using the rules, this is represented by

$$
12\left[\frac{1}{(2\pi)^{\frac{3}{2}}}\frac{1}{(2E_{\boldsymbol{q}})^{\frac{1}{2}}}\mathrm{e}^{\mathrm{i}q\cdot z}\right]\left[\int \frac{\mathrm{d}^4 k}{(2\pi)^4}\frac{\mathrm{i}\,\mathrm{e}^{-\mathrm{i}k\cdot(z-z)}}{k^2-m^2+\mathrm{i}\epsilon}\right]\left[\frac{1}{(2\pi)^{\frac{3}{2}}}\frac{1}{(2E_{\boldsymbol{p}})^{\frac{1}{2}}}\mathrm{e}^{-\mathrm{i}p\cdot z}\right]. \quad (19.15)
$$

Remember that to compute the contribution of this term to the S-matrix element we will need to integrate the variable z over all space. That is, on including the normalization factors, we have that $\mathcal{A}$, which is what we'll call the contribution to the amplitude $\langle q|\hat{S}^{(1)}|p\rangle$ from the above contraction, is given by

$$
\begin{aligned}
\mathcal{A} &= 12\times (2\pi)^3 (2E_{\boldsymbol{q}})^{\frac{1}{2}}(2E_{\boldsymbol{p}})^{\frac{1}{2}}\frac{(-\mathrm{i}\lambda)}{4!}\int \mathrm{d}^4 z\left[\frac{1}{(2\pi)^{\frac{3}{2}}}\frac{1}{(2E_{\boldsymbol{q}})^{\frac{1}{2}}}\mathrm{e}^{\mathrm{i}q\cdot z}\right.\\
&\quad \times \left.\frac{\mathrm{d}^4 k}{(2\pi)^4}\frac{\mathrm{i}\,\mathrm{e}^{-\mathrm{i}k\cdot(z-z)}}{k^2-m^2+\mathrm{i}\epsilon}\frac{1}{(2\pi)^{\frac{3}{2}}}\frac{1}{(2E_{\boldsymbol{p}})^{\frac{1}{2}}}\mathrm{e}^{-\mathrm{i}p\cdot z}\right] \\
&= \frac{(-\mathrm{i}\lambda)}{2}\int \mathrm{d}^4 z\,\frac{\mathrm{d}^4 k}{(2\pi)^4}\mathrm{e}^{\mathrm{i}q\cdot z}\frac{\mathrm{i}}{k^2-m^2+\mathrm{i}\epsilon}\mathrm{e}^{-\mathrm{i}p\cdot z}. \qquad (19.16)
\end{aligned}
$$

Notice that the normalization factors of (2π) and $(2E_{\boldsymbol{p}})$ have cancelled. This was one intention of the conventions for decorating the earlier equation with these factors.

We could, at this point, collect the terms together and prepare to integrate the position of the interactions over all spacetime. However, there's still one more part of the plan to carry out.

Step V

Make sense of a term by drawing a Feynman diagram. A Feynman diagram represents an amplitude in the expansion of the S-matrix. A particular term in the expansion will comprise a certain number of interactions. This number is the same as the order of expansion of the $\hat{S}$-operator. A second-order term has two interactions, a third-order three, etc. To represent the interaction in a diagram we draw the interaction vertices exactly as shown in Fig. 19.2. Each line segment (we will call these *legs*) emerging from an interaction vertex represents an uncontracted field operator $\hat{\phi}(z)$. The Wick contractions (that we carried out in step IV) join the legs of the interaction to each other [for contractions like $\overset{\frown}{\hat{\phi}(x)\hat{\phi}(y)}$] or to external particles (for contractions like $\overset{\frown}{\hat{\phi}(x)\hat{a}_{\boldsymbol{p}}^{\dagger}}$).

Here is a list of rules for drawing a diagram.

- Draw the interaction vertices and label them with their spacetime coordinates [Fig. 19.3(a)].

- Contractions between the initial state and a field, i.e. $\overset{\frown}{\hat{\phi}(x)\hat{a}_{\boldsymbol{p}}^{\dagger}}$, are drawn as incoming lines connecting to one of the legs of the vertex [Fig. 19.3(b)]. This corresponds to a real, on-mass-shell particle coming into the story.

- Propagators resulting from the field–field contractions $\overset{\frown}{\hat{\phi}(x)\hat{\phi}(y)}$ are drawn as lines linking the points [Fig. 19.3(c)]. We can think of these as virtual particles which are internal to the story the diagram is telling.

- Contractions between the final state and a field $\overset{\frown}{\hat{a}_{\boldsymbol{q}}\hat{\phi}(x)}$ are drawn as an outgoing lines [Fig. 19.3(d)]. These correspond to on-mass-shell outgoing particles.

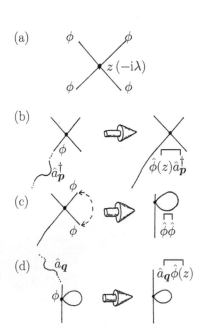

Fig. 19.3 Steps in drawing a diagram.

Example 19.4

Let's again consider the term

$$-\frac{12\mathrm{i}\lambda}{4!}\int \mathrm{d}^4z\langle 0|\overset{\frown}{\hat{a}_{\boldsymbol{q}}\hat{\phi}(z)}\,\overset{\frown}{\hat{\phi}(z)\hat{\phi}(z)}\,\overset{\frown}{\hat{\phi}(z)\hat{a}_{\boldsymbol{p}}^{\dagger}}|0\rangle. \tag{19.17}$$

The process of drawing a diagram is shown in Fig. 19.3.

- It's a first-order diagram (it contains one copy of $\mathcal{H}_\mathrm{I}$) so we draw one interaction vertex at position z [Fig. 19.3(a)].

- The contraction $\hat{\phi}(z)\hat{a}_{\boldsymbol{p}}^{\dagger}$ is represented by an incoming line that grabs one of the vertex legs [Fig. 19.3(b)].

- The contraction $\hat{\phi}(z)\hat{\phi}(z)$ ties together two vertex legs [Fig. 19.3(c)].

- The contraction $\hat{a}_{\boldsymbol{q}}\hat{\phi}(z)$ grabs the remaining vertex stub and forms a line leaving the diagram [Fig. 19.3(d)].

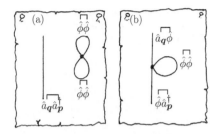

Fig. 19.4 First-order contributions in $\hat{\phi}^4$ theory. (a) A disconnected diagram with two connected parts. (b) Connected diagram with one loop.

The diagrams that result from our Wick expansion of the S-matrix to first order are shown in Fig. 19.4. The expression for eqn 19.11 is shown in Fig. 19.4(a), that for eqn 19.12 is shown in Fig. 19.4(b).

19.3 Anatomy of a diagram

The diagrams we've drawn so far could correspond to stories describing the interaction of particles if you think of time running upwards. A contraction gives you a full diagram. This may be made of many pieces. Figure 19.4(a) is a diagram made of two pieces, while Fig. 19.4(b) is a diagram made of a single piece. We call a piece of a diagram a *connected diagram*. Figure 19.4(a) is a *disconnected diagram* made up of two *connected diagrams*. Disconnected processes cannot influence each other (because they are not connected to each other!) and so physical intuition tells us that we will probably only have to consider connected diagrams (that expectation will turn out to be right on the money).[5] There's more vocabulary to be learnt in order to talk diagrams with the professionals:

- *External lines* have one end which appears not to be connected to anything. (Actually, they indicate a connection with the world exterior to the physical process.) Figure 19.4(b) or the left-hand piece of Fig. 19.4(a) are connected diagrams with external lines.

- A *vacuum diagram* has no external lines. The right-hand piece of Fig. 19.4(a) is a (connected) vacuum diagram. Vacuum diagrams don't affect transition probabilities because they are not connected to the incoming or outgoing particles; they only affect $\langle 0|\hat{S}|0\rangle$ (and therefore only contribute a phase $e^{i\phi}$ to transition amplitudes).

A particular connected diagram might contain:

- *Vertices* where lines join together. These represent interactions.

- External *incoming lines* (we draw these below the vertices). They represent on-mass-shell particles entering the process.

- External *outgoing lines* (we draw these above the vertices). They represent on-mass-shell particles leaving the process.

- *Internal lines* (joining two vertices). These represent virtual particles which are off-mass-shell and therefore exist internally within the diagram.

[5]The **cluster decomposition principle** guarantees that distant experiments yield uncorrelated results. An S-matrix constructed out of creation and annihilation operators turns out automatically to satisfy this principle, which therefore provides the deep reason why these operators are *required* in quantum field theory, rather than just being a convenience. The cluster decomposition principle may be justified for connected and disconnected Feynman diagrams using the linked cluster theorem, discussed in Chapter 22. See Weinberg, Chapter 4 for a detailed discussion.

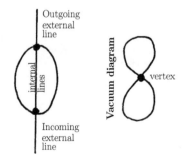

Fig. 19.5 The anatomy of a disconnected third-order diagram.

These various properties are illustrated in Fig. 19.5.

Why do we draw Feynman diagrams? Simple: knowing how the diagrams relate to the contractions means that we can simply draw diagrams for a particular interaction and, instead of going through all of the leg work of doing an expansion, we just write down the equation to which a diagram corresponds. If someone were now just to give us a Feynman diagram, how could we translate it into a term in the expansion? Here are the rules:

Remember that these are the rules for ϕ^4 theory. Other theories will have slightly different rules.

Feynman rules for ϕ^4 theory in position space

To calculate an amplitude in the S-matrix expansion, translate a Feynman diagram into equations as follows:

- Each vertex contributes a factor $-i\lambda$.
- Each internal line gives a propagation factor $\Delta(x-y)$, where x and y are the start and end points of the line.
- External lines contribute incoming $(-ip \cdot x)$ or outgoing $(+ip \cdot x)$ waves $e^{\pm ip \cdot x}$.
- Integrate the positions of the vertices over all spacetime.
- In order to get the right coefficient in front of the term divide by the symmetry factor D.

This last point is treated in the following section.

19.4   Symmetry factors

A potentially tricky thing is how to work out what number D we divide by in working out our Feynman diagram amplitude. The number arises from the number of ways there are to produce a certain diagram through contractions, divided by 4! (In fact the reason for the 4! is to simplify this number as much as possible.) Rather than deriving them we'll quote the result which the interested (and/or masochistic) reader can easily justify by working through the combinatorics.[6] The general rule is as follows: *if there are m ways of arranging vertices and propagators to give identical parts of a diagram (keeping the outer ends of external lines fixed and without cutting propagator lines) we get a factor $D_i = m$.* The symmetry factor is given by the product of all symmetry factors $D = \prod_i D_i$.

Two very useful special cases, which aren't immediately apparent from the general rule, are as follows:

- Every propagator with two ends joined to one vertex (a loop) gives a factor $D_i = 2$.
- Every pair of vertices directly joined by n propagators gives a factor $D_i = n!$

[6]Often the symmetry factors aren't required to understand the physics of what a diagram is telling you, but when combining several diagrams it's useful to know them.

Example 19.5

Figure 19.6 shows several diagrams to which we can apply the rules.

- (a) The bubble propagator has $D = 2$ since it contains a propagator with two ends joined to a vertex. Remember you are not allowed to move the ends of external lines, so there are no more factors to worry about.

- (b) The two-bubble propagator has $D = 4$. Each bubble gives a factor of 2, so $D = 2 \times 2$. The bubble propagators in (a) and (b) are examples of **self-energy diagrams**. They are basically free propagators with extra loops in, but don't interact with anything else. We will see later that these extra loops just change the constants in the free propagator, or to use the lingo of quantum field theory, they 'renormalize the free propagator'.

- (c) The double-bubble vacuum graph has two bubbles (or loops), each contributing a factor 2, but the bubbles can be swapped over (rotating the diagram 180° about a vertical axis passing through the middle of it), and we get the same diagram, giving us an extra factor $m = 2$. Therefore, we have $D = 8$.

- The Saturn diagram (d) has a pair of vertices joined by three lines, so $D = 3! = 6$.

- Diagram (e) has a factor of 2 from the bubble. It also has two vertices joined by two lines, contributing a factor 2!, so $D = 2 \times 2! = 4$.

- Diagram (f) has three vertices. We think of these as being grouped in two pairs (with the middle vertex counted twice). Each pair is joined together by two lines (and so each contributes a factor 2!). In total we therefore have $D = 2 \times 2! = 4$.

- Diagram (g) has a bubble, so $D = 2$.

- Diagram (h) has two bubbles but the top and bottom parts between the vertices can be swapped giving an extra factor of 2, so $D = 2 \times 2 \times 2 = 8$.

- Diagram (i) has a pair of vertices joined by two lines so $D = 2$.

Other theories will have slightly different rules for their symmetry factors.

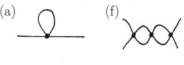

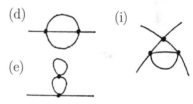
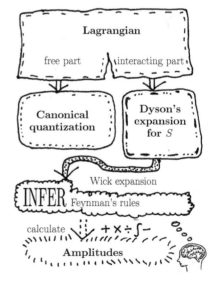

Fig. 19.6 Examples of diagrams in ϕ^4 theory.

The derivation of the Feynman rules represents the point where we can start calculating scattering amplitudes for physical processes. A large part of formulating a useful quantum field theory may be reduced to the process of deriving the Feynman rules following the path shown in Fig. 19.7.

19.5   Calculations in p-space

Calculations turn out to be far easier in momentum space, as illustrated by the following example: the second-order 'Saturn diagram'.

Fig. 19.7 The process of deriving the Feynman rules.

Example 19.6

Let's see how the Saturn diagram arises. Consider the second-order $O(\lambda^2)$ term in the expansion of $\langle q|\hat{S}|p\rangle$ (i.e. two copies of $\hat{\mathcal{H}}_I(x)$ in the expansion),

$$
\begin{aligned}
\hat{S}^{(2)} &= \frac{(-\mathrm{i})^2}{2!} T \int \mathrm{d}^4 y\, \mathrm{d}^4 w\, \hat{\mathcal{H}}_I(y)\hat{\mathcal{H}}_I(w) \\
&= \frac{(-\mathrm{i}\lambda)^2}{2!(4!)^2} T \int \mathrm{d}^4 y\, \mathrm{d}^4 w\, \hat{\phi}(y)\hat{\phi}(y)\hat{\phi}(y)\hat{\phi}(y)\hat{\phi}(w)\hat{\phi}(w)\hat{\phi}(w)\hat{\phi}(w). \quad (19.18)
\end{aligned}
$$

(a) 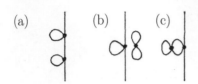 (b) (c)

(d) 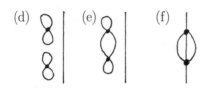 (e) (f)

Fig. 19.8   Examples of second-order ϕ^4 diagrams.

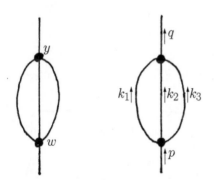

Fig. 19.9 (a) Position and (b) momentum space Feynman diagrams.

We now apply Wick's theorem and obtain the diagrams shown in Fig. 19.8. The one in which we're interested is Fig. 19.8(f), which comes from the term

$$\langle q|\hat{S}^{(2)}|p\rangle = (2\pi)^3 (2E_q)^{\frac{1}{2}} (2E_p)^{\frac{1}{2}}$$

$$\times \frac{(-i\lambda)^2}{D} \int d^4y d^4w \, \langle 0|\hat{a}_q \overset{\frown}{\hat{\phi}(y)} \hat{\phi}(y) \hat{\phi}(y) \hat{\phi}(w) \hat{\phi}(w) \hat{\phi}(w) \hat{a}_p^\dagger|0\rangle, \quad (19.19)$$

where $D = 6$ is the symmetry factor here. The amplitude for the Saturn Feynman diagram in Fig. 19.9(a) is given by

$$-\frac{\lambda^2}{6} \int d^4y d^4w \, e^{iq\cdot y} \Delta(y-w)\Delta(y-w)\Delta(y-w)e^{-ip\cdot w}. \quad (19.20)$$

Each spacetime propagator Δ gives us a factor of

$$\Delta(y-w) = \int \frac{d^4k}{(2\pi)^4} e^{-ik\cdot(y-w)} \frac{i}{k^2 - m^2 + i\epsilon}. \quad (19.21)$$

So, upon collecting all of the exponentials together, we have

$$\text{Saturn} = -\frac{\lambda^2}{6} \int \frac{d^4k_1}{(2\pi)^4} \frac{d^4k_2}{(2\pi)^4} \frac{d^4k_3}{(2\pi)^4} \frac{i}{(k_1^2 - m^2 + i\epsilon)} \frac{i}{(k_2^2 - m^2 + i\epsilon)} \frac{i}{(k_3^2 - m^2 + i\epsilon)}$$

$$\times \int d^4y d^4w \left(e^{iq\cdot y}e^{-i(k_1+k_2+k_3)\cdot y}\right) \left(e^{i(k_1+k_2+k_3)\cdot w}e^{-ip\cdot w}\right).$$

Next we notice that some of the integrals will give us delta functions, which simplify things considerably. Integrating y over all space gives us an additional factor of $\int d^4y \, e^{-i(k_1+k_2+k_3-q)\cdot y} = (2\pi)^4\delta^{(4)}(k_1+k_2+k_3-q)$ and we have

$$\text{Saturn} = -\frac{\lambda^2}{6} \int \frac{d^4k_1}{(2\pi)^4} \frac{d^4k_2}{(2\pi)^4} \frac{d^4k_3}{(2\pi)^4} \frac{i}{(k_1^2 - m^2 + i\epsilon)} \frac{i}{(k_2^2 - m^2 + i\epsilon)} \frac{i}{(k_3^2 - m^2 + i\epsilon)}$$

$$\times \int d^4w \left(e^{i(k_1+k_2+k_3)\cdot w}e^{-ip\cdot w}\right) (2\pi)^4\delta^{(4)}(k_1+k_2+k_3-q),$$

which means that we may set $k_1 = q - k_2 - k_3$ upon doing the k_1 momentum integral. We see that the integral has given us a constraint on the allowed values of the momenta. Translated into real life, this means that the sum of four-momenta at each vertex is zero. Energy-momentum is therefore conserved at each interaction vertex. So now we have

$$\text{Saturn} = -\frac{\lambda^2}{6} \int \frac{d^4k_2}{(2\pi)^4} \frac{d^4k_3}{(2\pi)^4} \frac{i}{[(q-k_2-k_3)^2 - m^2 + i\epsilon]} \frac{i}{(k_2^2 - m^2 + i\epsilon)} \frac{i}{(k_3^2 - m^2 + i\epsilon)}$$

$$\times \int d^4w \left(e^{iq\cdot w}e^{-ip\cdot w}\right).$$

The w integration gives $\int d^4w \, e^{i(q-p)\cdot w} = (2\pi)^4\delta^{(4)}(q-p)$. The overall energy-momentum is conserved and the entire diagram carries an overall energy-momentum conserving delta function around with it. We're left with two momentum integrals to do. Finally we arrive at the momentum space result:

$$\text{Saturn} = \frac{-\lambda^2}{6}(2\pi)^4\delta^{(4)}(q-p)\int \frac{d^4k_2}{(2\pi)^4}\frac{d^4k_3}{(2\pi)^4} \frac{i}{[(q-k_2-k_3)^2 - m^2 + i\epsilon]}$$

$$\times \frac{i}{(k_2^2 - m^2 + i\epsilon)} \frac{i}{(k_3^2 - m^2 + i\epsilon)}. \quad (19.22)$$

Although we've ended up with the same number of integrals as with the position space version, we've expunged all of the exponential factors. In addition, since the energy-momentum conserving delta function is common to all diagrams, we often choose not to carry it around, but to understand that energy-momentum is implicitly conserved (and reinstate the factor when it's needed in the mathematics!).

We may now write down a set of Feynman rules for ϕ^4 theory in momentum space:

Feynman rules for ϕ^4 theory in momentum space

- Each vertex contributes a factor $-i\lambda$ [Fig. 19.10(a)].
- Label each internal line with a momentum q flowing along it and describe it by a propagator $\frac{i}{q^2 - m^2 + i\epsilon}$ [Fig. 19.10(b)].
- Force the sum of each momentum coming into a vertex to be equal to the momentum leaving it.
- Integrate over unconstrained internal momenta with a measure $\frac{d^4 q}{(2\pi)^4}$.
- External lines contribute a factor 1 [Fig. 19.10(c)].
- Divide by the symmetry factor.
- Include an overall energy-momentum conserving delta function for each diagram.

(a)

$= -i\lambda$

(b)

$= \dfrac{i}{q^2 - m^2 + i\epsilon}$

(c)

$= 1$

$= 1$

Remember that the external lines don't contribute a propagator. In momentum space the external line parts are just $\hat{a}_{\boldsymbol{p}} \hat{\phi}(x) = 1$. Remember also that you only need to integrate over *unconstrained* momenta. That is, only integrate over momenta which are not already determined by the conditions that momentum is conserved both at the vertices and overall for a diagram. (If in doubt you can include a delta function for each vertex and integrate over all momenta.)

Fig. 19.10 Momentum space Feynman rules for ϕ^4 theory.

Example 19.7

Let's now apply our rules to some example Feynman diagrams. Some terms in the expansion of $\langle q|\hat{S}|p\rangle$ up to second order are shown in Fig. 19.11. Applying the rules to some of these yield the following for their contribution to $\langle q|\hat{S}|p\rangle$:

$$(b) \;=\; (2\pi)^4 \delta^{(4)}(q-p) \frac{(-i\lambda)}{2} \int \frac{d^4 k}{(2\pi)^4} \frac{i}{k^2 - m^2 + i\epsilon},$$

$$(f) \;=\; (2\pi)^4 \delta^{(4)}(q-p) \frac{(-i\lambda)}{2} \int \frac{d^4 k_3}{(2\pi)^4} \frac{i}{k_3^2 - m^2 + i\epsilon} \times$$
$$(2\pi)^4 \delta^{(4)}(0) \frac{(-i\lambda)}{8} \int \frac{d^4 k_1}{(2\pi)^4} \frac{d^4 k_2}{(2\pi)^4} \frac{i}{(k_1^2 - m^2 + i\epsilon)} \frac{i}{(k_2^2 - m^2 + i\epsilon)},$$

$$(g) \;=\; (2\pi)^4 \delta^{(4)}(q-p) \times$$
$$\frac{(-i\lambda)^2}{4} \int \frac{d^4 k_1}{(2\pi)^4} \frac{d^4 k_2}{(2\pi)^4} \frac{i}{(k_1^2 - m^2 + i\epsilon)} \frac{i}{(p^2 - m^2 + i\epsilon)} \frac{i}{(k_2^2 - m^2 + i\epsilon)},$$

$$(h) \;=\; (2\pi)^4 \delta^{(4)}(q-p) \times$$
$$\frac{(-i\lambda)^2}{6} \int \frac{d^4 k_1}{(2\pi)^4} \frac{d^4 k_2}{(2\pi)^4} \frac{i}{(k_1^2 - m^2 + i\epsilon)} \frac{i}{(k_2^2 - m^2 + i\epsilon)} \frac{i}{[(p - k_1 - k_2)^2 - m^2 + i\epsilon]}.$$

In summary: the amplitude $\mathcal{A} = {}^{\text{out}}\langle q|p\rangle^{\text{in}} = \langle q|\hat{S}|p\rangle$ may be written as a sum of diagrams. Each diagram stands for an integral. We call

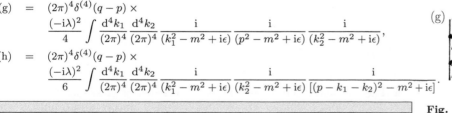
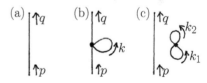
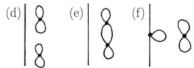

Fig. 19.11 Examples of Feynman diagrams for ϕ^4 theory, up to second order in the interaction strength, for the S-matrix element $\langle q|\hat{S}|p\rangle$.

diagrams which describe how interactions affect the amplitudes of single particles self-energy diagrams.

We haven't yet discussed how to finally do the integral to finally get the numbers out. This is because often the integrals give us divergent (i.e. infinite) results! Taming these fearsome infinities reveals a huge amount about the physics lying behind quantum field theory and we discuss that in Chapter 32. Some diagrams don't diverge, of course, and we examine some of these in Chapter 20. For now, we need some more experience getting used to how the expansions go.

19.6　A first look at scattering

We have looked at diagrams describing single particles. What about a process of two particles entering and then leaving? The amplitude we're after is

$$\langle q_1 q_2 | \hat{S} | p_2 p_1 \rangle = (2\pi)^6 (16 E_{\boldsymbol{p}_1} E_{\boldsymbol{p}_2} E_{\boldsymbol{q}_1} E_{\boldsymbol{q}_2})^{\frac{1}{2}} \langle 0 | \hat{a}_{\boldsymbol{q}_1} \hat{a}_{\boldsymbol{q}_2} \hat{S} \hat{a}^\dagger_{\boldsymbol{p}_2} \hat{a}^\dagger_{\boldsymbol{p}_1} | 0 \rangle, \tag{19.23}$$

describing particles entering in momentum states p_1 and p_2 and leaving in states q_1 and q_2. We could of course now start the process again and work through all of the contractions. However, this is now unnecessary since we have Feynman diagrams for the theory.

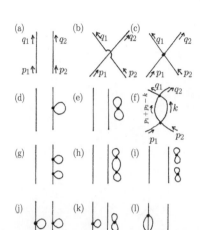

Fig. 19.12 Examples of Feynman diagrams for the S-matrix element $\langle q_1 q_2 | \hat{S} | p_2 p_1 \rangle$ up to second order in the interaction strength.

Example 19.8

Some diagrams contributing to the matrix element $\langle q_1 q_2 | \hat{S} | p_2 p_1 \rangle$ are shown in Fig. 19.12. The first two yield up

$$\text{Diagram (a)} \quad = \quad (2\pi)^6 \, (16 E_{\boldsymbol{p}_1} E_{\boldsymbol{p}_2} E_{\boldsymbol{q}_1} E_{\boldsymbol{q}_2})^{\frac{1}{2}} \, \delta^{(3)}(\boldsymbol{q}_1 - \boldsymbol{p}_1) \, \delta^{(3)}(\boldsymbol{q}_2 - \boldsymbol{p}_2),$$

$$\text{Diagram (b)} \quad = \quad (2\pi)^6 \, (16 E_{\boldsymbol{p}_1} E_{\boldsymbol{p}_2} E_{\boldsymbol{q}_1} E_{\boldsymbol{q}_2})^{\frac{1}{2}} \, \delta^{(3)}(\boldsymbol{q}_1 - \boldsymbol{p}_2) \, \delta^{(3)}(\boldsymbol{q}_2 - \boldsymbol{p}_1).$$

These clearly aren't very interesting since no interactions occur. They just arise from the zeroth-order part of the $\hat{S}$-operator, $\hat{S}^{(0)} = 1$. They don't contribute to scattering, and therefore aren't measured in an experiment. We'll ignore them.

Some more interesting examples of amplitudes are the following:

$$\text{Diagram (c)} \quad = \quad (2\pi)^4 \delta^{(4)}(q_1 + q_2 - p_1 - p_2)(-\mathrm{i}\lambda),$$

$$\text{Diagram (f)} \quad = \quad (2\pi)^4 \delta^{(4)}(q_1 + q_2 - p_1 - p_2) \frac{(-\mathrm{i}\lambda)^2}{2}$$

$$\times \quad \int \frac{\mathrm{d}^4 k}{(2\pi)^4} \frac{\mathrm{i}}{[k^2 - m^2 + \mathrm{i}\epsilon]} \frac{\mathrm{i}}{[(p_1 + p_2 - k)^2 - m^2 + \mathrm{i}\epsilon]}.$$

Chapter summary

- The S-matrix may be expanded in a five-step process.
- The terms in the expansion can be encoded in Feynman diagrams.
- It's easiest to work in momentum space.

Exercises

(19.1) Write down the momentum space amplitudes for the processes shown in the Feynman diagrams in Fig. 19.6.

(19.2) (a) Draw the interaction vertex for ϕ^3 theory, which is a scalar field theory with Lagrangian $\mathcal{L} = \frac{1}{2}(\partial_\mu \phi)^2 - \frac{m^2}{2}\phi^2 - \frac{\eta}{3!}\phi^3$.
(b) By expanding the S-matrix find the contributions to the amplitude $\langle q|\hat{S}|p \rangle$ up to second order in the interaction strength. Draw the corresponding Feynman diagrams. What are the symmetry factors?
(c) Draw the new, connected diagrams that contribute to $\langle q|\hat{S}|p \rangle$ when you consider the fourth-order contribution.
(d) Which connected diagrams contribute to the two-particle S-matrix element $\langle q_1 q_2|\hat{S}|p_2 p_1 \rangle$ at fourth order in the interaction.
For simplicity, just draw topologically distinct diagrams, ignoring new ones produced by permutations of external lines.

(19.3) Consider the ABA theory, defined by the Lagrangian

$$\mathcal{L} = \frac{1}{2}(\partial_\mu \phi_A)^2 - \frac{m_A^2}{2}\phi_A^2 \qquad (19.24)$$
$$+ \frac{1}{2}(\partial_\mu \phi_B)^2 - \frac{m_B^2}{2}\phi_B^2 - \frac{g}{2}\phi_A \phi_B \phi_A.$$

This theory describes the interaction of two scalar fields.

(a) Draw the interaction vertex using different styles of line for the different fields.
(b) Draw the Feynman diagrams that contribute to the scattering amplitude $\langle q_A|\hat{S}|p_A \rangle$ up to fourth order (here p_A refers to an A-particle in momentum state p). Write expressions for the amplitudes for these diagrams. You may ignore symmetry factors.
(c) Draw the Feynman diagrams that contribute to the scattering $\langle q_B|\hat{S}|p_B \rangle$ up to fourth order in the interaction. Write expressions for the amplitudes for these diagrams. Again, you may ignore symmetry factors.
(d) Draw Feynman diagrams that contribute to the A-particle scattering $\langle q_{A1} q_{A2}|\hat{S}|p_{A2} p_{A1} \rangle$ up to fourth order.
(e) What are the rules for the symmetry factors of this theory?

(19.4) (a) Show that the amplitude for the double bubble diagram [Fig. 19.6(c)] is given by

$$\mathcal{A} = \frac{-i\lambda}{8}\left[\int \frac{d^4 p}{(2\pi)^4} \frac{i}{p^2 - m^2 + i\epsilon} \right]^2 \int d^4 x.$$
$$(19.25)$$

Here the $\int d^4 x$ factor gives us a quantity proportional to the volume of the system $\mathcal{V}$ multiplied by the total time T. *This factor is infinite, but we will show later that this will not worry us.*
(b) Argue that this factor arises for all vacuum diagrams.

Scattering theory

And the LORD shall cause his glorious voice to be heard, and shall shew the lighting down of his arm, ... with scattering, and tempest, and hailstones.
Isaiah 30:30

One important application of quantum field theory is the calculation of scattering cross-sections. It is important since scattering cross-sections are measured in many experiments. In this chapter we examine scattering using Hideki Yukawa's $\psi^\dagger\psi\phi$ theory, which is an illuminating toy model describing the interactions of scalar fields. The joy of this model is that it bears a strong resemblance to quantum electrodynamics (QED), which describes electrodynamics to an astounding degree of accuracy. Specifically, in $\psi^\dagger\psi\phi$ theory the phion excitations in the ϕ-field take the role of (massive, scalar) photons and the psion excitations in the ψ-field describe (complex scalar, bosonic) electrons.

20.1 Another theory: Yukawa's $\psi^\dagger\psi\phi$ interactions

The theory describes a complex scalar field ψ and a real scalar field ϕ interacting. The full Lagrangian for this theory is

$$\mathcal{L} = \partial^\mu\psi^\dagger\partial_\mu\psi - m^2\psi^\dagger\psi + \frac{1}{2}(\partial_\mu\phi)^2 - \frac{1}{2}\mu^2\phi^2 - g\psi^\dagger\psi\phi. \qquad (20.1)$$

Here psions have mass m and phions have mass μ. This Lagrangian is the sum of the free scalar field and free complex scalar fields Lagrangian with an interaction part, shown in Fig. 20.1, given by $\mathcal{L}_I = -g\psi^\dagger\psi\phi$. We start by writing down the mode expansions of each of the free fields:

$$\hat{\psi}(x) = \int \frac{\mathrm{d}^3 p}{(2\pi)^{\frac{3}{2}}} \frac{1}{(2E_{\boldsymbol{p}})^{\frac{1}{2}}} \left(\hat{a}_{\boldsymbol{p}}\mathrm{e}^{-\mathrm{i}p\cdot x} + \hat{b}_{\boldsymbol{p}}^\dagger\mathrm{e}^{\mathrm{i}p\cdot x} \right),$$

$$\hat{\psi}^\dagger(x) = \int \frac{\mathrm{d}^3 p}{(2\pi)^{\frac{3}{2}}} \frac{1}{(2E_{\boldsymbol{p}})^{\frac{1}{2}}} \left(\hat{a}_{\boldsymbol{p}}^\dagger\mathrm{e}^{\mathrm{i}p\cdot x} + \hat{b}_{\boldsymbol{p}}\mathrm{e}^{-\mathrm{i}p\cdot x} \right),$$

$$\hat{\phi}(x) = \int \frac{\mathrm{d}^3 q}{(2\pi)^{\frac{3}{2}}} \frac{1}{(2\varepsilon_{\boldsymbol{q}})^{\frac{1}{2}}} \left(\hat{c}_{\boldsymbol{q}}\mathrm{e}^{-\mathrm{i}q\cdot x} + \hat{c}_{\boldsymbol{q}}^\dagger\mathrm{e}^{\mathrm{i}q\cdot x} \right), \qquad (20.2)$$

where $E_{\boldsymbol{p}} = (\boldsymbol{p}^2 + m^2)^{\frac{1}{2}}$ and $\varepsilon_{\boldsymbol{q}} = (\boldsymbol{q}^2 + \mu^2)^{\frac{1}{2}}$. Here the $\hat{a}$-operators describe the creation and annihilation of psions, the $\hat{b}$-operators describe the creation and annihilation of antpsions and the $\hat{c}$-operators

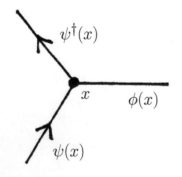

Fig. 20.1 The $\psi^\dagger\psi\phi$ interaction vertex.

create and destroy scalar phions. The interaction Hamiltonian to use in Dyson's equation is[1] $\hat{\mathcal{H}}_I(z) = g\hat{\psi}^\dagger(z)\hat{\psi}(z)\hat{\phi}(z)$, whose interaction diagram is shown in Fig. 20.1. We decorate the interaction diagram with an arrow pointing toward the interaction blob for the $\hat{\psi}$-field and an arrow pointing away for the $\hat{\psi}^\dagger$-field. The meaning of these arrows will be to show particle number flow, as discussed a little later.

To calculate S-matrix elements we'll work through our five-point plan. The most important new feature of this theory is that now we have the possibility of antiparticles in our interactions.

Step I: We decide what to calculate. The one psion in, one psion out amplitude is

$$\mathcal{A} = \langle q|\hat{S}|p\rangle = (2\pi)^3 (2E_q)^{\frac{1}{2}} (2E_p)^{\frac{1}{2}} \langle 0|\hat{a}_q \hat{S}\hat{a}_p^\dagger|0\rangle. \qquad (20.3)$$

Step II: We expand the $\hat{S}$-operator using Dyson's expansion

$$\begin{aligned}\hat{S} = {} & T\left[1 + (-ig)\int d^4z\, \hat{\psi}^\dagger(z)\hat{\psi}(z)\hat{\phi}(z) \right. \qquad (20.4)\\ & \left. + \frac{(-ig)^2}{2!}\int d^4y d^4w\, \left[\hat{\psi}^\dagger(y)\hat{\psi}(y)\hat{\phi}(y)\right]\left[\hat{\psi}^\dagger(w)\hat{\psi}(w)\hat{\phi}(w)\right] + \dots\right]\end{aligned}$$

Step III: The next stage is to plug in to get the amplitudes for the various processes we're going to calculate. All first-order terms will turn out to give zero (try it!) so we'll consider the second-order term

$$\mathcal{A}^{(2)} = \tfrac{(-ig)^2}{2!}(2\pi)^3(2E_p)^{\frac{1}{2}}(2E_q)^{\frac{1}{2}}\int d^4y d^4w\, \langle 0|T\hat{a}_q\hat{\psi}^\dagger(y)\hat{\psi}(y)\hat{\phi}(y)\hat{\psi}^\dagger(w)\hat{\psi}(w)\hat{\phi}(w)\hat{a}_p^\dagger|0\rangle. \qquad (20.5)$$

Step IV: We use Wick to digest the matrix elements. Here are the rules for the nonzero contractions for this theory:

$$\begin{aligned}\overline{\hat{\psi}(x)\hat{\psi}^\dagger(y)} &= \int \tfrac{d^4p}{(2\pi)^4}\tfrac{ie^{-ip\cdot(x-y)}}{p^2-m^2+i\epsilon}, & \overline{\hat{\phi}(x)\hat{\phi}(y)} &= \int \tfrac{d^4q}{(2\pi)^4}\tfrac{ie^{-iq\cdot(x-y)}}{q^2-\mu^2+i\epsilon},\\ \overline{\hat{a}_p\hat{\psi}^\dagger(x)} &= \tfrac{1}{(2\pi)^{\frac{3}{2}}}\tfrac{1}{(2E_p)^{\frac{1}{2}}}e^{ip\cdot x}, & \overline{\hat{\psi}(x)\hat{a}_p^\dagger} &= \tfrac{1}{(2\pi)^{\frac{3}{2}}}\tfrac{1}{(2E_p)^{\frac{1}{2}}}e^{-ip\cdot x},\\ \overline{\hat{b}_p\hat{\psi}(x)} &= \tfrac{1}{(2\pi)^{\frac{3}{2}}}\tfrac{1}{(2E_p)^{\frac{1}{2}}}e^{ip\cdot x}, & \overline{\hat{\psi}^\dagger(x)\hat{b}_p^\dagger} &= \tfrac{1}{(2\pi)^{\frac{3}{2}}}\tfrac{1}{(2E_p)^{\frac{1}{2}}}e^{-ip\cdot x},\\ \overline{\hat{c}_q\hat{\phi}(x)} &= \tfrac{1}{(2\pi)^{\frac{3}{2}}}\tfrac{1}{(2\varepsilon_q)^{\frac{1}{2}}}e^{iq\cdot x}, & \overline{\hat{\phi}(x)\hat{c}_q^\dagger} &= \tfrac{1}{(2\pi)^{\frac{3}{2}}}\tfrac{1}{(2\varepsilon_q)^{\frac{1}{2}}}e^{-iq\cdot x}.\end{aligned} \qquad (20.6)$$

All other contractions give zero. Finally, we'll agree to work in momentum space and note that the symmetry factor for all diagrams in $\hat{\psi}^\dagger\hat{\psi}\hat{\phi}$ theory is $D = 1$.

Example 20.1

Using these contractions let's see what sort of diagrams are thrown up. Consider the second-order term with the following contraction

$$\langle 0|\hat{a}_q\hat{\psi}^\dagger(y)\hat{\psi}(y)\hat{\phi}(y)\hat{\psi}^\dagger(w)\hat{\psi}(w)\hat{\phi}(w)\hat{a}_p^\dagger|0\rangle. \qquad (20.7)$$

[1]Remember that $\hat{\mathcal{H}}_I = -\hat{\mathcal{L}}_I$ so $\hat{\mathcal{L}}_I = -g\hat{\psi}^\dagger\hat{\psi}\hat{\phi}$ implies $\hat{\mathcal{H}}_I = g\hat{\psi}^\dagger\hat{\psi}\hat{\phi}$.

[2] This name was coined by Sidney Coleman. When the journal *Physical Review* objected to the name, Coleman suggested the alternative 'Spermion'. *Physical Review* relented.

(a)

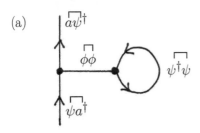

(b)

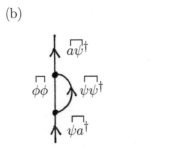

Fig. 20.2 (a) A tadpole diagram. (b) An oyster diagram.

(a)

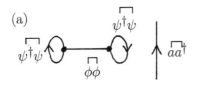

(b)

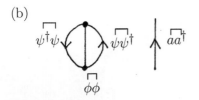

Fig. 20.3 Disconnected diagrams making second-order contributions.

This leads to the Feynman diagram shown in Fig. 20.2(a), which is known as a tadpole.[2] In a theory with no particles in the ground state this diagram gives zero.

Consider another set of contractions:

$$\langle 0|\hat{a}_{\boldsymbol{q}}\hat{\psi}^{\dagger}(y)\hat{\psi}(y)\hat{\phi}(y)\hat{\psi}^{\dagger}(w)\hat{\psi}(w)\hat{\phi}(w)\hat{a}_{\boldsymbol{p}}^{\dagger}|0\rangle. \tag{20.8}$$

This leads to the Feynman diagram shown in Fig. 20.2(b), known as an oyster, which when translated into equations gives an amplitude $\mathcal{A}_{\text{oyster}}$:

$$(-\mathrm{i}g)^2 \int \frac{\mathrm{d}^4k}{(2\pi)^4} \frac{\mathrm{i}}{(k^2-\mu^2+\mathrm{i}\epsilon)} \frac{\mathrm{i}}{[(p-k)^2-m^2+\mathrm{i}\epsilon]} (2\pi)^4\delta^{(4)}(q-p). \tag{20.9}$$

There are also two disconnected contributions involving vacuum diagrams, shown in Fig. 20.3.

There are a few new things to note about the Feynman diagrams in this theory. The first is that we include arrows on the psion and antipsion lines of our Feynman diagrams, but not the phi lines. Psions have arrows going in the direction of 'time' (that is, up the page), antipsions have arrows going against the 'time' direction. These arrows don't represent the direction of momentum. Actually arrows on the lines represent the (conventional) direction of Noether current J_{Nc} (i.e. conserved particle number flow). Motivated by the expression $\hat{Q}_{\text{Nc}} = \int \mathrm{d}^3p\,(\hat{n}_{\boldsymbol{p}}^{(a)} - \hat{n}_{\boldsymbol{p}}^{(b)})$ we say that incoming particles increase particle number and correspond to lines inwards on the diagram whilst incoming antiparticles reduce particle number resulting in outward going lines in the diagram. To avoid confusion, it sometimes helps to draw extra lines showing the directions of momenta (see, e.g. Fig. 20.6). Notice that this implies that an antiparticle has momentum in the opposite direction to its number flow.

20.2 Scattering in the $\psi^{\dagger}\psi\phi$ theory

So far we've only considered one particle coming in and one particle leaving. Now let's consider the process of two psions coming in, scattering, then two psions leaving. The S-matrix element we're after is

$$\langle q_1 q_2|\hat{S}|p_2 p_1\rangle = \langle 0|\hat{a}_{\boldsymbol{q}_1}\hat{a}_{\boldsymbol{q}_2}\hat{S}\hat{a}_{\boldsymbol{p}_2}^{\dagger}\hat{a}_{\boldsymbol{p}_1}^{\dagger}|0\rangle. \tag{20.10}$$

It's fairly obvious that all first-order terms are zero (since it's impossible to draw two-in two-out Feynman diagrams with only one of our interaction vertices). We'll consider the second-order terms.

Example 20.2

From the contraction

$$\langle 0|\hat{a}_{\boldsymbol{q}_1}\hat{a}_{\boldsymbol{q}_2}\hat{\psi}^{\dagger}(y)\hat{\psi}(y)\hat{\phi}(y)\hat{\psi}^{\dagger}(w)\hat{\psi}(w)\hat{\phi}(w)\hat{a}_{\boldsymbol{p}_2}^{\dagger}\hat{a}_{\boldsymbol{p}_1}^{\dagger}|0\rangle, \tag{20.11}$$

we have the diagram shown in Fig. 20.4(a). Two psions enter, one emits a force-carrying, virtual phion which collides with the second psion. The two psions then leave. This is known in particles physics as a *t*-channel process (following notation introduced by Stanley Mandelstam in 1958 in which the three possible processes were assigned the letters '*s*', '*t*' and '*u*').

A closely related process arises from the following contraction

$$\langle 0|\hat{a}_{\boldsymbol{q}_1}\hat{a}_{\boldsymbol{q}_2}\hat{\psi}^\dagger(y)\hat{\psi}(y)\hat{\phi}(y)\hat{\psi}^\dagger(w)\hat{\psi}(w)\hat{\phi}(w)\hat{a}^\dagger_{\boldsymbol{p}_2}\hat{a}^\dagger_{\boldsymbol{p}_1}|0\rangle, \qquad (20.12)$$

represented by the diagram in Fig. 20.4(b) and is known as a *u*-channel process. It's the same as the *t*-channel process, except that the initial and final particles have changed places after the interaction. To obtain the amplitude $\langle q_1 q_2|\hat{S}|p_2 p_1\rangle$ for the scattering of indistinguishable particles at second order add the amplitudes from the *t*- and *u*-channel processes.

What about the following process describing psions interacting with antipsions (denoted $|\bar{p}\rangle$)? The expression

$$\langle \bar{q}_1 q_2|\hat{S}|p_2\bar{p}_1\rangle = \langle 0|\hat{b}_{\boldsymbol{q}_1}\hat{a}_{\boldsymbol{q}_2}\hat{S}\hat{a}^\dagger_{\boldsymbol{p}_2}\hat{b}^\dagger_{\boldsymbol{p}_1}|0\rangle \qquad (20.13)$$

describes a psion and antipsion entering, interacting, then leaving. We can still have the contraction which leads to a process analogous to Fig. 20.4(a), although this will involve antipsion lines with arrows pointing against the direction of time (i.e. down the page) attached to one vertex. In addition, another interesting new contraction possibility occurs here. Look at

$$\langle 0|\hat{b}_{\boldsymbol{q}_1}\hat{a}_{\boldsymbol{q}_2}\hat{\psi}^\dagger(y)\hat{\psi}(y)\hat{\phi}(y)\hat{\psi}^\dagger(w)\hat{\psi}(w)\hat{\phi}(w)\hat{a}^\dagger_{\boldsymbol{p}_2}\hat{b}^\dagger_{\boldsymbol{p}_1}|0\rangle, \qquad (20.14)$$

which corresponds to the diagram in Fig. 20.4(c). This depicts a psion and an antipsion coming in and annihilating. They become a virtual phion which decays into a psion and antipsion. This is known in particle physics as an *s*-channel process.

Note that the *u*-channel process now involves the exchange of the particle and antiparticle, so becomes distinguishable from the *t*- and *s*-channel processes. It therefore contributes to $\langle q_1 \bar{q}_2|\hat{S}|p_2\bar{p}_1\rangle$, rather than $\langle \bar{q}_1 q_2|\hat{S}|p_2\bar{p}_1\rangle$. To obtain the amplitude $\langle \bar{q}_1 q_2|\hat{S}|p_2\bar{p}_1\rangle$ at second order we therefore need to add the amplitudes from the *t*- and *s*-channel processes only.

The idea behind all of this exploration is to write down a set of Feynman rules for this theory, which we now do.

Feynman rules for $\psi^\dagger\psi\phi$ theory

- Each vertex contributes a factor $-ig$ [Fig. 20.5(a)].
- For each phion internal line carrying momentum q include a propagator $\frac{i}{q^2-\mu^2+i\epsilon}$ [Fig. 20.5(b)]. For a psion internal line include a propagator $\frac{i}{q^2-m^2+i\epsilon}$.
- Integrate over all undetermined momenta.
- Incoming and outgoing lines contribute a factor 1 [Fig. 20.5(c-d)].
- All symmetry factors are 1.
- Include an overall energy-momentum conserving delta function for each diagram.

Stanley Mandelstam (1928–2016)

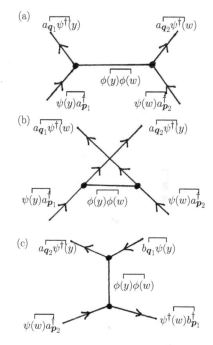

Fig. 20.4 (a) A *t*-channel scattering process. (b) A *u*-channel process. (c) An *s*-channel process involving particles scattering from antiparticles.

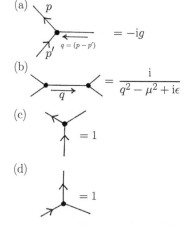

Fig. 20.5 Feynman rules for $\psi^\dagger\psi\phi$ theory.

(a)
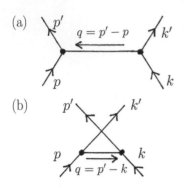

(b)
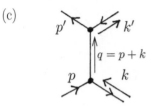

(c)

Fig. 20.6 The t-, u- and s-channel scattering processes in momentum space.

Example 20.3

The amplitude for the second-order scattering diagrams (including the overall energy-momentum conserving delta function) is

$$\mathcal{A}^{(2)} = (-\mathrm{i}g)^2 \frac{\mathrm{i}}{q^2 - \mu^2 + \mathrm{i}\epsilon}(2\pi)^4\,\delta^{(4)}\left(\sum p_\mathrm{f} - \sum p_\mathrm{i}\right). \quad (20.15)$$

Here we don't integrate over q, as it is completely determined by momentum conservation for these diagrams. Referring to Fig. 20.6 we have the following amplitudes:

- For the t-channel process $q^2 = (p'-p)^2 = t$ and we have

$$\mathcal{A}_t = (-\mathrm{i}g)^2 \frac{\mathrm{i}}{t - \mu^2 + \mathrm{i}\epsilon}(2\pi)^4\,\delta^{(4)}(p'+k'-p-k). \quad (20.16)$$

- For the u-channel process we have $q^2 = (p'-k)^2 = u$ and

$$\mathcal{A}_u = (-\mathrm{i}g)^2 \frac{\mathrm{i}}{u - \mu^2 + \mathrm{i}\epsilon}(2\pi)^4\,\delta^{(4)}(p'+k'-p-k). \quad (20.17)$$

- For the s-channel process we have $q^2 = (p+k)^2 = s$ and

$$\mathcal{A}_s = (-\mathrm{i}g)^2 \frac{\mathrm{i}}{s - \mu^2 + \mathrm{i}\epsilon}(2\pi)^4\,\delta^{(4)}(p'+k'-p-k). \quad (20.18)$$

20.3 The transition matrix and the invariant amplitude

One thing that one immediately notices about the expansion of the $\hat{S}$-operator is that the first term is unity. This means that the only nonzero amplitude from this term in the expansion has identical initial and final states. This is a rather dull result as far as scattering is involved (since nothing happens) so it's commonly removed by writing $\hat{S} = 1 + \mathrm{i}\hat{T}$, where the matrix elements of $\hat{T}$ are often called the transition or T-matrix. Another irritation is that diagrams carry the overall energy-momentum conserving delta function. This is often factored out to make the so-called invariant amplitude $\mathcal{M}$, defined for a two-in, two-out process as

$$\langle p_{1\mathrm{f}}p_{2\mathrm{f}}|\mathrm{i}\hat{T}|p_{2\mathrm{i}}p_{1\mathrm{i}}\rangle = (2\pi)^4\delta^{(4)}(p_{1\mathrm{f}} + p_{2\mathrm{f}} - p_{2\mathrm{i}} - p_{1\mathrm{i}})\mathrm{i}\mathcal{M}, \quad (20.19)$$

which at least prevents us from having to continuously write down the delta function.[3]

We're now ready to reveal a wonderful simplification. Remember that diagrams fall into two classes: connected diagrams and disconnected diagrams. It turns out that *only fully connected diagrams contribute to the T-matrix*, a point which makes good physical sense since disconnected diagrams correspond to physical processes which do not influence each other. We also note that there are a class of diagrams which have loops attached to the external legs. These are bad news and often lead to unpleasant and non-trivial infinities. The good news is that they don't contribute to the T-matrix either, so we can remove them.[4] We are left

---

[3]The factor of i is included so that the amplitudes match up with the results from the conventions of non-relativistic scattering theory.

[4]To remove them, we define the act of **amputation**, which involves chopping off all such carbuncles on the external legs.

with the appealing result that

$$\mathrm{i}\mathcal{M}(2\pi)^4\,\delta^{(4)}\left(\sum p_\mathrm{f} - \sum p_\mathrm{i}\right) = \sum \left(\begin{array}{c}\text{All connected, amputated Feynman}\\ \text{diagrams with incoming momentum } p_\mathrm{i}\\ \text{and outgoing momentum } p_\mathrm{f}\end{array}\right).$$

$$(20.20)$$

We now need to design an experiment which measures $\mathcal{M}$ and that turns out to be a scattering experiment.

20.4   The scattering cross-section

When firing one particle at another, we want to think about the amplitude of scattering of our incoming particle in terms of how big the other particle appears to be in cross-sectional area terms. Imagine driving at night, where your car headlights give you a beam of particles with which you probe the inky blackness. A sheep shows up more in your headlights than a field mouse, simply because it's bigger and scatters more photons [Fig. 20.7(a)]. Similarly a cow will scatter more light if seen from the side than from the front, because it offers a larger cross-sectional area in the former case [Fig. 20.7(b)]. However it's not just about area – white sheep are more visible than black sheep – but we fold this all in to our definition of the scattering cross-section σ in which we measure the rate R of a scattering process occurring as $R = \sigma L$, where L is the luminosity[5] of our incoming beam of particles.

Scattering can occur in all directions, and so in each bit of solid angle $\mathrm{d}\Omega$ there will be a bit of scattering $\mathrm{d}\sigma$. A detector usually only covers a range of solid angle, so it is usual to think in terms of the **differential cross-section**[6] defined by $\mathrm{d}\sigma/\mathrm{d}\Omega$. Fermi's golden rule allows us to relate the scattering cross-section to the modulus squared of a matrix element. After taking proper account of all the normalization factors one can show,[7] for example, that for a scattering process involving two equal-mass particles scattering off each other then

$$\frac{\mathrm{d}\sigma}{\mathrm{d}\Omega} = \frac{|\mathcal{M}|^2}{64\pi^2 E_{\mathrm{CM}}^2},\qquad(20.21)$$

where E_{CM} is the total energy in the centre of mass frame.

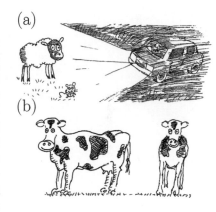

Fig. 20.7   Scattering cross-sections. (a) $\sigma_{\text{sheep}} > \sigma_{\text{field mouse}}$. (b) $\sigma_{\text{cow, side}} > \sigma_{\text{cow, front}}$.

[5]Luminosity is a quantity with dimensions $\text{Time}^{-1} \times \text{Area}^{-1}$ to be evaluated for the case under study.

[6]The relationship between σ and $\mathrm{d}\sigma/\mathrm{d}\Omega$ is given by

$$\int_0^{2\pi} \mathrm{d}\phi \int_{-1}^{1} \mathrm{d}(\cos\theta)\frac{\mathrm{d}\sigma}{\mathrm{d}\Omega} = \sigma.$$

[7]This result is discussed in Peskin and Schroeder, Chapter 4.

Example 20.4

We're finally going to work out some scattering amplitudes and cross-sections for $\psi^\dagger\psi\phi$ theory. We start with the scattering of two distinguishable psion particles. In this case only the t-channel diagram [Fig. 20.6(a)] contributes. Let's further simplify by considering only a non-relativistic case. This means that we can take the four-momentum $p = (E_{\boldsymbol{p}}, \boldsymbol{p})$ to be $p \approx (m, \boldsymbol{p})$. We have

$$\begin{array}{ll} p_{1\mathrm{i}} \approx (m, \boldsymbol{p}), & p_{2\mathrm{i}} \approx (m, \boldsymbol{k}),\\ p_{1\mathrm{f}} \approx (m, \boldsymbol{p}'), & p_{2\mathrm{f}} \approx (m, \boldsymbol{k}'), \end{array}\qquad(20.22)$$

so t may be written $t = q^2 = (p_{1f} - p_{1i})^2 \approx -|\boldsymbol{p}' - \boldsymbol{p}|^2 = -|\boldsymbol{q}|^2$, where $\boldsymbol{q} = \boldsymbol{p}' - \boldsymbol{p}$ is the three-momentum transfer. The scattering amplitude is given by evaluating the t-channel Feynman diagram yielding

$$i\mathcal{M} = \frac{ig^2}{|\boldsymbol{q}|^2 + \mu^2},$$
(20.23)

where we ignore the $i\epsilon$, since it's unnecessary here. Plugging into our expression for the cross-section and noting that $E_{\mathrm{CM}} = 2m$, we obtain

$$\frac{\mathrm{d}\sigma}{\mathrm{d}\Omega} = \frac{1}{256\pi^2 m^2}\left(\frac{g^2}{|\boldsymbol{q}|^2 + \mu^2}\right)^2.$$
(20.24)

Max Born (1882–1970)

We have a measurable prediction from the theory. Interestingly this calculation can also be done in non-relativistic quantum mechanics using Born's approximation, which says that the amplitude for scattering from a potential $V(\boldsymbol{r})$ is $\langle p'|i\hat{T}|p\rangle = -i\tilde{V}(\boldsymbol{q})(2\pi)\delta(E_{p'} - E_p)$, where $\tilde{V}(\boldsymbol{q}) = \langle p'|\hat{V}(\boldsymbol{r})|p\rangle = \int \mathrm{d}^3r\, \mathrm{e}^{i(\boldsymbol{p}-\boldsymbol{p}')\cdot\boldsymbol{r}} V(\boldsymbol{r})$. Comparing with our quantum field theory result, we see that Born says that our scattering potential is

$$\tilde{V}(\boldsymbol{q}) = \frac{-g^2}{|\boldsymbol{q}|^2 + \mu^2}.$$
(20.25)

What function has a Fourier transform that looks like this? The answer is

$$V(\boldsymbol{r}) = -\frac{g^2}{4\pi|\boldsymbol{r}|}\mathrm{e}^{-\mu|\boldsymbol{r}|};$$
(20.26)

[8] See Chapter 17.

the potential dies away over a distance $1/\mu$. But this is just Yukawa's potential![8] This is the reason why the $\psi^\dagger\psi\phi$ theory is often called Yukawa theory.

[9] Readers interested in fermions, photons and quantum electrodynamics may now proceed to Chapter 36. However, the main route of the book will be different.

We now have enough formalism to start calculating amplitudes relevant in real systems.[9] However, in the next chapter we will take a break from quantum mechanics and discuss the seemingly unrelated subject of statistical physics. This subject has an unexpected connection to quantum field theory that will provide some powerful new insights.

Chapter summary

- Yukawa's theory involves an interaction term $\hat{\mathcal{H}}_{\mathrm{I}} = g\hat{\psi}^\dagger\hat{\psi}\hat{\phi}$.
- We have introduced two new diagrams: tadpole and oyster.
- The scattering matrix and cross-section can be related to the invariant amplitude $\mathcal{M}$. The only contribution to $\mathcal{M}$ comes from connected, amputated Feynman diagrams.

Exercises

(20.1) Verify the rules for the contractions in eqn 20.6.

(20.2) (a) Verify that the Fourier transform of $V(\boldsymbol{r}) = -\frac{g^2}{4\pi|\boldsymbol{r}|}\mathrm{e}^{-\mu|\boldsymbol{r}|}$ is $\tilde{V}(\boldsymbol{q}) = \frac{-g^2}{|\boldsymbol{q}|^2+\mu^2}$.

(b) By taking an appropriate limit, find the form of the Fourier transform of the Coulomb potential.

Part V

Interlude: wisdom from statistical physics

Statistical physics is the study of large assemblies of atoms or particles and allows one to extract thermodynamic quantities through a process of averaging. Typically you write down the partition function Z for the system and manipulate it to generate the quantities you want. In quantum field theory there is an analogous process which involves a generating function $Z[J]$ which can be processed to bring forth Green's functions.

- A rapid crash course in basic statistical physics is given in Chapter 21, showing how the partition function can be used to manufacture any desired thermodynamic quantity or correlation function.

- Chapter 22 develops these ideas for quantum field theory, introducing the generating functional $Z[J]$ and linking this with the S-matrix via the Gell-Mann–Low theorem and showing how these concepts are connected with diagrams.

21

Statistical physics: a crash course

Relax, you know more than you think you do.
Benjamin Spock (1903–1998), *Baby and Child Care*

21.1 Statistical mechanics in a nutshell

Consider a system with a Hamiltonian $\hat{H}_0$ that has energy eigenvalues E_λ. The probability of the system being in some particular state $|\lambda\rangle$ with energy E_λ at a temperature T is given by the **Gibbs distribution**

$$p_\lambda = \frac{e^{-\beta E_\lambda}}{Z}, \tag{21.1}$$

where $\beta = 1/k_B T$ and Z is known as the **partition function** which is defined as a sum over states $Z = \sum_\lambda e^{-\beta E_\lambda}$, or equivalently

$$Z = \sum_\lambda \langle \lambda | e^{-\beta \hat{H}_0} | \lambda \rangle = \text{Tr}\left[e^{-\beta \hat{H}_0} \right]. \tag{21.2}$$

The term Z is needed to guarantee that $\sum_\lambda p_\lambda = 1$, which is a requirement of any probability distribution. However, the partition function Z is more important than a mere normalization constant. It turns out that all of the information about the system is contained in Z and for this reason Z is also known as a **generating function**.

Consider a one-dimensional magnet which is made up of a lattice of N sites, each one labelled by an index i, decorated with spins (shown in Fig. 21.1). The spins have magnitude $S = \frac{1}{2}$ and can point up ($S_z = +\frac{1}{2}$) or down ($S_z = -\frac{1}{2}$).[1] We define the operator-valued field $\hat{\phi}_i$. This is the same sort of field as we had before: we input a lattice point position i and get out an operator which we use to act on a state of the system. In this case the operator is $\hat{S}_{zi}$ which we understand to act only on the spin in the ith position.

$i = 1 \quad 2 \quad 3 \quad 4 \quad 5 \quad 6 \quad 7 \quad 8 \quad 9 \quad 10$

Fig. 21.1 A simple model of a one-dimensional magnet.

[1]This, of course, is shorthand. We mean that the spin states are eigenstates of the $\hat{S}_z$ operator, which has eigenvalues $S_z = \pm\frac{1}{2}$.

Example 21.1

A state-vector $|A\rangle$ lists the values of each of the spins in our one-dimensional magnet, e.g. $|A\rangle = |\uparrow\uparrow\downarrow\downarrow\downarrow \ldots\rangle$. We input a position (e.g. $i = 4$) and obtain an operator corresponding to that position $\hat{\phi}_4$. Acting on the state $|A\rangle$ with $\hat{\phi}_4$ yields

$$\hat{\phi}_4 |\uparrow\uparrow\downarrow\downarrow\downarrow \ldots\rangle = -\frac{1}{2}|\uparrow\uparrow\downarrow\downarrow\downarrow \ldots\rangle, \tag{21.3}$$

since the spin at position $i = 4$ has eigenvalue $S_{z4} = -\frac{1}{2}$.

Acting on $|\lambda\rangle$ with $\hat{\phi}_i$ yields $S_{zi}^{(\lambda)}|\lambda\rangle$, where $S_{zi}^{(\lambda)}$ is the eigenvalue of the operator $\hat{S}_z$ applied to the ith spin for the state $|\lambda\rangle$. As usual, the expectation value of the $\hat{\phi}_i$ operator for the state $|\lambda\rangle$ is $\langle\lambda|\hat{\phi}_i|\lambda\rangle$. At a temperature T we might want to know the thermal expectation value of the ith spin, denoted $\langle\hat{\phi}_i\rangle_t$ which is found by summing over the expectation values weighted by the probabilities of finding a state of energy E_λ thus:

$$
\begin{aligned}
\langle\hat{\phi}_i\rangle_t = \sum_\lambda S_{zi}^{(\lambda)} p_\lambda &= \frac{1}{Z}\sum_\lambda S_{zi}^{(\lambda)} \mathrm{e}^{-\beta E_\lambda} \\
&= \frac{1}{Z}\sum_\lambda \langle\lambda|\hat{\phi}_i|\lambda\rangle \mathrm{e}^{-\beta E_\lambda} \\
&= \frac{1}{Z}\sum_\lambda \langle\lambda|\hat{\phi}_i \mathrm{e}^{-\beta\hat{H}_0}|\lambda\rangle. \\
&= \frac{\mathrm{Tr}\left[\hat{\phi}_i \mathrm{e}^{-\beta\hat{H}_0}\right]}{Z}.
\end{aligned} \tag{21.4}
$$

The quantity $\hat{\rho} = \mathrm{e}^{-\beta\hat{H}_0}/Z$ is known as the probability density operator. Its matrix elements are known collectively as the density matrix. We conclude that to find the thermal expectation value of an operator in statistical physics we need to multiply it by the density matrix and then take a trace over all of the states. That's pretty much all there is to statistical physics.

In the absence of a magnetic field, a system of non-interacting spins at some particular temperature is expected to have $\langle\hat{\phi}_i\rangle_t = 0$ for all i. This reflects the fact that spin i has equal probability to be found with $S_z = 1/2$ or $-1/2$. If however a system shows spin order (more usually called magnetic order) then we would expect $\langle\phi_i\rangle_t \neq 0$ for all i.

21.2 Sources in statistical physics

There's a neater way of getting the thermal average of the field $\langle\hat{\phi}_i\rangle_t$. It involves adding to the Hamiltonian $\hat{H}_0$ a fictional term $\hat{H}_s = -\frac{1}{\beta}\sum_k J_k\hat{\phi}_k$. Such a term, which involves coupling a field J_k to the spin at position k, is known as a **source term** for reasons which will come apparent later.[2] With the inclusion of the source term, the partition function is now

$$
Z(J) = \mathrm{Tr}\left[\mathrm{e}^{-\beta\hat{H}}\right] = \mathrm{Tr}\left[\mathrm{e}^{-\beta\hat{H}_0 + \sum_k J_k\hat{\phi}_k}\right]. \tag{21.5}
$$

The point of this procedure is that we can now use the partition function $Z(J)$ to work out the thermal average $\langle\hat{\phi}_i\rangle_t$ by differentiating $Z(J)$ with respect to J_i, and evaluating this at $J = 0$ and dividing by $Z(J = 0)$. That is to say

$$
\langle\hat{\phi}_i\rangle_t = \frac{1}{Z(J=0)}\frac{\partial Z(J)}{\partial J_i}\bigg|_{J=0} = \frac{\mathrm{Tr}\left[\hat{\phi}_i \mathrm{e}^{-\beta\hat{H}_0}\right]}{Z(J=0)}, \tag{21.6}
$$

as we had before.

[2]The source term for a magnet is actually proportional to the magnetic field B_i, i.e. the value of the B-field evaluated at position i.

But that's not all. We can also use this philosophy to calculate **correlation functions**. A useful question in statistical physics is to ask about paired averages. For example, if we know the value of the spin at position i, we'd like to know the probability that the spin at position j is in the same state. The quantity we're after is $G_{ij} = \langle \hat{\phi}_i \hat{\phi}_j \rangle_t$. This tells us about the degree of correlation between spins in different parts of the system. If the spins are completely random, then $G_{ij} = 0$, whereas if they are completely aligned then $G_{ij} = \frac{1}{4}$ (because the spin eigenvalues are $\pm\frac{1}{2}$). More interesting is when the correlation is partial and then we might expect to find $G_{ij} = \frac{1}{4}$ when $|i-j|$ is small but $G_{ij} \to 0$ as $|i-j| \to \infty$. Sometimes it is interesting to look for deviations from **order**, which is the situation when $\langle \hat{\phi}_i \rangle_t \neq 0$. In that case, one can examine the **connected correlation function**

$$G_{cij} = \langle \hat{\phi}_i \hat{\phi}_j \rangle_t - \langle \hat{\phi}_i \rangle_t \langle \hat{\phi}_j \rangle_t. \tag{21.7}$$

Our notation makes these correlation functions look rather like cousins of the Green's functions that we've been considering previously. Some example correlations we can find include

$$
\begin{aligned}
G_{ij} &= \langle \hat{\phi}_i \hat{\phi}_j \rangle_t = \frac{1}{Z(0)} \frac{\partial^2 Z(J)}{\partial J_i \partial J_j}\bigg|_{J=0}, \\
G_{ijk} &= \langle \hat{\phi}_i \hat{\phi}_j \hat{\phi}_k \rangle_t = \frac{1}{Z(0)} \frac{\partial^3 Z(J)}{\partial J_i \partial J_j \partial J_k}\bigg|_{J=0},
\end{aligned} \tag{21.8}
$$

and in general

$$\langle \hat{\phi}_{i_1}...\hat{\phi}_{i_n} \rangle_t = \frac{1}{Z(0)} \frac{\partial^n Z(J)}{\partial J_{i_1}...\partial J_{i_n}}\bigg|_{J=0}. \tag{21.9}$$

The moral of this story is that by linearly coupling an operator-valued field $\hat{\phi}_i$ to a source J_i our partition function becomes a generating function. Once we've found the generating function we gain access to the average value of $\hat{\phi}_i$ and a vast set of related quantities.

21.3   A look ahead

There is a close analogy between statistical physics, as exemplified by the magnet, and quantum field theory. In fact everything we're looked at in this chapter has an analogue in field theory. In going to field theory we'll take a continuum limit, turning the lattice label i into the position x so that $\hat{\phi}_i \to \hat{\phi}(x)$. The correlation functions of statistical physics become propagators in field theory and our goal is to find these.

This is all very nice and certainly deserves further examination,[3] but the point of this chapter is to tell you that to reach the goal of finding propagators for a field theory we can use a generating functional $Z[J]$ similar to that used in statistical physics. The generating functional for fields will allow us, through (functional) differentiation, to find all of the

[3]See Chapter 25.

	Statistical physics (on a lattice)	Field theory (continuum)
Source	J_i	$J(x)$
Generating function	$Z(J)$	$Z[J(x)]$
Green's functions	$G^{(n)}_{i_1...i_n} = \langle \hat{\phi}_{i_1}...\hat{\phi}_{i_n} \rangle_t$	$G^{(n)}(x_1,...,x_n) = \langle \Omega|T\hat{\phi}(x_1)...\hat{\phi}(x_n)|\Omega \rangle$
Differentiation recipe	$G^{(n)}_{i_1...i_n} = \frac{1}{Z(J=0)} \frac{\partial^n Z(J)}{\partial J_{i_1}...\partial J_{i_n}}\Big|_{J=0}$	$G^{(n)}(x_1,...,x_n) = \frac{1}{i^n}\frac{1}{Z[J=0]}\frac{\delta^n Z[J]}{\delta J(x_1)...\delta J(x_n)}\Big|_{J=0}$
Order	$\langle \hat{\phi}_i \rangle_t \neq 0$	$\langle \Omega|\hat{\phi}(x)|\Omega \rangle \neq 0$

Table 21.1 Analogies between quantities in statistical physics on a lattice and those in quantum field theory.

propagators. In equations, this can be written as

$$
\begin{aligned}
G(x_1,...,x_n) &= \langle \Omega|T\hat{\phi}(x_1)...\hat{\phi}(x_n)|\Omega \rangle \\
&= \frac{1}{i^n}\frac{1}{Z[J=0]}\frac{\delta^n Z[J]}{\delta J(x_1)...\delta J(x_n)}\Big|_{J=0}. \quad (21.10)
\end{aligned}
$$

One of the marvellous things about field theory is that even when we can't solve a theory exactly, the generating functional $Z[J]$ and the Green's functions $G^{(n)}(x_1,...,x_n)$ can be expressed in terms of Feynman diagrams. We will examine this in the next chapter.

Chapter summary

- Correlation functions in statistical mechanics can be extracted by differentiating the partition function Z with respect to the source term J. This approach carries over into quantum field theory because Green's functions can be extracted from a generating functional $Z[J]$. Some analogies between statistical mechanics and quantum field theory are tabulated in Table 21.1.

Exercises

(21.1) Define the density operator as $\hat{\rho} = \frac{e^{-\beta\hat{H}}}{Z}$. Using the rule that thermal averages are given by $\langle \hat{A} \rangle_t = \text{Tr}\left[\hat{\rho}\hat{A}\right]$, show that for a quantum oscillator, with Hamiltonian $\hat{H} = \omega \hat{a}^\dagger \hat{a}$, the thermal average

of the number of excitations is given by

$$
\langle \hat{n} \rangle_t = \langle \hat{a}^\dagger \hat{a} \rangle = \frac{1}{e^{\beta\omega}-1}. \quad (21.11)
$$

(21.2) *Response functions* or *susceptibilities* are ubiquitous in physics. Consider the forced quantum oscillator

$$L = \frac{1}{2}m\dot{x}(t)^2 - \frac{1}{2}m\omega^2 x(t)^2 + f(t)x(t), \quad (21.12)$$

and take the definition of the response function to be

$$\langle\psi(t)|\hat{x}(t)|\psi(t)\rangle = \int_{-\infty}^{\infty} dt' \chi(t-t')f(t'). \quad (21.13)$$

(a) Using the interaction representation and treating $\hat{H}' = -f(t)\hat{x}(t)$ as the interaction part, show that, to first order in the force function $f_I(t)$, we have

$$|\psi_I(t)\rangle = |0\rangle + i\int_{-\infty}^{t} dt' f_I(t')\hat{x}_I(t')|0\rangle. \quad (21.14)$$

(b) Using the previous result, show that

$$\chi(t-t') = i\theta(t-t')\langle 0| \left[\hat{x}_I(t), \hat{x}_I(t')\right] |0\rangle. \quad (21.15)$$

(c) Using the fact that $\hat{x}_I(t) = \left(\frac{1}{2m\omega}\right)^{\frac{1}{2}} \left(\hat{a}e^{-i\omega t} + \hat{a}^\dagger e^{i\omega t}\right)$, calculate the response function for the quantum oscillator at temperature $T = 0$.

(d) Using the fact that at nonzero T we have $\langle\hat{n}\rangle_t = (e^{\beta\omega} - 1)^{-1}$, calculate the response function for the quantum oscillator at nonzero T.

(e) Compare the behaviour of the response function with that of the correlation function S defined as

$$S = \frac{1}{2}\langle\{\hat{x}_I(t'), \hat{x}_I(t)\}\rangle = \frac{1}{2}\langle\hat{x}_I(t')\hat{x}_I(t) + \hat{x}_I(t)\hat{x}_I(t')\rangle. \quad (21.16)$$

(21.3) *Diffusion problems are common in thermodynamics. Here we meet the diffusion equation. The diffusion equation for the number density of particles $n(\boldsymbol{x}, t)$ is*

$$\frac{\partial n(\boldsymbol{x}, t)}{\partial t} - D\boldsymbol{\nabla}^2 n(\boldsymbol{x}, t) = 0. \quad (21.17)$$

Consider a point source at time $t = 0$ and position $\boldsymbol{x} = \boldsymbol{y}$, giving us the boundary conditions $G(\boldsymbol{x} - \boldsymbol{y}, t = 0) = \delta^{(3)}(\boldsymbol{x} - \boldsymbol{y})$. Show that, for times $t > 0$ the Green's function for this equation is given by

$$\tilde{G}(\omega, \boldsymbol{q}) = \frac{1}{-i\omega + D|\boldsymbol{q}|^2}, \quad (21.18)$$

and identify the position of the pole in the complex ω plane.

Hint: one way to do this problem which makes the boundary conditions clear is to take a Laplace transformation in time and a Fourier transform in space. See Chaikin and Lubensky, Chapter 7, for a discussion.

The generating functional for fields

22

Since the finished product cannot be served to the table as it is, this isn't so much a recipe, a method or a list of ingredients: it is a way of life, for if there is fresh stock made every week, a reputation as a good cook will follow.
Len Deighton (1929–), *Action Cook Book*

In this chapter we will present a general method of finding Green's functions based on the philosophy of generating functionals. We will explain the meaning of the generating functional of quantum field theory, $Z[J]$, and show that all Green's functions may be extracted from it through differentiation. This will allow us to relate together the S-matrix and Green's functions, linking two of the most important topics in quantum field theory.

Although this chapter contains a lot of formal material the ideas are quite simple (and have much in common with the world of statistical physics). In fact the rules that emerge for perturbation theory allow us to write generating functionals and Green's functions as sums of Feynman diagrams, relating formal objects such as $G(x, y)$ and $Z[J]$ to a series of cartoons.

22.1 How to find Green's functions

All of the information about a field theory is available in the form of Green's functions. Happily there's an easy way to work out Green's functions which is based on a trick which enables us to lump all of the information about a quantum field theory in a single entity called the **generating functional** or $Z[J]$. The generating functional is replete with all that juicy data in a zipped up and compressed form and there's a simple procedure for extracting it. This is identical to the case of statistical physics, where all correlation functions were obtainable from the partition function $Z(J)$.

Just as in statistical physics we make use of the philosophy of sources. In order to make a generating functional we'll add a source function $J(x)$ to our Lagrangian density, which couples linearly[1] with the field $\phi(x)$:

$$\mathcal{L}[\phi(x)] \quad \rightarrow \quad \mathcal{L}[\phi(x)] + J(x)\phi(x). \qquad (22.1)$$

The reason why $J(x)$ is referred to as a source is that in quantum field theory this term will generate particle excitations in the field $\phi(x)$. By

---

[1]Note that by *adding* $J\phi$ to $\mathcal{L}$, we will end up *subtracting* $J\phi$ from $\mathcal{H}$.

turning on a source we are, essentially, grabbing the field and shaking excitations out.

We now define the generating functional for quantum fields as

$$Z[J] \;=\; \langle\Omega|\hat{U}(\infty,-\infty)|\Omega\rangle_J, \tag{22.2}$$

where $\hat{U}$ is the time-evolution operator for the full Hamiltonian $\hat{H}$ of the theory and the J subscript on the right-hand side reads 'in the presence of a source J' when said out loud. We also have that $\hat{H}|\Omega\rangle = 0$. That is $|\Omega\rangle$ is the physical ground state of the system and the ground state energy is defined to be zero. Turning eqn 22.2 into words we see that the generating functional $Z[J]$ tells us the amplitude

$$Z[J] = \left\langle \begin{array}{c} \text{no particles} \\ \text{at } x^0 = \infty \end{array} \middle| \begin{array}{c} \text{no particles} \\ \text{at } y^0 = -\infty \end{array} \right\rangle_J, \tag{22.3}$$

that is, in the presence of a source, we start with no particles and end with no particles. We could call $Z[J]$ the 'no particle propagator'.

As we have defined it, $Z[J]$ allows two classes of process apart from the rather trivial option of nothing happening at all: (i) those where a particle is created by the source J and later absorbed by J; (ii) those where a particle spontaneously appears and then disappears. The latter class, corresponding to $Z[J=0]$, is described by vacuum diagrams and needs to be removed by normalization in order for the generating functional to work. We therefore define a normalized generating functional:

$$\mathcal{Z}[J] = \frac{Z[J]}{Z[J=0]}. \tag{22.4}$$

This guarantees that $\mathcal{Z}[J=0] = 1$ or, in other words, that the amplitude to start and end with no particles in the absence of sources is unity.

We now need to find a way of calculating $Z[J]$. Writing this in terms of the time-evolution operator as $Z[J] = \langle\Omega|\hat{U}(\infty,-\infty)|\Omega\rangle_J$ turns this into a problem of finding $\hat{U}(\infty,-\infty)$. A clue is provided by the way in which we worked out the S-matrix in the last few chapters. In the interaction picture, the time-evolution operator $\hat{U}_{\mathrm{I}}(\infty,-\infty)$ is known as the $\hat{S}$-operator and the Dyson expansion gives us a way to calculate $\hat{S}$ via $\hat{S} = T\mathrm{e}^{-\mathrm{i}\int \mathrm{d}^4x\,\hat{\mathcal{H}}_{\mathrm{I}}}$. Things aren't so simple here though, since $\hat{U}(\infty,-\infty)$ in eqn 22.2 is defined in the Heisenberg picture. The problem is that we don't have free fields in the Heisenberg picture so we can't use Wick's theorem to grind down long strings of operators (since Wick's theorem only works on free fields). However we can still write a Dyson-like equation for the time evolution in terms of the Heisenberg fields $\hat{\phi}_{\mathrm{H}}$. If we treat the source term like we would an interaction by saying $\hat{\mathcal{H}}_{\mathrm{I}} = -J\hat{\phi}_{\mathrm{H}}$, then the time-evolution operator is given by Dyson's expression $U(\infty,-\infty) = T\mathrm{e}^{-\mathrm{i}\int \mathrm{d}^4x\,\hat{\mathcal{H}}_{\mathrm{I}}}$. Splitting this up in this way will only generate processes involving the sources and not the vacuum diagrams, and so will result in the normalized generating functional. We can therefore write

$$\mathcal{Z}[J] = \langle\Omega|T\mathrm{e}^{\mathrm{i}\int \mathrm{d}^4x\,J(x)\hat{\phi}_{\mathrm{H}}(x)}|\Omega\rangle, \tag{22.5}$$

where the fields $\hat{\phi}_{\mathrm{H}}(x)$ evolve according to the full Hamiltonian (rather than a free Hamiltonian H_0 used in the interaction picture) and we use the interacting ground state $|\Omega\rangle$ rather than the free ground state $|0\rangle$. Again we stress this is all still permissible, as long as we remember that we wouldn't be able to use Wick's theorem. We expand $\mathcal{Z}[J]$ to obtain

$$\mathcal{Z}[J] = 1 + \sum_{n=1}^{\infty} \frac{\mathrm{i}^n}{n!} \int \mathrm{d}^4 x_1 ... \mathrm{d}^4 x_n \, J(x_1)...J(x_n) \langle \Omega | T \hat{\phi}_{\mathrm{H}}(x_1)...\hat{\phi}_{\mathrm{H}}(x_n) | \Omega \rangle.$$
(22.6)

Recalling that Green's functions are defined by

$$G^{(n)}(x_1, ..., x_n) = \langle \Omega | T \hat{\phi}_{\mathrm{H}}(x_1)...\hat{\phi}_{\mathrm{H}}(x_n) | \Omega \rangle,$$
(22.7)

we notice that we can get access to them through a manipulation of the generating function $\mathcal{Z}[J]$ by differentiating eqn 22.6 as follows:

$$G^{(n)}(x_1, ..., x_n) = \frac{1}{\mathrm{i}^n} \left. \frac{\delta^n \mathcal{Z}[J]}{\delta J(x_1)...\delta J(x_n)} \right|_{J=0}.$$
(22.8)

That is, we get the Green's functions (which contain all of the information about a system) by differentiating a single functional $\mathcal{Z}[J]$. This method of doing quantum field theory is illustrated in Fig. 22.1. In terms of the unnormalized generating functional $Z[J]$ we have the equivalent rule

$$G^{(n)}(x_1, ..., x_n) = \frac{1}{\mathrm{i}^n} \frac{1}{Z[J=0]} \left. \frac{\delta^n Z[J]}{\delta J(x_1)...\delta J(x_n)} \right|_{J=0},$$
(22.9)

which closely resembles the version from statistical physics.

We've done what we set out to do: we have found a single functional that contains all of the information about the system, which can be used to find Green's functions. All we need to do now is find an automated way of calculating $Z[J]$ for any theory. It turns out that there are two ways: the first is by relating $Z[J]$ to the S-matrix; the second uses Feynman's path integral. In this chapter we pursue the first approach, leaving the second for later (Chapter 24).

22.2   Linking things up with the Gell-Mann–Low theorem

We'll examine the first method: calculating $Z[J]$ from the S-matrix. We stress that the S-matrix is defined for the interaction picture, so we'll need an equation for $Z[J]$ given explicitly in terms of interaction picture fields. We'll pluck the answer from the air and then prove we've plucked correctly. We propose a form for the normalized generating function $\mathcal{Z}[J]$ which looks like:

$$\mathcal{Z}[J] = \frac{Z[J]}{Z[0]} = \frac{\langle 0 | T e^{-\mathrm{i} \int \mathrm{d}^4 x [\hat{\mathcal{H}}_{\mathrm{I}} - J(x)\hat{\phi}_{\mathrm{I}}(x)]} | 0 \rangle}{\langle 0 | T e^{-\mathrm{i} \int \mathrm{d}^4 x \hat{\mathcal{H}}_{\mathrm{I}}} | 0 \rangle},$$
(22.10)

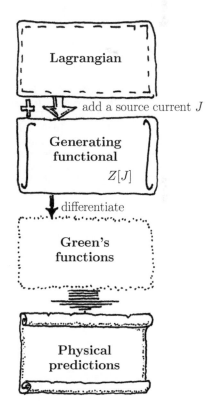

Fig. 22.1 Doing quantum field theory using generating functionals

which looks like a good guess because it has the form of $Z[J]$ in eqn 22.5, but this time writing the interaction Hamiltonian in the form $\hat{\mathcal{H}}_I - J\hat{\phi}_I$. This isn't quite in terms of an expression for the S-matrix yet as we have the source term in the numerator. Using the rule that you differentiate n times, divide by i^n and set $J = 0$ to yield a Green's function, we have the Green's function

$$G^{(n)}(x_1, ..., x_n) = \frac{\langle 0|T\hat{\phi}_I(x_1)...\hat{\phi}_I(x_n)e^{-i\int d^4x\hat{\mathcal{H}}_I}|0\rangle}{\langle 0|Te^{-i\int d^4x\hat{\mathcal{H}}_I}|0\rangle}, \qquad (22.11)$$

and this can be written in terms of the $\hat{S}$-operator as

$$G^{(n)}(x_1, ..., x_n) = \frac{\langle 0|T\hat{\phi}_I(x_1)...\hat{\phi}_I(x_n)\hat{S}|0\rangle}{\langle 0|\hat{S}|0\rangle}. \qquad (22.12)$$

All that's left is now is to prove that we've made a correct guess, so that

$$\langle\Omega|T\hat{\phi}_H(x_1)...\hat{\phi}_H(x_n)|\Omega\rangle = \frac{\langle 0|T\hat{\phi}_I(x_1)...\hat{\phi}_I(x_n)\hat{S}|0\rangle}{\langle 0|\hat{S}|0\rangle}. \qquad (22.13)$$

This result turns out to be true and is known as the **Gell-Mann–Low theorem**.[2] Let's be completely clear about what eqn 22.13 is saying. First examine the left-hand side:

- $|\Omega\rangle$ is the interacting ground state, defined by the full Hamiltonian as $\hat{H}|\Omega\rangle = 0$. $\hat{\phi}_H(x)$ are Heisenberg field operators. They time-evolve according to the full Hamiltonian $\hat{H}$.

Next consider the right-hand side:

- Since we've turned off the source term now, the Hamiltonian can be split up into free (solvable) and interacting (usually unsolvable) parts $\hat{H} = \hat{H}_0 + \hat{H}'$. The right-hand side is therefore written in terms of the interaction picture. $|0\rangle$ is the free ground state, defined such that $\hat{H}_0|0\rangle = 0$. Quantities $\hat{\phi}_I(x)$ are interaction picture field operators. These time-evolve according to the free part of the Hamiltonian $\hat{H}_0$.

[2]The proof of the theorem can be found in the book by P. Coleman. It was first proved by Gell-Mann and Low in 1951. Murray Gell-Mann (1929–) received the 1969 Nobel Prize in Physics and, borrowing a term from *Finnegans Wake*, gave the name to 'quarks'. Francis Low (1921–2007) worked with Gell-Mann at Princeton, and spent most of his later career at MIT.

22.3   How to calculate Green's functions with diagrams

We've seen how to get a Green's function from the generating functional of a theory. It's simply a matter of differentiating. It's natural to be curious about what happens if we try to directly relate the Green's functions to Feynman diagrams. The Gell-Mann–Low theorem is a quotient:

$$G^{(n)}(x_1, ..., x_n) = \frac{\langle 0|T\hat{\phi}_I(x_1)...\hat{\phi}_I(x_n)\hat{S}|0\rangle}{\langle 0|\hat{S}|0\rangle}. \qquad (22.14)$$

Presumably there might be some cancellation between the Feynman diagrams thrown up in the numerator against those coming from the denominator. It turns out that this cancellation does indeed take place and leaves us with a wonderfully simple result:

$$G^{(n)} = \langle\Omega|T\hat{\phi}_{\mathrm{H}}(x_1)...\hat{\phi}_{\mathrm{H}}(x_n)|\Omega\rangle = \sum \left(\begin{array}{c} \text{All connected diagrams} \\ \text{with } n \text{ external lines} \end{array} \right).$$

(22.15)

Note that for $n > 2$ this sum excludes vacuum diagrams but includes those diagrams formed from products of diagram parts which are connected to the external lines. For example for $n = 4$, we include products of two diagrams each with two external lines.

Why is this? If we start with the Gell-Mann–Low formula (eqn 22.14), then the numerator factorizes into

$$\langle 0|T\hat{\phi}_{\mathrm{I}}(x_1)...\hat{\phi}_{\mathrm{I}}(x_n)\hat{S}|0\rangle = \sum \left(\begin{array}{c} \text{Connected diagrams} \\ \text{with } n \text{ external lines} \end{array} \right) \times \exp\left[\sum \left(\begin{array}{c} \text{Connected} \\ \text{vacuum diagrams} \end{array} \right) \right],$$

and the denominator is simply $\langle 0|\hat{S}|0\rangle = \exp\left[\sum \left(\begin{array}{c} \text{Connected} \\ \text{vacuum diagrams} \end{array} \right) \right]$,

so we are able to show that

$$\begin{aligned} G^{(n)}(x_1, ..., x_n) &= \langle\Omega|T\hat{\phi}_{\mathrm{H}}(x_1)...\hat{\phi}_{\mathrm{H}}(x_n)|\Omega\rangle \\ &= \frac{\sum \left(\begin{array}{c} \text{Connected diagrams} \\ \text{with } n \text{ external lines} \end{array} \right) \times \exp\left[\sum \left(\begin{array}{c} \text{Connected} \\ \text{vacuum diagrams} \end{array} \right) \right]}{\exp\left[\sum \left(\begin{array}{c} \text{Connected} \\ \text{vacuum diagrams} \end{array} \right) \right]}, \end{aligned}$$

and that proves eqn 22.15, the most important result of this chapter.

Example 22.1

Let's see how this unfolds for the single-particle propagator $G^{(2)}(x_1, x_2)$. A typical term will result in a disconnected diagram made up of a connected diagram with external lines and several connected vacuum diagrams. The value of the full diagram is the product of all of the connected diagrams. We label each connected vacuum diagram by some number i. There may be more than one copy of each connected vacuum diagram: we'll call the number of copies n_i. Call the value of a connected vacuum diagram V_i. Put all of this together and write the value of a full, disconnected diagram, representing one particular contraction, as

$$\left(\begin{array}{c} \text{Connected diagram} \\ \text{with external lines} \end{array} \right) \times \prod_i \frac{1}{n_i!}(V_i)^{n_i},$$

(22.16)

where we need a factor $\frac{1}{n_i!}$ because we can interchange the n_i copies of the connected diagram i. The sum of all diagrams is then given by

$$\sum_{\substack{\text{Connected diagrams} \\ \text{with external lines}}} \sum_{(\text{all } \{n_i\})} \left(\begin{array}{c} \text{Connected diagram} \\ \text{with external lines} \end{array} \right) \times \prod_i \frac{1}{n_i!}(V_i)^{n_i},$$

(22.17)

where, to generate all possible disconnected diagrams, we have summed over all possible connected diagrams with external lines and over all ordered sets of integers $\{n_i\}$. To break this down we may factorize out the sum over connected diagrams with external lines:

$$\left[\sum \left(\begin{array}{c} \text{Connected diagram} \\ \text{with external lines} \end{array} \right) \right] \times \sum_{(\text{all}\{n_i\})} \prod_i \frac{1}{n_i!}(V_i)^{n_i}.$$

(22.18)

Then the rest of the expression may be factorized thus:

$$\left[\sum\left(\begin{array}{c}\text{Connected diagram}\\\text{with external lines}\end{array}\right)\right] \times \left(\sum_{n_1}\frac{1}{n_1!}(V_1)^{n_1}\right)\left(\sum_{n_2}\frac{1}{n_2!}(V_2)^{n_2}\right)\cdots$$

$$= \left[\sum\left(\begin{array}{c}\text{Connected diagram}\\\text{with external lines}\end{array}\right)\right] \times \prod_i\left(\sum_{n_i}\frac{1}{n_i!}(V_i)^{n_i}\right)$$

$$= \left[\sum\left(\begin{array}{c}\text{Connected diagram}\\\text{with external lines}\end{array}\right)\right] \times \prod_i e^{V_i}$$

$$= \left[\sum\left(\begin{array}{c}\text{Connected diagram}\\\text{with external lines}\end{array}\right)\right] \times e^{\sum_i V_i}. \tag{22.19}$$

Now, as long as we know the Feynman rules for a theory, we can forget about the S-matrix and the interaction picture. We've reduced the whole perturbation theory to one expression. To get the n-point Green's function we just sum all Feynman diagrams with n external lines. Amazingly simple.

22.4   More facts about diagrams

We've seen that $G^{(2)}(x_1, x_2)$, which is the amplitude for a particle to start at x_2 and propagate to x_1, may be written as a sum of diagrams describing all the ways in which a particle can start at x_2 and finish up at x_1. We ask next, is there a way of expressing $Z[J]$ in terms of diagrams? Since $Z[J]$ describes a situation where we have sources (and sinks) of particles, but that we start at $t = -\infty$ with no particles and end up at $t = \infty$ with no particles, we might expect to be able to express $Z[J]$ as a sum of diagrams describing ways of starting and ending without particles. This is indeed the case and we have

$$Z[J] = \langle\Omega(\infty)|\Omega(-\infty)\rangle_J = \langle 0|\hat{S}|0\rangle_J \tag{22.20}$$

$$= \sum\left(\begin{array}{c}\text{Disconnected vacuum and}\\\text{source-to-source diagrams}\end{array}\right).$$

The sum here is made up of all of the ways of starting and ending with no particles, which corresponds to the sum of all disconnected vacuum and source-to-source diagrams, of which there are potentially a very great number. Fortunately there's an even simpler way of stating this result, which is:

$$Z[J] = e^{\sum\left(\begin{array}{c}\textit{Connected}\text{ vacuum and}\\\text{source-to-source diagrams}\end{array}\right)}. \tag{22.21}$$

[3]The linked-cluster theorem, which equates the right-hand sides of eqns 22.20 and 22.21, can be stated more memorably as

$$\sum\left(\begin{array}{c}\text{all}\\\text{diagrams}\end{array}\right) = e^{\sum\left(\begin{array}{c}\text{connected}\\\text{diagrams}\end{array}\right)}.$$

This simplification is based on the **linked-cluster theorem**,[3] which we now show follows from more combinatoric arguments, very similar to those used earlier. We call a vacuum or source-to-source diagram

V_i. Now, in general, the amplitude for a general, possibly disconnected diagram is given by $\prod_i \frac{(V_i)^{n_i}}{n_i!}$ and the generating functional is given by the expression

$$Z = \prod_i \sum_{n_i=0}^{\infty} \frac{(V_i)^{n_i}}{n_i!} = e^{\sum_i V_i},\qquad(22.22)$$

proving the theorem. (Notice that all of the leg work was done in Example 22.1!)

Example 22.2

We can use the linked-cluster theorem to simplify the expression for the normalized generating functional $\mathcal{Z}[J] = Z[J]/Z[0]$. If we set $J = 0$, we turn off the sources and we have

$$Z[0] = e^{\sum(\text{Connected vacuum diagrams})}.\qquad(22.23)$$

We write

$$Z[J] = e^{\sum\left(\begin{array}{c}\text{Connected source-}\\\text{to-source diagrams}\end{array}\right)} e^{\sum(\text{Connected vacuum diagrams})}.\qquad(22.24)$$

The normalized generating functional is then

$$\mathcal{Z}[J] = \frac{Z[J]}{Z[0]} = e^{\sum\left(\begin{array}{c}\text{Connected source-}\\\text{to-source diagrams}\end{array}\right)},\qquad(22.25)$$

allowing us to give more meaning to the normalized functional $\mathcal{Z}[J]$.

Notice that we derive most of the results on the way so that we never have to use them. (One assumes this is a bit like learning knife fighting.) The end result for a particular theory is that the propagator is represented by a sum of connected diagrams. Of course, the details of what the diagrams look like come from a one-off, nitty-gritty calculation. We'll see a few more examples of these in the following chapters. Finally we'll need to do the integrals. This will lead into even stranger territory.

Chapter summary

- Green's functions are derived from generating functionals. This approach shows that $G^{(n)}$ is simply the sum of all diagrams with n external lines.
- The generating functional $Z[J]$ is the exponential of the sum of all connected source-to-source and vacuum diagrams.

Exercises

(22.1) Consider the forced quantum oscillator

$$L = \frac{1}{2}m\dot{x}(t)^2 - \frac{1}{2}m\omega^2 x(t)^2 + f(t)x(t), \quad (22.26)$$

where $f(t)$ is a constant force f_0 which acts only for a finite length of time: $0 \leq t \leq T$. We can use the S-matrix to work out the generating functional for this theory, by treating $f(t)x(t)$ as an interaction term.

(a) Explain why we may write $\langle 0|\hat{S}|0\rangle = Z[f(t)]$ for this theory.

(b) By expanding the $\hat{S}$-matrix and show that $Z[f(t)] = e^{(\text{Dumbbell})}$, where (Dumbbell) is the Feynman diagram shown in Fig. 22.2(a) whose amplitude is

$$(\text{Dumbbell}) = \frac{(-\text{i})^2}{2}\int dt dt' \, f(t)\langle 0|T\hat{x}(t)\hat{x}(t')|0\rangle f(t').$$
$$(22.27)$$

(c) Working in frequency space, show that the Dumbbell may be written

$$(\text{Dumbbell}) = -\frac{1}{2m}\int \frac{d\nu}{2\pi}\tilde{f}(-\nu)\frac{\text{i}}{\nu^2 - \omega^2 + \text{i}\epsilon}\tilde{f}(\nu),$$
$$(22.28)$$

where $\tilde{f}(\nu) = \frac{\text{i}f_0}{\nu}\left(1 - e^{\text{i}\nu T}\right)$.

(d) What does the quantity $|Z[f(t)]|^2$ describe physically? Using the identity $\frac{1}{E+\text{i}\epsilon} = \mathcal{P}\frac{1}{E} - \text{i}\pi\delta(E)$, where $\mathcal{P}$ is the principal part (see Appendix B), show that this quantity is given by

$$|Z[f(t)]|^2 = \exp\left(-\frac{2f_0^2}{m\omega^3}\sin^2\frac{\omega T}{2}\right). \quad (22.29)$$

(22.2) Now consider scalar field theory in the presence of a source field

$$\mathcal{L} = \frac{1}{2}[\partial_\mu \phi(x)]^2 - \frac{m^2}{2}\phi(x)^2 + gJ(x)\phi(x). \quad (22.30)$$

Notice the resemblance between this theory and the forced oscillator.

(a) Show that the amplitude for the Dumbbell diagram in this theory [Fig. 22.2(b)] is given by

$$\frac{(-\text{i}g)^2}{2}\int d^4x d^4y \int \frac{d^4p}{(2\pi)^4} J(x)\frac{\text{i}e^{-\text{i}p\cdot(x-y)}}{p^2 - m^2 + \text{i}\epsilon}J(y).$$
$$(22.31)$$

(b) Consider a source field describing two static sources localized at x_1 and x_2:

$$J(x) = \delta^{(3)}(x - x_1) + \delta^{(3)}(x - x_2). \quad (22.32)$$

Show that the interaction between the two sources can be written

$$(\text{Dumbbell})_{12} \propto -\text{i}g^2\left(\int dx^0 dy^0 \frac{dp^0}{2\pi}e^{-\text{i}p^0(x^0-y^0)}\right)$$
$$\times \int \frac{d^3p}{(2\pi)^3}\frac{e^{\text{i}p\cdot(x_1-x_2)}}{p^2 - m^2 + \text{i}\epsilon},$$

where the self-interaction of the sources is ignored.

(c) Show further that

$$(\text{Dumbbell})_{12} \propto \text{i}g^2\int dx^0 \int \frac{d^3p}{(2\pi)^3}\frac{e^{\text{i}p\cdot(x_1-x_2)}}{p^2 + m^2}.$$
$$(22.33)$$

(d) Finally, using the fact that $e^{-\text{i}ET} = \langle 0|\hat{S}|0\rangle = e^{(\text{Dumbbell})}$, where T is the time over which the sources act (see Exercise 43.5 if in doubt), show that the interaction energy E between two spatially separate sources is given by

$$E \propto -g^2\int \frac{d^3p}{(2\pi)^3}\frac{e^{\text{i}p\cdot(x_1-x_2)}}{p^2 + m^2}. \quad (22.34)$$

The important feature here is that the energy is negative. The two sources therefore attract each other by exchanging a virtual particle of mass m.
(e) Compare the previous result to the corresponding one for a (spin-1) vector field dealt with in Chapter 24. Is the interaction attractive or repulsive? A graviton is a spin-2 object. What would you expect for the interaction of sources in a gravitational field. See Feynman's Lectures on Gravitation, Chapter 3, for a discussion.

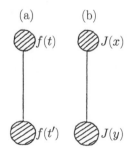

(a) (b)

Fig. 22.2 (a) Dumbbell diagram for Exercise 22.1. (b) Dumbbell diagram for Exercise 22.2.

Part VI

Path integrals

In this part we use a reformulation of quantum field theory which is due to Richard Feynman, the so-called path integral approach. Quantities of interest in quantum field theory, such as the propagator, can be calculated using a functional integral, which is an integral over all possible paths of the particle from one spacetime point to another.

- We describe Feynman's path integral approach in Chapter 23 and demonstrate how to evaluate certain Gaussian integrals, using the simple harmonic oscillator as an example.

- These ideas are then applied in Chapter 24 to working out functional integrals and thus calculating Green's functions.

- There is a rather subtle connection between quantum field theory and statistical physics, and this is explored in Chapter 25, where the concepts of *imaginary time* and the *Wick rotation* are introduced.

- An important concept is symmetry breaking and this is treated in Chapter 26. This idea is introduced in the context of phase transitions and we show how to describe symmetry breaking with a Lagrangian. Breaking a continuous symmetry leads to *Goldstone modes* and we also show the consequence of symmetry breaking in a gauge theory, thereby introducing the Higgs mechanism.

- Chapter 27 describes *coherent states* which are eigenstates of the annihilation operator. These states do not have a definite number of particles but they can be made to have a well-defined phase and are used to describe laser fields and superfluids.

- We introduce a new type of mathematical object in Chapter 28: *Grassmann numbers*. These anticommute and can be used to describe fermions. We show how to construct a fermion path integral using Grassmann numbers.

23
Path integrals: I said to him, 'You're crazy'

Thirty-one years ago Dick Feynman told me about his 'sum over histories' version of quantum mechanics. 'The electron does anything it likes', he said. 'It goes in any direction at any speed, forward and backward in time, however it likes, and then you add up the amplitudes and it gives you the wave-function.' I said to him, 'You're crazy'. But he wasn't.
F. J. Dyson (1923–)

In this chapter, we introduce the path integral formulation of quantum mechanics.

23.1 How to do quantum mechanics using path integrals

Quantum mechanics is all about probability amplitudes. One way to do quantum mechanics is to ask a question like 'what's the amplitude that a particle starts at point q_a at time t_a and ends up at point q_b at time t_b?' The amplitude for the process is, as we've discussed, known as the propagator. Previously, we found the propagator using a conventional approach to quantum mechanics espoused by Schrödinger and Heisenberg. In this chapter we'll find the propagator using Richard Feynman's alternative approach. This will then be applied to quantum fields.

The propagator depends on the trajectory that, for whatever reason, the particle takes in getting from A to B. Some possible paths are shown in Fig. 23.1. In classical mechanics, we'd know what to do: write down a Lagrangian L; plug that Lagrangian into the Euler–Lagrange equation and thus generate the equations of motion, which we'd then (at least try to) solve. The trajectory that we'd calculate would be the one that minimizes the action

$$S = \int_{t_A}^{t_B} dt \, L[q(t)], \qquad (23.1)$$

just as we discussed in Chapter 1.

How does this work in quantum mechanics? Richard Feynman's suggestion was that in getting from A to B a particle will take *every possible trajectory*. We'll say that again. A particle will take every single possible trajectory, forward and backward in time, zig-zagging, looping, whatever. To get the quantum amplitude Feynman says that each

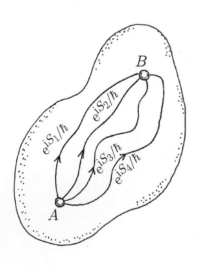

Fig. 23.1 Some possible paths with their amplitudes. A particle will take all of them.

trajectory contributes a complex factor $\mathrm{e}^{\mathrm{i}S/\hbar}$, where S is the action describing that trajectory, and we add up the contributions to get the amplitude. Crazy!

Let's see if we can justify this approach of getting the amplitude by summing over $\mathrm{e}^{\mathrm{i}S/\hbar}$ terms associated with each trajectory. This is also helpfully known as the **sum over histories** approach because the present wave function is obtained by summing over all possible past trajectories. The propagator amplitude G to get from spacetime point A $[\equiv (t_{\mathrm{a}}, q_{\mathrm{a}})]$ to point B $[\equiv (t_{\mathrm{b}}, q_{\mathrm{b}})]$ may be written as $\langle q_{\mathrm{b}}|\hat{U}(t_{\mathrm{b}}, t_{\mathrm{a}})|q_{\mathrm{a}}\rangle$, where $|q\rangle$ is an eigenstate of position and $\hat{U}(t_{\mathrm{b}}, t_{\mathrm{a}})$ is the time-evolution operator $\hat{U}(t_{\mathrm{b}}, t_{\mathrm{a}}) = \mathrm{e}^{-\mathrm{i}\hat{H}(t_{\mathrm{b}}-t_{\mathrm{a}})}$ (where we've returned to our usual units employing $\hbar = 1$). We have then

$$G = \langle q_{\mathrm{b}}|\mathrm{e}^{-\mathrm{i}\hat{H}(t_{\mathrm{b}}-t_{\mathrm{a}})}|q_{\mathrm{a}}\rangle. \qquad (23.2)$$

To deal with this we split the trajectory into N infinitesimal steps of length Δt as shown in Fig. 23.2. This is known as **time-slicing**. The fact that $\hat{U}(t)$ is a unitary operator allows us to time-slice. Unitarity lets us say $\hat{U}(t_{\mathrm{b}} - t_{\mathrm{a}}) = \hat{U}(t_{\mathrm{b}} - t_x)\hat{U}(t_x - t_{\mathrm{a}})$ and so time-slicing $\hat{U}(t) = \mathrm{e}^{-\mathrm{i}\hat{H}(t_{\mathrm{b}}-t_{\mathrm{a}})}$ in this way yields

$$\begin{aligned} G &= \langle q_{\mathrm{b}}|(\mathrm{e}^{-\mathrm{i}\hat{H}\Delta t})^N|q_{\mathrm{a}}\rangle \\ &= \langle q_{\mathrm{b}}|\mathrm{e}^{-\mathrm{i}\hat{H}\Delta t}\dots\mathrm{e}^{-\mathrm{i}\hat{H}\Delta t}\dots\mathrm{e}^{-\mathrm{i}\hat{H}\Delta t}|q_{\mathrm{a}}\rangle. \end{aligned} \qquad (23.3)$$

Notice that each of these mini time-evolution operators moves the particle for time Δt between two positions q_n and q_{n+1} (say) as shown in Fig. 23.2.

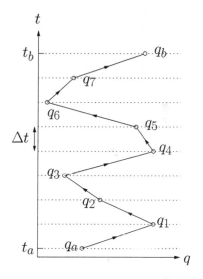

Fig. 23.2 Time slicing a trajectory from $(t_{\mathrm{a}}, q_{\mathrm{a}})$ to $(t_{\mathrm{b}}, q_{\mathrm{b}})$.

Example 23.1

Remember that in quantum mechanics we often make use of the completeness of a set of states. For the position states q_n, we have $\int \mathrm{d}q_n|q_n\rangle\langle q_n| = 1$. The trick to get any further with G is to insert a 'resolution of the identity' (also known as a *fat unity*) in between each mini time-evolution operator. We'll need $N-1$ fat unities to do this. Inserting these in (working from right to left) we obtain

$$\begin{aligned} G &= \langle q_{\mathrm{b}}|\mathrm{e}^{-\mathrm{i}\hat{H}\Delta t}\left[\int \mathrm{d}q_{N-1}|q_{N-1}\rangle\langle q_{N-1}|\right]\mathrm{e}^{-\mathrm{i}\hat{H}\Delta t}\dots \\ &\dots\mathrm{e}^{-\mathrm{i}\hat{H}\Delta t}\left[\int \mathrm{d}q_{n+1}|q_{n+1}\rangle\langle q_{n+1}|\right]\mathrm{e}^{-\mathrm{i}\hat{H}\Delta t}\left[\int \mathrm{d}q_n|q_n\rangle\langle q_n|\right]\mathrm{e}^{-\mathrm{i}\hat{H}\Delta t}\dots \\ &\dots\mathrm{e}^{-\mathrm{i}\hat{H}\Delta t}\left[\int \mathrm{d}q_1|q_1\rangle\langle q_1|\right]\mathrm{e}^{-\mathrm{i}\hat{H}\Delta t}|q_{\mathrm{a}}\rangle. \end{aligned} \qquad (23.4)$$

Notice what's happening here. In each resolution we're varying the q_n's independently and making new trajectories as we do so. In Fig. 23.2 this corresponds to moving the blobs along each line independently to some new position to make a new trajectory. We take a snapshot of the trajectory and then move the blobs again. The path integral is done by adding up all of the different trajectories that you make. In this way we add all of the possible paths between q_a and q_b. A little rearranging reveals that we have

$$G = \int \mathrm{d}q_1 \mathrm{d}q_2\dots\mathrm{d}q_{N-1}\,\langle q_{\mathrm{b}}|\mathrm{e}^{-\mathrm{i}\hat{H}\Delta t}|q_{N-1}\rangle\dots\langle q_{n+1}|\mathrm{e}^{-\mathrm{i}\hat{H}\Delta t}|q_n\rangle\dots\langle q_1|\mathrm{e}^{-\mathrm{i}\hat{H}\Delta t}|q_{\mathrm{a}}\rangle. \qquad (23.5)$$

Between the two amplitudes at the end we have a string of mini-propagators $G_n = \langle q_{n+1}|e^{-i\hat{H}\Delta t}|q_n\rangle$. Let's look at one of the typical mini-propagators:

$$
\begin{aligned}
G_n &= \langle q_{n+1}|e^{-i\hat{H}\Delta t}|q_n\rangle \\
&= \langle q_{n+1}|e^{-i\left[\frac{\hat{p}^2}{2m}+\hat{V}(q)\right]\Delta t}|q_n\rangle,
\end{aligned}
\tag{23.6}
$$

where we've written $\hat{H} = \hat{p}^2/2m + \hat{V}(q)$. To evaluate the mini-propagator we need to replace the operators by eigenvalues. The potential term acts[1] on the position state as follows: $e^{-i\hat{V}(q)\Delta t}|q_n\rangle = |q_n\rangle e^{-iV(q_n)\Delta t}$ [we've moved the state $|q_n\rangle$ past the potential operator $\hat{V}(q)$, replacing the operator with its eigenvalue $V(q_n)$]. We have then

$$
G_n = \langle q_{n+1}|e^{-i\frac{\hat{p}^2}{2m}\Delta t}|q_n\rangle e^{-iV(q_n)\Delta t}.
\tag{23.7}
$$

Next we need to know how $\hat{p}$ acts on a position state. Since $|q_n\rangle$ isn't an eigenstate of $\hat{p}$ (since $[\hat{q},\hat{p}] = i$) we just expand $|q_n\rangle$ in terms of momentum eigenstates in the usual way $|q_\mathrm{n}\rangle = \int dp|p\rangle\langle p|q_n\rangle = \int dp|p\rangle e^{-ipq_n}/(2\pi)^{\frac{1}{2}}$. Replacing $|q_n\rangle$ in this way gives us

$$
\begin{aligned}
G_n &= \int \frac{dp}{(2\pi)^{\frac{1}{2}}} \langle q_{n+1}|e^{-i\frac{\hat{p}^2}{2m}\Delta t}|p\rangle e^{-ipq_n} e^{-iV(q_n)\Delta t} \\
&= \int \frac{dp}{(2\pi)^{\frac{1}{2}}} \langle q_{n+1}|p\rangle e^{-i\frac{p^2}{2m}\Delta t} e^{-ipq_n} e^{-iV(q_n)\Delta t} \\
&= \int \frac{dp}{(2\pi)} e^{ipq_{n+1}} e^{-i\frac{p^2}{2m}\Delta t} e^{-ipq_n} e^{-iV(q_n)\Delta t} \\
&= \int \frac{dp}{(2\pi)} e^{-i\frac{p^2}{2m}\Delta t + ip(q_{n+1}-q_n)} e^{-iV(q_n)\Delta t}.
\end{aligned}
\tag{23.8}
$$

We've now removed all of the operators in favour of numbers and are left with an integral over p to do. This is a Gaussian integral which can be done exactly (see the next section[2]). The result is

$$
G_n = \left(\frac{-im}{2\pi\Delta t}\right)^{\frac{1}{2}} e^{\frac{im}{2}\frac{(q_{n+1}-q_n)^2}{\Delta t}} e^{-iV(q_n)\Delta t}.
\tag{23.9}
$$

Inserting back into eqn 23.5, and writing the factor $(-im/2\pi\Delta t)^{\frac{1}{2}} = \xi^{-1}$, we find that the propagator amplitude G can be written as a product of these mini-propagators

$$
G = \prod_{n=1}^{N-1} \int \frac{dq_n}{\xi} e^{\frac{im}{2}\frac{(q_{n+1}-q_n)^2}{(\Delta t)^2}\Delta t} e^{-iV(q_n)\Delta t}.
\tag{23.10}
$$

Lastly we take the limit $N \to \infty$, $\Delta t \to 0$. This corresponds to upping the contrast on our grid in Fig. 23.2, turning the jagged trajectories into smooth curves. Taking these limits turns $\frac{(q_{n+1}-q_n)^2}{(\Delta t)^2}$ into $\dot{q}^2$ and $\sum_n \Delta t \to \int dt$.

We obtain our answer for the total amplitude

$$
G = \int \mathcal{D}\left[q(t)\right] e^{i\int dt\left[\frac{m\dot{q}^2}{2}-V(q)\right]},
\tag{23.11}
$$

where we define the **integration measure** $\int \mathcal{D}[q(t)]$ as

$$
\int \mathcal{D}\left[q(t)\right] = \lim_{N\to\infty} \prod_{n=1}^{N-1} \int \frac{dq_n}{\xi}.
\tag{23.12}
$$

Equation 23.11 is an example of a **functional integral**, a sum over all possible trajectories encoded as a vast multiple integral over all time-sliced coordinates q_n, and in fact as we let $N \to \infty$ the functional integral

---

[1]There's some sleight of hand here. What we're saying is that $e^{-i(\hat{T}+\hat{V})\Delta t} = e^{-i\hat{T}\Delta t}e^{-i\hat{V}\Delta t} + O(\Delta t^2)$, and since Δt is assumed small, all is well. In general if two operators have a commutation relation $[\hat{A},\hat{B}] = \hat{C}$ then $e^{\hat{A}+\hat{B}} = e^{\hat{A}}e^{\hat{B}}e^{-\hat{C}/2}$. Here we take $\hat{A} = -i\hat{T}\Delta t$, $\hat{B} = -i\hat{V}\Delta t$ and so $\hat{C} = -(\Delta t)^2[\hat{T},\hat{V}]$ and thus $e^{-\hat{C}/2} \approx 1 + O[(\Delta t)^2]$.

[2]The result we need is that $\int_{-\infty}^{\infty} dx\, e^{-\frac{ax^2}{2}+bx} = \sqrt{\frac{2\pi}{a}} e^{\frac{b^2}{2a}}$, with $a = i\Delta t/m$ and $b = i(q_{n+1}-q_n)$.

becomes an infinitely multiple integral. The curly $\mathcal{D}$ hides the guts of the functional integral: the instruction to integrate over all possible trajectories.

Noticing that the Lagrangian is $L = m\dot{q}^2/2 - V(q)$ and restoring $\hbar$, we recognize

$$G = \int \mathcal{D}[q(t)] \, e^{\frac{i}{\hbar} \int dt \, L[q(t)]} = \int \mathcal{D}[q(t)] \, e^{iS/\hbar}, \qquad (23.13)$$

which is what we set out to justify. To recap: we calculate an amplitude by adding up contributions $e^{iS/\hbar}$ from each possible trajectory, of which there are an infinite number.[3]

What does the path integral tell us about quantum amplitudes? Although we add up all of the paths, each path is weighted by a phase factor, and we expect that there's some degree of cancellation between the contribution from different paths. Also, as we approach the classical world (mathematically achieved by sending $\hbar \to 0$) we should recover Lagrange's result that the trajectory the particle takes is the one for which the action is stationary, i.e. $\delta S/\delta q(t) = 0$. This can be understood by considering a particular trajectory with action S, typically large compared to $\hbar$, so that its contribution is weighted by $e^{iS/\hbar}$ (the phase $S/\hbar$ can be anywhere on the Argand clock). A close neighbouring trajectory will have a different action S', also large compared to $\hbar$, and if $\delta S = (S' - S) \gg \hbar$ it is likely to have a completely different phase. It doesn't take much imagination to realize that a bunch of such trajectories will all have essentially random phases and will cancel, giving no net contribution. In contrast, trajectories sufficiently close to the stationary value that Lagrange would have predicted (some are shown in Fig. 23.3) all have similar actions and thus similar phases, giving a strong in-phase contribution. Therefore, the sum is dominated by the stationary result. The quantum mechanical effects, known as quantum corrections or **quantum fluctuations**, are given, for the most part, by the trajectories close to the classical path of least action.[4]

23.2 The Gaussian integral

We've outlined a new way to think about quantum mechanics, but it's not worth much if we can't actually do a path integral. In this section we'll build up a mathematical armoury to deal with evaluating path integrals. It turns out that it's all related to a simple Gaussian integral and we'll build it up via a series of successive steps of generalization.

Step 0: The basic Gaussian integral around which this section is based is given by the standard integral[5]

$$\int_{-\infty}^{\infty} dx \, e^{-x^2} = \sqrt{\pi}. \qquad (23.14)$$

Step 1: The first generalization involves rescaling $x^2 \to ax^2/2$ to get

$$\int_{-\infty}^{\infty} dx \, e^{-\frac{ax^2}{2}} = \sqrt{\frac{2\pi}{a}}. \qquad (23.15)$$

[3]Doing the sum relies on the shady business of defining an integration measure $\int \mathcal{D}[q(t)]$ which doesn't look very well behaved in the limit $N \to \infty$. Although this is a legitimate concern, we'll see that we'll always be able to manoeuvre ourselves out of the trouble caused by $\int \mathcal{D}[q(t)]$.

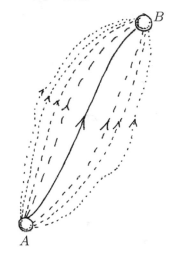

Fig. 23.3 The classical path is shown as a solid line, alternative paths allowed by quantum mechanics are shown as dashed lines.

[4]An approach to path integral calculations where we find the classical trajectory and then find the approximate form of the quantum corrections, is known as the stationary phase approximation and is used in Chapter 50.

[5]Write $I = \int_{-\infty}^{\infty} dx \, e^{-x^2}$ and $I = \int_{-\infty}^{\infty} dy \, e^{-y^2}$, then multiply the two equivalent integrals together to obtain

$$I^2 = \int_{x=-\infty}^{\infty} \int_{y=-\infty}^{\infty} dx dy \, e^{-(x^2+y^2)}$$

$$= \int_{\theta=0}^{2\pi} d\theta \int_{r=0}^{\infty} dr \, re^{-r^2},$$

where in the last equality we've switched to polar coordinates. These are standard integrals, yielding $I^2 = (2\pi) \times \frac{1}{2} = \pi$ and hence $I = \sqrt{\pi}$.

Step 2: A very important variation, which turns out to be the useful one for evaluating path integrals, is the integral:

$$\int_{-\infty}^{\infty} \mathrm{d}x\, e^{-\frac{ax^2}{2}+bx} = \sqrt{\frac{2\pi}{a}} e^{\frac{b^2}{2a}}, \qquad (23.16)$$

which is also not too difficult to prove, as shown below.

Example 23.2

Equation 23.16 is proved by 'completing the square'. The maximum of $P(x) = -\frac{ax^2}{2} + bx$ is $P_{\max} = \frac{b^2}{2a}$ at $x = \frac{b}{a}$. We rewrite

$$
\begin{aligned}
P(x) - P_{\max} &= -\frac{ax^2}{2} + bx - \frac{b^2}{2a} \\
&= -\frac{a}{2}\left(x - \frac{b}{a}\right)^2.
\end{aligned} \qquad (23.17)
$$

If we now change variables $y = (x - \frac{b}{a})$ so that $\mathrm{d}x = \mathrm{d}y$, then eqn 23.16 becomes

$$
\begin{aligned}
\int_{-\infty}^{\infty} \mathrm{d}x\, e^{-\frac{ax^2}{2}+bx} &= \int_{-\infty}^{\infty} \mathrm{d}x\, e^{-\frac{a}{2}\left(x-\frac{b}{a}\right)^2 + \frac{b^2}{2a}} \\
&= e^{\frac{b^2}{2a}} \int_{-\infty}^{\infty} \mathrm{d}y\, e^{-\frac{ay^2}{2}} = \sqrt{\frac{2\pi}{a}} e^{\frac{b^2}{2a}}. \qquad (23.18)
\end{aligned}
$$

There are some things to note about this equation for later use. Most of them are obvious, but important.

- Since x was our integration variable it doesn't appear on the right-hand side. We say that we have 'integrated out' x.
- We are left with an expression which has the structure

$$\int \mathrm{d}x\, e^{-x\frac{a}{2}x + bx} = \sqrt{\frac{2\pi}{a}} e^{\frac{1}{2}\left(b\frac{1}{a}b\right)}, \qquad (23.19)$$

where on the left-hand side a touches two x's and b touches one, while on the right-hand side the two b's are linked by the $\frac{1}{a}$.

We'll see this structure again a little later.

Step 3: In order to be able to integrate over functions like $q(t)$, we have to consider an intermediate step: integrating over vectors. This is because a function can be thought of as an infinite dimensional vector. We start with an N-dimensional vector $\mathbf{x}$, with components[6] x_j and consider the integral

$$\mathcal{I} = \int \mathrm{d}x_1 \mathrm{d}x_2 \ldots \mathrm{d}x_N\, e^{-\frac{1}{2}x_i A_{ij} x_j} = \int \mathrm{d}^N x\, e^{-\frac{1}{2}\mathbf{x}^{\mathrm{T}} \mathbf{A}\mathbf{x}}, \qquad (23.20)$$

where $\mathbf{A}$ is an $N \times N$, real, symmetric matrix.[7]

[6]For simplicity here we label ordinary vectors in Euclidean space with indices in the down position and assume a sum over repeated indices. We will return to our usual vector notation in the next chapter.

[7]This is nothing to worry about as we're used to doing multiple dimensional integrals over volumes or surfaces. Remember that the tricky thing about these integrals is the dependence of one component on another. Things are a lot simpler if we can separate the components out. That's exactly what we're going to do here.

Example 23.3

To evaluate the integral we make an orthogonal transformation $\mathbf{A} = \mathbf{O}^{\mathrm{T}}\mathbf{D}\mathbf{O}$, where we choose our matrices $\mathbf{O}$ in such a way as to make $\mathbf{D}$ a diagonal matrix. Our integral is then

$$\mathcal{I} = \int d^N x \, e^{-\frac{1}{2}\mathbf{x}^{\mathrm{T}}\mathbf{O}^{\mathrm{T}}\mathbf{D}\mathbf{O}\mathbf{x}}. \tag{23.21}$$

Next make the transformation $\mathbf{O}\mathbf{x} = \mathbf{y}$, and since the matrix $\mathbf{O}$ is orthogonal, the Jacobian of this transformation is 1 so $\int d^N x = \int d^N y$, and we get

$$\mathcal{I} = \int d^N y \, e^{-\frac{1}{2}\mathbf{y}^{\mathrm{T}}\mathbf{D}\mathbf{y}}. \tag{23.22}$$

As $\mathbf{D}$ is diagonal, we have $\mathbf{y}^{\mathrm{T}}\mathbf{D}\mathbf{y} = \sum_i D_{ii}(y_i)^2$, which allows us to separate out our multidimensional integral.

$$
\begin{aligned}
\int d^N y \, e^{-\frac{1}{2}\mathbf{y}^{\mathrm{T}}\mathbf{D}\mathbf{y}} &= \int dy_1 \, e^{-\frac{D_{11}(y_1)^2}{2}} \int dy_2 \, e^{-\frac{D_{22}(y_2)^2}{2}} \ldots \int dy_N \, e^{-\frac{D_{NN}(y_N)^2}{2}} \\
&= \left(\frac{2\pi}{D_{11}}\right)^{\frac{1}{2}} \left(\frac{2\pi}{D_{22}}\right)^{\frac{1}{2}} \ldots \left(\frac{2\pi}{D_{NN}}\right)^{\frac{1}{2}} \\
&= \prod_i^N \left(\frac{2\pi}{D_{ii}}\right)^{\frac{1}{2}} = \left(\frac{(2\pi)^N}{\det \mathbf{D}}\right)^{\frac{1}{2}} = \left(\frac{(2\pi)^N}{\det \mathbf{A}}\right)^{\frac{1}{2}}, \tag{23.23}
\end{aligned}
$$

where we've used $\prod_i D_{ii} = \det \mathbf{D} = \det \mathbf{A}$.

Step 4: We're finally in a position to generalize our important result, eqn 23.19, to the multidimensional case. We want to evaluate the integral

$$\mathcal{K} = \int d^N x \, e^{-\frac{1}{2}\mathbf{x}^{\mathrm{T}}\mathbf{A}\mathbf{x}+\mathbf{b}^{\mathrm{T}}\mathbf{x}}, \tag{23.24}$$

where $\mathbf{b}$ is an N-dimensional vector.

Example 23.4

By analogy with Example 23.2 we consider $P(\mathbf{x}) = -\frac{1}{2}\mathbf{x}^{\mathrm{T}}\mathbf{A}\mathbf{x} + \mathbf{b}^{\mathrm{T}}\mathbf{x}$. We can easily find its minimum (remembering that A_{ij} is symmetric and using $\frac{\partial x_i}{\partial x_j} = \delta_{ij}$):

$$
\begin{aligned}
\frac{\partial P}{\partial x^k} &= -\frac{1}{2}\delta_{ik}A_{ij}x_j - \frac{1}{2}x_i A_{ij}\delta_{jk} + b_j \delta_{jk} \\
&= -A_{kj}x_j + b_k \\
&= -(\mathbf{A}\mathbf{x} - \mathbf{b})_k = 0. \tag{23.25}
\end{aligned}
$$

This gives us a minimum of $P_{\min} = \frac{\mathbf{b}^{\mathrm{T}}\mathbf{A}^{-1}\mathbf{b}}{2}$ at $\mathbf{x} = \mathbf{A}^{-1}\mathbf{b}$. So again, just as in eqn 23.17, we can write

$$
\begin{aligned}
P(\mathbf{x}) - P_{\min} &= -\frac{1}{2}\mathbf{x}^{\mathrm{T}}\mathbf{A}\mathbf{x} + \mathbf{b}^{\mathrm{T}}\mathbf{x} - \frac{\mathbf{b}^{\mathrm{T}}\mathbf{A}^{-1}\mathbf{b}}{2} \\
&= -\frac{1}{2}(\mathbf{x} - \mathbf{A}^{-1}\mathbf{b})^{\mathrm{T}}\mathbf{A}(\mathbf{x} - \mathbf{A}^{-1}\mathbf{b}). \tag{23.26}
\end{aligned}
$$

This all looks familiar, of course. Again, we can perform a transformation and integrate over our new variable $\mathbf{y} = (\mathbf{x} - \mathbf{A}^{-1}\mathbf{b})$. We'll obtain a factor of the form $e^{\frac{1}{2}\mathbf{b}^{\mathrm{T}}\mathbf{A}^{-1}\mathbf{b}}$ emerging at the front.

The previous example leads us to

$$\mathcal{K} = \int \mathrm{d}^N x \, e^{-\frac{1}{2}\mathbf{x}^\mathrm{T}\mathbf{A}\mathbf{x}+\mathbf{b}^\mathrm{T}\mathbf{x}} = \left(\frac{(2\pi)^N}{\det\mathbf{A}}\right)^{\frac{1}{2}} e^{\frac{1}{2}\mathbf{b}^\mathrm{T}\mathbf{A}^{-1}\mathbf{b}}. \qquad (23.27)$$

Again we note that we've integrated out $\mathbf{x}$, and while $\mathbf{A}$ touches two $\mathbf{x}$'s on the left-hand side, $\mathbf{A}^{-1}$ takes $\mathbf{b}$ to $\mathbf{b}^\mathrm{T}$.

Step 5: Functions can be perfectly well described as infinite dimensional vectors, and so our next generalization is the functional integral

$$\mathcal{Q} = \int \mathcal{D}[f(x)] \, e^{-\frac{1}{2}\int \mathrm{d}x \mathrm{d}y \, f(x)A(x,y)f(y)+\int \mathrm{d}x \, b(x)f(x)}. \qquad (23.28)$$

To deal with this we'll revert to the discrete world by turning the functions into N-dimensional vectors; we'll write down the solution and then upgrade the result back into the world of functions.

Example 23.5

We convert $A(x,y) \to \mathbf{A}$, $f(x) \to \mathbf{x}$ and so on to obtain

$$\int \mathcal{D}[f(x)] \, e^{-\frac{1}{2}\int \mathrm{d}x\mathrm{d}y \, f(x)A(x,y)f(y)+\int \mathrm{d}x \, b(x)f(x)} \to \int \mathrm{d}^N x \, e^{-\frac{1}{2}\mathbf{x}^\mathrm{T}\mathbf{A}\mathbf{x}+\mathbf{b}^\mathrm{T}\mathbf{x}}$$

$$= \left(\frac{(2\pi)^N}{\det\mathbf{A}}\right)^{\frac{1}{2}} e^{\frac{1}{2}\mathbf{b}^\mathrm{T}\mathbf{A}^{-1}\mathbf{b}} \quad \text{(and then returning to functions)}$$

$$\to \left(\frac{(2\pi)^N}{\det A(x,y)}\right)^{1/2} e^{\frac{1}{2}\int \mathrm{d}x\mathrm{d}y \, b(x)A^{-1}(x,y)b(y)}. \qquad (23.29)$$

We have the final result that

$$\begin{aligned} \mathcal{Q} &= \int \mathcal{D}[f(x)] \, e^{-\frac{1}{2}\int \mathrm{d}x\mathrm{d}y \, f(x)A(x,y)f(y)+\int \mathrm{d}x \, b(x)f(x)} \\ &= B \left[\det A(x,y)\right]^{-1/2} e^{\frac{1}{2}\int \mathrm{d}x\mathrm{d}y \, b(x)A^{-1}(x,y)b(y)}, \qquad (23.30) \end{aligned}$$

where we've introduced a constant B to soak up the factor of $(2\pi)^{\frac{N}{2}}$ (which is a potential embarrassment in the limit $N \to \infty$).

The inverse function $A^{-1}(x,y)$ is that function that satisfies

$$\int \mathrm{d}z \, A(x,z) \, A^{-1}(z,y) = \delta(x-y). \qquad (23.31)$$

That's just another way of saying that $A^{-1}(x,y)$ is the Green's function for the operator $A(x,y)$. It's been a long slog, but we can now do one sort of functional integral,[8] as long as we can find the determinant $\det A(x,y)$ and find the Green's function of $A(x,y)$. In the next section, at long last, we'll finally do a functional integral!

---

[8]The key path integral equation (eqn 23.30), which we'll use most frequently in later chapters, may be summarized as

$$\int \mathcal{D}[f] \, e^{-\frac{1}{2}\int\int fAf+\int bf} = N e^{\frac{1}{2}\int bA^{-1}b},$$

where N is a constant.

23.3   The propagator for the simple harmonic oscillator

An illuminating example of the path integral approach is the simple harmonic oscillator. We're going to find the probability amplitude that if, at $t_a = 0$, we put a particle in the minimum of a harmonic potential located at $q = 0$, when we come back at time $t_b = T$ the particle will still be at $q = 0$, where we originally put it. We need, therefore, to compute the propagator

$$
\begin{aligned}
&G(q_b = 0, t_b = T, q_a = 0, t_a = 0) \\
&= \left\langle \begin{array}{c|c} \text{Particle ends at} & \text{Particle starts at} \\ q_b = 0, t_b = T & q_a = 0, t_a = 0 \end{array} \right\rangle \\
&= \int_{q_a = 0, t_a = 0}^{q_b = 0, t_b = T} \mathcal{D}[q(t)]\, e^{i \int_0^T dt\, L[q(t)]}.
\end{aligned} \tag{23.32}
$$

Let's do the integral. The Lagrangian for a simple harmonic oscillator is $L = \frac{m\dot{q}(t)^2}{2} - \frac{m\omega_0^2 q(t)^2}{2}$, where $\omega_0 = (K/m)^{1/2}$. The path integral is therefore

$$
G = \int \mathcal{D}[q(t)]\, e^{\frac{im}{2} \int dt \left[\left(\frac{dq(t)}{dt} \right)^2 - \omega_0^2 q(t)^2 \right]}. \tag{23.33}
$$

Unfortunately this is not in the form of the one integral we know how to do (that looks like $\int \mathcal{D}q\, e^{-\frac{1}{2}\int q\hat{A}q + \int bq}$). Can we massage the simple-harmonic oscillator into this form? We can and the trick, which is based on integrating by parts, is one that we'll use repeatedly for path integrals.

Example 23.6

The action in the argument of the exponential part of the path integral is

$$
S = \int dt \left[\frac{m\dot{q}(t)^2}{2} - \frac{m\omega_0^2 q(t)^2}{2} \right], \tag{23.34}
$$

and the trick just involves integrating the first term, which represents the kinetic energy, by parts

$$
I = \int dt \left(\frac{\partial q(t)}{\partial t} \right)^2 = \int dt \left(\frac{\partial q(t)}{\partial t} \right) \left(\frac{\partial q(t)}{\partial t} \right), \tag{23.35}
$$

giving

$$
I = \left[q(t) \frac{\partial q(t)}{\partial t} \right]_{t=0}^{T} - \int dt\, q(t) \frac{\partial^2}{\partial t^2} q(t). \tag{23.36}
$$

The boundary conditions ($q = 0$ at the beginning and end of the trajectory) wipe out the first term and the second term has the form '$q\hat{C}q$', where $\hat{C}$ is an operator.

The use of our neat trick gives the functional integral

$$
G(0, T, 0, 0) = \int \mathcal{D}[q(t)]\, e^{\frac{i}{2} \int dt\, q(t)\hat{C}q(t)}, \tag{23.37}
$$

with $\hat{C} = m\left(-\frac{\partial^2}{\partial t^2} - \omega_0^2\right)$. We can do this integral using a slightly modified form of the result in eqn 23.30. Setting $A(x,y) = -i\hat{C}$ and $b(x) = 0$ in eqn 23.30, we discretize the integral to obtain

$$G \to \int dq_1 dq_2 \ldots dq_N\, e^{\frac{i}{2}\sum_{ij} q_i \hat{C}_{ij} q_j} = \left(\frac{(2\pi i)^N}{\det \hat{\mathbf{C}}}\right)^{\frac{1}{2}}. \qquad (23.38)$$

Again this fills us with a sense of dread since the result is quite frightening with its factor $(2\pi i)^N$ that won't look pretty as $N \to \infty$. Undaunted, we'll assume that when we pop back into the world of functions the answer will at least be that $\to G \propto [\det \hat{C}(x,y)]^{-1/2}$. Admittedly, the determinant of the differential operator $\hat{C}(x,y)$ might also send a chill down the spine. To give such an object meaning, we can find the eigenvalues (and eigenfunctions) of the operator and use the fact that the determinant is given by the product of eigenvalues. Our answer is then

$$G(0,T,0,0) = B \det\left[m\left(-\frac{\partial^2}{\partial t^2} - \omega_0^2\right)\right]^{-\frac{1}{2}}, \qquad (23.39)$$

where B is some constant coming from the integration measure that we'll worry about shortly.

Example 23.7

The next job is to work out the determinant. This we do with the trick described above that makes use of the fact that the determinant of a matrix is equal to the product of its eigenvalues. Finding the eigenfunctions of the operator $\hat{C}$ isn't too tough either. You can check that $m\left(-\frac{\partial^2}{\partial t^2} - \omega_0^2\right)q(t) = \lambda_n q(t)$ has eigenfunctions $q(t) = c\sin(u_n t)$, with eigenvalues $\lambda_n = m(u_n^2 - \omega_0^2)$, of which there are an infinite number. Remembering that our boundary conditions are that $q = 0$ at $t = 0$ and $q = 0$ at $t = T$ we need $u_n = n\pi/T$ for everything to work. Putting this all together, we have an answer

$$G(0,T,0,0) = B\left\{\prod_{n=1}^{\infty} m\left[\left(\frac{n\pi}{T}\right)^2 - \omega_0^2\right]\right\}^{-\frac{1}{2}}. \qquad (23.40)$$

At this stage we notice that the answer isn't very useful in its present form. This is because the infinite product looks like it gives infinity whenever $\omega_0 T = n\pi$ and we don't know what B is yet.

A sanity check is useful here. When there's no spring constant $\omega_0 = 0$, there's no harmonic potential and we have a free particle. In that case we have

$$G_{\omega_0=0}(0,T,0,0) = B\left\{\prod_{n=1}^{\infty} m\left(\frac{n\pi}{T}\right)^2\right\}^{-\frac{1}{2}}, \qquad (23.41)$$

but we already know (from Example 16.5) that for a free particle $G_{\text{free}}(0,T,0,0) = \left(\frac{-im}{2\pi T}\right)^{\frac{1}{2}}$. This will be enough to provide the proportionality constant B for our problem.

We multiply eqn 23.40 by $G_{\text{free}}/G_{\omega_0=0} = 1$:

$$
\begin{aligned}
G(0,T,0,0) &= G(0,T,0,0)\frac{G_{\text{free}}(0,T,0,0)}{G_{\omega_0=0}(0,T,0,0)} \\[2mm]
&= \frac{B\left\{\prod_{n=1}^{\infty} m\left[\left(\frac{n\pi}{T}\right)^2 - \omega_0^2\right]\right\}^{-\frac{1}{2}}}{B\left\{\prod_{n=1}^{\infty} m\left(\frac{n\pi}{T}\right)^2\right\}^{-\frac{1}{2}}}\left(\frac{-\mathrm{i}m}{2\pi T}\right)^{\frac{1}{2}} \\[2mm]
&= \left\{\prod_{n=1}^{\infty}\left[1 - \left(\frac{\omega_0 T}{n\pi}\right)^2\right]\right\}^{-\frac{1}{2}}\left(\frac{-\mathrm{i}m}{2\pi T}\right)^{\frac{1}{2}},\qquad (23.42)
\end{aligned}
$$

which removes all of the dangerous terms. Finally we use the identity $\prod_{n=1}^{\infty}[1 - (x/n\pi)^2]^{-1} = x/\sin x$ and write

$$
G(0,T,0,0) = \left(\frac{\omega_0 T}{\sin \omega_0 T}\right)^{\frac{1}{2}}\left(\frac{-\mathrm{i}m}{2\pi T}\right)^{\frac{1}{2}}.\qquad (23.43)
$$

The final answer is that

$$
G(0,T,0,0) = \left(\frac{-\mathrm{i}m\omega_0}{2\pi \sin \omega_0 T}\right)^{\frac{1}{2}},\qquad (23.44)
$$

and the probability density $|G|^2$ derived from this is shown in Fig. 23.4. Interestingly, the probability density peaks strongly for $T = n\pi/\omega_0$, where n is an integer, making it very likely that the particle will be found at the origin at these times. Perhaps this shouldn't come as too much of a surprise: we know that classical particle oscillates back and forth in an harmonic potential. From the path integral point of view, the peaks in probability density result from the constructive interference of all of the possible particle paths that start and end at the origin in a time T; while the minima result from a large degree of cancellation of such path sections. It's as if the paths at the origin go in and out of focus as a function of time!

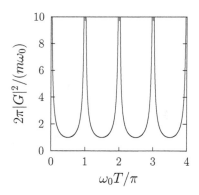

Fig. 23.4 The (scaled) probability density $|G(0,T,0,0)|^2$ as a function of time for a particle that starts at the origin of a harmonic oscillator potential to be found there a time T later.

Chapter summary

- Feynman's path integral approach involves finding amplitudes by adding up contributions from all possible trajectories. Each trajectory makes a contribution $\exp(\mathrm{i}S/\hbar)$.
- One type of functional integral has been evaluated and yields the result $\int \mathcal{D}[f]\mathrm{e}^{-\frac{1}{2}\int fAf + \int bf} = \left(\frac{(2\pi)^N}{\det A}\right)^{\frac{1}{2}}\mathrm{e}^{\frac{1}{2}\int bA^{-1}b}$.
- The simple harmonic oscillator provides an example showing the infinities that have to be tamed in a typical calculation.

Exercises

(23.1) *A back of the envelope derivation of the most important result of the chapter.*

Start with a Lagrangian $L = \frac{1}{2}x\hat{A}x + bx$, where $\hat{A}$ is an operator. We will be brash and treat $\hat{A}$ as a number. Use the Euler–Lagrange equation to find x and show that the Lagrangian may be expressed equivalently as $L = -b\frac{1}{2\hat{A}}b$.

(23.2) *The path integral derivation of Wick's theorem.*

(a) Calculate $\int_{-\infty}^{\infty} dx\, x^2 e^{-\frac{1}{2}ax^2}$.

(b) Define

$$\langle x^n \rangle = \frac{\int_{-\infty}^{\infty} dx\, x^n e^{-\frac{1}{2}ax^2}}{\int_{-\infty}^{\infty} dx\, e^{-\frac{1}{2}ax^2}}. \qquad (23.45)$$

Calculate (i) $\langle x^2 \rangle$, (ii) $\langle x^4 \rangle$ and (iii) $\langle x^n \rangle$.

(c) Make sense of the result for $\langle x^n \rangle$ diagrammatically, representing each factor of $\frac{1}{a}$ as a line linking factors of x.

(d) Now consider the N-dimensional vector $\mathbf{x}$ and the integral

$$\begin{aligned} \mathcal{K} &= \int dx_1 \ldots dx_N\, e^{-\frac{1}{2}\mathbf{x}^T \mathbf{A}\mathbf{x} + \mathbf{b}^T \mathbf{x}} \\ &= \left(\frac{(2\pi)^N}{\det \mathbf{A}}\right)^{\frac{1}{2}} e^{\frac{1}{2}\mathbf{b}^T \mathbf{A}^{-1}\mathbf{b}}. \qquad (23.46) \end{aligned}$$

By differentiating with respect to components of the vector $\mathbf{b}$, and then setting $\mathbf{b} = 0$, show that

$$\langle x_i x_j \rangle = \frac{\int dx_1 \ldots dx_N\, x_i x_j e^{-\frac{1}{2}\mathbf{x}^T \mathbf{A}\mathbf{x}}}{\int dx_1 \ldots dx_N\, e^{-\frac{1}{2}\mathbf{x}^T \mathbf{A}\mathbf{x}}} = (\mathbf{A}^{-1})_{ij}. \qquad (23.47)$$

(e) Using these results, argue that

$$\begin{aligned} \langle x_i x_j x_k x_l \rangle &= (\mathbf{A}^{-1})_{ij}(\mathbf{A}^{-1})_{kl} + (\mathbf{A}^{-1})_{ik}(\mathbf{A}^{-1})_{jl} \\ &\quad + (\mathbf{A}^{-1})_{il}(\mathbf{A}^{-1})_{jk}. \qquad (23.48) \end{aligned}$$

(f) Write down an expression for the general case $\langle x_i \ldots x_z \rangle$.

This is the basis of Wick's theorem in the path integral approach.

(23.3) Consider the forced quantum oscillator from Exercise 22.1

$$L = \frac{1}{2}m\dot{x}(t)^2 - \frac{1}{2}m\omega^2 x(t)^2 + f(t)x(t). \qquad (23.49)$$

Take $f(t)$ to be a constant force f_0 which acts only for a finite length of time: $0 \le t \le T$.

(a) Show that the amplitude for a particle that is in the ground state at $t = 0$ to be in the ground state at $t = T$ is given by

$$\mathcal{A} = e^{-\frac{1}{2}\int dt' dt f(t)G(t,t')f(t')}, \qquad (23.50)$$

where

$$G(t,t') = \frac{\theta(t - t')e^{-i\omega(t-t')} + \theta(t' - t)e^{i\omega(t-t')}}{2m\omega}. \qquad (23.51)$$

(b) Carry out the integral (being careful of the limits) and show that the amplitude is given by

$$\mathcal{A} = \exp\left[\frac{if_0^2}{2m\omega^2}\left(T - \frac{\sin\omega T}{\omega} + i\frac{2}{\omega}\sin^2\frac{\omega T}{2}\right)\right]. \qquad (23.52)$$

(c) What is the probability for the oscillator to still be in its ground state at $t = T$? Compare this with the result of Exercise 22.1.

(d) What is the physical meaning of the imaginary part of the argument in the exponential?

Field integrals

24

Life is not a walk across a field
Russian proverb, quoted by Boris Pasternak (1890–1960) in his poem *Hamlet*

In this chapter we will apply the functional integrals of the last chapter to quantum fields. Remarkably, the generating functional $Z[J]$ for quantum fields can be written as a functional integral. In this chapter we'll find that integral. Using functional integrals can be seen as an alternative to canonical quantization and the $\check{S}$-operator (as shown in Fig. 24.1) although it's best to have the two approaches available since both have their strengths and weaknesses.

24.1 The functional integral for fields

In the previous chapter we calculated a propagator amplitude for a particle by doing a functional integral. This involved summing all possible trajectories for a single particle travelling between two spacetime points. Our result was that

$$G(x, t_x, y, t_y) = \int \mathcal{D}[q(t)] \, e^{i \int_{t_x}^{t_y} dt \, L[q(t)]}. \qquad (24.1)$$

However, we know that single-particle quantum mechanics isn't enough to describe reality: we need fields. We therefore want to find a functional integral that sums all possible *field configurations* that can exist between the two spacetime points. The bad news is that this integral won't yield a propagator. In the next example we'll pursue a method of integrating over field configurations and in the following section we will show (the good news) how this leads directly to the generating functional.

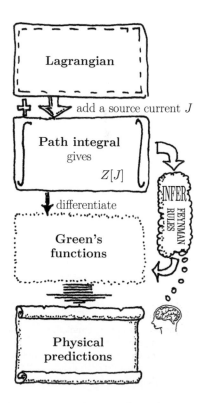

Fig. 24.1 Using functional integrals to do quantum field theory.

Example 24.1

The jump between integrating over particle trajectories to integrating over field configurations is quite straightforward. We can still write down the $e^{iS/\hbar}$ factor for fields, but now the action S is the integral of the Lagrangian density over spacetime: $S = \int d^4x \, \mathcal{L}[\phi(x)]$. Note that the fields appearing here are classical Heisenberg fields. Their time dependence is given by the full Hamiltonian and they haven't been through a canonical quantization machine so have no commutation relations. A functional integral over fields may then be written as

$$\int \mathcal{D}[\phi(x)] \, e^{i \int_{-\infty}^{\infty} d^4x \, \mathcal{L}[\phi(x)]}. \qquad (24.2)$$

The difference with the path integral that we had before is that the dynamic variable is now the field $\phi(x)$ rather than the trajectory $q(t)$ and the integral is therefore over functions of position and time rather than just the functions of time. Despite this difference, we do the integral in the same way as before: we discretize. Previously we just sliced up time, now we slice up spacetime. We obtain a multiple integral over N spacetime variables.

We know how to do the integral for the special, but important, case that the Lagrangian can be worked into the form $L[\boldsymbol{q}] = \frac{1}{2}\boldsymbol{q}^{\mathrm{T}}\boldsymbol{A}\boldsymbol{q} + \mathrm{i}\boldsymbol{b}^{\mathrm{T}}\boldsymbol{q}$. Based on the observation that

Here's the recipe to revert back from continuous functions to a discrete lattice:

- The field $\phi(x)$ evaluated at a spacetime point x becomes an N-component vector q_j, where $j = 1,\ldots,N$ labels a point in spacetime.

- Differential operators become real symmetric matrices. Things like $\frac{\partial\phi}{\partial x}$ become $\frac{1}{l}(q_i - q_j) = A_{ij}q_j$.

- Integrals become sums: $\int \mathrm{d}x$ becomes $l\sum_j$.

$$\int \mathrm{d}x\, e^{\frac{\mathrm{i}}{2}ax^2 + \mathrm{i}bx} = \left(\frac{2\pi\mathrm{i}}{a}\right)^{\frac{1}{2}} e^{-\frac{\mathrm{i}b^2}{2a}}, \tag{24.3}$$

we can read off that the answer is

$$\mathcal{P} = \int \mathrm{d}^N q\, e^{\frac{\mathrm{i}}{2}\mathbf{q}^{\mathrm{T}}\mathbf{A}\mathbf{q}+\mathrm{i}\mathbf{b}^{\mathrm{T}}\mathbf{q}} = \left(\frac{(2\pi\mathrm{i})^N}{\det\mathbf{A}}\right)^{\frac{1}{2}} e^{-\frac{\mathrm{i}}{2}\mathbf{b}^{\mathrm{T}}\mathbf{A}^{-1}\mathbf{b}}. \tag{24.4}$$

Our important result for a calculable functional integral over fields is that

$$\int \mathcal{D}[\phi(x)]\, e^{\frac{\mathrm{i}}{2}\int \mathrm{d}^4x\,\mathrm{d}^4y\,\phi(x)A(x,y)\phi(y)+\mathrm{i}\int \mathrm{d}^4x\,b(x)\phi(x)}$$
$$= B\left[\det A(x,y)\right]^{-\frac{1}{2}} e^{-\frac{\mathrm{i}}{2}\int \mathrm{d}^4x\,\mathrm{d}^4y\,b(x)A^{-1}(x,y)b(y)}, \tag{24.5}$$

where we've hidden the factor $(2\pi\mathrm{i})^{\frac{N}{2}}$ in the (divergent) constant B as we did before.

24.2   Which field integrals should you do?

We're now at the stage where we'd like to have a go at doing a functional integral over some fields. Before getting too excited about all of this, we need to decide which integrals are worth doing. What are we trying to learn? Which fields do we start at, and which do we end with?

We want access to all of the Green's functions for a theory, since these contain all of the physics. If we only want to work out one thing, then it should be the generating functional $Z[J]$. We saw before that this functional contained all of the Green's functions and therefore all of the information for the theory. Remember that the generating functional $Z[J]$ tells us the amplitude

$$Z[J] = \left\langle \begin{array}{c} \text{no particles} \\ \text{at } x^0 = \infty \end{array} \middle| \begin{array}{c} \text{no particles} \\ \text{at } y^0 = -\infty \end{array} \right\rangle_J, \tag{24.6}$$

that is, in the presence of a source, we start with no particles and end with no particles.[1] The remarkable answer turns out to be that the generating functional can be written in terms of the functional integral as[2]

$$Z[J] = \int \mathcal{D}[\phi(x)]\, e^{\mathrm{i}\int \mathrm{d}^4x\,\{\mathcal{L}[\phi(x)]+J(x)\phi(x)\}}. \tag{24.7}$$

In order to go about calculating $Z[J]$, we start by splitting the Lagrangian up into two parts $\mathcal{L} = \mathcal{L}_0 + \mathcal{L}_\mathrm{I}$. As usual, $\mathcal{L}_0$ is the free part: it's the part that can be cast in the quadratic form $\frac{1}{2}\phi(x)\hat{A}\phi(x)$ and is therefore solvable by our one known field integral. The other term $\mathcal{L}_\mathrm{I}$ is the interaction part: that is, the part which isn't solvable by canonical

[1]Also remember that our goal is actually the Green's functions. To get to these we can take functional derivatives of the generating functional $\mathrm{i}^n G^{(n)}(x_1,\ldots,x_n) = \frac{1}{Z[J]}\frac{\delta^n Z[J]}{\delta J(x_1)\ldots\delta J(x_n)}\big|_{J=0}$ (eqn 21.10).

[2]Note that this is the unnormalized generating functional. Recall that to obtain the normalized generating functional we can simply divide through by $Z[J=0]$ since $\mathcal{Z}[J] = Z[J]/Z[0]$.

quantization. Written in terms of functional integrals, the generating functional is therefore

$$Z[J] = \int \mathcal{D}[\phi(x)]\, e^{i \int d^4x\, (\mathcal{L}_0 + \mathcal{L}_I + J\phi)}. \tag{24.8}$$

Now let's try to do the integral.

24.3   The generating functional for scalar fields

We'll calculate a generating functional for the free scalar field. Following the custom that quantities involving free fields are given a subscript '0' this will be denoted $Z_0[J]$. The free Lagrangian is given by

$$\mathcal{L}_0 = \frac{1}{2}(\partial_\mu \phi)^2 - \frac{m^2}{2}\phi^2. \tag{24.9}$$

The generating functional $Z_0[J]$ for the free scalar field is

$$Z_0[J] = \int \mathcal{D}\phi\, e^{\frac{i}{2} \int d^4x \{(\partial_\mu \phi)^2 - m^2\phi^2\} + i \int d^4x\, J\phi}. \tag{24.10}$$

Note that in eqn 24.10 and later we've suppressed the spacetime variable x to avoid unnecessary clutter.

As we've said above, we can do the integral if we massage $Z_0[J]$ into the form of eqn 24.4. We again invoke the neat trick of doing by parts the integral of the kinetic energy part of the Lagrangian density in the argument of the exponential ($e^{\frac{i}{2} \int d^4x\, (\partial_\mu \phi)^2}$):

$$\int d^4x\, (\partial_\mu \phi)^2 = [\phi(\partial_\mu \phi)]_{-\infty}^{\infty} - \int d^4x\, \phi\, \partial^2\phi, \tag{24.11}$$

where we assume that the first term disappears at the boundary, that is the field dies off as we head out to spacetime infinity. This trick enables us to make the replacement $\int d^4x\, (\partial_\mu \phi)^2 \to -\int d^4x\, \phi\, \partial^2\phi$ and allows us to rewrite the functional integral as

$$Z_0[J] = \int \mathcal{D}\phi\, e^{\frac{i}{2} \int d^4x\, \phi[-(\partial^2 + m^2)]\phi + i \int d^4x\, J\phi}, \tag{24.12}$$

which is in the form of the one field integral that we know how to do.[3] We have that $\hat{A} = -(\partial^2 + m^2)$ and our answer is that

$$Z_0[J] = B \left[\det \hat{A}(x,y)\right]^{-\frac{1}{2}} e^{-\frac{1}{2} \int d^4x d^4y\, J(x)[i\hat{A}^{-1}(x,y)]J(y)}, \tag{24.13}$$

where B is a (potentially infinite) constant determined by the integration measure. The determinant also looks worryingly infinite!

Normalization will save us here. The normalized generating functional is $\mathcal{Z}_0[J] = Z[J]/Z[J=0]$, where $Z_0[J=0] = B \det \left[\hat{A}(x,y)\right]^{-\frac{1}{2}}$. This removes two divergent quantities (B and the determinant) at a stroke.

[3]We will see that this is telling us that a Lagrangian that can be massaged into this form may be quantized using functional integration. The family of quadratic and bilinear Lagrangians for which this is true are exactly those that may be diagonalized by canonical quantization and are conventionally called non-interacting.

Recall that the normalization ensures that the amplitude to start and end with no particles, in the absence of a source, is unity.

Finally, we recognize $\hat{A}(x,y) = -(\partial^2 + m^2)$ as an operator giving rise to the Klein–Gordon equation of motion and note that its inverse determines the free scalar propagator via

$$i\hat{A}^{-1}(x,y) = \Delta(x,y) = \int \frac{\mathrm{d}^4 p}{(2\pi)^4} \frac{\mathrm{i}e^{-\mathrm{i}p\cdot(x-y)}}{p^2 - m^2 + \mathrm{i}\epsilon}. \tag{24.14}$$

Example 24.2

To check this, we recall that the inverse of an operator satisfies $\hat{A}\hat{A}^{-1} = I$, which in the continuum limit becomes

$$-(\partial_x^2 + m^2)\hat{A}^{-1} = \delta^{(4)}(x-y), \tag{24.15}$$

where the ∂_x^2 operator acts on the x variable. Next we note that we defined $\Delta(x,y)$ via

$$-(\partial_x^2 + m^2)\Delta(x,y) = \mathrm{i}\delta^{(4)}(x-y), \tag{24.16}$$

so we can identify $\mathrm{i}\hat{A}^{-1} = \Delta(x,y)$, that is, the Green's function of the equation of motion, also known as the Feynman propagator.

Putting all of this together gives us a normalized generating functional for the free scalar field of

Notice that the field ϕ doesn't appear on the right-hand side of eqn (24.17). We say that we've 'integrated out' the field $\phi(x)$.

$$\begin{aligned} \mathcal{Z}_0[J] &= \frac{\int \mathcal{D}\phi\, e^{\frac{1}{2}\int \mathrm{d}^4 x\, \phi\{-(\partial^2+m^2)\}\phi + \mathrm{i}\int \mathrm{d}^4 x\, J\phi}}{\int \mathcal{D}\phi\, e^{\frac{1}{2}\int \mathrm{d}^4 x\, \phi\{-(\partial^2+m^2)\}\phi}} \\ &= e^{-\frac{1}{2}\int \mathrm{d}^4 x \mathrm{d}^4 y\, J(x)\Delta(x-y)J(y)}. \end{aligned} \tag{24.17}$$

Example 24.3

We now justify that eqn 24.17 is the same generating functional for the free scalar field that we defined in Chapter 22. The Chapter 22 definition of the generating functional may be written

$$\mathcal{Z}[J] = \sum_{n=0}^{\infty} \frac{\mathrm{i}^n}{n!} \int \mathrm{d}^4 x_1 ... \mathrm{d}^4 x_n\, J(x_1)...J(x_n)\langle\Omega|T\hat{\phi}_\mathrm{H}(x_1)...\hat{\phi}_\mathrm{H}(x_n)|\Omega\rangle. \tag{24.18}$$

Note that for a non-interacting field we have $\hat{\phi}_\mathrm{H} = e^{\mathrm{i}\hat{H}_0 t}\hat{\phi}e^{-\mathrm{i}\hat{H}_0 t} = \hat{\phi}_\mathrm{I}$ and so we can use Wick's theorem (which applies to freely evolving interaction picture fields only) to expand any time-ordered products. They fall apart into a sum over products of free propagators. For example, the $n = 4$ term is

$$\langle 0|T\hat{\phi}_\mathrm{I}(x_1)\hat{\phi}_\mathrm{I}(x_2)\hat{\phi}_\mathrm{I}(x_3)\hat{\phi}_\mathrm{I}(x_4)|0\rangle = \Delta(x_1-x_2)\Delta(x_3-x_4) + \Delta(x_1-x_3)\Delta(x_2-x_4) + \Delta(x_1-x_4)\Delta(x_2-x_3).$$

Renaming variables, we see that each of the three terms will therefore contribute a factor $[\int J(x_1)\Delta(x_1-x_2)J(x_2)]^2$. This enables us to write the fourth-order term as

$$\begin{aligned} &\frac{\mathrm{i}^4}{4!}\int \mathrm{d}^4 x_1...\mathrm{d}^4 x_4\, J(x_1)...J(x_4)\langle\Omega|T\hat{\phi}_\mathrm{H}(x_1)...\hat{\phi}_\mathrm{H}(x_4)|\Omega\rangle \\ &= \frac{1}{2!}\left(-\frac{1}{2}\int \mathrm{d}^4 x_1 \mathrm{d}^4 x_2\, J(x_1)\Delta(x_1-x_2)J(x_2)\right)^2. \end{aligned} \tag{24.19}$$

Repeating this process for other values of n generates the corresponding terms in the exponential and the answer indeed reduces to eqn 24.17.

The point of finding the generating function is to give us access to propagators. Specifically we have for free fields that the propagator is given, in terms of the normalized generating functional, by

$$
\begin{aligned}
G_0^{(n)}(x_1, ..., x_n) &= \frac{1}{\mathrm{i}^n} \frac{\delta^n \mathcal{Z}_0[J]}{\delta J(x_1)...\delta J(x_n)}\bigg|_{J=0} \\
&= \frac{1}{\mathrm{i}^n} \frac{1}{\mathcal{Z}_0[J=0]} \frac{\delta^n \mathcal{Z}_0[J]}{\delta J(x_1)...\delta J(x_n)}\bigg|_{J=0} . \quad (24.20)
\end{aligned}
$$

We'll evaluate this in two different ways for the single-particle propagator $G_0(x, y)$. Differentiating the expression for the functional integral $\mathcal{Z}_0[J]$ with respect to the J's gives us

$$
G_0(x, y) = \frac{\int \mathcal{D}\phi \, \phi(x)\phi(y)\mathrm{e}^{\mathrm{i}\int \mathrm{d}^4x \mathcal{L}_0[\phi]}}{\int \mathcal{D}\phi \, \mathrm{e}^{\mathrm{i}\int \mathrm{d}^4x \mathcal{L}_0[\phi]}}, \quad (24.21)
$$

while differentiating the expression for the normalized generating functional $\mathcal{Z}_0[J] = \mathrm{e}^{-\frac{1}{2}\int \mathrm{d}^4x \mathrm{d}^4y \, J(x)\Delta(x,y)J(y)}$ gives us the expected answer $G_0(x, y) = \Delta(x, y)$. Thus the Green's function, which we wrote down as $\langle \Omega | T \hat{\phi}_\mathrm{H}(x)\hat{\phi}_\mathrm{H}(y) | \Omega \rangle$, is obtained by integrating two fields weighted by a factor $\mathrm{e}^{\mathrm{i}\int \mathrm{d}^4x \mathcal{L}_0[\phi]}$ which tells us how the amplitudes of different field trajectories are distributed.[4]

Example 24.4

We'll apply what we've learnt to the interesting case of the free massive vector field theory from Chapter 13 described by

$$
\begin{aligned}
\mathcal{L} &= -\frac{1}{4}(F_{\mu\nu}F^{\mu\nu}) + \frac{1}{2}m^2 A_\mu A^\mu \\
&= -\frac{1}{2}(\partial_\mu A_\nu \partial^\mu A^\nu - \partial_\mu A_\nu \partial^\nu A^\mu) + \frac{1}{2}m^2 A^\mu A_\mu \quad (24.23)
\end{aligned}
$$

(see Exercise 13.4 if in doubt). The only complication with this is the different components of the A^μ field. Note that in this case the result of the one functional integral that we can do is written

$$
\mathcal{Z}_0[J] = \frac{\int \mathcal{D}A \, \mathrm{e}^{\frac{\mathrm{i}}{2}\int \mathrm{d}^4x \, A^\mu \hat{K}_{\mu\nu}A^\nu + \mathrm{i}\int \mathrm{d}^4x \, J_\mu A^\mu}}{\int \mathcal{D}A \, \mathrm{e}^{\frac{\mathrm{i}}{2}\int \mathrm{d}^4x \, A^\mu \hat{K}_{\mu\nu}A^\nu}} = \mathrm{e}^{-\frac{1}{2}\int \mathrm{d}^4x \mathrm{d}^4y \, J^\mu(x)[\mathrm{i}\hat{K}_{\mu\nu}^{-1}]J^\nu(y)} .
$$

$$(24.24)$$

We can, therefore, only do our functional integral if the Lagrangian is in the form $\frac{1}{2}A^\mu \hat{K}_{\mu\nu}A^\nu$, where $\hat{K}_{\mu\nu}$ is a differential operator (with an inverse). Just as before, the way to achieve this is, as usual, to integrate $\mathcal{L}$ by parts. Of the three terms in the Lagrangian, the final one is already in the correct form. Integrating the first one, and disregarding the boundary term, we get $\frac{1}{2}A_\mu \partial^2 A^\mu = \frac{1}{2}A^\mu \partial^2 g_{\mu\nu}A^\nu$. The second term yields $-\frac{1}{2}A^\mu \partial_\mu \partial_\nu A^\nu$. The result[5] is that

$$
\frac{1}{2}A^\mu \hat{K}_{\mu\nu}A^\nu = \frac{1}{2}A^\mu \left[(\partial^2 + m^2)g_{\mu\nu} - \partial_\mu \partial_\nu\right]A^\nu. \quad (24.25)
$$

We therefore do the integral

$$
\begin{aligned}
\mathcal{Z}_0[J] &= \frac{\int \mathcal{D}A \, \mathrm{e}^{\frac{\mathrm{i}}{2}\int \mathrm{d}^4x \, A^\mu[(\partial^2+m^2)g_{\mu\nu}-\partial_\mu\partial_\nu]A^\nu + \mathrm{i}\int \mathrm{d}^4x \, J_\mu A^\mu}}{\int \mathcal{D}A \, \mathrm{e}^{\frac{\mathrm{i}}{2}\int \mathrm{d}^4x \, A^\mu[(\partial^2+m^2)g_{\mu\nu}-\partial_\mu\partial_\nu]A^\nu}} \\
&= \mathrm{e}^{-\frac{1}{2}\int \mathrm{d}^4x \mathrm{d}^4y \, J^\mu(x)[\mathrm{i}\hat{K}_{\mu\nu}^{-1}]J^\nu(y)}, \quad (24.26)
\end{aligned}
$$

[4]This is just as we have in classical statistical physics, where the analogous expression for the average of two fields is

$$
\langle \phi_1 \phi_2 \rangle_\mathrm{t} = \frac{\sum_{ij} \phi_i \phi_j \mathrm{e}^{-\beta H(\phi_i, \phi_j)}}{\sum_{ij} \mathrm{e}^{-\beta H(\phi_i, \phi_j)}} \quad (24.22)
$$

where the sum is over all configurations of ϕ_i and ϕ_j weighted by the Boltzmann factor and divided by the normalizing partition function.

[5]You are invited to confirm this result in Exercise 24.1.

where $i\hat{K}_{\mu\nu}^{-1}$ is the free particle propagator for the theory $G_{0\mu\nu}(x,y)$. This already tells us something very interesting. Namely, that we have access to the free propagator for a theory if we can work out $i\hat{K}_{\mu\nu}^{-1}$. This is often an easier method than working out, for example, $G_{0\mu\nu}(x,y) = \langle 0|T\hat{A}_\mu(x)\hat{A}_\nu^\dagger(y)|0\rangle$ as we would have had to have done here using a mode expansion if we were still using canonical quantization. Following this through, we can therefore obtain the free particle propagator from the following equation (with the indices all included):

$$\left[(\partial^2 + m^2)g^{\mu\nu} - \partial^\mu\partial^\nu\right] G_{0\nu\lambda}(x,y) = ig_\lambda^\mu\delta^{(4)}(x-y), \tag{24.27}$$

and the easiest way of doing this is to take a Fourier transform, yielding

$$[-(p^2 - m^2)g^{\mu\nu} + p^\mu p^\nu]\tilde{G}_{0\nu\lambda}(p) = ig_\lambda^\mu, \tag{24.28}$$

whose solution may be shown to be

$$\tilde{G}_{0\nu\lambda}(p) = \frac{-i\left(g_{\nu\lambda} - p_\nu p_\lambda/m^2\right)}{p^2 - m^2}, \tag{24.29}$$

which (with the addition of $i\epsilon$ in the denominator) is the massive vector field propagator. Interestingly, this has the form of the scalar propagator which we met previously, but this time with the transverse projection tensor on top, containing all of the indices.

Chapter summary

- The normalized generating functional for free scalar fields is $\mathcal{Z}_0[J] = e^{-\frac{1}{2}\int d^4x d^4y\ J(x)\Delta(x-y)J(y)}$.

Exercises

(24.1) (a) Verify eqn 24.25.

(b) Show that eqn 24.29 solves eqn 24.28.

(24.2) Consider the ϕ^4 Lagrangian $\mathcal{L} = \frac{1}{2}(\partial_\mu\phi)^2 - \frac{m^2}{2}\phi^2 - \frac{g}{8}\phi^4$. Shift the Lagrangian by adding a new field σ via

$$\mathcal{L}' = \mathcal{L} + \frac{1}{2g}\left(\sigma - \frac{g}{2}\phi^2\right)^2. \tag{24.30}$$

(a) By performing a functional integral over the field σ, show that σ doesn't change the dynamics of the theory.

(b) Confirm that σ has no effect on the dynamics of the theory by finding the Euler–Lagrange equation for σ and showing that it has no time derivatives. *This means that it can only provide a constraint and may be eliminated, much like the component A^0 in the massive vector field in Chapter 13.*

(c) Contrast the Feynman diagrams of ϕ^4 theory with those of the shifted theory involving σ.

The introduction of a new field to shift the Lagrangian in this way is known as a Hubbard–Stratonovich transformation and is very useful in removing the inconvenient ϕ^4 term from a theory.

(24.3) *In this problem we motivate the perturbation expansion for the functional integral.*

We want to do the integral

$$Z(J) = \int dx\, e^{-\frac{1}{2}Ax^2 - \frac{\lambda}{4!}x^4 + Jx}. \tag{24.31}$$

(a) Show that this integral may be recast as

$$Z(J) = \left[\sum_{n=0}^\infty \frac{1}{n!}\left(-\frac{\lambda}{4!}\frac{\partial^4}{\partial J^4}\right)^n\right]\int dx\, e^{-\frac{1}{2}Ax^2 + Jx}, \tag{24.32}$$

which has the advantage that we've effectively removed the interaction from inside the integral.
(b) Re-exponentiate the series in the first square bracket to give a differential operator acting on an integral and then do the integral to obtain the result

$$Z(J) = \left[e^{-\frac{\lambda}{4!}\frac{\partial^4}{\partial J^4}} \right] \left[\left(\frac{2\pi}{A}\right)^{\frac{1}{2}} e^{\frac{J^2}{2A}} \right]. \quad (24.33)$$

We now have a perturbation series that can be expanded by operating on the free generating function $Z_0(J) = \left(\frac{2\pi}{A}\right)^{\frac{1}{2}} e^{\frac{J^2}{2A}}$.

(24.4) By analogy with the previous problem, the generating functional for ϕ^4 theory is given by

$$Z[J] = \left[e^{-\frac{i\lambda}{4!}\int d^4 z \frac{\delta^4}{\delta J(z)^4}} \right] Z_0[J], \quad (24.34)$$

where $Z_0[J]$ is the normalized generating functional for the free scalar field. In this problem we investigate the first-order term in the expansion of $Z[J]$, given by

$$Z_1[J] = \left[-\frac{i\lambda}{4!}\int d^4 z \frac{\delta^4}{\delta J(z)^4} \right] e^{-\frac{1}{2}\int d^4 x d^4 y\, J(x)\Delta(x-y)J(y)}. \quad (24.35)$$

We will act on $Z_0[J] = e^{-\frac{1}{2}\int J\Delta J}$ with $\frac{\delta}{\delta J(z)}$ four times. *This will rely on some functional differentiation from Chapter 1.*
(a) Verify that hitting $Z_0[J]$ once gives us

$$\frac{\delta Z_0[J]}{\delta J(z)} = \left[-\int d^4 y\, \Delta(z-y)J(y) \right] Z_0[J]; \quad (24.36)$$

(b) the second time:

$$\frac{\delta^2 Z_0[J]}{\delta J(z)^2} = \left[-\Delta(z-z) + \left\{ \int d^4 y\, \Delta(z-y)J(y) \right\}^2 \right] Z_0[J]; \quad (24.37)$$

(c) the third:

$$\frac{\delta^3 Z_0[J]}{\delta J(z)^3} = \left[3\Delta(z-z)\left\{ \int d^4 y\, \Delta(z-y)J(y) \right\} - \left\{ \int d^4 y\, \Delta(z-y)J(y) \right\}^3 \right] Z_0[J]; \quad (24.38)$$

(d) and finally the fourth:

$$\frac{\delta^4 Z_0[J]}{\delta J(z)^4} = \left[3\Delta(z-z)^2 \right. \quad (24.39)$$

$$-6\Delta(z-z)\left\{ \int d^4 y\, \Delta(z-y)J(y) \right\}^2$$

$$\left. + \left\{ \int d^4 y\, \Delta(z-y)J(y) \right\}^4 \right] Z_0[J].$$

(e) Expand the brackets, multiply by $-i\lambda/4!$ and integrate $\int d^4 z$ to end up with the final result that $Z_1[J]$ is given by

$$Z_1[J] = -i\lambda \left[\frac{1}{8}\int d^4 z\, \Delta(z-z)^2 \right. \quad (24.40)$$

$$-\frac{1}{4}\left\{ \int d^4 z d^4 y_1 d^4 y_2 \Delta(z-z) \right.$$

$$\times \Delta(z-y_1)J(y_1)\Delta(z-y_2)J(y_2) \Big\}$$

$$+\frac{1}{4!}\left\{ \int d^4 z d^4 y_1 d^4 y_2 d^4 y_3 d^4 y_4\, \Delta(z-y_1) \right.$$

$$\times J(y_1)\Delta(z-y_2)J(y_2)\Delta(z-y_3)J(y_3)$$

$$\left. \times \Delta(z-y_4)J(y_4) \Big\} \right] Z_0[J].$$

We see that this has given us terms in J of order $O(J) = 0$, 2 and 4.
(f) Interpret this result by drawing Feynman diagrams for each term. *This procedure can be used to work out the Feynman rules for a theory. For more detail see Ryder, Chapter 6.*

25

Statistical field theory

It is the mark of a truly intelligent person to be moved by statistics.
George Bernard Shaw (1856–1950)

We've seen how taking the derivatives of the generating functional $Z[J] = \int \mathcal{D}\phi \, e^{i \int d^4 x (\mathcal{L}[\phi] + J\phi)}$ allows us access to the all-important Green's functions of a quantum field theory via

$$\langle \Omega | T \hat{\phi}(x_1)...\hat{\phi}(x_n) | \Omega \rangle = \frac{1}{i^n} \frac{1}{Z[J=0]} \frac{\delta^n Z[J]}{\delta J(x_1)...\delta J(x_n)} \bigg|_{J=0}. \quad (25.1)$$

This is very similar indeed to the state of affairs in statistical mechanics that we examined in Chapter 21. There we found that we had access to all of the interesting correlation functions via derivatives of the partition function

$$\langle \hat{\phi}_{i_1}...\hat{\phi}_{i_n} \rangle_t = \frac{1}{Z(J=0)} \frac{\partial^n Z(J)}{\partial J_{i_1}...\partial J_{i_n}} \bigg|_{J=0}. \quad (25.2)$$

In fact, our motivation for searching for a quantum field generating functional was simply the ease with which calculations in statistical physics can be made with a partition function. The similarity between the two disciplines is not just a superficial resemblance but reflects something much more fundamental. The crux of the matter is that the Green's functions of quantum field theory are related to time-evolution operators, which (roughly speaking) look like a complex exponential $e^{-iEt/\hbar}$, with E an energy and t the time. The probabilities (and density matrices) of statistical physics are based on the Boltzmann factor, given (roughly) by the real exponential $e^{-E/k_B T}$. The fundamental elements of the two theories are therefore both exponentials, albeit one is real and the other is complex. This similarity can be exploited by making the substitution

$$\frac{it}{\hbar} \to \frac{1}{k_B T}, \quad (25.3)$$

[1] Usually we employ units where $k_B = \hbar = 1$ so that $\beta = 1/T = it$.

and the claim is that this will map a quantum field theory on to a statistical field theory. The advantage of this mapping is that it allows us to transfer results from one subject directly to the other, and vice versa. Within the framework of the mapping, inverse temperature[1] $\beta = 1/k_B T$ in statistical physics behaves like **imaginary time** in a quantum field theory. It turns out that the concept of imaginary time is a remarkably powerful one with uses across quantum and statistical field theory. We examine the imaginary time formalism in the next section.

25.1 Wick rotation and Euclidean space

Imaginary time, assigned the symbol $\tau = it$, is defined, not simply by multiplying time by i, but by making a rotation in the complex plane by an angle[2] $\pi/2$ (Fig 25.1). This transformation is known as a **Wick rotation** and leads to a very nice feature for spacetime four-vectors. By rotating the time-like part of our vectors we essentially remove the annoyance of the Minkowski metric $(+,-,-,-)$ and replace it (to within an overall minus sign) by the more civilized $(+,+,+,+)$. This means that all four components of our vectors are treated equally, in the same way that all spatial components are treated equally when doing Euclidean geometry in three dimensions. We refer to this four-dimensional space as **Euclidean space**.

Example 25.1

Let's work through the consequences of this transformation. To make the Wick rotation to Euclidean space we will rotate the time coordinate by $-\pi/2$, defining $x_0 = -i\tau$. This means that

$$x^2 = x_0^2 - |\boldsymbol{x}|^2 = -(\tau^2 + |\boldsymbol{x}|^2) = -x_{\rm E}^2, \tag{25.4}$$

where $x_{\rm E}^2 = \tau^2 + \boldsymbol{x}^2$. Note that it also implies that

$$d^4x = -id^4x_{\rm E} = -id\tau\,d^3x. \tag{25.5}$$

We still want things like Fourier transforms to work in Euclidean space. The requirement that the exponential factor $e^{ip\cdot x}$ will not blow up requires us to rotate the energy part of the momentum vector p_0 by $\pi/2$ in the complex plane. We therefore define $p_0 = i\omega$, giving

$$p^2 = p_0^2 - |\boldsymbol{p}|^2 = -(\omega^2 + |\boldsymbol{p}|^2) = -p_{\rm E}^2. \tag{25.6}$$

We can also write the element of four-volume in momentum space as

$$d^4p = id^4p_{\rm E} = id\omega\,d^3p, \tag{25.7}$$

and the inner product between p and x as

$$p \cdot x = p_0 x_0 - \boldsymbol{p} \cdot \boldsymbol{x} = \omega\tau - \boldsymbol{p} \cdot \boldsymbol{x}. \tag{25.8}$$

We also consider the derivatives. We have that $\partial_0 = i\partial_\tau$ and

$$\partial^2 = \partial_0^2 - \boldsymbol{\nabla}^2 = -(\partial_\tau^2 + \boldsymbol{\nabla}^2) = -\partial_{\rm E}^2, \tag{25.9}$$

where $\partial_{\rm E}^2 = \partial_\tau^2 + \boldsymbol{\nabla}^2$. Finally we need to note that

$$(\partial_\mu\phi)^2 = -\left[(\partial_\tau\phi)^2 + (\boldsymbol{\nabla}\phi)^2\right] = -(\partial_{{\rm E}\mu}\phi)^2. \tag{25.10}$$

---

[2]Positive angles correspond to anti-clockwise rotations in the complex plane.

Euclid of Alexandria was a Greek mathematician who lived around 300 BC. He is best known for his *Elements*, the textbook which laid the foundations for geometry and has influenced generations mathematicians for more than two thousand years.

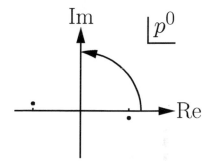

Fig. 25.1 The Wick rotation involves a rotation of $-\pi/2$ in the complex plane. We also show the poles of the Feynman propagator, demonstrating that Wick rotating of the contour of integration from the real axis (integrating from $-\infty$ to ∞ as we do in calculating Feynman amplitudes) to the imaginary axis (integrating $-i\infty$ to $i\infty$) is permissible as we don't cross any singularities.

To summarize the previous example:

The rules of Euclidean space

$$
\begin{array}{lll}
p_0 = i\omega & p^2 = -(\omega^2 + \boldsymbol{p}^2) = -p_{\rm E}^2 & d^4p = id\omega\,d^3p = id^4p_{\rm E} \\
x_0 = -i\tau & x^2 = -(\tau^2 + \boldsymbol{x}^2) = -x_{\rm E}^2 & d^4x = -id\tau\,d^3x = -id^4x_{\rm E} \\
\partial_0 = i\partial_\tau & \partial^2 = -(\partial_\tau^2 + \boldsymbol{\nabla}^2) = -\partial_{\rm E}^2 & (\partial_\mu\phi)^2 = -\left[(\partial_\tau\phi)^2 + (\boldsymbol{\nabla}\phi)^2\right]
\end{array}
$$

$$\tag{25.11}$$

Example 25.2

Our goal is to see the effect of the Wick rotation on the path integral. We start with the ϕ^4 theory with sources for which we have

$$iS = i \int d^4x \left[\frac{1}{2}(\partial_\mu \phi)^2 - \frac{m^2}{2}\phi^2 - \frac{\lambda}{4!}\phi^4 + J\phi \right],\quad (25.12)$$

[3]Note that we often write $\mathcal{L}_E[\phi] = \frac{1}{2}(\partial_\tau \phi)^2 + U[\phi]$, where $U[\phi]\left[= \frac{1}{2}(\boldsymbol{\nabla}\phi)^2 + \frac{m^2}{2}\phi^2 + \frac{\lambda}{4!}\phi^4\right]$ is the potential energy density.

Making the rotation results in the expression[3]

$$iS = -\int d^4x_E \left[\frac{1}{2}(\partial_\tau \phi)^2 + \frac{1}{2}(\boldsymbol{\nabla}\phi)^2 + \frac{m^2}{2}\phi^2 + \frac{\lambda}{4!}\phi^4 - J\phi \right]. \quad (25.13)$$

Defining

$$S_E = \int d^4x_E \left(\mathcal{L}_E[\phi] - J\phi \right) = \int d^4x_E \left[\frac{1}{2}(\partial_\tau \phi)^2 + \frac{1}{2}(\boldsymbol{\nabla}\phi)^2 + \frac{m^2}{2}\phi^2 + \frac{\lambda}{4!}\phi^4 - J\phi \right], \quad (25.14)$$

we write a Euclidean path integral

$$Z[J] = \int \mathcal{D}\phi \, e^{-S_E} = \int \mathcal{D}\phi \, e^{-\int d^4x_E \{\mathcal{L}_E[\phi] - J(x)\phi(x)\}}. \quad (25.15)$$

This is a useful mathematical object as it stands since it makes the path integral somewhat more respectable. As we shall see in the next example, it removes the ambiguity in defining the propagator which forced us to introduce those factors of $i\epsilon$.

Example 25.3

This Wick rotation may be carried out on some of our most useful equations to express them in Euclidean space. The Feynman propagator for free scalar fields in Minkowski space is

$$\Delta(x) = \int \frac{d^4p}{(2\pi)^4} \frac{ie^{-ip\cdot x}}{p^2 - m^2 + i\epsilon}, \quad (25.16)$$

which is the solution to the equation

$$(\partial^2 + m^2)\Delta(x-y) = -i\delta^{(4)}(x-y). \quad (25.17)$$

[4]We drop the subscript 'E' on position and momentum variables in Euclidean space from this point.

In Euclidean space[4] this becomes

$$\Delta_E(x) = \int \frac{d^4p}{(2\pi)^4} \frac{e^{-ip\cdot x}}{p^2 + m^2}. \quad (25.18)$$

In the last equation we've dropped the $i\epsilon$, since it is no longer required in this equation which has no poles (see also Fig. 25.1). This is the solution to the equation

$$(-\partial_E^2 + m^2)\Delta_E(x-y) = \delta^{(4)}(x-y). \quad (25.19)$$

That this is the case may be shown by transforming from Minkowski to Euclidean space. We make the changes

$$\partial^2 \to -\partial_E^2, \quad \Delta(x) \to \Delta_E(x). \quad (25.20)$$

To see how the delta function transforms note that its definition is

$$\delta(x^0)\delta^{(3)}(\boldsymbol{x}) = \int \frac{d^4p}{(2\pi)^4} e^{-ip\cdot x} \to i\int \frac{d\omega d^3p}{(2\pi)^4} e^{-ip\cdot x} = i\delta(\tau)\delta^{(3)}(\boldsymbol{x}), (25.21)$$

which may be substituted to give the desired result.

25.2   The partition function

We will now attempt to express the partition function as a path integral and will find that it involves the Euclidean generating functional. The partition function is given by $Z = \mathrm{Tr}\left[\mathrm{e}^{-\beta \hat{H}}\right] = \sum_\lambda \langle \lambda | \mathrm{e}^{-\beta \hat{H}} | \lambda \rangle$. We know how to express a similar object as a path integral. In Chapter 23 we found

$$\langle q_\mathrm{B} | \mathrm{e}^{-\mathrm{i}\hat{H} t_\mathrm{B}} | q_\mathrm{A} \rangle = \int \mathcal{D}[q(t)]\, \mathrm{e}^{\mathrm{i}\int_0^{t_\mathrm{B}} \mathrm{d}t L[q(t)]}, \qquad (25.22)$$

which tells us that to evolve the state at A into that at B we add up all of the possible trajectories that start at A at $t = 0$ and end at B at $t = t_\mathrm{B}$. Let's cook up a similar description of $\sum_\lambda \langle \lambda | \mathrm{e}^{-\beta \hat{H}} | \lambda \rangle$.

First set $|q_\mathrm{A}\rangle = |q_\mathrm{B}\rangle = |\lambda\rangle$, sum over λ and make the replacement $t_\mathrm{B} \to -\mathrm{i}\beta$. This turns the left-hand side into the desired $\sum_\lambda \langle \lambda | \mathrm{e}^{-\beta \hat{H}} | \lambda \rangle$, where the consequences of the sum will be discussed below. Now for the right-hand side. The replacement $t_\mathrm{B} \to -\mathrm{i}\beta$ provides the integral in the exponent with the new limits $\int_{t=0}^{t=-\mathrm{i}\beta}$. To give this meaning we make the Wick rotation $t \to -\mathrm{i}\tau$ and we obtain

$$\sum_\lambda \langle \lambda | \mathrm{e}^{-\beta \hat{H}} | \lambda \rangle = \sum_\lambda \int \mathcal{D}[q(\tau)]\, \mathrm{e}^{-\int_0^\beta \mathrm{d}\tau L_\mathrm{E}[q(\tau)]}, \qquad (25.23)$$

where $L_\mathrm{E}[q(\tau)] = \frac{m}{2}\left(\frac{\mathrm{d}q}{\mathrm{d}\tau}\right)^2 + V(q)$ is the Euclidean version of the Lagrangian for a single particle.

The question is, how does the sum on the left affect the trajectories over which we'll need to integrate? If we think of $\mathrm{e}^{-\beta \hat{H}}$ as an evolution operator, then we're taking a state $|\lambda\rangle$, evolving τ from 0 to β and requiring that we end up at the same state. This means that we must have $q(\tau = 0) = q(\tau = \beta)$. But that's not all. If we imagine τ as extending from $-\infty$ to ∞ then the boundary conditions are periodic: after a period β, the state returns to where it started. The sum on the left means that we sum over all configurations subject to these periodic boundary conditions. A quotable conclusion is that 'statistical physics takes place in periodic, imaginary time'.

This approach can be simply extended to fields. If $\hat{H}$ is the Hamiltonian describing what's happening to fields in three-dimensional space then we have a partition function given by the field integral

$$Z = \mathrm{Tr}\left[\mathrm{e}^{-\beta \hat{H}}\right] = \int_\mathrm{PBC} \mathcal{D}\phi\, \mathrm{e}^{-\int_0^\beta \mathrm{d}\tau \int \mathrm{d}^3 x \mathcal{L}_\mathrm{E}[\phi(x)]}, \qquad (25.24)$$

where the periodic boundary conditions (illustrated in Fig. 25.2 and denoted 'PBC') are now $\phi(0, \boldsymbol{x}) = \phi(\beta, \boldsymbol{x})$ and $\mathcal{L}_\mathrm{E}[\phi]$ is the Euclidean Lagrangian density.[5] Identifying $\mathrm{d}\tau\, \mathrm{d}^3 x$ with $\mathrm{d}^4 x_\mathrm{E}$, we see that this is the Wick rotated version of the functional integral.

We made the replacement $x_0 \to -\mathrm{i}\tau$ and so in momentum space we must make the replacement $p_0 \to \mathrm{i}\omega$. In fact, to guarantee the periodic boundary conditions are encoded in our functional integral we choose to work with the Fourier transforms of the fields. Since statistical physics

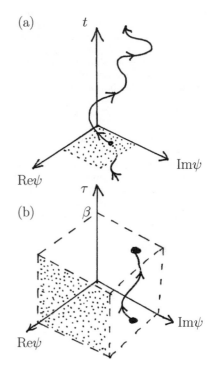

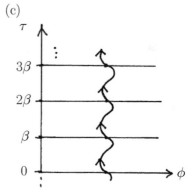

Fig. 25.2 (a) The trajectory of a complex scalar field ψ in time t, where $-\infty \le t \le \infty$. (b) In statistical physics the field is defined only from imaginary time $\tau = 0$ to $\tau = \beta$. For Bose fields like $\psi(\tau)$ or the real scalar field $\phi(\tau)$ we impose the boundary condition $\phi(0) = \phi(\beta)$ to give a picture of periodic boundary conditions shown in (c).

[5]Note, however, that these boundary conditions are appropriate for Bose fields only. For Fermi fields we must impose the antiperiodic boundary condition that $\psi(0, \boldsymbol{x}) = -\psi(\beta, \boldsymbol{x})$.

[6]The rather awkward normalization is chosen to make the field $\phi_n(\boldsymbol{p})$ dimensionless. This is done with malice aforethought in order to guarantee that the expressions which will result make sense in describing a statistical system. An important expression required to manipulate these fields is $\int \mathrm{d}\tau \mathrm{d}^3 x\, \mathrm{e}^{\mathrm{i}\omega_n\tau} \mathrm{e}^{-\mathrm{i}\boldsymbol{p}\cdot\boldsymbol{x}} = \beta\mathcal{V}\delta_{\boldsymbol{p},0}\delta_{\omega_n,0}$.

[7]This is another condition only appropriate for Bose fields. To impose the antiperiodic boundary conditions for Fermi fields we require that the Matsubara frequencies obey $\omega_n = \frac{(2n+1)\pi}{\beta}$.

Takeo Matsubara (1921–2014)

involves systems in boxes and explicit factors of volume we will put the system in such a box of volume $\mathcal{V}$ which discretizes momentum space. The Fourier representation is then given by[6]

$$\phi(\tau,\boldsymbol{x}) = \sqrt{\frac{\beta}{\mathcal{V}}} \sum_{n,\boldsymbol{p}} \mathrm{e}^{-\mathrm{i}\omega_n\tau + \mathrm{i}\boldsymbol{p}\cdot\boldsymbol{x}} \tilde{\phi}_n(\boldsymbol{p}), \qquad (25.25)$$

where the frequencies (quantized by the boundary conditions) must satisfy $\omega_n = \frac{2\pi n}{\beta}$ with n an integer to ensure the periodicity.[7] The ω_n's are known as **Matsubara frequencies** and in our calculations the integral over imaginary time will become a sum over these frequencies.

We may now state the rules of statistical field theory required to find the partition function.

The rules of statistical field theory

To find the partition function for scalar field theory follow the following recipe

- Write down the Euclidean Lagrangian $\mathcal{L}_\mathrm{E} = \frac{1}{2}(\partial_\tau\phi)^2 + U[\phi]$.
- The partition function is given by the functional integral of the Euclidean action, with periodic boundary conditions (PBCs)

$$Z[J] = \int_\mathrm{PBC} \mathcal{D}\phi\, \mathrm{e}^{-\int \mathrm{d}\tau \mathrm{d}^3 x\, (\mathcal{L}_\mathrm{E}[\phi] - J\phi)}, \qquad (25.26)$$

  where ϕ is a function of τ and $\boldsymbol{x}$.
- Impose the PBCs with the expansion

$$\phi(\tau,\boldsymbol{x}) = \sqrt{\frac{\beta}{\mathcal{V}}} \sum_{n,\boldsymbol{p}} \mathrm{e}^{-\mathrm{i}\omega_n\tau + \mathrm{i}\boldsymbol{p}\cdot\boldsymbol{x}} \tilde{\phi}_n(\boldsymbol{p}), \qquad (25.27)$$

  where $\omega_n = 2\pi n/\beta$.
- Note that we now use

$$\int \mathrm{d}\tau\, \mathrm{d}^3 x\, \mathrm{e}^{\mathrm{i}\omega_n\tau - \mathrm{i}\boldsymbol{p}\cdot\boldsymbol{x}} = \beta\mathcal{V}\delta_{\boldsymbol{p},0}\delta_{\omega_n,0}. \qquad (25.28)$$

Example 25.4

Let's examine the familiar example of free scalar field theory in more detail. The Euclidean Lagrangian is given by

$$\mathcal{L}_\mathrm{E} = \frac{1}{2}(\partial_\tau\phi)^2 + \frac{1}{2}(\boldsymbol{\nabla}\phi)^2 + \frac{m^2}{2}\phi^2. \qquad (25.29)$$

This leads to a partition function

$$Z[J] = \int_\mathrm{PBC} \mathcal{D}\phi\, \mathrm{e}^{-\int \mathrm{d}\tau \mathrm{d}^3 x \left[\frac{1}{2}(\partial_\tau\phi)^2 + \frac{1}{2}(\boldsymbol{\nabla}\phi)^2 + \frac{m^2}{2}\phi^2 - J\phi\right]}. \qquad (25.30)$$

We will calculate some thermodynamic properties of this theory. We therefore set $J = 0$ and insert the Fourier expansion of the fields into the argument of the exponential. We obtain an argument of

$$\frac{-\beta}{2\mathcal{V}} \int d\tau d^3x \sum_{n,m,\boldsymbol{p},\boldsymbol{q}} \tilde{\phi}_m(\boldsymbol{q}) e^{-i\omega_m \tau + i\boldsymbol{q}\cdot\boldsymbol{x}} [(-i\omega_n)(-i\omega_m) + (-i\boldsymbol{p})\cdot(-i\boldsymbol{q}) + m^2] \tilde{\phi}_n(\boldsymbol{p}) e^{-i\omega_n \tau + i\boldsymbol{p}\cdot\boldsymbol{x}}$$

$$= \frac{-\beta}{2\mathcal{V}} \sum_{n,m,\boldsymbol{p},\boldsymbol{q}} \tilde{\phi}_m(\boldsymbol{q})(-\omega_n\omega_m - \boldsymbol{p}\cdot\boldsymbol{q} + m^2)\tilde{\phi}_n(\boldsymbol{p})\beta\mathcal{V}\delta_{\boldsymbol{p},-\boldsymbol{q}}\delta_{\omega_m,-\omega_n}$$

$$= -\frac{\beta^2}{2} \sum_{n,\boldsymbol{p}} \tilde{\phi}_n(-\boldsymbol{p})(\omega_n^2 + \boldsymbol{p}^2 + m^2)\tilde{\phi}_n(\boldsymbol{p}). \tag{25.31}$$

This results in a partition function

$$Z[J = 0] = \int_{\text{PBC}} \mathcal{D}\tilde{\phi} \, e^{-\frac{\beta^2}{2}\sum_{n\boldsymbol{p}} \tilde{\phi}_n(-\boldsymbol{p})(\omega_n^2 + \boldsymbol{p}^2 + m^2)\tilde{\phi}_n(\boldsymbol{p})}. \tag{25.32}$$

This is, of course, in the form of our favourite (and only) tractable functional integral. The solution (for $J = 0$) is that

$$Z[J = 0] = N \det \left[\beta^2(\omega_n^2 + \boldsymbol{p}^2 + m^2)\right]^{-\frac{1}{2}}. \tag{25.33}$$

This is diagonal in $\boldsymbol{p}$, allowing us to write $\ln[Z(J = 0)]$ as the expression $-\frac{1}{2}\sum_{n,\boldsymbol{p}} \ln \left[\beta^2(\omega_n^2 + \boldsymbol{p}^2 + m^2)\right] +$const. The sum over n can be done (using a couple of tricks described in the book by Kapusta and Gale) which allows the replacement

$$\sum_n \ln \left[\beta^2(\omega_n^2 + E_{\boldsymbol{p}}^2)\right] = \beta E_{\boldsymbol{p}} + 2\ln\left(1 - e^{-\beta E_{\boldsymbol{p}}}\right) + \text{const.}, \tag{25.34}$$

where $E_{\boldsymbol{p}} = \sqrt{\boldsymbol{p}^2 + m^2}$. To obtain the thermodynamic limit we make our usual replacement for summing over closely spaced levels in phase space, $\sum_{\boldsymbol{p}} \to \mathcal{V}\int \frac{d^3p}{(2\pi)^3}$, leading to a result

$$\ln Z = \mathcal{V} \int \frac{d^3p}{(2\pi)^3} \left[-\frac{1}{2}\beta E_{\boldsymbol{p}} - \ln\left(1 - e^{-\beta E_{\boldsymbol{p}}}\right)\right]. \tag{25.35}$$

We then have a Helmholtz energy $F = \frac{\mathcal{V}}{\beta} \int \frac{d^3p}{(2\pi)^3} \left[\frac{1}{2}\beta E_{\boldsymbol{p}} + \ln\left(1 - e^{-\beta E_{\boldsymbol{p}}}\right)\right]$.

Notice that the zero-point energy leads to a ground-state energy given by $E_0 = -\frac{\partial}{\partial\beta} \ln Z = \frac{\mathcal{V}}{2} \int \frac{d^3p}{(2\pi)^3} E_{\boldsymbol{p}}$ [which is just like the $\frac{1}{2}\hbar\omega$ from $(n + \frac{1}{2}\hbar\omega)$] and this is associated with a pressure $P_0 = -\frac{\partial F}{\partial\mathcal{V}} = \frac{E_0}{\mathcal{V}}$. Our meaningful energies and pressures will be expressed relative to these vacuum values. Finally we obtain an energy, expressed relative to the (arbitrary) ground state energy of

$$E - E_0 = \mathcal{V} \int \frac{d^3p}{(2\pi)^3} \frac{E_{\boldsymbol{p}}}{e^{\beta E_{\boldsymbol{p}}} - 1}, \tag{25.36}$$

and a pressure

$$P - P_0 = \frac{1}{\beta} \int \frac{d^3p}{(2\pi)^3} \ln\left(1 - e^{-\beta E_{\boldsymbol{p}}}\right). \tag{25.37}$$

Although we have not learnt anything we couldn't have extracted with simpler methods it is, as usual, heartening that the more sophisticated machinery produces the same results. The power of statistical field theory comes in extracting predictions from unsolvable theories such as ϕ^4 theory, to which we now turn.

25.3 Perturbation theory and Feynman rules

The joy of statistical field theory is that the rules we have already learnt which allow us to encode perturbation theory in terms of Feynman diagrams carry over wholesale to statistical field theory. This is not magic;

we have made every effort to construct things this way! We found before that the generating functional could be expressed in terms of Feynman diagrams as $Z[J] = \mathrm{e}^{\Sigma \left(\begin{smallmatrix} \text{Connected source-to-source} \\ \text{and vacuum diagrams} \end{smallmatrix} \right)}$. Since the results of statistical field theory depend on $\ln Z$ rather than Z we use the neat result that $\ln Z[J] = \sum \left(\begin{smallmatrix} \text{Connected source-to-source} \\ \text{and vacuum diagrams} \end{smallmatrix} \right)$. Remember that the sources are included in order that we can extract Green's functions. If we are just interested in the thermodynamic equilibrium properties we can turn off the sources and we have the key result

$$\ln Z[J=0] = \sum (\text{Connected vacuum diagrams}). \qquad (25.38)$$

In statistical physics we often find ourselves in a situation where the average value of the operator $\langle \hat{\phi}(x) \rangle_t \neq 0$. An example is the ordered magnet encountered in the next chapter. In that case, more useful in determining the fluctuations in the fields is the connected correlation function

$$G_c(x, y) = \langle \hat{\phi}(x) \hat{\phi}(y) \rangle_t - \langle \hat{\phi}(x) \rangle_t \langle \hat{\phi}(y) \rangle_t. \qquad (25.39)$$

[8]See Exercise 25.2.

This may also be derived[8] directly from $\ln Z$ too:

$$
\begin{aligned}
G_c(x, y) &= \langle \hat{\phi}(x) \hat{\phi}(y) \rangle_t - \langle \hat{\phi}(x) \rangle_t \langle \hat{\phi}(y) \rangle_t = \frac{\partial^2}{\partial J(x) \partial J(y)} \ln Z[J] \Big|_{J=0} \\
&= \sum \left(\begin{smallmatrix} \text{All connected diagrams with two} \\ \text{external legs} \end{smallmatrix} \right). \qquad (25.40)
\end{aligned}
$$

Since there are far fewer connected diagrams than disconnected diagrams, this is a wonderfully labour-saving simplification. This also explains why $G_c(x, y)$ is known as the connected correlation function. Notice that this is the same result we had in eqn 22.15, where we insisted on normalizing the generating functional.

We already know how to draw diagrams, so the question we must ask is how the calculations encoded by these diagrams differ from those we've seen before. Luckily, all of the Feynman rules and momentum space tricks we've built up can be employed with little modification. In the case of scalar field theory, the propagator in momentum space $\mathrm{i}/(p_0^2 - \boldsymbol{p}^2 - m^2 + \mathrm{i}\epsilon)$ becomes the Euclidean space version $1/(\omega_n^2 + \boldsymbol{p}^2 + m^2)$. Integrals over all momenta will now involve a sum over Matsubara frequencies in place of the integral over p_0. The conversion rule is that

$$\int \frac{\mathrm{d}^4 p}{(2\pi)^4} \to \frac{1}{\beta} \sum_n \int \frac{\mathrm{d}^3 p}{(2\pi)^3}. \qquad (25.41)$$

As long as we also have $(2\pi)\delta(p_n^0 - p_m^0) \to \beta \delta_{\omega_n, \omega_m}$ then all is well.

As an example, we now state the Feynman rules for translating diagrams into thermodynamic quantities for ϕ^4 statistical field theory:

> **The Feynman rules for ϕ^4 theory for $T \neq 0$**
>
> To calculate the amplitude of a diagram in statistical field theory
>
> - Each internal propagator line contributes $\dfrac{1}{\omega_n^2 + \boldsymbol{p}^2 + m^2}$.
> - A factor $-\lambda$ results from each vertex.
> - Integrate over all unconstrained internal momenta with a measure $\frac{1}{\beta}\sum_n \int \frac{\mathrm{d}^3 p}{(2\pi)^3}$.
> - Divide by the symmetry factor.
> - An overall energy-momentum conserving delta function $(2\pi)^3 \delta^{(3)}(\boldsymbol{p}_{\text{in}} - \boldsymbol{p}_{\text{out}})\beta\delta_{\omega_n,\omega_m}$ is understood for each diagram. For vacuum diagrams, this results in a factor $\beta(2\pi)^3 \delta^3(0) \to \beta\mathcal{V}$ in the thermodynamic limit.

For practice, we now evaluate a few diagrams in statistical field theory.

(a)

(b)

Fig. 25.3 (a) The double-bubble diagram is the first-order correction to $\ln Z[J = 0]$ in ϕ^4 theory. (b) The line-bubble diagram is the first-order correction to the propagator.

Example 25.5

The first-order correction to $\ln Z[J = 0]$ is given by the double-bubble diagram shown in Fig. 25.3(a). The symmetry factor for this one is $D = 8$, so it makes a contribution to $\ln Z$ of

$$\ln Z_{(1)} = -\frac{\lambda \mathcal{V}\beta}{8}\left[\frac{1}{\beta}\sum_n \int \frac{\mathrm{d}^3 p}{(2\pi)^3}\frac{1}{\omega_n^2 + \boldsymbol{p}^2 + m^2}\right]^2. \tag{25.42}$$

The first-order correction to the propagator $\tilde{G}_{(1)}(k, q)$ is the line-bubble shown in Fig. 25.3(b). The symmetry factor is $D = 2$ and the amplitude for this diagram is given by[9]

$$\tilde{G}_{(1)}(k, q) = \frac{1}{\omega_m^2 + \boldsymbol{k}^2 + m^2}\left\{\frac{-\lambda}{2}\left[\frac{1}{\beta}\sum_n \int \frac{\mathrm{d}^3 p}{(2\pi)^3}\frac{1}{\omega_n^2 + \boldsymbol{p}^2 + m^2}\right]\right\}$$
$$\times \frac{1}{\omega_l^2 + \boldsymbol{q}^2 + m^2}(2\pi)^3\delta^{(3)}(\boldsymbol{k} - \boldsymbol{q})\beta\delta_{\omega_m,\omega_l}. \tag{25.43}$$

[9]Here the incoming line carries four-momentum $q = (\omega_l, \boldsymbol{q})$ and outgoing line four-momentum $k = (\omega_m, \boldsymbol{k})$.

Example 25.6

We finish this chapter by examining some of the consequences of these corrections on the thermodynamics of the scalar field with ϕ^4 interactions. The loop integral and Matsubara sum in brackets in eqns 25.42 and 25.43 can be done.[10] In fact, it may be shown that it falls apart into two pieces:

$$\int \frac{\mathrm{d}^4 p}{(2\pi)^4}\frac{1}{(p^0)^2 + E_{\boldsymbol{p}}^2} + \int \frac{\mathrm{d}^3 p}{(2\pi)^3}\frac{1}{E_{\boldsymbol{p}}}\left(\frac{1}{e^{\beta E_{\boldsymbol{p}}} - 1}\right). \tag{25.44}$$

[10]See Kapusta and Gale, Chapter 3.

The first term is independent of temperature and will not concern us here. The second term provides the first-order thermodynamic correction to $\ln Z$ and to the particle energies.

The resulting contribution to the temperature-dependent part of $\ln Z$ is

$$\ln Z_{(1)} = -\frac{\lambda\beta\mathcal{V}}{8}\left[\int \frac{\mathrm{d}^3 p}{(2\pi)^3}\frac{1}{E_{\boldsymbol{p}}}\left(\frac{1}{e^{\beta E_{\boldsymbol{p}}} - 1}\right)\right]^2. \tag{25.45}$$

[11] See Exercise 25.3.

This will give a contribution to the pressure that varies[11] as T^4 in the limit that $m \to 0$.

Turning to the line-bubble correction to the propagator: we note that in Chapter 16 it was suggested that the stuff that fits between the two external lines of the propagator, known as the self-energy, gives a correction to the particle dispersion $E_{\boldsymbol{p}}$. (In Chapter 33 we will see that this is true.) For our present purposes we claim that the correction to $E_{\boldsymbol{p}}$ is provided by the part in curly brackets in eqn 25.43. In particular, it is straightforward to show that in the limit that $m \to 0$ the correction to $E_{\boldsymbol{p}}$ varies as λT^2.

Chapter summary

- Statistical field theory is obtained from quantum field theory by a Wick rotation, so that (temperature)$^{-1}$ behaves like imaginary time.

- The diagrammatic techniques developed for quantum fields can then be used to solve problems in statistical physics.

Exercises

(25.1) (a) What are the dimensions of the field in ϕ^4 theory? Show that the normalization in eqn 25.25 makes the field $\tilde{\phi}_n(\boldsymbol{p})$ dimensionless.
(b) What normalization would be required for non-relativistic particles? What about Dirac particles?
(c) Investigate the consequences of choosing a different normalization for ϕ^4 theory.

(25.2) Verify that $G_c(x, y)$ may be obtained by differentiating $\ln Z[J]$ as suggested in the text.

(25.3) Verify the T dependence of the corrections to ϕ^4 theory claimed in the chapter, in the limit $m \to 0$.

(25.4) (a) After Wick rotating the Schrödinger equation and investigating the consequences, show that wave functions evolve according to $\psi(\tau) = e^{-\hat{H}\tau}\psi(0)$ and the Heisenberg equation of motion becomes $\frac{\partial \hat{A}_{\mathrm{H}}}{\partial \tau} = \left[\hat{H}, \hat{A}_{\mathrm{H}}\right]$.
(b) For a system described by a Hamiltonian $\hat{H} = \omega \hat{c}^\dagger \hat{c}$ show that the operators evolve according to $\hat{c}(\tau) = e^{-\omega\tau}\hat{c}$ and $\hat{c}^\dagger(\tau) = e^{\omega\tau}\hat{c}^\dagger$.

(25.5) Using a Wick rotation of $\int \frac{\mathrm{d}E}{2\pi} \tilde{G}(E) e^{-iE(t_x - t_y)} = \int \frac{\mathrm{d}E}{2\pi} i e^{-iE(t_x - t_y)}/(E - E_{\boldsymbol{p}} + i\epsilon)$, suggest a propagator for describing non-relativistic electrons at nonzero temperature. *By convention the propaga-*

tor for $T \neq 0$ is defined as minus this quantity, preventing a proliferation of minus signs.

(25.6) Define the imaginary time free propagator for the quantum oscillator as

$$G(\tau) = -\langle T\hat{x}(\tau)\hat{x}(0)\rangle_{\mathrm{t}}, \qquad (25.46)$$

where T is the imaginary time ordering symbol and the average is a thermal one. *Vitally important here is that $\langle \hat{a}^\dagger \hat{a}\rangle_{\mathrm{t}} = \langle \hat{n}\rangle_{\mathrm{t}} = (e^{\beta\omega} - 1)^{-1}$ rather than zero!*
(a) Show that

$$G(\tau) = -\frac{1}{2m\omega}\left[\theta(\tau)(\langle\hat{n}\rangle_{\mathrm{t}} + 1)e^{-\omega\tau} + \theta(\tau)\langle\hat{n}\rangle_{\mathrm{t}}e^{\omega\tau} + \theta(-\tau)(\langle\hat{n}\rangle_{\mathrm{t}} + 1)e^{\omega\tau} + \theta(-\tau)\langle\hat{n}\rangle_{\mathrm{t}}e^{-\omega\tau}\right]. \qquad (25.47)$$

(b) Using the definition

$$\tilde{G}(i\nu_n) = \int_0^\beta \mathrm{d}\tau\, G(\tau) e^{i\nu_n\tau}, \qquad (25.48)$$

where ν_n is a Matsubara frequency satisfying $\nu_n = \frac{2\pi n}{\beta}$, show that the propagator may be expressed as

$$\tilde{G}(i\nu_n) = \frac{1}{2m\omega}\left[\frac{1}{i\nu_n - \omega} - \frac{1}{i\nu_n + \omega}\right] = \frac{1/m}{(i\nu_n)^2 - \omega^2}. \qquad (25.49)$$

Broken symmetry

26

On the earth the broken arcs; in the heaven, a perfect round.
Robert Browning (1812–1889)

In this chapter we turn to the profoundly important topic of broken symmetry. It's hard to overestimate the influence this concept has had on condensed matter and particle physics and on quantum field theory in general. We will start by discussing the arena where these ideas were born: Lev Landau's theory of the statistical physics of phase transitions. We then turn to classical field theory and will see how broken symmetry manifests itself in a system described by a Lagrangian.

26.1 Landau theory

Our discussion of Landau's theory of phase transitions begins with a simple observation about magnets. We imagine a magnet whose spins can point either up or down. It's well known that at high temperatures each spin is equally likely to be found up or down. The magnetization, that is, the spatial average of the magnetic moment, is zero. This system is shown in Fig. 26.1(a). This system has a symmetry: turn all of the spins through 180° [as in Fig. 26.1(b)] and the magnetization is still zero. Of course, each individual moment is pointing in the opposite direction, but the number pointing in the upwards direction is still half of the total and the magnetization is still zero. The symmetry here is a global one: we rotate *all* of the spins through the same angle, here 180°.

It is found experimentally that upon cooling the system through a critical temperature T_c, the system undergoes a **phase transition** and the magnetization M becomes nonzero as all of the spins line up along a single direction, as shown in Fig. 26.2(a). The direction along which the spins align could *either* be all in the up direction *or* the down direction. If we rotate each spin of the aligned system through 180° [Fig. 26.2(b)] M is obviously reversed. We say that the system has broken (or lowered) its symmetry in the ordered phase.

One puzzling feature of this story is the reason why the system chose to point all of the spins in the one direction rather than the other. After all, there's nothing in the Hamiltonian which describes the system that distinguishes between up and down. The original symmetry of the system appears to have *spontaneously* broken. The same thing happens in the Euler strut shown in Fig. 26.3. A weight is balanced on top of an elastic strut. If the weight is large enough the strut will buckle. The

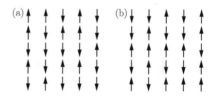

Fig. 26.1 The magnet at $T > T_c$. The magnetization, or average moment, is zero in (a) and, after rotation of each spin through 180°, is still zero in (b).

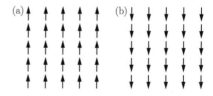

Fig. 26.2 For temperatures $T < T_c$ the spins align along a single direction. The magnetization in (a) is different in (b), where each spin has been rotated by 180°.

[1] Of course, in real life, some very small perturbation or fluctuation will have tipped the system towards its choice of ground state.

[2] Landau's scheme is a form of *mean-field theory*, where the magnetization is described by a uniform field M. We will meet mean field theories throughout this book, most notably in Chapter 43.

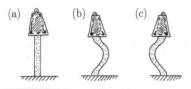

Fig. 26.3 The Euler strut. (a) This has a vertical axis of symmetry and can buckle either (b) one way or (c) the other, in each case breaking symmetry.

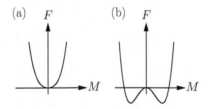

Fig. 26.4 The Landau free energy for (a) $T > T_c$ with a minimum at $M = 0$ and (b) for $T < T_c$ with minima at $\pm M_0$.

[3] This means that any slight perturbation of the system which has been prepared in a $M = 0$ state will propel it into one, or other, of the new minima that have emerged at $\pm M_0$.

[4] This classification is due to Philip Anderson and described in detail in his classic *Basic Notions of Condensed Matter Physics*.

buckling can be *either* to the left *or* to the right. There is nothing in the underlying physics of the strut and weight that allows one to predict which way it will go.[1] The result is that the ground state does not have the symmetry of the Hamiltonian describing the system. For the magnet this Hamiltonian could be $\hat{H} = -J \sum_i \hat{S}_i^z \hat{S}_{i+1}^z$, where the up direction is along z. This has no bias either favouring the spins pointing up or favouring them pointing down.

Lev Landau had a scheme for thinking about phase transitions involving a breaking of symmetry.[2] Since equilibrium in thermodynamic systems relies on both minimizing the internal energy U and maximizing the entropy S of a system, Landau considered the free energy $F = U - TS$. To find the equilibrium state of the system we need to minimize F. The free energy is a function of an **order parameter**, namely some field describing the system whose thermal average is zero in the $T > T_c$ unbroken symmetry state and nonzero in the $T < T_c$ broken symmetry state. For a magnet, the order parameter is simply the magnetization field M. With joyful and deliberate ignorance of any microscopic description, Landau wrote $F(M)$ as a power series

$$F = F_0 + aM^2 + bM^4 + \dots, \tag{26.1}$$

where a and b are parameters which are independent of M, but may in principle depend on the temperature. This free energy has the symmetry $M \to -M$, that is, reversing the magnetization doesn't affect the energy. If a is positive we have the free energy shown in Fig. 26.4(a), which is minimized at $M = 0$, which is the correct prediction for the high-temperature regime. If, however the parameter a is negative then we have the free energy shown in Fig. 26.4(b), which is known, by some, as Lifshitz's buttocks. This has two minima at nonzero values of magnetization. These minima correspond to the spins all aligning up ($M = +M_0$) or all aligning down ($M = -M_0$). The previous minimum $M = 0$ is now at a position of metastable equilibrium.[3] If we take $a = a_0(T - T_c)$ (with a_0 a constant) and b to be T-independent, then clearly a is positive for $T > T_c$ and negative for $T < T_c$, and so we predict a phase transition at a temperature $T = T_c$.

Example 26.1

We can find the minima of F straightforwardly for $T < T_c$. We set

$$\frac{\partial F}{\partial M} = 2aM + 4bM^3 = 0, \tag{26.2}$$

from which we conclude that the minima occur at

$$M_0^2 = -\frac{a}{2b}. \tag{26.3}$$

For $T < T_c$ we have two minima at $M = \pm M_0$. The system will choose one of these as its new ground state and the symmetry will be broken. When a symmetry is broken in a many-particle system there are four features[4] that one should look out for.

- **Phase transitions** We saw that in Landau's example, the parameter a in the free energy was temperature dependent. At a temperature T_c, at which a changes sign, a phase transition takes place. The transition separates two distinct states of different symmetry. The low-temperature phase has lost some symmetry, more precisely it is missing a symmetry element.[5]

- **New excitations** Our philosophy has been that every particle is an excitation of the vacuum of a system. When a symmetry is broken we end up with a new vacuum (e.g. a vacuum with $M = -M_0$). The fact that the vacuum is different means that the particle spectrum should be expected to be different to that of the unbroken symmetry state (such as $M = 0$ in our example). We will see that new particles known as Goldstone modes can emerge upon symmetry breaking.[6]

- **Rigidity** Any attempt to deform the field in the broken symmetry state results in new forces emerging. Examples of rigidity include phase stiffness in superconductors, spin stiffness in magnets and the mechanical strength of crystalline solids.

- **Defects** These result from the fact that the symmetry may be broken in different ways in different regions of the system, and are topological in nature. An example is a domain wall in a ferromagnet. These are described in Chapter 29.

26.2 Breaking symmetry with a Lagrangian

Let us now apply the arguments developed for magnets to our Lagrangian treatment of (classical) field theory.[7] Here instead of following the ground state magnetization, we're interested in the ground state value of $\phi(x)$ or (returning briefly to quantum mechanics) the expectation value of the field operator $\langle\Omega|\hat{\phi}(x)|\Omega\rangle$, which is the order parameter in field theory. We start with a scalar field theory

$$\mathcal{L} = \frac{1}{2}(\partial_\mu\phi)^2 - U(\phi), \tag{26.4}$$

where we've split off all of the potential energy terms and called them[8] $U(\phi)$. In the familiar case of ϕ^4 theory we have that

$$U(\phi) = \frac{\mu^2}{2}\phi^2 + \frac{\lambda}{4!}\phi^4. \tag{26.5}$$

This function resembles the free energy in our magnet example above. It admits the global symmetry $\phi(x) \to -\phi(x)$. Assuming μ^2 is positive we have a potential with a minimum at $\phi = 0$ as shown in Fig. 26.5(a). The minimum corresponds to a quantum mechanical ground state, also known as the vacuum, of $\langle\Omega|\hat{\phi}(x)|\Omega\rangle = 0$. As we've seen earlier, the excitations of this field are phions with mass $m = \mu$.

[5]In fact, you don't need to regard the parameters in a model as being dependent on anything: in the Lagrangian examples below we won't, and then the phase transition is not such a useful concept. However, the link between phase transitions and symmetry breaking is a very powerful one that unifies our understanding of an enormous range of phenomena in Nature.

[6]See Section 26.3.

[7]Classical field theory will be enough to show us the main features of broken symmetry. See S. Coleman and Peskin and Schroeder for more details on the quantum mechanics of broken symmetry.

[8]More strictly $U(\phi)$ is the potential energy density. It should not be confused with the internal energy, which is also conventionally called U.

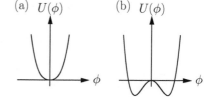

Fig. 26.5 Equation 26.5 for (a) $\mu^2 > 0$ with a minimum at $\phi = 0$ and (b) for $\mu^2 < 0$ with minima at $\pm\sqrt{6\mu^2/\lambda}$.

[9]From $\partial U/\partial \phi = 0$ we find

$$0 = -\mu^2\phi + \frac{\lambda}{3!}\phi^3,$$

from which we deduce stationary points at $(0, \pm\sqrt{6\mu^2/\lambda})$. The second derivative,

$$\frac{\partial^2 U}{\partial \phi^2} = -\mu^2 + \frac{\lambda\phi^2}{2},$$

is negative (and equals $-\mu^2$) for $\phi = 0$ and is positive (and equals $+2\mu^2$) for the minima at $\phi_0 = \pm\sqrt{6\mu^2/\lambda}$.

[10]For the ordinary scalar field Lagrangian, the factor multiplying the term quadratic in field is $m^2/2$, with m the mass of the particle excitations in the theory.

[11]In the quantum mechanical version we say that the ground state does not have the symmetry of the Hamiltonian.

Now we turn to a very interesting possibility: what if we swap the sign in front of μ^2? In that case we have $U(\phi) = -\frac{\mu^2}{2}\phi^2 + \frac{\lambda}{4!}\phi^4$ and again have a potential that looks like Fig. 26.5(b). The minima[9] of the potential now occur at

$$\phi_0 = \pm\left(\frac{6\mu^2}{\lambda}\right)^{\frac{1}{2}}, \tag{26.6}$$

so we have the choice of two new vacua. The system will spontaneously choose one and the symmetry $\phi_0 \to -\phi_0$ of the ground state is broken.

Next we ask if the new vacuum has the same excitations (a.k.a. particles) as those of the unbroken symmetry state.

Example 26.2

To do this we choose a vacuum (for definiteness let's choose $+\phi_0$) and we expand the field around this new minimum. The Taylor expansion gives us

$$\begin{aligned}
U(\phi - \phi_0) &= U(\phi_0) + \left(\frac{\partial U}{\partial \phi}\right)_{\phi_0}(\phi - \phi_0) + \frac{1}{2!}\left(\frac{\partial^2 U}{\partial \phi^2}\right)_{\phi_0}(\phi - \phi_0)^2 + \dots \\
&= U(\phi_0) + \mu^2(\phi - \phi_0)^2 + \dots, \tag{26.7}
\end{aligned}$$

and since $U(\phi_0)$ is an ignorable constant, we can write the Lagrangian in terms of $\phi' = \phi - \phi_0$ as

$$\mathcal{L} = \frac{1}{2}(\partial\phi')^2 - \mu^2\phi'^2 + O(\phi'^3). \tag{26.8}$$

By comparison with our original scalar field theory,[10] we find from eqn 26.8 that the mass of the phion excitations in the broken symmetry state is now no longer $m = \mu$ but has become $m = \sqrt{2}\mu$.

An important point to note is that the Lagrangian does not 'break symmetry' itself. Indeed our Lagrangian, just like the Landau free energy for the magnet, is invariant with respect to the transformation $\phi \to -\phi$. The breaking of symmetry is a property of the *ground state* (or *vacuum*) of the system.[11]

To conclude, the effect of spontaneous symmetry breaking is (i) for ϕ or (again, briefly returning to quantum mechanics) the vacuum expectation value $\langle\Omega|\hat{\phi}(x)|\Omega\rangle$ to acquire a non-vanishing, constant amplitude ϕ_0, in this case $\phi_0 = \left(\frac{6\mu^2}{\lambda}\right)^{1/2}$; (ii) the particle excitations of the theory are the same as before, but now have a mass $\sqrt{2}\mu$.

26.3   Breaking a continuous symmetry: Goldstone modes

The scalar field theory had a discrete global symmetry ($\phi \to -\phi$). More dramatic and interesting still is the effect of symmetry breaking for a theory with a continuous global symmetry. A two-component field with the sign of its mass term flipped looks like

$$\mathcal{L} = \frac{1}{2}\left[(\partial_\mu \phi_1)^2 + (\partial_\mu \phi_2)^2\right] + \frac{\mu^2}{2}\left(\phi_1^2 + \phi_2^2\right) - \frac{\lambda}{4!}\left(\phi_1^2 + \phi_2^2\right)^2. \quad (26.9)$$

This has a global $SO(2)$ symmetry: it is symmetric with respect to rotations in the (internal) $\phi_1(x)$-$\phi_2(x)$ plane (where the same rotation takes place at all points in spacetime). The potential $U(\phi_1, \phi_2)$ for this theory is sketched in Fig. 26.6(b) and looks a lot like the bottom of a wine bottle [as shown in Fig. 26.6(a)]. There are an infinite number of potential minima, and these are found to lie on a circle [Fig. 26.6(c)] whose equation can easily be found[12] to be $\phi_1^2 + \phi_2^2 = \frac{6\mu^2}{\lambda}$.

Let's break the symmetry by supposing the system chooses a particular vacuum. Each is as good as any other, so for computational simplicity, let's choose $(\phi_1, \phi_2) = \left(+\sqrt{\frac{6\mu^2}{\lambda}}, 0\right)$ and expand around this broken symmetry ground state using variables $\phi_1' = \phi_1 - \sqrt{\frac{6\mu^2}{\lambda}}$ and $\phi_2' = \phi_2$. This is illustrated in Fig. 26.7.

Example 26.3

With the potential $U(\phi_1, \phi_2)$ given by

$$U(\phi_1, \phi_2) = -\frac{\mu^2}{2}(\phi_1^2 + \phi_2^2) + \frac{\lambda}{4!}(\phi_1^2 + \phi_2^2)^2, \quad (26.10)$$

Taylor expanding around the minimum $(\phi_1, \phi_2) = \left(\sqrt{\frac{6\mu^2}{\lambda}}, 0\right)$ gives $\partial^2 U/\partial\phi_1^2 = 2\mu^2$ and $\partial^2 U/\partial\phi_2^2 = 0$ and so we obtain (ignoring constant terms and truncating the expansion at order ϕ^2)

$$\mathcal{L} = \frac{1}{2}\left[(\partial\phi_1')^2 + (\partial\phi_2')^2\right] - \mu^2(\phi_1')^2 + O(\phi'^3). \quad (26.11)$$

Equation 26.11 shows that the particles of the ϕ_1' field now have mass $m = \sqrt{2}\mu$, but that there is no quadratic term for the ϕ_2' term: thus the particles of the ϕ_2' field are completely massless! This seemingly magical result is less surprising if we look again at the potential landscape shown in Fig. 26.7. We can see immediately that excitations in the ϕ_1 direction will cost energy since such excitations have to climb up the walls of the potential. However, a small movement in the ϕ_2 direction corresponds to rolling round the gutter. Such motion has no opposing force as it involves no change in potential energy and therefore costs no energy, exactly what we would expect for a massless excitation. In other words, we still have particles in the ϕ_2 field, it's just that their dispersion starts at the origin, rather than from some offset determined by the mass. In the language of condensed matter physics one would say that the excitations in the ϕ_2 field are gapless, rather than massless, but it amounts to the same thing.

[12]To quickly show this, you can write $U(x) = -\frac{\mu^2}{2}x + \frac{\lambda}{4!}x^2$ where $x = \phi_1^2 + \phi_2^2$, and then $\partial U/\partial x = 0$ yields the result.

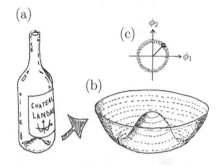

Fig. 26.6 (a) The potential for the $SO(2)$ symmetry breaking looks like the bottom of a punted wine bottle. (b) There is a maximum at the point $\phi_1 = \phi_2 = 0$, but surrounding this there is a set of minima which lie on a circle. (c) The circle of minima are shown on a ϕ_1-ϕ_2 plot (this is therefore viewing the surface sketched in (b) from 'above'). The symmetry can then be broken by choosing a particular point in the circle of minima and setting this to be the ground state. We can then examine small deviations away from that point.

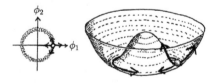

Fig. 26.7 Breaking symmetry at the position $(\phi_1, \phi_2) = \left(+\sqrt{\frac{6\mu^2}{\lambda}}, 0\right)$. There are two possible excitations: the particle can oscillate up and down in the ϕ_1 direction (a massive excitation) or trundle round in the gutter (a massless excitation).

> This vanishing of the mass is a manifestation of **Goldstone's theorem**, which says that breaking a continuous symmetry always results in a massless excitation, known as a **Goldstone mode**. The massless particle associated with this excitation is known as a **Goldstone boson**.

Jeffrey Goldstone (1933–). The bosons had previously been discovered by Yoichiro Nambu (1921–2015) in the context of superconductivity.

[13] See chapter 12, although note that there we were dealing with a non-relativistic theory and wrote $\Psi(x) = \sqrt{\rho(x)}e^{i\theta(x)}$. Here, in dealing with the relativistic theory we write $\psi(x) = \varrho(x)e^{i\theta(x)}$ and so $\int d^3x\,|\varrho(x)|^2$ [rather than $\int d^3x\,\rho(x)$] gives the number density of particles.

Example 26.4

Perhaps more illuminating is the complex scalar field theory. In this theory we'll break global $U(1)$ symmetry. Of course $U(1)$ is isomorphic to $SO(2)$, so we won't have anything too different from the previous example. The $U(1)$ theory, with a positive mass term, reads

$$\mathcal{L} = (\partial^\mu\psi)^\dagger(\partial_\mu\psi) + \mu^2\psi^\dagger\psi - \lambda(\psi^\dagger\psi)^2. \tag{26.12}$$

This theory enjoys a *global* $U(1)$ symmetry: a transformation $\psi \to \psi e^{i\alpha}$ has no effect on the Lagrangian. We switch to 'polar coordinates'[13] via $\psi(x) = \varrho(x)e^{i\theta(x)}$ and this theory reads

$$\mathcal{L} = (\partial_\mu\varrho)^2 + \varrho^2(\partial_\mu\theta)^2 + \mu^2\varrho^2 - \lambda\varrho^4, \tag{26.13}$$

where the two fields are now called $\varrho(x)$ and $\theta(x)$. The symmetry transformation for this version is $\varrho \to \varrho$ and $\theta \to \theta + \alpha$.

Let's break this symmetry. The potential part is $U = -\mu^2\varrho^2 + \lambda\varrho^4$ whose minima are on a circle of radius $\varrho = \sqrt{\frac{\mu^2}{2\lambda}}$. We set $\varrho_0 = \sqrt{\mu^2/2\lambda}$ and (arbitrarily, but with convenience in mind) choose a minimum by selecting $\theta_0 = 0$, then expand around the minimum with $\varrho' = \varrho - \varrho_0$ and $\theta' = \theta - \theta_0$, to obtain

$$\begin{aligned}\mathcal{L} = \ &\left(\tfrac{\mu^2}{2\lambda}\right)(\partial_\mu\theta')^2 &&(\theta'\text{-field terms})\\ &+(\partial_\mu\varrho')^2 - 2\mu^2\varrho'^2 - 4\left(\tfrac{\mu^2\lambda}{2}\right)^{\frac{1}{2}}\varrho'^3 - \lambda\varrho'^4 &&(\varrho'\text{-field terms})\\ &+\left[\varrho'^2 + \left(\tfrac{2\mu^2}{\lambda}\right)^{\frac{1}{2}}\varrho'\right](\partial_\mu\theta')^2 + \dots &&(\text{interaction terms}),\end{aligned} \tag{26.14}$$

where constant terms have been dropped and terms have been arranged according to the contributions of the two individual fields and then the interactions between them. As might be expected, there is no term in θ'^2 and so the θ'-field is massless. As before, this is because it costs nothing to roll in the gutter. The mass of the ϱ'-field has been shifted to $m = \sqrt{2}\mu$.

[14] Philip Anderson (1923–), Francois Englert (1932–), Robert Brout (1928–2011), Peter Higgs (1929–), Tom Kibble (1932–2016), Gerald Guralnik (1936–2014) and Carl Hagan (1937–). Higgs and Englert shared the 2013 Nobel Prize in physics for their discovery (Englert and Brout worked together, but Brout died before the prize was awarded). Anderson (who won the Nobel prize in 1977 for other work) had a non-relativistic version in 1962 which was inspired by superconductivity, and in 1964 all the others (Brout and Englert published first, closely followed by Higgs) produced relativistic treatments of what we now call the Higgs mechanism (though only Higgs mentioned the massive boson).

26.4 Breaking a symmetry in a gauge theory

An extraordinary feature that emerges upon breaking a symmetry in a gauge theory was spotted by Anderson, Higgs, Kibble, Guralnik, Englert and Brout and is famously known as the **Higgs mechanism**.[14] As we've said previously (see Chapter 14), a Lagrangian which has a local symmetry contains gauge fields. The simplest example is the gauged complex scalar field theory, whose Lagrangian (with a flipped signed for the mass term) is given by

$$\mathcal{L} = (\partial^\mu\psi^\dagger - iqA^\mu\psi^\dagger)(\partial_\mu\psi + iqA_\mu\psi) + \mu^2\psi^\dagger\psi - \lambda(\psi^\dagger\psi)^2 - \frac{1}{4}F_{\mu\nu}F^{\mu\nu}, \tag{26.15}$$

which is symmetric under the local transformation $\psi \to \psi e^{i\alpha(x)}$ as long as we also transform $A_\mu \to A_\mu - \frac{1}{q}\partial_\mu \alpha(x)$. This Lagrangian describes a world made up of two sorts of oppositely charged, massive scalar particles with energies $E_{\boldsymbol{p}} = (\boldsymbol{p}^2 + \mu^2)^{\frac{1}{2}}$ and two sorts of transversely polarized photons with energies $E_{\boldsymbol{p}} = |\boldsymbol{p}|$, which are, of course, massless.

Now we'll see what happens when we break symmetry in a system with a local symmetry. Working in polars, the ground state has the field $\psi(x) = \varrho(x)e^{i\theta(x)}$ and takes on a unique phase angle $\theta(x) = \theta_0$ for all x. We are no longer permitted to change the phase of the ground state at different values of x (local symmetry) or indeed change the phase of the ground state for the entire system (global symmetry). The broken symmetry ground state therefore breaks global symmetry and, as a consequence, local symmetry too.

Next we turn our attention to the excitations. Remember that the Lagrangian is gauge invariant so we are able to make gauge transformations to simplify the physics as much as possible. In the next example we show that the gauge transformation we make to elucidate the particle spectrum of the broken symmetry system reveals something quite exciting.

Example 26.5

In polar coordinates, the second bracketed term in eqn 26.15 can be written

$$\partial_\mu \psi + iqA_\mu \psi = (\partial_\mu \varrho)e^{i\theta} + i(\partial_\mu \theta + qA_\mu)\varrho e^{i\theta}. \tag{26.16}$$

Thus we notice that A_μ enters the theory as

$$A_\mu + \frac{1}{q}\partial_\mu \theta \equiv C_\mu, \tag{26.17}$$

and so we will replace A_μ by C_μ, thereby simplifying the term $(\partial^\mu \psi^\dagger - iqA^\mu \psi^\dagger)(\partial_\mu \psi + iqA_\mu \psi) = (\partial_\mu \varrho)^2 + \varrho^2 q^2 C_\mu C^\mu$. The replacement is a gauge invariant one since $F_{\mu\nu} = \partial_\mu A_\nu - \partial_\nu A_\mu = \partial_\mu C_\nu - \partial_\nu C_\mu$. The Lagrangian in eqn 26.15 can now be written in terms of ϱ and C_μ fields as

$$\mathcal{L} = (\partial_\mu \varrho)^2 + \varrho^2 q^2 C^2 + \mu^2 \varrho^2 - \lambda \varrho^4 - \frac{1}{4}F^{\mu\nu}F_{\mu\nu}, \tag{26.18}$$

where we use the shorthand $C^2 \equiv C^\mu C_\mu$.

Now to break the symmetry! The minima of the potential are again on the circle at $\varrho = \sqrt{\frac{\mu^2}{2\lambda}}$. We choose the symmetry to be broken with $\varrho_0 = \sqrt{\frac{\mu^2}{2\lambda}}$ and $\theta_0 = 0$. The expansion that reveals the excitations above the ground state is most easily done by expanding in terms of a field χ, defined as $\frac{\chi}{\sqrt{2}} = \varrho - \varrho_0$. Ignoring constant terms we obtain

$$\begin{aligned}
\mathcal{L} = \; & \frac{1}{2}(\partial_\mu \chi)^2 - \mu^2 \chi^2 - \sqrt{\lambda}\mu\chi^3 - \frac{\lambda}{4}\chi^4 \\
& - \frac{1}{4}F_{\mu\nu}F^{\mu\nu} + \frac{M^2}{2}C^2 \\
& + q^2 \left(\frac{\mu^2}{\lambda}\right)^{\frac{1}{2}} \chi C^2 + \frac{1}{2}q^2 \chi^2 C^2 + \dots,
\end{aligned} \tag{26.19}$$

where $M = q\sqrt{\frac{\mu^2}{\lambda}}$.

As usual, we have the contribution that derives from the radial field (here called χ), which has a mass $\sqrt{2}\mu$. However, there is a big surprise here in the second line of eqn 26.19: the theory also now contains a *massive* vector field C_μ, whose particles have mass M. But what of the θ field? It has completely disappeared! The excitations of the θ field, which were massless in the global symmetry breaking version, have disappeared and have been replaced by those of a massive vector field[15] $C_\mu(x)$. As Sidney Coleman put it, it's as if the massless photon field $A_\mu(x)$ has eaten the Goldstone bosons from the θ field and grown massive, changing its name to $C_\mu(x)$. The theory now describes one sort of scalar particle, which is an excitation in the $\chi(x)$ field with energy $E_{\boldsymbol{p}} = (\boldsymbol{p}^2 + 2\mu^2)^{\frac{1}{2}}$, and three sorts of massive vector particle, which are excitations in the $C_\mu(x)$ field with energies $E_{\boldsymbol{p}} = \left[\boldsymbol{p}^2 + \left(\frac{q^2\mu^2}{\lambda}\right)\right]^{\frac{1}{2}}$.

One might now worry that the loss of the Goldstone bosons means that we have lost some of the precious degrees of freedom in the mathematical description of the fields. This worry is unfounded because upon symmetry breaking we have the change

$$\begin{pmatrix} 2 \times \text{ massive scalar particles} \\ 2 \times \text{ massless photon particles} \end{pmatrix} \rightarrow \begin{pmatrix} 1 \times \text{ massive scalar particles} \\ 3 \times \text{ massive vector particles} \end{pmatrix},$$

(26.20)

that is, four types of particle are excitable before and after we break symmetry.

So where did the Goldstone boson go? The answer becomes clear when we reexamine what we did to remove it: we made the change of variables $A_\mu + \frac{1}{q}\partial_\mu\theta = C_\mu$, which is a gauge transformation! If the Goldstone particle can be removed with a simple gauge transformation then it can't have been there in the first place. In other words, it must be **pure gauge**.[16] Another way of describing the Higgs mechanism, therefore, is the removal by gauge transformation of all Goldstone modes.

This feature of symmetry breaking, where the Goldstone mode may be removed and turned massive through combination with a gauge field, has become one of the most important in modern physics. Historically it told particle physicists, puzzled by that apparent lack of Goldstone particles in Nature, that symmetry breaking allows another way. We will return to this undeniably important topic in Chapter 46.

26.5   Order in reduced dimensions

Field theory allows us to ask questions about systems which have less (or more!) than the ordinary three spatial dimensions of our world. One famous result is the **Coleman–Mermin–Wagner theorem**, which tells us that symmetry breaking is impossible in two or fewer spatial dimensions in systems with an order parameter with a continuous symmetry.

We'll show that this is the case for complex scalar field theory with spontaneous symmetry breaking.[17] The idea of the proof is to ask what the fluctuations look like at one point in space. We get this by evaluating the Euclidean (two-point) correlation function at the origin. We want[18]

[15]Recall from Chapter 13 that the Lagrangian for a massive vector field is $\mathcal{L} = -\frac{1}{4}F^{\mu\nu}F_{\mu\nu} + \frac{m^2}{2}A^\mu A_\mu$.

[16]The notion of a pure gauge will appear again in our treatment of topological effects in Chapter 29.

Sidney Coleman (1937–2007)
N. David Mermin (1935–)
Herbert Wagner (1935–)

[17]We use the theory of a complex scalar field $\psi(x)$ since there a continuous $U(1)$ internal symmetry is broken, leading to the emergence of a Goldstone mode. The presence of the Goldstone mode is crucial here.

[18]Strictly we should write the integration measure as $\int \mathcal{D}\psi\mathcal{D}\psi^\dagger$, but we will abbreviate it here to save on clutter.

$$G(0,0) = \langle |\hat{\psi}(0)|^2 \rangle = \frac{1}{Z} \int \mathcal{D}\psi \, |\psi(0)|^2 \mathrm{e}^{-S_{\mathrm{E}}[\phi]}$$

$$= \lim_{x \to 0} \frac{1}{Z} \int \mathcal{D}\psi \, \psi(x)\psi^\dagger(0) \mathrm{e}^{-S_{\mathrm{E}}[\phi]}. \quad (26.21)$$

We can evaluate this since we know that the propagator is given in d-dimensional Euclidean space by

$$G(x,y) = \langle \hat{\psi}(x)\hat{\psi}^\dagger(y) \rangle = \frac{1}{Z} \int \mathcal{D}\psi \, \psi(x)\psi^\dagger(y)\mathrm{e}^{-S_{\mathrm{E}}[\phi]} = \int \frac{\mathrm{d}^d p}{(2\pi)^d} \frac{\mathrm{e}^{-i p \cdot (x-y)}}{p^2 + m^2}. \quad (26.22)$$

If the system breaks symmetry then the propagating particles will be Goldstone particles. These are massless, so give rise to a fluctuation with $m = 0$ and result in a correlation function

$$G(0,0) = \int \frac{\mathrm{d}^d p}{(2\pi)^d} \frac{1}{p^2}, \quad (26.23)$$

whose integrand contains a singularity at $p = 0$. For $d > 2$ the singularity is integrable and the fluctuations in the field are well behaved. For $d \leq 2$ the singularity is not integrable and the fluctuations diverge. The divergent influence of Goldstone bosons breaks up the order in the system. We conclude that symmetry breaking is impossible in our complex scalar field theory for $d \leq 2$.

Chapter summary

- In many phase transitions, symmetry is broken below T_c and the ground state possesses only a subset of the symmetry of the Hamiltonian.

- Breaking a continuous symmetry results in a massless excitation, known as a Goldstone mode or Goldstone boson. These are not stable in two or fewer dimensions and symmetry breaking is then not possible (Coleman–Mermin–Wagner theorem).

- In a gauge theory, the Goldstone mode can be removed and becomes massive through combination with a gauge field (the Higgs mechanism).

A simple example of the Coleman–Mermin–Wagner theorem can be found in the behaviour of a ferromagnet. Magnons (quantized spin waves) are bosons and cost energy $E_{\boldsymbol{p}} = \alpha \boldsymbol{p}^2$ (in the low $\boldsymbol{p}$ limit) where $\alpha > 0$ is a constant. Thus the number of magnons excited at temperature T is proportional to

$$\int \frac{\mathrm{d}|\boldsymbol{p}| \, |\boldsymbol{p}|^{d-1}}{\mathrm{e}^{\beta E_{\boldsymbol{p}}} - 1}.$$

The integrand is proportional to $|\boldsymbol{p}|^{d-3}$ at low momentum (assuming $T \neq 0$), and therefore diverges when $d \leq 2$. Thus long-range order in a ferromagnet is wiped out by spin waves at all nonzero temperatures in two or fewer dimensions. This argument uses no quantum field theory, and holds only in this specific case (the analogous argument is more elaborate for antiferromagnets, see the book by Auerbach) but it is apparent that the mathematical origin is the same as that used for $G(0,0)$ in eqn 26.23, namely that an integral of this form diverges when $d \leq 2$.

Exercises

(26.1) *In this problem we prove that if the vacuum is invariant under an internal symmetry then a multiplet structure (that is, a different field with the same energy) will emerge.*

Consider creation operators for two types of particles $\hat{\phi}_A^\dagger$ and $\hat{\phi}_B^\dagger$ with the property

$$\left[\hat{Q}_N, \hat{\phi}_A^\dagger\right] = \hat{\phi}_B^\dagger, \qquad (26.24)$$

for some generator of a symmetry group, such that $[\hat{Q}_N, \hat{H}] = 0$. We impose the condition

$$\hat{Q}_N|0\rangle = 0. \qquad (26.25)$$

(a) Show that $e^{i\alpha\hat{Q}_N}|0\rangle = |0\rangle$.
(b) Show also that

$$E_A = E_B, \qquad (26.26)$$

where $\hat{H}\hat{\phi}_A^\dagger|0\rangle = E_A\hat{\phi}_A^\dagger|0\rangle$ and $\hat{H}\hat{\phi}_B^\dagger|0\rangle = E_B\hat{\phi}_B^\dagger|0\rangle$. Note that symmetry breaking occurs in the contrasting case that $\hat{Q}_N|0\rangle \neq 0$.

(26.2) The Fabri–Picasso theorem demonstrates that if a Lagrangian is invariant under an internal symmetry with charge operator $\hat{Q}_N = \int d^3x\,\hat{J}^0$, then there are two possibilities: either
(i) (as above) we have $\hat{Q}_N|0\rangle = 0$; or
(ii) $\hat{Q}_N|0\rangle$ has an infinite norm.
Here we prove this fact.
(a) By using the translation operator $e^{i\hat{p}\cdot a}$ show that

$$\langle 0|\hat{J}_N^0(x)\hat{Q}_N|0\rangle = \langle 0|\hat{J}_N^0(0)\hat{Q}_N|0\rangle. \qquad (26.27)$$

(b) By considering the matrix element $\langle 0|\hat{Q}_N\hat{Q}_N|0\rangle$ along with the result of (a), show that either $\hat{Q}_N|0\rangle = 0$ or $\hat{Q}_N|0\rangle$ has an infinite norm.
Note that if $\hat{Q}_N|0\rangle$ has an infinite norm, then it does not exist in the same space as $|0\rangle$. This is the case in spontaneous symmetry breakdown. In this case, the state $\hat{Q}_N|0\rangle$ is not zero, but another possible vacuum state. *See Aitchison and Hey, Chapter 17 for more details.*

(26.3) *We'll work through a famous proof of Goldstone's theorem. This can be found in many books.* Start with a theory with a continuous symmetry with charge $\hat{Q}_N$, where $\hat{Q}_N|0\rangle \neq 0$. Consider a field $\hat{\phi}(y)$ which is not invariant under $\hat{Q}_N$ such that

$$\left[\hat{Q}_N, \hat{\phi}(y)\right] = \hat{\psi}(y), \qquad (26.28)$$

where $\hat{\psi}$ is some other field. We are going to examine $\langle 0|\hat{\psi}(0)|0\rangle$, which we assume takes on a nonzero value when symmetry is broken.
(a) Show that

$$\frac{\partial}{\partial x^0}\langle 0|\hat{\psi}(0)|0\rangle = -\int d\boldsymbol{S}\cdot\langle 0|\left[\hat{\boldsymbol{J}}_N(x), \hat{\phi}(0)\right]|0\rangle, \qquad (26.29)$$

where $\boldsymbol{J}_N$ is the space-like part of the Noether current. *Hint: You should assume $\partial_\mu\hat{J}_N^\mu = 0$.*
Since the commutator in the previous equation contains local operators potentially separated by

a large space-like interval, we conclude that $\langle 0|\hat{\psi}(0)|0\rangle$ is independent of time.
(b) Insert a resolution of the identity to show

$$\int d^3x \sum_n \left[\langle 0|\hat{J}_N^0(0)|n\rangle\langle n|\hat{\phi}(0)|0\rangle e^{-ip_n\cdot x}\right.$$
$$\left. - \langle 0|\hat{\phi}(0)|n\rangle\langle n|\hat{J}_N^0(0)|0\rangle e^{ip_n\cdot x}\right] \neq 0. \qquad (26.30)$$

(c) Evaluate the spatial integral to verify

$$\sum_n (2\pi)^3\delta^{(3)}(\boldsymbol{p}_n)\left[\langle 0|\hat{J}_N^0(0)|n\rangle\langle n|\hat{\phi}(0)|0\rangle\; e^{ip_n^0\cdot x^0}\right.$$
$$\left. - \langle 0|\hat{\phi}(0)|n\rangle\langle n|\hat{J}_N^0(0)|0\rangle e^{-ip_n^0\cdot x^0}\right] \neq 0. \qquad (26.31)$$

From (a) we conclude that this expression is independent of x^0.
(d) Argue that for $|n\rangle = |0\rangle$, eqn. 26.31 vanishes.
(e) Argue that if $|n\rangle$ describes a massive particle state then matrix element $\langle n|\hat{J}_N^0|0\rangle$ must vanish.
(f) Argue that if we want $\langle n|\hat{J}_N^0|0\rangle \neq 0$ then we require $p_n^0 \rightarrow 0$ as $\boldsymbol{p}_n \rightarrow 0$.
(g) This proves Goldstone's theorem: the states linked to the ground state via the Noether current are massless Goldstone modes. Convince yourself that this is so! *Again, see Aitchison and Hey for more details.*

(26.4) Consider another gauge theory: complex scalar electrodynamics with a symmetry breaking mass term, whose Lagrangian is

$$\mathcal{L} = -\frac{1}{4}F_{\mu\nu}F^{\mu\nu} + |D_\mu\psi|^2 - V(\psi), \qquad (26.32)$$

where $V(\psi) = -m^2(\psi^\dagger\psi) + \frac{\lambda}{2}(\psi^\dagger\psi)^2$ and $D_\mu = \partial_\mu + iqA_\mu$.
(a) Take the broken symmetry ground state to be $\psi_0 = \left(\frac{m^2}{\lambda}\right)^{\frac{1}{2}}$. Expand about this using $\psi = \psi_0 + \frac{1}{\sqrt{2}}(\phi_1 + i\phi_2)$ and show that the potential of the broken symmetry theory is

$$U(\phi_1, \phi_2) = -\frac{1}{2\lambda}m^4 + m^2\phi_1^2 + O(\phi_i^3), \qquad (26.33)$$

exactly as in the case without electromagnetism.
(b) Consider the kinetic energy term. Show that, on symmetry breaking, this becomes

$$|D_\mu\psi|^2 = \frac{1}{2}(\partial_\mu\phi_1)^2 + \frac{1}{2}(\partial_\mu\phi_2)^2 \qquad (26.34)$$
$$+ \sqrt{2}q\psi_0 A_\mu\partial^\mu\phi_2 + q^2\psi_0^2 A_\mu A^\mu + \dots$$

Interpret this result in terms of massive photons. *This example shows that the massive gauge particle in a broken symmetry system may be found by expanding the covariant derivative.*

Coherent states

27

'Tis all in pieces, all coherence gone
John Donne (1572–1631)

There exists a class of problems in which a macroscopic number of quanta all pile into the same momentum state. Examples include the laser, where we put a large number of photons in the same wave-vector state into a cavity, and the superfluid, where a macroscopic number of Bose particles sit in a zero-momentum state. The best way to describe such systems uses a set of basis states discovered by Schrödinger and christened **coherent states** by Roy Glauber. A coherent state is an eigenstate of an annihilation operator. In real space coherent states are Gaussian wavepackets. As the occupation of the coherent state becomes large, the wavepacket becomes more and more narrow, increasingly resembling an object with classical properties. In this chapter we examine the properties of these states. Having built up quantum field theory from an analogy between identical, non-interacting particles and harmonic oscillators we start with the coherent states of the harmonic oscillator.

Roy Glauber (1925–)

27.1 Coherent states of the harmonic oscillator

The harmonic oscillator has a Hamiltonian $\hat{H} = \omega\left(\hat{a}^\dagger \hat{a} + \frac{1}{2}\right)$. Recall from Chapter 2 that this was obtained from the Hamiltonian[1] $\hat{H} = \omega(\hat{P}^2 + \hat{Q}^2)/2$ by the introduction of creation and annihilation operators $\hat{a}$ and $\hat{a}^\dagger$. This produced energy eigenstates $|n\rangle$ containing n quanta and with energy eigenvalue $(n + \frac{1}{2})\omega$. Thus $\langle n|(\hat{P}^2 + \hat{Q}^2)|n\rangle = n + \frac{1}{2}$, but P and Q are individually not good quantum numbers.[2] For a classical harmonic oscillator, the position and momentum behave like $Q = Q_0 \cos(\omega t - \varphi)$ and $P = -P_0 \sin(\omega t - \varphi)$, so that both quantities oscillate but there is a definite 90° phase relationship between them.[3] A quantum mechanical state that possesses this phase relationship is what we are after, and so we are led to consider eigenstates of the operator combination $\hat{Q} + i\hat{P}$ in which this phase relationship is hard wired. Since $\hat{a} = (\hat{Q} + i\hat{P})/\sqrt{2}$, we thus seek eigenstates of $\hat{a}$. We therefore define a coherent state $|\alpha\rangle$ by

$$\hat{a}|\alpha\rangle = \alpha|\alpha\rangle, \qquad (27.1)$$

where $\hat{a}$ is the annihilation operator and α is the eigenvalue (which will

[1] In this chapter we use reduced variables $\hat{P}^2 = \frac{\hat{p}^2}{m\omega}$ and $\hat{Q}^2 = m\omega\hat{x}^2$, and hence

$$\hat{Q} = \frac{1}{\sqrt{2}}(\hat{a} + \hat{a}^\dagger), \quad \hat{P} = -\frac{i}{\sqrt{2}}(\hat{a} - \hat{a}^\dagger).$$

[2] In fact $\langle\hat{Q}\rangle = \langle\hat{P}\rangle = 0$ for energy eigenstates.

[3] This feature is not captured in our quantum mechanical treatment of the harmonic oscillator in Chapter 2, since energy eigenstates $|n\rangle$ are, by definition, stationary states and *individually* do not possess any signature of anything oscillating at angular frequency ω. Coherent states will achieve this by being *combinations* of energy eigenstates. Coherent combination will give rise to interference, and this will recover wavelike behaviour.

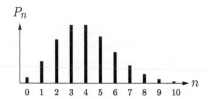

Fig. 27.1 The Poisson distribution $P_n = \mathrm{e}^{-|\alpha|^2}|\alpha|^{2n}/n! = \mathrm{e}^{-\langle n\rangle}\langle n\rangle^n/n!$ plotted for $|\alpha|^2 = \langle n\rangle = 4$. Note that a coherent state consists of a sum of harmonic oscillator states: the quantity P_n represents the probability of finding this coherent state in the nth harmonic oscillator state.

be a complex number).[4] Of course, a coherent state will be a sum of many energy eigenstates and so we may write $|\alpha\rangle = \sum_{n=0}^{\infty} c_n|n\rangle$. Substituting this into eqn 27.1 leads to the recursion relation $c_{n+1} = \alpha c_n/\sqrt{n+1}$, and hence[5]

$$
\begin{aligned}
|\alpha\rangle &= c_0\left(|0\rangle + \frac{\alpha}{\sqrt{1!}}|1\rangle + \frac{\alpha^2}{\sqrt{2!}}|2\rangle + \frac{\alpha^3}{\sqrt{3!}}|3\rangle + \ldots\right) \\
&= c_0\left(1 + \frac{\alpha}{1!}\hat{a}^\dagger + \frac{\alpha^2}{2!}(\hat{a}^\dagger)^2 + \frac{\alpha^3}{3!}(\hat{a}^\dagger)^3 + \ldots\right)|0\rangle, \quad (27.2)
\end{aligned}
$$

and normalization $\langle \alpha|\alpha\rangle = 1$ fixes $c_0 = \mathrm{e}^{-|\alpha|^2/2}$. Thus we can write the coherent state compactly as

$$|\alpha\rangle = \mathrm{e}^{-\frac{|\alpha|^2}{2}}\mathrm{e}^{\alpha\hat{a}^\dagger}|0\rangle. \qquad (27.3)$$

The amplitude to find $|\alpha\rangle$ in the state $|n\rangle$ is $c_n = \langle n|\alpha\rangle = \mathrm{e}^{-\frac{|\alpha|^2}{2}}\alpha^n/\sqrt{n!}$, corresponding to a probability density $P_n = |c_n|^2 = \mathrm{e}^{-|\alpha|^2}|\alpha|^{2n}/n!$ and this represents a Poisson distribution (see Fig. 27.1). In the next example, we establish some basic properties of this probability distribution of coherent states.

Example 27.1

We don't know the number of quanta in a coherent state but we can work out the average number[6]

$$\langle \hat{n}\rangle = \langle \alpha|\hat{a}^\dagger\hat{a}|\alpha\rangle = |\alpha|^2. \qquad (27.4)$$

We can calculate the uncertainty Δn in the number of quanta in a coherent state from $\Delta n = \sqrt{\langle \hat{n}^2\rangle - \langle n\rangle^2}$. Thus by working out

$$\langle \hat{n}^2\rangle = \langle \alpha|\hat{a}^\dagger\hat{a}\hat{a}^\dagger\hat{a}|\alpha\rangle = |\alpha|^4 + |\alpha|^2, \qquad (27.5)$$

we find that the uncertainty Δn is

$$\Delta n = |\alpha|, \qquad (27.6)$$

and therefore the fractional uncertainty is

$$\frac{\Delta n}{\langle \hat{n}\rangle} = \frac{|\alpha|}{|\alpha|^2} = \frac{1}{|\alpha|}. \qquad (27.7)$$

We conclude that although we don't know exactly how many quanta are in a coherent state, the fractional uncertainty in the number tends to zero as α (and hence the average occupation) tends to infinity.

Example 27.2

In this example we see how coherent states respond to an operator $\hat{U}(\theta) = \mathrm{e}^{-\mathrm{i}\theta\hat{n}}$, where θ is a real number and $\hat{n}$ is our usual number operator. Let's practise first on the state $|n\rangle$ (where $\hat{n}|n\rangle = n|n\rangle$). This is easily done when we realize that $\hat{U}(\theta)$ can be written as a power series and hence

$$\hat{U}(\theta)|n\rangle = \left(1 + (-\mathrm{i}\theta\hat{n}) + \frac{(-\mathrm{i}\theta\hat{n})^2}{2!} + \cdots\right)|n\rangle = \mathrm{e}^{-\mathrm{i}\theta n}|n\rangle, \qquad (27.8)$$

since any power of the operator $\hat{n}$ on the state $|n\rangle$ just gives that power of the number n. This now allows us to work out

$$\hat{U}(\theta)|\alpha\rangle = e^{-\frac{|\alpha|^2}{2}} \sum_{n=0}^{\infty} \frac{(\alpha e^{-i\theta})^n}{\sqrt{n!}}|n\rangle = |\alpha e^{-i\theta}\rangle. \qquad (27.9)$$

Thus the action of $\hat{U}(\theta)$ on the coherent state $|\alpha\rangle$ is to turn it into another coherent state with an eigenvalue multiplied by $e^{-i\theta}$.

This last result tells us that the time-dependence of the coherent state $|\alpha\rangle$ under the Hamiltonian $\hat{H} = (\hat{n} + \frac{1}{2})\omega$ is given by[7]

$$|\alpha(t)\rangle = e^{-i\hat{H}t}|\alpha(0)\rangle = e^{-i\omega t/2}|\alpha(0)e^{-i\omega t}\rangle. \qquad (27.10)$$

The exponential prefactor $e^{-i\omega t/2}$ is a simple phase factor originating from the zero-point energy, but the main message should be read from the state $|\alpha(0)e^{-i\omega t}\rangle$: coherent states remain coherent over time and the eigenvalue α executes circular motion around the Argand diagram at an angular speed ω. It looks as if we have recovered the coherent oscillatory behaviour we were after. But is there a way to visualize these states?

[7]Equation 27.10 should be contrasted with the analogous result for energy eigenstates of the harmonic oscillator. Equation 27.8 gives

$$|n(t)\rangle = e^{-i(n+\frac{1}{2})\omega t}|n(0)\rangle,$$

so in this case the only time-dependence in the state is found in the uninteresting phase factor.

27.2 What do coherent states look like?

To see what a state looks like as a function of some coordinate Q we project it along Q to form a wave function, via the amplitude $\langle Q|\alpha\rangle$.

Example 27.3

Let $|Q\rangle$ and $|P\rangle$ be the eigenstates of $\hat{Q}$ and $\hat{P}$. We write $\hat{a}|\alpha\rangle = \frac{1}{\sqrt{2}}\left(\hat{Q} + i\hat{P}\right)|\alpha\rangle = \alpha|\alpha\rangle$ and act on the right with a state $\langle Q|$ to obtain

$$\frac{1}{\sqrt{2}}\langle Q|\left(\hat{Q} + i\hat{P}\right)|\alpha\rangle = \alpha\langle Q|\alpha\rangle. \qquad (27.11)$$

Since $\hat{Q}|Q\rangle = Q|Q\rangle$ and $\hat{P} = -i\frac{\partial}{\partial Q}$, we can make the replacements $\langle Q|\hat{Q} = Q\langle Q|$ (since position is real) and $\langle Q|i\hat{P} = \frac{\partial}{\partial Q}\langle Q|$, from which we obtain the differential equation for the wave function:

$$\frac{\partial}{\partial Q}\langle Q|\alpha\rangle = -\left(Q - \sqrt{2}\alpha\right)\langle Q|\alpha\rangle. \qquad (27.12)$$

This equation has the normalized solution

$$\langle Q|\alpha\rangle = \frac{1}{\pi^{\frac{1}{4}}}e^{-(Q-\sqrt{2}\alpha)^2/2}, \qquad (27.13)$$

so the wave function is a Gaussian. This one is displaced by $\sqrt{2}\alpha = \langle \hat{Q}\rangle + i\langle\hat{P}\rangle$. You can also show that the wave function in momentum space is given by

$$\langle P|\alpha\rangle = \frac{1}{\pi^{\frac{1}{4}}}e^{-(P+i\sqrt{2}\alpha)^2/2}. \qquad (27.14)$$

Using the position and momentum wave functions we can work out the uncertainty in position and momentum. This is most easily done via the definitions

$$(\Delta Q)^2 = \langle\hat{Q}^2\rangle - \langle\hat{Q}\rangle^2, \quad (\Delta P)^2 = \langle\hat{P}^2\rangle - \langle\hat{P}\rangle^2. \qquad (27.15)$$

We find (Exercise 27.3) that

$$(\Delta P)^2 (\Delta Q)^2 = \frac{1}{4}. \tag{27.16}$$

Comparing this with the uncertainty principle $\Delta Q \Delta P \geq 1/2$ shows that the coherent states have the minimum possible value of uncertainty. This is the sense in which coherent states are the closest quantum mechanics allows us to come to classical objects which are localized in position and momentum space.

Finally we ask what these states do as a function of time. We have shown that the state $|\alpha(0)\rangle$ evolves to $|\alpha(t)\rangle$ (apart from an uninteresting phase factor) where $\alpha(t) = \alpha(0)e^{-i\omega t}$. An immediate consequence of this is that the expectation value of the position and momentum of our state may be calculated. Since we have $\langle \hat{a} \rangle = \langle \alpha|\hat{a}|\alpha \rangle = \alpha$ and $\langle \hat{a}^\dagger \rangle = \langle \alpha|\hat{a}^\dagger|\alpha \rangle = \alpha^*$, we also have

$$\langle \hat{Q} \rangle = \frac{\alpha + \alpha^*}{\sqrt{2}} \quad \text{and} \quad \langle \hat{P} \rangle = -i\frac{\alpha - \alpha^*}{\sqrt{2}}, \tag{27.17}$$

and we find that (writing $\alpha(0) = \alpha_0 e^{i\varphi}$)

$$\langle \hat{Q} \rangle = \sqrt{2}\alpha_0 \cos(\omega t - \varphi) \quad \text{and} \quad \langle \hat{P} \rangle = -\sqrt{2}\alpha_0 \sin(\omega t - \varphi), \tag{27.18}$$

meaning that the expectation values execute simple harmonic motion (see Fig. 27.2). These results are exactly as we expected for the classical harmonic oscillator at the beginning of the chapter.

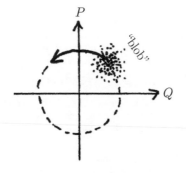

Fig. 27.2 A coherent state can be pictured as a 'blob' orbiting the origin in a plot of P against Q. Its mean position is given by $\langle \hat{Q} \rangle = \sqrt{2}\mathrm{Re}\,\alpha$ and $\langle \hat{P} \rangle = \sqrt{2}\mathrm{Im}\,\alpha$, where $\alpha(t) = \alpha(0)e^{-i\omega t}$. It is a blob, rather than a point, because there is uncertainty in both Q and P, given by $\Delta Q = \Delta P = \frac{1}{\sqrt{2}}$ (see Exercise 27.3).

27.3   Number, phase and the phase operator

Because α is a complex number, we can write $\alpha = |\alpha|e^{i\theta}$ and obtain

$$|\alpha\rangle = e^{-\frac{|\alpha|^2}{2}} \left(1 + \frac{|\alpha|e^{i\theta}\hat{a}^\dagger}{1!} + \frac{|\alpha|^2 e^{2i\theta}(\hat{a}^\dagger)^2}{2!} + \frac{|\alpha|^3 e^{3i\theta}(\hat{a}^\dagger)^3}{3!} + \dots\right)|0\rangle. \tag{27.19}$$

If we differentiate with respect to the phase θ we find that

$$-i\frac{\partial}{\partial \theta}|\alpha\rangle = \hat{n}|\alpha\rangle. \tag{27.20}$$

This implies that just as we have conjugate variables momentum and position linked via $\hat{p} = -i\frac{\partial}{\partial x}$, we identify conjugate variables n and θ. However, a closer examination of the phases of states shows that some care is needed in their interpretation. The definition of an operator that measures the phase of a system turns out to be rather problematical. One possibility suggested by Dirac is to make a phase operator via

$$\begin{aligned} \hat{a} &= e^{i\hat{\phi}}\sqrt{\hat{n}}, \\ \hat{a}^\dagger &= \sqrt{\hat{n}}e^{-i\hat{\phi}}. \end{aligned} \tag{27.21}$$

These already look rather odd in that it's unclear how to interpret the square root of the number operator. We'll examine the consequences of these definitions.

Example 27.4

The exponential operators can be rearranged to read

$$\begin{aligned} e^{i\hat{\phi}} &= \hat{a}(\hat{n})^{-\frac{1}{2}}, \\ e^{-i\hat{\phi}} &= (\hat{n})^{-\frac{1}{2}}\hat{a}^{\dagger}. \end{aligned} \tag{27.22}$$

Since we know that $\hat{a}^{\dagger}\hat{a} = \hat{n}$, then we must also have

$$e^{-i\hat{\phi}}e^{i\hat{\phi}} = \hat{n}^{-\frac{1}{2}}\hat{n}\hat{n}^{-\frac{1}{2}} = 1. \tag{27.23}$$

However, if we try to find $e^{i\hat{\phi}}e^{-i\hat{\phi}}$ we encounter a problem since

$$e^{i\hat{\phi}}e^{-i\hat{\phi}} = (\hat{n})^{-1}(\hat{n}+1) \neq 1, \tag{27.24}$$

which means that the exponential operator $e^{i\hat{\phi}}$ is not unitary. Since these exponential operators are not unitary, the phase operator $\hat{\phi}$ cannot be Hermitian[8] and so cannot be used to describe an observable. There's another problem. The phase operators yield a commutation relation (exercise)

$$\left[\hat{n}, e^{i\hat{\phi}}\right] = -e^{i\hat{\phi}}, \tag{27.25}$$

which implies[9] that we have

$$\left[\hat{n}, \hat{\phi}\right] = i. \tag{27.26}$$

This, in turn, gives us a number-phase uncertainty relation

$$\Delta n \Delta \phi \geq \frac{1}{2}. \tag{27.27}$$

This is a problem if taken literally, since if we take Δn to be small we must have $\Delta\phi$ greater than 2π. Dirac was so troubled by these inconsistencies that he left these operators out of the third edition of his seminal textbook *The Principles of Quantum Mechanics*.

[8]Remember that $\hat{U} = e^{i\hat{H}}$, where $\hat{U}$ is a unitary operator and $\hat{H}$ is a Hermitian operator.

[9]If $[\hat{A}, \hat{B}] = c$ then $[\hat{A}, e^{\lambda\hat{B}}] = \lambda c e^{\lambda\hat{B}}$. We take $\hat{A} = \hat{n}$, $\hat{B} = \hat{\phi}$ and $\lambda = i$. It follows that $c = i$.

One possible way forward was suggested by Peter Carruthers in the 1960s.[10] Here we regard the exponentials $e^{\pm i\phi}$ as a whole as the fundamental operators. For things to work out we need the exponentiated phase operators to obey the relations

$$\begin{aligned} \widehat{e^{i\phi}}|n\rangle &= |n-1\rangle, \\ \widehat{e^{-i\phi}}|n\rangle &= |n+1\rangle. \end{aligned} \tag{27.28}$$

We may then form Hermitian operators from the exponentials via the definitions

$$\begin{aligned} \widehat{\cos\phi} &= \frac{1}{2}\left(\widehat{e^{i\phi}} + \widehat{e^{-i\phi}}\right), \\ \widehat{\sin\phi} &= \frac{1}{2i}\left(\widehat{e^{i\phi}} - \widehat{e^{-i\phi}}\right). \end{aligned} \tag{27.29}$$

These operators, which do not commute with each other, have the properties

$$\begin{aligned} \left[\hat{n}, \widehat{\cos\phi}\right] &= -i\widehat{\sin\phi}, \\ \left[\hat{n}, \widehat{\sin\phi}\right] &= i\widehat{\cos\phi}. \end{aligned} \tag{27.30}$$

Peter Carruthers (1935–1997) is perhaps best known for his leadership of the theoretical division at Los Alamos National Laboratory from 1973–1980, overseeing a notable period of expansion.

[10]Actually by three students in his class. His problem set contained the following: *Apparently no one has investigated whether quantities $\hat{\phi}$ and $\sqrt{\hat{n}}$ 'defined' by equations (1) really exist. A bonus will be given for an answer to this question.* The bonus turned out to be a bottle of beer. See M.M. Nieto, arXiv:hep-th/9304036v1 for more details.

and uncertainty relations

$$\Delta n \Delta \cos \phi \;\geq\; \frac{1}{2}|\langle \widehat{\sin \phi} \rangle|,$$

$$\Delta n \Delta \sin \phi \;\geq\; \frac{1}{2}|\langle \widehat{\cos \phi} \rangle|. \qquad (27.31)$$

Carruthers' prescription then gives us a meaningful phase operator to play with, as long as we agree only to ask questions about sines and cosines of the phase. In the next section we examine some examples of coherent states.

27.4   Examples of coherent states

Example 27.5

A box containing an electromagnetic field (or in the jargon, a cavity containing photons) allows various normal modes. If we put a fixed number of photons into the box and have them all in a single mode labelled by a wave vector $\boldsymbol{k}$, then they will be in an eigenstate of $\hat{n}_{\boldsymbol{k}}$, which we will write as $|n_{\boldsymbol{k}}\rangle$. Since $\hat{n}_{\boldsymbol{k}}|n_{\boldsymbol{k}}\rangle = n_{\boldsymbol{k}}|n_{\boldsymbol{k}}\rangle$, we know precisely the number of photons (i.e. we know $n_{\boldsymbol{k}}$) and we have absolutely no uncertainty in this number ($\Delta n_{\boldsymbol{k}} = 0$). However, we have complete uncertainty in the phase of this state.

If instead we put photons, still all with the same wave vector $\boldsymbol{k}$, into a coherent state $|\alpha_{\boldsymbol{k}}\rangle$, we are now uncertain about how many photons we have ($\Delta n_{\boldsymbol{k}} = |\alpha_{\boldsymbol{k}}|$), by the uncertainty in phase[11] $\Delta \cos \phi_{\boldsymbol{k}} \to 0$ as $|\alpha_{\boldsymbol{k}}| \to \infty$. As we put more photons into this state ($\langle \hat{n}_{\boldsymbol{k}} \rangle = |\alpha_{\boldsymbol{k}}|^2$) then $|\alpha_{\boldsymbol{k}}|$ grows and the state becomes more phase coherent. This situation is a good model for what occurs inside a laser.

[11]It is possible to show that

$$\Delta \cos \phi_{\boldsymbol{k}} = \frac{\sin \phi_{\boldsymbol{k}}}{2|\alpha_{\boldsymbol{k}}|}.$$

A proof is given in the book by Loudon.

Example 27.6

Another important use of the coherent states is in describing the matter fields of superfluids and superconductors.[12] These phenomena are examined in detail a little later, but here we introduce a method of describing superfluids in terms of coherent states. Like the laser, the superfluid also contains a large number of quanta in the same quantum state. The particles are interacting bosons and are described by a coherent state

$$|\psi\rangle = |\alpha_{\boldsymbol{p}_0}\alpha_{\boldsymbol{p}_1}...\rangle. \qquad (27.32)$$

The coherent state $|\psi\rangle$ is an eigenstate of the field annihilation operator $\hat{\Psi}(\boldsymbol{x})$ given by

$$\hat{\Psi}(\boldsymbol{x}) = \frac{1}{\sqrt{\mathcal{V}}} \sum_{\boldsymbol{p}} \hat{a}_{\boldsymbol{p}} e^{i\boldsymbol{p}\cdot\boldsymbol{x}}, \qquad (27.33)$$

so that

$$\hat{\Psi}(\boldsymbol{x})|\psi\rangle = \psi(\boldsymbol{x})|\psi\rangle. \qquad (27.34)$$

The eigenvalue $\psi(\boldsymbol{x})$ can be written simply as

$$\psi(\boldsymbol{x}) = \frac{1}{\sqrt{\mathcal{V}}} \sum_{\boldsymbol{p}} \alpha_{\boldsymbol{p}} e^{i\boldsymbol{p}\cdot\boldsymbol{x}}, \qquad (27.35)$$

[12]Superfluids are described in Chapter 42 and superconductors in Chapter 44.

and is often called a **macroscopic wave function**, for reasons which will become apparent.[13]

In the superfluid a macroscopic number of particles reside in the $\boldsymbol{p} = 0$ state. We then have $\psi(\boldsymbol{x}) = \alpha_0/\sqrt{\mathcal{V}}$, a simple complex number which we can write as $\sqrt{n_0}\, e^{i\theta_0}$. Thus the expected *number density* of bosons is $|\alpha_0|^2/\mathcal{V} = n_0$ and since n_0 is very large (we have macroscopically occupied the coherent state), the uncertainty in the phase is vanishingly small.

We conclude that the macroscopic wave function $\psi_0(\boldsymbol{x})$ for a state where the $\boldsymbol{p}$ state is coherently occupied is best written

$$\psi_0(\boldsymbol{x}) = \sqrt{n_0}\, e^{i\theta_0}, \qquad (27.36)$$

where θ_0 is a constant field. We can therefore think of superfluid order as a spontaneous symmetry breaking of phase symmetry, resulting in the phase becoming fixed to the same value everywhere. This is examined in more detail in Chapter 42.

[13] Note that the expectation value $\langle\psi|\hat{\Psi}(\boldsymbol{x})|\psi\rangle = \psi(\boldsymbol{x}) = \frac{1}{\sqrt{\mathcal{V}}}\sum_{\boldsymbol{p}} \alpha_{\boldsymbol{p}} e^{i\boldsymbol{p}\cdot\boldsymbol{x}}$, which is significant since for occupation-number states (with fixed numbers of particles) we would expect

$$\begin{aligned}&\langle n_{\boldsymbol{p}_1} n_{\boldsymbol{p}_2}...|\hat{\Psi}(\boldsymbol{x})|n_{\boldsymbol{p}_1} n_{\boldsymbol{p}_2}...\rangle\\ &= \frac{1}{\sqrt{\mathcal{V}}}\sum_{\boldsymbol{p}}\langle n_{\boldsymbol{p}_1} n_{\boldsymbol{p}_2}...|\hat{a}_{\boldsymbol{p}}|n_{\boldsymbol{p}_1} n_{\boldsymbol{p}_2}...\rangle e^{i\boldsymbol{p}\cdot\boldsymbol{x}}\\ &= 0,\end{aligned}$$

since $\langle n_{\boldsymbol{p}_i}|\hat{a}_{\boldsymbol{p}_i}|n_{\boldsymbol{p}_i}\rangle = 0$ for all i. So clearly the interactions between bosons, which lead to a coherent state $|\psi\rangle$, are doing something very special. In fact, it will transpire that $\psi(\boldsymbol{x})$ is the order parameter for the ordered, coherent state of bosons, known as a superfluid and discussed in Chapter 42.

Chapter summary

- A coherent state is an eigenstate of the annihilation operator with eigenvalue α, a complex number.

- An occupation-number state contains a fixed number of particles but the phase is not well defined. A coherent state does not contain a fixed number of particles, but that number is Poisson distributed with mean $|\alpha|^2$. Its phase however can be well-defined, and its uncertainty decreases as $|\alpha|$ increases.

- Coherent states can be used to describe lasers and superfluids.

Exercises

(27.1) Verify that

$$\langle\alpha|\beta\rangle = e^{\alpha^*\beta - \frac{|\alpha|^2}{2} - \frac{|\beta|^2}{2}}. \qquad (27.37)$$

Compare this expression to the analogous one for energy eigenstates of the harmonic oscillator, $\langle m|n\rangle = \delta_{mn}$. Why is there a difference? Use these results to show that $\langle n|\hat{a}|n\rangle = 0$, but $\langle\alpha|\hat{a}|\alpha\rangle = \alpha$. Again, why the difference?

(27.2) Show that

$$\alpha = \frac{\langle\hat{Q}\rangle + i\langle\hat{P}\rangle}{\sqrt{2}}. \qquad (27.38)$$

(27.3) (a) Show that $\langle\hat{Q}\rangle = \sqrt{2}\,\mathrm{Re}\,\alpha$, $\langle\hat{P}\rangle = \sqrt{2}\,\mathrm{Im}\,\alpha$, $\langle\hat{Q}^2\rangle = \frac{1}{2} + 2(\mathrm{Re}\,\alpha)^2$ and $\langle\hat{P}^2\rangle = \frac{1}{2} + 2(\mathrm{Im}\,\alpha)^2$.
(b) Show that $\Delta Q = \Delta P = \frac{1}{\sqrt{2}}$.

(c) Using the previous results, show that the uncertainty relation for coherent states is indeed

$$(\Delta P)^2(\Delta Q)^2 = \frac{1}{4}. \qquad (27.39)$$

(27.4) Verify that the momentum space coherent state is given by

$$\langle P|\alpha\rangle = \frac{1}{\pi^{\frac{1}{4}}} e^{-(P + i\sqrt{2}\alpha)^2}. \qquad (27.40)$$

(27.5) Show that

$$\left(\hat{Q} - \langle Q\rangle\right)|\alpha\rangle = -i\left(\hat{P} - \langle P\rangle\right)|\alpha\rangle. \qquad (27.41)$$

One way of stating this result in words is that, for coherent states, the uncertainty is balanced between position and momentum.

(27.6) Show that
$$\left[\hat{n}, e^{i\hat{\phi}}\right] = -e^{i\hat{\phi}}. \qquad (27.42)$$

(27.7) Once again consider the forced quantum oscillator
$$L = \frac{1}{2}m\dot{x}(t)^2 - \frac{1}{2}m\omega^2 x(t)^2 + f(t)x(t), \qquad (27.43)$$

where $f(t)$ is the force which you should assume acts only for a finite length of time: $0 \leq t \leq T$. The evolution of $\langle \hat{x}(t)\rangle$ will be given by the sum of the time dependence in the absence of the force added to the response:
$$\hat{x}(t) = \hat{x}_0(t) + \int dt' \chi(t - t') f(t'). \qquad (27.44)$$

(a) Show that $\hat{x}(t)$ is given by
$$\hat{x}(t) = \left(\frac{1}{2m\omega}\right)^{\frac{1}{2}} \left(\hat{a}e^{-i\omega t} + \hat{a}^\dagger e^{i\omega t}\right) \qquad (27.45)$$
$$+ \int dt' \theta(t - t') \frac{if(t')}{2m\omega}\left(e^{-i\omega(t-t')} - e^{i\omega(t-t')}\right).$$

Hint: Use the definition of the susceptibility in Exercise 21.2.

(b) For times $t > T$ verify that this may be written
$$\hat{x}(t) = \left(\frac{1}{2m\omega}\right)^{\frac{1}{2}}\left[\left(\hat{a} + \frac{i}{(2m\omega)^{\frac{1}{2}}}\tilde{f}(\omega)\right)e^{-i\omega t}\right.$$
$$\left. + \left(\hat{a}^\dagger - \frac{i}{(2m\omega)^{\frac{1}{2}}}\tilde{f}(-\omega)\right)e^{i\omega t}\right]. (27.46)$$

(c) Argue that the Hamiltonian must take the form
$$\hat{H} = \omega\left(\hat{a}^\dagger - \frac{i}{(2m\omega)^{\frac{1}{2}}}\tilde{f}(-\omega)\right)\left(\hat{a} + \frac{i}{(2m\omega)^{\frac{1}{2}}}\tilde{f}(\omega)\right). \qquad (27.47)$$

(d) Show that this Hamiltonian is diagonalized by the coherent state $|\alpha\rangle$ which is the eigenstate of the $\hat{a}$ operator and find α.

(e) Show that the number of quanta emitted by the source is
$$\langle \hat{n}\rangle = \frac{|\tilde{f}(\omega)|^2}{2m\omega}. \qquad (27.48)$$

and find the energy imparted to the system by the source.

(f) Find an expression for $|\alpha\rangle$ in terms of the unforced ground state $|0\rangle$.

(27.8) Consider (yet again) the forced quantum oscillator from the previous question with the forcing part expressed as an interaction.

(a) Show that
$$-i\int dt \hat{H}_I(t)|0\rangle = \frac{i\tilde{f}(\omega)}{(2m\omega)^{\frac{1}{2}}}|1\rangle. \qquad (27.49)$$

(b) By considering the S-matrix, show that the amplitude for the source to create a state containing a single quantum is given by
$$\mathcal{A}_1 = \frac{i\tilde{f}(\omega)}{(2m\omega)^{\frac{1}{2}}}e^{(\text{Dumbbell})}, \qquad (27.50)$$

where (Dumbbell) denotes the Feynman diagram in Fig. 22.2.

(c) Extend the previous result to show that the amplitude for the source emitting n quanta is given by
$$\mathcal{A}_n = \frac{1}{\sqrt{n!}}\left(\frac{i\tilde{f}(\omega)}{(2m\omega)^{\frac{1}{2}}}\right)^n e^{(\text{Dumbbell})}. \qquad (27.51)$$

(d) Show that the probability of the source emitting n quanta is given by
$$P_n = \frac{|\alpha|^{2n}}{n!}e^{-|\alpha|^2}, \qquad (27.52)$$

where $|\alpha| = \frac{|\tilde{f}(\omega)|}{(2m\omega)^{\frac{1}{2}}}$. Comparing with the previous question, explain why this result was inevitable.

(27.9) A coherent state is an eigenstate of the annihilation operator $\hat{a}$. Why do we never bother with the eigenstates of the creation operator $\hat{a}^\dagger$?

Grassmann numbers: coherent states and the path integral for fermions

28

Thus far, our discussion of fields and coherent states has concentrated exclusively on bosons. We would also like to describe fermions with path integrals and coherent states. This turns out to be a non-trivial matter.[1]

Consider the path integral, for example. In calculating $Z[J]$ for Bose fields we have argued that we integrate over the classical action in the presence of sources $Z[J] = \int \mathcal{D}\phi\, e^{-i \int d^4 x (\mathcal{L} + J\phi)}$. Unlike the canonical formalism, the path integral formulation doesn't deal with operators and their commutation relations; just classical fields. (The quantum behaviour comes about through the constructive interference of the classical paths.) This formulation works for Bose fields because the ordinary numbers outputted by the functions over which we integrate *commute*, just like the quantum operators of the canonical formalism.

As we discussed in Chapter 3 the operator-valued fields that represent fermions in the canonical approach *anticommute*. This is the property that, for example

$$\hat{\psi}(x)\hat{\psi}(y) = -\hat{\psi}(y)\hat{\psi}(x), \qquad (28.1)$$

where $\hat{\psi}(x)$ is an operator that annihilates a fermion at spacetime point x. In order to be able to describe fermions with a path integral we need a new sort of number: a number that anticommutes.

Even before the advent of quantum mechanics, the properties of such numbers had already been examined by Hermann Grassmann. Although Grassmann's work was largely neglected during most of his life, the anticommuting numbers he invented are now known as **Grassmann numbers**. In this chapter we will examine the properties of Grassmann numbers. This will allow us to write down a coherent state for fermions and will allow us to write a path integral for fermions. The latter will be useful in our description of the quantum field of the electron.

[1]For this reason, Chapter 28 can be skipped on a first reading. Grassmann numbers will only be used in Section 38.2.

Hermann Grassmann (1809–1877)

28.1 Grassmann numbers

Grassmann numbers anticommute. This means that two Grassmann numbers η and ζ have the property that

$$\eta\zeta = -\zeta\eta. \qquad (28.2)$$

Immediately following from this is that $\eta^2 = \eta\eta = -\eta\eta = 0$. In words: the square of a Grassmann number is zero. This simplifies the algebra of Grassmann numbers. Because $\eta^2 = 0$ we can't have any terms in the Taylor expansion of a function with more than a single power of η, and so the most general function of η is then given by

$$f(\eta) = a + b\eta, \tag{28.3}$$

where a and b are real numbers.[2]

[2]Real numbers commute with Grassmann numbers.

We'd like to be able to do calculus with Grassmann numbers. We define the act of differentiating via

$$\frac{\partial}{\partial\eta}\eta = 1, \quad \frac{\partial}{\partial\eta}a = 0. \tag{28.4}$$

Note also that this derivative operator itself anticommutes, so if ζ is another Grassmann number we have $\frac{\partial}{\partial\eta}\zeta\eta = -\zeta\frac{\partial}{\partial\eta}\eta = -\zeta$.

Example 28.1

Differentiating the most general function of a Grassmann variable, we have

$$\frac{\partial f(\eta)}{\partial\eta} = b. \tag{28.5}$$

The most general function of two Grassmann variables is written

$$g(\eta, \zeta) = a + b\eta + c\zeta + e\eta\zeta. \tag{28.6}$$

We can differentiate this to show that

$$\frac{\partial}{\partial\eta}g(\eta, \zeta) = b + e\zeta, \quad \frac{\partial}{\partial\zeta}g(\eta, \zeta) = c - e\eta. \tag{28.7}$$

We would also like to integrate Grassmann numbers. We define integration over Grassmann variables via

$$\int d\eta\, \eta = 1, \quad \int d\eta = 0. \tag{28.8}$$

(Note again that the integration operator $d\eta$ is to be regarded as a Grassmann number.) As a result of the definitions, integration of our most general function $f(\eta)$, given in eqn 28.3, yields

$$\int d\eta\, f(\eta) = \int d\eta\, a + \int d\eta\, b\eta = b. \tag{28.9}$$

Notice that this is the same result that we obtained by differentiating. In fact it is a general result that integration and differentiation of Grassmann numbers yield exactly the same result!

Example 28.2

Let's check this assertion by integrating the function $g(\eta, \zeta)$. Integrating with respect to η, we have

$$\int d\eta\, (a + b\eta + c\zeta + e\eta\zeta) = b + e\zeta, \tag{28.10}$$

while integrating with respect to ζ gives us

$$\int d\zeta\, (a + b\eta + c\zeta + e\eta\zeta) = c - e\eta, \tag{28.11}$$

which is identical to what we had in eqn 28.7.

28.2 Coherent states for fermions

One use of Grassmann numbers is in defining a coherent state that described fermions. Taking the creation operator for a fermion to be $\hat{c}^\dagger$ we define a fermionic coherent state as $|\eta\rangle = \mathrm{e}^{-\eta\hat{c}^\dagger}|0\rangle$, where η, as usual, is a Grassmann number. We can immediately simplify this using $\eta^2 = 0$ and write

$$|\eta\rangle = |0\rangle - \eta|1\rangle. \tag{28.12}$$

In order for things to work out, we require that Grassmann numbers like η anticommute with fermion operators. For example, we require $\{\eta, \hat{c}\} = 0$.

Example 28.3

We will check that a coherent state defined this way is the eigenstate of the fermion annihilation operator $\hat{c}$. We have

$$\hat{c}|\eta\rangle = \hat{c}|0\rangle - \hat{c}\eta|1\rangle = 0 + \eta\hat{c}|1\rangle = \eta|0\rangle. \tag{28.13}$$

But

$$\eta|\eta\rangle = \eta|0\rangle - \eta^2|1\rangle = \eta|0\rangle, \tag{28.14}$$

and so $\hat{c}|\eta\rangle = \eta|\eta\rangle$ as required.

We can also define a state $\langle\bar{\eta}|\hat{c}^\dagger = \langle\bar{\eta}|\bar{\eta}$, where

$$\langle\bar{\eta}| = \langle 0| - \langle 1|\bar{\eta} = \langle 0| + \bar{\eta}\langle 1|. \tag{28.15}$$

Note that $\bar{\eta}$ is not the complex conjugate of η and $\langle\bar{\eta}|$ is not the adjoint of $|\eta\rangle$. With these definitions it follows[3] that the value of an inner product is

$$\langle\bar{\zeta}|\eta\rangle = \mathrm{e}^{\bar{\zeta}\eta}. \tag{28.16}$$

[3]See Exercise 28.1.

Finally, for these coherent states to be useful, we require that they form a complete set via the completeness relation

$$\int \mathrm{d}\eta\,\mathrm{d}\bar{\eta}\,\mathrm{e}^{\bar{\eta}\eta}|\bar{\eta}\rangle\langle\eta| = 1. \tag{28.17}$$

28.3 The path integral for fermions

We start by evaluating the Gaussian integral for coherent states[4]

$$\int \mathrm{d}\eta\,\mathrm{d}\bar{\eta}\,\mathrm{e}^{\bar{\eta}a\eta} = \int \mathrm{d}\eta\,\mathrm{d}\bar{\eta}\,(1 + \bar{\eta}a\eta) = \int \mathrm{d}\eta\,a\eta = a. \tag{28.18}$$

This carries over to the case of Grassmann-valued (N-component) vectors $\boldsymbol{\eta} = (\eta_1, \eta_2, ..., \eta_N)$ and $\bar{\boldsymbol{\eta}} = (\bar{\eta}_1, \bar{\eta}_2, ..., \bar{\eta}_N)$ and we find[5] that for a matrix $\mathbf{A}$:

$$\int \mathrm{d}^N\eta\,\mathrm{d}^N\bar{\eta}\,\mathrm{e}^{\bar{\boldsymbol{\eta}}\mathbf{A}\boldsymbol{\eta}} = \det \mathbf{A}, \tag{28.19}$$

where $\mathrm{d}^N\eta\,\mathrm{d}^N\bar{\eta} = \mathrm{d}\eta_1\mathrm{d}\bar{\eta}_1\mathrm{d}\eta_2\mathrm{d}\bar{\eta}_2...\mathrm{d}\eta_N\mathrm{d}\bar{\eta}_N$.

Finally, we can work out the most important integral of all. Starting with Grassmann numbers ψ and η, we will want to integrate an equation of the form

$$\int \mathrm{d}\psi\,\mathrm{d}\bar{\psi}\,\mathrm{e}^{\bar{\psi}K\psi + \bar{\eta}\psi + \bar{\psi}\eta}. \tag{28.20}$$

[4]Had these been ordinary c-numbers, we would have

$$\int \mathrm{d}z\,\mathrm{d}z^*\,\mathrm{e}^{-z^*az} = \frac{\pi}{a}, \quad (\mathrm{Re}\,a > 0),$$

where $\mathrm{d}z\,\mathrm{d}z^*$ represents the independent integration over real and imaginary parts of z. While the factor of π is neither here nor there, the important point to note is that the a comes out in the numerator for the Grassmann version.

[5]See Exercise 28.3.

[6]Recall that we originally solved the path integral problem $I = \int dx\, e^{-\frac{ax^2}{2}+bx}$ by completing the square, writing

$$-\frac{ax^2}{2} + bx = -\frac{a}{2}\left(x - \frac{b}{a}\right)^2 + \frac{b^2}{2a},$$

allowing us to conclude that $I = \sqrt{\frac{2\pi}{a}}e^{\frac{b^2}{2a}}$.

We play the usual trick of completing the square[6] and we find

$$\bar{\psi}K\psi + \bar{\eta}\psi + \bar{\psi}\eta = \left(\bar{\psi} + \bar{\eta}K^{-1}\right)K\left(\psi + K^{-1}\eta\right) - \bar{\eta}K^{-1}\eta, \quad (28.21)$$

which enables us to conclude that

$$\int d\psi d\bar{\psi}\; e^{\bar{\psi}K\psi + \bar{\eta}\psi + \bar{\psi}\eta} = Ce^{-\bar{\eta}K^{-1}\eta}, \quad (28.22)$$

where the number C (which will be proportional to a determinant for the vector case) will later be removed by normalization. (Remember that most of our problems only require part of the solution: the part in the exponential is important and the multiplying prefactor far less so.)

We can immediately extend this result to the functional integral we need. Our goal is to find the generating functional for fermions

$$Z[\bar{\eta}, \eta] = \int \mathcal{D}\psi \mathcal{D}\bar{\psi}\, e^{i\int d^4x\left[\mathcal{L}(\bar{\psi},\psi) + \bar{\eta}(x)\psi(x) + \bar{\psi}(x)\eta(x)\right]}. \quad (28.23)$$

[7]Recall that to normalize we define $\mathcal{Z}[\bar{\eta},\eta] = Z[\bar{\eta},\eta]/Z[0,0]$.

If necessary we massage the Lagrangian $\mathcal{L}(\bar{\psi},\psi)$ into the form $\bar{\psi}\hat{K}\psi$, with $\hat{K}$ a differential operator, allowing us to read off the answer that the normalized[7] generating functional is given by:

$$\mathcal{Z}[\bar{\eta},\eta] = e^{-i\int d^4x d^4y\, \bar{\eta}(x)\hat{K}^{-1}(x,y)\eta(y)}. \quad (28.24)$$

As before, we can then read off the propagator as $i\hat{K}^{-1}$ and the quantization process is complete!

Chapter summary

- Grassmann numbers anticommute and can be used to describe fermions.
- The fermion coherent state can be written $|\eta\rangle = |0\rangle - \eta|1\rangle$. Such states can be used to construct the fermion path integral.

Exercises

(28.1) From the definitions of the coherent states $|\eta\rangle$ and $\langle\bar{\zeta}|$, prove that $\langle\bar{\zeta}|\eta\rangle = e^{\bar{\zeta}\eta}$.

(28.2) Find (a) $\int d\bar{\eta}d\eta\,\eta\bar{\eta}$ and (b) $\int d\bar{\eta}d\eta\,\bar{\eta}\eta$.

(28.3) Show that $\int d^2\eta d^2\bar{\eta}\, e^{\bar{\eta}\mathbf{A}\eta} = \det\mathbf{A}$, where $\eta = \begin{pmatrix}\eta_1\\\eta_2\end{pmatrix}$, $\bar{\eta} = \begin{pmatrix}\bar{\eta}_1 & \bar{\eta}_2\end{pmatrix}$, $d^2\eta d^2\bar{\eta} = d\eta_1 d\bar{\eta}_1 d\eta_2 d\bar{\eta}_2$, and $\mathbf{A}$ is a 2×2 matrix.

(28.4) Prove that

$$\int d\eta d\bar{\eta}\, e^{\bar{\eta}\eta}|\bar{\eta}\rangle\langle\eta| = |0\rangle\langle 0| + |1\rangle\langle 1|. \quad (28.25)$$

and may therefore be taken as a resolution of the identity.

(28.5) Verify eqn 28.21.

Part VII

Topological ideas

In many cases in quantum field theory we are concerned with working out some integral over spacetime coordinates and therefore the precise geometry of spacetime is important. However, there are some problems in which the physics is entirely insensitive to the shape of spacetime, that is it has no way of sensing the spacetime metric. In this case, the theory will give an answer which is some kind of invariant which nevertheless does depend on some global property of the *topology* of the spacetime.

- In Chapter 29 we introduce some simple topological ideas and use them to describe simple topological objects which can exist in field theories. These include *kinks* and *vortices*.

- Topological field theories are discussed in Chapter 30 and we provide some concrete examples of applications of these ideas by discussing *anyons*, *fractional statistics* and *Chern–Simons* theories.

29

Topological objects

[1]Newton-John's grandfather was Max Born (1882–1970), one of the founding fathers of quantum mechanics. The quoted lyric is by John Farrar.

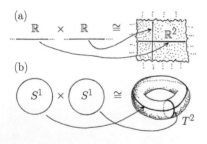

Fig. 29.1 A cup can be continuously deformed into a doughnut (a torus).

[2]This is why projections of the Earth's surface on maps usually involve a cut through one of the major oceans.

Fig. 29.2 A product space is constructed by attaching a copy of one space to every single point in another space. (a) A product space of the real line ($\mathbb{R}$) with itself produces the two-dimensional plane $\mathbb{R}^2$. (b) The product of two circles (S^1) produces the torus T^2. A cup can be continuously deformed into a doughnut (a torus).

You better shape up
Olivia Newton-John (1948–), *Grease*[1]

In this chapter we meet a new class of excitation in the field known as a **topological object**. These are configurations of the field which may exist in systems that show spontaneous symmetry breaking. As discussed in Chapter 26, when a symmetry is broken we should watch out for objects known as defects that reflect the possibility that a field can break symmetry in different ways in different regions of space. The objects we will encounter in this chapter are examples of such defects. We will discuss the two simplest topological objects: kinks and vortices. Both of these may be thought of as time-independent lumps of energy which are held together by interactions. An important additional feature of vortices is that they're not stable alone and we must add *gauge fields* in order for them to be realized. As always, the presence of gauge fields provides the theory with some particularly interesting properties. We begin with a crash course on basic topology.

29.1   What is topology?

The doughnut and the coffee mug are topologically equivalent because one can be continuously deformed into the other (Fig. 29.1). We imagine that everything is made of some kind of mouldable clay and so all objects can be pressed and prodded so that their shape changes. However, you are not allowed to puncture or heal, to add or remove holes, so that neighbouring points in the object remain neighbouring after the deformation. Topological spaces are given mathematical names: the real line is denoted $\mathbb{R}$, so the plane is $\mathbb{R}^2$ and the n-dimensional version is $\mathbb{R}^n$. One can define a **product space** in an obvious way, and so $\mathbb{R} \times \mathbb{R} = \mathbb{R}^2$ [Fig. 29.2(a)]. Similarly $\mathbb{R} \times \mathbb{R} \times \cdots \times \mathbb{R} = \mathbb{R}^n$. A segment of a $\mathbb{R}$ joined head to tail gives a circle, denoted S^1. The sphere is S^2. (Note that this is a two-dimensional space, because by 'sphere' mathematicians mean the surface of a ball, not its interior.) The sphere cannot be embedded in $\mathbb{R}^2$ (you can't fully represent a sphere on a piece of paper without cutting it[2]) but can be embedded in $\mathbb{R}^3$. (Similarly S^1 can't be embedded in $\mathbb{R}$.) The torus T^2 (back to the doughnut again) can be constructed using the product $T^2 = S^1 \times S^1$ [Fig. 29.2(b)].

Once we have a topological space, we can define a path through the space as a mapping $f(x)$ of the real-line segment $[a, b]$ (shorthand for

$a \leq x \leq b$) to the topological space. If $f(a) = f(b)$, the path becomes a loop. Paths can be continuously deformed into each other in an obvious way and one finds that in a particular topological space the set of possible loops can be divided up into a number of classes and these can be mapped onto a group (called the **fundamental group** and given the symbol π_1). For example, in $\mathbb{R}^n$ all loops are **contractible** (i.e. can be continuously deformed to a point) and so there is only one class of loop and π_1 is a trivial group just consisting of the identity element. It's more interesting if the space has a hole in it, so that the loops can be divided into classes, each one characterized by the number of times the loop winds round the hole. This integer is known as the **winding number**. Thus $\pi_1 = \mathbb{Z}$, the set of integers. Further examples are $\pi_1(S^1) = \mathbb{Z}$ and $\pi_1(T^2) = \mathbb{Z} \times \mathbb{Z}$ (see Fig. 29.3).

The crucial point about topological arguments is that they do not rely on any notion of the geometrical structure of space. Topological spaces are continuously deformable and therefore a topological description reflects the underlying 'knottiness' of the problem.

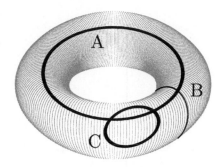

Fig. 29.3 Loops on a surface of a torus T^2 can be contractible (continuously deformable to a point), such as C, or non-contractible such as A and B (each of these winds round once). Any loop can be described using two winding numbers, the number of times the loop winds round like A and the number of times like B. In topological parlance: $\pi_1(T^2) = \mathbb{Z} \times \mathbb{Z}$.

Example 29.1

A thin solenoid (radius R, aligned along $\hat{e}_z$) inserted into the two-slit experiment (see Fig. 29.4) gives rise to a magnetic vector potential outside the solenoid equal to $A^\theta = BR^2/2r$ and inside $A^\theta = Br/2$. The magnetic field $\boldsymbol{B} = \boldsymbol{\nabla} \times \boldsymbol{A}$ is then equal to zero outside and $(0,0,B)$ inside the solenoid.

Even though $\boldsymbol{B} = 0$ outside the solenoid, the electronic wave function can still be affected. This is because a plane wave $\psi \propto e^{i\boldsymbol{p}\cdot\boldsymbol{r}/\hbar}$ and in the presence of $\boldsymbol{A}$ the momentum $\boldsymbol{p} \to \boldsymbol{p} - q\boldsymbol{A}$ and so over a trajectory ψ acquires an extra phase factor $e^{i\Delta\alpha}$ where $\Delta\alpha = -\frac{q}{\hbar}\int \boldsymbol{A}\cdot d\boldsymbol{r}$. This might not seem to matter because if you perform a gauge transformation $\boldsymbol{A} \to \boldsymbol{A} + \boldsymbol{\nabla}\chi$ then $\Delta\alpha = -\frac{q}{\hbar}\int (\boldsymbol{A}+\boldsymbol{\nabla}\chi)\cdot d\boldsymbol{r}$ and so $\Delta\alpha$ can be anything you want, all you need to do is to pick a particular form for χ. However, when we consider the two-slit experiment, we are interested in the phase difference $\Delta\delta$ between the two paths which is given by

$$\Delta\delta = \Delta\alpha_1 - \Delta\alpha_2 = -\frac{q}{\hbar}\int_{a\to 1\to b} \boldsymbol{A}\cdot d\boldsymbol{r} + \frac{q}{\hbar}\int_{a\to 2\to b} \boldsymbol{A}\cdot d\boldsymbol{r}$$
$$= \frac{q}{\hbar}\oint \boldsymbol{A}\cdot d\boldsymbol{r}, \tag{29.1}$$

where the final integral is carried out anticlockwise around the trajectories. However, we may use Stokes' theorem to write

$$\Delta\delta = \frac{q}{\hbar}\int \boldsymbol{\nabla}\times\boldsymbol{A}\cdot d\boldsymbol{S} = \frac{q}{\hbar}\int \boldsymbol{B}\cdot d\boldsymbol{S} = \frac{q}{\hbar}\Phi, \tag{29.2}$$

where Φ is the flux through the solenoid. This leads to a shift in the interference pattern on the screen and is observed experimentally. Gauge transformations make no difference here because $\oint \boldsymbol{\nabla}\chi\cdot d\boldsymbol{r} = 0$. Remarkably we expect a nonzero gauge-independent phase difference $\Delta\delta$ even though electrons only pass through regions without magnetic field. The effect is known as the **Aharonov–Bohm effect** and is topological. The electronic wave function is defined on the plane $\mathbb{R}^2$ minus the origin (where we place the flux), i.e. a sheet with a hole in it. Electromagnetism has $U(1)$ symmetry, which has the same topology as S^1 (the phase can be defined on a circle in the Argand diagram). Thus to define a phase everywhere on a sheet with a hole in it amounts to mapping S^1 on to a path around the hole. These mappings fall into disjoint classes (labelled by an integer winding number, because $\pi_1(S^1) = \mathbb{Z}$) and cannot be deformed into each other.

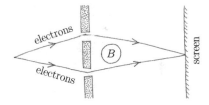

Fig. 29.4 The Aharonov–Bohm effect. Electrons diffract through two slits but their paths enclose a solenoid. The field inside the solenoid is B, but the electrons only pass through regions in which there is no magnetic field.

Yakir Aharonov (1932–)
David Bohm (1917–1992)

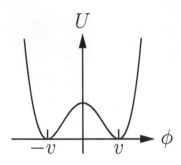

Fig. 29.5 The double well potential.

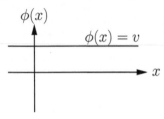

Fig. 29.6 The uniform ground state of the field in a system with broken symmetry.

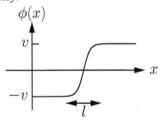

Fig. 29.7 The kink solution.

[3]Notice that the kink has energy inversely proportional to the coupling constant. This is significant since it implies that an expansion in powers of the coupling constant, of the sort we make in perturbation theory, would never predict such a field configuration. This is one of the hallmarks of topological objects: they are fundamentally non-perturbative.

29.2 Kinks

Consider the Lagrangian $\mathcal{L} = \frac{1}{2}(\partial_\mu \phi)^2 - U(\phi)$ in $(1+1)$-dimensional spacetime, with $U(\phi) = \frac{\lambda}{4}(v^2 - \phi^2)^2$. This is just our symmetry breaking Lagrangian for ϕ^4 theory as can be quickly seen by expanding the bracket and setting $v^2 = \frac{m^2}{\lambda}$. The potential, shown in Fig. 29.5, has two minima with $U(\phi) = 0$ at $\phi(x) = \pm v$. The vacuum of the theory is to be found at $\phi(x) = v$ or $\phi(x) = -v$. We know that one way to work out what sort of particles live in this potential is to expand about one of the vacua. The excitations are then predicted to be particles of mass $(2\lambda v^2)^{\frac{1}{2}}$. The point of this chapter is to tell you that objects other than particles can live in the potential too. These are stationary configurations of the field $\phi(x)$ whose energy density $\frac{1}{2}(\partial_i \phi)^2 + U(\phi)$ goes to zero at $x = \pm\infty$, but which do something non-trivial in between.

Let's examine a field which has zero energy density at $x = \pm\infty$. Such a field must be stationary and occupy a zero of the potential at $\pm\infty$. A boring way of achieving this is for $\phi(x) = v$ as shown in Fig. 29.6. The configuration $\phi(x) = -v$ is similar. This is just symmetry breaking: the system falls into a ground state of $\phi = v$ or $\phi = -v$. However, there is a more interesting configuration of the field. What if the field takes $\phi(-\infty) = -v$ and $\phi(\infty) = v$? Then we must have something like the object shown in Fig. 29.7. This configuration is known as a **kink** and involves the curious state of affairs of half of the field living in one vacuum and half in the other. Notice that the field has broken the symmetry in two different ways. At $x \to \infty$ the field takes on the value $\phi = v$, whereas for $x \to -\infty$ it takes on value $\phi = -v$. We might say that the fields at $\pm\infty$ live in difference vacua. To find out whether the kink is stable, we may evaluate its total energy. If the energy is finite then we're in luck as such configurations are allowed to exist.

Example 29.2

We will evaluate the energy of a kink. We take the kink to be stationary in time, so $\partial_0 \phi = 0$. We want to work out $E = \int \mathrm{d}x \left[\frac{1}{2}(\partial_1 \phi)^2 + U(\phi)\right]$. From the equations of motion we have that $\frac{\partial^2 \phi}{\partial x^2} = \frac{\partial U}{\partial \phi}$, which may be integrated to give $\frac{1}{2}\left(\frac{\partial \phi}{\partial x}\right)^2 = U(\phi)$. We therefore have an expression for the energy given by

$$
\begin{aligned}
E &= \int \mathrm{d}x \left[\frac{1}{2}\left(\frac{\partial \phi}{\partial x}\right)^2 + U(\phi)\right] \\
&= \int_{-\infty}^{\infty} \mathrm{d}x\, 2U(\phi) = \int_{-v}^{v} \mathrm{d}\phi \frac{\mathrm{d}x}{\mathrm{d}\phi} 2U(\phi) \\
&= \int_{-v}^{v} \mathrm{d}\phi \left[2U(\phi)\right]^{\frac{1}{2}}.
\end{aligned}
\tag{29.3}
$$

Plugging in $[2U(\phi)]^{\frac{1}{2}} = (\lambda/2)^{1/2}(v^2 - \phi^2)$, we obtain the energy[3]

$$
E = \frac{1}{\sqrt{2}} \frac{4m^3}{3\lambda}.
\tag{29.4}
$$

This is finite and the kink is therefore allowed to exist!

The kink has a finite energy and, by construction, is a time-independent field configuration. It is also of a finite size l (see Fig. 29.8), the extent of which is determined by a balance of kinetic and potential energy. The kinetic energy's contribution $\int dx \frac{1}{2}(\partial_1 \phi)^2 \approx l \times (v/l)^2$ tries to smear the kink out, but the potential contribution $\int dx\, U \approx \lambda v^4 l$ tries to limit the region over which the field changes. The minimum energy is then to be found when the size of the kink is $l \approx (\lambda v^2)^{-\frac{1}{2}} = 1/m$, where m is the mass of the conventional particles of the theory.

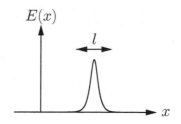

Fig. 29.8 The energy density of a kink is localized in an interval l.

Example 29.3

The domain wall in a ferromagnet has much in common with the kink described here. For the ferromagnet the vacua correspond to all spins pointing up ($\phi = v$) or down ($\phi = -v$). The kink corresponds a region where the fields change from up to down over a finite distance. This separates two magnetic **domains**. A domain is a region where the symmetry is broken in a particular way. It is, for example, part of the Universe where the vacuum corresponds to all spins pointing up. Domain walls therefore separate regions with different vacua.

Domain walls are real and their existence may be detected in magnets through a variety of means. In a magnet, domains allow a magnetic system of finite spatial extent to save energy by reducing the field density in free space. However, the domain wall might be expected to exist on the more general grounds discussed here.

Also interesting is an idea of Nambu's that in reaching its final state the Universe has broken various symmetries. This, he suggests, might imply that domains with different vacua may exists across the Universe.

We have a field configuration with finite energy above the ground state which is of finite spatial size. The theory is translationally invariant: the kink's centre can be anywhere. The theory is Lorentz invariant: the kink can be boosted to an arbitrary velocity. The kink is, therefore, very much like a particle. In addition, the kink is stable: any attempt to remove the kink involves lifting a (semi)infinite length of field from one of the potential minima to another. This costs an infinite amount of energy and is therefore impossible. A more mathematical statement of this is that the kink is not *deformable*. We cannot hold the ends of the kink tightly and then remove the part that crosses the axis in between. The one way of annihilating a kink would be to have an anti-kink: an object whose field crosses the axis in the opposite direction so that $\phi(-\infty) = v$ and $\phi(\infty) = -v$, as shown in Fig. 29.9. The kink and antikink pair is deformable. If we hold the ends of the field tightly we see that we can continuously change the part that crosses the axis to remove the parts that cross the axis. We end with a field lying in the vacuum at $\phi = -v$.

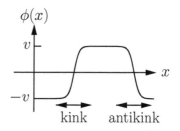

Fig. 29.9 A kink and antikink.

This stability of kinks is encoded in a description of its charge. A charge, remember, is designed to be a conserved quantity. A single kink should carry a topological kink-charge of $Q_\mathrm{T} = 1$; an antikink a kink-charge of $Q_\mathrm{T} = -1$. In order to come up with a sensible recipe for finding the kink-charge, we define the kink-current[4] as

$$J_\mathrm{T}^\mu = \frac{1}{2v}\varepsilon^{\mu\nu}\partial_\nu \phi, \qquad (29.5)$$

[4]Note that this is different to the Noether currents that we have considered so far. Its existence owes nothing to a symmetry of the Lagrangian.

[5]Remember that our convention is not to treat this as a tensor and so $\varepsilon^{\mu\nu} = \varepsilon_{\mu\nu}$.

[6]To summarize the argument so far, the theory

$$\mathcal{L} = \frac{1}{2}(\partial_\mu\phi)^2 - U(\phi),$$

with

$$U(\phi) = \frac{\lambda}{4}(v^2 - \phi^2)^2,$$

allows the existence of kinks. These are topological objects which behave rather like particles. The field at $x = \infty$ and $x = -\infty$ live in different vacua.

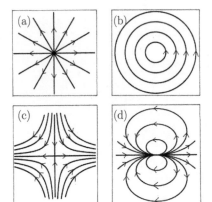

Fig. 29.10 Some vortices with winding number n equal to (a) 1; (b) 1; (c) -1; (d) 2.

[7]The phase φ can be chosen to globally alter the directions at infinity. So, for example, Fig. 29.10(a) corresponds to $\varphi = 0$, while Fig. 29.10(b) corresponds to $\varphi = \pi/2$. This can also be absorbed into the complex amplitude K.

where $\varepsilon^{\mu\nu}$ is the antisymmetry symbol, whose signs are determined by knowing that $\varepsilon^{01} = 1$. This leads to an expression for the charge of a kink of

$$\begin{aligned} Q_{\mathrm{T}} &= \int_{-\infty}^{\infty} dx\, J_{\mathrm{T}}^0 = \frac{1}{2v}\int_{-\infty}^{\infty} dx\, \frac{\partial\phi}{\partial x} \\ &= \frac{1}{2v}[\phi(\infty) - \phi(-\infty)] = 1, \end{aligned} \tag{29.6}$$

whereas an antikink has $Q_{\mathrm{T}} = -1$. Notice that the phions of the theory do not have a kink-charge since $Q_{\mathrm{T}} = \phi(\infty) - \phi(-\infty) = 0$. We call the kink-charge a **topological charge**, which explains the subscript. Its existence is independent of the geometry of spacetime. That geometry is encoded in $g^{\mu\nu}$, which is an instruction book telling us what watches and measuring rods measure in spacetime. In contrast the topological charge has its indices summed with the antisymmetric symbol[5] $\varepsilon^{\mu\nu}$. The dependence on $\varepsilon^{\mu\nu}$ rather than $g^{\mu\nu}$ turns out to be a general feature of topological objects and is examined in more detail in the next chapter.[6]

29.3 Vortices

Now we move to $(2+1)$-dimensional spacetime and examine the same Lagrangian. Our potential now resembles a Mexican hat, whose minima describe a circle in the ϕ_1-ϕ_2 plane. It will be convenient to work in polar (internal) coordinates (r, θ) and write $\phi(x) = \phi_1(x) + i\phi_2(x) \equiv r(x)e^{i\theta(x)}$. In the case of the kink we found a topological field whose two ends lived in different vacua at spatial $|\boldsymbol{x}| \to \infty$. By analogy we ask what a continuous field would look like whose elements live in different vacua at spatial infinity. Some examples of such configurations are shown in Fig. 29.10. These objects are known as vortices. They are described by a field whose form at infinity is written

$$\phi(\boldsymbol{x}) = Ke^{i[n\theta(\boldsymbol{x})+\varphi]} \quad (|\boldsymbol{x}| \to \infty), \tag{29.7}$$

where $\theta(\boldsymbol{x}) = \tan^{-1}(x^2/x^1)$ and φ is a constant, arbitrary phase.[7] Notice how this equation relates the direction in which the field ϕ points in the internal complex plane to the angle θ that tells us where we are in coordinate space. Here n is known as the **winding number**; it tells us how many times the field winds around as we travel around the circle at spatial infinity.

Example 29.4

We may evaluate the energy of a vortex. For a static field configuration we have a Hamiltonian

$$\mathcal{H} = \frac{1}{2}\boldsymbol{\nabla}\phi^\dagger \cdot \boldsymbol{\nabla}\phi + U(\phi), \tag{29.8}$$

and we take $U(\phi) = \left(|K|^2 - \phi^\dagger\phi\right)^2$, so that $U(\phi) = 0$ on the boundary at infinity. The gradient of the vortex in cylindrical (spatial) coordinates is

$$\boldsymbol{\nabla}\phi = \frac{1}{r}(inKe^{in\theta})\hat{\boldsymbol{e}}_\theta. \tag{29.9}$$

The core of the vortex looks frighteningly singular, but we will assume it is ultimately well behaved and call the core energy $E_{\text{core}}(a)$, where a is the core size. We evaluate the energy of the rest of the configuration at distances larger than $r = a$ and find

$$E = E_{\text{core}} + \int_a^\infty dr d\theta\, r\mathcal{H} = E_{\text{core}} + \pi n^2 |K|^2 \int_a^\infty dr \frac{1}{r}. \qquad (29.10)$$

This is a logarithmically divergent energy. We conclude that a single vortex is not a stable object.

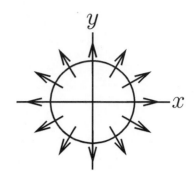

Fig. 29.11 Fields at large distances for an $n = 1$ vortex.

The single vortex is unstable. This is understandable if we look at the form of the field at large distances (Figs. 29.11 and 29.12). It is still swirly as $r \to \infty$. Actually this is predicted by a general theorem due to G.H. Derrick, which says that *time-independent topological objects are impossible for our theory in more than one spatial dimension.*

To stabilize the vortex we must tamper with the Lagrangian. The way to do this is to add a gauge field. Recall that adding a gauge field (known as 'gauging the theory') is done by introducing a covariant derivative as follows:

$$D_\mu \phi = \partial_\mu \phi + iqA_\mu \phi. \qquad (29.11)$$

In order to cancel the divergence of the vortex we're going to choose our gauge field in such a way that it cancels the divergent part of the vortex energy, which resulted from the derivative of the field configuration. We therefore want the covariant derivative $D_\mu \phi$ to vanish at infinity. A gauge field that does this is one whose limit as $r \to \infty$ tends to $\boldsymbol{A}(r,\theta) = \frac{1}{q}\boldsymbol{\nabla}(n\theta)$.

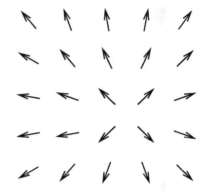

Fig. 29.12 The vortex field configuration for an $n = 1$ vortex.

[8]Recall that $\boldsymbol{A} = A^i = -A_i$ in following the sign changes here.

Example 29.5

The components of the proposed gauge field at infinity are[8]

$$A_r \to 0, \quad A_\theta \to -\frac{n}{qr} \quad \text{as } r \to \infty. \qquad (29.12)$$

As $r \to \infty$ the components of the covariant derivative become

$$D_r \phi = 0, \quad D_\theta \phi = \frac{1}{r}\frac{\partial \phi}{\partial \theta} + iqA_\theta \phi = 0 \quad \text{as } r \to \infty. \qquad (29.13)$$

This shows that the covariant derivative (which represents the 'kinetic energy' of the field) vanishes at infinity. Since it was this term that diverged for the ungauged theory, we see that $\boldsymbol{A}$ has saved the day.

The gauge field should also be expected to make a contribution to the Lagrangian of the form $-\frac{1}{4}F_{\mu\nu}F^{\mu\nu}$. However, this is not the case here. Usually we have a gauge field which can be changed by a gauge transformation which adds a contribution $\partial_\mu \chi$. In our case we notice that at infinity $\boldsymbol{A} = \boldsymbol{\nabla}\chi$, where $\chi = \frac{1}{q}n\theta$. That is, our gauge field is entirely formed from a gauge transformation and is therefore another example of the **pure gauge** which we first saw in Chapter 26. This means that $F_{\mu\nu} = 0$ and the gauge field cannot result in a nonzero $-\frac{1}{4}F^2$ contribution to the Lagrangian. This is also a relief, since such a contribution might also threaten to blow up, ruining the stability of the vortex.

What then is the contribution of the gauge field to the physics? The answer is found by carrying out the integral $\oint \boldsymbol{A} \cdot \mathrm{d}\boldsymbol{l}$ around a circle at infinity. From Stokes' theorem we know what this means physically: the integral is $\oint \boldsymbol{A} \cdot \mathrm{d}\boldsymbol{l} = \int \boldsymbol{B} \cdot \mathrm{d}\boldsymbol{S} = \Phi$, that is the magnetic flux through the circle. Carrying out this integral we obtain

$$\Phi = \oint \boldsymbol{A} \cdot \mathrm{d}\boldsymbol{l} = \oint A^\theta r \, \mathrm{d}\theta = \int \mathrm{d}\theta \, \frac{n}{q} = \frac{2\pi n}{q}. \tag{29.14}$$

We see that vortices carry quantized magnetic flux!

The next step in this line of argument to examine a topological object in (3+1)-dimensional spacetime. Remarkably, this object turns out to be a form of a magnetic monopole and is examined in Chapter 49. In the next chapter we turn our attention to another aspect of topology and describe a field theory that can be legitimately described as topological.

Chapter summary

- Topology relates the structure of objects that are preserved under continuous deformations involving stretching but no cutting or gluing.
- Topological objects include kinks and vortices and can be characterized by a topological charge.

Exercises

(29.1) Show that the antikink has kink-charge $Q_\mathrm{T} = -1$ and that two kinks and three antikinks have a total charge of $Q_\mathrm{T} = -1$.

(29.2) This is the first time we have used the antisymmetric symbol $\varepsilon^{\mu\nu}$ in anger. *We regard the up and down versions as identical since we are not using it as a tensor.* Let's examine its properties. Take $\varepsilon^{01} = 1$ and $\varepsilon^{012} = 1$ and show the following:
(a) $\varepsilon^{ij}\varepsilon^{ij} = 2$.
(b) $\varepsilon^{ij}\varepsilon^{in} = \delta_{jn}$.
(c) $\varepsilon^{ij}\varepsilon^{mn} = \delta_{jn}\delta_{im} - \delta_{in}\delta_{jm}$.
(d) $\varepsilon^{ijk}\varepsilon^{imn} = \delta_{jm}\delta_{kn} - \delta_{jn}\delta_{km}$.

(29.3) A simple model of a domain wall in a magnet says that the interaction energy of two neighbouring (classical) spins of magnitude S is given by $E = -JS^2 \cos\theta$ where θ is the relative angle between the spins and J is a constant.
(a) Show that the energy of a domain wall sepa-

rating a domain of spins up and a domain of spins down costs energy

$$U = \frac{1}{2} J S^2 \frac{\pi^2}{N}, \tag{29.15}$$

where N is the number of spins in the wall.
(b) Show that this picture will not lead to a stable wall.
(c) In reality spin–orbit effects lead to an energy cost for not having spins pointing along certain spatial directions. In this case these lead to a contribution to the energy of $\frac{NK}{2}$. Show that the wall is now stable and find its size.

(29.4) (a) Draw vortices with winding numbers $n = -2$ and $n = 3$.
(b) Draw the fields resulting from vortices with $n = 1$ and $n = -1$ sharing the same region of space. This situation is discussed in Chapter 34.

Topological field theory

*Imagine the perplexity of a man outside time and space, who
has lost his watch, his measuring rod and his tuning fork.*
Alfred Jarry (1873–1907), *Exploits and Opinions of Doctor
Faustroll, Pataphysician*

So far we have met topological objects that can exist in scalar field theories with a broken symmetry. In this chapter we will take the study
of topology further and cook up an inherently **topological field theory**. This theory will apply to the special case of (2+1)-dimensional
spacetime. We will see that physics in the flatland of (2+1) dimensions
has some remarkable properties whose origin may be traced back to
topological considerations.

We start our discussion by examining the sorts of particle statistics
that apply in two spatial dimensions. We will see that flatland supports
anyons: particles that are neither bosons nor fermions. We will then
go on to formulate a field theory that not only captures anyon statistics,
but also describes some curious new electrodynamics.

30.1 Fractional statistics à la Wilczek: the strange case of anyons

We are quite used to the idea of bosons and fermions as entities distinguished by their behaviour upon exchange of identical particles. Specifically, changing the labels on two identical particles results in a change
of the wave function according to the rule

$$\psi(x_1, x_2) = \pm \psi(x_2, x_1), \tag{30.1}$$

where the $+$ sign applies to bosons and the $-$ sign to fermions.

For the case of two-dimensional space, the exchange of particles needs
more careful attention. In fact, we need to think more precisely about
what the exchange of particles actually involves. We do not simply
make the two particles vanish and then reappear in different positions.
Rather we imagine the process in which particles are moved around
each other in real space. Let's identify two moving-around processes.
We start with two identical particles at positions x_1 and x_2 and identify
two fundamentally distinct ways of exchanging them. The result of
processes of type A is to move $x_1 \to x_1$ and $x_2 \to x_2$. Some examples
are shown in Fig. 30.1(a) and (b). In this sort of process the particles

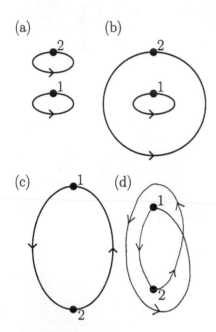

Fig. 30.1 Examples of ways of exchanging particles. (a) and (b) are type A processes, where the particles end up at the same positions. In case (b) one particle loops around the other once in the exchanging process. (c) and (d) are type B processes where the particles exchange positions. In case (d) one particle loops once around the other during the exchange.

Frank Wilczek (1951–) was a co-recipient of the Nobel Prize in Physics in 2004 for their discovery of asymptotic freedom. His popular book *The Lightness of Being* (2008) is a very readable account of modern ideas in particle physics.

end up where they were originally, although they may move around each other. The result of type B processes is to move $x_1 \to x_2$ and $x_2 \to x_1$ [Fig. 30.1(c) and (d)]. Again particles may move around each other several times before settling at their final positions [as in Fig. 30.1(d)]. This is particle exchange, albeit in a more tangible form than the usual magic trick method.

We ask what the relative quantum mechanical phase difference is between processes of type A and type B. The key parameter to consider is the angle that one particle is moved around the other. This is where topology comes in: given a set of particle paths, we may smoothly distort the paths of the particles, but we may not change the number of times particles wrap around each other without introducing singularities in the particle paths. Processes of type A involves rotating particle 2 around particle 1 by angle $\phi = 2\pi p$, where the winding number p takes an integer value (including zero). Processes of type B involves rotations of $\phi = \pi(2p + 1)$. Each value of p describes a topologically distinct process.

We suppose that these topologically distinct processes make a multiplicative contribution to the total wave function (or, in field theory, to the path integral) of $\Phi(\phi)$ which we expect to be pure phase. If we carry out a sequential string of these processes then we require that the angles add, whilst the wave functions should multiply. That is, we need $\Phi(\phi_1 + \phi_2) = \Phi(\phi_1)\Phi(\phi_2)$ which implies that $\Phi(\phi) = e^{i\eta\phi}$ where η is a parameter which, crucially, doesn't need to be an integer.

Now we compare our real life exchange to the old-fashioned definition in eqn 30.1. If we carry out the exchange $(x_1, x_2) \to (x_2, x_1)$ [shown in Fig. 30.1(c)] then the formal definition of exchange embodied in eqn 30.1 tells us that the wave function should be identical (for bosons) or pick up a minus sign (fermions). However, the realistic version of exchange merely tells us that $\phi = \pi$, resulting in a phase factor of $\Phi(\phi) = e^{i\eta\pi}$. The two versions of exchange are only identical for the special cases that (a) we have $\eta =$ even integer, when we recover the expected exchange behaviour for bosons or (b) we have $\eta =$ odd integer, when we recover fermion exchange. However, this analysis shows that there are many more possible values of η in two dimensions. It doesn't have to be an integer! We are therefore not tied simply to bosons and fermions, the freedom to choose η means we can have any exchange statistics. Particles with such statistics were dubbed **anyons** by Frank Wilczek, who has made important contributions to elucidating the study of these particles and their 'fractional statistics'.

Notice that having only two spatial dimensions to play with was vital to the argument. In three spatial dimensions all of the type A processes are topologically identical since they are all deformable into paths where the particles don't move. Similarly all of the type B processes are topologically identical and may be reduced to a simple exchange of particles. This reduction occurs because the extra dimension allows us to move the paths past each other in the third dimension, shrinking all loops to zero.

30.2 Chern–Simons theory

We now turn to the problem of writing down a Lagrangian that can be described as topological. What do we mean by this? Our usual Lagrangians have relied on the use of the metric tensor $g_{\mu\nu}$, which is an object that tells us how to measure displacements in space and time. It provides a set of instructions for taking the scalar product of vectors through a contraction of indices. In contrast, in a topological theory the contraction of indices is carried out through the use [in (2+1) dimensions] of the antisymmetric symbol $\varepsilon^{\mu\nu\lambda}$. Topological theories are therefore blind to the details of the watches and measuring rods that are contained in $g_{\mu\nu}$. Their content is dependent purely on the topology of the manifold[1] in which we're working. The first consequence of this feature is a shock when we try to derive the Hamiltonian from a topological Lagrangian: the Hamiltonian vanishes!

[1]A **manifold** is a topological space that locally resembles Euclidean space.

Example 30.1

To see this we note that a general method of finding the energy-momentum tensor $T^{\mu\nu}$ is to vary the action with respect to the components of the metric. For an action S_{top} derived by integrating a topological Lagrangian we have

$$T^{\mu\nu} = \frac{-2}{\sqrt{-\det g}} \frac{\delta S_{\text{top}}}{\delta g_{\mu\nu}}. \tag{30.2}$$

Since topological theories don't include $g_{\mu\nu}$ at all then we must have $T^{\mu\nu} = 0$. Since the Hamiltonian is the 00th component of this tensor then we conclude that $H = 0$.

A consequence of this is that all of the states of a topological Hamiltonian have energy $E = 0$. So there is a significant degeneracy in the system. Unfortunately it is rather difficult to determine how many states we have since this depends on the topology of the manifold.

We will now describe a topological term that may be part of a Lagrangian. This is known as a **Chern–Simons** Lagrangian and applies to the case of (2+1)-dimensional spacetime. It is written[2]

$$\mathcal{L} = -\frac{1}{2}\kappa\varepsilon^{\mu\nu\lambda}a_\mu\partial_\nu a_\lambda, \tag{30.3}$$

where a_μ is a $U(1)$ gauge field and κ is a constant. Chern–Simons terms are easy to spot since they have the form $\mathcal{L} = \varepsilon a\partial a$. We will see that Chern–Simons theory contrasts in many respects with the other $U(1)$ gauge theory with which we're familiar: electromagnetism.

Chern–Simons theories were first introduced into field theory by Edward Witten (1951–) based on work in differential geometry. They are named after Shiing-Shen Chern (1911–2004), mathematician, and James Harris Simons (1938–), mathematician, hedge fund manager and philanthropist.

[2]Notice that the Chern–Simons theory will work in (2+1)-dimensional spacetime but not in (3+1) dimensions. Try writing it in (3+1) dimensions and you'll see that the indices won't match up.

Example 30.2

The Chern–Simons Lagrangian gives rise to a gauge invariant action. Actually the gauge invariance is slightly more subtle than the usual version seen in electromagnetism. To show this, we make the gauge transformation $a_\mu \to a_\mu + \partial_\mu\chi$, and we obtain

$$\varepsilon^{\mu\nu\lambda}a_\mu\partial_\nu a_\lambda \to \varepsilon^{\mu\nu\lambda}a_\mu\partial_\nu a_\lambda + \varepsilon^{\mu\nu\lambda}\partial_\mu\chi\partial_\nu a_\lambda. \tag{30.4}$$

Here we have used the fact that $\varepsilon^{\mu\nu\lambda}\partial_\mu\partial_\nu\chi = 0$, which follows from the antisymmetry of $\varepsilon^{\mu\nu\lambda}$. The resulting change in the action (that is, the second term in the above equation) may be written as a total derivative

$$\delta S = \int \mathrm{d}^3x\, \varepsilon^{\mu\nu\lambda}\partial_\mu(\chi\partial_\nu a_\lambda). \tag{30.5}$$

This vanishes under the assumption that we may drop boundary terms. We conclude that the Chern–Simons Lagrangian is gauge invariant, as long as we discard the boundary terms.

The next thing to note is that if we plug this term into the Euler–Lagrange equations we obtain an equation of motion for the a_μ fields as

$$\kappa\varepsilon^{\mu\nu\lambda}\partial_\nu a_\lambda = 0, \tag{30.6}$$

or defining a field strength tensor $f_{\mu\nu} = \partial_\mu a_\nu - \partial_\nu a_\mu$ we have an equation of motion $\frac{1}{2}\kappa\varepsilon^{\mu\nu\lambda}f_{\nu\lambda} = 0$ or

$$f_{\mu\nu} = 0. \tag{30.7}$$

This looks very unexciting! Unlike electromagnetism, for which, in the absence of sources, we have an equation of motion $\partial_\mu F^{\mu\nu} = 0$ which supports plane waves, the Chern–Simons theory seems to have no dynamics of its own. Nevertheless, the Chern–Simons theory will turn out to be interesting, but we just need to couple it to another field.

Therefore, we now couple our Chern–Simons field a_μ to a source J^μ which is the conserved current of some other field and write

$$\mathcal{L} = -\frac{1}{2}\kappa\varepsilon^{\mu\nu\lambda}a_\mu\partial_\nu a_\lambda + a_\mu J^\mu, \tag{30.8}$$

and this leads to an equation of motion

$$J^\mu = \kappa\varepsilon^{\mu\nu\lambda}\partial_\nu a_\lambda = \frac{1}{2}\kappa\varepsilon^{\mu\nu\lambda}f_{\nu\lambda}. \tag{30.9}$$

Clearly this expression for current isn't a consequence of Noether's theorem; rather, it occurs as a result of our coupling a source to the Chern–Simons gauge field by hand. The Chern–Simons term therefore provides a *constraint* on the current.

In order to understand the properties of the current we now point out an odd, but undeniably relevant feature of (2+1)-dimensional spacetime. This feature stems from the fact that, in two spatial dimensions, the cross product results in a pseudoscalar, rather than a pseudovector. The result is that the magnetic field in (2+1)-dimensional spacetime is a pseudoscalar: $B = \varepsilon^{ij}\partial_i A^j$.

Example 30.3

To see this we note that in three dimensions we have the definition of the cross product in component form $V^i = \varepsilon^{ijk}A^j B^k$. In two dimensions we lose an index from our antisymmetric symbol and we have $S = \varepsilon^{ij}A^i B^j$. We run out of indices, so the result of the cross product is a pseudoscalar as claimed.

Now turning to the components of the current $J^\mu = (\rho, \boldsymbol{J})$ predicted from eqn 30.9, we find

$$\rho = \kappa(\partial_1 a_2 - \partial_2 a_1), \quad J^1 = \kappa(-\partial_0 a_2 + \partial_2 a_0), \quad J^2 = \kappa(\partial_0 a_1 - \partial_1 a_0).$$
$$(30.10)$$

Although the field a_μ does not describe electromagnetism, we may use the language of electromagnetism and call $b = \partial_2 a_1 - \partial_1 a_2$ the a_μ magnetic field b and call $\partial_0 a_i - \partial_i a_0$ the ith component of the a_μ electric field e^i. We therefore have

$$\rho = -\kappa b, \quad J^i = -\kappa \varepsilon^{ij} e^j, \qquad (30.11)$$

where $\varepsilon^{12} = 1$. The second of these equations tells us that an a_μ electric field in the y-direction gives a source current in the x-direction. The meaning of the first equation is made clear if we integrate over the two-dimensional space to find that the charge Q of the source field is related to the flux of the a_μ magnetic field by

$$Q = -\kappa \int \mathrm{d}^2 x \, b, \qquad (30.12)$$

or

$$\left(\begin{array}{c} \text{Charge of} \\ \text{source field} \end{array} \right) \propto \left(\begin{array}{c} \text{Flux of} \\ a_\mu \text{ field} \end{array} \right). \qquad (30.13)$$

Thus we conclude the following:

Chern–Simons theory ties a_μ-flux to the charge of the source field.

30.3   Fractional statistics from Chern–Simons theory

The fact that external source charge couples to a_μ-flux has an influence on the statistics we find on the exchange of particles. Recall that in (2+1) dimensions moving one particle completely around another gives rise to a phase contribution $\mathrm{e}^{2\pi i \eta}$, where η is an odd integer for fermions and an even integer for bosons.

Now recall the Aharonov–Bohm effect from Chapter 29. Moving one charged particle completely around a source of flux gives rise to a phase contribution $\mathrm{e}^{iq\Phi}$, where q is the particle charge and Φ is the flux. Using our Chern–Simons result that $q = -\kappa\Phi$, we obtain a prediction of the Aharonov–Bohm phase of $\mathrm{e}^{-i\frac{q^2}{\kappa}}$. However, particle exchange (which tells us about statistics) is slightly different to this in that it involves moving one particle around the other through an angle π. Interpreting the above Aharonov–Bohm process as *two* lots of particle exchange (and therefore contributing a phase $2\pi\eta$) we obtain a result for the statistics of our particles,[3] namely that

$$2\pi\eta = -\frac{q^2}{\kappa}, \qquad (30.14)$$

[3] In fact, there's another tricky factor of two here. One is tempted to say that since all particles carry both charge and flux then moving one charge around a flux also moves a flux around a charge, doubling the effect compared to what we have written. A more careful treatment confirms our naive picture of a single charge moving around a single flux and gives eqn 30.14. See Wen, Chapter 7 for further discussion.

and hence $\eta = -\frac{q^2}{2\pi\kappa}$ is not necessarily an integer. This implies that the particles exhibit fractional statistics.

This chapter has explained how to construct a topological Lagrangian. It is based on gauge fields; it supports excitations with fractional statistics; its charged excitations carry flux. In Chapter 45 we'll examine a phase of matter which is described by a Chern–Simons theory: this is the fractional quantum Hall fluid.

Chapter summary

- Topological field theories such as Chern–Simons theory are built from products of gauge theories summed together using the antisymmetric symbol $\varepsilon^{\mu\nu\lambda}$ in place of a metric tensor. These theories provide constraints on other fields to which they are coupled.

- Chern–Simons theory attaches flux to the charge of the source field.

- Chern–Simons theories may produce fractional excitations.

Exercises

(30.1) (i) Verify that $\varepsilon^{\mu\nu\lambda}\partial_\mu\partial_\nu\chi = 0$.
(ii) Verify eqn 30.4.

(30.2) Show that the Chern–Simons term won't work in (3+1)-dimensional spacetime. Suggest a form for (4+1)-dimensional spacetime.

(30.3) Derive the equations of motion for the Chern–Simons fields coupled to a source in eqn 30.9 using the Euler–Lagrange equations.

(30.4) The propagator for Chern–Simons theory may be found in a manner very similar to that used in electromagnetism in Chapter 39. Starting with Chern–Simons theory we add a gauge fixing term $-\frac{1}{2\xi}(\partial_\mu a^\mu)^2$ to obtain

$$\mathcal{L} = -\frac{\kappa}{2}\varepsilon^{\mu\nu\lambda}a_\mu\partial_\nu a_\lambda - \frac{1}{2\xi}(\partial_\mu a^\mu)^2 + J^\mu a_\mu. \quad (30.15)$$

(a) Show that the equations of motion for the field a_μ are given by

$$J^\mu = \left[\kappa\varepsilon^{\mu\nu\lambda}\partial_\nu - \frac{1}{\xi}\partial^\mu\partial^\lambda\right]a_\lambda = M^{\mu\lambda}a_\lambda. \quad (30.16)$$

(b) Show that the inverse of the equations of motion is given, in momentum space, by the matrix

$$(M^{-1})_{\lambda\sigma} = -\frac{i\varepsilon_{\lambda\nu\sigma}p^\nu}{\kappa p^2} + \xi\frac{p_\lambda p_\sigma}{p^4}, \quad (30.17)$$

which provides the propagator for the theory via $i(M^{-1})_{\lambda\sigma}$.

(c) Show that the Chern–Simons term therefore leads to a term in the Lagrangian that may be written

$$\mathcal{L} = -\frac{1}{2}J^\mu\left(\frac{\varepsilon_{\mu\nu\lambda}\partial^\nu}{\kappa\partial^2}\right)J^\lambda. \quad (30.18)$$

This is known as the Hopf term.

(30.5) Consider *Chern–Simons electromagnetism* described by a Lagrangian

$$\mathcal{L} = -\frac{1}{4}f_{\mu\nu}f^{\mu\nu} - \frac{\kappa}{2}\varepsilon^{\mu\nu\lambda}a_\mu\partial_\nu a_\lambda - \frac{1}{2\xi}(\partial_\mu a^\mu)^2, \quad (30.19)$$

where $f_{\mu\nu} = \partial_\mu a_\nu - \partial_\nu a_\mu$.
(a) Find the equations of motion for this theory.
(b) Show that the inverse of the equations of motion is given by

$$(M^{-1})_{\lambda\sigma} = \frac{p^2 g_{\lambda\sigma} - p_\lambda p_\sigma + i\kappa\varepsilon_{\lambda\nu\sigma}p^\nu}{p^2(p^2 - \kappa^2)} + \xi\frac{p_\lambda p_\sigma}{p^4}. \quad (30.20)$$

What is the significance of the pole in the propagator at $p^2 = \kappa^2$?

Part VIII

Renormalization: taming the infinite

This part is structured as follows:

- Interactions in quantum field theories change the properties of the particles, producing dressed particles, also known as quasiparticles. These are examined in Chapter 31 which also gives an outline of Landau's Fermi liquid picture.

- Quantum field theory has some uncomfortable divergences and Chapter 32 shows how these can be tamed using *renormalization* using additional *counterterms*.

- These ideas are applied to deriving propagators and making Feynman diagrams in Chapter 33. This allows us to write down a Green's function called the *self-energy* which describes the change in the mass of a particle due to the interactions, and also a vertex function which describes screening.

- The *renormalization group* is introduced in Chapter 34. Examples treated include asymptotic freedom, Anderson localization, and the Kosterlitz–Thouless transition.

- Chapter 35 treats the ferromagnetic transition as a further example of the renormalization group procedure and shows how critical exponents can be extracted.

31 Renormalization, quasiparticles and the Fermi surface

This chapter focuses on how the behaviour of particles is affected by interactions between them. We will see that the passage from a non-interacting theory to an interacting theory involves not only a propensity for particles to scatter from each other but, more dramatically, *changes in the properties of particles themselves, such as their mass and charge*. We will say that the values of these properties are **renormalized**. Furthermore, we will find that interactions even change what we mean by a 'particle'. The process of renormalization may be imagined as a particle dressing itself in interactions.

A rather familiar example of interactions changing the apparent properties of particles is the screening of a positive charge embedded in a metal. The positive charge builds up a cloud of electron density around it. Distant electrons do not feel its full charge because the nearby electron density screens the charge, reducing its apparent value. A second example is the effective mass of an electron in a crystal. The fact that a crystal contains immobile ion cores which interact electrostatically with the electron means that the apparent mass of the electron is not that of the bare electron in a vacuum, but instead an effective mass m^*.

We will discover that analogous processes occur in all interacting quantum field theories. Of course the difference between condensed matter physics and a 'fundamental' theory such as quantum electrodynamics is that, for the former, we may measure the mass and charge of electrons both inside the crystal and in vacuo. In contrast, we cannot remove electrons from the vacuum of quantum electrodynamics to see the difference in their properties caused by interactions. However, the framework in which we understand the fate of interacting particles is the same in both cases and it is to this set of ideas which we now turn.

31.1 Recap: interacting and non-interacting theories

Our simplest field theories describe particles that don't interact. We call these free or non-interacting theories.[1]

---

[1] The simple scalar field theory with Lagrangian $\mathcal{L} = \frac{1}{2}(\partial_\mu \phi)^2 - \frac{m^2}{2}\phi^2$ is an example of a free field theory.

Example 31.1

Here's a recap of the properties of a non-interacting theory. For a theory described by a free Hamiltonian $\hat{H}_0$, we define freely evolving field operators and a vacuum ground state. These may be defined through

$$\hat{H}_0|0\rangle = 0, \quad \hat{\boldsymbol{p}}|0\rangle = 0, \quad \langle 0|\hat{\phi}(x)|0\rangle = 0. \tag{31.1}$$

Particles are excitations of the vacuum. Creation and annihilation operators create single-particle modes, one at a time, labelled by their momenta. An example is $\hat{a}_{\boldsymbol{p}}^{\dagger}|0\rangle = |\boldsymbol{p}\rangle$. The amplitude for a field operator to create a (relativistically normalized) single particle with momentum p is given by

$$\langle p|\hat{\phi}^{\dagger}(x)|0\rangle = e^{ip\cdot x}. \tag{31.2}$$

Finally, to talk about particles we need a propagator. This describes a single particle, inserted into the system at y and removed at x, and which, for a non-interacting theory, will not affect or be affected by any other particles in the system. The Feynman propagator describing the evolution of a free, single particle is given by

$$\tilde{G}_0(p) = \frac{i}{p^2 - m^2 + i\epsilon}. \tag{31.3}$$

If we didn't know the mass of the particle before (from its dispersion relation $E_{\boldsymbol{p}}^2 - \boldsymbol{p}^2 = m^2$) then we could read it off the propagator as the pole of the Green's function.

So much for the world of non-interacting particles. What happens when particles interact? Consider a Hamiltonian written as a sum of non-interacting and interacting parts

$$\hat{H} = \hat{H}_0 + \lambda\hat{H}'. \tag{31.4}$$

In order to examine the interacting system, we will imagine what happens to a free theory described by $\hat{H}_0$ when we slowly turn on an interaction $\hat{H}'$. To do this we multiply $\hat{H}'$ by a function $\lambda(T)$ which starts vanishingly small and slowly approaches unity as we turn up some parameter T, as shown in Fig. 31.1. As λ grows, our system gradually becomes a bubbling cauldron of interactions: the new eigenstates of the full Hamiltonian $\hat{H}$ (called $|\boldsymbol{p}_\lambda\rangle$) are different to those of $\hat{H}_0$ (called $|\boldsymbol{p}\rangle$). The interacting ground state $|\Omega\rangle$ is different from the non-interacting ground state $|0\rangle$ and the dispersion of the particles is also altered. If we put a test particle into this system it will interact with particles and antiparticles, pulling them out of the vacuum. These particles and antiparticles may also interact with still more particles and antiparticles, doing all manner of complicated things. Our original particle may be lost in all of this havoc and it might be doubted that we can even identify single particles in this system any more. We may have to abandon our quantum field theory which, after all, is based around the notion of creating and annihilating single particles. We are led to ask: are there such things as single-particle excitations in an interacting system? The point of this chapter is to show that there are! They are called **dressed particles** or **quasiparticles**.

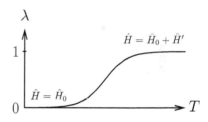

Fig. 31.1 Adiabatic turning on of $\hat{H}'$. The full Hamiltonian is given by $\hat{H} = \hat{H}_0 + \lambda(T)\hat{H}'$. This thought experiment allows us to examine what happens to particles in an interacting theory.

> Quasiparticles are the excitations in interacting systems. They resemble free particles, but have different masses and interact differently with their environment. The process of free particles turning into dressed particles is called **renormalization**.

31.2   Quasiparticles

We imagine that as the interaction is turned on the particle begins its transformation from a single non-interacting particle into a quasiparticle. We say that the particle becomes dressed by interactions and becomes a quasiparticle. The process of renormalization is therefore embodied in the equation

$$(\text{Quasiparticle}) = (\text{real particle}) + (\text{interactions}), \qquad (31.5)$$

where the word quasiparticle may be substituted by 'dressed particle' or 'renormalized particle', depending on the context.

Let's examine what happens to a free particle as the interaction is turned on. How do we know if it's possible for an interacting quasiparticle to exist with the same momentum once we've turned on the interaction? To answer this, we act on the interacting vacuum with the particle creation operator $\hat{\phi}^\dagger(x) \propto \int_{\boldsymbol{p}} \hat{a}_{\boldsymbol{p}}^\dagger e^{i p \cdot x} + \hat{a}_{\boldsymbol{p}} e^{-i p \cdot x}$ that we had for the non-interacting system. As we turn on the interaction the ground state is no longer $|0\rangle$, it's now $|\Omega\rangle$ (where $\hat{H}|\Omega\rangle = 0$). The excited states are called $|\boldsymbol{p}_\lambda\rangle$, where λ tells us that we have interactions. The key to understanding dressed particles is that the state $\hat{a}_{\boldsymbol{p}}^\dagger|\Omega\rangle$ created by our field operator won't generally contain one single-particle excitation. In fact, our operator $\hat{a}_{\boldsymbol{p}}^\dagger$ is a bull in a china shop: there's nothing to stop the free field operator producing several excitations all at once!

To make a single-particle excitation with momentum $\boldsymbol{p}$ in the free theory we acted on the free ground state with an operator $\hat{a}_{\boldsymbol{p}}^\dagger$ and said $|\boldsymbol{p}\rangle = \hat{a}_{\boldsymbol{p}}^\dagger|0\rangle$. To make a single particle with momentum $\boldsymbol{p}$ in the interacting theory we might act on the interacting ground state $|\Omega\rangle$ with a different operator $|\boldsymbol{p}_\lambda\rangle = \hat{q}_{\boldsymbol{p}}^\dagger|\Omega\rangle$. This all makes sense, in that we need to have the right tools for the job, that is, the right operators for the ground states, and we make the expected number of particles (i.e. one). Unfortunately we don't know what $\hat{q}_{\boldsymbol{p}}^\dagger$ is and so can't use it on the system. All we have is the single-particle creation operators $\hat{a}_{\boldsymbol{p}}^\dagger$ for the non-interacting states.

But what happens if we use the wrong operator on the interacting ground state? If we apply the operators for the non-interacting theory $\hat{a}_{\boldsymbol{p}}^\dagger$ to the interacting ground state $|\Omega\rangle$ we make a superposition of states. One might be the state we intend to make $|\boldsymbol{p}_\lambda\rangle$, but others will be multiparticle states whose momenta each add up to $\boldsymbol{p}$. That is

$$\hat{a}_{\boldsymbol{p}}^\dagger|\Omega\rangle = |\boldsymbol{p}_\lambda\rangle\langle\boldsymbol{p}_\lambda|\hat{a}_{\boldsymbol{p}}^\dagger|\Omega\rangle + \sum(\text{multiparticle parts}). \qquad (31.6)$$

For example, $\hat{a}_{\boldsymbol{p}}^{\dagger}|\Omega\rangle$, might contain a multiparticle state[2] with two particles and an antiparticle like this: $\hat{q}_{\boldsymbol{p}_1}^{\dagger}\hat{q}_{\boldsymbol{p}_2}^{\dagger}\hat{q}_{\boldsymbol{p}_1+\boldsymbol{p}_2-\boldsymbol{p}}|\Omega\rangle$.

In fact, it can be more complicated still, since we may only *approximately* create a state $|\boldsymbol{p}_\lambda\rangle$ (along with multiparticle parts) and in actual fact create an ultimately unstable, narrow wave packet of width $\Gamma_{\boldsymbol{p}}$ resembling this state. In particle physics this is known as a **resonance**. In this case the energy of the particle may have an imaginary part $\Gamma_{\boldsymbol{p}}$. A complex energy $E_{\boldsymbol{p}}+\mathrm{i}\Gamma_{\boldsymbol{p}}$ will cause a particle to time evolve according to $\mathrm{e}^{\mathrm{i}E_{\boldsymbol{p}}t}\mathrm{e}^{-\Gamma_{\boldsymbol{p}}t}$ or, in words, the particle is unstable to decay and has a lifetime[3] $(2\Gamma_{\boldsymbol{p}})^{-1}$. After a time that's long compared to $(2\Gamma_{\boldsymbol{p}})^{-1}$ we can't expect there to be a quasiparticle any more.

With all of these particles flying about and possibly decaying how can we say whether we have a single particle? We call the amplitude for making the desired single-particle part the **quasiparticle weight** $Z_{\boldsymbol{p}}^{\frac{1}{2}}$, where $Z_{\boldsymbol{p}}^{\frac{1}{2}}=\langle p_\lambda|\hat{a}_{\boldsymbol{p}}^{\dagger}|\Omega\rangle$, and where $|p_\lambda\rangle$ is a relativistically normalized momentum state. We say that we have a quasiparticle if $Z_{\boldsymbol{p}}^{\frac{1}{2}}\neq 0$. In the interacting theory, we can state that the amplitude in position space for creating a single particle with a momentum $\boldsymbol{p}$ is

$$\langle p_\lambda|\hat{\phi}^{\dagger}(x)|\Omega\rangle = Z_{\boldsymbol{p}}^{\frac{1}{2}}\mathrm{e}^{\mathrm{i}p\cdot x}\mathrm{e}^{-\Gamma_{\boldsymbol{p}}t}. \tag{31.7}$$

In fact, to meaningfully describe a particle as a quasiparticle we will require that $E_{\boldsymbol{p}}>\Gamma_{\boldsymbol{p}}$, that is, its energy must be greater than its decay rate. Comparing with the analogous result obtained for non-interacting scalar field theory we conclude that, when $\Gamma_{\boldsymbol{p}}$ is small, we can describe particles in an interacting theory by multiplying[4] our single-particle wave functions by the factor $Z_{\boldsymbol{p}}^{\frac{1}{2}}$.

It's important to note that the energies of interacting single particles are different from those of free particles. In the (relativistic) scalar field theory we're describing[5] the interactions change the mass of the single-particle excitation from m (often called the bare mass) to m_λ, giving an energy for the state $|\boldsymbol{p}_\lambda\rangle$ of $(\boldsymbol{p}^2+m_\lambda^2)^{\frac{1}{2}}$. Since m_λ is the mass of a single particle measured in our experiments, it's often called the **physical mass** and given the symbol m_P. The following table shows quantites for both non-interacting and interacting theories.

Non-interacting:	$|\boldsymbol{p}\rangle=\hat{a}_{\boldsymbol{p}}^{\dagger}|0\rangle$	$\langle p|\hat{\phi}^{\dagger}(x)|0\rangle=\mathrm{e}^{\mathrm{i}p\cdot x}$	m
Interacting:	$|\boldsymbol{p}_\lambda\rangle=\hat{q}_{\boldsymbol{p}}^{\dagger}|\Omega\rangle$	$\langle p_\lambda|\hat{\phi}^{\dagger}(x)|\Omega\rangle=Z_{\boldsymbol{p}}^{\frac{1}{2}}\mathrm{e}^{\mathrm{i}p\cdot x-\Gamma_{\boldsymbol{p}}t}$	m_P

31.3   The propagator for a dressed particle

For serious calculations we need to identify a propagator for the interacting system via our definition

$$G(x,y)=\langle\Omega|T\hat{\phi}_\mathrm{H}(x)\hat{\phi}_\mathrm{H}^{\dagger}(y)|\Omega\rangle. \tag{31.8}$$

[2]Note that, in a metal, these multiparticle parts may be thought of in terms of the emission of electron–hole pairs. See Chapter 43.

[3]Where the factor of two comes from considering the square modulus of the particle wave function.

[4]In the jargon, one says 'renormalizing' rather than 'multiplying'.

[5]In which the free particle dispersion is given by $E_{\boldsymbol{p}}=(\boldsymbol{p}^2+m^2)^{\frac{1}{2}}$.

[6]This can be achieved simply by sliding a resolution of the identity between the field operators in eqn 31.8. This resolution will be made up from the exact eigenstates $|p_\lambda\rangle$ of the full Hamiltonian $\hat{H}$. For details of the calculation, see Peskin and Schroeder, Chapter 7.

It can be shown, without the aid of perturbation theory, what the general form for the interacting propagator G should be.[6] The propagator, which is built out of free operators $\hat{a}_p$ and $\hat{a}_p^\dagger$, will not only describe the propagation of single-particle states, but also describe the fate of the multiparticle states, and takes the form

$$\tilde{G}(p) = \frac{iZ_p}{p^2 - m_P^2 + i\Gamma_p} + \left(\begin{array}{c} \text{multiparticle} \\ \text{parts} \end{array} \right), \qquad (31.9)$$

where, as before, $Z_p^{\frac{1}{2}} = \langle p_\lambda | \hat{\phi}^\dagger(0) | \Omega \rangle$ is the amplitude for creating a single-particle state $|p_\lambda\rangle$ out of the interacting vacuum and m_P is the mass of the state $|p_\lambda\rangle$. Notice the role of Γ_p here: it is the inverse quasiparticle lifetime in the place of the usual infinitesimal factor ϵ.

In most cases involving long-lived elementary particles Z_p is independent of the three-momentum p and is simply denoted Z. As we shall see, a dependence on p is more relevant for metals in solids.

Example 31.2

One way to write the full propagator that makes contact with experiment uses the **spectral density function** $\rho(M^2)$. For example, for scalar field theory we rewrite the propagator in terms of the free propagator[7] $\Delta(x, y, M^2)$ which is now made a function of particle mass as well as the spacetime points x and y:

$$G(x,y) = \int_0^\infty \frac{dM^2}{(2\pi)} \rho(M^2) \Delta(x, y, M^2). \qquad (31.10)$$

[7]Recall that the free scalar field propagator $\Delta(x, y, M^2)$ can be written in momentum space as

$$\Delta(p, M^2) = \frac{i}{p^2 - M^2 + i\epsilon}.$$

Suppose we have a spectral density function

$$\rho(M^2) = (2\pi)\delta(M^2 - m_P^2)Z + \left(\begin{array}{c} \text{multiparticle} \\ \text{parts} \end{array} \right). \qquad (31.11)$$

This is shown in Fig. 31.2(a). It is made up of a single-particle peak at mass $M \equiv m_P$: the physical mass of a single particle in the interacting theory. It also contains a multiparticle continuum which starts around $4m_P^2$, which is the threshold for making two real particles in the centre of mass frame. This sort of spectral function is the one usually seen in particle physics problems where the creation of more than one particle is due to pair production. Plugging in this spectral density we obtain

$$\tilde{G}(p) = \frac{iZ}{p^2 - m_P^2 + i\epsilon} + \int_{\approx 4m_P^2}^\infty \frac{dM^2}{2\pi} \rho(M^2) \frac{i}{p^2 - M^2 + i\epsilon}. \qquad (31.12)$$

Notice that the lifetime of the particle is ϵ^{-1}, which is effectively infinite since the particle pole is infinitesimally close to the real axis. We expect these particles to be long-lived and stable.

Now suppose we have the spectral density in Fig. 31.2(b), which has a quasiparticle peak with width $2\Gamma_p$. This leads to a propagator

$$\tilde{G}(p) = \frac{iZ_p}{p^2 - m_P^2 + i\Gamma_p} + \left(\begin{array}{c} \text{multiparticle} \\ \text{parts} \end{array} \right), \qquad (31.13)$$

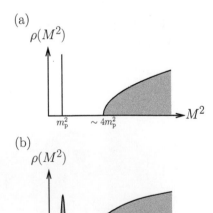

(a)
$\rho(M^2)$

m_P^2 $\sim 4m_P^2$ M^2

(b)
$\rho(M^2)$

m_P^2 M^2

Fig. 31.2 (a) The spectral function showing a single-particle peak. (b) The spectral function showing a quasiparticle peak with lifetime $(2\Gamma_p)^{-1}$.

where the quasiparticle now has a finite lifetime $(2\Gamma_p)^{-1}$ (and we allow the possibility that Z is a function of p). We see that our requirement that $E_p > \Gamma_p$ for a meaningful quasiparticle tells us that the width of the peak in the spectral weight should be narrow compared to the distance of the peak from the origin. A spectral weight with a broad quasiparticle peak is often seen in condensed matter physics where there is usually a large number of very closely spaced energy levels. As a result the quasiparticle is a wave packet spread over (potentially very many) levels which eventually decays away.

In addition to the quasiparticle having a finite lifetime it can also be shown that the multiparticle part usually undergoes some degree of destructive interference and decays after a time that is fairly small compared to Γ_p^{-1}, leaving the lone quasiparticle amongst the ruins!

Once we obtain the full propagator $\tilde{G}(p)$ we can immediately read off the mass energy of the excited states m_{P} from the position of the pole in the function, the residue at that pole gives us $iZ_{\boldsymbol{p}}$, the lifetime is determined from $\Gamma_{\boldsymbol{p}}$ and, of course, we can use $\tilde{G}(p)$ to calculate other quantities too, such as scattering amplitudes.

31.4 Elementary quasiparticles in a metal

The concept of dressed particles is especially important in condensed matter physics. Although, in reality, the condensed matter physicist usually deals with a system made up from a large number of very strongly interacting, real particles (such as electrons in a metal), renormalization involves changing our point of view and describing the system in terms of a small number of weakly interacting or non-interacting, fictitious particle-like excitations. Despite the fact that the ground state of a system, such as a crystal for example, isn't an empty box[8] we regard it as a vacuum, in that it contains no excitations. A weakly excited state of the whole system (that is, one at low temperature) can then be described in terms of a small number of these excitations. These are often also called the **elementary excitations** of the system.[9]

A distinction can be made between two types of elementary excitation. The first are called **collective excitations** and correspond to particle-like excitations that occur through motion of *all* of the constituent parts of the underlying system. An example is the chain of coupled masses we considered in Chapter 2. There we described the excited states of the system, which involved the motion of all of the masses, as particle-like entities called phonons.[10]

The second sort of elementary excitation is the quasiparticle, a single-particle excitation from the non-interacting system dressed in interactions. An example is the single-particle excitation of an electron gas. In a non-interacting electron gas [Fig. 31.3(a)] we have single-particle excitations above the ground state called electrons and holes[11] [Fig. 31.3(b)]. As we turn on the interactions these excited electrons and holes become quasielectrons and quasiholes [Fig. 31.3(c)] .

Example 31.3

Let's look at the case of electron quasiparticles in more detail. We start with the non-interacting version of a metal: the Fermi gas at $T = 0$. This comprises a box of N electrons, stacked up in momentum eigenstates $|\boldsymbol{p}\rangle$ up to the Fermi level $|\boldsymbol{p}_{\mathrm{F}}\rangle$. (For simplicity we regard the electrons as spinless and, in that case, the momentum states may contain either 1 or 0 electrons.) The momentum distribution $n_{\boldsymbol{p}}^{(0)}$ for the Fermi gas is shown in Fig. 31.4(a). This is the ground state of this non-interacting system, which we call $|0\rangle$. The particle dispersion is given by $E_{\boldsymbol{p}}^{(0)} = \boldsymbol{p}^2/2m_{\mathrm{e}}$, where m_{e} is the free electron mass. Note that near the Fermi level the dispersion may be written $E_{\boldsymbol{p}}^{(0)} = v_{\mathrm{F}}(|\boldsymbol{p}| - p_{\mathrm{F}})$, where

$$v_{\mathrm{F}} = \boldsymbol{\nabla}_{\boldsymbol{p}} E_{\boldsymbol{p}}|_{|\boldsymbol{p}|=p_{\mathrm{F}}} = \frac{p_{\mathrm{F}}}{m_{\mathrm{e}}}, \qquad (31.14)$$

[8]Far from it, it usually comprises $N \approx 10^{23}$ particles of one sort or another.

[9]The approach of describing a system in terms of small numbers of excitations works in condensed matter physics because we are most often interested in the behaviour of a solid at low temperature and this behaviour is dominated by the occupation of only those energy levels of the system very close to the ground state.

[10]Notice that if we turn off the interactions in the system (in this case the springs between the masses) then these excitations cease to exist. Another example of a collective excitation is the plasmon, which is a collective excitation of all of the electrons in a metal. We will examine these in Chapter 43.

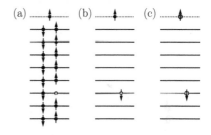

Fig. 31.3 (a) The non-interacting electron gas with a single electron above the Fermi level. (b) The particle excitations in the non-interacting electron gas. (c) The quasiparticles in the interacting electron gas [identical to (b) but with renormalized masses].

[11]Remember that in the picture discussed here, excitations only exist above the ground state. The ground state here consists simply of electron states filled up with electrons all the way to the Fermi level. In the non-interacting system an electron above the Fermi level is an excitation, which becomes a quasiparticle upon turning on the interactions. An electron below the Fermi sea is not an excitation and does not become a quasiparticle. A contrasting philosophy, where *all* electrons become quasiparticles, is Landau's Fermi liquid theory, which is examined later in the chapter.

[12]The Fermi surface is spherical here so the Fermi velocity only depends on the magnitude of $\boldsymbol{p}_F$ and we write $|\boldsymbol{p}_F| = p_F$.

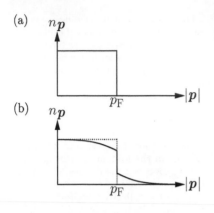

(a) $n_{\boldsymbol{p}}$

p_F $|\boldsymbol{p}|$

(b) $n_{\boldsymbol{p}}$

p_F $|\boldsymbol{p}|$

Fig. 31.4 (a) The momentum distribution of the Fermi gas. (b) The momentum distribution of the interacting Fermi system. A discontinuity of size Z_{p_F} exists at the Fermi surface.

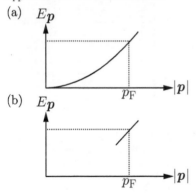

(a) $E_{\boldsymbol{p}}$

p_F $|\boldsymbol{p}|$

(b) $E_{\boldsymbol{p}}$

p_F $|\boldsymbol{p}|$

Fig. 31.5 (a) The dispersion of the free particles in a Fermi gas. (b) Near the Fermi momentum the dispersion is given by $E_{\boldsymbol{p}} = (|\boldsymbol{p}| - p_F)v_F$, where $v_F = p_F/m^*$ and m^* is the effective mass of the particles.

Philip Anderson is arguably the most influential condensed matter physicist since Landau. Not only is he the discoverer/creator of a variety of theoretical descriptions of electronic and magnetic phenomena (among many other things); his elucidation of the structure underlying the subject pervades current condensed matter physics. (Even the name 'condensed matter physics' is one of his contributions!)

is known as the Fermi velocity[12] [see Fig. 31.5(b)].

We now turn on interactions between the electrons. The interacting ground state of the system is $|\Omega\rangle$, the eigenstates of the system are now called $|\boldsymbol{p}_\lambda\rangle$, and the energies of these states are now $E_{\boldsymbol{p}}$. Near the Fermi level we again have $E_{\boldsymbol{p}} = v_F(|\boldsymbol{p}| - p_F)$, but now $v_F = p_F/m^*$, where we call m^* the effective mass. Following from our assertion that we may describe the weakly excited states of a system as quasiparticles, we expect that those states that lie close to the Fermi level p_F will have nonzero values of $Z_{\boldsymbol{p}}^{\frac{1}{2}}$ and may therefore be described as containing quasiparticles. As will be shown in Chapter 43 the free propagator for particles in a metal is given by

$$\tilde{G}_0(p) = \frac{i\theta(|\boldsymbol{p}| - p_F)}{E - E_{\boldsymbol{p}}^{(0)} + i\epsilon} + \frac{i\theta(p_F - |\boldsymbol{p}|)}{E - E_{\boldsymbol{p}}^{(0)} - i\epsilon}. \tag{31.15}$$

The interacting version becomes

$$\tilde{G}(p) = \frac{iZ_{\boldsymbol{p}}\theta(|\boldsymbol{p}| - p_F)}{E - E_{\boldsymbol{p}} + i\Gamma_{\boldsymbol{p}}} + \frac{iZ_{\boldsymbol{p}}\theta(p_F - |\boldsymbol{p}|)}{E - E_{\boldsymbol{p}} - i\Gamma_{\boldsymbol{p}}} + \left(\begin{array}{c}\text{Multiparticle}\\\text{parts}\end{array}\right). \tag{31.16}$$

It will be useful in what follows to have the propagator in three-momentum space and the time domain. This is achieved by doing a Fourier transform (see Appendix B) with the result

$$G(\boldsymbol{p}, t) = Z_{\boldsymbol{p}}\left[\theta(t)\theta(|\boldsymbol{p}| - p_F)e^{-iE_{\boldsymbol{p}}t}e^{-\Gamma_{\boldsymbol{p}}t} - \theta(-t)\theta(p_F - |\boldsymbol{p}|)e^{-iE_{\boldsymbol{p}}t}e^{\Gamma_{\boldsymbol{p}}t}\right]$$
$$+ \left(\begin{array}{c}\text{Multiparticle}\\\text{parts}\end{array}\right). \tag{31.17}$$

We may use the result of the last example to test whether we really have quasiparticles in a real-life metal by making a direct measurement of the quasiparticle weight $Z_{\boldsymbol{p}}$. To measure $Z_{\boldsymbol{p}}$ we make a measurement of the momentum distribution $n_{\boldsymbol{p}}$ of the weakly excited system (that is, when it contains some quasiparticles), given by $n_{\boldsymbol{p}} = \langle \hat{a}_{\boldsymbol{p}}^\dagger \hat{a}_{\boldsymbol{p}} \rangle$ which may be related to the propagator via (see Exercise 31.1.)

$$n_{\boldsymbol{p}} = -\lim_{t \to 0^-} G(\boldsymbol{p}, t), \tag{31.18}$$

giving us a prediction

$$n_{\boldsymbol{p}} = Z_{\boldsymbol{p}}\theta(p_F - |\boldsymbol{p}|) - \left(\begin{array}{c}\text{Multiparticle}\\\text{parts}\end{array}\right). \tag{31.19}$$

Assuming that the multiparticle contribution varies smoothly, we conclude that the interacting metal should have a discontinuity in its momentum distribution $n_{\boldsymbol{p}}$ at its Fermi surface of size $Z_{|\boldsymbol{p}|=p_F}$, as shown in Fig. 31.4(b). This prediction can be tested by performing Compton scattering experiments on metals. The results are found to be in good agreement with the theoretical predictions.

31.5 The Landau Fermi liquid

In general, I believe that the attitude is at least justifiable that the instructive use of quasi-particles by the founders of solid state physics and by Landau's school is hardly less rigorous than the sophisticated many-body theory, and perhaps it is more foolproof, because it has less appearance of being rigorous.

P. W. Anderson (1923–), *Concepts in Solids*

A rather different way of understanding metals in terms of fictitious particles was formulated by Lev Landau[13] and is known as **Fermi liquid theory**. It has been so useful in understanding metals that it is little exaggeration to claim that it has become the standard model of the metal.[14] Landau's theory is phenomenological, but appealingly intuitive since it involves a description of a strongly interacting metal as being almost identical to the non-interacting Fermi gas! Moreover Fermi liquid theory allows the complete description of an interacting system in terms of a small number of parameters, while avoiding all of the complexities of perturbation theory. Central to the model is the fact that Landau describes an interacting metal in terms of quasiparticles; but these are different to the field theory quasiparticles described thus far in this chapter. We will call the former Landau quasiparticles to avoid confusion.

In Landau's picture we pay particular attention to the process of 'turning on' the interaction between the non-interacting electrons of the Fermi gas. Landau assumed that if we very slowly turn on the interaction the system evolves continuously from Fermi gas to Fermi liquid, with each single-particle momentum eigenstate of the gas evolving into a single-particle momentum eigenstate of the liquid. This *adiabatic turning on* may be shown in ordinary quantum mechanics to lead to a very small amplitude for transitions out of the level, providing the density of final states is small, as it is for electrons within the Fermi sea as a result of the Pauli principle.

Vital to Landau's Fermi liquid concept is the notion that *all* of the the states $|\boldsymbol{p}\rangle$ occupied by an electron in the gas with $n_{\boldsymbol{p}}^{(0)} = 1$ become single-particle eigenstates $|\boldsymbol{p}\rangle$ in the liquid occupied by a Landau quasiparticle. The Landau quasiparticles in the interacting ground state therefore also have momentum distribution $n_{\boldsymbol{p}}^{(0)}$. We say that there is a one-to-one correspondence between the free particles and Landau quasiparticles. On turning on the interaction the Landau quasiparticles take on an effective mass m^* which parametrizes the change in energy of the eigenstates due to the effect of the field from the other quasiparticles.

Despite the change in energy, there should be no ambiguity in the identity of an occupied eigenstate after turning on the interaction and therefore the energy levels of the states shouldn't cross during the turning on process. That is to say, upon turning on the interaction we must have the process shown in Fig. 31.6(a) rather than that in Fig. 31.6(b). We might ask whether this non-crossing of levels is a realistic proposition. Recall that when two eigenstates are related by a symmetry transformation which leaves the Hamiltonian invariant we obtain a degeneracy in energy. If we turn on a perturbing potential V which leads to a nonzero matrix element δ between the two states we have a Hamiltonian $H = \begin{pmatrix} E & \delta \\ \delta & E \end{pmatrix}$ which leads to a splitting in energy of 2δ. It's as if the energy levels repel each other by virtue of a matrix element δ existing between them, which prevents them from ever crossing. The consequence of this for the metal is that, as we turn on the interaction,

[13]Lev Landau (1908–1968) was a Soviet physicist and one of the greatest scientists of the twentieth century who made contributions in many areas, including phase transitions, magnetism, superconductivity, superfluidity, plasma physics and neutrinos. He was also the model for the principal character, Viktor Shtrum, in Vassily Grossman's novel *Life and Fate*.

[14]The concept of a Fermi liquid is a high successful phenomenological theory of interacting electrons. It is a bit of a departure from our main theme of field theories of interacting systems, but it is too useful to omit and more advanced treatments than the one given here demonstrate that the machinery of diagrammatic perturbation theory can indeed be employed to derive important results in Fermi liquid theory.

(a)

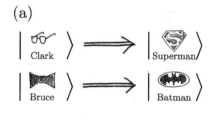

(b)

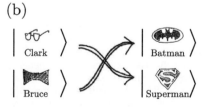

Fig. 31.6 Turning on interactions changes the states but the energy levels shouldn't cross during the turning on process. Thus we can have (a) but not (b).

the levels repel each other when they get too close and never cross. This is shown in Fig. 31.7. However, this repulsion does not occur when the matrix element between two levels is zero. This is the case when the levels have different symmetries and then there is nothing stopping the levels from crossing.[15] Since in the absence of a phase transition the slow turning-on of interactions provides a one-to-one correspondence between Landau quasiparticles and free particles, the theory is said to possess **adiabatic continuity**.

Notice how different the Landau quasiparticles are to our field theory idea of quasiparticles. In field theory the ground state contains no quasiparticles; in Fermi liquid theory it contains as many Landau quasiparticles as we had electrons in the Fermi gas, guaranteeing charge conservation, provided Landau quasiparticles carry the same charge as electrons. Landau quasiparticles do not, therefore, rely on the renormalization constant $Z_{\boldsymbol{p}}$ to guarantee their existence as our field theory quasiparticles do.

An insight comes when, instead of starting with a ground state Fermi gas, we start with a ground state Fermi gas with an additional electron added with $|\boldsymbol{p}'| > p_{\mathrm{F}}$ and, again, turn on the interaction. We would hope that the excited electron evolves into a quasiparticle in a momentum eigenstate $|\boldsymbol{p}'\rangle$. However, unlike the electrons within the Fermi sea, the extra electron potentially has rather a lot of phase space to explore as we turn on the interactions. Through interactions with the other electrons it may therefore have a considerable amplitude to scatter out of its state $|\boldsymbol{p}'\rangle$ and so the excited quasiparticle will have a finite lifetime $(2\Gamma_{\boldsymbol{p}'})^{-1}$ against scattering into other states and decaying.[16] The phase space for this scattering, restricted by the Pauli principle, is such that those states near the Fermi momentum have the smallest probabilities for scattering and therefore survive for longest in quasiparticle momentum states. This is examined in the exercises with the result that the decay rate varies as $\Gamma_{\boldsymbol{p}} \sim (|\boldsymbol{p}| - p_{\mathrm{F}})^2$. We conclude that really it is meaningful to speak of the properties of a Landau quasiparticle only near the Fermi surface in a metal where $\Gamma_{\boldsymbol{p}}$ is small.

Finally, if we lift a Landau quasiparticle out of a level below p_{F} in the ground state and promote it to a level with momentum $\Delta\boldsymbol{p}$ above p_{F}, we simultaneously create an unoccupied quasiparticle state or *hole*[17] with momentum $\Delta\boldsymbol{p}$ below p_{F}. Since $\Delta\boldsymbol{p} = |\boldsymbol{p}| - p_{\mathrm{F}}$, the quasielectrons and quasiholes will both have decay rates $\Gamma_{\boldsymbol{p}} \propto (|\boldsymbol{p}| - p_{\mathrm{F}})^2$, so it is only meaningful to discuss their properties if their momenta are close to p_{F}.

To summarize the philosophy of the Fermi liquid:

> The Fermi liquid ground state contains Landau quasiparticles stacked up to the Fermi level. Landau quasiparticles have the same charge as ordinary electrons but their mass takes on a renormalized value m^*.
> A quasiparticle excitation outside the Fermi sea (or, equivalently, a quasihole within the sea) will have a decay rate $\Gamma_{\boldsymbol{p}} \propto (|\boldsymbol{p}| - p_{\mathrm{F}})^2$.

A cartoon of the Landau Fermi liquid theory is shown in Fig. 31.8.

[15]This is exactly the case at a phase transition when, as we've seen, the system finds a new ground state with a lower symmetry than its previous ground state. The requirement that the levels don't cross is therefore a requirement that the evolution of the system doesn't involve a change in thermodynamic phase.

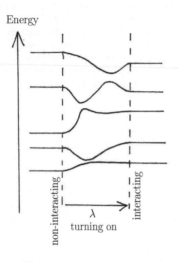

Energy

non-interacting

interacting

λ

turning on

Fig. 31.7 Evolution of single-particle energy levels as interactions are turned on. Notice how they repel when they draw close to each other and never cross.

[16]Notice that this sets a limit on the speed at which we may turn on the interactions. Although we would like to turn them on infinitely slowly, any slower than $\Gamma_{\boldsymbol{p}'}^{-1}$ and our excited quasiparticle loses its momentum and, consequently, its identity. For a general quantum liquid to exist it must be possible to select a turning on time that allows the ground state to evolve adiabatically, but also retains the low-lying quasiparticles.

[17]Holes will be examined in Chapter 43.

Where the Fermi liquid and quantum field theories coincide is in the description of the weakly excited states of the metal. The excited Landau quasiparticles have exactly the same properties as our quantum field theory quasiparticles and the argument about their lifetime applies to both. Near the Fermi surface the energy of an excited quasiparticle may be written $E_{\boldsymbol{p}} \approx \frac{p_{\mathrm{F}}}{m^*}(|\boldsymbol{p}| - p_{\mathrm{F}})$ and its decay rate $\Gamma_{\boldsymbol{p}} \propto (|\boldsymbol{p}| - p_{\mathrm{F}})^2$ We see that for small $(|\boldsymbol{p}| - p_{\mathrm{F}})$ the real part of the quasiparticle energy will be larger than the imaginary part, or $E_{\boldsymbol{p}} > \Gamma_{\boldsymbol{p}}$, and a meaningful quasiparticle exists in the metal.

We now complete our description of the Landau Fermi liquid with a discussion of the total energy of a weakly excited state of a metal. In the spirit of many of Landau's arguments, in which any ignorance of microscopic details is no barrier to formulating a meaningful power series expansion, we write the energy of the Fermi liquid in the limit of a low density of quasiparticles as

$$E = E_{\mathrm{g}} + \sum_{\boldsymbol{p}} (E_{\boldsymbol{p}}^{(0)} - \mu)\, \delta n_{\boldsymbol{p}} + \frac{1}{2} \sum_{\boldsymbol{p}\boldsymbol{p}'} f_{\boldsymbol{p}\boldsymbol{p}'}\, \delta n_{\boldsymbol{p}}\, \delta n_{\boldsymbol{p}'} + \cdots, \qquad (31.20)$$

where $\delta n_{\boldsymbol{p}} = n_{\boldsymbol{p}} - n_{\boldsymbol{p}}^{(0)}$ is the difference between a distribution with excitations and the ground state distribution[18] and $E_{\mathrm{g}} = \sum_{\boldsymbol{p}} E_{\boldsymbol{p}}^{(0)} n_{\boldsymbol{p}}^{(0)}$ is the energy of the ground state. The term linear in $\delta n_{\boldsymbol{p}}$ describes the excitation of isolated quasiparticles of energy $E_{\boldsymbol{p}}^{(0)}$ $\mu = \partial E/\partial n_{\boldsymbol{p}}$. The second-order coefficient[19] $f_{\boldsymbol{p}\boldsymbol{p}'} = \partial^2 E/\partial n_{\boldsymbol{p}} \partial n_{\boldsymbol{p}'}$ describes the contribution from quasiparticle–quasiparticle scattering (i.e. from interactions), and can be described via an interaction Hamiltonian $\mathcal{H}_{\mathrm{I}} = \frac{1}{2} \sum_{\boldsymbol{p}\boldsymbol{p}'} f_{\boldsymbol{p}\boldsymbol{p}'}\, \delta n_{\boldsymbol{p}}\, \delta n_{\boldsymbol{p}'}$.

Example 31.4

To understand the scattering process in more detail, one should include spin and this amounts to replacing $f_{\boldsymbol{p}\boldsymbol{p}'}$ by $f_{\boldsymbol{p}\sigma;\boldsymbol{p}'\sigma'}$. One can show[20] that for spin-conserving interactions this quantity can be written as

$$f_{\boldsymbol{p}\sigma;\boldsymbol{p}'\sigma'} = f^{\mathrm{s}}(\cos\theta) + f^{\mathrm{a}}(\cos\theta)\boldsymbol{\sigma} \cdot \boldsymbol{\sigma}', \qquad (31.21)$$

where $\cos\theta = \frac{\boldsymbol{p} \cdot \boldsymbol{p}'}{|\boldsymbol{p}||\boldsymbol{p}'|}$. The functions f^{s} and f^{a} can be expanded in terms of Legendre polynomials so that

$$f^{s,a}(\cos\theta) = \frac{1}{g(E_{\mathrm{F}})} \sum_{\ell=0}^{\infty} P_\ell(\cos\theta) F_\ell^{\mathrm{s,a}}, \qquad (31.22)$$

where F_ℓ^{s} and F_ℓ^{a} are **Landau parameters** and $g(E_{\mathrm{F}})$ is the density of states at the Fermi level.[21] These expressions allow some of the key properties of the Landau Fermi liquid to be deduced. For example, one can show that the effective mass is $m^* = m(1+F_1^{\mathrm{s}})$ and the spin susceptibility is $\chi = \mu_0 \mu_{\mathrm{B}}^2 g(E_{\mathrm{F}})/(1+F_0^{\mathrm{a}})$. Many of the predictions of Landau's Fermi liquid theory may be shown (after much hard graft!) to coincide with those of quantum field theory and, owing to its comparative ease of use, it is still regarded as the best way of understanding the results of experiment on numerous condensed matter systems.

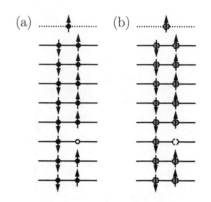

Fig. 31.8 A cartoon of the Landau Fermi liquid. (a) The non-interacting Fermi gas with a single electron promoted into an excited state (indicated as a dotted line). (b) In the Landau Fermi liquid the interactions are turned on so that the electrons become quasiparticles and the holes become quasiholes. There is a one-to-one correspondence between the quasiparticles and quasiholes in the interacting liquid and the electrons and holes in the non-interacting picture.

[18]This implies that it is neither $n_{\boldsymbol{p}}$ nor $n_{\boldsymbol{p}}^{(0)}$ that is the crucial quantity, but rather it is their difference $\delta n_{\boldsymbol{p}}$, which tells us the number of excitations in the excited state. This is fortunate since we know neither $n_{\boldsymbol{p}}$ nor $n_{\boldsymbol{p}}^{(0)}$ with much accuracy, but we can find $\delta n_{\boldsymbol{p}}$.

[19]The partial differentials in the expressions for $E_{\boldsymbol{p}}^{(0)} - \mu$ and $f_{\boldsymbol{p}\boldsymbol{p}'}$ are evaluated about the ground state configurations, i.e. a frozen Fermi sea.

[20]Further details and a fuller account may be found in the book by P. Coleman.

[21]See Chapter 43.

We have seen in this chapter that renormalization is an essential step in understanding what happens to particles in an interacting quantum field theory. In fact, as shown in Fig. 31.9, we could summarize the steps to make a working theory as: (i) write a Lagrangian; (ii) quantize the free part; (iii) derive Feynman rules for interactions; (iv) renormalize. However, renormalization is often presented as a method of eliminating nonsensical infinities from theories. The elimination of these infinities is the subject of the next chapter. It should be remembered throughout that renormalization is necessary for all interacting theories, independent of whether or not we have infinities.

Chapter summary

- Turning on the interactions in a theory changes the properties of the particles. We say that they dress themselves in interactions.
- The interacting propagator is

$$\tilde{G}(p) = \frac{iZ_{\boldsymbol{p}}}{p^2 - m_{\mathrm{P}}^2 + i\Gamma_{\boldsymbol{p}}} + (\text{multiparticle parts}),$$

where $Z_{\boldsymbol{p}}$ is the quasiparticle weight

- In Landau's Fermi liquid theory there is a one-to-one correspondence between electrons in the non-interacting theory and quasiparticles in the interacting theory. The effect of interactions is to renormalize physical quantities, such as the effective mass, via the Landau parameters F_{ℓ}^{s} and F_{ℓ}^{a}.

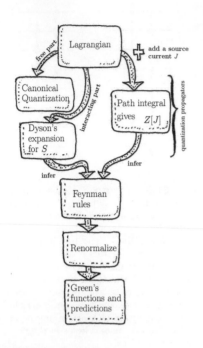

Fig. 31.9 The process of doing quantum field theory includes the necessary step of renormalizing.

Exercises

(31.1) Verify that, for fermions,

$$n_{\boldsymbol{p}} = -\lim_{t \to 0-} G(\boldsymbol{p}, t). \qquad (31.23)$$

(31.2) Consider an electron with energy $E_1 \geq E_{\mathrm{F}}$ scattering with an electron with energy $E_2 \leq E_{\mathrm{F}}$ at $T = 0$. In order for this to occur we must have final electron state $E_3 \geq E_{\mathrm{F}}$ and $E_4 \geq E_{\mathrm{F}}$.
(a) Show that this implies that the lifetime of an electron with $E_1 = E_{\mathrm{F}}$ is infinite.
(b) If E_1 is a little different to E_{F}, why does the scattering rate go as $(E_1 - E_{\mathrm{F}})^2$?
(c) For $T \neq 0$ argue that we expect a scattering

rate $\frac{1}{\tau} = a(E_1 - E_{\mathrm{F}})^2 + b(k_{\mathrm{B}}T)^2$, where a and b are constants.

(31.3) (a) Using the relation

$$\frac{1}{x_0 + i\epsilon} = \frac{\mathcal{P}}{x_0} - i\pi\delta(x_0), \qquad (31.24)$$

(discussed in Appendix B) identify the delta function part of the propagator $\frac{iZ_{\boldsymbol{p}}}{E - E_{\boldsymbol{p}} + i\epsilon}$.
(b) For the propagator $\frac{iZ_{\boldsymbol{p}}}{E - E_{\boldsymbol{p}} + i\Gamma_{\boldsymbol{p}}}$, show that the full width at half maximum of the peak is given by $2\Gamma_{\boldsymbol{p}}$.

Renormalization: the problem and its solution

32

It turns out that the obvious generalization of this idiotically simple manipulation gets rid of all of the infinities for any field theory with polynomial interactions, to any order in perturbation theory.
Sidney Coleman (1937–2007)

In the last chapter, we saw how interactions dress a particle and change its propagator. The dressed particles were quite different to the bare particles that are revealed after canonical quantization of the free theory. This might mean that our perturbation theory, which involved expansions in the bare mass m and coupling constant λ, is at risk. It is: it turns out that we've been doing our perturbation theory about the wrong masses and coupling constants! The solution to this problem involves making a shift so that we're expanding in physical masses and coupling constants. The procedure also has a bonus. It solves a big problem in our perturbation theory: the problem of divergent amplitudes.

In this chapter we approach the problem by starting with the divergences in our perturbation theory. We will look at why we obtain infinities in our calculations and suggest an idiotically simple mathematical workaround. The physical content of this workaround will turn out to shift the parameters of the theory from the wrong ones (m and λ) to the right ones (m_P and λ_P).

32.1 The problem is divergences

We've looked at calculating amplitudes and propagators for several interacting theories. Perhaps you're curious (or suspicious) as to why we haven't done the integrals at the end of the calculations. The reason is that many of them diverge. Understanding the source of the divergences relies purely on the following facts (with a finite and positive):

$$\int_a^\infty \mathrm{d}x\, x^n = \left[\frac{x^{n+1}}{n+1}\right]_a^\infty \text{ diverges for } n \geq 0, \qquad (32.1)$$

$$\int_a^\infty \frac{\mathrm{d}x}{x} = [\ln x]_a^\infty \text{ diverges}, \qquad (32.2)$$

$$\int_a^\infty \frac{\mathrm{d}x}{x^m} = \left[\frac{x^{-m+1}}{-m+1}\right]_a^\infty = \frac{a^{-m+1}}{m-1} \text{ for } m > 1. \qquad (32.3)$$

In the first integral the divergence arises because there are more powers of x in the numerator than the denominator. The second integral, where the number of powers of x is the same on top and bottom, is called **logarithmically divergent**, for rather obvious reasons. The last example is convergent.

Let us return to ϕ^4 theory, which is described by a Lagrangian

$$\mathcal{L} = \frac{1}{2}(\partial_\mu \phi)^2 - \frac{m^2}{2}\phi^2 - \frac{\lambda}{4!}\phi^4. \tag{32.4}$$

We ended our analysis of this theory in Chapter 19 with a perturbation expansion encoded in Feynman diagrams. Consider the amplitude for two-particle scattering, which is equal to the sum of all connected, amputated Feynman diagrams with four external legs. Up to second order in the expansion (i.e. drawing all diagrams with one or two interaction vertices) we obtain the diagrams shown in Fig. 32.1. The main message of this section is that the diagrams with loops in them [(b),(c) and (d)] are divergent. This is to say, with the upper limit of the integral set to infinity we get infinite and therefore nonsensical answers. This is a disaster.

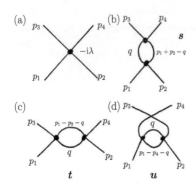

Fig. 32.1 The expansion of four-point diagrams for two-particle scattering up to second order in the interaction strength. Diagrams (b)–(d) define the processes labelled s, t and u, respectively.

Example 32.1

Let's see how this unfolds. We've encountered the first diagram before. The result is

$$i\mathcal{M}_a = -i\lambda. \tag{32.5}$$

For the subsequent diagrams, the integral we'll need is

$$\int_0^\Lambda \frac{d^4q}{(2\pi)^4} \frac{i}{q^2 - m^2 + i\epsilon} \frac{i}{(p-q)^2 - m^2 + i\epsilon} = -4ia\ln\left(\frac{\Lambda}{p}\right), \tag{32.6}$$

where a is some numerical constant whose exact value isn't important to us. Notice that the limits of this integral are zero and Λ. The parameter Λ is a large momentum cut-off, and we want to send $\Lambda \to \infty$. By counting powers we see that there are four powers of momentum on top and four below, telling us that the integral $\sim \int \frac{d^4q}{q^4}$ is logarithmically divergent. In fact, for the diagrams in Figs. 32.1(b)–(d) we obtain the amplitudes:

$$i\mathcal{M}_b = ia\lambda^2\left\{\ln\Lambda^2 - \ln\left[(p_1+p_2)^2\right]\right\} = ia\lambda^2\left\{\ln\Lambda^2 - \ln s\right\}, \tag{32.7}$$

$$i\mathcal{M}_c = ia\lambda^2\left\{\ln\Lambda^2 - \ln\left[(p_1-p_3)^2\right]\right\} = ia\lambda^2\left\{\ln\Lambda^2 - \ln t\right\}, \tag{32.8}$$

$$i\mathcal{M}_d = ia\lambda^2\left\{\ln\Lambda^2 - \ln\left[(p_1-p_4)^2\right]\right\} = ia\lambda^2\left\{\ln\Lambda^2 - \ln u\right\}. \tag{32.9}$$

All of these tend to ∞ as $\Lambda \to \infty$. The total amplitude for two-particle scattering is given by the sum of these diagrams, giving

$$\begin{aligned} i\mathcal{M} &= i\left(\mathcal{M}_a + \mathcal{M}_b + \mathcal{M}_c + \mathcal{M}_d\right) \\ &= -i\lambda + ia\lambda^2\left\{3\ln\Lambda^2 - \ln s - \ln t - \ln u\right\}, \end{aligned} \tag{32.10}$$

and so we need to find a way to tame the $ia\lambda^2\left\{3\ln\Lambda^2\right\} \propto \ln\Lambda$ term.

We can avoid the divergence in these integrals if we make the pragmatic choice to only integrate up to a large, but finite, value of the cut-off Λ. This amounts to deliberately ignoring any fine-scale details in the fields below a length scale $\approx 1/\Lambda$. This is frequently what is done in condensed matter physics, where we usually ignore all detail smaller than the size of an atom. If we apply our theory to fundamental particles it's not quite so obvious why there should be a smallest scale but it might represent some graininess in spacetime.[1]

For now, let's see if we can live (uncomfortably) with $\Lambda \neq \infty$. Introducing finite Λ has an immediate consequence: the amplitudes we calculate using Feynman diagrams will depend on Λ. This is a serious problem: if a prediction for something we want to measure depends on Λ then we have an arbitrary constant which will need to be given some value. One strategy to avoid this is to add terms to the Lagrangian that, when we do the perturbation expansion up to some order (second order say), will remove the dependence of amplitudes on Λ. Doing this will clean up the theory up to second order. Although we should still expect the higher order terms to diverge (i.e. blow up as $\Lambda \to \infty$), we'll at least have healed the theory to the extent that we can obtain a prediction of physical behaviour that is valid up to second order.

[1] Actually, the idea of a length scale like this emerges very naturally from a way of looking at the Universe called the renormalization group, which we discuss in Chapter 34.

32.2 The solution is counterterms

The terms that are added to the Lagrangian to remove the dependence on Λ, and hence the divergences, are called **counterterms**. We saw that for ϕ^4 theory at second order the divergent part of the amplitude looked like $6ia\lambda^2 \ln \Lambda$; we therefore change the Lagrangian by adding a counterterm $-(6a\lambda^2 \ln \Lambda)\phi^4/4!$ thus:

$$\mathcal{L} \to \mathcal{L} - \left(\frac{6a\lambda^2 \ln \Lambda}{4!}\right)\phi^4. \qquad (32.11)$$

We now have a Lagrangian

$$\mathcal{L} = \frac{1}{2}(\partial_\mu \phi)^2 - \frac{m^2}{2}\phi^2 - \frac{\lambda}{4!}\phi^4 + \frac{C^{(2)}}{4!}\phi^4, \qquad (32.12)$$

where the counterterm $C^{(2)} = -6a\lambda^2 \ln \Lambda$ and we write the superscript (2) to remind ourselves that we're only getting rid of divergences at second order.

Now we start again with our new Lagrangian. We canonically quantize, use the Dyson expansion and figure out the Feynman rules and then sum the new diagrams. With the counterterm, the Dyson expansion of the $\hat{S}$-operator is now just

$$\hat{S} = Te^{-i\int d^4x \frac{1}{4!}\left(\lambda\hat{\phi}^4 - C^{(2)}\hat{\phi}^4\right)}. \qquad (32.13)$$

Every order of the expansion therefore also includes a contribution from the counterterm, whose Feynman diagram is shown in Fig. 32.2. Since the counterterm is proportional to ϕ^4 it has the same behaviour as the $\frac{\lambda}{4!}\phi^4$ term. It therefore represents a vertex which, instead of carrying a factor $-i\lambda$ carries a factor $iC^{(2)}$.

We calculate amplitudes to second order in λ and (we only need to) include one diagram involving a counterterm and we obtain

$$i\mathcal{M}^{(2)} = -i\lambda + ia\lambda^2 \left[3\ln\Lambda^2 - \ln s - \ln t - \ln u\right] + iC^{(2)}, \qquad (32.14)$$

and substituting $C^{(2)} = -6a\lambda^2 \ln\Lambda$ yields

$$i\mathcal{M}^{(2)} = -i\lambda - ia\lambda^2 \left(\ln s + \ln t + \ln u\right). \qquad (32.15)$$

Fig. 32.2 The C-counterterm.

This amplitude depends on the momenta of the incoming and outgoing particles (which is no problem) but crucially it doesn't depend on Λ. We've done what we set out to do! Predictions for the theory will now not depend on an arbitrary cut-off Λ, and hence won't yield infinity when we send $\Lambda \to \infty$.

32.3   How to tame an integral

We now have a plan, which is to start calculating using the Lagrangian and then add a counterterm when we see a divergence. We then work out the coefficient $C^{(n)}$ in front of the counterterm and all is set. Quite apart from the as yet unexamined physical consequences of this extra term, we might worry that this process will turn out to be ultimately futile. What if you calculate to second order as above, but at third order not only do we need a counterterm $C^{(3)}\phi^4$ but also a new counterterm $D^{(3)}\phi^6$ to swallow up some extra divergent term that emerges? What if it gets even worse and, at 27th order in the expansion, we need to add 513 new types of counterterm to cancel all of the divergences and then, at 28th order, we find we need 752 new counterterms? A theory might never stop absorbing new types of counterterm!

Amazingly, it often turns out that we only need to add a small number of types of counterterm (three, for example, in QED) and that these cancel divergences to all orders of perturbation theory! We still need to know the right constants (like $C^{(n)}$) for each order of perturbation theory in order to cancel the divergences, but these will turn out to be numbers that we can calculate. Theories that have the property that only a finite number of counterterms are needed to cancel all divergences are said to be **renormalizable theories**. This is the miracle of renormalization.

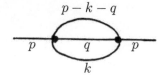

Fig. 32.3 The Saturn diagram.

Example 32.2

To get a feel for this we'll look at the Saturn diagram, shown in Fig. 32.3. Written out in full the amplitude is

$$i\mathcal{M} = \frac{(-i\lambda)^2}{6} \int_0^\Lambda \frac{\mathrm{d}^4q}{(2\pi)^4} \frac{\mathrm{d}^4k}{(2\pi)^4} \frac{i}{q^2 - m^2 + i\epsilon} \frac{i}{k^2 - m^2 + i\epsilon} \frac{i}{(p - q - k)^2 - m^2 + i\epsilon}.$$
$$(32.16)$$

These integrals are getting worse! If we're going to systematically renormalize any theory we'll need a system. The system involves writing the integral as a polynomial. Usefully there's a key fact that will help us: *Feynman diagrams can be expanded as a Taylor series in the external momentum.* The polynomial representing the diagram will have divergent coefficients, reflecting the divergence of the integral. We'll pick our counterterms to cancel the divergences in the coefficients. Happily, this results in counterterms of the right type to cancel the divergences in a particular integral.

The integral in eqn 32.16 may be written in a series about $p = 0$ as

$$I = \alpha + \beta p^2 + \gamma p^4 + \ldots, \tag{32.17}$$

where we don't have any odd terms because of the symmetry $\phi \equiv -\phi$ in the original Lagrangian.

Here's how to find the counterterms:

- The Saturn diagram integral is quadratically divergent. It has eight powers of momentum on top and six on the bottom. If we set the external momentum p equal to zero in our expansion then $I = \alpha$ and we conclude that α diverges quadratically. This tells us that we need one counterterm that diverges quadratically.

- If we differentiate the Saturn integral twice with respect to p then it becomes logarithmically divergent, since we have reduced the number of momenta on top by two. If we differentiate the series expansion I twice and set $p = 0$ we get $I'' = 2\beta$ and we conclude that β diverges logarithmically. We therefore need another counterterm which is logarithmically dependent on the cut-off which will have a Feynman rule giving rise to a p^2 factor.

- Differentiating the integral two more times makes it convergent so we don't need any more counterterms.

We need two counterterms to cancel the divergences in the Saturn integral I: one with coefficient A quadratically dependent on Λ and one with coefficient B which is logarithmically dependent. How do we make sure that the B counterterm will give a Feynman amplitude that is multiplied by p^2 and that A will not be multiplied by anything? The answer is to introduce counterterms into the Lagrangian that look like

$$\mathcal{L}_{\text{ct}} = \frac{A}{2}\phi^2 + \frac{B}{2}(\partial_\mu \phi)^2, \tag{32.18}$$

remembering that in momentum space $[\partial_\mu \phi(x)]^2 \rightarrow \tilde{\phi}(-p)p^2\tilde{\phi}(p)$ we have $B(\partial_\mu \phi)^2 \rightarrow p^2 B$ and $A \rightarrow A$. Adding the counterterms gives the full Lagrangian for renormalized ϕ^4 theory:

$$\mathcal{L} = \frac{1}{2}(\partial_\mu \phi)^2 - \frac{m^2}{2}\phi^2 - \frac{\lambda}{4!}\phi^4 + \frac{A}{2}\phi^2 + \frac{B}{2}(\partial_\mu \phi)^2 + \frac{C}{4!}\phi^4. \tag{32.19}$$

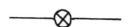

Fig. 32.4 The A and B counterterm diagram

Notice that all of the counterterms have the same form as the terms in the original equation; they just have different coefficients.

These counterterms all have Feynman diagrams and rules attached to them. We saw before that the $C^{(n)}$-counterterm diagram corresponds to a rule $iC^{(n)}$. The $B^{(n)}$ counterterm makes a contribution $i(p^2 B^{(n)})$, where p is the momentum along the line and the A counterterm contributes $iA^{(n)}$. The A and B terms are usually represented by the single diagram shown in Fig. 32.4, which has a rule $i(B^{(n)}p^2 + A^{(n)})$.

To summarize, we have the Feynman rules for renormalized ϕ^4 theory given in the box. These eliminate all infinities.

> **Feynman rules for renormalized ϕ^4 theory**
>
> - A factor $\frac{i}{p^2 - m^2 + i\epsilon}$ for each propagator.
> - A factor $-i\lambda$ for each interaction.
> - Add sufficient counterterm diagrams to cancel all infinities.
> - A factor $i\left(B^{(n)}p^2 + A^{(n)}\right)$ for each counterterm propagator, where n is the order of the diagram.
> - A factor $iC^{(n)}$ for each interaction counterterm, where n is the order of the diagram.
> - All other rules regarding integrating, symmetry factors and overall energy momentum conserving delta functions are identical as for the case of unrenormalized perturbation theory.

32.4   What counterterms mean

We now turn to the question of the physics behind the counterterms. We've added terms to the Lagrangian which surely alters the physics we're studying. What we are going to find is that the counterterms shift the parameters from fictional ones to the real-life ones, simultaneously removing infinities and forcing our theory to describe real life. To recap: we started with an interacting theory,

$$\mathcal{L} \;=\; \frac{1}{2}(\partial_\mu \phi)^2 - \frac{m^2}{2}\phi^2 - \frac{\lambda}{4!}\phi^4, \tag{32.20}$$

with the mass and coupling constant m and λ respectively. We'll call eqn 32.20 the unrenormalized Lagrangian. The excitations made by the field operators when the coupling $\lambda = 0$ have mass m. However, when we turn on the coupling this Lagrangian gives rise to revolting infinities unless we cut off the integrals at some momentum Λ.

To fix the divergence encountered as $\Lambda \to \infty$ we find we need to renormalize the theory. As described in the previous chapter, renormalization involves accepting that we're dealing with dressed particles, which have a mass m_P and coupling λ_P. These particles only involve a fraction $Z^{\frac{1}{2}}$ of the fields. This realization doesn't remove the infinities, however. To do that we need to include counterterms in our Lagrangian. The renormalized Lagrangian is

$$\begin{aligned}
\mathcal{L}' \;=\;& \frac{1}{2}(\partial_\mu \phi_r)^2 - \frac{m_P^2}{2}\phi_r^2 - \frac{\lambda_P}{4!}\phi_r^4 \\
& + \frac{B}{2}(\partial_\mu \phi_r)^2 + \frac{A}{2}\phi_r^2 + \frac{C}{4!}\phi_r^4,
\end{aligned} \tag{32.21}$$

[2]Note that $\phi_r(x)$ are known as **renormalized fields**. We assume Z is independent of three-momentum here.

where we have rescaled the fields using[2] $\phi = \sqrt{Z}\phi_r(x)$ and called the mass m_P and coupling constant λ_P respectively. This theory won't give us nonsensical, infinite results for $\Lambda \to \infty$.

Example 32.3

To see the consequence of renormalization, we now collect the coefficients of the renormalized Lagrangian

$$\mathcal{L} = \frac{1+B}{2}(\partial_\mu \phi_r)^2 - \frac{(m_P^2 - A)}{2}\phi_r^2 - \frac{(\lambda_P - C)}{4!}\phi_r^4, \qquad (32.22)$$

which suggests that the counterterms represent shifts in the parameters in the Lagrangian. Writing $-A = \delta m^2$, $-C = \delta\lambda$ and $B = \delta Z$ we have

$$\mathcal{L} = \frac{1+\delta Z}{2Z}(\partial_\mu \phi)^2 - \frac{(m_P^2 + \delta m^2)}{2Z}\phi^2 - \frac{(\lambda_P + \delta\lambda)}{4!Z^2}\phi^4, \qquad (32.23)$$

where we've restored the original, unrenormalized fields. From this, we read off that we can relate our original and renormalized Lagrangians through

$$\phi = \sqrt{Z}\phi_r, \quad Z = 1 + \delta Z, \quad m^2 = \frac{(m_P^2 + \delta m^2)}{Z}, \quad \lambda = \frac{(\lambda_P + \delta\lambda)}{Z^2}. \qquad (32.24)$$

This tells us that renormalization is simply an exercise in shifting parameters.

To truly see what's going on it may make more sense now to replay the argument backwards. We start with the unrenormalized Lagrangian and shift from parameters m and λ to parameters m_P and λ_P. We start by renormalizing the fields to obtain

$$\begin{aligned}\mathcal{L} &= \frac{1}{2}(\partial_\mu \phi)^2 - \frac{m^2}{2}\phi^2 - \frac{\lambda}{4!}\phi^4 \\ &= \frac{Z}{2}(\partial_\mu \phi_r)^2 - \frac{Zm^2}{2}\phi_r^2 - \frac{Z^2\lambda}{4!}\phi_r^4. \end{aligned} \qquad (32.25)$$

We then use the shifts

$$Z = 1 + \delta Z, \quad Zm^2 = m_P^2 + \delta m^2, \quad Z^2\lambda = \lambda_P + \delta\lambda, \qquad (32.26)$$

giving us

$$\mathcal{L} - \frac{1}{2}(\partial_\mu \phi_r)^2 - \frac{m_P^2}{2}\phi_r^2 - \frac{\lambda_P}{4!}\psi_r^4 + \frac{\delta Z}{2}(\partial_\mu \phi_r)^2 - \frac{\delta m^2}{2}\phi_r^2 - \frac{\delta\lambda}{4!}\phi_r^4, \qquad (32.27)$$

allowing us to identify the counterterms A, B and C as the shifts in the parameters $-\delta m^2$, δZ and $-\delta\lambda$.

We started the whole venture of quantum field theory by writing down a simple Lagrangian with parameters we thought would tell us about the masses and couplings of real particles in Nature. It turns out we were wrong. We were doing the *wrong* perturbation theory expanding a series in terms of the *wrong* constant mass m and *wrong* coupling λ. We therefore had asked a nonsensical question and got some infinite (and therefore nonsensical) answers. Actually we should have been expanding our series in terms of a mass m_P and coupling λ_P. The price we pay for making this shift from wrong to right variables is counterterms. In a sense, we're not really 'adding counterterms' at all; we're really taking a bare Lagrangian and making a shift of variables from wrong ones to the right ones and the shifts required are the counterterms. *Renormalization is not, therefore, an exercise in hiding infinities, it's an exercise in making a theory describe real life.* For ϕ^4 theory, the counterterms are divergent, meaning that these shifts may be infinite! This implies that m and λ, the bare mass and coupling are infinite quantities which are shifted by infinite amounts upon dressing themselves by interactions.[3]

There's one last, but very important, point to make. What number do we take for m_P and λ_P? What are the right masses and coupling we should expand around? The answer is that they're the ones Nature's given us! For this reason, m_P and λ_P are known as the 'physical' parameters, in contrast to the 'bare' parameters m and λ.

[3]This explains why it's often claimed that the bare charge of the electron is infinite. The spontaneous appearance of electron-positron pairs screens the charge leading to the finite charge we encounter in Nature.

32.5   Making renormalization even simpler

We'd like a method to avoid having to analyse each diagram separately as we did for the Saturn diagram above. Is there a method which can, once and for all, tell us how divergent a diagram or class of diagrams is? There is and it's based on simple dimensional analysis.

Define the superficial degree of divergence[4] of an integral arising from a Feynman diagram in ϕ^4 theory to be

$$D = \left(\begin{array}{c} \text{Powers of momentum} \\ \text{in numerator} \end{array} \right) - \left(\begin{array}{c} \text{Powers of momentum} \\ \text{in denominator} \end{array} \right).$$

(32.28)

If $D > 0$ the integrals diverge, if $D = 0$ they logarithmically diverge and if $D < 0$ they don't diverge. For a diagram with B_E external lines, the degree of divergence in ϕ^4 is given by $D = 4 - B_\text{E}$.

[4]The clumsy name, 'superficial' degree of divergence, arises because the actual degree of divergence can be complicated in gauge theories and some pathological cases, see Peskin and Schroeder, page 316.

Example 32.4

We can prove the above theorem for predicting whether diagrams diverge. In addition to the previously defined symbols

- V is the number of vertices,
- L is the number of loops,
- B_I is the number of internal lines.

Each loop brings with it an integral with four powers of momentum. Each internal line brings a propagator which brings -2 powers of momentum. Therefore

$$D = 4L - 2B_\text{I}.$$

(32.29)

To get the number of loops: L is the number of momenta we integrate over and each internal line gives a momentum. The number of loops is less than B_I though. Momentum-conserving delta functions eat up integrals and there are V of them (one for each vertex, since we must conserve momentum at every vertex). However, one of the delta functions conserves momentum for the entire diagram, so doesn't eat an integral. The number of loops L is therefore equal to the number of internal lines B_I, minus $(V - 1)$ the number of vertices, with one removed for overall momentum conservation:

$$L = B_\text{I} - (V - 1).$$

(32.30)

Each vertex has four lines coming out of it. Each external line comes out of one vertex and each internal line connects two vertices. Therefore

$$4V = B_\text{E} + 2B_\text{I}.$$

(32.31)

Inserting eqn 32.30 and eqn 32.31 into eqn 32.29 we get $D = 4 - B_\text{E}$ and the theorem is proved.

The great significance of this theorem is that it shows that the three counterterms we have identified are the only ones that will ever appear. The fact that $D = 4 - B_\text{E}$ means that the only diagrams that diverge (i.e. with $D \geq 0$) have ≤ 4 external legs. There are no allowable diagrams with one or three legs and we've considered the divergences for the two and four leg diagrams. We'll never need any more.

32.6 Which theories are renormalizable?

Renormalizable theories are those in which a finite number of counterterms cancel all divergences. Unfortunately, there are theories for which this property does not hold true. An example is Fermi's theory of weak interactions between fermions, whose Lagrangian is given by

$$\mathcal{L} = \bar{\psi}(\hat{\not{p}} - m)\psi + G(\bar{\psi}\psi)^2, \tag{32.32}$$

where ψ is the fermion field.[5] The superficial degree of divergence of this theory is given by

$$D = 4 - \frac{3}{2}F_{\mathrm{E}} + 2V, \tag{32.33}$$

where F_{E} is the number of external Fermi lines in the diagram. This has the unfortunate feature that D depends on V, the number of interaction vertices. If we consider fermion–fermion scattering, for which $F_{\mathrm{E}} = 4$ then for $V > 1$ we have divergent diagrams, which become more divergent as V gets larger. We would need new counterterms at every order of perturbation theory.

In fact, whether a theory is renormalizable or not may be read off simply from the dimensions of the interaction coupling constant. The rules are

- A *super-renormalizable theory* has only a finite number of superficially divergent diagrams. Its coupling constant has positive mass dimension.

- A *renormalizable theory* has a finite number of superficially divergent diagrams, however, divergences occur at all orders of perturbation theory. Its coupling constant is dimensionless.

- A *non-renormalizable theory* has diagrams that are all divergent at a sufficiently high order of perturbation theory. These theories have a negative mass dimension.

In our units the action $S = \int \mathrm{d}^4 x \, \mathcal{L}$ must be dimensionless since e^{iS} appears in the path integral. We therefore need $\mathcal{L}$ to have dimensions $[\mathrm{Length}]^{-4}$. Also note that, using these units, mass and energy have dimensions $[\mathrm{Length}]^{-1}$. Our ϕ^4 theory has a Lagrangian $\mathcal{L} = \frac{1}{2}(\partial_\mu \phi)^2 - \frac{m^2}{2}\phi^2 - \frac{\lambda}{4!}\phi^4$ from which we conclude that $[\phi] = [\mathrm{Mass}] = [\mathrm{Length}]^{-1}$ and λ is dimensionless. The ϕ^4 theory is therefore renormalizable. On the other hand, looking at the mass term in Fermi's theory we can conclude that $[\psi] = [\mathrm{Mass}]^{\frac{3}{2}}$ and that, therefore, $[G] = [\mathrm{Mass}]^{-2}$. The theory is non-renormalizable.

The reason for the rule may be seen if we ask about the momentum dependence of scattering in Fermi's theory for small external momenta. The lowest order contribution will vary as $\mathcal{M}_1 \sim G$. The next order is G^2 and so that the units agree with $\mathcal{M}_1$ we require $\mathcal{M}_2 \sim G^2 p^2$ or $\sim G^2 \Lambda^2$, when we integrate. This is divergent when we send $\Lambda \to \infty$. Things only get worse as we go to higher orders. This increase in powers of momentum will occur for any theory with a coupling which has

[5]The meaning of the cross through the momentum operator p and the bar over ψ will be explained in Chapter 36.

negative mass dimension. The disaster clearly doesn't occur for theories with dimensionless couplings, where no extra momentum dependence is required. For super-renormalizable theories, we need momentum factors raised to negative powers, improving convergence at large momenta.

Finally we note that this argument also allows us to see that non-renormalizable interactions in Fermi's theory only cause trouble when Λ is larger than $\approx G^{-1}$. That is, if we're interested in physics at much lower energies than G^{-1} then we never explore momenta between G^{-1} and infinity and so the presence of the non-renormalizable term in the Lagrangian wouldn't cause us trouble. It may be that all of the theories of Nature that, at one time, we believed to be renormalizable (including quantum electrodynamics and the electroweak theory) actually contain non-renormalizable terms in their true Lagrangians. It's just that all of our experience is limited to energies where we don't notice them since their coupling constants G_i are so small that our experiments cannot reach the energies of G_i^{-1}. In this way of looking at the world, our theories of Nature are low-energy, **effective theories**, which will eventually break down at high enough energies. The true theory of Nature may not, therefore, be a quantum field theory at all...

Chapter summary

> • Quantum field theory contains divergences and the solution to the problem is renormalization. Counterterms can be added to the Lagrangian to remove the divergences and these counterterms have the effect of shifting parameters in the original Lagrangian.

Exercises

(32.1) Consider a theory described by the Lagrangian
$\mathcal{L} = \frac{1}{2}(\partial_\mu \phi)^2 - (m^2/2)\phi^2 - (g/3!)\phi^3$.

(a) Write a renormalized Lagrangian and determine the relationships between the bare and renormalized parameters.

(b) Working in $(5+1)$-dimensional spacetime, determine the superficial degree of divergence of the two diagrams shown in Fig. 32.5.

(c) Suggest Feynman rules for the counterterms in the renormalized theory.

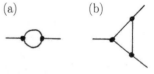

Fig. 32.5 Two divergent diagrams occurring in ϕ^3 theory.

Renormalization in action: propagators and Feynman diagrams

Now we are going to see yet another way in which the whole world is built out of recursion ... We are going to see that particles are - in a certain sense which can only be defined rigorously in relativistic quantum mechanics - nested inside each other in a way which can be described recursively, perhaps even by a sort of 'grammar'.

Douglas Hofstadter (1945–), *Gödel, Escher, Bach*

Some of the most useful tools in quantum field theory are propagators and Feynman diagrams. We've seen in the previous two chapters how renormalization affects single particles and Lagrangians. Now we'll examine how renormalization affects Feynman diagrams and propagators. This approach provides a particularly vivid picture of how a particle dresses itself in interactions. The main features of the propagator approach to renormalization is that the dressing-up process is implemented with two new Green's functions.[1] The first results from the single particle interacting with the vacuum as it propagates through spacetime. These interactions gives rise to a Green's function we call the **self-energy**. The self-energy describes the changes to the particle's mass caused by interactions. The second new Green's function comes from the virtual fluctuations screening the interactions between particles. The screening is described by the **vertex function** $\tilde{\Gamma}$. This screening changes the coupling constant of the theory.

[1] The beauty of describing renormalization in terms of Green's functions is that each of these new Green's functions may be written as a sum of Feynman diagrams.

The notation is unfortunate, but please try not to confuse the vertex function $\tilde{\Gamma}$ and the quasiparticle decay rate $\Gamma_{\boldsymbol{p}}$.

33.1 How interactions change the propagator in perturbation theory

We will discuss the usual example of a scalar field theory with ϕ^4 interactions, described by a Lagrangian

$$\mathcal{L} = \frac{1}{2}(\partial_\mu \phi)^2 - \frac{m^2}{2}\phi^2 - \frac{\lambda}{4!}\phi^4, \qquad (33.1)$$

where we've written the theory in terms of bare fields and parameters.

When there are no interactions present, the amplitude for a particle with momentum p to propagate between two points is given by $\tilde{G}_0(p) =$

(a)   (b)

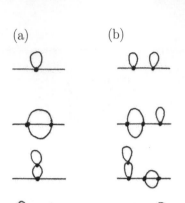

Fig. 33.1 (a) Some 1PI diagrams from ϕ^4 theory. They can't be turned into two meaningful diagrams by cutting one propagator line. (b) Diagrams that aren't 1PI. All can be turned into legitimate 1PI diagrams by cutting at some point along the horizontal line.

[2]Note that the 1PI self-energy does *not* include an overall energy-momentum conserving δ-function. We also *amputate* the external lines, that is, we don't include propagators for these lines.

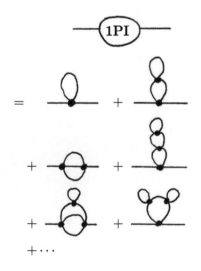

Fig. 33.2 Some contributions to the 1PI self-energy $-i\Sigma(p)$ in ϕ^4 theory.

$i/(p^2 - m^2 + i\epsilon)$. We saw previously (without using perturbation theory at all) that when you turn on interactions particles become dressed particles described by a propagator

$$\tilde{G}(p) = \frac{iZ_{\boldsymbol{p}}}{p^2 - m_{\mathrm{P}}^2 + i\Gamma_{\boldsymbol{p}}} + \left(\begin{array}{c} \text{Multiparticle} \\ \text{parts} \end{array} \right), \qquad (33.2)$$

where $Z_{\boldsymbol{p}}^{1/2} = \langle p_\lambda | \hat{\phi}^\dagger(0) | \Omega \rangle$ tells you how much particle we have in the momentum state $|p_\lambda\rangle$, $\Gamma_{\boldsymbol{p}}$ is the particle decay rate and m_{P} is the physical particle's exact mass. Notice that the pole is found at $p^2 = m_{\mathrm{P}}^2$ and the residue of the pole is $iZ_{\boldsymbol{p}}$. We now take a propagator for non-interacting particles with bare mass m and examine how it changes when we add an interaction to our theory as a perturbation. Since we know that the exact answer is given by eqn 33.2, we know how to extract the exact mass m_{P} of the particle from any propagator we calculate with perturbation theory: we simply look for the pole. We also know how to find the quasiparticle weight $Z_{\boldsymbol{p}}$: we find the residue at the pole. We therefore have a very important rule:

> The physical mass of a particle in an interacting theory is given by the position of the pole in the propagator. The quasiparticle weight is found from the residue at the pole.

In perturbation theory the propagator can be found by adding up Feynman diagrams. That is

$$\tilde{G}(p) = \sum \left(\begin{array}{c} \text{All connected Feynman diagrams} \\ \text{with two external legs} \end{array} \right). \qquad (33.3)$$

One helpful way to think of the propagator is to picture it as being made up of the two external legs and everything we can slot in between those two legs. What we can slot in is called the **self-energy**, and we describe that using what is called a **1-part irreducible diagram** (or **1PI diagram** for short), which is defined as *a connected diagram which can't be disconnected by cutting one internal propagator line*. Some examples of 1PI diagrams in ϕ^4 theory are shown in Fig. 33.1(a) and some examples of diagrams that aren't 1PI are shown in Fig. 33.1(b). These latter diagrams contain **cutlines**, and so fall apart into disconnected pieces if you attack one of the cutlines with a pair of scissors. In contrast, the connectedness of an 1PI diagram is immune to a single scissor attack on one of their internal lines. They essentially represent the smallest non-trivial Feynman diagram and a basic building block of more complex diagrams. They are, if you like, the guts of the self-energy.

The sum of 1PI diagrams with two external legs forms a Green's function known as the **1PI self-energy** $\tilde{\Sigma}(p)$, given by[2]

$$-i\tilde{\Sigma}(p) = \sum \left(\begin{array}{c} \text{All amputated 1PI diagrams} \\ \text{with two external lines} \end{array} \right). \qquad (33.4)$$

Some contributions to $-i\tilde{\Sigma}(p)$ in ϕ^4 theory are shown in Fig. 33.2. In the following example, we will explore how the propagator $\tilde{G}(p)$ can be expressed in terms of the 1PI self-energy $\tilde{\Sigma}(p)$.

Example 33.1

The definition of $\tilde{\Sigma}$ leads to one of the neatest tricks in perturbation theory. We can write the interacting propagator as a series of encounters with an interaction represented by $\tilde{\Sigma}$. This is the story of how the propagator dresses itself: it's the amplitude for propagating with no interactions, added to the amplitude for propagation being interrupted by one interaction with $\tilde{\Sigma}$, added to the amplitude for two interactions with $\tilde{\Sigma}$, added to... The series, shown in Fig. 33.3, is written as follows:

$$
\begin{aligned}
\tilde{G}(p) \;=\;& \frac{i}{p^2 - m^2} + \frac{i}{p^2 - m^2}\left[-i\tilde{\Sigma}(p)\right]\frac{i}{p^2 - m^2} \\
& + \frac{i}{p^2 - m^2}\left[-i\tilde{\Sigma}(p)\right]\frac{i}{p^2 - m^2}\left[-i\tilde{\Sigma}(p)\right]\frac{i}{p^2 - m^2} + \ldots \quad (33.5)
\end{aligned}
$$

We now treat this as a geometric series and sum:

$$
\begin{aligned}
\tilde{G}(p) \;=\;& \frac{i}{p^2 - m^2}\left\{ 1 + \left[-i\tilde{\Sigma}(p)\right]\frac{i}{p^2 - m^2} \right.\\
& \left. + \left[-i\tilde{\Sigma}(p)\right]\frac{i}{p^2 - m^2}\left[-i\tilde{\Sigma}(p)\right]\frac{i}{p^2 - m^2} + \ldots\right\} \\
=\;& \frac{i}{p^2 - m^2}\left\{ \frac{1}{1 - \frac{\tilde{\Sigma}(p)}{p^2-m^2}} \right\} \\
=\;& \frac{i}{p^2 - m^2 - \tilde{\Sigma}(p) + i\epsilon}, \quad (33.6)
\end{aligned}
$$

where, in the last line, we've reinserted the $i\epsilon$ from the free propagator. The sum can, more amusingly, be done in terms of diagrams as shown in Fig. 33.4. Equation 33.6 is another instance of Dyson's equation, which we met in Chapter 16. The form of the equation makes it look a lot like the free propagator, albeit with an extra self-energy term in the denominator.

To find the mass-energy m_P of the physical particles we just follow our renormalization rule: we look for the position of the pole. It is found using the equation

$$
p^2 - m^2 - \mathrm{Re}\left(\tilde{\Sigma}(p)\right) = 0, \quad (33.7)
$$

when $p^2 = m_\mathrm{P}^2$. That is to say that the physical mass of the particles is given by

$$
m_\mathrm{P}^2 = m^2 + \mathrm{Re}\left(\tilde{\Sigma}(p^2 = m_\mathrm{P}^2)\right), \quad (33.8)
$$

which tells us that the real part of the self-energy tells us the shift in energy caused by interactions. The latter equation is known as a renormalization condition.[3]

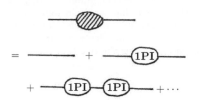

Fig. 33.3 Diagrammatic version of eqn 33.5.

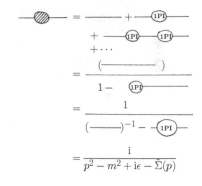

Fig. 33.4 Diagrammatic version of eqn 33.6

[3]It allows us, at the level of Feynman diagrams and perturbation theory, to renormalize a theory. By the same token, the decay rate of the physical particle is found from $\mathrm{Im}\left(\tilde{\Sigma}(p^2 = m_\mathrm{P}^2)\right)$.

33.2   The role of counterterms: renormalization conditions

But what of counterterms in this way of looking at things? Thus far we've summed all of the 1PI self-energy diagrams to make an object $\tilde{\Sigma}(p)$. In the absence of counterterms, the role of the self-energy $\tilde{\Sigma}(p)$ is to shift the mass from m to m_P. However, we have ignored the fact that many of the diagrams that contribute to $\tilde{\Sigma}(p)$ will be divergent, making the shift in mass seem dangerously infinite! One way to interpret this is to say that m must be infinite and requires an infinite shift to bring it down to the physical value m_P. However all this talk of infinity should

[4]In the theory of the electron gas, where we do not encounter divergences to the extent that we do in some other theories, the first option is the logical one. In quantum electrodynamics the second option is the one to choose.

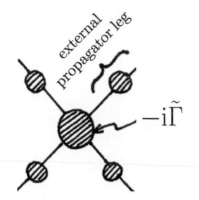

Fig. 33.5 In ϕ^4 theory the vertex function is contained in the Greens function shown here, corresponding to $\langle 0|T\hat{\phi}^4\hat{S}|0\rangle / \langle 0|\hat{S}|0\rangle$, or equivalently $\tilde{G}(p_1)\tilde{G}(p_2)\big[-\mathrm{i}\tilde{\Gamma}(p_1,p_2,p_3)\big]\tilde{G}(p_3)\tilde{G}(p_4)$. Diagrammatically, the vertex part $-\mathrm{i}\tilde{\Gamma}$, the central blob in the diagram above, may be extracted by amputating the external propagator legs.

[5]We could also define a 1PI vertex function, which is useful for many applications, but we won't need it here.

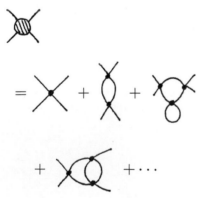

Fig. 33.6 Some contributions to the vertex function $-\mathrm{i}\tilde{\Gamma}$ in ϕ^4 theory. The vertex function is represented as a shaded blob with four external legs.

make us nervous and we would be better to remove divergences as we did in the previous chapter.

To remove the divergences in $\tilde{\Sigma}(p)$ we should really sum all of the self-energy diagrams along with counterterms in order to cancel infinities. Recall that the effect of including counterterms was to shift the mass of the theory from the unphysical fictional value of m to the measured and reliable value m_{P}. Using the 1PI self-energy $\tilde{\Sigma}(p)$, we won't need to shift the mass at all! That is, when we include counterterms we require the renormalization condition

$$\mathrm{Re}\left(\tilde{\Sigma}(p^2 = m_{\mathrm{P}}^2)\right) = 0. \qquad (33.9)$$

We see that there are two ways of doing business. We may use the renormalization condition eqn 33.8, which neglects counterterms with the result that the shifts from fictional to real masses are potentially infinite. On the other hand we can use the renormalization condition in eqn 33.9 which allows us to start with the mass set to the correct value and we ensure that it doesn't change as we do our perturbation theory.[4]

33.3   The vertex function

The next thing to do is to work out how to renormalize the coupling constant. Physically, this is changed because vacuum fluctuations screen the interactions between two particles. The interaction vertex in ϕ^4 theory has the Feynman rule

$$(\text{Vertex}) = (2\pi)^4 \delta^{(4)}(p_4 + p_3 - p_2 - p_1)(-\mathrm{i}\lambda), \qquad (33.10)$$

that is, we get a factor of $-\mathrm{i}\lambda$ for every vertex and the δ-function ensures that momentum is conserved. The vertex is plugged in between four external lines to tell us about two particle scattering.

We define the **vertex function** $\tilde{\Gamma}$ for ϕ^4 theory by[5]

$$-\mathrm{i}\tilde{\Gamma} = \sum \left(\begin{array}{c}\text{All connected four-point diagrams}\\ \text{with external legs amputated}\end{array}\right). \qquad (33.11)$$

This is shown diagrammatically in Fig. 33.5. (Note that $-\mathrm{i}\tilde{\Gamma}$ does not include an overall energy-momentum conserving δ-function.) Like the ϕ^4 interaction vertex, the vertex function may be plugged in between four propagators. It is designed to tell us how interactions (involving virtual particles) affect how real particles interact with each other. Some examples of contributions to $-\mathrm{i}\tilde{\Gamma}$ are shown in Fig. 33.6.

Example 33.2

For two-particle scattering $\tilde{\Gamma}$ can be obtained from the amplitude for the two-particle scattering process. To first order in the coupling, the four-point vertex function is given by a single interaction vertex: $-\mathrm{i}\tilde{\Gamma} = -\mathrm{i}\lambda$. To second order in the coupling we eliminated divergences using a counterterm (see Chapter 32) and saw that the two-particle scattering amplitude was given by

$$-\mathrm{i}\tilde{\Gamma}(p_1, p_2, p_3) = -\mathrm{i}\lambda - \mathrm{i}a\lambda^2 \left(\ln s + \ln t + \ln u\right), \qquad (33.12)$$

that is, it depends on the momenta of the incoming and outgoing particles [parametrized by $s = (p_1 + p_2)^2$, $t = (p_1 - p_3)^2$ and $u = (p_2 - p_3)^2$].

The last example demonstrates that the exact vertex function $\tilde{\Gamma}(p_1, p_2, p_3)$ is a function of the incoming and outgoing momenta. (Of course momentum conservation means we only need give three of the four momenta in the problem.) In order to define a renormalization condition for $\tilde{\Gamma}$ we need to choose the values of these momenta at which to fix $\tilde{\Gamma}$ to some value. However, unlike mass which is unambiguously defined, there is no unique definition of the coupling constant. We are therefore free to choose a renormalization condition at our convenience. We finally write the definition of the physical coupling constant:

$$-i\lambda_P = -i\tilde{\Gamma}(p_1, p_2, p_3) = \sum \begin{pmatrix} \text{Amputated, connected diagrams with} \\ \text{four external legs} \\ \text{and momenta } p_1, \, p_2 \text{ and } p_3 \end{pmatrix},$$
$$(33.13)$$

where the values of p_1, p_2 and p_3 that we choose are called the *renormalization point*. The sign is chosen so that to first order $\lambda_P = \lambda$.

In Chapter 41 we will use the analogous Green's functions $\tilde{\Sigma}$ and $\tilde{\Gamma}$ to assess how electromagnetic interactions change the properties of photons and electrons in quantum electrodynamics. In Chapter 43 we will examine the renormalization of the theory of metals in terms of the self-energy of electrons and the photons that mediate the Coulomb force that acts between them.

Example 33.3

We can use the definition of λ_P to calculate the amplitude for two-particle scattering up to second order in unrenormalized perturbation theory (i.e. where we haven't used counterterms to remove the cut-off). If the physical coupling constant λ_P is defined to be that measured at a renormalization point s_0, t_0, u_0 then, to second order, we have that

$$-i\lambda_P = -i\tilde{\Gamma}^{(2)}(s_0, t_0, u_0) = -i\lambda + ia\lambda^2 \left(3\ln\Lambda^2 - \ln s_0 - \ln t_0 - \ln u_0\right). \quad (33.14)$$

The two-particle scattering amplitude is given, for particles with momenta s, t and u, by

$$i\mathcal{M} = -i\lambda + ia\lambda^2 \left(3\ln\Lambda^2 - \ln s - \ln t - \ln u\right). \quad (33.15)$$

At the level of approximation to which we're working we may use eqn 33.14 to write

$$-i\lambda = -i\lambda_P - ia\lambda_P^2 \left(3\ln\Lambda^2 - \ln s_0 - \ln t_0 - \ln u_0\right) + O(\lambda^3), \quad (33.16)$$

which allows us to eliminate λ (and Λ) from eqn 33.15 to give the answer

$$i\mathcal{M} = -i\lambda_P - ia\lambda_P^2 \left[\ln\left(\frac{s}{s_0}\right) + \ln\left(\frac{t}{t_0}\right) + \ln\left(\frac{u}{u_0}\right)\right] + O(\lambda^3). \quad (33.17)$$

We see that we obtain the amplitude in terms of the physical coupling constant λ_P, where momenta are measured relative to the renormalization point. This argument may be repeated using renormalized perturbation theory (i.e. with the use of counterterms, which eliminate all $\ln\Lambda$ terms) to obtain an identical answer, but without the troubling business of carrying around the (potentially infinite) Λ-dependent term.

Chapter summary

- The self-energy $\tilde{\Sigma}$ can be related to the sum of all 1-part irreducible (1PI) diagrams with two external legs.
- In ϕ^4 theory, the vertex function $\tilde{\Gamma}$ can be related to the sum of all four-point diagrams with external legs amputated.

Exercises

(33.1) To find the quasiparticle weight we can expand the self-energy in a Taylor series. We'll do this about the interesting point $p^2 = m_P^2$.

$$\tilde{\Sigma}(p^2) \approx \tilde{\Sigma}(m_P^2) + (p^2 - m_P^2) \frac{d\tilde{\Sigma}(p^2)}{dp^2}\bigg|_{p^2=m_P^2} + \dots$$
(33.18)

(a) Using this expansion show that

$$\tilde{G}(p) \approx \frac{i}{(p^2 - m_P^2)\left[1 - \frac{d\tilde{\Sigma}(p^2)}{dp^2}\big|_{p^2=m_P^2}\right]}. \quad (33.19)$$

(b) Use this to show, to the order to which we're working, that

$$Z \approx 1 + \frac{d\tilde{\Sigma}(p^2)}{dp^2}\bigg|_{p^2=m_P^2}. \quad (33.20)$$

This gives us a scheme for calculating Z, the quasiparticle weight, in terms of the quasiparticle self-energy.

(33.2) (a) Draw diagrams showing the contributions to the self-energy $-i\tilde{\Sigma}(p)$ of psions for $\psi^\dagger\psi\phi$ theory up to fourth order in the interaction.

(b) Draw the contributions to the vertex function $-i\tilde{\Gamma}$ in $\psi^\dagger\psi\phi$ theory up to third order in the interaction.

Here the vertex function includes all amputated insertions with one psion line, one antipsion line and one phion line.

(33.3) Repeat the argument in Example 33.3 using renormalized perturbation theory (i.e. including the use of counterterms to remove the momentum cut-off Λ).

(33.4) Consider the action derived for a one-dimensional lattice in $(1+1)$-dimensional spacetime

$$S = \frac{1}{2}\sum_p \int \frac{d\omega}{(2\pi)} \tilde{\phi}_{-p}(-\omega)\left(\omega^2 - \omega_0^2 + \omega_0^2 \cos pa\right)\tilde{\phi}_p(\omega).$$
(33.21)

We will treat the final term in the bracket as a perturbation.

(a) Identify the free propagator and the interaction vertex.

(b) Write the full propagator as a sum to infinity involving the free propagator and interaction term.

(c) Show that, on carrying out the sum, the expected full propagator is recovered.

(33.5) Consider a model of an atom made up of two energy levels $(E_2 > E_1 > 0)$ occupied by a single fermion and described by a Hamiltonian

$$\hat{H} = E_1\hat{c}_1^\dagger\hat{c}_1 + E_2\hat{c}_2^\dagger\hat{c}_2 + V(\hat{c}_1^\dagger\hat{c}_2 + \hat{c}_2^\dagger\hat{c}_1). \quad (33.22)$$

(a) Treat the non-diagonal term as a perturbation and find the free propagators $\langle 0|T\hat{c}_1(t)\hat{c}_1^\dagger(t')|0\rangle$ and $\langle 0|T\hat{c}_2(t)\hat{c}_2^\dagger(t')|0\rangle$ for electrons in level $|1\rangle$ and $|2\rangle$ respectively.

(b) Now consider the interaction term. Find the Feynman rule for the processes that contribute to the self-energy and invent appropriate Feynman diagrams to describe an electron in the system.

Hint: the potential here is time independent, so the process to consider involves the electron changing states and immediately changing back.

(c) Find the self-energy of the electron in each state and show that the interaction causes a shift in the energy of the particle in level $|1\rangle$ by an amount

$$\Delta E_1 = -\frac{V^2}{E_2 - E_1}. \quad (33.23)$$

(d) Compare this result with (i) the exact solution and (ii) with second-order perturbation theory.

*(e) *Harder* In the time-dependent version of the problem the interaction part is given by

$$H' = \int \frac{d^3q}{(2\pi)^3} V(\omega_q)(\hat{a}_q e^{-i\omega_q t} + \hat{a}_q^\dagger e^{i\omega_q t})(\hat{c}_1^\dagger \hat{c}_2 + \hat{c}_2^\dagger \hat{c}_1).$$

$$(33.24)$$

By evaluating the self-energy of an electron in each of the levels, show that the decay rates of the levels are given by

$$2\Gamma_1 = 2\pi V(\omega_0)^2 \langle n(\omega_0) \rangle \int \frac{d^3q}{(2\pi)^3} \delta(q^0 - \omega_0),$$

$$2\Gamma_2 = 2\pi V(\omega_0)^2 [1 + \langle n(\omega_0) \rangle] \int \frac{d^3q}{(2\pi)^3} \delta(q^0 - \omega_0),$$

where $\omega_0 = E_2 - E_1$.

34 The renormalization group

[1]The renormalization group is a bit of a misnomer as it is not really a group. The name arises from the study of how a system behaves under rescaling transformations and such transformations do of course form a group. However, the 'blurring' that occurs when we rescale and then integrate up to a cut-off, thereby removing the fine structure (and this is the very essence of the renormalization group procedure) is not invertible (the fine details are lost and you can't put them back). Thus the transformations consisting of rescaling and integrating up to a cut-off do not form a mathematical group because the inverse transformation does not exist.

Kenneth Wilson (1936–2013) was awarded the Nobel Prize in Physics in 1982 in recognition of 'his theory for critical phenomena in connection with phase transitions', a theory that grew out of his work on the renormalization group.

Everything is in motion.
Everything flows.
Everything is vibrating.
William Hazlitt (1778–1830)

A question we keep asking in physics is 'when is a theory valid'. In the context of quantum field theory an answer is provided by the philosophy of the renormalization group.[1] This is a way of looking at the world pioneered by several physicists, but most notably Kenneth Wilson. It allows us to make sense of why a renormalized quantum field theory describes Nature.

34.1 The problem

Theories are described by Lagrangians. These are the sum of several terms, each of which is some combination of fields and their derivatives, multiplied by a so-called coupling constant. For ϕ^4 theory, we have the Lagrangian

$$\mathcal{L} = \frac{1}{2}(\partial_\mu \phi)^2 - \frac{m^2}{2}\phi^2 - \frac{\lambda}{4!}\phi^4, \qquad (34.1)$$

where m and λ are the coupling constants. Calculations that start with a Lagrangian like this soon encounter disaster in the form of divergent (i.e. infinite) integrals. That is, we take our Lagrangian, canonically quantize the free part, treat the interacting part as a perturbation to define an $\hat{S}$-operator using Dyson's equation and then start working out amplitudes in the form of S-matrix elements which lead to Feynman rules. If a Feynman diagram involves a loop, we often get a divergence. To cure a divergence, we can introduce a finite cut-off in momentum Λ in such a way that we integrate up to Λ rather than infinity. In the previous chapters our strategy was then to renormalize, that is to say that we remove the mention of Λ from things that we calculate. This involves expressing amplitudes in terms of physical coupling constants, which are the ones we measure at some agreed point in momentum space. Our definitions of coupling constants depend on this point in momentum space.

Kenneth Wilson had another strategy: rather than hiding the cut-off, we live with it. In Wilson's view, to properly define the theory, we need to admit that we do our integrals up to a maximum momentum Λ, which we are entirely free to choose. The arbitrary choice of Λ immediately raises the question: how do we know what value to choose? One answer is that we want the length scale Λ^{-1} to be far smaller than the length[2] scale p^{-1} of any of the physics in which we're interested. Consider, for example, a gas of atoms in a box. If we're interested in sound waves, which are oscillations in the number density of the constituents of the gas involving many millions of atoms, then the scale in which we're interested in p^{-1} is of the order of centimetres and we could take Λ^{-1} to be a few microns. If we're interested in the electron cloud of an atom in the gas, then we're interested in a length scale p^{-1} of the order of the size of an atom and we could take Λ^{-1} to be the size of an atomic nucleus. The size of Λ^{-1} that we choose therefore changes depending on the scale p^{-1} of the physics that we're analysing.

The renormalization group is a machine that tells us how the predictions of the theory change as we alter the scale of interest. It gives us an equation telling us how a coupling 'constant' changes with the length or momentum scale in which we're interested. To summarize this chapter:

> The goal of renormalization group analysis is to discover how the coupling constants change with the scale of interest.

[2]Remember that in units where $\hbar = c = 1$, mass m, energy E and momentum p have inverse units to length.

Example 34.1

In general, this goal may be achieved by examining how your Lagrangian behaves when the cut-off Λ is systematically varied. This approach will be outlined later. It is, however, sometimes possible to leapfrog the systematic method to get an equation for the coupling constant in terms of the scale of interest. We'll follow such an approach here first. Recall how we defined the renormalized coupling constant in ϕ^4 theory. We said that λ_P was the measured scattering amplitude when the particles had squared momenta s_0, t_0 and u_0, which is to say, the coupling constant is defined at a particular point in momentum space. How do we choose this point? A pragmatist would say that you choose it to be a convenient energy scale for the problem you're addressing. If we're dealing with electrons then we could set $s_0 = t_0 = u_0 = \mu^2$ where μ is close to the electron mass. If we're dealing with pions, we'd use μ close to the pion mass and so on. However, this 'idiot-proof' scheme could easily be challenged if an ingenious idiot started thinking about electron physics but set μ equal to the pion mass.

As we saw in Chapter 33 the physical coupling constant is given to second order by

$$-i\lambda_P(s_0, t_0, u_0) = -i\lambda + ia\lambda^2 \left(3\ln\Lambda^2 - \ln s_0 - \ln t_0 - \ln u_0\right). \quad (34.2)$$

This allowed us to write the two-particle scattering amplitude to second order:

$$i\mathcal{M} = -i\lambda_P(\mu) + ia[\lambda_P(\mu)]^2 \left[\ln\left(\frac{\mu^2}{s}\right) + \ln\left(\frac{\mu^2}{t}\right) + \ln\left(\frac{\mu^2}{u}\right)\right], \quad (34.3)$$

which is to say that it's given by the physical coupling constant added to a logarithmic correction. This correction is small since we chose μ^2 to be of order s_0, t_0 and u_0.

If, on the other hand, we idiotically choose a renormalization point μ'^2 to be much larger than s_0, t_0 and u_0, then the logarithmic term could be very large compared to the first term. Then we would have

$$i\mathcal{M} = -i\lambda_P(\mu') + ia[\lambda_P(\mu')]^2 \left[\ln\left(\frac{\mu'^2}{s}\right) + \ln\left(\frac{\mu'^2}{t}\right) + \ln\left(\frac{\mu'^2}{u}\right)\right], \quad (34.4)$$

where the second term is large compared to the first. Subtracting, we can find an expression for $\lambda_P(\mu')$ in terms of $\lambda_P(\mu)$:

$$\lambda_P(\mu') = \lambda_P(\mu) + 6a[\lambda_P(\mu)]^2 \ln\left(\frac{\mu'}{\mu}\right) + O(\lambda_P^3), \quad (34.5)$$

Remember that $[\text{Mass}] \equiv [\text{Length}]^{-1}$.

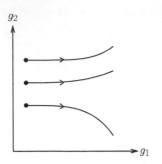

Fig. 34.1 The renormalization group flow in a configuration space defined by two coupling constants g_1 and g_2.

[3]Alternatively there is nothing stopping us examining the theory at short length scales in which case we could set $b < 1$.

[4]This is, of course, a different 'Gell-Mann–Low equation' to that described in Chapter 22. See M. Gell-Mann and F. Low, Phys. Rev. **84**, 350 (1951)

[5]Sometimes one can write

$$\beta(g) = \frac{\mathrm{d}g}{\mathrm{d}b}$$

instead. Since b divides Λ, the rescaling is more usefully modelled by looking at how g changes with $\ln b$. But, as we shall see in this chapter, different definitions can, and are, made.

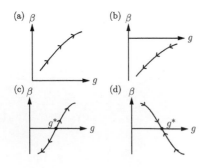

Fig. 34.2 Four examples of flows in a configuration space defined by a single coupling constant g, visualized by plotting $\beta(g) = \frac{\mathrm{d}g}{\mathrm{d}b}$. Arrows show the direction that the length scale $\ell \to \infty$.

which is expressed in differential form thus:

$$\mu \frac{\mathrm{d}}{\mathrm{d}\mu} \lambda_{\mathrm{P}}(\mu) = 6a\lambda_{\mathrm{P}}(\mu)^2 + O(\lambda_{\mathrm{P}}^3). \tag{34.6}$$

So in this case, we have a simple equation that tells us how the coupling λ changes with the momentum scale μ.

If we are interested in high energies, we can examine how λ behaves as μ, representing the scale of interest, gets larger and larger, that is, as $\mu \to \infty$. We call this the **ultraviolet behaviour**. Equally well, we might be interested in the physics at very large length scale, which would involve asking how λ changed as $\mu \to 0$. This is known as the **infrared behaviour**, since it corresponds to the limit of small energies.

34.2   Flows in parameter space

A theory is characterized by $\{g_i\}$, the set of coupling constants that measure the strength of the various interactions. (For ϕ^4 theory we have $g_2 = m$ and $g_4 = \lambda$.) Each coupling constant is a function of the cut-off Λ, so we can write the ith coupling constant as $g_i(\Lambda)$. We are going to examine what happens as we look at different length scales, which is achieved by changing the value of the cut-off. If we reduce the cut-off by a factor b, then we want to see how the coupling constants change and so we will look at the transformation

$$g_i(\Lambda) \to g_i(\Lambda/b). \tag{34.7}$$

If we are interested in the limit of large length scales (corresponding to what is measured in condensed matter physics) then we choose the direction of positive flow by fixing $b > 1$. This is known as **coarse graining**.[3] A theory can be described as a point in the multidimensional space formed by all of the $\{g_i\}$, i.e. by the point $(g_1, g_2, \cdots)$ in that space. The rescaling of the cut-off causes the point to travel through this multidimensional space along a **renormalization group trajectory**, and this is called a **renormalization group flow** (see Fig. 34.1). We will sometimes find fixed points in this space; these remain invariant under coarse graining and correspond to *scale-invariant* Lagrangians. We will see that such Lagrangians are very important in determining the physics described by the theory.

Example 34.2

A very simple example of renormalization group flow occurs for a theory when there is only a single coupling constant g. A graph with only one axis is hard to illustrate (much like the problem with the sound of one hand clapping) so to illustrate the flow it is conventional to plot the function

$$\beta(g) = \frac{\mathrm{d}g}{\mathrm{d}\ln b}, \tag{34.8}$$

which is known as a **Gell-Mann–Low equation**[4] or simply a **flow equation**.[5]

Note that:

- If β is always positive then g will be blasted out to infinity, as shown in Fig. 34.2(a). If β is always negative then g will be sucked backwards to $g = 0$ as shown in Fig. 34.2(b).

- If at position $g = g^*$ we have $\beta(g^*) = 0$ then g is stuck and will remain at g^* for evermore. We call g^* a **fixed point**.

- Even if a fixed point exists the system may never get stuck there. If the velocity function β looks like Fig. 34.2(c) then if $g > g^*$ the system will end up at ∞, while $g < g^*$ will be drawn back to $g = 0$: here g^* is known as a repulsive fixed point.

- If the function $\beta(g)$ looks like Fig. 34.2(d) then the system will be attracted towards g^* regardless of its initial condition. There are no prizes for guessing that under these circumstances g^* is known as an attractive fixed point.

Fig. 34.3 Removing the largest Fourier components results in a loss of detail.

34.3   The renormalization group method

Now that we know what it's useful for, we'd like to know how the renormalization group method works in more detail. We start with the functional integral, written in Euclidean space

$$Z(\Lambda) = \int_{\Lambda} \mathcal{D}\phi \, e^{-\int d^d x \mathcal{L}[\phi]}, \qquad (34.9)$$

where Λ is an instruction to integrate over configurations of $\phi(x)$ containing Fourier components up to Λ.

We will use the renormalization group method to examine the physics as we increase the length scale of interest. The method involves three steps. In **step I** we're going to *remove the largest Fourier components of momentum*. This is achieved by doing that part of the functional integral that involves the largest Fourier components in momentum space. The removal of these components is equivalent to losing one's glasses. A certain level of fine detail is lost and we have a simpler looking field (Fig. 34.3). We then compare the theory to what we started with. Since we now seem to be comparing apples (the original integral) and oranges (the partly integrated integral) we need to scale up the partly integrated integral so that it is defined over the same momentum range as the original. To do this involves (**step II**) relabelling the momenta and then we (**step III**) scale the size of the fields so that the Lagrangian resembles what we had before. The beauty of the procedure is that, in successful cases, it will not change the form of the Lagrangian, only the coupling constants. We then imagine repeating the procedure over and over again (losing successive pairs of spare glasses) and extract how the couplings change as we partially integrate the functional integral.

We now demonstrate the procedure, which is illustrated in Fig. 34.4. First we need to **set the scene**. We choose to work in momentum space, where we divide the fields $\tilde{\phi}(p)$ into those parts that vary slowly with p [called $\tilde{\phi}_s(p)$] and those parts [called $\tilde{\phi}_f(p)$] that vary quickly (see Fig. 34.5). We define the fast Fourier components as those in the momentum shell $\Lambda/b \leq |p| \leq \Lambda$, where $b > 1$ is a scaling factor, which

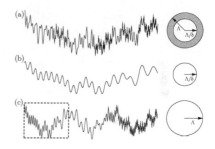

Fig. 34.4 The renormalization group process in momentum space. (a) The field $\phi(x)$ before the process. (b) The result of integrating out the largest momentum states (step I) is to remove the high spatial frequency components. (c) The field after rescaling (steps II and III). Note that the region in the dashed box is identical to the entire trace in (b).

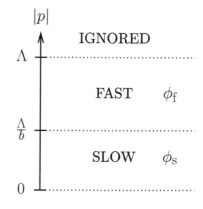

Fig. 34.5 Components in momentum space are divided up into slowly varying and fast varying, with any component with $|p| > \Lambda$ being ignored.

[6]We therefore have

$$\tilde{\phi}(p) = \tilde{\phi}_f(p) \quad \text{for } \Lambda/b \le p \le \Lambda,$$
$$\tilde{\phi}(p) = \tilde{\phi}_s(p) \quad \text{for } 0 \le p \le \Lambda/b.$$

will be central to our analysis. The slow Fourier components of the fields are then those with momenta[6] $0 \le p \le \Lambda/b$. The functional integral becomes

$$Z(\Lambda) = \int_{\Lambda/b} \mathcal{D}\phi_s \, e^{-\int d^d x \mathcal{L}[\phi_s]} \int_{\Lambda/b}^{\Lambda} \mathcal{D}\phi_f \, e^{-\int d^d x \mathcal{L}_I[\phi_s, \phi_f]}. \quad (34.10)$$

We're now ready to start the renormalization group procedure.

Step I: This is the difficult step. We need to integrate over the quickly varying fields. This is usually impossible, so we need an approximation scheme involving perturbation theory. Assuming we can do this, we'll obtain an action in which the remaining slow fields and coupling constants are different to what they were before. If we can (somehow) do the integral over the fast varying part ϕ_f we define the answer to be

$$\int_{\Lambda/b}^{\Lambda} \mathcal{D}\phi_f \, e^{-\int d^d x \mathcal{L}_I[\phi_s, \phi_f]} = e^{-\int d^d x \, \delta\mathcal{L}[\phi_s]}, \quad (34.11)$$

leading to the result for our integral:

$$Z(\Lambda) = \int_{\Lambda/b} \mathcal{D}\phi_s \, e^{-\int d^d x \, (\mathcal{L}[\phi_s] + \delta\mathcal{L}[\phi_s])}, \quad (34.12)$$

which only features the slow fields.

Step II: We relabel the momenta. We do this by expressing our theory in terms of scaled momenta $p' = pb$. Since $b > 1$ this has the effect of stretching the momentum space out over the range we had originally, as shown in Fig. 34.6. This is seen by examining the limit of the integral, which is $p'/b = \Lambda/b$ or $p' = \Lambda$.

Step III: Finally we scale the fields. We select the term that we imagine will be most important in determining the scaling behaviour and require it to be *invariant* with the scaling of step II. We use this to rescale the field $\tilde{\phi}(p) = \tilde{\phi}(p'/b) = b^{d-d_\phi} \tilde{\phi}'(p')$, where d_ϕ is chosen to leave the important term invariant.[7]

Now that the procedure has been carried out once, we repeat it an infinite number of times. In fact, we only need to carry out the procedure once and then infer the consequence of repeating it. That is, we do the approximate integral which tells us how all of the coupling constants change with scale. We then *imagine* a continuous flow of the coupling constant caused by repeating the renormalization group method again and again.

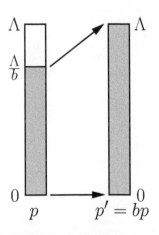

Fig. 34.6 Rescaling the momentum.

[7]The subject is notorious for its arcane terminology, and d_ϕ is known as the **anomalous dimension**.

[8]Sending ℓ to infinity corresponds to looking at the infrared behaviour of the couplings (since large length is the same as small energy). We could send $\ell \to 0$, if we want the ultraviolet behaviour.

By removing the fastest varying fields we are looking at the physics at lower momentum, or equivalently at longer length scale. We can therefore ask how the physics changes as we send the length scale of interest ℓ to infinity. One way to do this is to set $b = e^\ell$ and search for flow equations of the form

$$\frac{dg_i}{d\ell} = \beta_i(g_1, g_2, ..., g_N). \quad (34.13)$$

This is another form of the Gell-Mann–Low equation.[8]

The renormalization group is most famously applied to phase transitions, an example we will defer until the next chapter. Here we focus on three examples which illustrate other features of the renormalization group.

34.4   Application 1: asymptotic freedom

Perturbation theory is carried out in terms of series expansions in coupling constants. The renormalization group teaches us that coupling constants change depending on the region in energy-momentum space in which we're working. If we work in a region of momentum space where the coupling constants are small, then perturbation theory will work well. If, however, we work in a region where the coupling constants are large then we can expect perturbation theory to fail to yield sensible results. The coupling 'constant' in quantum electrodynamics is the electronic charge. As we'll see in Chapter 41, to first order, the β function describing the evolution of e is written[9] $\mu\frac{\mathrm{d}|e|}{\mathrm{d}\mu} = |e|^3/12\pi^2$, where μ is the energy scale of interest. The important thing to note is that the right-hand side is positive. This means that as we flow to large energy-momentum transfer (by increasing the mass-energy scale μ), the coupling constant gets larger and larger. Eventually, when the momentum involved in the processes we're calculating implies a very large mass-energy scale, perturbation theory ceases to be sufficient to capture the physics. Of course, running this process backwards (corresponding to examining processes involving small momentum transfer and large length scale) the coupling gets smaller and smaller and perturbation theory captures the physics better and better. This confirms the expected result that perturbation theory works best for processes involving small momenta and energy and gets worse as the energy gets larger. It also confirms the physical picture that electric charge is largest close to a charged source (that is at small length scales), and dies away at large distances.[10]

A surprising result was discovered for a class of theories, which includes the famous non-abelian Yang–Mills theory that will be examined in Chapter 46. In this case, β is negative! Therefore, if we flow to large momentum transfer, the coupling constants become small and perturbation theory at low orders is a very good approximation. However, now if we reverse the process, allowing the momentum transfer to become small, then g gets very large and perturbation theory no longer gives good results. Theories of this sort are said to have **asymptotic freedom**:[11] they are most strongly interacting at low energy scales, but at large scales they are weakly interacting (and therefore act like nearly free theories).[12]

The importance of asymptotic freedom is that it allows us to explain (among other things) the physics of the strong interaction. Under the normal (i.e. low energy) circumstances of much of the Universe, the quarks in a proton are very tightly bound. However, it was found exper-

[9]Notice again that slightly different definitions of the flow are used depending on context. This is nothing to worry about since the idea is always the same: we want to know how the coupling changes with scale.

[10]This occurs because of screening by virtual charged particle–antiparticle pairs. See Section 41.1 and Fig. 41.6.

[11]Frank Wilczek now regrets coining this term, which was first suggested by Sidney Coleman, and wishes he had used 'charge without charge' instead. The idea is that closer and closer to a quark the effective colour charge 'asymptotically' approaches zero. Having zero colour charge means complete 'freedom' from its interaction. Playing the argument in reverse, at large distances away from the intrinsically weak colour charge of the quark, the cloud of virtual particles give rise to **anti-screening** and this greatly enhances the effective colour charge and results in the strong interaction being strong.

[12]One might think of masses connected by springs as showing a sort of asymptotic freedom. At small length scales they interact weakly; as the length scale is increased and the springs become stretched they interact more strongly. This is why you cannot isolate quarks and gluons as free particles, but only find them in bound states inside mesons (quark, antiquark) or baryons (three quarks).

imentally in deep inelastic scattering experiments that quarks impacted by very highly energetic electrons behave as if they're free (or, at least that the coupling between quarks is very small). This is entirely the circumstance described by theories that show asymptotic freedom. The strong interaction is described by a theory known as quantum chromodynamics, which is a non-abelian gauge theory with an internal $SU(3)$ symmetry which gives rise to a conserved charge known as colour.[13] The theory shows asymptotic freedom. We introduce non-abelian gauge theories in Chapter 46.

[13]The virtual gluon excitations in the vacuum carry colour charge and serve to reinforce the field, rather than screen it. This effect is sometimes called antiscreening (see above) but is essentially another way of describing asymptotic freedom.

34.5   Application 2: Anderson localization

Crystalline metals conduct electricity, but what happens when you introduce more and more disorder? Philip Anderson realized in 1958 that beyond a critical amount of disorder, the diffusive motion that results from impurity scattering would completely stop and the electrons would become localized, residing in bound states rather than in the delocalized band states of the pure metal. This process is called **Anderson localization**. Thus we can imagine making a metal more and more impure by implanting specks of dirt in it and at some point we expect to cross from metallic behaviour to insulating behaviour, as Anderson localization sets in. This tipping point in behaviour is known as the *mobility edge*.

[14]See Exercise 34.1.

Let's think about the scaling properties of conductivity, since that will be important in a renormalization group treatment. A three-dimensional metal of volume L^3 conducts electrons. It has a small resistance R and a large conductance $G = 1/R$. How does the conductance vary with the length of a piece of metal? As examined in the exercises[14] $G(L) = \sigma L$, where σ is the conductivity. In fact, for a d-dimensional solid, $G(L) \propto L^{d-2}$. In contrast, an insulator has a small conductance, given by $G(L) \propto \mathrm{e}^{-L/\xi}$, where ξ is a length scale determined by microscopic considerations. What we really want to know is that if we start with a material with some fixed amount of dirt in it, do we expect it to be a conductor or an insulator? The renormalization group provides the answer! For this problem, David Thouless showed that the coupling constant of interest turns out to be the conductance G. We define our β function as[15]

David Thouless (1934–)

[15]This is another example of how the β function is defined in various ways to suit the problem in hand.

$$\beta(g) = \frac{\mathrm{d}\ln g}{\mathrm{d}\ln L} = \frac{L}{g}\frac{\mathrm{d}g}{\mathrm{d}L}, \qquad (34.14)$$

where the dimensionless conductance is defined as $g = \hbar G(L)/e^2$. Differentiating our expressions, we obtain a prediction for the scaling in the metallic and insulating regions for a d-dimensional solid:

$$\beta \approx \begin{cases} (d-2) & \text{metallic (large } g) \\ \ln g & \text{insulating (small } g). \end{cases} \qquad (34.15)$$

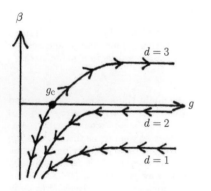

Fig. 34.7 The renormalization group flows for Anderson localization. Arrows show the flow for long length scale.

Patching these limits together we get the flow shown in Fig. 34.7. Notice the fixed point at g_{c}, which will turn out to represent the mobility edge.

The flow diagram predicts the behaviour of the system as we look at larger and larger length scales. This is useful as the limit of large length scales is what our measurements of electrical resistance probe.

A material containing a particular amount of dirt will be an insulator or metal, depending on whether the flows takes it to the left (small g) or right (larger g) respectively.[16] We see that, for $d = 3$, we have the behaviour we predicted in our naive picture above: if the material has a conductance above that at the mobility edge (the fixed point in the flow known as g_c) the material is destined to be a metal. If it has so much dirt that $g < g_c$ it will be an insulator. The mobility edge is represented by a *repulsive* fixed point. The surprise comes when we look in $d = 2$ and $d = 1$ dimensions. In these cases there is no fixed point and hence, no mobility edge! If the system has any dirt in it at all it *must* flow to $g = 0$ and will be an insulator. This is a profound statement!

[16]Note that a flow implying $g \to \infty$ doesn't mean that we will measure this behaviour in a real system. We would expect some other piece of physics, not included in our Lagrangian will cut off such flows eventually.

34.6   Application 3: the Kosterlitz–Thouless transition

For any theory of a magnet in two spatial dimensions with a continuous symmetry, the Coleman–Mermin–Wagner theorem (of Chapter 26) predicts that there can be no symmetry breaking phase transition to an ordered magnetic state. However, a phase transition of a different sort is predicted by a remarkable renormalization group analysis. This is a *topological* phase transition, which separates two phases of matter containing rather different topological objects.

We work in two spatial dimensions, and examine vortices, which are the topological objects that can exist in (2+1)-dimensional spacetime. As discussed in Chapter 29, lone vortices are unstable[17] at $T = 0$. An example of such a vortex is shown in Fig. 34.8. However, it's a different story at high temperatures. This is because, rather than minimizing the energy to find the thermodynamic equilibrium state of the system, we must minimize the free energy $F = U - TS$, and so at high temperature there is a need to maximize entropy. The presence of a vortex gives the system some entropy, since its centre may be placed at any one of a large number of lattice sites and this provides a way of minimizing the free energy. The entropic term in F is weighted by the temperature, making the entropic advantage of vortices dominant at high T. We are led to the conclusion that thermodynamics favours the proliferation of vortices at high temperature.

The question is, what happens at low temperatures? This is the question that Kosterlitz, Thouless and Berezinskii addressed. Typically of condensed matter applications of the renormalization group, we ask what happens to the system of vortices as we look at successively longer length scales. The renormalization group analysis of the two-dimensional vortex problem therefore involves asking what happens to the important coupling constants as we send the length scale of interest to infinity.

J. Michael Kosterlitz.   The transition was simultaneously discovered by Vadim L. Berezinskii in the Soviet Union.

[17]To recap: this is because the spin field is swirly at spatial infinity and this costs the system an infinite amount of energy for a single vortex.

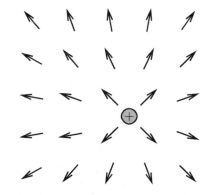

Fig. 34.8 A vortex has infinite energy, due to the swirliness at spatial infinity.

[19]The reason for writing the action this way is to prevent the integral from diverging at the centre of the vortex, where the field looks to wind an infinite amount.

[20]See Exercise 34.2.

[21]More generally J arises from the **rigidity** of the system. Rigidity is a general feature of ordered systems, reflecting the energy cost associated with deforming the order.

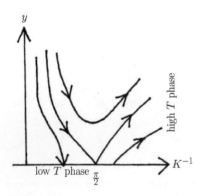

Fig. 34.9 Flow diagram for the Kosterlitz–Thouless transition. The horizontal axis is $K^{-1} = T/J$.

[22]See Altland and Simons, Section 8.6 for details of the calculation.

Example 34.3

We model this system with a complex scalar field given in polar coordinates $\phi(\boldsymbol{x}) = e^{i\theta(\boldsymbol{x})}$. At a point in space $\boldsymbol{x}$ we see the field as an arrow of unit length pointing along a direction given by an angle $\theta(\boldsymbol{x})$. Such a picture is known as the two-dimensional XY model[18] since the arrows are constrained to the two-dimensional plane, where their direction may vary continuously. The Euclidean action of a gas of vortices is given by[19]

$$S = S^{\text{core}}(a) + \frac{K}{2} \int_a^L \mathrm{d}^2 x \, |\boldsymbol{\nabla}\phi(\boldsymbol{x})|^2, \qquad (34.16)$$

where the first term represents the energy of the core region of the vortex, which has size a and the second term represents the energy cost of having a spin field that varies with position outside the core in a system with sides of length L. The constant $K = J/T$ where J is known as the stiffness of the spin system and T is the temperature. We may plug in our ansatz for the field at a vortex, given by $\phi(\boldsymbol{x}) = e^{i\theta(\boldsymbol{x})}$, where $\theta(\boldsymbol{x}) + \varphi = \tan^{-1}(x^2/x^1)$ for $|\boldsymbol{x}| \to \infty$, just as we had in Chapter 29. We integrate to find an action[20]

$$S = S^{\text{core}}(a) + \pi K \ln(L/a). \qquad (34.17)$$

What are the coupling constants that we change as the system flows to larger length scales?

- The *spin stiffness* J is one of the coupling constants and tells us how much energy cost we pay for having the spins describe some non-uniform configuration. This is the parameter that causes a single vortex to cost infinite energy.[21] At high temperature, where spins are presumably disordered, we do not expect any contribution of the spin stiffness to the action determining the physics of the system over long distances. At low temperature J will be decisive in determining the configuration of spins. We thus need to consider the ratio $K = J/T$ which reflects the balance between J and T and therefore we use K (actually K^{-1}) as one of our variables.

- The other coupling reflects the energy cost of a vortex core divided by the temperature. It is known as the *fugacity* and is given by $y = e^{-S^{\text{core}}(a)}$ (note that the Euclidean action is effectively a ratio of energy and temperature in statistical mechanics, see Chapter 25). The fugacity effectively tells us how the system feels the presence of the vortices. A small fugacity tells us that we have a configuration with small, energetic and compact vortex cores [with large $S^{\text{core}}(a)$] and that the spins don't swirl much at infinity; a large fugacity tells us that we have less action in the cores and a significant energy cost from swirling fields at large distances owing to the existence of free vortices. The energy cost here is, as usual for statistical mechanics problems, measured with respect to the energy scale set by $k_{\text{B}}T$.

The behaviour of these two couplings predicted by the renormalization group[22] is shown in Fig. 34.9. Following the arrows (which show the flow to large length scales) we see that most of the trajectories point to $K^{-1}, y \to \infty$. This is the expected behaviour in the high-temperature phase. The fact that $y \to \infty$ tells us that we have free vortices present making a non-uniform texture of spins at infinity. In this phase entropy wins out and J is ultimately an unimportant consideration in determining the thermodynamics, and so $K = J/T$ flows to zero.

The high-temperature phase, however, is not the only fate of the system. We see that in the bottom-left of the plot there are flows that lead to fixed points on the K^{-1} axis, with $y = 0$. This is the low-temperature phase of the system, where stable configurations of vortices are possible, even at $T = 0$. The fact that y flows to zero means that the vortex configurations must be quite compact, so that the cores look less and less significant at larger and larger length scales and that the spin texture at infinity must be uniform. We also have that $K \neq 0$, meaning that the system has rigidity. Since rigidity is one of the inevitable side-effects of *magnetic order*, this tells us that the low-temperature phase involves some (possibly exotic) type of ordering.

How can this be? The key to understanding the low-temperature phase is the fact that it involves *bound states of vortices and antivortices*. As shown in Fig. 34.10, an antivortex has spins that wind in the opposite direction to a vortex. A picture of the low-temperature phase is shown in Fig. 34.11 where we see that bound pairs of vortices and antivortices are the stable field configurations. The stability of these pairs at low temperature is evident if we look at the fields at a large distance from a vortex–antivortex pair as shown in Fig. 34.12. The clockwise swirl of the vortex at infinity is cancelled by the anticlockwise swirl of the antivortex, making $y \to 0$. This means that in a system with nonzero J, the pair costs a finite amount of energy.

We can even come up with an estimate of the transition temperature between the low and high temperature phases. Consider the stability of the high-temperature phase, containing a single vortex. The energy of the vortex is given (from eqn 34.17) by $\pi J \ln(L/a)$. The size of the entire system is L and the size of a vortex core is a. We may therefore fit a vortex into the system in L^2/a^2 ways, giving an entropy $k_B \ln \Omega = k_B \ln(L/a)^2$. The free energy is then $F = (\pi J - 2k_B T) \ln(L/a)$. The only way of preventing this free energy from becoming very large and positive is for $2k_B T > \pi J$. Below $T \approx \pi J/2k_B$ the system is unstable to the formation of bound vortex–antivortex pairs. Of course, this is identical to the fixed point in the renormalization-group flow diagram of Fig. 34.9 at $K^{-1} = \frac{\pi}{2}$ (apart, of course, from the factor k_B that was set to one in our hair-shirt quantum field treatment). The transition is often also called the *vortex unbinding transition*, describing the breakup of the bound states on warming through the transition region.

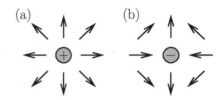

Fig. 34.10 (a) A vortex. (b) An antivortex.

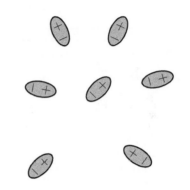

Fig. 34.11 Schematic of the low-temperature phase, viewed at long distance. The +s represent vortices, the −s antivortices. They form a gas of dipole-like pairs.

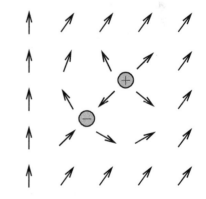

Fig. 34.12 A vortex dipole made from a vortex and antivortex in the same region of space. Notice that the fields may now be made uniform at infinity.

Chapter summary

- The renormalization group is a method for studying how coupling constants change with rescaling.
- This method provides insight into many physical phenomena, including asymptotic freedom, Anderson localization, and the Kosterlitz–Thouless transition.

Exercises

(34.1) *We will show using dimensional analysis that the conductance of a d-dimensional solid varies as* $G(L) \propto L^{(d-2)}$.
(a) For $d = 3$ consider a cubic sample with sides of length L and show $G = \sigma L$, where σ is the conductivity.
(b) For $d = 2$ consider a flat plate of thickness a and other sides of length L and show $G = \sigma a$.
(c) For $d = 1$ show $G = \sigma a^2 / L$.

(34.2) By considering the ansatz in the text verify that the action of a vortex is given by

$$S = S^{\text{core}}(a) + \pi K \ln(L/a). \qquad (34.18)$$

If instead the phase winds n times round a vortex, show that

$$S = S^{\text{core}}(a) + \pi K n^2 \ln(L/a). \qquad (34.19)$$

Ferromagnetism: a renormalization group tutorial

In this chapter we are going to study how the coupling constants in a Euclidean ϕ^4 theory vary with increasing length scale. This will tell us about the physics of phase transitions. To make things concrete, we will discuss the magnetic transition separating a paramagnetic phase and a ferromagnetic phase.[1] Perhaps the most wonderful feature of a phase transition is **universality**. This is the notion that some of the key properties of a phase transition depend only on the dimensionality of a system and dimensionality of its order parameter field. The details of the microscopic interactions between the constituent parts, whether the transition is ferromagnetic or superconducting, or whether it involves polymers or the early Universe, are all irrelevant. Thus focussing on the ferromagnetic transition will nevertheless give results that are applicable to any system described by the same Lagrangian.

[1]We have already discussed the mean-field treatment of this in Chapter 26. In this chapter we will include the possibility of fluctuations in the fields.

35.1 Background: critical phenomena and scaling

Magnetic phase transitions may be observed with several experimental techniques. Some thermodynamic properties of a magnet in which we might be interested include the magnetization M (which is the order parameter of the ferromagnet system); the magnetic susceptibility $\chi = \lim_{B \to 0} \frac{\mu_0 M}{B}$, which tells us how much magnetization we can achieve through applying a field B; and the heat capacity C, which tells us how the energy of the system varies with temperature. To these thermodynamic properties we can add the **correlation length** ξ, which tells us how well correlated the *fluctuations* are in space (Fig. 35.1).

The correlation length is related to the correlation functions, which are the Green's functions of Euclidean space. As described in Chapter 25, a slight modification of the usual correlation function $G(x, y) = \langle \hat{\phi}(x)\hat{\phi}(y) \rangle_t$ is useful when studying phase transitions. This is the **connected correlation function** $G_c(x, y)$, given by

$$G_c(x, y) = \langle \hat{\phi}(x)\hat{\phi}(y) \rangle_t - \langle \hat{\phi}(x) \rangle_t \langle \hat{\phi}(y) \rangle_t, \qquad (35.1)$$

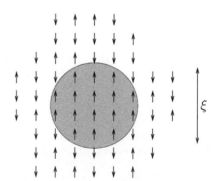

Fig. 35.1 The correlation length in a paramagnet showing a ferromagnetic fluctuation of size ξ.

and differs from $G(x, y)$ by the subtraction of a product of thermally averaged fields $\langle \hat{\phi}(x) \rangle_t \langle \hat{\phi}(y) \rangle_t$. At temperatures above the transition temperature T_c we have $\langle \hat{\phi}(x) \rangle_t = 0$ and the definition is identical to the usual form of the correlation function. At temperatures below T_c, subtracting off the mean-fields allows us to just see the fluctuations away from average behaviour. For example, in a magnet above T_c, despite the spins being disordered they may save energy by pointing in the same direction, and will tend to do so within the correlation length ξ. The correlation between spins at position x and position y decreases with increasing $|x - y|$ because thermal fluctuations have a tendency to randomize the spins. Thus we find that[2]

$$G_c(x, y) \sim e^{-|x-y|/\xi}, \qquad (35.2)$$

for $|x - y| \gg \xi$. The subtraction of the average fields in our definition of G_c means that this is also the behaviour below T_c and represents the fluctuations away from perfect order.

At temperatures close to the ordering temperature T_c it is found experimentally that the physical properties of a magnet described above follow *power law* behaviour[3] as a function of temperature, applied magnetic field or distance. This motivates the definition of **critical exponents** to describe the phase transition. Defining the reduced temperature as $t = (T - T_c)/T_c$, the critical exponents $\alpha, \beta, \gamma, \delta, \nu, \xi$ and η are related to physical properties as follows:

- Heat capacity: $C \sim |t|^{-\alpha}$,
- Magnetization: $M \sim (-t)^{\beta}$, for $B \to 0$, $T < T_c$,
- Magnetic susceptibility: $\chi \sim |t|^{-\gamma}$,
- Field dependence of χ at $T = T_c$: $\chi \sim |B|^{1/\delta}$,
- Correlation length: $\xi \sim |t|^{-\nu}$,
- The correlation function $G(r)$ behaves like

$$G(r) \sim \begin{cases} \frac{1}{|r|^{d-2+\eta}} & |r| \ll \xi \\ e^{-\frac{|r|}{\xi}} & |r| \gg \xi, \end{cases} \qquad (35.3)$$

where r is distance and d is the dimensionality of the system.

A renormalization group analysis of the ferromagnet relies on varying the length scale at which we examine the physics. Well before the ascent of the renormalization group, Benjamin Widom suggested that the critical behaviour that is observed experimentally may be explained if the free energy[4] of the system has a particular form under such a scaling of the length scale. The idea is that when rescaling lengths by a factor b (so that a length $L \to bL$), we should expect

$$f(t, h) = b^{-d} f(t', h'), \qquad (35.4)$$

where the factor b^{-d} is because of the L^{-d} factor in f. Widom assumed that $t' = b^{y_t} t$ and $h' = b^{y_h} h$, where y_h and y_t are exponents yet to be

[2]Closer to the disturbance the correlations are found to decay more slowly, as described below.

[3]Unlike, for example, exponential functions such as $e^{-x/\xi}$, power laws are free from any dependence on a length scale. This scale-free property is a hallmark of critical behaviour, where the characteristic fluctuations occur over all length scales.

[4]In these calculations, one studies the **reduced free energy** f (dividing F by both volume $V = L^d$ and temperature T) so that $f = F/(L^d T)$. We also define reduced temperature $t = |T - T_c|/T_c$ and reduced field $h = B/T$.

determined. Thus the Widom hypothesis is

$$f(t, h) = b^{-d} f(b^{y_t} t, b^{y_h} h). \qquad (35.5)$$

We may use the fact that b is an arbitrary parameter and choose a value of b such that $b^{y_t} t = 1$. This reduces the number of free parameters by one, and we obtain

$$f(t, h) = t^{d/y_t} f(1, h/t^{y_h/y_t}), \qquad (35.6)$$

which is Widom's hypothesis, written in reduced variables. Using this result, and also expressions relating the free energy to physical properties such as magnetization, heat capacity, magnetic susceptibility, etc., one can show that the critical exponents can be expressed in terms of the y's as follows:[5]

$$\alpha = 2 - \frac{d}{y_t}, \quad \beta = \frac{d - y_h}{y_t}, \quad \gamma = \frac{2y_h - d}{y_t},$$
$$\delta = \frac{y_h}{d - y_h}, \quad \nu = \frac{1}{y_t}, \quad \eta = 2 + d - 2y_h.$$

[5]See Exercise 35.1.

Our task is now to use the renormalization group to find y_t and y_h. We find these quantities by noting how the quantities t and h flow on renormalization. Since we have the predictions $t \to b^{y_t} t$ and $h \to b^{y_h} h$, we should be able to simply read off the values of y_t and y_h. If we can do this then we have access to all of the critical exponents, which we can compare with experiment.

35.2 The ferromagnetic transition and critical phenomena

Let's feed a theory of the ferromagnet through the renormalization group machine and extract some physics. A continuum field theory of the ferromagnet is given by the **Landau–Ginzburg** model. This model is equivalent to (time-independent) d-dimensional Euclidean ϕ^4 theory, which has an action[6]

$$S_E = \int d^d x \, \mathcal{L}_E = \int d^d x \left[\frac{1}{2} (\nabla \phi)^2 + \frac{m^2}{2} \phi^2 + \frac{\lambda}{4!} \phi^4 \right]. \qquad (35.7)$$

Vitaly Ginzburg (1916–2009)

[6]We can choose d to suit whatever dimensionality of system we need. Ordinarily we imagine materials having $d = 3$, but it is possible in condensed matter physics to find materials which effectively have $d = 2$ and $d = 1$.

Example 35.1

We may see what the Landau–Ginzburg model predicts in *mean-field theory* where we take ϕ to be uniform across the whole system.[7] We then have a free energy

$$F = \frac{m^2}{2} \Phi^2 + \frac{\lambda}{4!} \Phi^4, \qquad (35.8)$$

where $\Phi = \int d^d x \, \phi(x)$. If $\lambda = 0$ then the minimum free energy occurs at $\Phi = 0$ and we have a paramagnet. If $\lambda > 0$ and $m^2 = a(T - T_c) < 0$ then we have minima at $\Phi = [6a(T_c - T)/\lambda]^{\frac{1}{2}} \neq 0$, giving a ferromagnet. However, mean-field theory is blind to fluctuations. We will show with the more sophisticated approach of the renormalization group that even if $m^2 < 0$ and $\lambda > 0$ a phase transition is not guaranteed due to the influence of these wobbles in the field.

[7]Note that this approach recreates the Landau theory of phase transitions used in Chapter 26.

As a warm-up exercise we'll set the coupling $\lambda = 0$ and carry out the renormalization group procedure for the free theory.[8]

Example 35.2

We choose to work in momentum space, so we write the action in terms of fields in momentum space in the usual manner:

$$S_{\mathrm{E}} = \frac{1}{2} \int \frac{\mathrm{d}^d p}{(2\pi)^d} \, \tilde{\phi}(-p) \left(p^2 + m^2\right) \tilde{\phi}(p). \qquad (35.9)$$

To set the scene, we split the fields up into slow and fast via $\phi = \phi_{\mathrm{f}} + \phi_{\mathrm{s}}$ where

$$\begin{aligned} \tilde{\phi}(p) &= \tilde{\phi}_{\mathrm{f}}(p) \quad \text{for } \Lambda/b \leq p \leq \Lambda, \\ \tilde{\phi}(p) &= \tilde{\phi}_{\mathrm{s}}(p) \quad \text{for } 0 \leq p \leq \Lambda/b. \end{aligned} \qquad (35.10)$$

The action in eqn 35.9 is diagonal in p, which means that slow and fast fields will not be mixed up in the calculation of S. Let's go through the steps with this free theory.

Step I involves integrating over fast momenta. This just results in a constant contribution to the action, which we are free to ignore. We therefore have the integral

$$Z = \int_{\Lambda/b} \mathcal{D}\tilde{\phi}_{\mathrm{s}} \, \mathrm{e}^{-S_{\mathrm{E}}[\phi_s]} = \int_{\Lambda/b} \mathcal{D}\tilde{\phi}_{\mathrm{s}} \, \mathrm{e}^{-\frac{1}{2} \int \frac{\mathrm{d}^d p}{(2\pi)^d} \tilde{\phi}_{\mathrm{s}}(-p)(p^2 + m^2)\tilde{\phi}_{\mathrm{s}}(p)}. \qquad (35.11)$$

Step II: We scale the momenta with $p'/b = p$ to obtain an action

$$S_{\mathrm{E}} = \frac{b^{-d}}{2} \int_0^{\Lambda} \frac{\mathrm{d}^d p'}{(2\pi)^d} \, \tilde{\phi}_{\mathrm{s}}(-p'/b) \left(b^{-2} p'^2 + m^2\right) \tilde{\phi}_{\mathrm{s}}(p'/b). \qquad (35.12)$$

Step III: We scale the fields via $\tilde{\phi}_{\mathrm{s}}(p'/b) = b^{d-d_\phi} \tilde{\phi}'(p')$ to obtain

$$S_{\mathrm{E}} = \frac{b^{2(d-d_\phi)}b^{-d}}{2} \int_0^{\Lambda} \frac{\mathrm{d}^d p'}{(2\pi)^d} \, \tilde{\phi}'(-p') \left(b^{-2} p'^2 + m^2\right) \tilde{\phi}'(p'). \qquad (35.13)$$

Fixing the constant d_ϕ so that the gradient term $\tilde{\phi}(-p)p^2\tilde{\phi}(p)$ is invariant, we have $d - 2d_\phi - 2 = 0$ or $d_\phi = (d-2)/2$ and

$$S_{\mathrm{E}} = \frac{1}{2} \int_0^{\Lambda} \frac{\mathrm{d}^d p'}{(2\pi)^d} \, \tilde{\phi}'(-p') \left(p'^2 + m^2 b^2\right) \tilde{\phi}'(p'). \qquad (35.14)$$

The result of our renormalization procedure is that after scaling, we find that we have a mass term $m'^2 = m^2 b^2$. That is, since $b > 1$, the coupling constant m^2 gets larger with each rescaling.

Since $b > 1$, the fact that $m'^2 = m^2 b^2$ means that m^2 is a **relevant variable** and $\hat{\phi}^2$ is known as a relevant operator. This means that it gets larger with each renormalization. (In contrast a variable that becomes smaller at each rescaling is known as an **irrelevant variable**.) Identifying m^2 from our action with the reduced temperature t, we expect that each rescaling changes m^2 by a factor b^{y_t}. From this analysis we can conclude that $y_t = 2$ when $\lambda = 0$.

In order to find y_h we need to consider the application of an external magnetic field. Still holding $\lambda = 0$ we write an action that includes the influence of the field as

$$S_{\mathrm{E}} = \int \mathrm{d}^d x \left[\frac{1}{2} (\boldsymbol{\nabla}\phi)^2 + \frac{m^2}{2} \phi^2 - h\phi \right]. \qquad (35.15)$$

On rescaling the field-dependent term will become $-hb^{d-d_\phi} \tilde{\phi}'(p')$. Substituting for d_ϕ as before we have $h' = hb^{\frac{d+2}{2}}$, and so $y_h = (d+2)/2$. This allows us to write the exponents predicted from this $\lambda = 0$ theory.

Now let's turn on the interaction. Since we don't know how to integrate the functional integral for the ϕ^4 theory it should come as little surprise that neither can we do the integral over the fast modes for ϕ^4 theory. We therefore have to treat $\mathcal{L}_I = \frac{\lambda}{4!}\phi^4$ as a perturbation and do the integral approximately. Our previous method for working out approximations to the generating functional $Z[J]$ using Feynman diagrams may be adapted to get an approximate answer for the result of removing the fast modes. The only difference in calculating the Feynman diagrams for the renormalization group analysis is that we integrate over internal momenta *only over the fast modes, up to the cut-off.*

In order to work out the lowest order corrections to the couplings we consider diagrammatic corrections to the theory in order of how many loops they contain. The lowest order diagrammatic corrections to the theory therefore involve diagrams with a single loop that correct the self-energy (which determines m^2) and the phion–phion vertex function (which determines λ). These one-loop diagrams are shown in Fig. 35.2.

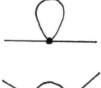

Example 35.3

We give the results of the important computations here.

The first diagram in Fig. 35.2 corresponds to an amplitude $-\frac{\lambda}{2}\int \frac{\mathrm{d}^d p}{(2\pi)^d}\frac{1}{p^2+m^2}$, which yields a contribution $\delta S^{(2)}$ to the quadratic part of the action given[9] by

$$\delta S^{(2)} = \frac{\lambda}{2}\int_{\Lambda/b}^{\Lambda}\frac{\mathrm{d}^d p}{(2\pi)^d}\frac{1}{p^2+m^2}\int \mathrm{d}^d x\,\frac{\phi_s^2}{2}. \tag{35.16}$$

Doing **I**: the integral over fast modes, **II**: rescaling the momentum and **III**: rescaling the fields, we obtain a contribution to the quadratic term

$$\delta S^{(2)} = b^2\left[\frac{\lambda\Omega_d}{2(d-2)}(1-b^{2-d}) - \frac{m^2\lambda\Omega_d}{2(d-4)}(1-b^{4-d})\right]\int \mathrm{d}^d x\,\frac{\phi_s^2}{2}, \tag{35.17}$$

which we give in terms of Ω_d, which is the volume of a unit sphere in d dimensions. Measured in units of 2π this volume is given by $\Omega_d = \frac{1}{(2\pi)^d}\frac{2\pi^{d/2}}{\Gamma(d/2)}$, where $\Gamma(z) = \int_0^\infty \mathrm{d}t\,\mathrm{e}^{-t}t^{z-1}$ is the Gamma function. The usual procedure in these calculations is to work in four-dimensional spacetime where the integrals converge and $\Omega_{d=4} = 1/8\pi^2$. To see the effects in other dimensions we set $d = 4 - \epsilon$, and expand in the small parameter ϵ. To examine the real-life case of three dimensions we then set $\epsilon = 1$, well outside the presumed range of applicability of this approach. Amazingly, this procedure describes reality very well!

Turning to the second diagram in Fig. 35.2, it turns out that we only need to evaluate the amplitude for zero external momentum,[10] which means both internal propagators carry the same momentum. We then obtain a contribution $\delta S^{(4)}$ to the quartic term in the action

$$
\begin{aligned}
\delta S^{(4)} &= \frac{3\lambda^2}{2}\int_{\Lambda/b}^{\Lambda}\frac{\mathrm{d}^d q}{(2\pi)^d}\frac{1}{(q^2+m^2)^2}\int \mathrm{d}^d x\,\frac{\phi_s^4}{4!} \\
&= b^{4-d}\frac{3\lambda^2\Omega_d}{2(d-4)}(1-b^{4-d})\int \mathrm{d}^d x\,\frac{\phi_s^4}{4!}.
\end{aligned} \tag{35.18}
$$

The result of this expansion is that we must shift the variables according to

$$
\begin{aligned}
m'^2 &= b^2\left[m^2 + \frac{\lambda\Omega_d}{2(d-2)}(1-b^{2-d}) - \frac{m^2\lambda\Omega_d}{2(d-4)}(1-b^{4-d})\right], \\
\lambda' &= b^{4-d}\left[\lambda - \frac{3\lambda^2\Omega_d}{2(d-4)}(1-b^{4-d})\right].
\end{aligned} \tag{35.19}
$$

Fig. 35.2 One-loop Feynman diagrams in ϕ^4 theory.

[9]Although both of our diagrams have symmetry factor 2, there is a subtlety here in working out the prefactors reflecting the fact that there is more than one equivalent diagram for each loop. As discussed in Binney, Dowrick, Fisher and Newman, the correct prefactor for a diagram with p loops is $\frac{2p!}{2^p 2p}$.

[10]See Altland and Simons, Section 8.4 for a discussion.

Now to extract the physics. We're looking at a system of condensed matter, so we're interested in a scaling that makes the length scale longer. We therefore set $b = e^l$. Since, as stated in the example, the integrals only converge for $d = 4$, we set $d = 4 - \epsilon$ and expand in the parameter ϵ. Evaluating the expressions to leading order in ϵ, using $\Omega_{4-\epsilon} \approx \Omega_4 = 1/8\pi^2$, we obtain the Gell-Mann–Low equations

$$\frac{\mathrm{d}m'^2}{\mathrm{d}l} = 2m^2 + \frac{\lambda}{16\pi^2}(1 - m^2),$$

$$\frac{\mathrm{d}\lambda'}{\mathrm{d}l} = \epsilon\lambda - \frac{3\lambda^2}{16\pi^2}. \tag{35.20}$$

As shown in Fig. 35.3, these equations have fixed points at $(m^2, \lambda) = (0, 0)$ and $\left(-\frac{\epsilon}{6}, \frac{16\pi^2}{3}\epsilon\right)$. By expanding the flow around the fixed points we can use this information to predict the behaviour of the ferromagnet.

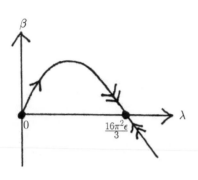

Fig. 35.3 The β function for the behaviour of the coupling constant λ.

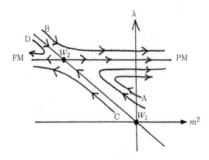

Fig. 35.4 Renormalization group flow for a ferromagnet. FM labels 'ferromagnet', PM labels 'paramagnet'.

[11] The predictions of the previous example are seen in the flows close to W_2, where we see λ decrease or increase depending, respectively, on whether we start above (flow B) or below (flow C) the value $\lambda^* = 16\pi^2\epsilon/3$. Flows A and D are clearly more complex and rely on the magnitude and behaviour of m^2.

Example 35.4

To see the behaviour near the fixed point we employ a first-order expansion to come up with matrix equations predicting the flow. A Taylor expansion of $f(x)$ about a fixed point x^* yields $f(x) = f(x^*) + f'(x^*)(x - x^*)$, but since $f(x^*) = 0$, by definition, we simply need to evaluate the derivatives at the fixed points. Our linearized flow equations take the form

$$\begin{pmatrix} \mathrm{d}m'^2/\mathrm{d}l \\ \mathrm{d}\lambda'/\mathrm{d}l \end{pmatrix} = \begin{pmatrix} \frac{\partial}{\partial m^2}\left(\frac{\mathrm{d}m'^2}{\mathrm{d}l}\right) & \frac{\partial}{\partial\lambda}\left(\frac{\mathrm{d}m'^2}{\mathrm{d}l}\right) \\ \frac{\partial}{\partial m^2}\left(\frac{\mathrm{d}\lambda'}{\mathrm{d}l}\right) & \frac{\partial}{\partial\lambda}\left(\frac{\mathrm{d}\lambda'}{\mathrm{d}l}\right) \end{pmatrix} \begin{pmatrix} m^2 - m^{*2} \\ \lambda - \lambda^* \end{pmatrix}, \tag{35.21}$$

with the elements evaluated at each of the fixed points.

For our ferromagnet, the matrices for the two sets of fixed points are

$$W_1 = \begin{pmatrix} 2 & \frac{1}{16\pi^2} \\ 0 & \epsilon \end{pmatrix}, \quad W_2 = \begin{pmatrix} 2 - \frac{\epsilon}{3} & \frac{1}{16\pi^2} \\ 0 & -\epsilon \end{pmatrix}. \tag{35.22}$$

Let's examine how these equations predict the behaviour of λ. For the fixed point at $(0, 0)$ (corresponding to W_1) we have $\lambda' = \epsilon\lambda l$. If $\epsilon > 0$ (that is, we are examining the physics in less than four dimensions since $d = 4 - \epsilon$) then λ increases as we look at larger scales. In that case λ is a relevant variable and will be important to the physics. Moving on to the other fixed point $(-\epsilon/6, 16\pi^2\epsilon/3)$, we see that close to $\lambda^* = 16\pi^2\epsilon/3$ we have $\lambda' = -\epsilon(\lambda - \lambda^*)l$. In this case λ is relevant (i.e. gets larger on rescaling) for $\lambda < \lambda^*$ but irrelevant (i.e. gets smaller on rescaling) for $\lambda > \lambda^*$. We will see the influence of this on the flow of the system next.

This information, along with an analogous analysis of how m^2 behaves, allows us to draw the phase diagram of the ferromagnet, as shown[11] in Fig. 35.4 for $\epsilon > 0$. The fixed point at W_1 is repulsive and that at W_2 is mixed: some flows are directed towards it, some away from it. Let's return to the physics of the system: at high temperatures we have a disordered, paramagnetic phase; at low temperatures, we suspect from our mean-field analysis that there exists an ordered, ferromagnetic phase. Relating this to our model, we write $m^2 = a(T - T_c)$, so flows that send m^2 to large, positive values predict that the system will end in the paramagnetic phase. Conversely, flows that send m^2 to negative values predict that the system becomes ferromagnetic. By examining the flows we ask what influence the fluctuations have on the phase transition. Recall that our mean-field analysis makes a prediction that any system that starts with a negative m^2 will be a ferromagnet and any system that starts with a positive m^2 will be paramagnetic.

The result of our renormalization group analysis is that flows starting with positive values of m^2 are always driven towards the paramagnetic phase, whatever the value of λ, in agreement with the mean-field treatment. However, starting with a negative value of m^2 doesn't guarantee a ferromagnetic ground state! The fact that we have fluctuations, whose strength is given by λ, may force the system to be a paramagnet. This is the case for the flow B in Fig. 35.4, where starting with a significant enough λ is enough to drag the system towards paramagnetism. On the other hand, flows starting with sufficiently small λ and sufficiently negative m^2 (such as paths C and D) eventually flow off to the left, pushing the system towards ferromagnetism.

Example 35.5

So much for our qualitative analysis of the phase diagram. Let's extract y_t and y_h and make some quantitative predictions regarding the thermodynamics of the ferromagnet. One of the uses of the linearized flow matrices is that they may be used to read off the scaling constant y_t. Since we have set $b = e^l$, we have that each rescaling goes as $m'^2 = e^{l y_t} m^2$. Close to the fixed point, where our linearized equations are valid this implies that $\Delta m'^2 = y_t(m^2 - m^{*2})\Delta l$. We can therefore read off from the matrix describing the flow around our nonzero fixed point that $y_t = 2 - \frac{\epsilon}{3}$. The one-loop correction doesn't change y_h, which continues to be given by $y_h = (d+2)/2 = (6-\epsilon)/2$ just as it was from the analysis of the $\lambda = 0$ case.

With these two pieces of information, we obtain all six of the critical parameters:

$$\alpha = \tfrac{\epsilon}{6}, \quad \beta = \tfrac{1}{2} - \tfrac{\epsilon}{6}, \quad \gamma = 1 + \tfrac{\epsilon}{6}, \quad \delta = 3 + \epsilon, \quad \nu = \tfrac{1}{2} + \tfrac{\epsilon}{12}, \quad \eta = 0. \quad (35.23)$$

These results can be used to make an estimate of the critical exponents of the three-dimensional Ising model, as shown in the following table.[12]

	α	β	γ	δ	ν	η
$\epsilon = 0$ ($d = 4$, mean field)	0	$\tfrac{1}{2}$	1	3	$\tfrac{1}{2}$	0
$\epsilon = 1$ $O(\epsilon)$	0.167	0.333	1.167	4	0.583	0
$\epsilon = 1$ $O(\epsilon^2)$	0.077	0.340	1.244	4.462	0.626	0.019
3D Ising	0.110	0.327	1.237	4.789	0.630	0.036
2D Ising (exact)	0	$\tfrac{1}{8}$	$\tfrac{7}{4}$	15	1	$\tfrac{1}{4}$

The first line of this table gives the mean-field results, exact for $d = 4$ and so can be obtained from eqn 35.23 by setting $\epsilon = 0$. The second line gives a crude, but surprisingly accurate, estimate for $d = 3$ by setting $\epsilon = 1$. One can derive expressions for the critical exponents to second order in ϵ [e.g. $\beta = \tfrac{1}{2} - \tfrac{\epsilon}{6} + \tfrac{\epsilon^2}{162} + O(\epsilon^3)$] and these yield the estimates in the third line of the table. These approach the currently accepted values[13] (fourth line) which agree well with experimental values. In the last line we list the exact results for the two-dimensional Ising model for comparison.

[12]For a discussion of the comparison see the books by Kaku, Zinn-Justin and Binney et al. See also Peskin and Schroeder, Chapter 13.

[13]Taken from A. Pelissetto and E. Vicari, Phys. Rep. **368**, 549 (2002), which reviews a large number of theoretical studies using various techniques and also summarises experimental results.

Chapter summary

- This chapter has provided a renormalization group analysis of the ferromagnetic transition following Widom's hypothesis for the scaling of the free energy.

Exercises

(35.1) (a) Using Widom's hypothesis in the form $f(t,h) = t^{\frac{d}{y_t}} \tilde{f}(h/t^{\frac{y_h}{y_t}})$, show that critical parameters are given in terms of the scaling parameters as follows

$$\alpha = 2 - \frac{d}{y_t}, \quad \beta = \frac{d - y_h}{y_t}, \quad \gamma = \frac{2y_h - d}{y_t}. \quad (35.24)$$

(b) We now redefine the scaling function slightly so that it reads $f(t,h) = h^{\frac{d}{y_h}} \tilde{g}(h/t^{\frac{y_h}{y_t}})$ where $g(z) = z^{-\frac{d}{y_h}} f(z)$. Use this form of the function to show that $\delta = \frac{y_h}{d - y_h}$.

(c) From these results prove the following relations:

$$\begin{array}{ll} \alpha + 2\beta + \gamma = 2 & \text{Rushbrooke's law,} \\ \alpha + \beta(\delta + 1) = 2 & \text{Griffith's law.} \end{array} \quad (35.25)$$

(d) Assuming the correlation function behaves as $G_c(x,t) = f\left(xt^{\frac{2-\alpha}{d}}\right)/x^{d-2+\eta}$, argue that we re-

quire $\nu d = 2 - \alpha$ (known as Josephson's law) and use this to show $\nu = \frac{1}{y_t}$.

(e) Finally use Fisher's law $(2 - \eta)\nu = \gamma$ to find η.

(35.2) (a) Starting from eqns 35.19, verify the Gell-Mann–Low equations in eqn 35.20.
Hint: let $b = 1 + \rho$ with ρ small. Then $\ln b \approx \rho$ and one may expand to order ρ to capture the flow equations.
(b) Verify the positions of the fixed points.
(c) Verify the matrices in 35.22 and the behaviour of the flow near the fixed points.

(35.3) What is the behaviour of λ for the case of dimensions $d > 4$?

Part IX

Putting a spin on QFT

This part extends our treatment of quantum field theory to spinors and allows us to treat the electron spin. These ideas set us on a path that inevitably arrives at quantum electrodynamics.

- The *Dirac equation* for an electron is introduced in Chapter 36. We describe the γ-matrices necessary for formulating this and also the concepts of *chirality* and *helicity*. The basis states are called *spinors* and we describe two types: *Dirac spinors* and *Weyl spinors*.

- We show how spinors transform under rotations and boosts in Chapter 37. We also demonstrate that the *parity* operator turns a left-handed spinor into a right-handed spinor and vice versa.

- In Chapter 38 we apply the machinery of quantization to the Dirac equation and derive the Noether current. This allows us to write down the *fermion propagator*.

- Arguably the most important achievement of quantum field theory is *quantum electrodynamics* (QED), which describes the interaction between electrons and photons, and this theory is presented in outline in Chapter 39.

- Three examples of ideas presented in the previous chapter are fleshed out in Chapter 40 via three examples. These are *Rutherford scattering*, the *Mott formula* and *Compton scattering*. We briefly discuss the useful feature of *crossing symmetry*.

- Some important consequences of QED arise when you consider processes involving electron–positron loops. These include the renormalization of the electron charge due to vacuum polarization and the deviation of the g-factor from $g = 2$. These ideas are explored in Chapter 41.

The Dirac equation

After seeing *2001: A Space Odyssey*, Paul Dirac was so impressed that, two days later, he attended three consecutive screenings of the film in the cinema. His favourite television programme was the Cher show.

[1]This is the subject examined in the two chapters following this one

I have an equation, do you have one too?
Paul Dirac (1902–1984), on meeting Richard Feynman

Our story so far: as a single-particle equation of motion, the Klein–Gordon equation has some serious problems. The first is negative energy states and the second is negative probabilities. Paul Dirac sees that the root of these problems lies in the fact that the Klein–Gordon equation contains a second-order time derivative as part of ∂^2. While staring into the fire in the senior common room of St John's College, Cambridge, Dirac realized that there's a way to construct an equation of motion that's first order in the time derivative.

In this chapter we examine Dirac's equation. We will treat it as an equation of motion for single particles and examine the form of the solutions and how features of these solutions describe the physical properties of real particles. Although our experience with the Klein–Gordon equation has taught us that, ultimately, a single-particle treatment of relativistic quantum mechanics will not provide an adequate description of Nature, the experience gained with the solutions of Dirac's equation will carry over to the world of Dirac fields[1] whose excitations are the fermions we observe in our Universe.

36.1 The Dirac equation

In order to avoid all of the problems with the Klein–Gordon equation, you might wonder whether we could take the square root of the guts of that equation $(\partial^2 + m^2)$ and then apply it to a wave function. A naive guess might be that if we could treat things like ∂^2 like ordinary algebraic symbols then we could write

$$(\partial^2 + m^2) = (\sqrt{\partial^2} + \mathrm{i}m)(\sqrt{\partial^2} - \mathrm{i}m). \tag{36.1}$$

Unfortunately, $\sqrt{\partial^2}$ isn't defined. How do we take the square root of the ∂^2 operator? Dirac found the answer was to define a new four-vector γ^μ, whose components have the properties

$$(\gamma^0)^2 = 1, \ (\gamma^1)^2 = -1, \ (\gamma^2)^2 = -1, \ (\gamma^3)^2 = -1. \tag{36.2}$$

Furthermore, these components aren't ordinary numbers; they anticommute. That is to say that for $\mu \neq \nu$ they have the intriguing property

$\gamma^\mu\gamma^\nu + \gamma^\nu\gamma^\mu = \{\gamma^\mu, \gamma^\nu\} = 0$. Taken with eqn 36.2, the properties of these γ's may be summarized in the compact form[2]

$$\{\gamma^\mu, \gamma^\nu\} = 2g^{\mu\nu}. \tag{36.3}$$

We use the gammas to cope with the square root in $\sqrt{\partial^2}$, via a strange looking new symbol $\not{a} = \gamma^\mu a_\mu$, which is pronounced 'a-slash'. We can then say that

$$\not{\partial}^2 = (\gamma^\mu\partial_\mu)^2 = \left(\gamma^0\frac{\partial}{\partial t} + \gamma^1\frac{\partial}{\partial x} + \gamma^2\frac{\partial}{\partial y} + \gamma^3\frac{\partial}{\partial z}\right)^2. \tag{36.4}$$

Expanding this bracket and using the all-important anticommutation relations we find that $\not{\partial}^2 = \partial^2$. Substituting $\not{\partial}^2$ for ∂^2 in the Klein–Gordon equation does indeed allow us to factorize

$$(\partial^2 + m^2) = (\not{\partial}^2 + m^2) = (\not{\partial} - \mathrm{i}m)(\not{\partial} + \mathrm{i}m). \tag{36.5}$$

Now let's just take the plus-sign bracket and treat it as an operator acting on the wave function $\psi(x)$,

$$(\not{\partial} + \mathrm{i}m)\psi(x) = 0, \tag{36.6}$$

and this may be tidied up to give us the famous **Dirac equation**:

$$(\mathrm{i}\gamma^\mu\partial_\mu - m)\psi(x) = 0. \tag{36.7}$$

We note first that the Dirac equation may also be written as $(\not{p} - m)\psi = 0$, where $\not{p} = \gamma^\mu(\mathrm{i}\partial_\mu)$ and we recognize the energy-momentum operator[3] $\hat{p}_\mu = \mathrm{i}\partial_\mu$.

We'll soon see that the dispersion relation of the particles described by the Dirac equation of motion continues to be $E_{\boldsymbol{p}}^2 = (\boldsymbol{p}^2 + m^2)$, which still admits negative energy states.[4]

To get to Dirac's equation, we have had to introduce the mysterious, anticommuting γ^μ objects. These will have a profound influence on the physics described by the Dirac equation because the simplest way to satisfy the commutation relations is to represent the γ's as four-by-four matrices. This means that the wave function $\psi(x)$ described by the Dirac equation is a special four-component wave function. Why four components? What could they represent? To make progress, we'll use an explicit form for the matrices and examine some illuminating special cases.

36.2 Massless particles: left- and right-handed wave functions

There is no unique way of writing the γ^μ matrices and so we have a choice about how to represent them. One set of matrices that satisfies

[2]Equation 36.3 defines what is known as a **Clifford algebra**, named after William Kingdon Clifford (1845–1879). See Penrose, Chapter 11 for more details.

[3]Recall that $\hat{p}_\mu = \mathrm{i}\partial_\mu = \mathrm{i}(\frac{\partial}{\partial t}, \boldsymbol{\nabla})$.

[4]It does successfully eliminate the negative probability problem. Of course, we've already found a way to deal with the negative energy states of the Klein–Gordon equation: Feynman's prescription of interpreting them as positive energy antiparticles, so the negative energy states are not a deal-breaker. In fact, the Dirac equation will turn out to provide the equation of motion for a type of *field* whose excitations are fermions such as the electron. As such, the results from this temporary foray into wave mechanics should be treated as provisional.

the commutation relations is

$$\gamma^0 = \begin{pmatrix} 0 & 0 & 1 & 0 \\ 0 & 0 & 0 & 1 \\ 1 & 0 & 0 & 0 \\ 0 & 1 & 0 & 0 \end{pmatrix}, \quad \gamma^1 = \begin{pmatrix} 0 & 0 & 0 & 1 \\ 0 & 0 & 1 & 0 \\ 0 & -1 & 0 & 0 \\ -1 & 0 & 0 & 0 \end{pmatrix},$$

$$\gamma^2 = \begin{pmatrix} 0 & 0 & 0 & -i \\ 0 & 0 & i & 0 \\ 0 & i & 0 & 0 \\ -i & 0 & 0 & 0 \end{pmatrix}, \quad \gamma^3 = \begin{pmatrix} 0 & 0 & 1 & 0 \\ 0 & 0 & 0 & -1 \\ -1 & 0 & 0 & 0 \\ 0 & 1 & 0 & 0 \end{pmatrix}. \tag{36.8}$$

[5] This is sometimes called the **Weyl representation**.

Writing the γ's in this form is known as the **chiral representation**,[5] for reasons that will become apparent. Writing out lots of four-by-four matrices is a tedious business, so to save on writing it's useful to write these four-by-four matrices out as two-by-two matrices, where each element is itself a two-by-two matrix.

Example 36.1

There are several ways to simplify the notation. Defining $I = \begin{pmatrix} 1 & 0 \\ 0 & 1 \end{pmatrix}$, we notice that we can express eqn 36.8 in a 2×2 form:

$$\gamma^0 = \begin{pmatrix} 0 & I \\ I & 0 \end{pmatrix}, \quad \gamma^1 = \begin{pmatrix} 0 & \sigma_x \\ -\sigma_x & 0 \end{pmatrix},$$

$$\gamma^2 = \begin{pmatrix} 0 & \sigma_y \\ -\sigma_y & 0 \end{pmatrix}, \quad \gamma^3 = \begin{pmatrix} 0 & \sigma_z \\ -\sigma_z & 0 \end{pmatrix}, \tag{36.9}$$

where the Pauli matrices have their usual definitions.

It turns out that γ^μ transforms as a four-vector. Motivated by this, we write $\gamma^\mu = (\gamma^0, \gamma^1, \gamma^2, \gamma^3) = (\gamma^0, \boldsymbol{\gamma})$ and then

$$\gamma^0 = \begin{pmatrix} 0 & I \\ I & 0 \end{pmatrix}, \quad \boldsymbol{\gamma} = \begin{pmatrix} 0 & \boldsymbol{\sigma} \\ -\boldsymbol{\sigma} & 0 \end{pmatrix}, \tag{36.10}$$

where $\boldsymbol{\sigma} = (\sigma^1, \sigma^2, \sigma^3)$. Finally, in a desire to write as little as possible, we define $\sigma^\mu = (I, \boldsymbol{\sigma})$ and $\bar{\sigma}^\mu = (I, -\boldsymbol{\sigma})$ to obtain[6]

$$\gamma^\mu = \begin{pmatrix} 0 & \sigma^\mu \\ \bar{\sigma}^\mu & 0 \end{pmatrix}. \tag{36.11}$$

[6] If, at any point in the future, any of this notation seems overly compact, then it may sometimes be helpful to write things out in full.

Substituting these γ's into the Dirac equation gives

$$(\gamma_0 \hat{p}^0 - \boldsymbol{\gamma} \cdot \hat{\boldsymbol{p}} - m)\psi(x) = 0, \tag{36.12}$$

where $\hat{p}^0 = i\partial_0$ and $\hat{\boldsymbol{p}} = -i\boldsymbol{\nabla}$. We'll make use of a handy two-by-two matrix form for the equations, where the four-component wave function ψ is written $\begin{pmatrix} \psi_L \\ \psi_R \end{pmatrix}$, where ψ_L and ψ_R are, themselves, two-component column matrices.[7] In this form the Dirac equation reads

[7] The meaning of these subscripts will be explained shortly.

$$\left[\begin{pmatrix} 0 & \hat{p}^0 \\ \hat{p}^0 & 0 \end{pmatrix} - \begin{pmatrix} 0 & \boldsymbol{\sigma} \cdot \hat{\boldsymbol{p}} \\ -\boldsymbol{\sigma} \cdot \hat{\boldsymbol{p}} & 0 \end{pmatrix} - \begin{pmatrix} m & 0 \\ 0 & m \end{pmatrix} \right] \begin{pmatrix} \psi_L \\ \psi_R \end{pmatrix} = 0, \tag{36.13}$$

which can be rewritten as

$$(\hat{p}_0 - \boldsymbol{\sigma} \cdot \hat{\boldsymbol{p}})\psi_{\mathrm{R}} = m\psi_{\mathrm{L}},$$
$$(\hat{p}_0 + \boldsymbol{\sigma} \cdot \hat{\boldsymbol{p}})\psi_{\mathrm{L}} = m\psi_{\mathrm{R}}. \qquad (36.14)$$

Now let's further simplify things by considering the special case of a massless particle (which obviously has $m = 0$). We then obtain

$$(\hat{p}^0 - \boldsymbol{\sigma} \cdot \hat{\boldsymbol{p}})\psi_{\mathrm{R}} = 0,$$
$$(\hat{p}^0 + \boldsymbol{\sigma} \cdot \hat{\boldsymbol{p}})\psi_{\mathrm{L}} = 0, \qquad (36.15)$$

which implies that a four-component eigenstate $\psi = \begin{pmatrix} \psi_{\mathrm{L}} \\ \psi_{\mathrm{R}} \end{pmatrix}$ splits into two two-component pieces, ψ_{L} and ψ_{R}, that don't get mixed up by the equations of motion. It's as if the Dirac equation is telling us that we have two sorts of massless Dirac particles in Nature: **left-handed particles** described by wave functions ψ_{L} and **right-handed particles** described by wave functions ψ_{R}. In the chiral representation, left-handed functions live in the upper two slots of the Dirac wave function and right-handed functions live in the lower slots.

Example 36.2

We need to be slightly more precise with our definition of left- and right-handed wave functions. To do this we define the **chirality** operator as

$$\gamma^5 = i\gamma^0\gamma^1\gamma^2\gamma^3$$
$$= \begin{pmatrix} -I & 0 \\ 0 & I \end{pmatrix}, \qquad (36.16)$$

where the final equality holds in the chiral representation only. A wave function $\begin{pmatrix} \psi_{\mathrm{L}} \\ 0 \end{pmatrix}$ is an eigenstate of γ^5 with eigenvalue -1. An entirely right-handed wave function $\begin{pmatrix} 0 \\ \psi_{\mathrm{R}} \end{pmatrix}$ is then an eigenstate with eigenvalue $+1$. The Dirac equation therefore predicts the existence of two types of Dirac wave function in the Universe: left-handed functions with chirality -1 and right-handed functions with chirality $+1$.

If we want to extract the left- and right-handed parts of a wave function we may define projection operators via

$$\hat{P}_{\mathrm{L}} = \frac{1-\gamma^5}{2}, \quad \hat{P}_{\mathrm{R}} = \frac{1+\gamma^5}{2}, \qquad (36.17)$$

which will project out the desired part.

We conclude that since ψ_{L} and ψ_{R} are eigenstates of the massless Dirac equation, then a free and massless left-handed particle may never change into a right-handed particle. Moreover, using the fact that, for massless particles, the eigenvalue of $\hat{p}^0$ is $E_{\boldsymbol{p}} = |\boldsymbol{p}|$, our eqns 36.15 give us

$$\frac{\boldsymbol{\sigma} \cdot \hat{\boldsymbol{p}}}{|\boldsymbol{p}|}\psi_{\mathrm{R}} = \psi_{\mathrm{R}}, \qquad (36.18)$$

$$\frac{\boldsymbol{\sigma} \cdot \hat{\boldsymbol{p}}}{|\boldsymbol{p}|}\psi_{\mathrm{L}} = -\psi_{\mathrm{L}}. \qquad (36.19)$$

This means that the massless states ψ_R and ψ_L, defined as being chirality eigenstates, are also eigenstates of the **helicity** operator

$$\hat{h} = \frac{\boldsymbol{\sigma} \cdot \hat{\boldsymbol{p}}}{|\boldsymbol{p}|}, \tag{36.20}$$

with eigenvalues $+1$ and -1 respectively.

From this discussion it looks rather like chirality and helicity, while being represented by different operators, tell us the same thing. However, one needs to be careful here: *helicity and chirality are identical for massless particles, but are, in general, not the same.* Clearly helicity tells us whether the particle's spin and momentum are parallel or antiparallel. Helicity therefore depends on the frame in which it is measured. In the case of massive particles it is possible to describe a particle with positive helicity and then to boost to a frame where the particle's momentum is reversed but its spin is unchanged, thereby reversing its helicity.

If we now consider massive particles again then the equations of motion revert back to eqn 36.14. This shows us that massive Dirac particles involve both left- and right-handed wave functions coupled by the particle's mass. We can think of massive Dirac particles oscillating back and forth in time between left- and right-handed at a rate determined by their mass. This is most easily seen by looking at massive Dirac particles at rest, where we have $i\partial_0 \psi_R = m\psi_L$ and $i\partial_0 \psi_L = m\psi_R$.

Example 36.3

We'll find the dispersion relation for massive Dirac particles. Assuming that ψ_L and ψ_R are eigenstates of momentum and energy, we may eliminate ψ_L from eqns. 36.14, to give us

$$(p^0 + \boldsymbol{\sigma} \cdot \boldsymbol{p})(p^0 - \boldsymbol{\sigma} \cdot \boldsymbol{p})\psi_R = m^2 \psi_R. \tag{36.21}$$

Multiplying this out gives

$$[(p^0)^2 - \boldsymbol{p}^2] = m^2, \tag{36.22}$$

which demonstrates that $E_{\boldsymbol{p}} = \pm(\boldsymbol{p}^2 + m^2)^{\frac{1}{2}}$ and we still have negative energy states as claimed above. Another way of saying this is that the eigenvalue of p^0 could be taken to be positive or negative and we still get the same dispersion.

---

[8]Of course we know that this is a bad way of thinking about them since physical antiparticles have positive energy. We apologize for the brief lapse into bad habits.

Finally we turn to antiparticle solutions of the Dirac equation. These are the solutions which appear to have negative values of energy.[8] Writing energies as $E = -|p_0|$, we find, for massless antiparticles, that

$$\left. \begin{array}{l} (-|p^0| - \boldsymbol{\sigma} \cdot \hat{\boldsymbol{p}})\psi_R = 0 \\ (-|p^0| + \boldsymbol{\sigma} \cdot \hat{\boldsymbol{p}})\psi_L = 0 \end{array} \right\} \text{antiparticles}, \tag{36.23}$$

which implies

$$\left. \begin{array}{l} \frac{\boldsymbol{\sigma} \cdot \hat{\boldsymbol{p}}}{|\boldsymbol{p}|}\psi_R = -\psi_R \\ \frac{\boldsymbol{\sigma} \cdot \hat{\boldsymbol{p}}}{|\boldsymbol{p}|}\psi_L = \psi_L \end{array} \right\} \text{antiparticles}. \tag{36.24}$$

This is to say that right-handed antiparticles have negative helicity and left-handed antiparticles have positive helicity. Note that this is the other way round compared to the particle case.

To summarize our progress so far. We've discovered that the four-component solutions to the Dirac wave function have left-handed parts in the upper two slots and right-handed parts in the lower two slots and can be written

$$\psi(x) = \left(\begin{array}{c} \psi_{\mathrm{L}}(x) \\ \psi_{\mathrm{R}}(x) \end{array} \right). \tag{36.25}$$

The four-component wave functions are known as **Dirac spinors** and are quite different to other objects that we have so far encountered such as scalars and four-vectors. The two-component objects $\psi_{\mathrm{L}}(x)$ and $\psi_{\mathrm{R}}(x)$ are themselves known as **Weyl spinors**. Spinors are best defined in terms of their transformation properties, which we turn to in the next chapter. In the meantime we will examine them further as solutions to Dirac's equation.

Hermann Weyl (1885–1955)

36.3 Dirac and Weyl spinors

We now investigate what's hidden away in $\psi_{\mathrm{L}}(x)$ and $\psi_{\mathrm{R}}(x)$. Let's work on the positive and negative energy solutions separately. Since physical particles and antiparticles really have positive energies, we'll conform to convention and call these the positive and negative *frequency* solutions. The positive frequency solutions will describe Fermi particles, while the negative frequency solutions will describe Fermi antiparticles. It's easy to show that a set of solutions to the Dirac equation may be written

$$\text{Particles: } u(p)\mathrm{e}^{-ip\cdot x} = \left(\begin{array}{c} u_{\mathrm{L}}(p) \\ u_{\mathrm{R}}(p) \end{array} \right) \mathrm{e}^{-ip\cdot x},$$

$$\text{Antiparticles: } v(p)\mathrm{e}^{ip\cdot x} = \left(\begin{array}{c} v_{\mathrm{L}}(p) \\ v_{\mathrm{R}}(p) \end{array} \right) \mathrm{e}^{ip\cdot x}. \tag{36.26}$$

Note that the negative frequency solutions are defined such that $p^0 = E > 0$ so that antiparticles will have positive energies. The four-component objects[9] $u(p)$ and $v(p)$ are momentum space Dirac spinors and themselves satisfy the momentum space Dirac equation:

$$(\not{p} - m)u(p) = 0,$$
$$(-\not{p} - m)v(p) = 0. \tag{36.27}$$

The left-handed [$u_{\mathrm{L}}(p)$ and $v_{\mathrm{L}}(p)$] and right-handed [$u_{\mathrm{R}}(p)$ and $v_{\mathrm{R}}(p)$] parts are momentum space Weyl spinors.

We start by considering the particle solution at rest where $p^\mu = (m, 0)$. In this case the Dirac eqn 36.27 becomes

$$\left(\begin{array}{cc} -m & m \\ m & -m \end{array} \right) \left(\begin{array}{c} u_{\mathrm{L}}(p^0) \\ u_{\mathrm{R}}(p^0) \end{array} \right) = 0, \tag{36.28}$$

with solutions

$$u(p^0) \equiv \left(\begin{array}{c} u_{\mathrm{L}}(p^0) \\ u_{\mathrm{R}}(p^0) \end{array} \right) = \sqrt{m} \left(\begin{array}{c} \xi \\ \xi \end{array} \right), \tag{36.29}$$

[9] Just as a polarization vector $\epsilon^\mu(p)$ carries the spin information for the photon, the spinors $u(p)$ and $v(p)$ carry the spin information of the fermion.

[10]The factor $\sqrt{m}$ is included in eqn 36.29 for later convenience. It results in $\bar{u}u = 2m$ where $\bar{u} \equiv u^\dagger \gamma^0$ is defined later (eqn 36.41). For massless particles one includes a factor $\sqrt{E_p}$ instead. See Peskin and Schroeder, Chapter 3, for more details.

where $\xi = \begin{pmatrix} \xi_1 \\ \xi_2 \end{pmatrix}$ is a two-component column vector that we will choose to normalize[10] according to $\xi^\dagger \xi = 1$. Note that ξ is also often called a spinor. The argument may be repeated for antiparticle solutions with the result

$$-\begin{pmatrix} m & m \\ m & m \end{pmatrix}\begin{pmatrix} v_L(p^0) \\ v_R(p^0) \end{pmatrix} = 0, \tag{36.30}$$

leading to

$$v(p^0) = \sqrt{m}\begin{pmatrix} \eta \\ -\eta \end{pmatrix}, \tag{36.31}$$

where we write $\eta = \begin{pmatrix} \eta_1 \\ \eta_2 \end{pmatrix}$ for the antiparticle spinor. Notice that the solutions to the Dirac equation have two degrees of freedom. These are the two components of ξ (for the positive frequency solutions) or η (for the negative frequency solutions). It turns out that it is ξ and η that tell us about the **spin** of the particle.

We know the form of the operator for the spin of a particle at rest is given by $\hat{S} = \frac{1}{2}\boldsymbol{\sigma}$. The object on which this operator acts is η or ξ. A particle with spin up along the z-axis therefore has $\xi = \begin{pmatrix} 1 \\ 0 \end{pmatrix}$ and one with spin down along the z-axis has $\xi = \begin{pmatrix} 0 \\ 1 \end{pmatrix}$. We can use these states as a basis to represent arbitrary spin states.

What about antiparticles? We define an antiparticle at rest with *physical* spin up along the z-direction to have $\eta = \begin{pmatrix} 0 \\ 1 \end{pmatrix}$, so that $\hat{S}_z\eta = -\frac{1}{2}\eta$. An antiparticle with physical spin down along z has $S_z = \frac{1}{2}$ and $\eta = \begin{pmatrix} 1 \\ 0 \end{pmatrix}$ so that $\hat{S}_z\eta = \frac{1}{2}\eta$. Admittedly this is rather confusing! The properties of particles and antiparticles are summarized in the table at the end of the chapter.

Example 36.4

A negative helicity antiparticle with momentum directed along the positive z-axis must have $\sigma\eta = +\eta$. It must also, therefore, have $\eta = \begin{pmatrix} 1 \\ 0 \end{pmatrix}$, which corresponds to physical spin down.

In eqn 36.32 we are again using the notation $\sigma = (I, \boldsymbol{\sigma})$ and $\bar{\sigma} = (I, -\boldsymbol{\sigma})$ and we understand that in taking the square root of a matrix we take the positive root of each eigenvalue.

What about the spinor for a particle in the general case of particles that aren't at rest? We quote the result here (and will justify it in the next chapter). For a particle or antiparticle with momentum p^μ the spinors are given by

$$u(p) = \begin{pmatrix} \sqrt{p\cdot\sigma}\,\xi \\ \sqrt{p\cdot\bar{\sigma}}\,\xi \end{pmatrix}, \quad v(p) = \begin{pmatrix} \sqrt{p\cdot\sigma}\,\eta \\ -\sqrt{p\cdot\bar{\sigma}}\,\eta \end{pmatrix}. \tag{36.32}$$

Example 36.5

We will evaluate the spinor for a particle with $p^\mu = (E_p, 0, 0, |\boldsymbol{p}|)$ and $\xi = \begin{pmatrix} 0 \\ 1 \end{pmatrix}$.
Evaluating the left-handed part $u_L(p) = \sqrt{p \cdot \sigma}\,\xi$ which lives in the upper two slots of the spinor, we have

$$\sqrt{p \cdot \sigma}\,\xi = \begin{pmatrix} E_p - |\boldsymbol{p}| & 0 \\ 0 & E_p + |\boldsymbol{p}| \end{pmatrix}^{\frac{1}{2}} \begin{pmatrix} 0 \\ 1 \end{pmatrix} = \begin{pmatrix} \sqrt{E_p - |\boldsymbol{p}|} & 0 \\ 0 & \sqrt{E_p + |\boldsymbol{p}|} \end{pmatrix} \begin{pmatrix} 0 \\ 1 \end{pmatrix},$$
(36.33)

which gives $u_L(p) = \sqrt{p \cdot \sigma}\,\xi = \sqrt{E_p + |\boldsymbol{p}|}\begin{pmatrix} 0 \\ 1 \end{pmatrix}$.

The analogous result for the right-handed, lower slots of $u(p)$ follows by a similar argument with the result that $u_R(p) = \sqrt{p \cdot \bar\sigma}\,\xi = \sqrt{E_p - |\boldsymbol{p}|}\begin{pmatrix} 0 \\ 1 \end{pmatrix}$.

We obtain the final result for the spinor

$$u(p) = \begin{pmatrix} \sqrt{E_p + |\boldsymbol{p}|}\begin{pmatrix} 0 \\ 1 \end{pmatrix} \\ \sqrt{E_p - |\boldsymbol{p}|}\begin{pmatrix} 0 \\ 1 \end{pmatrix} \end{pmatrix}.$$
(36.34)

The general expression for the spinors in eqn 36.32 allows us to explore the helicity and chirality of particles and antiparticles in more detail in the next example.

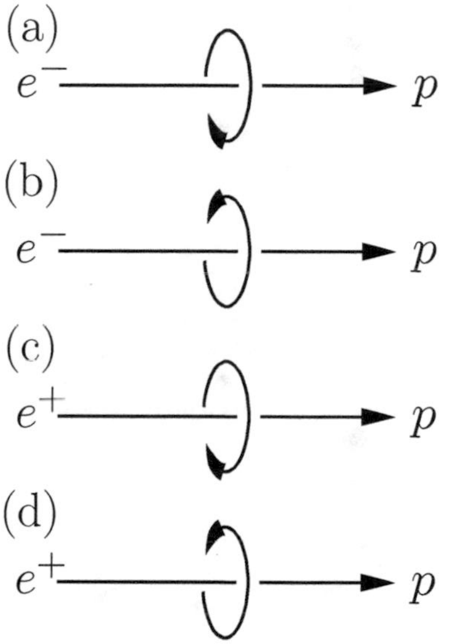

Example 36.6

We will look at particles and antiparticles in the ultra-relativistic limit, obtained by giving the particles a big boost. Consider a particle with spin up along the z-axis and momentum $|\boldsymbol{p}|$ along $+z$. This particle, shown in Fig. 36.1(a), has helicity $h = 1$, corresponding to physical spin parallel to the momentum. This is described by a spinor

$$u(p) = \begin{pmatrix} u_L(p) \\ u_R(p) \end{pmatrix} = \begin{pmatrix} \sqrt{E_p - \boldsymbol{p}\cdot\sigma}\begin{pmatrix} 1 \\ 0 \end{pmatrix} \\ \sqrt{E_p + \boldsymbol{p}\cdot\sigma}\begin{pmatrix} 1 \\ 0 \end{pmatrix} \end{pmatrix}.$$
(36.35)

This leads us to

$$u(p) = \begin{pmatrix} \sqrt{E_p - |\boldsymbol{p}|}\begin{pmatrix} 1 \\ 0 \end{pmatrix} \\ \sqrt{E_p + |\boldsymbol{p}|}\begin{pmatrix} 1 \\ 0 \end{pmatrix} \end{pmatrix} \xrightarrow[\text{boost}]{\text{big}} \sqrt{2E_p}\begin{pmatrix} 0 \\ 0 \\ 1 \\ 0 \end{pmatrix},$$
(36.36)

which has $u_L = 0$ and $u_R = \sqrt{2E_p}\begin{pmatrix} 1 \\ 0 \end{pmatrix}$.

We can carry out the same analysis for a negative helicity particle [Fig. 36.1(b)], which has spin down along the z-axis and momentum along $+z$. This has a spinor

$$u(p) = \begin{pmatrix} \sqrt{E_p - \boldsymbol{p}\cdot\sigma}\begin{pmatrix} 0 \\ 1 \end{pmatrix} \\ \sqrt{E_p + \boldsymbol{p}\cdot\sigma}\begin{pmatrix} 0 \\ 1 \end{pmatrix} \end{pmatrix} = \begin{pmatrix} \sqrt{E_p + |\boldsymbol{p}|}\begin{pmatrix} 0 \\ 1 \end{pmatrix} \\ \sqrt{E_p - |\boldsymbol{p}|}\begin{pmatrix} 0 \\ 1 \end{pmatrix} \end{pmatrix} \xrightarrow[\text{boost}]{\text{big}} \sqrt{2E_p}\begin{pmatrix} 0 \\ 1 \\ 0 \\ 0 \end{pmatrix},$$
(36.37)

which has $u_L = \sqrt{2E_p}\begin{pmatrix} 0 \\ 1 \end{pmatrix}$ and $u_R = 0$.

Fig. 36.1 Helicity states of particles and antiparticles. (a) A highly relativistic electron with spin up along the z-axis and momentum along z, which has $h = +1$. (b) The same with spin down, giving helicity $h = -1$. (c) A highly relativistic positron with physical spin up along the z-axis and momentum along z, giving a helicity $h = +1$ (d) The same with physical spin down along z and $h = -1$.

Now for the antiparticle states. Again we take momentum to be along $+z$. An antiparticle with physical spin up along the z-axis [Fig. 36.1(c)] has $\begin{pmatrix} 0 \\ 1 \end{pmatrix}$. It has a helicity of $+1$ (opposite to what a particle would have) and has a spinor

$$v(p) = \begin{pmatrix} \sqrt{E_{\boldsymbol{p}} - \boldsymbol{p}\cdot\boldsymbol{\sigma}}\begin{pmatrix} 0 \\ 1 \end{pmatrix} \\ -\sqrt{E_{\boldsymbol{p}} + \boldsymbol{p}\cdot\boldsymbol{\sigma}}\begin{pmatrix} 0 \\ 1 \end{pmatrix} \end{pmatrix} = \begin{pmatrix} 0 \\ \sqrt{E_{\boldsymbol{p}} + |\boldsymbol{p}|} \\ 0 \\ -\sqrt{E_{\boldsymbol{p}} - |\boldsymbol{p}|} \end{pmatrix} \xrightarrow[\text{boost}]{\text{big}} \sqrt{2E_{\boldsymbol{p}}}\begin{pmatrix} 0 \\ 1 \\ 0 \\ 0 \end{pmatrix}.$$
(36.38)

This has $v_{\mathrm{L}} = \sqrt{2E_{\boldsymbol{p}}}\begin{pmatrix} 0 \\ 1 \end{pmatrix}$ and $v_{\mathrm{R}} = 0$.

Finally, an antiparticle with spin down along the z-axis and momentum along $+z$ [Fig. 36.1(d)] has helicity -1 and a spinor

$$v(p) = \begin{pmatrix} \sqrt{E_{\boldsymbol{p}} - \boldsymbol{p}\cdot\boldsymbol{\sigma}}\begin{pmatrix} 1 \\ 0 \end{pmatrix} \\ -\sqrt{E_{\boldsymbol{p}} + \boldsymbol{p}\cdot\boldsymbol{\sigma}}\begin{pmatrix} 1 \\ 0 \end{pmatrix} \end{pmatrix} = \begin{pmatrix} \sqrt{E_{\boldsymbol{p}} - |\boldsymbol{p}|} \\ 0 \\ -\sqrt{E_{\boldsymbol{p}} + |\boldsymbol{p}|} \\ 0 \end{pmatrix} \xrightarrow[\text{boost}]{\text{big}} -\sqrt{2E_{\boldsymbol{p}}}\begin{pmatrix} 0 \\ 0 \\ 1 \\ 0 \end{pmatrix},$$
(36.39)

which has $v_{\mathrm{L}} = 0$ and $v_{\mathrm{R}} = -\sqrt{2E_{\boldsymbol{p}}}\begin{pmatrix} 1 \\ 0 \end{pmatrix}$.

Notice that highly relativistic particles with positive helicity have right-handed spinors, while relativistic antiparticles with positive helicity have left-handed spinors.

Imagine that there was an interaction that only coupled to left-handed particles. In the ultra-relativistic limit, where chirality and helicity become identical, we've demonstrated that only negative helicity particles and positive helicity antiparticles would interact. Amazingly, this is indeed the case for the **weak interaction**, which describes a field $W_\mu(x)$ which only couples to left-handed Dirac particles![11] This means that only a left-handed fermion may emit a $W^\pm$ particle. For a massive electron, which oscillates between being left- and right-handed, it may only emit a W^- particle (turning into a neutrino as it does so) when it is left-handed as shown in Fig. 36.2.

To summarize this section: the information content of a spinor tells us about the spin of a particle through the object ξ (or η for an antiparticle) which has two degrees of freedom. If we know ξ and the momentum of the particle then all four components of the spinor (two left-handed parts and two right-handed parts) are completely determined.

36.4   Basis states for superpositions

We turn to the orthogonality of the Dirac spinors. This property is useful if we are to use the spinors as a basis in order to express arbitrary wave functions as linear superpositions which contain spinor parts. Spinors carry information about the spin of the particle, so we want one spinor to be orthogonal to another if they describe different spin states. (The orthogonality of momentum states will be handled, as usual, by the $e^{-ip\cdot x}$ factors.)

Now, it's straightforward to show[12] that

$$u^\dagger(p)u(p) = 2E_{\boldsymbol{p}}\xi^\dagger\xi, \quad v^\dagger(p)v(p) = 2E_{\boldsymbol{p}}\eta^\dagger\eta.$$
(36.40)

[11]Neutrinos only exist as left-handed particles. They are (at least approximately) massless, which means that only negative helicity neutrinos and positive helicity antineutrinos have ever been observed.

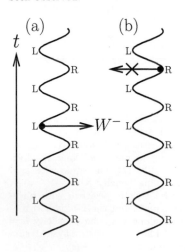

Fig. 36.2 Parity violation in the weak interaction. The electron oscillates between left- and right-handed chiralities. It may only emit a W^- particle when it is left-handed.

[12]See Exercise 36.3.

However this isn't going to be very useful. The reason is that these expressions are not Lorentz invariant: the combinations $u^\dagger u$ and $v^\dagger v$ are proportional to $E_{\boldsymbol{p}}$, which is frame-dependent. To make a Lorentz invariant quantity we define the adjoint or *bar* of a spinor as

$$\bar{u}(p) = u^\dagger(p)\gamma^0. \tag{36.41}$$

This is useful as a Lorentz invariant quantity can now be formed via

$$\bar{u}(p)u(p) = 2m\xi^\dagger\xi. \tag{36.42}$$

Now for the orthogonality: let's also choose a basis in which to describe the spin of a particle. We usually choose the basis $\xi^1 = \begin{pmatrix} 1 \\ 0 \end{pmatrix}$ and $\xi^2 = \begin{pmatrix} 0 \\ 1 \end{pmatrix}$ such that $\xi^{s\dagger}\xi^r = \delta^{sr}$. The result of all of this is that we can now write an expression for a Dirac spinor which is suitable for use as a basis state as

$$u^s(p) = \begin{pmatrix} \sqrt{p \cdot \sigma}\,\xi^s \\ \sqrt{p \cdot \bar{\sigma}}\,\xi^s \end{pmatrix}, \tag{36.43}$$

where $s = 1$ or 2 and we have

$$\bar{u}^s(p)u^r(p) = 2m\xi^{s\dagger}\xi^r = 2m\delta^{sr}. \tag{36.44}$$

We are now able to express an arbitrary particle solution of the Dirac equation as an integral over momentum and a sum over spin states as follows:

$$\psi^-(x) = \int \frac{\mathrm{d}^3p}{(2\pi)^{\frac{3}{2}}} \frac{1}{(2E_{\boldsymbol{p}})^{\frac{1}{2}}} \sum_{s=1}^{2} a_{s\boldsymbol{p}} u^s(p) e^{-\mathrm{i}p\cdot x}, \tag{36.45}$$

where $a_{s\boldsymbol{p}}$ is an amplitude here.

In the case of the antiparticle spinor we pick up a minus sign:

$$\bar{v}^s(p)v^r(p) = -2m\eta^{s\dagger}\eta^r = -2m\delta^{sr}, \tag{36.46}$$

and the antiparticle wave function is written

$$\psi^+(x) = \int \frac{\mathrm{d}^3p}{(2\pi)^{\frac{3}{2}}} \frac{1}{(2E_{\boldsymbol{p}})^{\frac{1}{2}}} \sum_{s=1}^{2} b_{s\boldsymbol{p}}^* v^s(p) e^{\mathrm{i}p\cdot x}, \tag{36.47}$$

where the components have amplitude $b_{s\boldsymbol{p}}^*$. Finally we note that the u's and v's are orthogonal:

$$\bar{u}^r(p)v^s(p) = \bar{v}^r(p)u^s(p) = 0. \tag{36.48}$$

We summarize all of our definitions in the master table below.

	Spinor	physical spin up	physical spin down	normalization
particle	$\begin{pmatrix} \sqrt{p \cdot \sigma}\,\xi \\ \sqrt{p \cdot \bar{\sigma}}\,\xi \end{pmatrix}$	$\begin{pmatrix} 1 \\ 0 \end{pmatrix}$	$\begin{pmatrix} 0 \\ 1 \end{pmatrix}$	$\bar{u}^s(p)u^r(p) = 2m\delta^{sr}$
antiparticle	$\begin{pmatrix} \sqrt{p \cdot \sigma}\,\eta \\ -\sqrt{p \cdot \bar{\sigma}}\,\eta \end{pmatrix}$	$\begin{pmatrix} 0 \\ 1 \end{pmatrix}$	$\begin{pmatrix} 1 \\ 0 \end{pmatrix}$	$\bar{v}^s(p)v^r(p) = -2m\delta^{sr}$

So why care about these spin sums? We'll see a little later that when we actually descend from our ivory tower to calculate something that people actually measure a certain amount of averaging is going to be required. This is because the incoming and outgoing polarizations of our particles are usually unknown. In that case the usual instruction is to *average over initial states and sum over final states*. This will lead to sums over the spin states where relaxations like $\sum_s u^s(p)\bar{u}^s(p) = \not{p} + m$ come in very useful.

Example 36.7

One of the most useful expressions in subsequent calculations is the sum

$$\sum_{s=1}^{2} u^s(p)\bar{u}^s(p) = \sum_s \begin{pmatrix} \sqrt{p\cdot\sigma}\,\xi^s \\ \sqrt{p\cdot\bar{\sigma}}\,\xi^s \end{pmatrix} \begin{pmatrix} \xi^{s\dagger}\sqrt{p\cdot\sigma} & \xi^{s\dagger}\sqrt{p\cdot\sigma} \end{pmatrix}. \tag{36.49}$$

We use the fact that

$$\sum_{s=1}^{2} \xi^s \xi^{s\dagger} = \begin{pmatrix} 1 \\ 0 \end{pmatrix}\begin{pmatrix} 1 & 0 \end{pmatrix} + \begin{pmatrix} 0 \\ 1 \end{pmatrix}\begin{pmatrix} 0 & 1 \end{pmatrix} = \begin{pmatrix} 1 & 0 \\ 0 & 1 \end{pmatrix}, \tag{36.50}$$

giving us a four-by-four matrix

$$\sum_{s=1}^{2} u^s(p)\bar{u}^s(p) = \begin{pmatrix} \sqrt{p\cdot\sigma}\sqrt{p\cdot\bar{\sigma}} & \sqrt{p\cdot\sigma}\sqrt{p\cdot\sigma} \\ \sqrt{p\cdot\bar{\sigma}}\sqrt{p\cdot\bar{\sigma}} & \sqrt{p\cdot\bar{\sigma}}\sqrt{p\cdot\sigma} \end{pmatrix} = \begin{pmatrix} m & p\cdot\sigma \\ p\cdot\bar{\sigma} & m \end{pmatrix}. \tag{36.51}$$

Finally we use our γ matrices to write this four-by-four object as

$$\sum_s u^s(p)\bar{u}^s(p) = \gamma\cdot p + m. \tag{36.52}$$

It's left as an exercise to prove the analogous expression for antiparticles

$$\sum_s v^s(p)\bar{v}^s(p) = \gamma\cdot p - m. \tag{36.53}$$

36.5   The non-relativistic limit of the Dirac equation

It is often stated that the Dirac equation is required to describe quantum mechanical spin. This isn't quite true as there's another equation that does this job perfectly well at low energy: the **Pauli equation** given by

$$\hat{H}\psi = \frac{(\boldsymbol{\sigma}\cdot\hat{\boldsymbol{p}})^2}{2m}\psi. \tag{36.54}$$

The Pauli equation emerges naturally as the non-relativistic limit of the Dirac equation.

Example 36.8

As in Chapter 12, we extract the non-relativistic limit by factorizing out the large mass contribution through a wave function $\psi_L(t,\boldsymbol{x}) = \phi_L(t,\boldsymbol{x})e^{-imt}$. From Dirac's equation $(\not{p} - m)\psi = 0$ we obtain

$$\begin{pmatrix} -m & \hat{E} - \boldsymbol{\sigma}\cdot\hat{\boldsymbol{p}} \\ \hat{E} + \boldsymbol{\sigma}\cdot\hat{\boldsymbol{p}} & -m \end{pmatrix}\begin{pmatrix} \phi_L(t,\boldsymbol{x})e^{-imt} \\ \phi_R(t,\boldsymbol{x})e^{-imt} \end{pmatrix} = 0, \tag{36.55}$$

where $\hat{E} = i\partial_0$. Noting that $\hat{E}\phi_a(t,\boldsymbol{x})e^{-imt} = e^{-imt}(m + \hat{E})\phi_a(t,\boldsymbol{x})$ (where $a = $ L or R) we obtain

$$\begin{aligned} -m\phi_L + (m + \hat{E} - \boldsymbol{\sigma}\cdot\hat{\boldsymbol{p}})\phi_R &= 0, \\ (m + \hat{E} + \boldsymbol{\sigma}\cdot\hat{\boldsymbol{p}})\phi_L - m\phi_R &= 0. \end{aligned} \tag{36.56}$$

Rewriting the second of these equations we produce

$$\phi_{\mathrm{R}} = \left(1 + \frac{\hat{E}}{m} + \frac{\boldsymbol{\sigma} \cdot \hat{\boldsymbol{p}}}{m}\right) \phi_{\mathrm{L}}, \tag{36.57}$$

allowing us to eliminate ϕ_{R} from the first equation, so that

$$\left(2\hat{E} + \frac{\hat{E}^2}{m} - \frac{(\boldsymbol{\sigma} \cdot \hat{\boldsymbol{p}})^2}{m}\right) \phi_{\mathrm{L}} = 0. \tag{36.58}$$

Recognizing that the $\hat{E}^2$ term may be neglected in comparison with the other terms at low energy, we end up with Pauli's equation

$$\hat{E}\phi_{\mathrm{L}} = \frac{(\boldsymbol{\sigma} \cdot \hat{\boldsymbol{p}})^2}{2m}\phi_{\mathrm{L}}. \tag{36.59}$$

The same equation applies for ϕ_{R} too, as can easily be shown.[13]

[13] See Exercise 36.8.

Example 36.9

The previous example gives us enough ammunition to show that the Pauli equation (and by extension, the Dirac equation) implies that the g-factor of the electron is $g = 2$. We may include the electromagnetic interaction in the Pauli equation by making the replacements[14] $\hat{\boldsymbol{p}} \to \hat{\boldsymbol{p}} - q\boldsymbol{A}$ and $\hat{E} \to \hat{E} - qA^0$. The result is

$$(\hat{E} - qA^0)\phi = \frac{[\boldsymbol{\sigma} \cdot (\hat{\boldsymbol{p}} - q\boldsymbol{A})]^2}{2m}\phi. \tag{36.60}$$

[14] This is equivalent to the replacement $\partial_\mu \to D_\mu$ that we examined in Chapter 14.

Multiplying out the right-hand side and using the identity $\sigma^i\sigma^j = \delta^{ij} + i\varepsilon^{ijk}\sigma^k$ we find[15]

$$(\hat{E} - qA^0)\phi = \frac{(\hat{\boldsymbol{p}} - q\boldsymbol{A})^2}{2m}\phi - \frac{q}{2m}\boldsymbol{\sigma} \cdot \boldsymbol{B}\phi, \tag{36.61}$$

where we have identified $\boldsymbol{B} = \boldsymbol{\nabla} \times \boldsymbol{A}$. The expected interaction Hamiltonian of a magnetic moment with an external $\boldsymbol{B}$-field may be written

$$\hat{H} = -\hat{\boldsymbol{\mu}} \cdot \boldsymbol{B} \equiv -g\frac{q}{2m}\hat{\boldsymbol{S}} \cdot \boldsymbol{B}. \tag{36.62}$$

[15] This is made slightly tricky by the fact that $\hat{\boldsymbol{p}}$ and $\boldsymbol{A}$ don't commute. If in doubt, write $\hat{\boldsymbol{p}} = -i\boldsymbol{\nabla}$ and remember that the differential operator acts on everything to its right.

Comparing this with $-\frac{q}{2m}\boldsymbol{B} \cdot \boldsymbol{\sigma}$ from eqn 36.61 and identifying the spin operator as $\hat{\boldsymbol{S}} = \frac{1}{2}\boldsymbol{\sigma}$ we conclude that $g = 2$.

The prediction that $g = 2$ (rather than $g = 1$) was historically an important one in confirming the validity of the Dirac equation. However, it is found experimentally that $g = 2.002319304....$ The reason for this discrepancy is that it is not appropriate to treat Dirac's equation as the equation of motion for a single particle as we have done here. In fact, the Dirac equation is correctly interpreted as providing the equation of motion for fermionic quantum fields. As we shall see in Chapter 41 this will lead to a prediction that agrees with experiment to a truly remarkable extent.

In this chapter we've encountered a new object, the spinor, to add to the scalar and vector objects that we've considered before. We have, however, treated the Dirac equation as a single-particle equation. We know that in reality particles are excitations in quantum fields. In order to be able to define spinor fields we need to know the all-important transformation properties of these spinor fields. In the next chapter we turn to the Lorentz transformation properties of these objects.

Chapter summary

- The Dirac equation $(\not{p} - m)\psi = 0$ results from an attempt to take the square root of the Klein–Gordon equation. The four components are required to represent spin and reflect symmetry under the parity operation.

- The operator '$\hat{p}$-slash' is $\not{p} = \gamma^{\mu}(\mathrm{i}\partial_{\mu})$ where γ_{μ} are the gamma matrices. In the chiral representation $\gamma^{\mu} = \begin{pmatrix} 0 & \sigma^{\mu} \\ \bar{\sigma}^{\mu} & 0 \end{pmatrix}$ where $\sigma^{\mu} = (I, \boldsymbol{\sigma})$, $\bar{\sigma}^{\mu} = (I, -\boldsymbol{\sigma})$ and $\gamma^{5} = \mathrm{i}\gamma^{0}\gamma^{1}\gamma^{2}\gamma^{3}$.

- Massless solutions of the Dirac equation are also the eigenstates of the chirality operator, and are known as left- and right-handed solutions. Left- or right-handed states with mass are not eigenstates of the Dirac equation because the mass term couples them. We can think of Dirac particles as oscillating in time between being left- and right-handed.

- There are positive and negative energy eigenstates of the Dirac equation. These give rise to particles and antiparticles.

- The non-relativistic limit of the Dirac equation yields the Pauli equation and predicts $g = 2$.

Exercises

(36.1) *An illustration of the reason for anticommutation and spin*

(a) Show that the Dirac equation can be recast in the form

$$\mathrm{i}\frac{\partial\psi}{\partial t} = \hat{H}_{\mathrm{D}}\psi, \tag{36.63}$$

where $\hat{H}_{\mathrm{D}} = \boldsymbol{\alpha} \cdot \hat{\boldsymbol{p}} + \beta m$ and find $\boldsymbol{\alpha}$ and β in terms of the γ matrices.

(b) Evaluate $\hat{H}_{\mathrm{D}}^{2}$ and show that for a Klein–Gordon dispersion to result we must have:

(i) that the α^{i} and β objects all anticommute with each other; and

(ii) $(\alpha^{i})^{2} = (\beta)^{2} = 1$.

This provides some justification for the anticommutation relations we imposed on the γs.

(c) Prove the following commutation relations

(i) $\left[\hat{H}, \hat{L}^{i}\right] = \mathrm{i}(\hat{\boldsymbol{p}} \times \boldsymbol{\alpha})^{i}$ where $\hat{\boldsymbol{L}} = \hat{\boldsymbol{x}} \times \hat{\boldsymbol{p}}$.

(ii) $\left[\hat{H}, \hat{S}^{i}\right] = -\mathrm{i}(\hat{\boldsymbol{p}} \times \boldsymbol{\alpha})^{i}$ where $\hat{\boldsymbol{S}} = \frac{1}{2}\boldsymbol{\Sigma}$ and we define $\boldsymbol{\Sigma} = \frac{\mathrm{i}}{2}\boldsymbol{\gamma} \times \boldsymbol{\gamma}$.

This tells us that the Dirac Hamiltonian doesn't commute with the orbital angular momentum operator $\hat{\boldsymbol{L}}$, mandating the existence of another form of angular momentum (i.e. the spin $\hat{\boldsymbol{S}}$) so that $\hat{H}$ commutes with the sum $\hat{\boldsymbol{J}} = \hat{\boldsymbol{L}} + \hat{\boldsymbol{S}}$.

(36.2) (a) Using the facts that $\gamma^{i\dagger} = -\gamma^{i}$ and $\gamma^{0\dagger} = \gamma^{0}$, take the Hermitian conjugate of the Dirac equation to show that we obtain the non-covariant form:

$$-\mathrm{i}\partial_{0}\psi^{\dagger}\gamma^{0} + \mathrm{i}\partial_{i}\psi^{\dagger}\gamma^{i} - m\psi^{\dagger} = 0. \tag{36.64}$$

(b) Show that to restore covariance we may multiply through by γ^{0} which results in

$$\mathrm{i}\partial_{\mu}\bar{\psi}\gamma^{\mu} + m\bar{\psi} = 0. \tag{36.65}$$

(36.3) (a) Verify eqn 36.40.

(b) Verify eqn 36.42.

(c) Verify eqn 36.46.

(36.4) The γ^{μ} matrices and spinors don't just have one form. Any form that obeys the anticommutation relations will also work. Consider, for example, the unitary transformation matrix

$$U = \frac{1}{\sqrt{2}} \begin{pmatrix} 1 & -1 \\ 1 & 1 \end{pmatrix}. \qquad (36.66)$$

Use this to transform the Dirac wave functions $\psi \to \hat{U}^\dagger \psi$ and show that for the resulting wave function to satisfy the Dirac equation the γ matrices have the explicit form

$$\gamma^0 = \begin{pmatrix} 1 & 0 \\ 0 & -1 \end{pmatrix}, \quad \gamma^i = \begin{pmatrix} 0 & \sigma^i \\ -\sigma^i & 0 \end{pmatrix}. \qquad (36.67)$$

This is known as the standard *or* Dirac–Pauli *representation.*

(36.5) Show that in (1+1)-dimensional spacetime, the requirement $\{\gamma^\mu, \gamma^\nu\} = 2g^{\mu\nu}$ is satisfied by $\gamma^0 = \sigma^2$ and $\gamma^1 = i\sigma^1$.

(36.6) Prove the following identities:
(a) $\{\gamma^\mu, \gamma^5\} = 0$.
(b) $(\gamma^\mu)^\dagger = \gamma^0 \gamma^\mu \gamma^0$.

(c) $(\gamma^5)^\dagger = \gamma^5$.
(d) $(\gamma^5)^2 = 1$.
(e) $(\gamma^5 \gamma^\mu)^\dagger = \gamma^0 \gamma^5 \gamma^\mu \gamma^0$.

(36.7) (a) Show that

$$i\bar{u}(p')\sigma^{\mu\nu}(p'-p)_\nu u(p) \qquad (36.68)$$
$$= \frac{1}{2}\bar{u}(p')\left[-\gamma^\mu(\not{p}'-m) + (m-\not{p})\gamma^\mu\right]u(p),$$

where $\sigma^{\mu\nu} = \frac{i}{2}[\gamma^\mu, \gamma^\nu]$.
(b) Show $\gamma^\mu \not{p} = 2p^\mu - \not{p}\gamma^\mu$.
(c) Use these two results to prove the **Gordon identity**

$$\bar{u}(p')\gamma^\mu u(p) = \bar{u}(p')\left[\frac{p'^\mu + p^\mu}{2m} + \frac{i\sigma^{\mu\nu}q_\nu}{2m}\right]u(p), \qquad (36.69)$$

where $q^\mu = (p'-p)^\mu$. This identity will be used in Chapter 41.

(36.8) Take the non-relativistic limit of the Dirac equation and find the equation of motion for ϕ_R.

37 How to transform a spinor

[1]The *Wisconsin State Journal*'s humorous writer 'Roundy' began his article as follows: *I've been hearing about a fellow they have up at the U. this spring – a mathematical physicist, or something, they call him – who is pushing Sir Isaac Newton, Einstein and all the others off the front page. So I thought I better go up and interview him for the benefit of the State Journal readers, same as I do all the other top notchers.* The article seems almost too good to be true, and unfortunately it seems that it is in fact a fake. Joseph Coughlin (nicknamed Roundy because of his rotund appearance) certainly existed but, as detailed in Graham Farmelo's biography of Dirac *The Strangest Man*, the article is most probably a spoof. Dirac however kept a copy of it.

[2]Recall that we define $\tanh \phi^i = \beta^i$.

[3]The commutation relations are

$$\begin{aligned} \left[J^i, J^j\right] &= \mathrm{i}\varepsilon^{ijk}J^k, \\ \left[J^i, K^j\right] &= \mathrm{i}\varepsilon^{ijk}K^k, \\ \left[K^i, K^j\right] &= -\mathrm{i}\varepsilon^{ijk}J^k. \end{aligned}$$

'Do you ever run across a fellow that even you can't understand?'
'Yes,' says he.
'This will make a great reading for the boys down at the office,' says I. 'Do you mind releasing to me who he is?'
'Weyl,' says he.
The interview came to a sudden end just then, for the doctor pulled out his watch and I dodged and jumped for the door. But he let loose a smile as we parted and I knew that all the time he had been talking to me he was solving some problem that no one else could touch. But if that fellow Professor Weyl ever lectures in this town again I sure am going to take a try at understanding him! A fellow ought to test his intelligence once in a while.
Roundy interviews Professor Dirac (1929), *Wisconsin State Journal*[1]

The humble electron cannot be described by either the scalar or vector field theories that we have discussed thus far. The electron, and indeed all spin-$\frac{1}{2}$ fermions, are excitations in a **spinor field**. The different sorts of field are classified by how they transform under rotations and Lorentz transformations. We know how scalars transform: by definition they don't change. We also know how to Lorentz transform and rotate vectors. In this chapter we discuss how to transform a spinor.

37.1 Spinors aren't vectors

Recall that a general Lorentz transformation and rotation can be written as a single matrix:

$$D(\boldsymbol{\theta}, \boldsymbol{\phi}) = \mathrm{e}^{-\mathrm{i}\boldsymbol{J}\cdot\boldsymbol{\theta} + \mathrm{i}\boldsymbol{K}\cdot\boldsymbol{\phi}}, \tag{37.1}$$

where $\boldsymbol{\theta}$ gives angle of rotation and $\boldsymbol{\phi}$ gives the velocity boost.[2] This general transformation may be applied to scalars, vectors or tensors by picking the appropriate form of the rotation generator $\boldsymbol{J}$ and boost generator $\boldsymbol{K}$, subject to their all-important commutation relations.[3] We've seen how rotations and boosts affect scalars and vectors, but how do they affect spinors? We have a problem in that our previous explicit expressions for $\boldsymbol{K}$ are only geared for vectors, which are objects whose rotations are described by the group $SO(3)$. We will see that spinors are rotated by the elements of a different group.

37.2 Rotating spinors

As Eugene Wigner proved, the rotations of spinors are generated by the elements of the group $SU(2)$. This is the group of two-by-two unitary matrices with unit determinant. The generators are the Pauli spin matrices whose commutation relations are $[\sigma^i, \sigma^j] = 2i\varepsilon^{ijk}\sigma^k$. This comes as no surprise since all of us learn at our mother's knee that two-component $S = 1/2$ spin states are acted on by an angular momentum operator $\hat{\boldsymbol{J}} = \frac{\boldsymbol{\sigma}}{2}$ and the angular momentum operator is the generator of rotations. The set of matrices that rotates two-component Weyl spinors is therefore

$$D(\boldsymbol{\theta}) = e^{-\frac{i}{2}\boldsymbol{\sigma}\cdot\boldsymbol{\theta}}. \tag{37.2}$$

These operators rotate the left-handed Weyl spinors $\psi_{\mathrm{L}}(p)$ and the right-handed Weyl spinors $\psi_{\mathrm{R}}(p)$ in exactly the same way, so the four-component Dirac spinor is rotated by the matrix

$$\begin{pmatrix} \psi_{\mathrm{L}} \\ \psi_{\mathrm{R}} \end{pmatrix} \rightarrow \begin{pmatrix} e^{-\frac{i}{2}\boldsymbol{\sigma}\cdot\boldsymbol{\theta}} & 0 \\ 0 & e^{-\frac{i}{2}\boldsymbol{\sigma}\cdot\boldsymbol{\theta}} \end{pmatrix} \begin{pmatrix} \psi_{\mathrm{L}} \\ \psi_{\mathrm{R}} \end{pmatrix}. \tag{37.3}$$

Example 37.1

With this knowledge we can calculate the 2×2 matrix that rotates Weyl spinors around a particular axis. We'll try the rotation by an angle θ^1 about the x-direction, whose matrix is[4]

$$D(\theta^1) = e^{-\frac{i}{2}\sigma^1\theta^1} = I\cos\frac{\theta^1}{2} - i\sigma^1\sin\frac{\theta^1}{2}. \tag{37.4}$$

To rotate four-component Dirac spinors we simply act on both the left- and right-handed spinors with this 2×2 matrix.

---

[4]Recall eqn 15.23.

37.3 Boosting spinors

Next we examine how to boost spinors by finding an explicit form for the boost generators $\boldsymbol{K}$. Since we know that spinors are rotated by the Pauli matrices via $\boldsymbol{J} = \frac{1}{2}\boldsymbol{\sigma}$ and we know the commutation relations of the generators $\boldsymbol{K}$ with themselves and $\boldsymbol{J}$, we guess that $\boldsymbol{K} \propto \boldsymbol{\sigma}$. We find that *two* possible choices of proportionality constants satisfy the commutation relations. It is easily checked that both

$$\boldsymbol{K} = \pm\frac{i\boldsymbol{\sigma}}{2} \tag{37.5}$$

will work. It seems, therefore, that there must be two sorts of spinors: those boosted by $\boldsymbol{K} = +\frac{i\boldsymbol{\sigma}}{2}$ and those boosted by $\boldsymbol{K} = -\frac{i\boldsymbol{\sigma}}{2}$ and the two choices of sign correspond to two distinct representations of the group. At rest, these two sorts of spinors should be indistinguishable. This is indeed correct and the two sorts of spinors are exactly the left- and right-handed Weyl spinors, ψ_{L} and ψ_{R}, that fell out of the Dirac equation in the previous chapter.

Example 37.2

To show that the Weyl spinors are transformed by different representations of the group we should be able to show that each representation obeys the Lie algebra of the generators of $SU(2)$, which is $[C^i, C^j] = i\varepsilon^{ijk}C^k$. This is possible if we define

$$J_{\pm} = \frac{J \pm iK}{2}, \tag{37.6}$$

since then the commutation relations read

$$\left[J_+^i, J_+^j\right] = i\varepsilon^{ijk}J_+^k, \tag{37.7}$$

$$\left[J_-^i, J_-^j\right] = i\varepsilon^{ijk}J_-^k. \tag{37.8}$$

Our new generators J_+ and J_- now have independent Lie algebras and each separately generates the rotations and boosts for a type of Weyl spinor. The simplest non-trivial case of this algebra corresponds to one or other of the new generators being zero, which gives the rules

$$\begin{aligned} J &= -iK \quad \text{for} \quad J_+ = 0, \\ J &= iK \quad \text{for} \quad J_- = 0. \end{aligned} \tag{37.9}$$

We then have generators $J = \sigma/2, K = i\sigma/2$ for the Weyl spinor ψ_L and generators $J = \sigma/2, K = -i\sigma/2$ for the Weyl spinor ψ_R, leading to our final answer for the transformations appropriate for Dirac spinors:

$$\begin{pmatrix} \psi_L \\ \psi_R \end{pmatrix} \to \begin{pmatrix} D_L(\boldsymbol{\theta}, \boldsymbol{\phi}) & 0 \\ 0 & D_R(\boldsymbol{\theta}, \boldsymbol{\phi}) \end{pmatrix} \begin{pmatrix} \psi_L \\ \psi_R \end{pmatrix}, \tag{37.10}$$

where

$$\begin{aligned} D_L(\boldsymbol{\theta}, \boldsymbol{\phi}) &= e^{\frac{\sigma}{2}\cdot(-i\boldsymbol{\theta}-\boldsymbol{\phi})}, \\ D_R(\boldsymbol{\theta}, \boldsymbol{\phi}) &= e^{\frac{\sigma}{2}\cdot(-i\boldsymbol{\theta}+\boldsymbol{\phi})}. \end{aligned} \tag{37.11}$$

After all of this formalism, we're left with the answer to our question of how to Lorentz boost a Dirac spinor. The Dirac wave function, which is made up of two spinors stacked on top of each other, is boosted according to $\psi \to e^{iK\cdot\phi}\psi$, or

$$\begin{pmatrix} \psi_L \\ \psi_R \end{pmatrix} \to \begin{pmatrix} e^{-\frac{1}{2}\sigma\cdot\phi} & 0 \\ 0 & e^{\frac{1}{2}\sigma\cdot\phi} \end{pmatrix} \begin{pmatrix} \psi_L \\ \psi_R \end{pmatrix}. \tag{37.12}$$

Example 37.3

We can get some practice by boosting the spinors along the z-direction.

$$\begin{pmatrix} \psi_L \\ \psi_R \end{pmatrix} \to \begin{pmatrix} e^{-\frac{1}{2}\sigma^3\phi^3} & 0 \\ 0 & e^{\frac{1}{2}\sigma^3\phi^3} \end{pmatrix} \begin{pmatrix} \psi_L \\ \psi_R \end{pmatrix}. \tag{37.13}$$

The elements of the boost matrix are given by

$$\begin{aligned} e^{\pm\frac{1}{2}\sigma^3\phi^3} &= 1 \pm \frac{\sigma^3\phi^3}{2} + \frac{1}{2!}\left(\frac{\sigma^3\phi^3}{2}\right)^2 \pm \dots \\ &= I\cosh(\phi^3/2) \pm \sigma^3\sinh(\phi^3/2). \end{aligned} \tag{37.14}$$

So the boost is

$$\begin{pmatrix} \psi_L \\ \psi_R \end{pmatrix} \to \begin{pmatrix} I\cosh\frac{\phi^3}{2} - \sigma^3\sinh\frac{\phi^3}{2} & 0 \\ 0 & I\cosh\frac{\phi^3}{2} + \sigma^3\sinh\frac{\phi^3}{2} \end{pmatrix} \begin{pmatrix} \psi_L \\ \psi_R \end{pmatrix}. \tag{37.15}$$

Reverting to 4×4 matrices we see that the components are boosted by a matrix

$$
\begin{pmatrix}
e^{-\frac{\phi^3}{2}} & 0 & 0 & 0 \\
0 & e^{\frac{\phi^3}{2}} & 0 & 0 \\
0 & 0 & e^{\frac{\phi^3}{2}} & 0 \\
0 & 0 & 0 & e^{-\frac{\phi^3}{2}}
\end{pmatrix}
\tag{37.16}
$$

Because $e^{\pm x} = \cosh x \pm \sinh x$ we can write $e^{\pm\frac{\phi^3}{2}} = \sqrt{\cosh \phi^3 \pm \sinh \phi^3}$. Identifying $\cosh \phi^3 = \gamma = E/m$ and $\sinh \phi^3 = \beta_z \gamma = |\boldsymbol{p}|/m$ we see that a boost along z is enacted by a matrix

$$
\frac{1}{\sqrt{m}}
\begin{pmatrix}
\sqrt{E - |\boldsymbol{p}|} & 0 & 0 & 0 \\
0 & \sqrt{E + |\boldsymbol{p}|} & 0 & 0 \\
0 & 0 & \sqrt{E + |\boldsymbol{p}|} & 0 \\
0 & 0 & 0 & \sqrt{E - |\boldsymbol{p}|}
\end{pmatrix}.
\tag{37.17}
$$

We found previously that the Dirac wave function for a particle at rest is given by $\begin{pmatrix} u_{\mathrm{L}} \\ u_{\mathrm{R}} \end{pmatrix} = \sqrt{m} \begin{pmatrix} \xi \\ \xi \end{pmatrix}$. Putting everything together we deduce that a boost in the z-direction has the following effect

$$
\begin{pmatrix}
u_{\mathrm{L}}^1 \\
u_{\mathrm{L}}^2 \\
u_{\mathrm{R}}^1 \\
u_{\mathrm{R}}^2
\end{pmatrix}
= \sqrt{m}
\begin{pmatrix}
\xi^1 \\
\xi^2 \\
\xi^1 \\
\xi^2
\end{pmatrix}
\rightarrow
\begin{pmatrix}
\sqrt{E - |\boldsymbol{p}|}\,\xi^1 \\
\sqrt{E + |\boldsymbol{p}|}\,\xi^2 \\
\sqrt{E + |\boldsymbol{p}|}\,\xi^1 \\
\sqrt{E - |\boldsymbol{p}|}\,\xi^2
\end{pmatrix}.
\tag{37.18}
$$

This provides some justification for the expression

$$
u(p) = \begin{pmatrix} \sqrt{p \cdot \sigma}\,\xi \\ \sqrt{p \cdot \bar{\sigma}}\,\xi \end{pmatrix},
\tag{37.19}
$$

as we claimed in the previous chapter.

37.4   Why are there four components in the Dirac equation?

The Dirac equation is written in terms of a four-component Dirac spinor ψ. These Dirac spinors have two degrees of freedom: the two components of ξ (or η). So why do we have four components in the Dirac equation?

We found that for the case of massless particles the Dirac spinors fell apart into two: left-handed Weyl spinors and right-handed Weyl spinors. Each of these spinors carried the same spin information in ξ (or η). The question is, then, why does the world need left- and right-handed fields. We will answer this question by asking what fundamental operation turns left-handed spinors into right-handed spinors? The answer is parity.

Many of the properties of fermions are invariant with respect to parity and so we require that Dirac particles have the same properties in a parity reversed world. (The weak interaction violates parity, but we ignore that for now.) The parity operation P does the following: it maps $x \to -x$ and $v \to -v$ and so it results in the changes to the generators:

$$
\mathsf{P}^{-1} \boldsymbol{K} \mathsf{P} = -\boldsymbol{K}, \quad \mathsf{P}^{-1} \boldsymbol{J} \mathsf{P} = +\boldsymbol{J}
\tag{37.20}
$$

(recall that we say that $\boldsymbol{K}$ is a vector and $\boldsymbol{J}$ is a pseudovector). The parity operation therefore turns a left-handed spinor into a right-handed spinor

$$\mathsf{P}\psi_\mathrm{L} \to \psi_\mathrm{R}, \quad \mathsf{P}\psi_\mathrm{R} \to \psi_\mathrm{L}. \qquad (37.21)$$

An explicit form for the parity operator, suitable for use on Dirac spinors, is $\mathsf{P} \equiv \gamma^0 = \begin{pmatrix} 0 & I \\ I & 0 \end{pmatrix}$.

As we have stressed, the Dirac wave functions are not actually single-particle wave functions: they represent fields. In the next chapter we will quantize these fields to make them operator-valued, so that when we input a position in spacetime we output a field operator. Dirac particles are excitations in these quantum fields.

Chapter summary

- We have derived the transformations for rotating and boosting spinors.
- Parity, enacted with the γ^0 matrix, swaps left- and right-handed parts of the wave function.

Exercises

(37.1) (a) Verify eqn 37.4.
(b) Show that the rotation matrix

$$D(\theta^3) = \mathrm{e}^{-\frac{\mathrm{i}}{2}\sigma^3\theta^3} = \begin{pmatrix} \mathrm{e}^{-\frac{\mathrm{i}\theta^3}{2}} & 0 \\ 0 & \mathrm{e}^{\frac{\mathrm{i}\theta^3}{2}} \end{pmatrix}. \qquad (37.22)$$

(37.2) Show that both choices of sign $\boldsymbol{K} = \pm\frac{\mathrm{i}\boldsymbol{\sigma}}{2}$ obey the necessary commutation relations.

(37.3) *A one-line derivation of the Dirac equation.*
(a) Given that the left- and right-handed parts of the Dirac spinor for a fermion at rest are identical, explain why we may write $(\gamma^0 - 1)u(p^0) = 0$.
(b) Prove that $\mathrm{e}^{\mathrm{i}\boldsymbol{K}\cdot\boldsymbol{\phi}}\gamma^0\mathrm{e}^{-\mathrm{i}\boldsymbol{K}\cdot\boldsymbol{\phi}} = \displaystyle{\not{p}}/m$.
(c) Use the result proven in (b) to boost

$$(\gamma^0 - 1)u(p^0) = 0,$$

and show that you recover the Dirac equation.

The quantum Dirac field

38

After examining the Dirac equation as an equation of motion for single particles and investigating the transformation properties of the spinor solutions we now turn to quantum fields that obey Dirac's equation as their equation of motion. We will approach the problem with a variety of the weapons we have developed in previous chapters. We start with canonical quantization. We will write down a Lagrangian density whose equations of motion are the Dirac equation. We will then quantize using the rules. The quantized field that we will produce is of fundamental importance since its excitations describe fermions such as the quarks and leptons of particle physics and various excitations in solids such as quasielectrons in metals. In order to carry out the sort of calculations that can make contact with experiments, such as scattering, we will need a propagator and Feynman rules. We will obtain the propagator with the path integral. Finally the properties of Noether's current will motivate us to write down the Lagrangian for perhaps the most successful of all quantum field theories: quantum electrodynamics.

38.1 Canonical quantization and Noether current

To quantize a field we start by writing down a Lagrangian density describing classical fields. It seems rather wrong-headed to write down a classical equation for a field that describes fermions. Although we can make a model of bosonic excitations like phonons out of masses and springs there is no such obvious classical analogue for fermions. Undaunted, we will write down the Lagrangian by choosing a form that when inserted into the Euler–Lagrange equation outputs Dirac's equation. The Lagrangian that leads to the Dirac equation is given by

$$\mathcal{L} = \bar{\psi}(\mathrm{i}\slashed{\partial} - m)\psi = \bar{\psi}(\mathrm{i}\gamma^\mu\partial_\mu - m)\psi, \tag{38.1}$$

where $\bar{\psi} = \psi^\dagger\gamma^0$. Just as free scalar field theory describes fields whose equation of motion is the Klein–Gordon equation, this Lagrangian describes fermion matter fields whose equation of motion is the Dirac equation.

We can use this Lagrangian to canonically quantize the Dirac field. This field has two components, just like the complex scalar field. The momentum conjugate to ψ is easily calculated using the ordinary method. We find that

$$\Pi_\psi^\mu = \frac{\delta\mathcal{L}}{\delta(\partial_\mu\psi)} = i\bar\psi\gamma^\mu, \tag{38.2}$$

and $\Pi_{\bar\psi}^\mu = 0$. We obtain a momentum field

$$\Pi_\psi^0 = i\bar\psi\gamma^0 = i\psi^\dagger. \tag{38.3}$$

The Hamiltonian follows in the usual fashion from $\mathcal{H} = \Pi_\psi^0\partial_0\psi - \mathcal{L}$, yielding

$$\mathcal{H} = \psi^\dagger(-i\gamma^0\boldsymbol{\gamma}\cdot\boldsymbol{\nabla} + m\gamma^0)\psi. \tag{38.4}$$

However, we also know another piece of information, namely the Dirac equation itself which simplifies things here. The Dirac equation tells us that $i\gamma^0\partial_0\psi = (-i\boldsymbol{\gamma}\cdot\boldsymbol{\nabla}+m)\psi$. This allows us to rewrite the Hamiltonian as

$$\mathcal{H} = \psi^\dagger i\partial_0\psi. \tag{38.5}$$

This looks much simpler! Next we make the fields quantum mechanical by imposing equal-time *anticommutation* relations:[1]

$$\{\hat\psi_a(t,\boldsymbol{x}),\hat\psi_b^\dagger(t,\boldsymbol{y})\} = \delta^{(3)}(\boldsymbol{x}-\boldsymbol{y})\delta_{ab}, \tag{38.6}$$

$$\{\hat\psi_a(t,\boldsymbol{x}),\hat\psi_b(t,\boldsymbol{y})\} = \{\hat\psi_a^\dagger(t,\boldsymbol{x}),\hat\psi_b^\dagger(t,\boldsymbol{y})\} = 0, \tag{38.7}$$

where a and b label the spinor components of the four-component field operators $\hat\psi(x)$. The mode expansions of these quantum fields are given by

$$\hat\psi(x) = \int\frac{d^3p}{(2\pi)^{\frac{3}{2}}}\frac{1}{(2E_{\boldsymbol{p}})^{\frac{1}{2}}}\sum_{s=1}^2\left(u^s(p)\hat a_{s\boldsymbol{p}}e^{-ip\cdot x} + v^s(p)\hat b_{s\boldsymbol{p}}^\dagger e^{ip\cdot x}\right),$$

$$\hat{\bar\psi}(x) = \int\frac{d^3p}{(2\pi)^{\frac{3}{2}}}\frac{1}{(2E_{\boldsymbol{p}})^{\frac{1}{2}}}\sum_{s=1}^2\left(\bar u^s(p)\hat a_{s\boldsymbol{p}}^\dagger e^{ip\cdot x} + \bar v^s(p)\hat b_{s\boldsymbol{p}}e^{-ip\cdot x}\right). \tag{38.8}$$

The creation and annihilation operators must themselves obey the anti-commutation relations

$$\{\hat a_{s\boldsymbol{p}},\hat a_{r\boldsymbol{q}}^\dagger\} = \{\hat b_{s\boldsymbol{p}},\hat b_{r\boldsymbol{q}}^\dagger\} = \delta^{(3)}(\boldsymbol{p}-\boldsymbol{q})\delta_{sr}. \tag{38.9}$$

Now it simply remains to insert the mode expansion into the Hamiltonian and normal order to obtain a quantized field theory.

[1] We also put hats on ψ and $\psi^\dagger$ to make the point that they are now operators.

Example 38.1

We obtain

$$\hat H = \int d^3x\frac{d^3p\,d^3q}{(2\pi)^3(2E_{\boldsymbol{p}}2E_{\boldsymbol{q}})^{\frac{1}{2}}}\sum_{s,r}\left(\bar u^s(p)\hat a_{s\boldsymbol{p}}^\dagger e^{ip\cdot x} + \bar v^s(p)\hat b_{s\boldsymbol{p}}e^{-ip\cdot x}\right)$$

$$\times\gamma^0 E_{\boldsymbol{q}}\left(u^r(q)\hat a_{r\boldsymbol{q}}e^{-iq\cdot x} - v^r(q)\hat b_{r\boldsymbol{q}}^\dagger e^{iq\cdot x}\right)$$

$$= \sum_{s,r}\int\frac{d^3p}{2}\left(u^{s\dagger}(p)\hat a_{s\boldsymbol{p}}^\dagger u^r(p)\hat a_{r\boldsymbol{p}} - v^{s\dagger}(p)\hat b_{s\boldsymbol{p}}v^r(p)\hat b_{r\boldsymbol{p}}^\dagger\right). \tag{38.10}$$

Lastly, we can use the properties of the normalized spinors to do the spin sum, since we know that $u^{s\dagger}u^r = 2E_{\boldsymbol{p}}\delta^{sr}$.

The (normal ordered) Hamiltonian becomes

$$\hat{H} = \int \mathrm{d}^3 p \sum_{s=1}^{2} E_{\boldsymbol{p}} \left(\hat{a}_{s\boldsymbol{p}}^\dagger \hat{a}_{s\boldsymbol{p}} + \hat{b}_{s\boldsymbol{p}}^\dagger \hat{b}_{s\boldsymbol{p}} \right). \qquad (38.11)$$

We see that the energy is given by the sum of the energies of the particles and antiparticles, exactly as we should expect.

The Lagrangian also has a global $U(1)$ symmetry, meaning that the Lagrangian is invariant with respect to the change $\psi(x) \to \psi(x)\mathrm{e}^{\mathrm{i}\alpha}$, so we can use Noether's theorem to find the conserved current and charge.

Example 38.2

With $\psi \to \psi \mathrm{e}^{\mathrm{i}\alpha}$ we also have $\bar{\psi} \to \bar{\psi}\mathrm{e}^{-\mathrm{i}\alpha}$. Since the transformation is global we also can write $\partial_\mu \psi \to \mathrm{e}^{\mathrm{i}\alpha}\partial_\mu \psi$. Putting this all together we get

$$\begin{aligned} \mathcal{L} &= \bar{\psi}(\mathrm{i}\slashed{\partial} - m)\psi \\ &\to \bar{\psi}\mathrm{e}^{-\mathrm{i}\alpha}(\mathrm{i}\slashed{\partial} - m)\psi \mathrm{e}^{\mathrm{i}\alpha} = \bar{\psi}(\mathrm{i}\slashed{\partial} - m)\psi. \end{aligned} \qquad (38.12)$$

The use of Noether's theorem goes through similarly to the case for the complex scalar field, except that we have

$$\Pi_\psi^\mu = \bar{\psi}\mathrm{i}\gamma^\mu, \quad \Pi_{\bar{\psi}}^\mu = 0, \qquad (38.13)$$

and $D\psi = \mathrm{i}\psi$. So, upon normal ordering and swapping signs to give the conventional number current direction, we obtain a Noether current operator

$$\hat{J}_{\mathrm{Nc}}^\mu = \hat{\bar{\psi}}\gamma^\mu \hat{\psi}. \qquad (38.14)$$

We can use this to work out the conserved $U(1)$ charge which, upon substitution of the mode expansion, yields up

$$\hat{Q}_{\mathrm{Nc}} = \int \mathrm{d}^3 p \sum_s \left(\hat{a}_{s\boldsymbol{p}}^\dagger \hat{a}_{s\boldsymbol{p}} - \hat{b}_{s\boldsymbol{p}}^\dagger \hat{b}_{s\boldsymbol{p}} \right). \qquad (38.15)$$

The conserved charge is given by the number of particles minus the number of antiparticles, as expected. Once again, this is just like the complex scalar field case.

38.2 The fermion propagator

The propagator for fermions can be worked out by calculating $G_0(x,y) = \langle 0|T\hat{\psi}(x)\hat{\bar{\psi}}(y)|0\rangle$. However, we can also use the path integral. For fermions we treat the classical fields ψ and $\bar{\psi}$ that appear in the path integral as Grassmann numbers.[2] In terms of the fields ψ and $\bar{\psi}$, the generating functional is written

$$Z[\eta,\bar{\eta}] = \int \mathcal{D}\psi \mathcal{D}\bar{\psi}\, \mathrm{e}^{\mathrm{i}\int \mathrm{d}^4 x \left[\bar{\psi}(x)\left(\mathrm{i}\slashed{\partial}-m\right)\psi(x)+\bar{\eta}(x)\psi(x)+\bar{\psi}(x)\eta(x)\right]}, \qquad (38.16)$$

Noether's theorem: $U(1)$ internal symmetry

$$\begin{aligned} D\psi &= \mathrm{i}\psi & D\psi^\dagger &= -\mathrm{i}\psi^\dagger \\ \Pi_\psi^\mu &= \mathrm{i}\bar{\psi}\gamma^\mu & \Pi_{\bar{\psi}}^\mu &= 0 \\ D\mathcal{L} &= 0 & W^\mu &= 0 \\ J_{\mathrm{Nc}}^0 &= \psi^\dagger \psi & J_{\mathrm{Nc}}^\mu &= \bar{\psi}\gamma^\mu \psi \\ \hat{Q}_{\mathrm{Nc}} &= \int \mathrm{d}^3 p \sum_{s=1}^{2}(\hat{n}_{s\boldsymbol{p}}^{(a)} - \hat{n}_{s\boldsymbol{p}}^{(b)}) \end{aligned}$$

See Exercise 38.4.

[2]See Chapter 28 for an introduction to Grassmann numbers.

where we notice that both $\psi(x)$ and $\bar{\psi}(x)$ must be coupled to separate source fields $\bar{\eta}(x)$ and $\eta(x)$ respectively. The leg-work for this calculation was carried out in Chapter 28 and so we note that the result is

$$Z[\bar{\eta}, \eta] = C e^{-\mathrm{i} \int \mathrm{d}^4 x \mathrm{d}^4 y \, \bar{\eta}(x)(\mathrm{i}\not\partial - m)^{-1}\eta(y)}, \qquad (38.17)$$

where C is proportional to a determinant. As usual, we normalize by dividing through by $Z[0,0]$ to obtain the normalized generating functional

$$\mathcal{Z}[\bar{\eta}, \eta] = e^{-\int \mathrm{d}^4 x \mathrm{d}^4 y \, \bar{\eta}(x)\mathrm{i}S(x-y)\eta(y)}, \qquad (38.18)$$

where $S(x - y) = (\mathrm{i}\not\partial - m)^{-1}$. The propagator is then the solution of the equation

$$(\mathrm{i}\not\partial - m)\mathrm{i}S(x) = \mathrm{i}\delta^{(4)}(x). \qquad (38.19)$$

The solution for the fermion propagator may be read off from this[3] as

$$G_0(x, y) = \mathrm{i}S(x - y) = \int \frac{\mathrm{d}^4 p}{(2\pi)^4} \frac{\mathrm{i}e^{-\mathrm{i}p\cdot(x-y)}}{\not{p} - m + \mathrm{i}\epsilon}. \qquad (38.20)$$

It's important to note that the fermion propagator $G_0(x, y)$ is a four-by-four matrix. This is revealed in momentum space if we multiply top and bottom by $(\not{p} + m)$ to give

$$
\begin{aligned}
\tilde{G}_0(p) &= \frac{\mathrm{i}}{\not{p} - m} = \frac{\mathrm{i}(\not{p} + m)}{p^2 - m^2 + \mathrm{i}\epsilon} \\
&= \frac{\mathrm{i}}{p^2 - m^2 + \mathrm{i}\epsilon} \begin{pmatrix} m & p^0 - \boldsymbol{p} \cdot \boldsymbol{\sigma} \\ p^0 + \boldsymbol{p} \cdot \boldsymbol{\sigma} & m \end{pmatrix}, \quad (38.21)
\end{aligned}
$$

where the latter equation is given in chiral representation. An interpretation of the four-by-four fermion propagator is revealed if we consider the roles of the left- and right-handed parts of the Dirac spinor. The fermion propagator G_0 is formed from the combination $\hat{\psi}(x)\hat{\bar{\psi}}(y)$ which, after substituting $\psi = \begin{pmatrix} \psi_\mathrm{L} \\ \psi_\mathrm{R} \end{pmatrix}$, has the form

$$\begin{pmatrix} \psi_\mathrm{L}(x)\psi_\mathrm{R}^\dagger(y) & \psi_\mathrm{L}(x)\psi_\mathrm{L}^\dagger(y) \\ \psi_\mathrm{R}(x)\psi_\mathrm{R}^\dagger(y) & \psi_\mathrm{R}(x)\psi_\mathrm{L}^\dagger(y) \end{pmatrix}. \qquad (38.22)$$

The top-left slot deals with a right-handed part entering and a left-handed part leaving. In contrast, the top-right slots tells us about a left-handed part entering and a left-handed part leaving. Recalling that a particle propagator is formed from a series $G = G_0 + G_0 V G_0 + G_0 V G_0 V G_0 + G_0 V G_0 V G_0 V G_0 + \ldots$, we may multiply the propagator matrices and obtain the matrix represented in Fig. 38.1. We see that the top-right slot still describes a left-handed part entering and a left-handed part leaving, but now expresses this as a superposition of all of the possible oscillations between left- and right-handed parts that result in the left-handed part leaving. The other slots of the matrix represent the other combinations of chiralities entering and leaving. In this sense

[3]Note that there is a difference of a minus sign in the equations defining the scalar and fermion propagators in terms of the Green's function of the equations of motion.

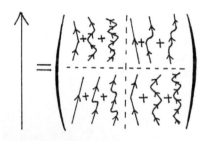

Fig. 38.1 The fermion propagator interpreted in terms of left- and right-handed spinors.

we may think of a massive Dirac particle propagating while oscillating between left- and right-handed.

With this in mind, our problem is solved since we have found the propagator for free fermions. This enables us to start solving scattering problems through the use of Feynman diagrams and it is the Feynman rules for fermions that we turn to next.

38.3 Feynman rules and scattering

The generating functional $Z[\bar{\eta}, \eta]$ can be used to find the Feynman rules for fermions. We won't do this, but simply list the rules in momentum space (see also Fig. 38.2).

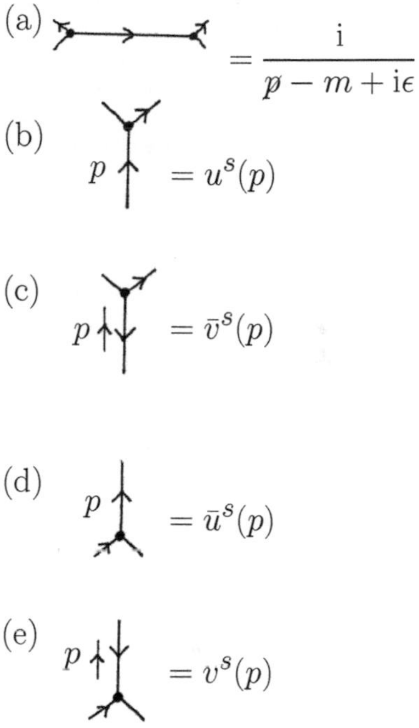

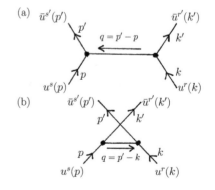

Fig. 38.2 Feynman rules for fermions.

> ## Feynman rules for fermions
>
> - A factor $i/(\not{p} - m + i\epsilon)$ for each internal fermion line.
> - A factor $u^s(p)$ for an incoming fermion with momentum p and spin s.
> - A factor $\bar{v}^s(p)$ for an incoming antifermion with momentum p and spin s.
> - A $\bar{u}^s(p)$ for an outgoing fermion.
> - A $v^s(p)$ for an outgoing antifermion.
> - Trace over the matrix product arising from a fermion loop.
> - Add minus sign factors reflecting fermion statistics, the most common of which is a factor -1 for every fermion loop.

We will now get some practice with the use of these rules.

Example 38.3

Let's use our rules to work out some Feynman diagrams for a theory of scalar phions interacting with electrons, via the interaction $\mathcal{L}_I = -g\bar{\psi}\psi\phi$. This is very much like the example of the complex scalar field interaction in Chapter 20, except that we need to consider the complication of the spinor polarizations.

We can start with electron–electron scattering mediated by the t-channel exchange of a virtual phion with mass m_ϕ, as shown in Fig. 38.3(a). Applying the rules, we obtain an amplitude

$$i\mathcal{M}_1 = (-ig)^2 \bar{u}^{s'}(p')u^s(p)\frac{i}{(p'-p)^2 - m_\phi^2}\bar{u}^{r'}(k')u^r(k), \qquad (38.23)$$

where we have ignored an overall energy-momentum conserving delta function. Next we consider electron–electron u-channel exchange scattering shown in Fig 38.3(b). Swapping the lines over corresponds to swapping operators in the S-matrix element, so this diagram comes with an extra minus sign. Applying the rules, we obtain

$$i\mathcal{M}_2 = -(-ig)^2 \bar{u}^{r'}(k')u^s(p)\frac{i}{(p'-k)^2 - m_\phi^2}\bar{u}^{s'}(p')u^r(k). \qquad (38.24)$$

We'll get a little further by considering the non-relativistic limit for distinguishable particles. This means that only the t-channel diagram contributes. For non-relativistic particles, we have the approximate properties:

$$p \approx (m_e, \boldsymbol{p}) \quad p' \approx (m_e, \boldsymbol{p'}) \quad k \approx (m_e, \boldsymbol{k}) \quad k' \approx (m_e, \boldsymbol{k'}) \qquad (38.25)$$

Fig. 38.3 (a) Electron–electron t-channel exchange. (b) Electron–electron u-channel exchange.

and, for spinors, that $u^s(p) = \sqrt{m_e}\begin{pmatrix} \xi^s \\ \xi^s \end{pmatrix}$. The spinor products become

$$\bar{u}^{s'}(p)u^s(p) = 2m_e\xi^{s'\,\dagger}\xi^s = 2m_e\delta_{s's}, \qquad (38.26)$$

showing that the scattering can't change the spin. Putting these contributions together yields a scattering amplitude[4]

$$i\mathcal{M} = \frac{4ig^2m_e^2}{(\boldsymbol{p}-\boldsymbol{p}')^2 + m_\phi^2}\delta_{s's}\delta_{r'r}. \qquad (38.27)$$

[4]This may be compared to the version in Chapter 20, which only differs in normalization and the absence of spin conserving Kronecker deltas. Here, as before, the Yukawa potential is attractive. In the next chapter we will recover the expected result that the electromagnetic interaction scattering like charges is repulsive.

We will see more examples of scattering in the next two chapters.

38.4   Local symmetry and a gauge theory for fermions

Our final job in this chapter will be to conjure up a gauge theory for fermions. Why do this? We know that electrons interact with electromagnetic fields (since they carry electric charge), and so we therefore plan to extend the gauge field Lagrangian of electromagnetism $\mathcal{L} = -\frac{1}{4}F_{\mu\nu}F^{\mu\nu}$ to include fermions and an interaction term telling us how photons and electrons interact. The gauge principle that we used in Chapter 14 allows us to do this since the act of promoting the Dirac Lagrangian to a gauge theory by making it locally $U(1)$ symmetric will provide us with an interaction term through the minimal coupling prescription.

As we've seen, the Dirac Lagrangian is invariant with respect to global $U(1)$ transformations, which is to say that the internal transformation $\psi \to \psi e^{i\alpha}$ leaves the Lagrangian invariant. The result of this is a conserved number current J_{Nc}. To make a gauge theory we will promote the global $U(1)$ symmetry to a local symmetry exactly as we have done before in Chapter 14. We simply replace our transformation with $\psi \to \psi e^{i\alpha(x)}\psi$ and introduce a gauge field via a covariant derivative $D_\mu = \partial_\mu + iqA_\mu(x)$. In order for the gauge field to ensure a local symmetry, it must itself transform as $A_\mu(x) \to A_\mu(x) - \frac{1}{q}\partial_\mu\alpha(x)$.

[5]We write $\not{D} = \gamma^\mu D_\mu = \gamma^\mu(\partial_\mu + iqA_\mu)$.

The Dirac Lagrangian, fixed up to have a local $U(1)$ invariance, is then given by[5]

$$\mathcal{L} = \bar{\psi}(i\not{D} - m)\psi. \qquad (38.28)$$

Inserting our covariant derivative, we have $\mathcal{L} = \bar{\psi}\left(i\gamma^\mu[\partial_\mu + iqA_\mu] - m\right)\psi$ or

$$\mathcal{L} = \bar{\psi}\left(i\not{\partial} - m\right)\psi - q\bar{\psi}\not{A}\psi. \qquad (38.29)$$

[6] It is also worth noting that the electromagnetic field tensor $F^{\mu\nu}$ may be generated by evaluating the commutator $[D^\mu, D^\nu] = iqF^{\mu\nu}$, as examined in Exercise 46.3.

Lo and behold, we are told how the gauge field $A_\mu(x)$ interacts with our fermion fields $\psi(x)$ and $\bar{\psi}(x)$: it's via the interaction term $\mathcal{L}_{\text{I}} = -q\bar{\psi}\gamma^\mu A_\mu\psi$. This method of determining the form of the interaction merely by considering the consequences of the substitution $p_\mu \to p_\mu - qA_\mu$ is another example of the minimal coupling prescription we saw in Chapter 14.[6]

In classical electromagnetism the conserved currents J_{em}^μ are the sources of the electromagnetic fields. These fields interact with the electric charges q, telling them how to move. This is encoded in an interaction between the electromagnetic current density $J_{\text{em}}^\mu(x)$ and the field $A^\mu(x)$ giving a contribution to the Lagrangian $\mathcal{L}_{\text{I}} = -J_{\text{em}}^\mu(x)A_\mu(x)$, where J_{em}^μ is a conserved current obeying $\partial_\mu J_{\text{em}}^\mu = 0$. Our minimal coupling prescription motivates us to make the step of identifying the fermion Noether current $J_{\text{Nc}}^\mu(x) = \bar{\psi}\gamma^\mu\psi$ with the electromagnetic current via $J_{\text{em}}^\mu = qJ_{\text{Nc}}^\mu$, where q is the electromagnetic charge. This constrains the source currents that may be coupled to the electromagnetic field and guarantees that the source current is a conserved quantity. This philosophy is illustrated in Fig. 38.4.

We end this chapter with a landmark equation. We will write down perhaps the most successful theory in modern physics: Quantum electrodynamics or QED. The QED Lagrangian includes the contributions from the gauge field of electromagnetism (eqn 5.49) and from the locally gauge invariant Dirac Lagrangian describing fermions and their interactions (eqn 38.28) giving

$$\mathcal{L} = -\frac{1}{4}F_{\mu\nu}F^{\mu\nu} + \bar{\psi}\left(\mathrm{i}\gamma^\mu\partial_\mu - m\right)\psi - q\bar{\psi}\gamma^\mu A_\mu\psi. \tag{38.30}$$

In the next chapter we will examine the predictions of this theory.

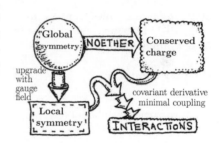

Fig. 38.4 The minimal coupling prescription in QED.

Chapter summary

- The Lagrangian that leads to the Dirac equation is $\mathcal{L} = \bar{\psi}(\mathrm{i}\slashed{\partial} - m)\psi$ where $\bar{\psi} = \psi^\dagger\gamma^0$ and can be quantized. The Noether current is $J_{\text{Nc}}^\mu = \bar{\psi}\gamma^\mu\psi$.
- The fermion propagator is $\mathrm{i}/(\slashed{p} - m)$.
- In order to have local gauge invariance, minimal coupling tells us that we must add to $\mathcal{L}$ a term $\mathcal{L}_{\text{I}} = -q\bar{\psi}\gamma^\mu A_\mu\psi$.

Exercises

(38.1) Try to canonically quantize the Dirac Lagrangian using commutation relations rather than anticommutation relations for the fields. You should get a minus sign in front of the $\hat{b}_p^\dagger\hat{b}_p$ term in the Hamiltonian. Explain why this is a catastrophe.
See Peskin and Schroeder, Chapter 3 for help.

(38.2) Verify eqn. (38.15).

(38.3) Show that the massless Dirac Lagrangian is invariant with respect to the global chiral $U(1)$ transformation $\psi(x) \to \mathrm{e}^{\mathrm{i}\alpha\gamma^5}\psi(x)$. Show that the Noether current is $J_{\text{Nc}}^\mu(x) = \bar{\psi}(x)\gamma^\mu\gamma^5\psi(x)$.

(38.4) Calculate the free fermion propagator by inserting the mode expansion into the equation $G_0(x, y) = \langle 0|T\hat{\psi}(x)\hat{\bar{\psi}}(y)|0\rangle$. You will need to use the spin summation rules from Chapter 36.

39

A rough guide to quantum electrodynamics

Sin-Itiro Tomonaga (1906-1979).

Quantum electrodynamics (QED) is the quantum field theory that describes the interaction of light and matter. Arguably one of the greatest intellectual achievements ever made, its discovery was due to several people but notably Tomonaga, Schwinger and Feynman, who shared the Nobel Prize in 1965.

Since QED describes an *interaction* it will come as little surprise that this is not a solvable theory and we must rely on perturbation theory and Feynman diagrams to understand the results of experiments. In this chapter we take the view that the Feynman diagrams *are* the theory and will therefore concentrate on finding the propagator for the photon and the Feynman rules for the photon–fermion interaction. One complication is that electrodynamics is a gauge invariant theory so we must expend some effort in ensuring that our dealings with the photon preserve this invariance. We will find that this can be done and, on the way, calculate our first amplitude in QED!

39.1 Quantum light and the photon propagator

[1]See eqn 24.29. Note that here we follow the convention, introduced in Chapter 17, that the free photon propagator is called $\tilde{D}_{0\mu\nu}(k)$.

Our task is to find the propagator for photons. We might guess that it's given by the $m \to 0$ limit of the propagator for the massive vector field we found[1] in Chapter 24, which is to say

$$\tilde{D}_{0\mu\nu}(k) = \lim_{m \to 0} \frac{\mathrm{i}(-g_{\mu\nu} + k_\mu k_\nu/m^2)}{k^2 - m^2 + \mathrm{i}\epsilon}. \tag{39.1}$$

However there's a problem in taking this limit: the (longitudinally projecting) term $k_\mu k_\nu/m^2$ threatens to blow up for photons on the mass shell, which have $k^2 = m^2 \to 0$. We will show later that this disaster does not occur and that gauge invariance allows the term $k_\mu k_\nu/m^2$ to be removed. Accepting this step for now, we are left with the photon propagator, given by

$$\tilde{D}_{0\mu\nu}(k) = \frac{-\mathrm{i}g_{\mu\nu}}{k^2 + \mathrm{i}\epsilon}. \tag{39.2}$$

[2]All photons that we detect actually interact with electrons in detectors such as the eye. They must *all* then, in some sense, be virtual! How can this be? We know that particles that are off-shell have the range over which they can propagate limited by the extent to which they're off-shell. If we see photons that have travelled from distant stars they have to be pretty close to being on-shell. We've seen before that when a particle is on-shell we hit the pole of the particle's propagator. Therefore photons from Andromeda, visible on a moonless night, must be so close to the pole that there can't be any observable effects from being off-shell.

As usual the propagator describes the contribution from a virtual particle, in this case the photon. By virtual we mean that the energy $E_{\boldsymbol{k}} \neq |\boldsymbol{k}|$, where $\boldsymbol{k}$ is the three-momentum of the photon.[2]

It's worth checking that our propagator correctly predicts the mechanism through which photons mediate the interaction between charges. Let's consider the fate of two electric charges interacting via the exchange of a photon, as shown in Fig. 39.1. Since we're interested in the photons we'll take the charges as forming electric currents J^μ. This has the advantage that the fermion and vertex parts of the amplitude are contained[3] in the current J^μ, leaving us free to concentrate on the physics of the photon.

We therefore consider the amplitude

$$\mathcal{A} = J_a^\mu \left(\frac{-\mathrm{i}g_{\mu\nu}}{k^2} \right) J_b^\nu, \qquad (39.3)$$

where we've left out the iϵ, since its presence won't be important for what follows.

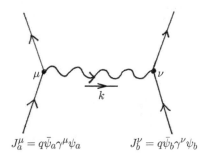

$$J_a^\mu = q\bar{\psi}_a \gamma^\mu \psi_a \qquad J_b^\nu = q\bar{\psi}_b \gamma^\nu \psi_b$$

Fig. 39.1 The exchange of a virtual photon.

Example 39.1

If we work in a frame where $k^\mu = (k^0, 0, 0, k^3)$, then the amplitude looks like

$$\mathcal{A} = -\frac{\mathrm{i}\left(J_a^0 J_b^0 - \boldsymbol{J}_a \cdot \boldsymbol{J}_b \right)}{(k^0)^2 - (k^3)^2}. \qquad (39.4)$$

Current conservation implies that $k_\mu J^\mu = 0$, so we have $k^0 J^0 - k^3 J^3 = 0$, which allows us to eliminate the component J^3 to yield

$$\mathcal{A} = \mathrm{i}\frac{J_a^0 J_b^0}{(k^3)^2} + \mathrm{i}\frac{J_a^1 J_b^1 + J_a^2 J_b^2}{(k^0)^2 - (k^3)^2}. \qquad (39.5)$$

Now for some interpretation.[4] The first term is in terms of J^0, the charge density. If we (inverse) Fourier transform this quantity we obtain an instantaneously acting Coulomb potential, which is repulsive between like charges

$$\int \frac{\mathrm{d}^4 k}{(2\pi)^4}\, \mathrm{e}^{-\mathrm{i}k \cdot x} \frac{J_a^0 J_b^0}{\boldsymbol{k}^2} \propto \frac{q^2}{4\pi|\boldsymbol{r}|} \delta(t_a - t_b). \qquad (39.6)$$

Don't worry about the instantaneousness of the term. It only looks unphysically instantaneous because we've split up the propagator in a non-covariant manner. Moreover, this is the term which dominates in the non-relativistic regime. This Coulomb interaction is of course the basis of much of condensed matter physics.

We argued above that the photons we observe are those very close to the pole. For our case of photons propagating along the z- (or 3-) direction, we look at the residue of the second term and we see that there seem to be two sorts of photon: those that couple J^1 currents and those that couple J^2 currents. These are the two physical transverse photon polarizations.

39.2 Feynman rules and a first QED process

We now turn to the interaction of the photon with a fermion. We saw in the previous chapter[5] that the result is an interaction term which expressed in Hamiltonian form is

$$\hat{\mathcal{H}}_\mathrm{I} = q\bar{\hat{\psi}}\gamma^\mu \hat{\psi}\hat{A}_\mu. \qquad (39.7)$$

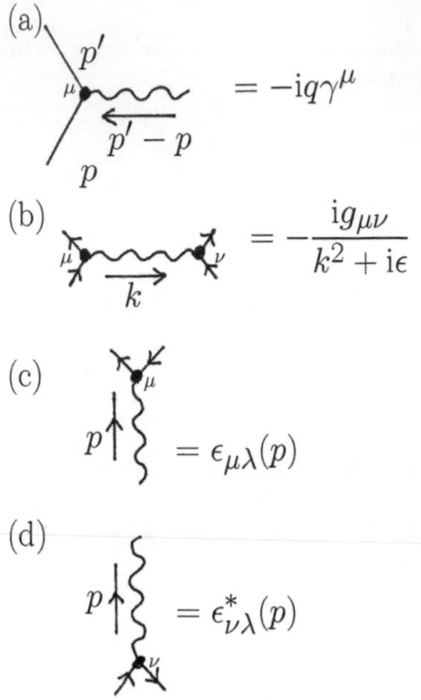

(a)

$$= -iq\gamma^\mu$$

(b)

$$= -\frac{ig_{\mu\nu}}{k^2 + i\epsilon}$$

(c)

$$= \epsilon_{\mu\lambda}(p)$$

(d)

$$= \epsilon^*_{\nu\lambda}(p)$$

Fig. 39.2 The QED Feynman rules.

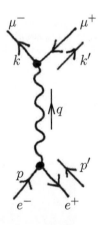

Fig. 39.3 A first QED process: $e^+e^- \rightarrow \mu^+\mu^-$.

Our treatment here follows Peskin and Schroeder.

[6]Remember that relativistic, right-handed fermions have $h = +1$ while right-handed antifermions have $h = -1$.

[7]See Chapter 36 if in doubt. We also assume here that $E = E_{\boldsymbol{p}} \approx E_{\boldsymbol{p}'}$.

The diagrammatic version of this interaction is shown in Fig. 39.2(a). This one simple vertex, whose translation in words is 'fermions can emit or absorb virtual photons', governs the entire theory.

We now have the ingredients to state the Feynman rules for quantum electrodynamics:

Feynman rules for QED

- Use all of the rules given previously for fermions.
- The interaction vertex contributes $-iq\gamma^\mu$, where $q(= Q|e|)$ is the charge [Fig. 39.2(a)].
- Every internal photon line contributes $\tilde{D}_{0\mu\nu}(k) = -ig_{\mu\nu}/(k^2+i\epsilon)$ [Fig. 39.2(b)].
- Incoming external photon lines contribute a polarization vector $\epsilon_{\mu\lambda}(p)$ [Fig. 39.2(c)].
- Outgoing external photon lines contribute $\epsilon^*_{\nu\lambda}(p)$ [Fig. 39.2(d)].

To get the feel of the rules, we will examine the case of $e^+e^- \rightarrow \mu^+\mu^-$, shown in the Feynman diagram in Fig. 39.3. Using the Feynman rules for fermions and photons, we obtain an invariant amplitude of

$$i\mathcal{M} = \bar{v}^{s'}(p')(-iQ|e|\gamma^\mu)u^s(p)\left(\frac{-ig_{\mu\nu}}{q^2}\right)\bar{u}^r(k)(-iQ|e|\gamma^\nu)v^{r'}(k'), \quad (39.8)$$

where we've written the charge as $Q|e|$. Note that the sign of the charge doesn't matter for many calculations and we will drop Q in what follows.

Example 39.2

We'll get some more practice manipulating spinors in these calculations by considering this process in the ultra-relativistic limit. We imagine the kinematics for the process to be those depicted in Fig. 39.4. Remember that in the ultra-relativistic limit particles in chirality eigenstates will simultaneously be in helicity eigenstates.

We'll start with a right-handed electron with initial momentum along $+z$. A relativistic, right-handed electron always has helicity $h = +1$ and so must have a spinor $\xi = \begin{pmatrix} 1 \\ 0 \end{pmatrix}$, corresponding to a physical spin-up along z. We will collide this electron with a right-handed positron with initial momentum along $-z$. A right-handed, highly relativistic *positron*[6] has $h = -1$, so its spinor is given by $\xi = \begin{pmatrix} 0 \\ 1 \end{pmatrix}$, corresponding to physical spin-up along z. Taking the limit of very large momenta[7] we have spinors for the incoming electrons given by

$$u(p) = \sqrt{2E}\begin{pmatrix} 0 \\ 0 \\ 1 \\ 0 \end{pmatrix} \quad v(p') = \sqrt{2E}\begin{pmatrix} 0 \\ 0 \\ 0 \\ -1 \end{pmatrix}. \quad (39.9)$$

To calculate our matrix element we need to evaluate products looking like $\bar{u}\gamma^\mu u = u^\dagger\gamma^0\gamma^\mu u$. It's therefore useful to know that

$$\gamma^0\gamma^\mu = \begin{pmatrix} 0 & 1 \\ 1 & 0 \end{pmatrix}\begin{pmatrix} 0 & \sigma^\mu \\ \bar{\sigma}^\mu & 0 \end{pmatrix} = \begin{pmatrix} \bar{\sigma}^\mu & 0 \\ 0 & \sigma^\mu \end{pmatrix}. \quad (39.10)$$

Plugging in our spinors and using this result gives us the result for the incoming interaction vertex that

$$\bar{v}(p')\gamma^\mu u(p) = 2E(0,0,0,-1)\begin{pmatrix} \bar{\sigma}^\mu & 0 \\ 0 & \sigma^\mu \end{pmatrix}\begin{pmatrix} 0 \\ 0 \\ 1 \\ 0 \end{pmatrix} = -2E\begin{pmatrix} 0 \\ 1 \\ i \\ 0 \end{pmatrix}. \quad (39.11)$$

So far, so good. Next we need to evaluate the outgoing part describing the muons. A useful insight here is that the quantity $\bar{v}(p')\gamma^\mu u(p)$ can be thought of as a four-vector describing the spin and momenta of the incoming electron states. We will take the inner product with a similar vector describing the outgoing muon states. We notice that the muon vector $\bar{u}(k)\gamma^\nu v(k')$ is almost the same four-vector as $\bar{v}(p')\gamma^\mu u(p)$ which we evaluated above for the incoming electrons. The difference is that it's complex-conjugated and rotated by an angle θ in the x-z plane. The complex conjugation is made easy with the identity[8] $[\bar{u}(p)\gamma^\mu u(k)]^* = \bar{u}(k)\gamma^\mu u(p)$ and we certainly know how to rotate a four-vector. Conjugating and rotating the result in eqn 39.11 therefore gives us

$$\begin{aligned} \bar{u}(k)\gamma^\mu v(k') &= [\bar{v}(k')\gamma^\mu u(k)]^* & \text{(Complex conjugation)} \\ &= [-2E(0,\cos\theta,i,-\sin\theta)]^* & \text{(Rotation about y by angle θ)} \\ &= -2E(0,\cos\theta,-i,-\sin\theta). \end{aligned}$$
$$(39.12)$$

Putting the amplitude together via the dot product $\bar{v}\gamma^\mu u g_{\mu\nu}\bar{u}\gamma^\nu v$ and including the photon propagator yields up the invariant amplitude

$$i\mathcal{M} = -i\frac{4e^2E^2}{q^2}(1+\cos\theta). \quad (39.13)$$

Noting that $q^2 = (2E)^2$, we obtain the simple result that $\mathcal{M}(e^-_{h+}e^+_{h-} \rightarrow \mu^-_{h+}\mu^+_{h-}) = -e^2(1+\cos\theta)$. (The subscript $h+$ means $h=1$; $h-$ means $h=-1$.)

This can be repeated for other combinations of chirality with the result that

$$\mathcal{M}(e^-_{h+}e^+_{h-} \rightarrow \mu^-_{h+}\mu^+_{h-}) = \mathcal{M}(e^-_{h-}e^+_{h+} \rightarrow \mu^-_{h-}\mu^+_{h+}) = -e^2(1+\cos\theta),$$
$$\mathcal{M}(e^-_{h+}e^+_{h-} \rightarrow \mu^-_{h-}\mu^+_{h+}) = \mathcal{M}(e^-_{h-}e^+_{h+} \rightarrow \mu^-_{h+}\mu^+_{h-}) = -e^2(1-\cos\theta), \quad (39.14)$$

and all other combinations yield an amplitude of zero.

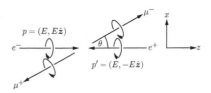

[8]See Exercise 39.3.

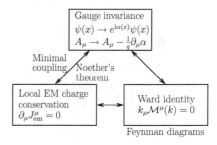

Fig. 39.5 Gauge invariance, the Ward identity and conservation of electromagnetic current are all intimately linked.

39.3 Gauge invariance in QED

Gauge invariance is central to our notion of QED. Global $U(1)$ symmetry of the Dirac Lagrangian guarantees the conservation of fermion number. Local $U(1)$ symmetry [guaranteed by the addition of the electromagnetic gauge field $A_\mu(x)$] allows us to identify the conserved fermion current with the conserved electromagnetic current and fixes the form of the QED interaction. Gauge invariance therefore leads to the conservation of electromagnetic current in QED. This is certainly central to the theory and must be maintained in our Feynman diagrams. Gauge invariance is indeed guaranteed in Feynman diagrams by the **Ward identity**. The symbiotic relationship between the concepts is shown in Fig. 39.5. The Ward identity comes in many forms[9] and the simplified one with which we'll be concerned is shown diagrammatically in Fig. 39.6. It says that *if* a sum of diagram parts contributing to an S-matrix element may be written $\mathcal{M}^\mu(k,p_1,p_2,...)$, where μ labels the vertex to which a photon line is attached, and that the photon line carries momentum k and the external lines are on mass shell, *then*

$$k_\mu\mathcal{M}^\mu(k,p_1,p_2,...) = 0. \quad (39.15)$$

$$k^\mu \cdot \left[\begin{array}{c} \\ \mu \quad k \\ \end{array}\right] = 0$$

S-matrix element

Fig. 39.6 The Ward identity. Dot the vector k^μ with the contribution to the S-matrix shown and you get zero.

John Ward (1924–2000). It has been said that his advances were used by others 'often without knowing it, and generally without quoting him' (M. Dunhill, *The Merton Record*, 1995).

[9]It is a special case of the Ward–Takahashi identities. See Peskin and Schroeder for its derivation.

Fig. 39.7 The half-dumbbell diagram.

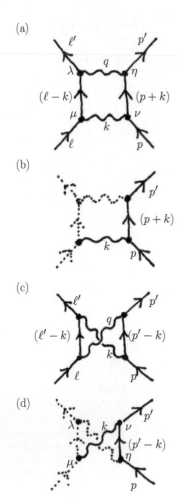

Fig. 39.8 Illustration of gauge invariance in the diagrammatic language used in Exercise 39.4.

This effectively kills the term $k_\mu k_\nu/k^2$ in the numerator of the photon propagator, since it always combines k_μ with the sort of vertex shown in Fig. 39.6.

We began our discussion of the photon propagator by arguing that the term $k_\mu k_\nu/k^2$ disappears from the numerator of our propagator. We conclude here that its disappearance is caused by gauge invariance and is encoded in Feynman diagrams by the Ward identity. A diagrammatic example of the Ward identity is examined in the exercises and a justification of the physics may be found in the next example.

Example 39.3

To see how gauge invariance leads to the disappearance of this term we naively couple a general source J^μ (not the conserved current of electromagnetism) to a massive vector field A_μ. We will calculate the amplitude for starting with no vector mesons and ending with one, represented by the half-dumbbell diagram shown in Fig 39.7. We can write the amplitude for creating a massive vector meson with any polarization as

$$\mathcal{A} \propto \sum_\lambda \epsilon^*_{\lambda\mu}(k)\tilde{J}^\mu(k), \qquad (39.16)$$

where the sum is over the polarizations. Assuming a basis of linear polarizations (i.e. so $\epsilon^*_{\lambda\mu} = \epsilon_{\lambda\mu}$), this leads to a probability

$$\begin{aligned} P &= |\mathcal{A}|^2 \propto \sum_\lambda \tilde{J}^\mu(k)\tilde{J}^{\nu\dagger}(k)\epsilon_{\lambda\mu}(k)\epsilon_{\lambda\nu}(k) \\ &= (-g_{\mu\nu} + k_\mu k_\nu/m^2)\tilde{J}^\mu(k)\tilde{J}^{\nu\dagger}(k), \qquad (39.17) \end{aligned}$$

where we've used the result for $\sum_\lambda \epsilon_{\lambda\mu}(k)\epsilon_{\lambda\nu}(k) = -P^{\mathrm{T}}_{\mu\nu}$ discussed in Chapter 13. Next we ask what this tells us about the source emitting a photon, by attempting to take a limit $m \to 0$. We immediately see that the $k_\mu k_\nu/m$ term blows up as $m \to 0$, which is the same problem we had with trying to apply the massive vector propagator to photons.

However, this won't cause a problem if $k_\mu \tilde{J}^\mu(k) = 0$, since this kills the troublesome term. Notice that this is current conservation $\partial_\mu J^\mu = 0$ written in momentum space! We conclude that the disappearance of the troublesome $k_\mu k_\nu/m^2$ term is due to electromagnetic current conservation, which is itself attributable to gauge invariance.

Looked at in another way, the disappearance of the $k_\mu k_\nu/k^2$ term allows us to add any multiple of this quantity back into the propagator. That is to say, vast helpings of a term that has no effect on the physics may be added for your convenience. A more general version of the photon propagator can therefore be given by

$$\tilde{D}_{0\mu\nu}(k) = \frac{-\mathrm{i}\left(g_{\mu\nu} + (1-\xi)k_\mu k_\nu/k^2\right)}{k^2 + \mathrm{i}\epsilon}. \qquad (39.18)$$

The simple choice $\xi = 1$ is known as Feynman gauge, while $\xi = 0$ is known as Landau gauge. As examined in the exercises, exactly this propagator follows from fixing the gauge in which we work in the Lagrangian itself. The freedom to choose ξ is therefore another manifestation of gauge invariance.

We have written down the Feynman rules for QED and examined a first process. In the next chapter we go further and look at the main (and measurable) application of QED calculations: calculating the scattering amplitudes for some fundamental processes.

Chapter summary

- The photon propagator has been derived and every internal photon line contributes a factor $-ig_{\mu\nu}/(k^2+i\epsilon)$, though alternative choices can be made in different gauges.
- We have presented the Feynman rules for QED.

Exercises

(39.1) *We will use the gauge fixing technique to find the QED propagator.*

(a) Show that the equations of motion for the electromagnetic field $A^\mu(x)$ written in momentum space are given by

$$- \left(k^2 g^{\mu\nu} - k^\mu k^\nu\right) \tilde{A}_\mu(k) = 0. \quad (39.19)$$

(b) We will attempt to find the inverse of these equations of motion. Explain why this must be of the form $(M^{-1})_{\nu\sigma} = C(k)g_{\nu\sigma} + H(k)k_\nu k_\sigma$.

(c) Show that defining the inverse via $M^{\mu\nu}(M^{-1})_{\nu\sigma} = g^\mu_\sigma$ demands the impossible condition

$$-k^2 C(k)g^\mu_\sigma + C(k)k^\mu k_\sigma = g^\mu_\sigma. \quad (39.20)$$

(d) We may fix this by adding a term to the Lagrangian that ensures we work in Lorentz gauge $\partial_\mu A^\mu = 0$. We therefore add a *gauge fixing* term and obtain a Lagrangian

$$\mathcal{L} = -\frac{1}{4}F^{\mu\nu}F_{\mu\nu} - \frac{1}{2\xi}(\partial_\mu A^\mu)^2, \quad (39.21)$$

where ξ is an arbitrary parameter. The point here is that the gauge fixing term will force the theory to be well behaved, but will contain the parameter ξ which (we claim) will not enter into any measurable quantity. Show that the equations of motion, written in momentum space, are now given by

$$\left(-k^2 g^{\mu\nu} + k^\mu k^\nu - \frac{1}{\xi}k^\mu k^\nu\right)\tilde{A}_\nu(k) = 0. \quad (39.22)$$

(e) Finally, verify that the photon propagator, defined as $i(M^{-1})_{\mu\nu}$, is given by

$$\tilde{D}_{0\mu\nu}(k) = -i\left(\frac{g_{\mu\nu} - (1-\xi)\frac{k_\mu k_\nu}{k^2}}{k^2 + i\epsilon}\right). \quad (39.23)$$

(39.2) Consider the reaction $e^+e^- \to \mu^+\mu^-$ again. Most experiments are done by firing unpolarized beams of electrons and positrons at each other. Since muon detectors are usually not sensitive to muon polarization we actually need to throw away the spin information. What we want is an *average over initial electron spin states and a sum over the final muon spin states*.

(a) Explain why this is the right thing to do and show that this prescription leads to a probability

$$\frac{1}{2}\sum_s \frac{1}{2}\sum_{s'}\sum_r\sum_{r'}|\mathcal{M}(s,s' \to r,r')|^2. \quad (39.24)$$

(b) Show that we obtain a scattering probability for unpolarized electrons and muons of $\frac{1}{4}\sum_{\text{spins}}|\mathcal{M}|^2 = e^4(1 + \cos^2\theta)$. *We will revisit this next chapter.*

(39.3) Verify $[\bar{u}(p)\gamma^\mu u(k)]^* = \bar{u}(k)\gamma^\mu u(p)$.

(39.4) We will examine how the term $k_\mu k_\nu/k^2$ is killed in a simple situation. Consider the Feynman diagram shown in Fig. 39.8(a) which contributes, via the S-matrix to the second-order correction to the amplitude for fermion–fermion scattering.

(a) Using the Feynman rules write down the amplitude for the whole process.

(b) Consider the bold part, shown in Fig. 39.8(b). This features a virtual electron propagator, with a photon line hitting it at the end labelled ν. Show that the amplitude $\mathcal{A}_{\text{part}}$ for this part of the diagram is

$$\mathcal{A}_{\text{part}} = \bar{u}(p')(-ie\gamma^\eta)\left(\frac{i}{\not{p}+\not{k}-m_e}\right)(-ie\gamma^\nu)$$
$$\times \left(\frac{-ig_{\mu\nu}+ik_\mu k_\nu/m_\gamma^2}{k^2+i\epsilon}\right)u(p), \quad (39.25)$$

where we've called the photon mass m_γ and the fermion mass m_e.

(c) Now just consider the dangerous part, which is proportional to $1/m_\gamma^2$. Show that the guts of this boil down to something proportional to

$$\mathcal{A}_{\text{guts}} = \bar{u}(p')\gamma^\eta\frac{k_\mu k_\nu/m_\gamma^2}{\not{p}+\not{k}-m_e}\gamma^\nu u(p). \quad (39.26)$$

(d) Show that this may be further simplified to

$$\mathcal{A}_{\text{guts}} = \frac{1}{m_\gamma^2}\frac{\bar{u}(p')\gamma^\eta k_\mu u(p)}{k^2+i\epsilon}. \quad (39.27)$$

Hint: rewrite the denominator as $\not{k} = (\not{p}+\not{k}-m_e)-(\not{p}-m_e)$ and use Dirac's equation in the form $\not{p}u(p)=m_e u(p)$.

(e) Now turn to the process shown in Fig. 39.8(c) which it is also necessary to consider when calculating the second-order correction to the fermion–fermion scattering amplitude. Just consider the part shown in Fig 39.8(d) (and note the change in labelling). Show that the dangerous guts of this part are given by

$$\mathcal{B}_{\text{guts}} = \bar{u}(p')\gamma^\nu\frac{k_\mu k_\nu/m_\gamma^2}{\not{p}'-\not{k}-m_e}\gamma^\eta u(p), \quad (39.28)$$

and that this may be reduced to

$$\mathcal{B}_{\text{guts}} = -\frac{1}{m_\gamma^2}\frac{\bar{u}(p')k_\mu\gamma^\eta u(p)}{k^2+i\epsilon}. \quad (39.29)$$

The point, then, is that we must consider both of these processes and the sum of the dangerous guts $\mathcal{A}_{\text{guts}}+\mathcal{B}_{\text{guts}}=0$, which means that the dangerous $k_\mu k_\nu/m_\gamma^2$ term disappears.

This argument is discussed in more depth in Zee, Chapter II.7.

QED scattering: three famous cross-sections

40

In this chapter we examine three famous scattering examples in QED: (1) Rutherford scattering, which led to the discovery of the atom; (2) Mott scattering, which is the relativistic version of Rutherford scattering and (3) Compton scattering, which demonstrates the particle-like properties of light. These examples will introduce a number of useful tricks which are employed freely in more advanced applications.

40.1 Example 1: Rutherford scattering

We'll examine the scattering of an electron from a heavy, charged object such as a point-like nucleus. This scattering of distinguishable particles is described by a Feynman diagram shown in Fig. 40.1(a). The nucleus is very heavy compared to the electron, so we treat the the process as involving an electron passing through a classical potential $A_{cl}^\mu(x)$. The Feynman rule for the vertex describing the electron interacting with the potential is $-iQ|e|\gamma_\mu \tilde{A}_{cl}^\mu(q)$ [Fig. 40.1(b)], that is, the electron (with $Q = -1$) interacts via the Fourier transform of the potential $A_{cl}^\mu(x)$. This is easy to understand: the interaction between an electron and a potential is $\mathcal{H}_I = Q|e|\bar\psi\gamma_\mu\psi A_{cl}^\mu$. The ψ brings an electron in and $\bar\psi$ takes it out, so the interaction vertex is proportional to what's left, which is $Q|e|\gamma_\mu A_{cl}^\mu$. For a static, positive electromagnetic potential we have $A_{cl}^0(\boldsymbol{r}) = \frac{Z|e|}{4\pi|\boldsymbol{r}|}$, and $\boldsymbol{A}_{cl}(\boldsymbol{r}) = 0$, so the potential we are after is the Fourier transform of the Coulomb interaction $\tilde{A}_{cl}^0(q) = \frac{Z|e|}{q^2}$. The amplitude for the diagram in Fig. 40.1(a) for a ($Q = -1$) electron to be scattered by a nuclear potential is given by

$$i\mathcal{M} = i\frac{Ze^2}{q^2}\bar{u}(p')\gamma^0 u(p). \tag{40.1}$$

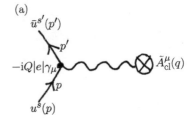

(a)

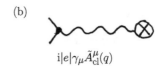

(b)

$i|e|\gamma_\mu \tilde{A}_{cl}^\mu(q)$

Fig. 40.1 (a) Scattering from a classical potential. (b) The Feynman rule for a $Q = -1$ electron to scatter from the potential.

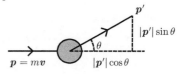

Fig. 40.2 Kinematics for Rutherford scattering.
[1]See Exercise 40.1.

Recall that in the non-relativistic limit we have $u(p) \approx \sqrt{m}\begin{pmatrix} \xi \\ \xi \end{pmatrix}$.

Example 40.1

The kinematics of the scattering is shown in Fig. 40.2. Calculating the transferred momentum gives us[1] $q^2 = -|\boldsymbol{q}|^2 = -4\boldsymbol{p}^2\sin^2\frac{\theta}{2}$. In this example we will work in the non-relativistic limit, for which we have that $\bar{u}(p')\gamma^0 u(p) = 2m\xi^\dagger\xi = 2m$. Putting everything together we obtain an amplitude

$$|\mathcal{M}|^2 = \frac{4Z^2e^4m^2}{16\boldsymbol{p}^4\sin^4(\theta/2)}. \tag{40.2}$$

[2]See, for example, Peskin and Schroeder, Chapters 4 and 5.

An alleged scientific discovery has no merit unless it can be explained to a barmaid.
Ernest Rutherford (1871–1937)

We collect here some useful tricks for evaluating expressions in QED.

Spinor trick 1:

$$[\bar{u}(f)\Gamma u(i)]^* = \left[\bar{u}(i)\gamma^0\Gamma^\dagger\gamma^0 u(f)\right]$$

Spinor trick 2:

$$\sum_s u^s(p)\bar{u}^s(p) = \not{p} + m$$

Trace trick 0:

$$\mathrm{Tr}(I) = 4$$

Trace trick 1:

$$\mathrm{Tr}(\text{odd number of } \gamma \text{ matrices}) = 0$$

Trace trick 2:

$$\mathrm{Tr}(\gamma^\mu\gamma^\nu) = 4g^{\mu\nu}$$

Trace trick 3:

$$\mathrm{Tr}(\gamma^\mu\gamma^\nu\gamma^\rho\gamma^\sigma)$$
$$= 4(g^{\mu\nu}g^{\rho\sigma} - g^{\mu\rho}g^{\nu\sigma} + g^{\mu\sigma}g^{\nu\rho})$$

Trace trick 4:

$$\gamma^\mu\not{d}\gamma_\mu = -2\not{d}$$

Photon trick:

$$\sum_{\text{polarizations}} \epsilon_\mu(p)\epsilon_\nu^*(p) \to -g_{\mu\nu}$$

See Peskin and Schroeder, Chapter 5, for the origin of the photon trick.

[3]Manipulations used in eqn 40.7 are:

Line 1: $u^s(p)_\beta\bar{u}^s(p)_\eta = [\not{p} + m]_{\beta\eta}$.

Line 2: Rearranging and using spinor trick 2 on primed variables.

Line 3: Identifying a trace.

Finally, we can relate our amplitude to something measurable. For scattering from a potential, the differential cross-section[2] is given by

$$\frac{d\sigma}{d\Omega} = \frac{|\mathcal{M}|^2}{(4\pi)^2}, \tag{40.3}$$

and so plugging in our expression for $|\mathcal{M}|^2$, we obtain

$$\frac{d\sigma}{d\Omega} = \frac{Z^2\alpha^2}{4m^2\boldsymbol{v}^4\sin^4(\theta/2)}, \tag{40.4}$$

which is Ernest Rutherford's celebrated result.

40.2   Example 2: Spin sums and the Mott formula

If we work through the previous example again, but this time without invoking the non-relativistic approximation, we obtain the relativistically correct version of Rutherford's result, which is known as the Mott formula. The complication with working with relativistic particles will be the manipulation of spinors.

Example 40.2

The scattering amplitude is still given by $i\mathcal{M} = i\frac{Ze^2}{q^2}\bar{u}^{s'}(p')\gamma^0 u^s(p)$, but here we've included the spin indices. We could work out the spinor part by brute force. However, this isn't usually necessary in real life. Experiments involve scattering initially unpolarized fermions and detecting the products of the scattering in detectors which aren't sensitive to polarization. We therefore average over the initial spin state and sum over the final spin states. We therefore want $\frac{1}{2}\sum_{s',s}|\mathcal{M}|^2$, whose spinor part is given by

$$\frac{1}{2}\sum_{s',s}|\bar{u}^{s'}(p')\gamma^0 u^s(p)|^2 = \frac{1}{2}\sum_{s',s}\left[\bar{u}^{s'}(p')\gamma^0 u^s(p)\right]\left[\bar{u}^{s'}(p')\gamma^0 u^s(p)\right]^*. \tag{40.5}$$

The calculations of these spin sums are quite straightforward, but do rely on the knowledge of a number of tricks (see sidenote). We will draw freely from them throughout this chapter.

Using spinor trick 1, we have $|\bar{u}(f)\gamma^0 u(i)|^2 = \left[\bar{u}(f)\gamma^0 u(i)\right]\left[\bar{u}(i)\gamma^0 u(f)\right]$. Using this, and writing out indices, our spinor term becomes

$$\frac{1}{2}\sum_{s,s'}\bar{u}^{s'}(p')_\alpha\gamma^0_{\alpha\beta}u^s(p)_\beta\bar{u}^s(p)_\eta\gamma^0_{\eta\lambda}u^{s'}(p')_\lambda, \tag{40.6}$$

where the explicit inclusion of subscripts allows us to rearrange the elements behind the sum sign (and we assume a sum over repeated indices, as usual). Next we use spinor trick 2 which says $\sum_s [u^s(p)\bar{u}^s(p)]_{\alpha\beta} = [\not{p} + m]_{\alpha\beta}$. This is used to perform the sum over the spin polarizations s and s' as follows[3]

$$\begin{aligned}\frac{1}{2}\sum_{s,s'}|\bar{u}^{s'}(p')\gamma^0 u^s(p)|^2 &= \frac{1}{2}\sum_{s'}\bar{u}^{s'}(p')_\alpha\gamma^0_{\alpha\beta}[\not{p}+m]_{\beta\eta}\gamma^0_{\eta\lambda}u^{s'}(p')_\lambda \\ &= \frac{1}{2}\gamma^0_{\alpha\beta}[\not{p}+m]_{\beta\eta}\gamma^0_{\eta\lambda}[\not{p}'+m]_{\lambda\alpha} \\ &= \frac{1}{2}\mathrm{Tr}\left[\gamma^0(\not{p}+m)\gamma^0(\not{p}'+m)\right].\end{aligned} \tag{40.7}$$

To recap, we've boiled down the sum over spinors into the trace of a number of matrices.

Next we note that the vector $\not{a} = \gamma^\mu a_\mu$ and we use some fun properties of the traces of products of γ matrices. Trace trick 1 says that traces of odd numbers of γs vanish which means we may write:

$$\frac{1}{2}\mathrm{Tr}\left[\gamma^0(\not{p}+m)\gamma^0(\not{p}'+m)\right] = \frac{1}{2}\left\{\mathrm{Tr}\left[\gamma^0\not{p}\gamma^0\not{p}'\right] + m^2\mathrm{Tr}\left[(\gamma^0)^2\right]\right\}. \tag{40.8}$$

It's fairly easy to work out $\mathrm{Tr}(\gamma^0\not{p}\gamma^0\not{p}')$ by plugging in explicit forms of the matrices, or alternatively we can use the trace trick 3, to obtain $4(E_{\boldsymbol{p}'}E_{\boldsymbol{p}}+\boldsymbol{p}'\cdot\boldsymbol{p})$. Trace trick 0 gives us $4m^2$ for the last term. Remembering that the scattering is elastic, we obtain

$$\frac{1}{2}\sum_{s,s'}|\bar{u}^{s'}(p')\gamma^0 u^s(p)|^2 = 2(E_{\boldsymbol{p}}^2 + \boldsymbol{p}\cdot\boldsymbol{p}' + m^2). \tag{40.9}$$

Referring back to the kinematics of the scattering, it's straightforward to show that $2(E_{\boldsymbol{p}}^2+\boldsymbol{p}^2\cos\theta+m^2) = 4E_{\boldsymbol{p}}^2(1-\beta\sin^2\theta/2)$, which combined with $|\boldsymbol{q}|^2 = 4\boldsymbol{p}^2\sin^2\theta/2$ allows us to complete the problem. We finally obtain

$$\frac{1}{2}\sum_{s,s'}|\mathcal{M}|^2 = \frac{Z^2e^4}{4\boldsymbol{p}^2\beta^2\sin^4\theta/2}(1-\beta^2\sin^2\theta/2). \tag{40.10}$$

We can utilize our cross-section equation once more to obtain

$$\frac{\mathrm{d}\sigma}{\mathrm{d}\Omega} = \frac{Z^2\alpha^2}{4\boldsymbol{p}^2\beta^2\sin^4\theta/2}(1-\beta^2\sin^2\theta/2), \tag{40.11}$$

which is **Mott's formula**.

I myself am neither an experimentalist nor a real mathematician; my theory stops at the Schrödinger equation. What I've done in this subject is to look at the evidence, do calculations on the back of an envelope and say to the theoretician 'if **you** *apply your techniques to this problem, this is how it will come out' and to the experimentalists just the same thing.*
Nevill Mott (1905–1996)

40.3 Example 3: Compton scattering

Compton scattering is the process $e^- + \gamma \to e^- + \gamma$. The two lowest order diagrams (which are second order in the interaction term) are shown in Fig. 40.3.

Arthur Compton (1892–1962)
With his wife and two sons, Dr. Compton lives in Chicago in a big brick house filled with souvenirs of their world tour. He does not know the taste of hard liquor, almost never smokes, always offers a cigaret to women visitors. He plays such a bang-up game of tennis that he sometimes has a hard time finding worthy opponents. Several times a month he puts in an evening of mandolin-playing with three friends. When his graduate students have finished an examination, he likes to dine them and take them to the theatre.
Time Magazine (1936).

Example 40.3

As practice in translating Feynman diagrams, we may translate the Compton diagrams into amplitudes. The first diagram yields (upon dropping Qs again)

$$\mathrm{i}\mathcal{M}_{s\text{-channel}} = \bar{u}(p')(\mathrm{i}|e|\gamma^\nu)\epsilon^*_{\nu\lambda'}(k')\frac{\mathrm{i}}{(\not{p}+\not{k})-m+\mathrm{i}\epsilon}\epsilon_{\mu\lambda}(k)(\mathrm{i}|e|\gamma^\mu)u(p), \tag{40.12}$$

and the second diagram gives

$$\mathrm{i}\mathcal{M}_{u\text{-channel}} = \bar{u}(p')(\mathrm{i}|e|\gamma^\nu)\epsilon_{\nu\lambda}(k)\frac{\mathrm{i}}{(\not{p}-\not{k}')-m+\mathrm{i}\epsilon}\epsilon^*_{\mu\lambda'}(k')(\mathrm{i}|e|\gamma^\mu)u(p). \tag{40.13}$$

These amplitudes may be evaluated to give the Klein–Nishina formula. However, the full derivation of the Klein–Nishina formula is rather lengthy and can be found in many books,[4] so we'll evaluate the s-channel contribution from the first diagram for the special case of a highly relativistic particle. (Recall that here 'relativistic' means we can ignore the mass of the electron.) We therefore have that $p^2 = 0$ and (as usual) $k^2 = 0$. We're interested in $|\mathcal{M}|^2$ averaged over initial spin states and photon polarizations and summed over final ones (we pick up a factor 1/4 in doing this). We therefore want (employing spinor trick 1[5] and dropping the photon polarization labels):

$$\frac{1}{4}\sum_{s,s'}\sum_{\text{polarizations}}\frac{e^4}{(p+k)^2}\left[\bar{u}^{s'}(p')\gamma^\nu\epsilon^*_\nu(k')(\not{p}+\not{k})\epsilon_\mu(k)\gamma^\mu u^s(p)\right]$$
$$\times\left[\bar{u}^s(p)\gamma^\sigma\epsilon^*_\sigma(k)(\not{p}+\not{k})\epsilon_\rho(k')\gamma^\rho u^{s'}(p')\right]. \tag{40.14}$$

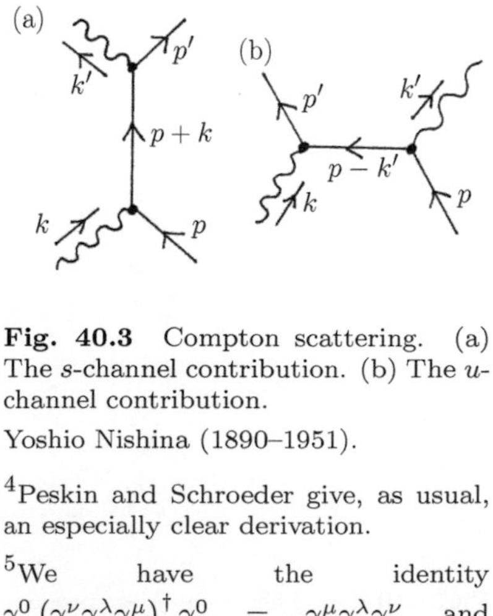

Fig. 40.3 Compton scattering. (a) The s-channel contribution. (b) The u-channel contribution.

Yoshio Nishina (1890–1951).

[4] Peskin and Schroeder give, as usual, an especially clear derivation.

[5] We have the identity $\gamma^0\left(\gamma^\nu\gamma^\lambda\gamma^\mu\right)^\dagger\gamma^0 = \gamma^\mu\gamma^\lambda\gamma^\nu$, and so $\gamma^0\left(\gamma^\nu\epsilon^*_\nu\not{p}\epsilon_\mu\gamma^\mu\right)^\dagger\gamma^0 = \gamma^\mu\epsilon^*_\mu\not{p}\epsilon_\nu\gamma^\nu$.

Successive lines of eqn 40.15 use the following ideas:

Line 1: Photon trick.

Line 2: Contracting indices and spinor trick 2.

Line 3: Using $\not{p}'^2 = p^2 = 0$ and trace trick 4.

Line 4: Trace trick 3.

Line 4: $u = (p' - k)^2 = -2p' \cdot k$, $s = (p + k)^2 = 2p \cdot k$.

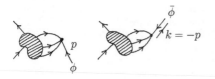

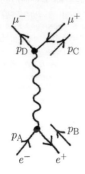

Fig. 40.4 Crossing symmetry. A Feynman diagram with an incoming particle may be manipulated to describe a new process involving an outgoing antiparticle with opposite charge and four-momentum.

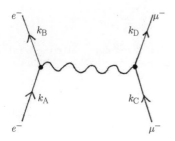

Fig. 40.5 The s-channel process $e^-e^+ \to \mu^-\mu^+$.

Fig. 40.6 The t-channel process $e^-\mu^- \to e^-\mu^-$ found by crossing the previous diagram.

Freely employing our array of tricks we have

$$
\begin{aligned}
\frac{1}{4} \sum |\mathcal{M}_{s\text{-channel}}|^2 &= \frac{e^4}{4s^2} \sum_{s,s'} \Big(g_{\mu\sigma} g_{\nu\rho} \bar{u}^{s'}(p') \gamma^\nu (\not{p} + \not{k}) \gamma^\mu u^s(p) \\
&\quad \times \bar{u}^s(p) \gamma^\sigma (\not{p} + \not{k}) \gamma^\rho u^{s'}(p') \Big) \\
&= \frac{e^4}{4s^2} \text{Tr} \left[\gamma^\nu (\not{p} + \not{k}) \gamma^\mu \not{p} \gamma_\mu (\not{p} + \not{k}) \gamma_\nu \not{p}' \right] \\
&= \frac{e^4}{s^2} \text{Tr} \left[\not{k} \not{p} \not{k} \not{p}' \right] \\
&= \frac{4e^4}{s^2} 2(p \cdot k)(p' \cdot k) = -\frac{2e^4 u}{s}.
\end{aligned} \tag{40.15}
$$

It turns out that the second diagram contributes $-\frac{2e^4 s}{u}$ and there is no interference term between the two. The electron spin and photon polarization averaged, squared amplitude for Compton scattering is then

$$
\frac{1}{4} \sum_{s,s'} \sum_{\text{polarizations}} |\mathcal{M}|^2 = -2e^4 \left(\frac{u}{s} + \frac{s}{u} \right). \tag{40.16}
$$

40.4   Crossing symmetry

As a final illustration of the utility of Feynman diagrams in describing scattering, we briefly note another very useful feature. Feynman's interpretation of antiparticles as particles travelling backward in time may be used to manipulate Feynman diagrams, allowing us access to several more physical amplitudes from the evaluation of a single Feynman amplitude.

The principle is shown in Fig. 40.4, where we see that if we start with a diagram with an incoming particle ϕ then we may flip the leg to create a valid process involving an outgoing antiparticle $\bar{\phi}$ with opposite charge and momentum. This process, known as a **crossing**, has the wonderful property that

$$
\mathcal{M}\left(\phi(p) + ... \to ...\right) = \mathcal{M}\left(... \to ... + \bar{\phi}(k)\right), \quad p = -k. \tag{40.17}
$$

Example 40.4

Consider the s-channel process $e^-e^+ \to \mu^-\mu^+$ examined in Chapter 39 and shown in Fig. 40.5. If we 'cross' the diagram by reversing the momentum direction of the incoming e^+ and outgoing μ^+ (changing the sign of their charge as we do so), then upon flipping the diagram over we obtain Fig. 40.6 describing the process $e^-\mu^- \to e^-\mu^-$. The amplitude for the resulting diagram is the same as the $\mathcal{M}$ we calculated in Chapter 39 if we make the replacements: $p_A \to k_A$, $p_B \to -k_B$, $p_C \to k_C$ and $p_D \to -k_D$. Note that we have turned an s-channel process into a t-channel process with our crossing, reflected by the fact that instead of reversing individual momentum we may simply replace $s \to t$ in the amplitude.

Crossing symmetries are clearly a very useful feature of the formalism because calculating $|\mathcal{M}|^2$ for one process immediately gives us access to $|\mathcal{M}|^2$ for all processes related by crossings. As we have only briefly touched on this topic here, we recommend you consult one of the standard texts before deploying crossings in anger. Peskin and Schroeder and Halzen and Martin both contain several examples.

Chapter summary

- This chapter has illustrated three simple scattering examples from QED: (1) Rutherford scattering; (2) the Mott formula; (3) Compton scattering.
- Crossing symmetry makes a Feynman amplitude for one process describe another process.

Exercises

(40.1) Verify that the transferred momentum in the Rutherford calculation is given by $q^2 = -4p^2 \sin^2 \frac{\theta}{2}$.

(40.2) Verify the Trace Tricks used in this chapter.

(40.3) (a) Calculate $\text{Tr}(\gamma^0 \not{p} \gamma^0 \not{p}')$ using explicit forms of the matrices.

(b) Repeat the calculation using Trace Trick 3.

(40.4) Consider the $e^- e^+ \rightarrow \mu^- \mu^+$ problem from the previous chapter. Using the tricks considered in this chapter show that

$$\frac{1}{4} \sum_{\text{spins}} |\mathcal{M}|^2 = \frac{e^4}{4q^4} \text{Tr}\left[(\not{p}' - m_e)\gamma^\mu(\not{p} + m_e)\gamma^\nu\right]$$
$$\times \text{Tr}\left[(\not{k} + m_\mu)\gamma_\mu(\not{k}' - m_\mu)\gamma_\nu\right] \quad (40.18)$$

(b) Taking all particles to be highly relativistic, show that

$$\frac{1}{4} \sum_{\text{spins}} |\mathcal{M}|^2 = \frac{8e^4}{q^4} \left[(p \cdot k)(p' \cdot k') + (p \cdot k')(p' \cdot k)\right]. \quad (40.19)$$

(c) Finally, use the kinematics discussed in Chapter 39 to show $\frac{1}{4}\sum_{\text{spins}}|\mathcal{M}|^2 = e^4(1 + \cos^2\theta)$.
See Peskin and Schroeder, Chapter 5 for help.

(40.5) Show that eqn 40.19, which describes the process $e^- e^+ \rightarrow \mu^- \mu^+$, may be written in terms of Mandelstam variables as

$$\frac{1}{4} \sum_{\text{spins}} |\mathcal{M}|^2 = \frac{8e^4}{s^2}\left[\left(\frac{t}{2}\right)^2 + \left(\frac{u}{2}\right)^2\right]. \quad (40.20)$$

Use a crossing to determine the analogous equation for the process $e^- \mu^- \rightarrow e^- \mu^-$.

(40.6) **Møller scattering** *is the process* $e^- e^- \rightarrow e^- e^-$.
(a) Two diagrams contribute to the Feynman amplitude for this process. Identify these and show that

$$\mathcal{M} = -\frac{e^2}{t}\bar{u}(k)\gamma^\mu u(p)\bar{u}(k')\gamma_\mu u(p')$$
$$+ \frac{e^2}{u}\bar{u}(k')\gamma^\mu u(p)\bar{u}(k)\gamma_\mu u(p') \quad (40.21)$$

(b) Show further that, in the ultra-relativistic limit,

$$\frac{1}{4} \sum_{\text{spins}} |\mathcal{M}|^2 = \frac{e^4}{4}\left\{\frac{1}{t^2}\text{Tr}\left[\not{k}\gamma^\mu \not{p}\gamma^\nu\right]\text{Tr}\left[\not{k}'\gamma_\mu \not{p}'\gamma_\nu\right]\right.$$
$$+ \frac{1}{u^2}\text{Tr}\left[\not{k}'\gamma^\mu \not{p}\gamma^\nu\right]\text{Tr}\left[\not{k}\gamma_\mu \not{p}'\gamma_\nu\right]$$
$$- \frac{1}{tu}\text{Tr}\left[\not{k}\gamma^\mu \not{p}\gamma^\nu \not{k}'\gamma_\mu \not{p}'\gamma_\nu\right]$$
$$\left. - \frac{1}{tu}\text{Tr}\left[\not{k}'\gamma^\mu \not{p}\gamma^\nu \not{k}\gamma_\mu \not{p}'\gamma_\nu\right]\right\}.$$

(c) Use a crossing symmetry to turn the previous equation into one describing the process $e^- e^+ \rightarrow e^- e^+$, which is known as **Bhabha scattering**.

<table>
</table>

41

The renormalization of QED and two great results

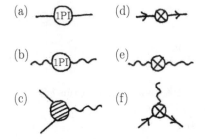

Fig. 41.1 Feynman diagrams for renormalized QED.

His laboratory is his ball-point pen.
Caption to a photograph of Julian Schwinger (1918–1994)
(holding a pen)

$$\frac{\alpha}{2\pi}$$

Engraved on Julian Schwinger's tombstone

Some really stunning consequences of QED are seen when we examine processes that contain electron–positron loops. The integrals we encounter when considering these processes contain divergences that we must tame with renormalization. In this chapter we will demonstrate two of the great results that follow from renormalizing QED. These are (1) the fact that the electronic charge is not a constant and (2) the fact that the g-factor of the electron is not quite $g = 2$.

To renormalize QED we calculate three Green's functions:

(a) Electron 1PI self-energy $\quad -i\tilde{\Sigma}(\not{p})$,
(b) Photon 1PI self-energy $\quad i\tilde{\Pi}^{\mu\nu}(q)$, $\qquad$ (41.1)
(c) Interaction vertex function $\quad -iQ|e|\tilde{\Gamma}^{\mu}(p, p')$.

These functions, each made of an infinite sum of Feynman diagrams, are themselves shown diagrammatically in Fig. 41.1(a–c). In calculating each of these Green's functions we will encounter contributions which lead to divergences. These divergences are fixed through the introduction of counterterms. QED is a renormalizable theory and only three counterterm diagrams, each corresponding to one of the Green's functions above, are required. The QED counterterms are shown in Fig. 41.1(d–f). They have Feynman rules:

(d) Electron self-energy $\quad$ (Bird on a wire) $\quad i(\not{p}B + A)$,
(e) Photon self-energy $\quad$ (Bird on a wave) $\quad i(g^{\mu\nu}q^2 - q^\mu q^\nu)C$,
(f) Interaction vertex $\quad$ (Electrocuted bird) $\quad iQ|e|\gamma^\mu D$.
$\qquad$ (41.2)

In order to renormalize the theory we impose conditions on each of the Green's functions which ensure that we are expanding in the correct masses and coupling constants. For QED the renormalization conditions are:

$$-i\tilde{\Sigma}(\not{p} = m) = 0 \qquad \text{fixes electron mass to } m,$$
$$i\tilde{\Pi}^{\mu\nu}(q = 0) = 0 \qquad \text{fixes photon mass to } 0, \qquad (41.3)$$
$$-iQ|e|\tilde{\Gamma}^{\mu}(p' - p = 0) = -iQ|e|\gamma^\mu \qquad \text{fixes charge to } Q|e|.$$

We will begin our tour of the consequences of renormalizing QED by examining the case of the photon propagator and show how renormalizing it explains the dielectric properties of the vacuum.

41.1 Renormalizing the photon propagator: dielectric vacuum

A dielectric is an insulator which may be polarized with an applied electric field. This is encoded in the dielectric constant of a material ϵ, which tells us how much the dielectric alters the electrostatic potential of a charge. Renormalized QED predicts that the interacting vacuum is a dielectric! The dielectric properties of the vacuum arise because interactions dress the bare photon in a cloud of electron–positron pairs. To see how this works we will examine the photon's self-energy. As in Chapter 33 we define the 1PI photon self-energy which is given by

$$i\tilde{\Pi}^{\mu\nu}(q) = \sum \left(\begin{array}{c} \text{All 1PI diagrams that can be} \\ \text{inserted between photon lines} \end{array} \right). \tag{41.4}$$

(This differs in sign from the 1PI self-energy Green's functions encountered previously.) Some examples of contributions to $i\tilde{\Pi}^{\mu\nu}(q)$ are shown in Fig. 41.2. Notice that these involve the photon removing electron–positron pairs from the vacuum in various ways and then returning them.

Looked at in terms of pictures, it's clear how the photon self-energy will renormalize the photon propagator. As shown in Fig. 41.3, we simply sum the contribution of the self-energy to infinity. The version in equations is made a little trickier by the presence of the indices, which reflect the vector-like nature of the photon field.[1]

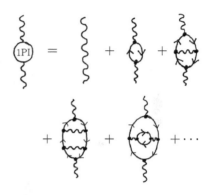

Fig. 41.2 Some contributions to the Green's function $i\tilde{\Pi}^{\mu\nu}(q)$.

---

[1] Remember that the massless nature of the photon field means that photons are not vector particles.

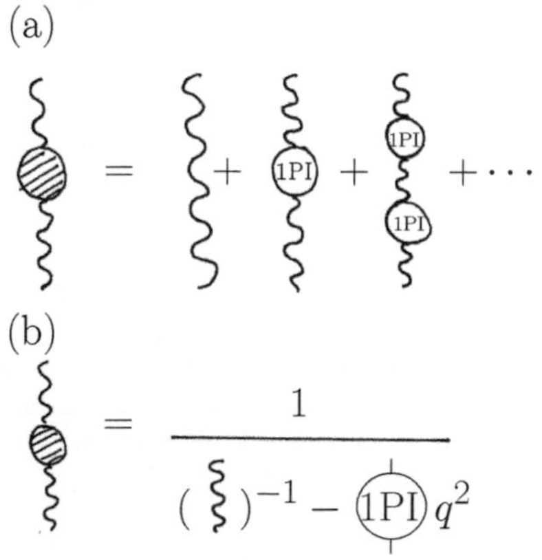

Example 41.1

In equations the sum is

$$\tilde{D}_{\mu\nu}(q) = \tilde{D}_{0\mu\nu}(q) + \tilde{D}_{0\mu\lambda}(q)[i\tilde{\Pi}^{\lambda\eta}(q)]\tilde{D}_{0\eta\nu}(q) + \ldots$$

$$= \frac{-ig_{\mu\nu}}{q^2} + \left(\frac{-ig_{\mu\lambda}}{q^2}\right) i\tilde{\Pi}^{\lambda\eta}(q) \left(\frac{-ig_{\eta\nu}}{q^2}\right) + \ldots$$

$$= \frac{-ig_{\mu\nu}}{q^2} + \left(\frac{-i}{q^2}\right) i\tilde{\Pi}_{\mu\nu}(q) \left(\frac{-i}{q^2}\right) + \left(\frac{-i}{q^2}\right) i\tilde{\Pi}^{\eta}_{\mu}(q) \left(\frac{-i}{q^2}\right) i\tilde{\Pi}_{\eta\nu}(q) \left(\frac{-i}{q^2}\right) + \ldots \tag{41.5}$$

Summing the series we obtain

$$\tilde{D}_{\mu\nu}(q) = -\frac{ig_{\mu\nu}}{q^2} + \frac{1}{q^2}\tilde{\Pi}_{\mu\eta}(q)\tilde{D}^{\eta}_{\nu}(q), \tag{41.6}$$

or

$$\left(q^2 g_{\mu\eta} - \tilde{\Pi}_{\mu\eta}(q)\right)\tilde{D}^{\eta}_{\nu}(q) = -ig_{\mu\nu}. \tag{41.7}$$

Things are simplified rather by the use of the Ward identity, which guarantees that $q_\nu \tilde{\Pi}^{\mu\nu}(q) = 0$. As a result, we can write the photon self-energy as

$$\tilde{\Pi}^{\mu\nu}(q) = (q^2 g^{\mu\nu} - q^\mu q^\nu)\tilde{\Pi}(q). \tag{41.8}$$

Substitution of this new form yields

$$q^2[1 - \tilde{\Pi}(q)]\tilde{D}_{\mu\nu}(q) + \tilde{\Pi}(q)q_\mu q_\eta \tilde{D}^{\eta}_{\nu}(q) = -ig_{\mu\nu}, \tag{41.9}$$

Fig. 41.3 Renormalization of the photon propagator.

whose solution is

$$\tilde{D}_{\mu\nu}(q) = \frac{-ig_{\mu\nu}}{q^2[1 - \tilde{\Pi}(q)]} + \frac{iq_\mu q_\nu \tilde{\Pi}(q)}{q^4[1 - \tilde{\Pi}(q)]}. \tag{41.10}$$

In practical calculations the $q_\mu q_\nu$ part never contributes to any observable amplitude. This is because $\tilde{D}_{\mu\nu}(q)$ is always coupled to conserved currents and we know that $q_\mu J_{\text{em}}^\mu = 0$.

The resulting renormalized photon propagator is given by

$$\tilde{D}_{\mu\nu}(q) = \frac{-ig_{\mu\nu}}{q^2[1 - \tilde{\Pi}(q)]}. \tag{41.11}$$

Note that we would expect that $\tilde{\Pi}(q)$, which contains loops, will diverge unless we use counterterms in our calculation of $\tilde{\Pi}(q)$.

Example 41.2

Let's see how to use the counterterm. Instead of summing over all 1PI insertions, we'll limit our calculation at a single electron–positron loop, which we'll call $i\tilde{\pi}^{\mu\nu}(q) = i(g^{\mu\nu}q^2 - q^\mu q^\nu)\tilde{\pi}(q)$. To prevent the loop causing any trouble we include a second-order counterterm which makes a contribution $i(g^{\mu\nu}q^2 - q^\mu q^\nu)C^{(2)}$. The total insertion in the photon line is shown in Fig. 41.4. Summing this insertion to all orders, we obtain

$$\tilde{D}_{\mu\nu}(q) = \frac{-ig_{\mu\nu}}{q^2\left\{1 - \left[\tilde{\pi}(q) + C^{(2)}\right]\right\}}. \tag{41.12}$$

The renormalization condition is that the photon self-energy $\tilde{\Pi}(q)$ vanishes at $q^2 = 0$, implying that $\tilde{\pi}(q = 0) = -C^{(2)}$ and the propagator becomes

$$\tilde{D}_{\mu\nu}(q) = \frac{-ig_{\mu\nu}}{q^2\left\{1 - \left[\tilde{\pi}(q) - \tilde{\pi}(0)\right]\right\}}. \tag{41.13}$$

This will be enough to remove the divergences from the calculation of $\tilde{D}_{\mu\nu}(q)$ to second order.

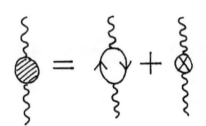

Fig. 41.4 The lowest order correction to the photon propagator and the counterterm that prevents its divergence.

The bare photon propagates while tearing electron–positron pairs from the vacuum. These processes have the effect of modifying the photon's amplitude for propagation between two points. More than that, we know that virtual photons (and remember that all photons are, to some extent, virtual photons) couple to a fermion line at both ends of their trajectory. If we write the vertex Feynman rule as $-iQ|e_0|\gamma^\mu$, we can bundle up the factors of $|e_0|$ with the photon propagator, so that we have[2] (considering $Q = 1$ for simplicity):

[2]Using the shorthand that e_0 is the unrenormalized charge and assuming that the counterterm contribution is accounted for by $\tilde{\Pi}(0)$.

$$\tilde{D}_{\mu\nu}(q) = \frac{-ie_0^2 g_{\mu\nu}}{q^2\left\{1 - \left[\tilde{\Pi}(q) - \tilde{\Pi}(0)\right]\right\}} = \frac{-ig_{\mu\nu}}{q^2}\left(\frac{e_0^2}{1 - \left[\tilde{\Pi}(q) - \tilde{\Pi}(0)\right]}\right). \tag{41.14}$$

The term in the round brackets plays the role of the square of the momentum-dependent electric charge $e(q)$. We could say that the real-life electric charge, renormalized by the interactions with the vacuum, is given by $|e| = |e_0|/\left\{1 - \left[\tilde{\Pi}(q) - \tilde{\Pi}(0)\right]\right\}^{\frac{1}{2}}$. The electric charge therefore depends on the momentum transferred by the virtual photon that mediates the electromagnetic force! A physical explanation for this feature of the dielectric properties of the vacuum is provided by the concept of **screening**, as we discuss below.

Example 41.3

In what follows we will require an expression for the one-loop amplitude in Fig. 41.5, which may be translated from diagram to equation, giving

$$i\tilde{\pi}(q) = (-1)(-ie_0)^2 \int \frac{d^4p}{(2\pi)^4} \text{Tr} \left(\gamma^\mu \frac{i}{\not{p} - m} \gamma^\nu \frac{i}{\not{p} + \not{q} - m} \right). \tag{41.15}$$

This integral can be done. With the inclusion of the counterterm it is found that $\tilde{\pi}(q) - \tilde{\pi}(0)$, the one-loop contribution to $\tilde{\Pi}(q)$, is given by

$$\tilde{\pi}(q) - \tilde{\pi}(0) = -\frac{e_0^2}{2\pi^2} \int_0^1 dx\, (x - x^2) \ln \left(\frac{m^2}{m^2 - (x - x^2)q^2} \right). \tag{41.16}$$

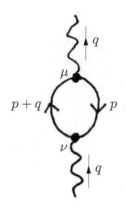

Fig. 41.5 The photon propagator with a single electron–hole loop insertion.

We've seen that the renormalized propagator is given, to second order in the interaction, by

$$\begin{aligned}
\tilde{D}_{\mu\nu}(q) &= \frac{-ig_{\mu\nu}}{q^2} \frac{e_0^2}{\{1 - [\tilde{\pi}(q) - \tilde{\pi}(0)]\}} \\
&\approx \frac{-ig_{\mu\nu}e_0^2}{q^2} [1 + \tilde{\pi}(q) - \tilde{\pi}(0)]. \tag{41.17}
\end{aligned}$$

To see the consequence of this on electrostatics, we need the static version of the propagator, obtained by taking $q^2 = -\boldsymbol{q}^2$ thus:

$$\tilde{\pi}(-\boldsymbol{q}) - \tilde{\pi}(0) = -\frac{e_0^2}{2\pi^2} \int_0^1 dx\, (x - x^2) \ln \left(\frac{m^2}{m^2 + (x - x^2)\boldsymbol{q}^2} \right). \tag{41.18}$$

Expanding this in the limit $\boldsymbol{q}^2 \ll m$ gives us

$$\tilde{D}_{\mu\nu}(\boldsymbol{q}) \approx \frac{ig_{\mu\nu}e_0^2}{\boldsymbol{q}^2} \left(1 + \frac{\alpha}{15\pi} \frac{\boldsymbol{q}^2}{m^2}, \right), \tag{41.19}$$

where $\alpha = e_0^2/4\pi$.

Recall from Chapter 20 our (Born approximation) interpretation of the momentum space propagator as proportional to a matrix element of a potential $V(\boldsymbol{r})$ via $\tilde{D}_{00}(\boldsymbol{q}) \propto - \int d^3r\, e^{i\boldsymbol{q}\cdot\boldsymbol{r}} V(\boldsymbol{r})$. Doing the inverse Fourier transform of $\tilde{D}_{00}(\boldsymbol{q})$ yields a potential due to the electron of

$$V(\boldsymbol{r}) = - \left\{ \frac{\alpha}{|\boldsymbol{r}|} + \frac{4\alpha^2}{15m^2} \delta^{(3)}(\boldsymbol{r}) \right\}. \tag{41.20}$$

The first term is Coulomb's potential for a point charge. The second term is the correction due to the screening of the electron charge that is provided by the virtual electron–hole pairs. The screening effect is visualized in Fig. 41.6, where we represent the pairs as effective dipoles. The screening is known as **vacuum polarization**.

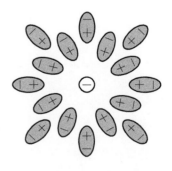

Fig. 41.6 Screening in QED.

Importantly, this effect can be measured! It causes a shift in the hydrogen energy levels of $\Delta E = -\frac{4\alpha^2}{15m^2}|\psi(0)|^2$, where $\psi(x)$ is the hydrogen wave function (which is only nonzero at the origin for $l = 0$ levels). This effect causes a shift in the $2S_{\frac{1}{2}}$ to $2P_{\frac{1}{2}}$ transition in hydrogen of -27 MHz which, despite making up only a small part of the famous Lamb shift (of $+1057$ MHz), has indeed been experimentally verified.[3] The effect of vacuum polarization is also very important in the quantum field theory of metals and will be discussed further in Chapter 43.

41.2 The renormalization group and the electric charge

Using what we know about vacuum polarization along with the renormalization group (of Chapter 34) we can work out how the electric charge changes, depending on the size of the momentum that the photon is carrying. That is, how the QED coupling depends on the energy scale. To do this we need to calculate the β-function, given here by $\beta = \mu\frac{d|e|}{d\mu}$, which tells us how the renormalized charge e depends on the choice of energy scale μ.

[3]Willis E. Lamb, Jr (1913–2008) measured the energy level shift in hydrogen in 1947. [See W. E. Lamb and R. C. Retherford, Phys. Rev. **72**, 241 (1947)]. Historically, this measurement was of immense importance in convincing physicists to take corrections due to virtual photons seriously. Shortly after the announcement of the experimental result, Hans Bethe (1906–2005) made a non-relativistic calculation of the shift caused by the self-energy of an electron in the electromagnetic field of a nucleus using an early form of renormalization. The agreement with experiment was good and the fully relativistic explanation soon followed from a number of physicists. See Schweber for the history and Weinberg, Chapter 14, for the full calculation of the Lamb shift.

Example 41.4

Using our equation for the renormalized electric charge we can expand for small $\hat{\pi}(q)$ to obtain

$$e^2(\mu) = \frac{e_0^2}{1 - (\tilde{\pi}(\mu) - \tilde{\pi}(0))} \approx e_0^2[1 + (\tilde{\pi}(\mu) - \tilde{\pi}(0))] + O(e^4), \quad (41.21)$$

or

$$|e| \approx |e_0|\left[1 + \frac{1}{2}(\tilde{\pi}(\mu) - \tilde{\pi}(0))\right]. \quad (41.22)$$

The β-function is then given by

$$\beta = \mu\frac{d|e|}{d\mu} = \frac{1}{2}\mu|e_0|\frac{d\tilde{\pi}(\mu)}{d\mu}. \quad (41.23)$$

Now we use eqn 41.16 which states that

$$\tilde{\pi}(\mu) - \tilde{\pi}(0) = -\frac{e_0^2}{2\pi^2}\int_0^1 dx\,(x - x^2)\ln\left[\frac{m^2}{m^2 - (x - x^2)\mu^2}\right]. \quad (41.24)$$

In the large energy scale limit ($\mu \gg m$) differentiation gets us

$$\frac{d\tilde{\pi}(\mu)}{d\mu} = \frac{e_0^2}{\pi^2\mu}\int_0^1 dx\,(x - x^2) = \frac{e_0^2}{6\pi^2\mu}. \quad (41.25)$$

Note that, to the order to which we're working, it's permissible to swap e_0 for e on the right-hand side in what follows.

We obtain a β-function

$$\beta = \mu\frac{d|e|}{d\mu} = +\frac{|e|^3}{12\pi^2}, \quad (41.26)$$

where we accentuate the $+$ sign, since that's the most important part. This shows that the electromagnetic coupling increases with increasing energy scale. Another way of saying this is that it increases with decreasing length scale. The electron appears to be more strongly charged as you get closer to it. This makes sense in the context of the last section, where we interpreted vacuum polarization as a screening process. As you get closer to the electron you penetrate the cloud of screening pairs and the charge seems to increase.

41.3 Vertex corrections and the electron *g*-factor

Other people publish to show you how to do it, but Julian Schwinger publishes to show you that only he can do it.
Anon, quoted from S. Schweber, *QED and the Men who Made It*

The magnetic moment operator of the electron may be written

$$\hat{\boldsymbol{\mu}} = g\left(\frac{Q|e|}{2m}\right)\hat{\boldsymbol{S}}, \tag{41.27}$$

where $Q = -1$ for the electron. The Dirac equation predicts[4] that the electron *g*-factor of the electron is exactly 2. In this section we will make the link between renormalized QED and the *g*-factor. This will lead to Julian Schwinger's famous prediction that the first-order correction to the QED interaction vertex causes a shift from $g = 2$ to $g = 2 + \alpha/\pi$. That is to say that the fact that electrons can emit virtual photons changes the form of the electron–photon interaction in a measurable manner.

The Green's function for the electron–photon vertex can be written $-iQ|e|\tilde{\Gamma}^\mu(p, p')$ and its diagram is shown in Fig. 41.7. It is defined as[5]

$$-iQ|e|\tilde{\Gamma}^\mu(p, p') = \sum \left(\begin{array}{c}\text{All amputated insertions with one} \\ \text{incoming fermion line, one outgoing} \\ \text{fermion line and one photon line.}\end{array}\right). \tag{41.28}$$

Some examples are shown in Fig. 41.8. The vertex function can be read as describing an off mass-shell photon decaying into an electron–positron pair. Since the photon has[6] $J^P = 1^-$, there are two possible configurations of the pair. It can have $L = 0$ and $S = 1$, or $L = 2$ and $S = 1$. This fact is reflected in the form in which we write the vertex function as a sum of two terms.

$$\tilde{\Gamma}^\mu(p, p') = \gamma^\mu F_1(q) + \frac{i\sigma^{\mu\nu}q_\nu}{2m} F_2(q), \tag{41.29}$$

where $q^\mu = p'^\mu - p^\mu$ and where $\sigma^{\mu\nu} = \frac{i}{2}[\gamma^\mu, \gamma^\nu]$. The function $F_1(q)$ is known as the Dirac form factor[7] and $F_2(q)$ is known as the Pauli form factor. For vanishing q we need only consider the first-order contribution to $\tilde{\Gamma}^\mu = \gamma^\mu$, which is the first diagram in Fig. 41.8. We read off that for $q \to 0$ we have $F_1(0) = 1$ and $F_2(0) = 0$.

[4]See Example 36.9.

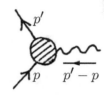

Fig. 41.7 The vertex Green's function $-iQ|e|\Gamma^\mu(p, p')$.

[5]The overal sign is chosen because the first-order interaction is given by $-iQ|e|\gamma^\mu$ so to first-order $\tilde{\Gamma}^\mu = \gamma^\mu$.

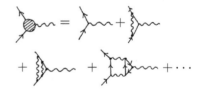

Fig. 41.8 Contributions to the Green's function $-iQ|e|\tilde{\Gamma}^\mu$. The contribution to first order is simply given by the interaction vertex $-i|e|\gamma^\mu$.

[6]In this notation we list the total angular momentum J and the intrinsic parity P of the photon.

[7]In fact, $F_1(q)$ may be thought of as a Fourier transform of the charge distribution and has the property that $F_1(0) = 1$ at all orders of perturbation theory.

We will now sketch out the determination of the g-factor of the electron in a manner intended 'to show you how to do it'. Since the calculation is quite lengthy and involved, we will break it down into a number of steps.

Step I: Consider the diagram in Fig. 41.9 showing the interaction vertex coupled to a classical source $A^{\text{cl}}_\mu(x)$, giving a Feynman rule contribution of $-\mathrm{i}Q|e|\tilde{\Gamma}^\mu \tilde{A}^{\text{cl}}_\mu(q)$, where q^μ is the momentum of the photon. Coupling this to the spinor current of the incoming and outgoing electron, we have an amplitude

$$\mathrm{i}\mathcal{M} = -\mathrm{i}Q|e|\bar{u}(p')\tilde{\Gamma}^\mu(p,p')u(p)\tilde{A}^{\text{cl}}_\mu(q), \qquad (41.30)$$

where $q = p' - p$ and an energy momentum conserving delta function has been suppressed. Inserting eqn 41.29 for $\tilde{\Gamma}^\mu$ yields

$$\mathrm{i}\mathcal{M} = -\mathrm{i}Q|e|\bar{u}(p')\left[\gamma^\mu F_1(q) + \frac{\mathrm{i}\sigma^{\mu\nu}q_\nu}{2m}F_2(q)\right]u(p)\tilde{A}^{\text{cl}}_\mu(q). \qquad (41.31)$$

Step II: The next step makes the $\bar{u}'\gamma^\mu u$ in the first term much more complicated and uses an identity, known as the Gordon decomposition (proved in Exercise 36.7), given by

$$\bar{u}(p')\gamma^\mu u(p) = \bar{u}(p')\left[\frac{p'^\mu + p^\mu}{2m} + \frac{\mathrm{i}\sigma^{\mu\nu}q_\nu}{2m}\right]u(p). \qquad (41.32)$$

The reason for this step is that is eliminates γ^μ, pushing the spin-independent part into the first term and the spin-dependent part into the second. Substituting this into our Feynman amplitude yields

$$\begin{aligned}
\mathrm{i}\mathcal{M} = {} & -\mathrm{i}Q|e|\bar{u}(p')\left(\frac{p'^\mu + p^\mu}{2m}\right)u(p)\tilde{A}^{\text{cl}}_\mu(q)F_1(q) \\
& + \frac{Q|e|}{2m}\left[\bar{u}(p')\sigma^{\mu\nu}q_\nu u(p)\right]\tilde{A}^{\text{cl}}_\mu(q)\left\{F_1(q) + F_2(q)\right\}.
\end{aligned} \qquad (41.33)$$

We now have an expression for the amplitude which is separated into the spin-independent first term and a spin-dependent second term. The information on the g-factor is to be found in the second term.

Step III: Now we take the non-relativistic limit. This involves writing $u(p) \approx \sqrt{m}\begin{pmatrix} \xi \\ \xi \end{pmatrix}$. We also need to use the explicit forms for the components of $\sigma^{\mu\nu}$

$$\sigma^{0i} = \begin{pmatrix} -\mathrm{i}\sigma^i & 0 \\ 0 & \mathrm{i}\sigma^i \end{pmatrix} \quad \text{and} \quad \sigma^{ij} = \begin{pmatrix} \varepsilon^{ijk}\sigma^k & 0 \\ 0 & \varepsilon^{ijk}\sigma^k \end{pmatrix}. \qquad (41.34)$$

Using these we find that $\bar{u}'\sigma^{0i}u = 0$ and, more importantly, that $\bar{u}'\sigma^{ij}u = 2m\left[\xi'^\dagger\sigma^k\xi\right]\varepsilon^{ijk}$.

Putting this all together we have, for the spin-dependent part of the amplitude of our diagram in the limit $q \to 0$, that

$$\begin{aligned}
\text{Spin-dependent amplitude} &= (2m)\frac{Q|e|}{2m}\left[\xi'^\dagger\sigma^k\xi\right]\varepsilon^{ijk}q^j\tilde{A}^{\text{cl}i}(q)\left\{F_1(0) + F_2(0)\right\} \\
&= (2m)\frac{\mathrm{i}Q|e|}{2m}\left[\xi'^\dagger\sigma^k\xi\right]\tilde{B}^k(\boldsymbol{q})\left\{1 + F_2(0)\right\}, \qquad (41.35)
\end{aligned}$$

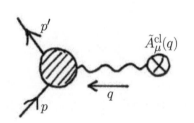

Fig. 41.9 The vertex function coupled to a classical source.

where we've assumed $A^{\text{cl}}(x) = (0, \boldsymbol{A}^{\text{cl}}(\boldsymbol{x}))$ and used the fact that, in momentum space, the components of the magnetic field are given by $i\varepsilon^{ijk}q^i\tilde{A}^{\text{cl}j}(\boldsymbol{q}) = \tilde{B}^k(\boldsymbol{q})$. We conclude that the spin-dependent part of the amplitude is proportional to $Q|e|\{1 + F_2(0)\}\langle\hat{\boldsymbol{S}}\rangle \cdot \boldsymbol{B}$, where the expectation value of the spin operator is $\langle\hat{\boldsymbol{S}}\rangle = \xi'^\dagger\boldsymbol{\sigma}\xi/2$.

We now have an amplitude, so we may extract a scattering potential using the Born approximation.

$$i\mathcal{M} = -i\langle f|\hat{V}(\boldsymbol{x})|i\rangle. \tag{41.36}$$

Since we are still using the relativistic normalization, we need to correct by dividing through by twice the energy of the electron that's scattered ($= 2m$). We obtain a real-space potential

$$V(\boldsymbol{x}) = -\frac{Q|e|}{m}\{1 + F_2(0)\}\langle\hat{\boldsymbol{S}}\rangle \cdot \boldsymbol{B}(\boldsymbol{x}). \tag{41.37}$$

Comparing this to the expected expression for the potential energy of a magnetic moment $\boldsymbol{\mu}$ in a magnetic field, namely

$$V(\boldsymbol{x}) = -\langle\hat{\boldsymbol{\mu}}\rangle \cdot \boldsymbol{B}(\boldsymbol{x}) = -\left[g\left(\frac{Q|e|}{2m}\right)\langle\hat{\boldsymbol{S}}\rangle\right] \cdot \boldsymbol{B}(\boldsymbol{x}), \tag{41.38}$$

we may read off that $g = 2[1 + F_2(0)]$.

So g is not exactly 2! There's a correction determined by $F_2(0)$. The correction can be calculated by evaluating the contributions to $\tilde{\Gamma}^\mu$.

Step IV: Let's look at the contributions to $-iQ|e|\tilde{\Gamma}^\mu$, order by order. To first order we have the bare vertex whose contribution (for $Q = -1$) is $i|e|\gamma^\mu$, so the first-order contribution to Γ^μ is γ^μ implying $F_1(0) = 1$ and $F_2(0) = 0$ as we said earlier. There are no second-order diagrams, but to third order the only diagram that contributes to $F_2(q)$ is that shown in Fig. 41.10.

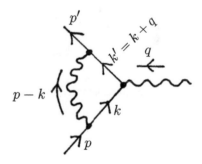

Fig. 41.10 The third-order diagram which contributes to $F_2(0)$.

Example 41.5

From the labelled diagram, shown in Fig. 41.10, we can write down the amplitude for the vertex as

$$\tilde{\Gamma}^\mu_{\text{diagram}} = \int \frac{\mathrm{d}^4k}{(2\pi)^4}\frac{-ig_{\nu\rho}}{(p-k)^2 + i\epsilon}\bar{u}(p')(i|e|\gamma^\nu)\frac{i}{\not{k}' - m + i\epsilon}\gamma^\mu\frac{i}{\not{k} - m + i\epsilon}(i|e|\gamma^\rho)u(p). \tag{41.39}$$

This is divergent, so we must also include a third-order counterterm. With the inclusion of the counterterm in Fig. 41.1(e) and two pages or so of Dirac algebra[8] one can derive a result for the Pauli form factor which is $F_2(0) = e^2/8\pi^2$.

The result: We evaluate $g = 2[1 + F_2(0)]$ with $F_2(0) = \frac{e^2}{8\pi^2}$ and find the famous result for the shift in the g-factor:

$$g = 2\left(1 + \frac{\alpha}{2\pi} + \dots\right). \tag{41.40}$$

This was also Julian Schwinger's result and one for which he justifiably gained a spontaneous round of applause when he revealed, in a talk at

[8]See Peskin and Schroeder, which is particularly informative on this subject.

Recall that in the units used here, the fine structure constant is given by $\alpha = e^2/4\pi$. Because $\alpha = 1/137.036\dots$ we have $g = 2.0023\dots$ In SI units, $\alpha = e^2/4\pi\epsilon_0\hbar c$.

[9]See Peskin and Schroeder, page 196, for a discussion of the remarkable agreement of QED with modern experimental tests.

the New York APS meeting in 1948, that his theory agreed precisely with the crucial experiments. Of course, higher order corrections to this and other QED processes may be evaluated and have led to QED becoming the most stringently tested theory in physics.[9]

Chapter summary

- Renormalizing the photon propagator shows that the electric charge is given by $|e| = |e_0|/\left\{1 - \left[\tilde{\Pi}(q) - \tilde{\Pi}(0)\right]\right\}^{\frac{1}{2}}$. The electric charge is screened by virtual electron–positron pairs (known as vacuum polarization) and this gives rise to the Lamb shift.
- The Dirac theory predicts $g = 2$, but including the electron–photon vertex $-iQ|e|\tilde{\Gamma}(p,p')$ (for an off mass-shell photon decaying into an electron–positron pair) yields $g = 2\left(1 + \frac{\alpha}{2\pi} + \cdots\right)$ which agrees extremely well with experiment.

Exercises

(41.1) Verify

$$\tilde{D}_{\mu\nu}(q) = \frac{-ig_{\mu\nu}}{q^2[1 - \tilde{\Pi}(q)]} + \frac{iq_\mu q_\nu \tilde{\Pi}(q)}{q^4[1 - \tilde{\Pi}(q)]}, \quad (41.41)$$

is the solution to eqn 41.9.

(41.2) (a) Verify

$$\tilde{D}_{\mu\nu}(\boldsymbol{q}) \approx \frac{ig_{\mu\nu}e_0^2}{\boldsymbol{q}^2}\left(1 + \frac{\alpha}{15\pi}\frac{q^2}{m^2}\right). \quad (41.42)$$

(b) Take the inverse Fourier transform and show you obtain the form in the text.

(41.3) The electron 1PI self-energy is defined as

$$-i\tilde{\Sigma} = \sum \left(\begin{array}{c}\text{All 1PI insertions with incoming}\\ \text{and outgoing electron lines}\end{array}\right). \quad (41.43)$$

Draw contributions to $-i\tilde{\Sigma}$ in QED up to fourth order in the interaction.

(41.4) The *Schwinger model* deals with quantum electrodynamics in (1+1)-dimensional spacetime and predicts a one-loop vacuum polarization

$$-i\tilde{\Pi}_{\mu\nu}(p) = -\frac{ie^2}{\pi p^2}(p_\mu p_\nu - p^2 g_{\mu\nu}). \quad (41.44)$$

Show that the photon takes on a mass as a result and determine its size.

(41.5) Verify eqn 41.25.

(41.6) (a) Show that

$$\sigma^{0i} = \begin{pmatrix} -i\sigma^i & 0 \\ 0 & i\sigma^i \end{pmatrix} \quad (41.45)$$

and

$$\sigma^{ij} = \begin{pmatrix} \varepsilon^{ijk}\sigma^k & 0 \\ 0 & \varepsilon^{ijk}\sigma^k \end{pmatrix}. \quad (41.46)$$

(b) Show that $\bar{u}\sigma^{ij}u = 2m\xi^\dagger \sigma^k \xi \varepsilon^{ijk}$.

Part X

Some applications from the world of condensed matter

Quantum field theory finds many applications in the field of condensed matter physics which is concerned with the properties of many-particle systems whose fundamental excitations can be regarded as particles. We have already seen that lattice vibrations can be treated as particles called phonons. In the case of metallic systems which contain large numbers of electrons, the interactions between electrons can be very important in determining the properties.

- Our first condensed matter physics example is the superfluid, and in Chapter 42 we show how an approximation due to Bogoliubov allows weakly interacting Bose systems to be treated. The quasiparticles are called bogolons and the ground state of the system can be treated as a coherent state. The Noether current is shown to arise from a gradient in the phase of this coherent state.

- Interactions between electrons in a metal are treated in Chapter 43 and we introduce the Cooper, Hartree and Fock terms in the perturbation expansion for the ground state energy shift. We explain how to model the excitations and study them using propagators and we introduce the random phase approximation.

- The BCS theory of superconductivity is presented in Chapter 44 and we show how to write down the BCS coherent state, derive the quasiparticles and explore the effect of broken symmetry. This allows us to explore the Higgs mechanism in a superconductor.

- The fractional quantum Hall effect is introduced in Chapter 45 and we explain how quasiparticles carrying fractional charge emerge in this theory.

A note on notation: In this part we conform to the conventions of the condensed matter literature by denoting the number of particles by N and define the number density of particles $n = N/\mathcal{V}$, where $\mathcal{V}$ is the volume. For consistency we write number operators and occupation numbers for the state with momentum $\boldsymbol{p}$ as $\hat{N}_{\boldsymbol{p}}$ and $N_{\boldsymbol{p}}$ respectively.

42 Superfluids

There is no doubt a special place in hell being reserved for me at this very moment for this mean trick, for the task is impossible.
Robert Laughlin (1950–), on asking his students to deduce superfluidity from first principles

A gas of weakly interacting, non-relativistic bosons is a problem where the methods of quantum field theory allow us to understand the ground state, the excitations and the breaking of symmetry. Not only is this an illustration of many of the methods we've discussed; it also applies to one of the most fascinating phenomena in condensed matter physics: superfluidity. A **superfluid** is a state of matter where momentum flows without dissipation. When ^{4}He is cooled below 2.17 K it becomes a superfluid and may flow through capillaries without resistance. It also exhibits the fountain effect whereby a beaker of superfluid spontaneously empties itself.

The dispersion of superfluid helium has been measured with inelastic neutron scattering and is shown in Fig. 42.1. At low momentum the dispersion is linear, at large momentum the energy goes as $\boldsymbol{p}^2/2m$ as expected for free particles. In the middle is a large minimum. We will discover that our gas of weakly interacting bosons captures the low and high momentum behaviour of the superfluid dispersion. We will also find that the form of the field that we obtain on spontaneous symmetry breaking necessarily gives us a superflow of momentum.

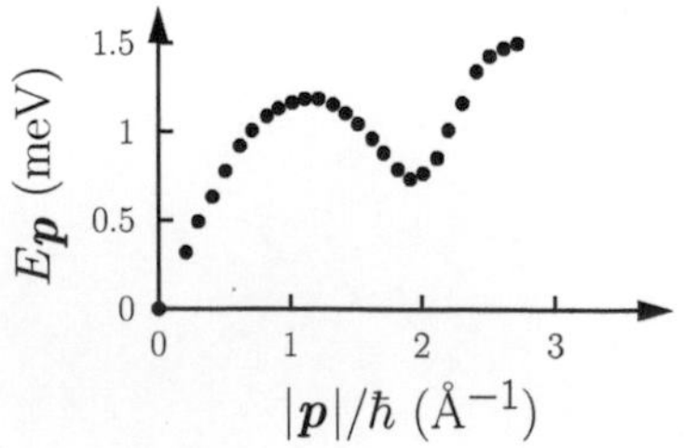

Fig. 42.1 The measured dispersion of superfluid helium. Data taken from A.B.D. Woods and R.A. Cowley, Can. J. Phys. **49**, 177 (1971).

42.1 Bogoliubov's hunting license

The problem revolves around the Hamiltonian for non-relativistic Bose particles interacting with a momentum-independent potential g. The Hamiltonian describing this state of affairs is

$$
\begin{aligned}
\hat{H} &= \frac{1}{2m} \int \mathrm{d}^3x\, \boldsymbol{\nabla}\hat{\phi}^\dagger(\boldsymbol{x}) \cdot \boldsymbol{\nabla}\hat{\phi}(\boldsymbol{x}) \\
&+ \frac{g}{2} \int \mathrm{d}^3x \mathrm{d}^3y\, \hat{\phi}^\dagger(\boldsymbol{x})\hat{\phi}^\dagger(\boldsymbol{y})\hat{\phi}(\boldsymbol{y})\hat{\phi}(\boldsymbol{x})\delta^{(3)}(\boldsymbol{x}-\boldsymbol{y}) \\
&= \sum_{\boldsymbol{p}} \frac{\boldsymbol{p}^2}{2m}\hat{a}_{\boldsymbol{p}}^\dagger \hat{a}_{\boldsymbol{p}} + \frac{g}{2\mathcal{V}} \sum_{\boldsymbol{k}\boldsymbol{p}\boldsymbol{q}} \hat{a}_{\boldsymbol{p}-\boldsymbol{q}}^\dagger \hat{a}_{\boldsymbol{k}+\boldsymbol{q}}^\dagger \hat{a}_{\boldsymbol{k}} \hat{a}_{\boldsymbol{p}},
\end{aligned} \tag{42.1}
$$

where the second form follows from putting the system in a box and inserting the mode expansion $\hat{\phi}(\boldsymbol{x}) = \frac{1}{\sqrt{\mathcal{V}}} \sum_{\boldsymbol{p}} \hat{a}_{\boldsymbol{p}} e^{i\boldsymbol{p}\cdot\boldsymbol{x}}$. As usual, we call the interacting ground state of this Hamiltonian $|\Omega\rangle$. This is intended to describe a very large number of interacting particles. For a macroscopic sample of matter in a laboratory this number might be around 10^{23}. Unfortunately, the interaction part of the Hamiltonian, given by $\hat{H}_I = \frac{g}{2\mathcal{V}} \sum_{\boldsymbol{kpq}} \hat{a}^\dagger_{\boldsymbol{p-q}} \hat{a}^\dagger_{\boldsymbol{k+q}} \hat{a}_{\boldsymbol{k}} \hat{a}_{\boldsymbol{p}}$, is too complicated to solve, so we need an approximation. Bogoliubov came up with one by thinking about the low-energy behaviour of the system.[1] Bogoliubov's assumption (based upon Bose–Einstein condensation) is that, if the system has very little energy, the number of particles N_0 in the $\boldsymbol{p}=0$ ground state of the Hamiltonian will be macroscopically large. The consequence of this is that instead of the exact relation $\hat{a}_{\boldsymbol{p}}|\Omega\rangle = \sqrt{N_0}|N_0-1\rangle$ (where $|N_0-1\rangle$ is a state achieved by removing one zero momentum particle from the ground state), we can take $\hat{a}_{\boldsymbol{p}=0}|\Omega\rangle \approx \sqrt{N_0}|\Omega\rangle$.

Bogoliubov used this as an approximation[2] to allow the replacement of operators $\hat{a}_{\boldsymbol{p}=0}$ and $\hat{a}^\dagger_{\boldsymbol{p}=0}$ with the number $\sqrt{N_0}$. He then broke down the sum over momentum states $\sum_{\boldsymbol{kpq}} \hat{a}^\dagger_{\boldsymbol{p-q}} \hat{a}^\dagger_{\boldsymbol{k+q}} \hat{a}_{\boldsymbol{k}} \hat{a}_{\boldsymbol{p}}$ by considering, in turn, the terms in the sum in which particular momentum subscripts are zero, and replacing that operator with the number $\sqrt{N_0}$. Since terms with odd numbers of operators will always give zero expectation value, we need only consider the cases of two or four of the indices being zero. The method then relies on sending the momentum index of pairs, or all four, of the operators to zero. Then we replace the operators $\hat{a}_0$ and $\hat{a}^\dagger_0$ with $\sqrt{N_0}$.

Nikolay Bogoliubov (1909–1982)

[1] For 'low-energy', you might read 'low-temperature' if you're imagining an experiment in a laboratory.

[2] In his words, this approximation provides a 'hunting license' to search for new phenomena.

Example 42.1

We take turns setting pairs of momentum subscripts to zero. Replacing pairs of subscripts yields six terms:

$$\boldsymbol{p}=0,\ \boldsymbol{k}=0,\qquad \hat{a}^\dagger_{-\boldsymbol{q}}\hat{a}^\dagger_{\boldsymbol{q}}\hat{a}_0\hat{a}_0 \to N_0\hat{a}^\dagger_{-\boldsymbol{q}}\hat{a}^\dagger_{\boldsymbol{q}},$$
$$\boldsymbol{p}=0,\ \boldsymbol{k+q}=0,\qquad \hat{a}^\dagger_{-\boldsymbol{q}}\hat{a}^\dagger_0\hat{a}_{\boldsymbol{k}}\hat{a}_0 \to N_0\hat{a}^\dagger_{\boldsymbol{k}}\hat{a}_{\boldsymbol{k}},$$
$$\boldsymbol{p}=0,\ \boldsymbol{p-q}=0,\qquad \hat{a}^\dagger_0\hat{a}^\dagger_{\boldsymbol{k}}\hat{a}_{\boldsymbol{k}}\hat{a}_0 \to N_0\hat{a}^\dagger_{\boldsymbol{k}}\hat{a}_{\boldsymbol{k}},$$
$$\boldsymbol{k}=0,\ \boldsymbol{k+q}=0,\qquad \hat{a}^\dagger_{\boldsymbol{p}}\hat{a}^\dagger_0\hat{a}_0\hat{a}_{\boldsymbol{p}} \to N_0\hat{a}^\dagger_{\boldsymbol{p}}\hat{a}_{\boldsymbol{p}},$$
$$\boldsymbol{k}=0,\ \boldsymbol{p-q}=0,\qquad \hat{a}^\dagger_0\hat{a}^\dagger_{\boldsymbol{q}}\hat{a}_0\hat{a}_{\boldsymbol{p}} \to N_0\hat{a}^\dagger_{\boldsymbol{p}}\hat{a}_{\boldsymbol{p}},$$
$$\boldsymbol{k+q}=0,\ \boldsymbol{p-q}=0,\qquad \hat{a}^\dagger_0\hat{a}^\dagger_0\hat{a}_{-\boldsymbol{q}}\hat{a}_{\boldsymbol{q}} \to N_0\hat{a}_{-\boldsymbol{q}}\hat{a}_{\boldsymbol{q}}. \qquad (42.2)$$

Lastly, we replace all four of the subscripts to get

$$\hat{a}^\dagger_0\hat{a}^\dagger_0\hat{a}_0\hat{a}_0 = N_0^2. \qquad (42.3)$$

With this approximation we obtain an approximate interaction Hamiltonian

$$\hat{H}_I \approx \frac{g}{2\mathcal{V}}\left[N_0^2 + 4N_0\sum_{\boldsymbol{p}\neq0}\hat{a}^\dagger_{\boldsymbol{p}}\hat{a}_{\boldsymbol{p}} + N_0\sum_{\boldsymbol{p}\neq0}(\hat{a}^\dagger_{\boldsymbol{p}}\hat{a}^\dagger_{-\boldsymbol{p}} + \hat{a}_{\boldsymbol{p}}\hat{a}_{-\boldsymbol{p}})\right]. \qquad (42.4)$$

Now we must prepare ourselves for a shock. This seemingly quite reasonable approximation of saying that there are a large number of particles

[3]Note that this global transformation doesn't depend on the momentum.

[4]This result is perhaps unsurprising in that our 'hunting license' approximation is to equate $|N-1\rangle$ with $|\Omega\rangle$, which shows a slightly cavalier attitude to particle number.

in the lowest momentum state has a dramatic consequence: it breaks a symmetry. Our original Hamiltonian is invariant under the global $U(1)$ transformation $\hat{\phi}(\boldsymbol{x}) \to \hat{\phi}(\boldsymbol{x})\mathrm{e}^{\mathrm{i}\alpha}$, or equivalently $\hat{a}_{\boldsymbol{p}} \to \hat{a}_{\boldsymbol{p}}\mathrm{e}^{\mathrm{i}\alpha}$. Looking at $\hat{H}_{\mathrm{I}}$, we see that the terms $\hat{a}_{\boldsymbol{p}}^{\dagger}\hat{a}_{-\boldsymbol{p}}^{\dagger}$ and $\hat{a}_{\boldsymbol{p}}\hat{a}_{-\boldsymbol{p}}$ both change if we make the latter transformation.[3] We have therefore lost the $U(1)$ symmetry and its conserved quantity, which (remembering back to our discussion of global $U(1)$ symmetry) was particle number.[4]

Since we've now lost the conservation of particle number, we don't know what value to assign to N_0. To remedy this we set the total number of particles N equal to the number in the $\boldsymbol{p}=0$ ground state, plus all the rest, that is:

$$N = N_0 + \sum_{\boldsymbol{p}\neq 0} \hat{a}_{\boldsymbol{p}}^{\dagger}\hat{a}_{\boldsymbol{p}}. \tag{42.5}$$

The result of these manipulations is to yield an effective Hamiltonian

$$\hat{H} = \sum_{\boldsymbol{p}\neq 0}\left(\frac{\boldsymbol{p}^2}{2m}+ng\right)\hat{a}_{\boldsymbol{p}}^{\dagger}\hat{a}_{\boldsymbol{p}} + \frac{1}{2}\sum_{\boldsymbol{p}\neq 0}ng(\hat{a}_{\boldsymbol{p}}^{\dagger}\hat{a}_{-\boldsymbol{p}}^{\dagger}+\hat{a}_{\boldsymbol{p}}\hat{a}_{-\boldsymbol{p}}), \tag{42.6}$$

where $n=N/\mathcal{V}$ and we've dropped the constant $\frac{1}{2}gn^2$ term.

The next problem is that the potential term in the Hamiltonian isn't diagonal. That is, it isn't expressed in terms of number operators of the form $\hat{b}_{\boldsymbol{q}}^{\dagger}\hat{b}_{\boldsymbol{q}}$. This may spell trouble unless there's a way to turn these non-diagonal objects into diagonal ones. That is, we want to turn objects like $\hat{a}_{\boldsymbol{p}}^{\dagger}\hat{a}_{-\boldsymbol{p}}^{\dagger}$ and $\hat{a}_{\boldsymbol{p}}\hat{a}_{-\boldsymbol{p}}$ into number operators.

42.2 Bogoliubov's transformation

The way to turn these non-diagonal bilinears into number operators is to transform to a new set of operators. The procedure that does this is a **Bogoliubov transformation** which defines a new set of operators $\hat{\alpha}_{\boldsymbol{p}}$ and $\hat{\alpha}_{\boldsymbol{p}}^{\dagger}$ via

$$\begin{pmatrix}\hat{a}_{\boldsymbol{p}}\\\hat{a}_{-\boldsymbol{p}}^{\dagger}\end{pmatrix}=\begin{pmatrix}u_{\boldsymbol{p}}&-v_{\boldsymbol{p}}\\-v_{\boldsymbol{p}}&u_{\boldsymbol{p}}\end{pmatrix}\begin{pmatrix}\hat{\alpha}_{\boldsymbol{p}}\\\hat{\alpha}_{-\boldsymbol{p}}^{\dagger}\end{pmatrix}, \tag{42.7}$$

and we will use these operators to diagonalize the Hamiltonian. The quantities $u_{\boldsymbol{p}}$ and $v_{\boldsymbol{p}}$ obey the following rules:

$$u_{\boldsymbol{p}}^2-v_{\boldsymbol{p}}^2=1,\quad u_{\boldsymbol{p}}^*=u_{\boldsymbol{p}},\quad v_{\boldsymbol{p}}^*=v_{\boldsymbol{p}}, \tag{42.8}$$

which are cunningly designed so that the new operators $\hat{\alpha}_{\boldsymbol{p}}$ and $\hat{\alpha}_{\boldsymbol{p}}^{\dagger}$ obey the same commutation relations as $\hat{a}_{\boldsymbol{p}}$ and $\hat{a}_{\boldsymbol{p}}^{\dagger}$:

$$[\hat{\alpha}_{\boldsymbol{p}},\hat{\alpha}_{\boldsymbol{q}}^{\dagger}]=\delta_{\boldsymbol{pq}},\quad [\hat{\alpha}_{\boldsymbol{p}},\hat{\alpha}_{\boldsymbol{q}}]=[\hat{\alpha}_{\boldsymbol{p}}^{\dagger},\hat{\alpha}_{\boldsymbol{q}}^{\dagger}]=0. \tag{42.9}$$

The physical significance of the new operators is that they describe a new sort of excitation: a new bosonic quasiparticle we will call a **bogolon**.

Example 42.2

We can show how the bogolon operator diagonalizes the Hamiltonian. The Hamiltonian can be written in matrix form as

$$\hat{H} = \sum_{p \neq 0} \left(\begin{array}{cc} \hat{a}_p^\dagger & \hat{a}_{-p} \end{array}\right) \left(\begin{array}{cc} \epsilon_p & \frac{1}{2}ng \\ \frac{1}{2}ng & 0 \end{array}\right) \left(\begin{array}{c} \hat{a}_p \\ \hat{a}_{-p}^\dagger \end{array}\right), \qquad (42.10)$$

where $\epsilon_p = \frac{p^2}{2m} + ng$. We're going to make the transformation given in eqn 42.7 and fix the constants u_p and v_p to make the resulting Hamiltonian matrix diagonal, which is to say

$$\hat{H} = \sum_{p \neq 0} \left(\begin{array}{cc} \hat{\alpha}_p^\dagger & \hat{\alpha}_{-p} \end{array}\right) \left(\begin{array}{cc} D_{11} & 0 \\ 0 & D_{22} \end{array}\right) \left(\begin{array}{c} \hat{\alpha}_p \\ \hat{\alpha}_{-p}^\dagger \end{array}\right). \qquad (42.11)$$

Assuming that we're successful in diagonalizing, how do we get the excitation energy, that is the constant in front of $\hat{\alpha}_p^\dagger \hat{\alpha}_p$? The trick is to look at the diagonalized form which is

$$
\begin{aligned}
\hat{H} &= \sum_{p \neq 0} \left[D_{11} \hat{\alpha}_p^\dagger \hat{\alpha}_p + D_{22} \hat{\alpha}_{-p} \hat{\alpha}_{-p}^\dagger \right] \\
&= \sum_{p \neq 0} \left[D_{11} \hat{\alpha}_p^\dagger \hat{\alpha}_p + D_{22} (1 + \hat{\alpha}_{-p}^\dagger \hat{\alpha}_{-p}) \right]. \qquad (42.12)
\end{aligned}
$$

The sum over $-p$'s can be re-indexed to be a sum over positive p's and we see that the constant in front of $\hat{\alpha}_p^\dagger \hat{\alpha}_p$ is $D_{11} + D_{22}$, otherwise known as the trace of the diagonal matrix. The trace is given by

$$E_p = \epsilon_p(u_p^2 + v_p^2) - 2ng u_p v_p. \qquad (42.13)$$

The condition for the off-diagonal elements to be zero is that

$$\frac{2u_p v_p}{u_p^2 + v_p^2} = \frac{ng}{\epsilon_p}. \qquad (42.14)$$

Substituting the condition allows us to eliminate u_p and v_p and we obtain[5] the answer given in eqn 42.15.

[5]You are invited to verify this in Exercise 42.1.

The diagonalized Hamiltonian is given by

$$\hat{H} = \sum_{p} E_p \hat{\alpha}_p^\dagger \hat{\alpha}_p, \quad \text{where} \quad E_p = \sqrt{\frac{p^2}{2m}\left(\frac{p^2}{2m} + 2ng\right)}. \qquad (42.15)$$

The point of diagonalizing the Hamiltonian is to tell us about the quasiparticle excitations from the ground state. These are the bogolons, which represent the motion of a large number of the original interacting bosons. The operator $\hat{\alpha}_p^\dagger$ creates a single bogolon. These bogolons don't interact with each other so the problem is solved.

Turning to the bogolon dispersion, shown in Fig. 42.2, we have that for small momenta the dispersion is linear and looks like

$$E_p = \left(\frac{ng}{m}\right)^{\frac{1}{2}} |p|. \qquad (42.16)$$

This linear dispersion occurs in another important problem: that of phonons. In a vibrating system phonon quasiparticles are collective excitations of a large number of atoms. The small-momentum bogolons

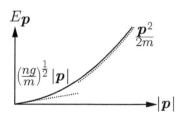

Fig. 42.2 The dispersion predicted by Bogoliubov's model. At low momentum the energy is linear in $|p|$, at large momentum it is quadratic.

are similar, and they may be thought of as representing the complicated, collective motion of a large number of bosons. Loosely speaking, the original bosons have a number density whose behaviour is wave-like and these waves are quantized into bogolons. At large momentum the dispersion becomes $E_{\boldsymbol{p}} = \boldsymbol{p}^2/2m$ telling us that large-momentum bogolons behave[6] like free particles with a mass m.

[6]We say that the particles are *ballistic*.

Finally we note that Bogoliubov's treatment does not predict the minimum seen in experimentally in Fig. 42.1. This owes its existence in liquid helium to the fact that the interactions in that system are very strong, not weak, as considered here. The minimum corresponds to excitations known as rotons, which represent a back-flow of Bose particles, rather like the motion of particles in a smoke ring.[7]

[7]See Feynman, *Statistical Physics*.

42.3 Superfluids and fields

The argument presented above, in terms of creation and annihilation operators, represents a quick and illuminating way of solving Bogoliubov's problem. An alternative approach is to attack the problem using the Lagrangian and the machinery of spontaneous symmetry breaking that we developed in Chapter 26.

We started our treatment with Bogoliubov's approximation $\hat{a}_{\boldsymbol{p}=0}|\Omega\rangle \approx \sqrt{N_0}|\Omega\rangle$. Actually we know that something very similar to this relation holds exactly, by definition, for a *coherent state*. If $|\Omega\rangle$ is a coherent state, we have $\hat{a}_{\boldsymbol{p}=0}|\Omega\rangle = \sqrt{N_0}\,e^{i\theta_0}|\Omega\rangle$. Note also that if this coherent state is macroscopically occupied then the uncertainty in the phase $\Delta\cos\theta$ tends to zero, so that the phase θ_0 is a well defined quantity. Thus Bogoliubov's 'approximation' holds exactly for a macroscopically occupied coherent state.

We can also understand how the occurrence of such a state necessarily implies broken symmetry. In real space (and ignoring time variation for now) the annihilation field $\hat{\Phi}(\boldsymbol{x}) = \frac{1}{\sqrt{\mathcal{V}}}\sum_{\boldsymbol{p}}\hat{a}_{\boldsymbol{p}}e^{i\boldsymbol{p}\cdot\boldsymbol{x}}$ has a coherent state eigenstate $|\phi(\boldsymbol{x})\rangle$ defined such that

$$\hat{\Phi}(\boldsymbol{x})|\phi(\boldsymbol{x})\rangle = \sqrt{\rho(\boldsymbol{x})}e^{i\theta(\boldsymbol{x})}|\phi(\boldsymbol{x})\rangle, \tag{42.17}$$

where $\rho(\boldsymbol{x})$ is a number density of particles. Substituting the mode expansion shows that $\hat{\Phi}(\boldsymbol{x})$ has the property, for the coherent ground state $|\Omega\rangle$, that

$$\begin{aligned}\langle\Omega|\hat{\Phi}(x)|\Omega\rangle &= \frac{1}{\sqrt{\mathcal{V}}}\sum_{\boldsymbol{p}}\langle\Omega|\hat{a}_{\boldsymbol{p}}|\Omega\rangle e^{i\boldsymbol{p}\cdot\boldsymbol{x}} \\ &= \frac{1}{\sqrt{\mathcal{V}}}\sum_{\boldsymbol{p}}\sqrt{N_{\boldsymbol{p}}}e^{i\theta_{\boldsymbol{p}}}e^{i\boldsymbol{p}\cdot\boldsymbol{x}},\end{aligned} \tag{42.18}$$

but since, for $|\Omega\rangle$, only the $\boldsymbol{p}=0$ state is occupied with any sizeable probability, this term dominates the sum and we have

$$\langle\Omega|\hat{\Phi}(\boldsymbol{x})|\Omega\rangle = \sqrt{n}\,e^{i\theta_0}, \tag{42.19}$$

where $\sqrt{n} = \sqrt{N_{\boldsymbol{p}=0}/\mathcal{V}}$. An important point here is that the field $\hat{\Phi}(\boldsymbol{x})$ has developed a ground state with a nonzero vacuum expectation value, which is a tell-tale sign of a broken symmetry state. In fact we take $\langle\Omega|\hat{\Phi}(\boldsymbol{x})|\Omega\rangle$ as the order parameter of the system. We also read off from eqn 42.17 that in the ground state $|\Omega\rangle$ the number operator expectation value is $\sqrt{\rho(\boldsymbol{x})} = \sqrt{n}$ across the sample and that the state's phase is $\theta(\boldsymbol{x}) = \theta_0$. We stress that while there is still uncertainty in the number n in a macroscopically occupied coherent state, the uncertainty in the phase θ_0 is zero. Thus, the breaking of symmetry involves fixing the phase to $\theta(\boldsymbol{x}) = \theta_0$ across the entire system. Clearly it is this picking of a unique phase that breaks the $U(1)$ symmetry of the system.[8]

How do we end up with a Lagrangian that describes non-relativistic bosons which is unstable to symmetry breaking? A system containing a fixed number of non-interacting, non-relativistic bosons has a Hamiltonian $\hat{H}_0 = (E_{\boldsymbol{p}} - \mu)\hat{a}_{\boldsymbol{p}}^\dagger\hat{a}_{\boldsymbol{p}}$, with $E_{\boldsymbol{p}} = \boldsymbol{p}^2/2m$ and where μ is the chemical potential. Translating back into position space fields we have a Lagrangian for the non-relativistic system, including chemical potential, given by

$$\mathcal{L} = i\Phi^\dagger\partial_0\Phi - \frac{1}{2m}\boldsymbol{\nabla}\Phi^\dagger\cdot\boldsymbol{\nabla}\Phi + \mu\Phi^\dagger\Phi. \qquad (42.20)$$

This is similar to the form we had in Chapter 12. The interaction is included by subtracting a term $\frac{g}{2}(\Phi^\dagger\Phi)^2$.

Example 42.3

The full, interacting Lagrangian is

$$\mathcal{L} = i\Phi^\dagger\partial_0\Phi - \frac{1}{2m}\boldsymbol{\nabla}\Phi^\dagger\cdot\boldsymbol{\nabla}\Phi + \mu\Phi^\dagger\Phi - \frac{g}{2}(\Phi^\dagger\Phi)^2, \qquad (42.21)$$

which, with its positive mass-like term $\mu\Phi^\dagger\Phi$, is unstable to spontaneous symmetry breaking. The Lagrangian may be helpfully rewritten (to within an ignorable constant) as

$$\mathcal{L} = i\Phi^\dagger\partial_0\Phi - \frac{1}{2m}\boldsymbol{\nabla}\Phi^\dagger\cdot\boldsymbol{\nabla}\Phi - \frac{g}{2}\left(n - \Phi^\dagger\Phi\right)^2, \qquad (42.22)$$

and remembering that the number density of particles is given by $\Phi^\dagger\Phi$, we interpret $n = \mu/g$ as the boson density of the ground state.

The Lagrangian for non-relativistic bosons with a short ranged repulsive interaction is therefore given by

$$\mathcal{L} = i\Phi^\dagger\partial_0\Phi - \frac{1}{2m}\boldsymbol{\nabla}\Phi^\dagger\cdot\boldsymbol{\nabla}\Phi - \frac{g}{2}(n - \Phi^\dagger\Phi)^2, \qquad (42.23)$$

which is unstable to symmetry-breaking. To see the consequences of broken symmetry we will use the polar coordinates of Chapter 12, so that the field looks like $\Phi(x) = \sqrt{\rho(x)}e^{i\theta(x)}$, and so is described by an amplitude-field part $\rho(x)$ and a phase-field part $\theta(x)$. Our Lagrangian becomes[9]

$$\mathcal{L} = -\rho\partial_0\theta - \frac{1}{2m}\left[\frac{1}{4\rho}(\boldsymbol{\nabla}\rho)^2 + \rho(\boldsymbol{\nabla}\theta)^2\right] - \frac{g}{2}(n - \rho)^2, \qquad (42.24)$$

[8]The breaking of symmetry is often represented pictorially by drawing a set of compass needles which depict $\theta(\boldsymbol{x})$: a phase angle which may be different at every point in space. The breaking of symmetry, and formation of a superfluid phase, corresponds to all of the needles lining up, as shown in Fig. 42.3(a). This picture of a superfluid is known as the XY model, where the name is motivated by the freedom of the needles to point in the x-y plane in the normal state.

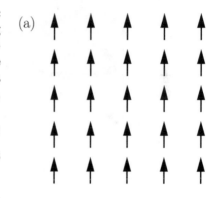
(a)

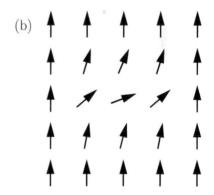
(b)

Fig. 42.3 (a) Order in the superfluid corresponds to a uniform phase angle across the entire system. This is represented on a lattice here. (b) A gradient in the phase $\boldsymbol{\nabla}\theta(x)$ results in a current in the superfluid.

[9]Note that $\Phi^\dagger = \sqrt{\rho}\,e^{-i\theta}$ and so

$$\partial_0\Phi = i\partial_0\theta\sqrt{\rho}\,e^{i\theta} + \frac{1}{2\sqrt{\rho}}e^{i\theta}\partial_0\rho,$$

resulting in

$$\Phi^\dagger\partial_0\Phi = i\rho\partial_0\theta + \frac{1}{2}\partial_0\rho.$$

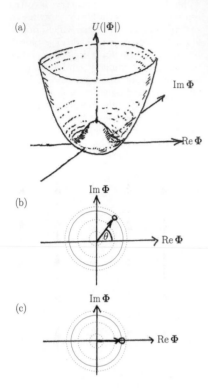

(a) $U(|\Phi|)$

Im Φ

Re Φ

(b)

Im Φ

θ

Re Φ

(c)

Im Φ

Re Φ

Fig. 42.4 (a) The Mexican hat potential in the superfluid problem. There is one of these at every point x in space. (b) The potential viewed from above as contours, with a minimum at $\rho = n$. Each value of θ is equivalent in energy. (c) Upon symmetry breaking the vacuum state picks a unique value of θ (here $\theta = 0$) at every point across the system.

[10]Particle physicists sometimes call this integrating out the high-energy particles.

[11]This is a Berry phase term which reveals some interesting physics in the presence of vortices. See Wen, Section 3.6.

where we have dropped the term $i\partial_0\rho/2$ as it is a total derivative and should therefore give a vanishing contribution to the action since our fields are defined to vanish at $\pm\infty$.

This Lagrangian has a Mexican hat potential (Fig. 42.4) which has a minimum at $\rho = n$ and is invariant with respect to *global* $U(1)$ transformations of phase. That is, we can swap $\theta(x) \to \theta(x) + \alpha$ and the Lagrangian is the same as long as α is the same at every point in spacetime. The possible ground states of the theory are therefore to be found at $\Phi(x) = \sqrt{\rho(x)}e^{i\theta(x)} = \sqrt{n}e^{i\theta_0}$, where θ_0 is the same at every point in space, but may itself take any value. We break the global $U(1)$ symmetry of the theory by fixing the phase of the ground state to be at a particular value, say $\theta_0 = 0$ as shown in Fig. 42.4(c). Breaking this continuous $U(1)$ phase symmetry will result in the emergence of a Goldstone mode, i.e. an excitation which costs vanishingly little energy to excite at low momentum.

To find the Goldstone mode we use a method much beloved of condensed matter physicists. We are going to remove the modes that cost a lot of energy.[10] This will leave behind a theory that only includes the low-energy excitations, which (we hope) will include the Goldstone modes. Staring at the form of the potential, we see that excitations that involve climbing the wall of the potential by changing ρ will cost more energy than those involving rolling around in the gutter (changing θ). We therefore examine the excitations that arise from departures from the ground state by considering an excited state field $\sqrt{\rho(x)} = \sqrt{n} + h$. This species of excited state just involve climbing the wall slightly from the ground state. We use these excited states to expand the Lagrangian, which yield (on dropping[11] the total time derivative $-n\partial_0\theta$):

$$\mathcal{L} = -\frac{1}{2m}(\boldsymbol{\nabla}h)^2 - 2gnh^2 - \left(2\sqrt{n}\partial_0\theta\right)h - \frac{n}{2m}(\boldsymbol{\nabla}\theta)^2 + \dots \quad (42.25)$$

To remove the energetic states we may plug a Lagrangian into a path integral and integrate out the variable describing these states. We will use this technique to remove the amplitude h from our Lagrangian.

Example 42.4

We'll make use of two neat functional integral tricks here. Remember that when dealing with the Lagrangian in a path integral, we can make the integrate-by-parts substitution to the quadratic parts:

$$(\partial_\mu\phi)^2 - m^2\phi^2 \to -\phi\left(\partial^2 + m^2\right)\phi, \quad (42.26)$$

which for our equation in terms of h requires

$$-\frac{1}{2m}(\boldsymbol{\nabla}h)^2 - 2gnh^2 \to -h\left(-\frac{1}{2m}\boldsymbol{\nabla}^2 + 2gn\right)h. \quad (42.27)$$

Next we use the one functional integral we can do, i.e. $\int \mathcal{D}\phi\, e^{\frac{1}{2}\int \phi\hat{K}\phi + i\int J\phi} = e^{-\frac{1}{2}\int J(i\hat{K}^{-1})J}$, which allows us to make the replacement

$$\frac{1}{2}\phi\left[-(\partial^2 + m^2)\right]\phi + J\phi \to \frac{1}{2}J(x)\frac{1}{\partial^2 + m^2}J(y). \quad (42.28)$$

Setting $\phi = -\sqrt{2}h$ and $J = \sqrt{2n}\partial_0\theta$, we have

$$h\left(-\frac{1}{2m}\nabla^2 + 2gn\right)h - (2\sqrt{n}\partial_0\theta)\,h \to \sqrt{n}\partial_0\theta\,\frac{1}{(-1/2m)\nabla^2 + 2gn}\,\sqrt{n}\partial_0\theta. \quad (42.29)$$

Our Lagrangian becomes

$$\mathcal{L} = n\partial_0\theta\,\frac{1}{2gn - \frac{1}{2m}\nabla^2}\,\partial_0\theta - \frac{n}{2m}(\nabla\theta)^2 + \dots \quad (42.30)$$

To obtain the low-energy, small-momentum behaviour we treat the energy density term $(1/2m)\nabla^2$ as small compared to the potential density term $2gn$ and we find a Lagrangian purely describing the low-energy physics

$$\mathcal{L} = \frac{1}{2g}\left(\frac{\partial\theta}{\partial t}\right)^2 - \frac{n}{2m}(\nabla\theta)^2. \quad (42.31)$$

This is merely a wave equation Lagrangian with a linear dispersion and a wave speed $c = \sqrt{\frac{gn}{m}}$. We have found our Goldstone mode: the θ field has no mass term, so the energy of excitations tends to zero as $p \to 0$. Our manipulations have resulted in a low-energy dispersion

$$E_{\boldsymbol{p}} = \left(\frac{ng}{m}\right)^{\frac{1}{2}}|\boldsymbol{p}|, \quad (42.32)$$

just as we had before. We've shown through another method that breaking the $U(1)$ symmetry gives the linearly dispersing mode at low momentum.

Noether's theorem: $U(1)$ internal symmetry for the low-energy theory

$$D\theta = 1$$
$$\Pi_\theta^0 = \tfrac{1}{g}\partial^0\theta \qquad \Pi_\theta^i = \tfrac{n}{m}\partial^i\theta$$
$$D\mathcal{L} = 0 \qquad W^\mu = 0$$
$$J_{\mathrm{Nc}}^0 = -\tfrac{1}{g}\partial_0\theta \qquad \boldsymbol{J}_{\mathrm{Nc}} = \tfrac{n}{m}\nabla\theta$$

42.4   The current in a superfluid

Let's examine the fate of Noether's symmetry current for the low-energy theory. Looking at the Lagrangian in eqn 42.31, we see that the $U(1)$ translation $\theta(x) \to \theta(x) + \alpha$ leaves the function invariant.[12] This is merely the $U(1)$ symmetry that is spontaneously broken in the ground state. Applying Noether's theorem to this symmetry we have that $D\theta(x) = 1$ and

$$\Pi_\theta^0 = \tfrac{1}{g}\partial_0\theta, \quad \Pi_\theta^i = \tfrac{n}{m}\partial^i\theta. \quad (42.33)$$

[12]See Chapter 12 for a reminder.

As a result, we have the conserved currents

$$J_{\mathrm{Nc}}^0(x) = -\tfrac{1}{g}\partial_0\theta(x), \quad \boldsymbol{J}_{\mathrm{Nc}}(x) = \tfrac{n}{m}\nabla\theta(x). \quad (42.34)$$

This final equation tells us that gradients in the phase in a superfluid [as shown in Fig. 42.3(b)] result in currents. The broken symmetry ground state of the superfluid has a uniform phase and if we deform the phase we set up a current in the system along the gradient. Notice that the current $\boldsymbol{J}_{\mathrm{Nc}}$ depends on the *strength* of the condensate (or vacuum expectation value) given by n. In the absence of symmetry breaking,

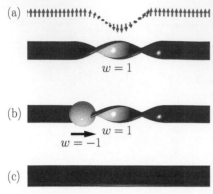

(a)

$w = 1$

(b)

$w = 1$

$w = -1$

(c)

Fig. 42.5 (a) The superfluid with a $w = 1$ phase twist. Any attempt to untwist the phase would tear the superfluid. (b) The same superfluid and a $w = -1$ vortex propagating to the right. (c) The propagation of the vortex unwinds the twist allowing the superfluid to dissipate momentum.

[13]We have considered annihilating our broad phase twist in Fig. 42.5(a) with a vortex in Fig. 42.5(b). The broadness of the phase twist and the localized nature of the vortex are irrelevant. What matters is their topological character, namely that one has $w = 1$ and the other has $w = -1$ so that they annihilate each other.

this is zero, so it makes sense that this is indeed the current that results from symmetry breaking.

What does this tell us about the dissipationless flow of momentum which, after all, is one of the most striking aspects of a superfluid? We will use a beautiful argument based on topology that demonstrates that a moving superfluid can't dissipate momentum. That is, it's a runaway train that can't slow down! The superfluid's broken symmetry ground state is a constant field $\Phi = \sqrt{n}e^{i\theta_0}$. We imagine putting our superfluid in a box of size L with periodic boundaries $\Phi(x = 0) = \Phi(L)$. Now we boost the superfluid so it has a momentum P. This changes the ground state $\Phi \rightarrow e^{iPx}\Phi = \sqrt{n}e^{i(\theta_0 + Px)}$. Notice that the boost effectively twists the phase of our field. The boost also twists the boundary conditions; instead of $\Phi(0) = \Phi(L)$, they become $\Phi(0)e^{iPL} = \Phi(L)$. This is shown in Fig. 42.5(a). To satisfy the periodic boundary conditions, we must have $PL = 2\pi w$, where w is an integer (known as the winding number, as discussed in Chapter 29).

The nonzero momentum state has more energy than the ground state. If the superfluid can slow down it will lower its energy. Now we ask, can we untwist the superfluid phase to remove this extra momentum? The answer is no, we can't. The nonzero momentum state involves an integer number of 2π rotations of the phase angle between the ends of the box. The angles at the ends of the box are locked to each other by the periodic boundary conditions, so removing the twists will lead to singularities in the phase field, costing enormous amounts of energy.

The only way we could remove a twist is to create a vortex, a localized phase twist. Such vortices carry integer winding numbers, so creating a vortex which travels across the box will remove a twist as shown in Fig. 42.5(b).[13] The creation of a vortex is a tunnelling process with a finite energy barrier. Ignoring such subtleties would lead us to believe that a moving superfluid is a perpetual motion machine. However, Nature is not so kind and allows a quantum tunnelling process which unwinds the condensate. As a result, the superfluid can, over a very long period (and even in the absence of excitations) dissipate some momentum.

Our model of weakly interacting bosons captures the small and large momentum behaviour of the dispersion, so explains at least some of the physics of the excited superfluid. One might question whether this dispersion is essential to the existence of the superfluid state. It is essential since it guarantees that the moving fluid can't dissipate too much momentum through its excitations. The reason is that a linear dispersion doesn't give rise to enough states for quasiparticles to scatter into and dissipate momentum. To see this consider the density of states $g(E)$ for a three-dimensional fluid of bosons. For the linear dispersion of a superfluid, this varies as $g(E) \propto E^2$, whereas for the quadratic dispersion of a normal particle, this varies as $g(E) \propto E^{\frac{1}{2}}$. As a result, there are fewer excited states available for a superfluid than a normal fluid at low values of energy, reducing the phase space for scattering events that will reduce momentum.

Example 42.5

It is also worth noting another beautiful argument by Landau that shows an upper bound on the critical velocity for superfluid flow is given by $(ng/m)^{\frac{1}{2}}$, a quantity which only exists due to the presence of *interactions*. In the simplest version of this argument, we imagine the superfluid as a body of mass M moving with velocity $\boldsymbol{v}$. The body creates an excitation, with momentum $\boldsymbol{p}$ and energy $E_{\boldsymbol{p}}$, with the result that the body's velocity changes to $\boldsymbol{v}'$. By conservation of momentum, we have $M\boldsymbol{v} = M\boldsymbol{v}' + \boldsymbol{p}$, from which we see that the body ends up with kinetic energy $\frac{M\boldsymbol{v}'^2}{2} = \frac{M\boldsymbol{v}^2}{2} - \boldsymbol{v} \cdot \boldsymbol{p} + \frac{p^2}{2M}$. The point is that the excitation cannot be created unless we have $\frac{M\boldsymbol{v}^2}{2} > \frac{M\boldsymbol{v}'^2}{2} + E_{\boldsymbol{p}}$, from which we conclude that

$$E_{\boldsymbol{p}} < \boldsymbol{v} \cdot \boldsymbol{p} - \frac{p^2}{2M} \approx \boldsymbol{v} \cdot \boldsymbol{p}, \qquad (42.35)$$

if M is assumed large. In other words, we have an upper bound on the superflow against creating excitations, given by $E_{\boldsymbol{p}}/|\boldsymbol{p}| = (ng/m)^{\frac{1}{2}}$. So while a non-interacting Bose gas (which has $g = 0$) will undergo a transition into a Bose–Einstein condensate at low temperatures, such a state will not exhibit a superflow as the critical velocity is zero.

Chapter summary

- Bogoliubov's approximation states that the number of particles in the $\boldsymbol{p} = 0$ ground state becomes macroscopically large, allowing the replacement of operators $\hat{a}_{\boldsymbol{p}=0}$ and $\hat{a}^{\dagger}_{\boldsymbol{p}=0}$ with the number $\sqrt{N_0}$.

- The effective Hamiltonian in a non-relativistic Bose gas can be diagonalized using a Bogoliubov transformation, and the quasi-particles are called bogolons.

- The low-energy excitations in a superfluid are Goldstone modes, the broken symmetry is the phase θ and the superfluid current is $\boldsymbol{J}_{\mathrm{Nc}} = \frac{n}{m}\boldsymbol{\nabla}\theta(x)$.

Exercises

(42.1) Verify eqn 42.15 using the method suggested in the text. You may find it useful to make the substitutions

$$\begin{aligned} u_{\boldsymbol{p}} &= \cosh\theta_{\boldsymbol{p}}, \\ v_{\boldsymbol{p}} &= \sinh\theta_{\boldsymbol{p}}. \end{aligned} \qquad (42.36)$$

(42.2) Verify the algebra leading to eqn 42.31.

(42.3) (a) Show that, in three dimensions, the density of states $g(E) = \mathrm{d}N(E)/\mathrm{d}E$ for a superfluid goes as $g(E) \propto E^2$.
(b) Show that for a Fermi gas $g(E) \propto E^{\frac{1}{2}}$.

The many-body problem and the metal

The sentries report Zulus to the south west. Thousands of them.
Colour Sergeant Bourne (Nigel Green), *Zulu* (1964)

The physics of large numbers of non-relativistic particles and their interactions is a branch of quantum field theory known as the many-body problem. In this chapter we'll examine the properties of a large number of non-relativistic fermions confined in a box, which is the basis of the description of electrons in solids. As we've seen, non-relativistic fermions scatter via the diagram shown in Fig. 43.1. The full formalism of Dirac spinors isn't required to understand this problem owing to the low energies involved in metals and so we can treat the particles as interacting via an instantaneous potential $V(\boldsymbol{x} - \boldsymbol{y}) = \sum_{\boldsymbol{q}} \mathrm{e}^{\mathrm{i}\boldsymbol{q}\cdot(\boldsymbol{x}-\boldsymbol{y})} \tilde{V}_{\boldsymbol{q}}$. Rather than starting with propagators and path integrals, we'll gain some intuition by treating the interaction with some very simple approximations.

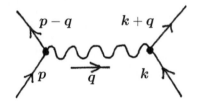

Fig. 43.1 The Coulomb vertex. Note that in this chapter the transferred momentum runs from left to right.

43.1 Mean-field theory

We start with the Hamiltonian

$$\hat{H} = \hat{H}_0 + \hat{V}, \tag{43.1}$$

where $\hat{H}_0 = \sum_{\boldsymbol{p}} \frac{\boldsymbol{p}^2}{2m} \hat{a}_{\boldsymbol{p}}^\dagger \hat{a}_{\boldsymbol{p}}$ and

$$\hat{V} = \frac{1}{2} \sum_{\boldsymbol{pkq}} \tilde{V}_{\boldsymbol{q}} \hat{a}_{\boldsymbol{p}-\boldsymbol{q}}^\dagger \hat{a}_{\boldsymbol{k}+\boldsymbol{q}}^\dagger \hat{a}_{\boldsymbol{k}} \hat{a}_{\boldsymbol{p}}, \tag{43.2}$$

where the anticommuting operators $\hat{a}_{\boldsymbol{p}}^\dagger$ and $\hat{a}_{\boldsymbol{p}}$ create and destroy fermions respectively. We take $\hat{H}_0$ as the dominant contribution and treat $\hat{V}$ as a perturbation. As usual we call $|0\rangle$ the ground state of H_0 with energy E_0 and $|\Omega\rangle$ the ground state of the full Hamiltonian of H with energy E.[1]

We need a way to deal with the perturbation provided by the complicated two-body potential $\hat{V}$. A good first step is to ask what the **mean-field** behaviour of the system is. Mean-field theory is a scheme of approximations that crops up all over physics and involves taking a system of many particles and asking how a single particle reacts to the average behaviour of all of the others. In the context of many-body

[1]It is worth noting that $|0\rangle$ doesn't mean that we have no particles in the ground state. Usually in condensed matter physics we study systems with a finite density of particles at zero temperature. The non-interacting ground state for a metal is a gas of electrons stacked up in energy up to the Fermi level p_{F}. So if we take spin into account then the ground state of a metal may be written $|0\rangle = \prod_{|\boldsymbol{p}|<p_{\mathrm{F}}} \hat{a}_{\boldsymbol{p}\uparrow}^\dagger \hat{a}_{\boldsymbol{p}\downarrow}^\dagger |\text{Empty box}\rangle$.

theory, we will interpret the mean-field approximation as a method of replacing pairs of operators with averages, or more correctly, with vacuum expectation values (VEVs), and we will later link this to the conventional idea of mean-field theory. The mean-field correction ΔE to the non-interacting ground state energy E_0 is found by taking a VEV of the perturbation, which is to say we approximate the expectation value of the troublesome four-operator term taken with the unperturbed ground state:[2] $\langle 0|\hat{a}_{p-q}^\dagger \hat{a}_{k+q}^\dagger \hat{a}_k \hat{a}_p|0\rangle$.

[2]This is exactly how we usually work out the energy shift caused by a perturbation in first-order perturbation theory, so represents nothing new. Novelty comes with the next step, where this VEV is boiled down to something more useful using a version of Wick's theorem.

Example 43.1

Wick's theorem reduces the four operators in eqn (43.2) into products of contractions of pairs of (normally ordered) operators.[3] We therefore apply

$$\hat{a}_{p-q}^\dagger \hat{a}_{k+q}^\dagger \hat{a}_k \hat{a}_p \rightarrow \prod N \begin{bmatrix} \text{all paired} \\ \text{contractions} \end{bmatrix}. \tag{43.3}$$

Wick's theorem on our string of four operators yields

$$\hat{a}_{p-q}^\dagger \hat{a}_{k+q}^\dagger \hat{a}_k \hat{a}_p = N\left[\hat{a}_{p-q}^\dagger \hat{a}_{k+q}^\dagger \hat{a}_k \hat{a}_p\right]$$

$$+ \quad \hat{a}_{p-q}^\dagger \hat{a}_{k+q}^\dagger\, N\left[\hat{a}_k \hat{a}_p\right] + N\left[\hat{a}_{p-q}^\dagger \hat{a}_{k+q}^\dagger\right] \hat{a}_k \hat{a}_p$$

$$+ \quad \hat{a}_{p-q}^\dagger \hat{a}_p\, N\left[\hat{a}_{k+q}^\dagger \hat{a}_k\right] + N\left[\hat{a}_{p-q}^\dagger \hat{a}_p\right] \hat{a}_{k+q}^\dagger \hat{a}_k$$

$$- \quad \hat{a}_{p-q}^\dagger \hat{a}_k\, N\left[\hat{a}_{k+q}^\dagger \hat{a}_p\right] - N\left[\hat{a}_{p-q}^\dagger \hat{a}_k\right] \hat{a}_{k+q}^\dagger \hat{a}_p$$

$$+ \quad \hat{a}_{p-q}^\dagger \hat{a}_{k+q}^\dagger \hat{a}_k \hat{a}_p + \hat{a}_{p-q}^\dagger \hat{a}_p \hat{a}_{k+q}^\dagger \hat{a}_k - \hat{a}_{p-q}^\dagger \hat{a}_k \hat{a}_{k+q}^\dagger \hat{a}_p. \tag{43.4}$$

Everything within the normal ordering signs is normally ordered as written, so these can be dropped. Mean-field theory involves replacing the contractions with averages taken over the ground state: $\langle 0|\hat{O}|0\rangle$.

[3]Since we're dealing with fermion operators in this chapter then we need to remember that swapping the order of two operators when we bring them together and make a VEV, gives us a minus sign.

The terms with uncontracted operators represent excitations, and so we ignore them for now, since we're concerned with the ground state properties. For the ground state energy shift $\langle \hat{V} \rangle$, we therefore only consider the completely contracted terms and this procedure therefore yields up the three terms:

[4]The Cooper term will be important when we come to deal with superconductors in the next chapter. We'll postpone discussing it until then and, for the moment, use the fact that under normal circumstances, the combination $\langle 0|\hat{a}_n^\dagger \hat{a}_m|0\rangle = 0$ for any n and m, and so we expect no contribution from C_0.

$$\frac{1}{2}\sum_{pkq} \tilde{V}_q \left[\underbrace{\langle \hat{a}_{p-q}^\dagger \hat{a}_{k+q}^\dagger \rangle \langle \hat{a}_k \hat{a}_p \rangle}_{\text{leads to } C_0} + \underbrace{\langle \hat{a}_{p-q}^\dagger \hat{a}_p \rangle \langle \hat{a}_{k+q}^\dagger \hat{a}_k \rangle}_{\text{leads to } D_0} \underbrace{-\langle \hat{a}_{p-q}^\dagger \hat{a}_k \rangle \langle \hat{a}_{k+q}^\dagger \hat{a}_p \rangle}_{\text{leads to } F_0} \right]$$
$$= C_0 + D_0 + F_0, \tag{43.5}$$

where we've shortened $\langle 0|\hat{O}|0\rangle$ to $\langle \hat{O} \rangle$ to save on clutter. We'll call C_0 the Cooper term,[4] D_0 is Hartree's direct term and F_0 Fock's exchange term.

Douglas Hartree (1897–1956)

Vladimir Fock (1898–1974)

Let's examine the Hartree term in more detail. It gives rise to a contribution to the ground state energy of

$$D_0 = \frac{1}{2}\sum_{pkq} \tilde{V}_q \langle \hat{a}_{p-q}^\dagger \hat{a}_p \rangle \langle \hat{a}_{k+q}^\dagger \hat{a}_k \rangle. \tag{43.6}$$

We know that $\langle 0|\hat{a}_r^\dagger \hat{a}_s|0\rangle = 0$ unless $r = s$, so we can immediately fix $q = 0$ in eqn 43.6. Recall that the $\tilde{V}_q$s are Fourier components of the real space potential $V(r)$. A wave with wave vector $q = 0$ is a constant (that is, it has an infinitely long wavelength), so the Hartree term evaluates the energy due to the constant part of the potential. Maybe this is what we would expect from a first-order guess! We obtain

$$D_0 = \frac{1}{2}\sum_{pk} \tilde{V}_{q=0}\langle \hat{a}_p^\dagger \hat{a}_p\rangle\langle \hat{a}_k^\dagger \hat{a}_k\rangle. \qquad (43.7)$$

Notice that we've reduced the VEVs of operator pairs to simple number operators and the Hartree term has become

$$D_0 = \frac{1}{2}\tilde{V}_{q=0}\left(\sum_p \langle 0|\hat{N}_p|0\rangle\right)^2. \qquad (43.8)$$

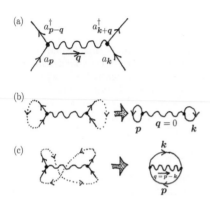

The mean-field treatment has a diagrammatic interpretation: we can take the interaction diagram and join the legs of the operators that we're averaging (Fig. 43.2). This turns out to be equivalent to the more usual ways of getting diagrams with the S-matrix or path integral approach. The Hartree term can therefore be represented in diagrammatic form as the double-headed tadpole shown in Fig. 43.2(b).

Another point to note is that for theories where there are no particles in the ground state $|0\rangle$ the tadpole [whose head corresponds to $\langle 0|\hat{a}_p^\dagger \hat{a}_p|0\rangle$, see Fig. 43.2(b)] gives zero. This isn't the case for nuclear or electronic matter however, where we have a finite number of fermions in the ground state.

Fig. 43.2 (a) The basic interaction diagram. (b) The double tadpole diagram giving the Direct (or Hartree) contribution to the ground state energy. (c) The double oyster diagram giving the Exchange (or Fock) contribution to the ground state energy.

Example 43.2

Let's see how the Hartree term looks in terms of position space fields. We examine the product $\langle \hat{a}_{p-q}^\dagger \hat{a}_p\rangle$. We have that $\hat{a}_p = \frac{1}{\sqrt{\mathcal{V}}}\int \mathrm{d}^3x\, \hat{\psi}(x)\mathrm{e}^{-i p\cdot x}$, which enables us to say

$$\langle \hat{a}_{p-q}^\dagger \hat{a}_p\rangle = \frac{1}{\mathcal{V}}\int \mathrm{d}^3x \mathrm{d}^3x'\, \langle \hat{\psi}^\dagger(x)\hat{\psi}(x')\rangle \mathrm{e}^{i(p-q)\cdot x}\mathrm{e}^{-i p\cdot x'}. \qquad (43.9)$$

This in turn allows us to write the Hartree energy as

$$D_0 = \frac{1}{2\mathcal{V}^2}\sum_{p,k,q}\int \mathrm{d}^3x \mathrm{d}^3x' \mathrm{d}^3y \mathrm{d}^3y'\, \tilde{V}_q\langle \hat{\psi}^\dagger(x)\hat{\psi}(x')\rangle\langle \hat{\psi}^\dagger(y)\hat{\psi}(y')\rangle \quad (43.10)$$
$$\times \mathrm{e}^{i(p-q)\cdot x}\mathrm{e}^{-i p\cdot x'}\mathrm{e}^{i(k+q)\cdot y}\mathrm{e}^{-i k\cdot y'}.$$

Doing the sums over p and k yields $\mathcal{V}\delta^{(3)}(x - x')$ and $\mathcal{V}\delta^{(3)}(y - y')$ respectively, which eat up two of the space integrals. We're left with

$$D_0 = \frac{1}{2}\sum_q \int \mathrm{d}^3x \mathrm{d}^3y\, \tilde{V}_q\langle \hat{\psi}^\dagger(x)\hat{\psi}(x)\rangle\langle \hat{\psi}^\dagger(y)\hat{\psi}(y)\rangle\mathrm{e}^{-i q\cdot(x-y)}. \quad (43.11)$$

Next we sum over q, which takes an inverse Fourier transform of the potential $\tilde{V}_q$, and we obtain

$$D_0 = \frac{1}{2}\int \mathrm{d}^3x \mathrm{d}^3y\, V(x - y)\langle \hat{\psi}^\dagger(x)\hat{\psi}(x)\rangle\langle \hat{\psi}^\dagger(y)\hat{\psi}(y)\rangle. \qquad (43.12)$$

This is just what one might have guessed classically for the energy of two charge distributions $\rho(x)$ and $\rho(y)$ interacting via a potential $V(x - y)$.

Of course, decompressing the notation here

$$\langle \hat{\psi}^\dagger(\boldsymbol{x})\hat{\psi}(\boldsymbol{x})\rangle = - \lim_{\substack{t' \to 0^- \\ \boldsymbol{x}' \to \boldsymbol{x}}} \langle 0|T\hat{\psi}(t', \boldsymbol{x}')\hat{\psi}^\dagger(0, \boldsymbol{x})|0\rangle, \qquad (43.13)$$

is related to a propagator.[5] The internal lines that form the double tadpole's two heads are just propagators, as we expect for a respectable Feynman diagram.

[5]See also eqn 31.18.

Next we examine the Fock term. This is represented by the double oyster diagram in Fig. 43.2(c) which corresponds to the equation

$$F_0 = -\frac{1}{2}\sum_{\boldsymbol{pkq}} \tilde{V}_{\boldsymbol{q}}\langle \hat{a}_{\boldsymbol{p}-\boldsymbol{q}}^\dagger \hat{a}_{\boldsymbol{k}}\rangle\langle \hat{a}_{\boldsymbol{k}+\boldsymbol{q}}^\dagger \hat{a}_{\boldsymbol{p}}\rangle. \qquad (43.14)$$

Example 43.3

Again, in order to make the VEVs nonzero, this gives us a condition on $\boldsymbol{q}$. This time we have that $\boldsymbol{p} - \boldsymbol{q} = \boldsymbol{k}$, leading to a result

$$
\begin{aligned}
F_0 &= -\frac{1}{2}\sum_{\boldsymbol{pk}} \tilde{V}_{\boldsymbol{p}-\boldsymbol{k}}\langle \hat{a}_{\boldsymbol{k}}^\dagger \hat{a}_{\boldsymbol{k}}\rangle\langle \hat{a}_{\boldsymbol{p}}^\dagger \hat{a}_{\boldsymbol{p}}\rangle \\
&= -\frac{1}{2}\sum_{\boldsymbol{pk}} \tilde{V}_{\boldsymbol{p}-\boldsymbol{k}}\langle 0|\hat{N}_{\boldsymbol{k}}|0\rangle\langle 0|\hat{N}_{\boldsymbol{p}}|0\rangle, \qquad (43.15)
\end{aligned}
$$

which is diagonal (that is, contains only number operators) but is clearly more complicated than what we had for the Hartree term. This is shown diagrammatically as the double oyster in Fig. 43.2(c). There is a lot of physics in the exchange term including, for example, magnetic ordering in metals, as examined in the exercises.

The position space version unfolds much as before, although now the momentum sums mix up $\boldsymbol{x}$'s and $\boldsymbol{y}$'s, giving us

$$F_0 = -\frac{1}{2}\int \mathrm{d}^3x\mathrm{d}^3y\, V(\boldsymbol{x} - \boldsymbol{y})\langle \hat{\psi}^\dagger(\boldsymbol{x})\hat{\psi}(\boldsymbol{y})\rangle\langle \hat{\psi}^\dagger(\boldsymbol{y})\hat{\psi}(\boldsymbol{x})\rangle. \qquad (43.16)$$

43.2   The Hartree–Fock ground state energy of a metal

We will use our mean-field results to work out an approximate value of the total ground state energy of a metal. This will involve potential energy contributions from the Hartree term and the Fock term and this is therefore known as the **Hartree–Fock** ground state energy.[6]

[6]The other contribution is from $\hat{H}_0$ and reflects fact that each electron has kinetic energy $E_{\boldsymbol{p}}^{(0)} = \frac{\boldsymbol{p}^2}{2m}$ and obeys the Pauli principle. Our approach will be to add these kinetic energies up by noting that the sums are carried out up to p_F and that states are so closely spaced that we may make the replacement $\sum_{|\boldsymbol{p}|<p_\mathrm{F}} \to \mathcal{V}\int_{|\boldsymbol{p}|<p_\mathrm{F}} \frac{\mathrm{d}^3 p}{(2\pi)^3}$.

Example 43.4

Recall that we put non-interacting electrons into a box, with one spin-up and one spin-down electron in each momentum state. The electrons stack up in energy and the upper-most occupied energy level is called the Fermi level. The number of occupied momentum states is

$$\sum_{|\boldsymbol{p}|<p_\mathrm{F}} \to \mathcal{V}\int_{|\boldsymbol{p}|<p_\mathrm{F}} \frac{\mathrm{d}^3 p}{(2\pi)^3} = \mathcal{V}\int_{|\boldsymbol{p}|=0}^{p_\mathrm{F}} (4\pi)\frac{\mathrm{d}|\boldsymbol{p}|}{(2\pi)^3}|\boldsymbol{p}|^2 = \frac{\mathcal{V}p_\mathrm{F}^3}{6\pi^2}, \qquad (43.17)$$

and with two spin states per momentum level, the total number of electron states is $N = \mathcal{V} p_{\mathrm{F}}^3/3\pi^2$. The electron density of the system $n = N/\mathcal{V}$ is related to the Fermi momentum p_{F} via $p_{\mathrm{F}} = (3\pi^2 n)^{\frac{1}{3}}$, and the Fermi energy is $E_{\mathrm{F}} = p_{\mathrm{F}}^2/(2m)$. The total energy of the box of non-interacting electrons is given by

$$ W_0 \;=\; 2\sum_{|\boldsymbol{p}|<p_{\mathrm{F}}} E_{\boldsymbol{p}}^{(0)} = 2\mathcal{V}\int_{|\boldsymbol{p}|<p_{\mathrm{F}}} \frac{\mathrm{d}^3 p}{(2\pi)^3}\frac{p^2}{2m} = \frac{3}{5} N E_{\mathrm{F}}, \qquad (43.18) $$

and so the kinetic energy per electron $W_0/N = \frac{3}{5}E_{\mathrm{F}}$. A popular unit for energy among theoreticians is the Rydberg (Ry), equal to $\hbar^2/(2m_e a_0^2)$ where a_0 is the Bohr radius. Writing the volume occupied per electron as $1/n = \frac{4}{3}\pi r^3$ where r is the average distance per particle, and using the dimensionless length $r_{\mathrm{s}} = r/a_0$, we can express the kinetic energy per electron as

$$ \frac{W_0}{N} = \frac{3}{5}E_{\mathrm{F}} = \frac{3}{5}\left(\frac{9\pi}{4}\right)^{2/3}\frac{1\,\mathrm{Ry}}{r_{\mathrm{s}}^2} \approx \frac{2.21}{r_{\mathrm{s}}^2}\frac{\mathrm{Rydbergs}}{\mathrm{electron}}. \qquad (43.19) $$

Now for the potential energy. This is not something we can calculate exactly, but our mean-field approach tells us that to first order we should add the Hartree term D_0 and Fock term F_0. However an immediate simplification is possible. Metals may be modelled as charge-neutral boxes full of electrons interacting via the Coulomb force. Although, in reality, metals contain positive ion cores and mobile electrons, this level of detail is not always necessary. Instead, we use a model known[7] as **jellium** in which the electrons are put in a box full of a homogenous, positively charged jelly in order to guarantee charge neutrality. This uniform positive charge exactly cancels the contribution from the Hartree term which originates from the uniform electron distribution. We therefore only need consider the Fock contribution to the energy given by $F_0 = -\frac{1}{2}\sum_{\boldsymbol{p}\boldsymbol{k}} \tilde{V}_{\boldsymbol{p}-\boldsymbol{k}}\langle \hat{N}_{\boldsymbol{p}}\rangle\langle \hat{N}_{\boldsymbol{k}}\rangle$. The ingredients of the sum are the ground state occupation numbers

$$ \langle \hat{N}_{\boldsymbol{p}}\rangle = \begin{cases} 1 & |\boldsymbol{p}| \le p_{\mathrm{F}} \\ 0 & |\boldsymbol{p}| > p_{\mathrm{F}}, \end{cases} \qquad (43.20) $$

and the Fourier transform of the interaction potential $\tilde{V}_{\boldsymbol{p}}$.

[7] Named by John Bardeen.

[8] Here we return to SI units, since these are often used in the many-body literature.

[9] The integral was evaluated in Chapter 17.

Example 43.5

The electrostatic potential energy between two electrons is given by[8] $V(\boldsymbol{x}-\boldsymbol{y}) = \frac{e^2}{4\pi\epsilon_0|\boldsymbol{x}-\boldsymbol{y}|}$. The Fourier transform can be most easily evaluated by working with the screened potential energy $V(\boldsymbol{r}) = (e^2/4\pi\epsilon_0)\mathrm{e}^{-\lambda|\boldsymbol{r}|}/|\boldsymbol{r}|$ and then sending $\lambda \to 0$. Thus[9]

$$ \tilde{V}_{\boldsymbol{q}} = \lim_{\lambda\to 0}\frac{1}{\mathcal{V}}\frac{e^2}{4\pi\epsilon_0}\int \mathrm{d}^3 r\,\frac{\mathrm{e}^{-\mathrm{i}\boldsymbol{q}\cdot\boldsymbol{r}}\mathrm{e}^{-\lambda|\boldsymbol{r}|}}{|\boldsymbol{r}|} = \lim_{\lambda\to 0}\frac{e^2}{\mathcal{V}\epsilon_0(\boldsymbol{q}^2+\lambda^2)} = \frac{e^2}{\mathcal{V}\epsilon_0\boldsymbol{q}^2}. \qquad (43.21) $$

Now to evaluate the Fock oyster term. It will be useful for what is to follow if we arrange the terms as follows (including a factor of 2 for spin):

$$F_0 = 2 \sum_{|\boldsymbol{p}|<p_\mathrm{F}} \frac{1}{2} \left[-\sum_{|\boldsymbol{k}|<p_\mathrm{F}} \tilde{V}_{\boldsymbol{p}-\boldsymbol{k}} \right]$$

$$= 2 \sum_{|\boldsymbol{p}|<p_\mathrm{F}} \frac{1}{2} \left[-\sum_{|\boldsymbol{k}|<p_\mathrm{F}} \frac{e^2}{V\epsilon_0} \frac{1}{|\boldsymbol{p}-\boldsymbol{k}|^2} \right]. \tag{43.22}$$

The part in the square bracket corresponds to the oyster diagram part shown in Fig. 43.3 and is known as the Fock self-energy[10] $\tilde{\Sigma}_{\boldsymbol{p}}^{(\mathrm{F})}$.

Example 43.6

The Fock self-energy may be evaluated as follows:[11]

$$\tilde{\Sigma}_{\boldsymbol{p}}^{(\mathrm{F})} = -\frac{e^2}{\epsilon_0} \int_{|\boldsymbol{k}|<p_\mathrm{F}} \frac{\mathrm{d}^3 k}{(2\pi)^3} \frac{1}{|\boldsymbol{p}-\boldsymbol{k}|^2} = -\frac{p_\mathrm{F}}{\pi} \left(\frac{e^2}{4\pi\epsilon_0} \right) F\left(\frac{|\boldsymbol{p}|}{p_\mathrm{F}} \right), \tag{43.23}$$

where

$$F(x) = 1 + \frac{1-x^2}{2x} \ln\left| \frac{1+x}{1-x} \right|. \tag{43.24}$$

The function $F(x)$ is shown in Fig. 43.4. It has an infinite slope at $x = 1$ corresponding, in our example, to the Fermi momentum.

The Fock energy is given (finally) by

$$F_0 = 2 \sum_{|\boldsymbol{p}|<p_\mathrm{F}} \frac{1}{2} \tilde{\Sigma}_{\boldsymbol{p}}^{(\mathrm{F})} = -V \frac{p_\mathrm{F}}{\pi} \left(\frac{e^2}{4\pi\epsilon_0} \right) \int_{|\boldsymbol{p}|<p_\mathrm{F}} \frac{\mathrm{d}^3 p}{(2\pi)^3} F\left(\frac{|\boldsymbol{p}|}{p_\mathrm{F}} \right)$$

$$= -\frac{3N p_\mathrm{F}}{2\pi} \left(\frac{e^2}{4\pi\epsilon_0} \right) \int_0^1 \mathrm{d}x \, x^2 F(x). \tag{43.25}$$

The integral yields $1/2$ and we obtain the potential energy per electron

$$\frac{F_0}{N} = -\frac{3p_\mathrm{F}}{4\pi} \frac{e^2}{4\pi\epsilon_0} = -\frac{0.916}{r_\mathrm{s}} \frac{\text{Rydbergs}}{\text{electron}}. \tag{43.26}$$

Putting this together with the kinetic energy yields the total Hartree–Fock energy per electron of jellium metal

$$\frac{E_\mathrm{HF}}{N} = \left(\frac{2.21}{r_\mathrm{s}^2} - \frac{0.916}{r_\mathrm{s}} \right) \frac{\text{Rydbergs}}{\text{electron}}. \tag{43.27}$$

Although this is as far as our mean-field theory will get us, this equation has the appearance of the start of a series in powers of r_s and we expect additional terms to add to this series. These terms were given the name **correlation energy**[12] by Eugene Wigner and Frederick Seitz. The next two terms were calculated by Gell-Mann and Brueckner in 1957 with the result

$$\frac{E}{N} = \left(\frac{2.21}{r_\mathrm{s}^2} - \frac{0.916}{r_\mathrm{s}} - 0.094 + 0.0622 \ln r_\mathrm{s} \right) \frac{\text{Rydbergs}}{\text{electron}}. \tag{43.28}$$

[10] We will see this again in the next section.

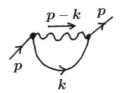

Fig. 43.3   The Fock contribution to the self-energy $\Sigma_{\boldsymbol{p}}^{(\mathrm{F})}$.

[11] Use has been made of the integral

$$\int_{|\boldsymbol{k}|<p_\mathrm{F}} \frac{\mathrm{d}^3 k}{|\boldsymbol{p}-\boldsymbol{k}|^2}$$
$$= \int_0^{p_\mathrm{F}} 2\pi k^2 \, \mathrm{d}k \int_0^\pi \frac{\sin\theta \, \mathrm{d}\theta}{k^2+p^2-2kp\cos\theta}$$
$$= \frac{2\pi}{p} \int_0^{p_\mathrm{F}} k \, \mathrm{d}k \ln\left| \frac{k+p}{k-p} \right|$$
$$= 2\pi p_\mathrm{F} \, F\left(\frac{|\boldsymbol{p}|}{p_\mathrm{F}} \right).$$

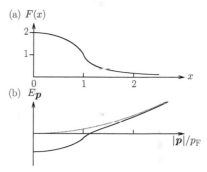

Fig. 43.4   (a) The function $F(x)$. (b) The energy of an electron in the Hartree–Fock approximation $E_{\boldsymbol{p}} = \frac{p^2}{2m} + \Sigma_{\boldsymbol{p}}^{\mathrm{F}}$.

[12] Feynman suggested it could be called the **stupidity energy**.

Eugene Wigner (1902–1995) made fundamental contributions to many areas of physics and mathematics. His sister was married to Dirac, who would often refer to her as 'Wigner's sister'.

Frederick Seitz (1911–2008) has arguably the best claim to be the founding father of solid state physics.

Keith Brueckner (1924–)

43.3   Excitations in the mean-field approximation

Next we examine the changes the interactions can make to the dispersion relations of single particles in the metal. For the moment we will return to the simple minded picture of a box of fermions, which does not include jellium's positive background.

The excitations of a quantum field theory are particles, which have a characteristic dispersion E_p. To examine the excitations of a field theory we want to manoeuvre our Hamiltonian into the form $\hat{H} = \sum_p E_p \hat{a}_p^\dagger \hat{a}_p$. So when we examine the fate of particles in mean-field theory we try to massage the two-particle interaction, described by the four operators in eqn 43.2, into one that describes a single particle interacting with an external field. The external field here is caused by the effect of all of the other particles, which (finally) explains the logic of naming this procedure a 'mean field' approximation. Recall from Chapter 4 that a single particle in an external potential is described by an operator $\tilde{V}_q \hat{a}_{p+q}^\dagger \hat{a}_p$, so we want to turn the tricky term $\hat{a}_{p-q}^\dagger \hat{a}_{k+q}^\dagger \hat{a}_k \hat{a}_p$ into a sum of terms that look like this. Actually we want to go further than this and have incoming and outgoing electrons residing in the same momentum state giving $\sum_p \tilde{V}_p^{\mathrm{MF}} \hat{a}_p^\dagger \hat{a}_p$, where $\tilde{V}_p^{\mathrm{MF}}$ is the effective mean-field potential through which the particle with momentum p passes. We do this with a recipe: we use Wick's theorem to generate terms containing the VEV of *two* of the four operators and treat this VEV part as an effective potential.

Example 43.7

Returning to our Wick expansion of the operator we identify the terms with one contraction and one pair of uncontracted operators. We obtain a sum of terms:

$$
\begin{aligned}
D &= \langle \hat{a}_{k+q}^\dagger \hat{a}_k \rangle \hat{a}_{p-q}^\dagger \hat{a}_p + \langle \hat{a}_{p-q}^\dagger \hat{a}_p \rangle \hat{a}_{k+q}^\dagger \hat{a}_k, \\
F &= -\langle \hat{a}_{k+q}^\dagger \hat{a}_p \rangle \hat{a}_{p-q}^\dagger \hat{a}_k - \langle \hat{a}_{p-q}^\dagger \hat{a}_k \rangle \hat{a}_{k+q}^\dagger \hat{a}_p, \\
C &= \langle \hat{a}_{k+q}^\dagger \hat{a}_{p-q}^\dagger \rangle \hat{a}_p \hat{a}_k + \langle \hat{a}_p \hat{a}_k \rangle \hat{a}_{k+q}^\dagger \hat{a}_{p-q}^\dagger.
\end{aligned}
\tag{43.29}
$$

We will continue to ignore the Cooper terms for now. The two terms in the D sum are identical after a re-indexing of the sums, and the same is true of the two terms in the F sum. We end up with

$$
\begin{aligned}
D &= 2\langle \hat{a}_{k+q}^\dagger \hat{a}_k \rangle \hat{a}_{p-q}^\dagger \hat{a}_p, \\
F &= -2\langle \hat{a}_{p-q}^\dagger \hat{a}_k \rangle \hat{a}_{k+q}^\dagger \hat{a}_p.
\end{aligned}
\tag{43.30}
$$

We now have a Hamiltonian that reads

$$
\hat{H} = \sum_p \frac{p^2}{2m} \hat{a}_p^\dagger \hat{a}_p + \sum_{qpk} \tilde{V}_q \langle \hat{a}_{k+q}^\dagger \hat{a}_k \rangle \hat{a}_{p-q}^\dagger \hat{a}_p - \sum_{qpk} \tilde{V}_q \langle \hat{a}_{p-q}^\dagger \hat{a}_k \rangle \hat{a}_{k+q}^\dagger \hat{a}_p.
\tag{43.31}
$$

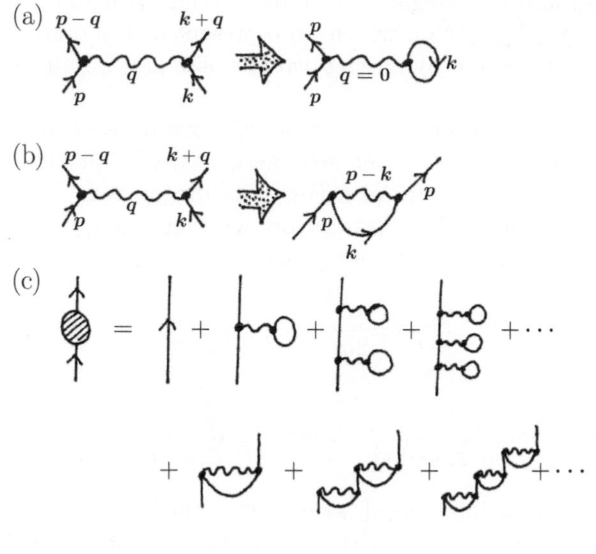

(a)

(b)

(c)

Fig. 43.5 (a) The tadpole, representing the first-order Hartree contribution to the single-particle energy. (b) The oyster giving the first-order Fock contribution to the single-particle energy. (c) The sum to infinity of oyster and tadpole diagrams.

Now we're in a position to examine the simplified version of the two-particle interaction. First, let's examine the direct term: exactly as before the VEV part fixes $q = 0$, giving a number operator and the expression

$$\hat{V}_{\text{direct}} = \sum_{pk} \tilde{V}_{q=0}\langle \hat{a}_k^\dagger \hat{a}_k\rangle \hat{a}_p^\dagger \hat{a}_p \qquad (43.32)$$

$$= \sum_p \left[\tilde{V}_{q=0}\sum_k \langle \hat{a}_k^\dagger \hat{a}_k\rangle \right] \hat{a}_p^\dagger \hat{a}_p.$$

Comparing with our equation for a single particle in an external potential, we see that it looks as if each of the particles is sailing through an effective potential $\tilde{V}_p^{\text{MF}} = \tilde{V}_{q=0}\sum_k \langle \hat{N}_k\rangle$. This is represented by the tadpole diagram in Fig. 43.5(a), which is generated by joining up two of the four legs. Notice that there are two ways of forming this diagram, re-creating the factor of two we had from Wick's expansion.

Next we examine the exchange term

$$\hat{V}_{\text{exchange}} = -\sum_{qpk} \tilde{V}_q \langle \hat{a}_{p-q}^\dagger \hat{a}_k\rangle \hat{a}_{k+q}^\dagger \hat{a}_p. \qquad (43.33)$$

In order to get a nonzero VEV, we just consider the terms with $p = k+q$. Then we obtain

$$\hat{V}_{\text{exchange}} = -\sum_p \left[\sum_k \tilde{V}_{p-k}\langle \hat{a}_k^\dagger \hat{a}_k\rangle \right] \hat{a}_p^\dagger \hat{a}_p. \qquad (43.34)$$

This corresponds to a single particle interacting with an effective potential $\hat{V}_p^{\mathrm{MF}} = \sum_k \tilde{V}_{p-k} \langle \hat{N}_k \rangle$, which we represent as the oyster diagram in Fig. 43.5(b), generated by joining two opposite legs from the interaction vertex.

Notice that the effective potentials we have derived are exactly the first-order contributions to the self-energy $\tilde{\Sigma}_p$ of a particle. They are each represented by a part of a Feynman diagram we can fit between the two external legs. We can therefore write the energy of the particles as a sum of kinetic and self-energy parts

$$E_p = \frac{p^2}{2m} + \tilde{\Sigma}_p^{(\mathrm{D})} + \tilde{\Sigma}_p^{(\mathrm{F})}, \tag{43.35}$$

where

$$\tilde{\Sigma}_p^{(\mathrm{D})} = \tilde{V}_{q=0} \sum_k \langle \hat{N}_k \rangle, \quad \tilde{\Sigma}_p^{(\mathrm{F})} = -\sum_k \tilde{V}_{p-k} \langle \hat{N}_k \rangle. \tag{43.36}$$

For the case of jellium metal, we note that the Hartree self-energy $\tilde{\Sigma}_p^{(\mathrm{D})}$ is cancelled by the positive background charge (just as previously) and the Fock term becomes $\tilde{\Sigma}_p^{(\mathrm{F})}$ from eqn 43.23, justifying its assignment in the previous section. The energy of electrons in the Hartree–Fock approximation is therefore given by

$$E_p = \frac{p^2}{2m} - \frac{p_\mathrm{F}}{\pi} \left(\frac{e^2}{4\pi\epsilon_0} \right) F \left(\frac{|p|}{p_\mathrm{F}} \right), \tag{43.37}$$

which is shown in Fig. 43.4(b). It is worth stressing at this point that this is not what is measured in a real metal! The Hartree–Fock approximation doesn't capture the physics very well and higher order corrections are necessary. This requires the full machinery of quantum field theory including propagators and Feynman rules and it is to this we now turn.

43.4 Electrons and holes

So far we have treated our metal as a box of electrons. For more serious calculations it is useful to distinguish two sorts of excitations. This is because a metal comprises electron states filled up to p_F and so the 'vacuum' for the metal is not empty. With this in mind we define new operators

$$\begin{aligned}
\hat{a}_p &= \theta(|p| - p_\mathrm{F})\hat{c}_p + \theta(p_\mathrm{F} - |p|)\hat{b}_p^\dagger, \\
\hat{a}_p^\dagger &= \theta(|p| - p_\mathrm{F})\hat{c}_p^\dagger + \theta(p_\mathrm{F} - |p|)\hat{b}_p,
\end{aligned} \tag{43.38}$$

with anticommutation relations

$$\{\hat{c}_p, \hat{c}_q^\dagger\} = \delta^{(3)}(p - q), \quad \{\hat{b}_p, \hat{b}_q^\dagger\} = \delta^{(3)}(p - q), \tag{43.39}$$

with all other anticommutators vanishing. We say that the $\hat{c}_p$ operators describe *electrons*, while the $\hat{b}_p$ operators describe **holes**.[13]

[13]Notice that, defined this way, electron operators only act for states above the Fermi surface $|p| > p_\mathrm{F}$ while hole operators only act for states below the Fermi surface $|p| < p_\mathrm{F}$. This, in turn, means that we describe the system as having electron excitations above the Fermi energy and hole excitations below the Fermi energy.

Defined in this way the kinetic energy of the system is given by

$$\hat{H}_0 = \sum_{|\boldsymbol{p}|<p_{\mathrm{F}}} E_{\boldsymbol{p}} + \sum_{|\boldsymbol{p}|>p_{\mathrm{F}}} E_{\boldsymbol{p}}\hat{c}_{\boldsymbol{p}}^{\dagger}\hat{c}_{\boldsymbol{p}} - \sum_{|\boldsymbol{p}|<p_{\mathrm{F}}} E_{\boldsymbol{p}}\hat{b}_{\boldsymbol{p}}^{\dagger}\hat{b}_{\boldsymbol{p}}. \qquad (43.40)$$

The first term is the ground state energy of all of the states, filled up to the Fermi level. The second term accounts for electron excitations. The final term accounts for hole excitations. Note that, with these definitions the holes make a negative contribution to the energy. Although there is a sense in which holes are antiparticles, they are not identical with positrons whose contribution to the total energy is always positive.

To perform calculations we need an electron–hole field operator, which is given by

$$\hat{\psi}(x) = \frac{1}{\sqrt{\mathcal{V}}} \sum_{\boldsymbol{p}} \left[\theta(|\boldsymbol{p}| - p_{\mathrm{F}})\hat{c}_{\boldsymbol{p}} + \theta(p_{\mathrm{F}} - |\boldsymbol{p}|)\hat{b}_{\boldsymbol{p}}^{\dagger} \right] \mathrm{e}^{-\mathrm{i}p\cdot x}. \qquad (43.41)$$

The use of this expansion to find the electron–hole propagator is examined in Exercise 43.4.

43.5 Finding the excitations with propagators

Our rather ad hoc mean-field procedure will not get us much further. Fortunately, we can use the technology of propagators to formalize the procedure, turning it into a perturbation series. This will have the advantage of showing how the full expansion of the four-operator term will go. Recall that, since we are dealing with an interacting theory we should think of our electrons and holes as quasiparticles. In the following example, we will recap the steps needed to make a perturbation expansion and re-create all of our results so far in a flash.[14]

[14]See Mahan, Chapter 2, P. Coleman, Chapter 8, or Abrikosov, Gorkov and Dzyaloshinski, Chapter 2, for the full story. Note also that we return to relativistic notation here, so that $x^{\mu} = (t, \boldsymbol{x})$, $p^{\mu} = (E, \boldsymbol{p})$, etc.

Example 43.8

The interaction Hamiltonian is $\mathcal{H}_{\mathrm{I}}(x, y) = \frac{1}{2}\psi^{\dagger}(\boldsymbol{x})\psi^{\dagger}(\boldsymbol{y})V(\boldsymbol{x}-\boldsymbol{y})\psi(\boldsymbol{y})\psi(\boldsymbol{x})\delta(x^0-y^0)$. This interaction gives us an $\hat{S}$-operator

$$\hat{S} = \mathrm{e}^{-\frac{\mathrm{i}}{2}\int \mathrm{d}^4x\mathrm{d}^4y\,\hat{\psi}^{\dagger}(\boldsymbol{x})\hat{\psi}^{\dagger}(\boldsymbol{y})V(\boldsymbol{x}-\boldsymbol{y})\hat{\psi}(\boldsymbol{y})\hat{\psi}(\boldsymbol{x})\delta(x^0-y^0)}. \qquad (43.42)$$

The basic building block of the perturbation expansion is the Feynman propagator, which we can obtain as the Green's function of the equation of motion. The equation of motion for non-relativistic electrons is the Schrödinger equation, so we want

$$\left(E_{\boldsymbol{p}} - \mathrm{i}\frac{\mathrm{d}}{\mathrm{d}t} \right) \tilde{G}_0(p) = -\mathrm{i}\delta(t), \qquad (43.43)$$

with $E_{\boldsymbol{p}} = \frac{\boldsymbol{p}^2}{2m}$. We obtain the propagator in momentum space as

$$\tilde{G}_0(p) = \frac{\mathrm{i}}{E - E_{\boldsymbol{p}} + \mathrm{i}\epsilon}, \qquad (43.44)$$

where we have added $\mathrm{i}\epsilon$ in the denominator to avoid integrals hitting the electron pole. This choice is discussed in Appendix B.

With the inclusion of holes, the free propagator for electrons in a metal is given by

$$\tilde{G}_0(p) = \frac{i\theta(|\boldsymbol{p}| - p_{\mathrm{F}})}{E - E_{\boldsymbol{p}} + i\epsilon} + \frac{i\theta(p_{\mathrm{F}} - |\boldsymbol{p}|)}{E - E_{\boldsymbol{p}} - i\epsilon}, \qquad (43.45)$$

which is a sum of electron and holes parts. This expression is similar to what we have for scalar particles, although with a differing sign. Since, for a particular choice of $\boldsymbol{p}$, we only need one of the terms in the expression for $\tilde{G}_0(p)$, a much simpler approach that is commonly employed is to write the electron–hole free propagator as a single expression

$$\tilde{G}_0(p) = \frac{i}{E - E_{\boldsymbol{p}} + i\delta_{\boldsymbol{p}}}, \qquad (43.46)$$

where $\delta_{\boldsymbol{p}} = +\epsilon$ for $|\boldsymbol{p}| > p_{\mathrm{F}}$ and $\delta_{\boldsymbol{p}} = -\epsilon$ for $|\boldsymbol{p}| < p_{\mathrm{F}}$.

Now that we have a propagator and an interaction, we're ready to write down the Feynman rules for a metal.

The Feynman rules for a metal

- A factor $\tilde{G}_0(p) = \frac{i}{E - E_{\boldsymbol{p}} + i\delta_{\boldsymbol{p}}}$ for each internal line.
- A factor $-i\tilde{V}_{\boldsymbol{q}}$ for each interaction vertex.
- A factor -1 for each closed fermion loop.
- Conserve energy-momentum at the vertices.
- Integrate over unconstrained energies and momenta with a measure $\mathcal{V} \int \frac{\mathrm{d}^4 p}{(2\pi)^4}$.
- A convergence factor e^{iE0^+} for non-propagating lines.

The final point is necessary for the two diagram parts shown in Fig. 43.6 and stems from the instantaneous nature of the interaction, which appears to allow lines to propagate instantaneously.[15] We're now ready to go and calculate some things with the Feynman rules.

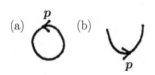

(a)   (b)

Fig. 43.6 Non-propagating lines.

[15]As shown in the examples below, it forces us to close the contour in our integrals in a particular manner, guaranteeing that we get a sensible answer.

43.6   Ground states and excitations

Ground state energies may be evaluated from perturbation theory by summing Feynman diagrams. As discussed in Exercise 43.5, this involves finding the amplitude $\langle 0|\hat{S}|0\rangle = e^{\Sigma(\text{Connected vacuum diagrams})}$, from which it follows that

$$-i\frac{E}{\mathcal{V}} = \frac{\Sigma\left(\begin{array}{c}\text{Connected} \\ \text{vacuum diagrams}\end{array}\right)}{\mathcal{V}T}. \qquad (43.47)$$

The lowest order vacuum diagrams are the double tadpole and oyster, just as we had with the mean-field approximation. The mean-field trick, which gave us the Hartree–Fock approximation, represents first-order terms in the perturbation expansion of the ground state energy. Higher order terms in the expansion give us diagrams such as those others shown in Fig. 43.7.

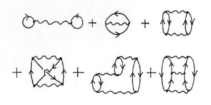

Fig. 43.7 Corrections to the ground state energy of a metal.

What are the particle energies? These can be accessed through the interacting propagator of the theory, which includes the effect of all self-energy contributions. Recall the exact result that

$$\tilde{G}(p) = \frac{iZ_{\boldsymbol{p}}}{E - E_{\boldsymbol{p}} + i\Gamma_{\boldsymbol{p}}} + \left(\begin{array}{c} \text{multiparticle} \\ \text{parts} \end{array}\right), \qquad (43.48)$$

so we should be able to read off the dispersion from the position of the pole in the denominator. To find the full propagator we use the prescription

$$\tilde{G}(p) = \sum \left(\begin{array}{c} \text{Connected diagrams with} \\ \text{two external legs} \end{array}\right). \qquad (43.49)$$

We will use Dyson's formula to formally sum the effect of self-energy processes to infinity and renormalize the propagator. As before, we define the 1PI self-energy as

$$-i\tilde{\Sigma}(p) = \sum \left(\begin{array}{c} \text{Amputated 1PI diagrams} \\ \text{with two external legs} \end{array}\right). \qquad (43.50)$$

The sum to infinity, shown for the example of Hartree–Fock theory in Fig. 43.5(c), yields Dyson's equation for the propagator

$$\tilde{G}(p) \;\; = \;\; \frac{1}{\tilde{G}_0(p)^{-1} + i\tilde{\Sigma}(p)} = \frac{i}{E - E_{\boldsymbol{p}} - \tilde{\Sigma}(p) + i\delta_{\boldsymbol{p}}}, \qquad (43.51)$$

so the energy of the excitations is $E = E_{\boldsymbol{p}} + \mathrm{Re}\left[\tilde{\Sigma}(E,\boldsymbol{p})\right]$. Some of the contributions to $\tilde{\Sigma}(p)$ are shown in Fig. 43.8.

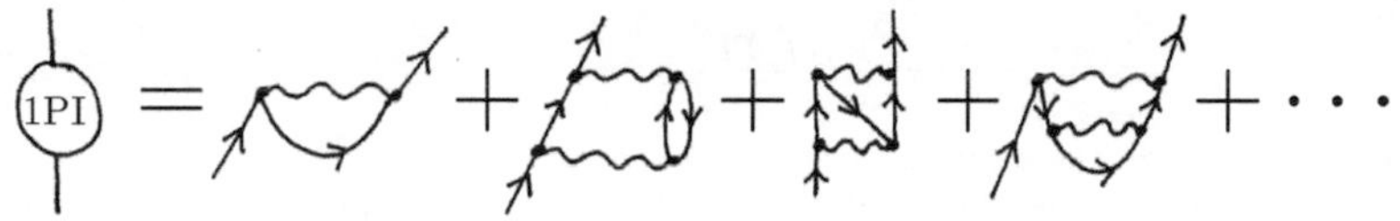

Fig. 43.8 Contributions to the electron 1PI self-energy of the metal.

Example 43.9

Let's examine the renormalization of the electron lines by first-order processes. The only self-energy terms containing one interaction wiggle are the tadpole and oyster diagrams shown in Fig. 43.5(a) and (b). This is, of course, identical to the mean-field result and demonstrates, again, that the mean-field approximation predicts the same contribution as first-order perturbation theory, namely the Hartree–Fock contributions. The joy of the propagator theory is that it tells us how to make sense of the self-energy contribution: we can use it to renormalize the propagator by summing to infinity.

In terms of propagators the diagrams correspond to

$$-i\tilde{\Sigma}(\text{Tadpole})_{\boldsymbol{p}} = (-1)\sum_{\boldsymbol{k}}\int\frac{\mathrm{d}E}{2\pi}(-i\tilde{V}_{\boldsymbol{q}=0})\tilde{G}_0(k)\mathrm{e}^{iE0^+},$$

$$-i\tilde{\Sigma}(\text{Oyster})_{\boldsymbol{p}} = \sum_{\boldsymbol{k}}\int\frac{\mathrm{d}E}{2\pi}(-i\tilde{V}_{\boldsymbol{p}-\boldsymbol{k}})\tilde{G}_0(k)\mathrm{e}^{iE0^+}. \qquad (43.52)$$

It's instructive to use the complex analysis of Appendix B to show the use of the e^{iE0^+} factors. Let's examine the integral

$$\int\frac{\mathrm{d}E}{2\pi}\tilde{G}_0(k)\mathrm{e}^{iE0^+}, \qquad (43.53)$$

whose pole structure is shown in Fig. 43.9(a). If we complete the contour in the upper half-plane [Fig. 43.9(b)] then we pick up all of the momentum state poles for $|\boldsymbol{k}| < p_{\mathrm{F}}$, completing in the lower half-plane picks up the momentum state poles for $|\boldsymbol{k}| > p_{\mathrm{F}}$. The convergence factor e^{iE0^+} makes the choice for us! If we complete in the lower half-plane (where imaginary E gets very large and negative) this factor will diverge. On the other hand, it will vanish for a contour completed in the upper half-plane in the limit that the contour becomes very large. We therefore pick up the poles in the upper half-plane for all $|\boldsymbol{k}| < p_{\mathrm{F}}$, each contributing residue i. We find

$$\int\frac{\mathrm{d}E}{2\pi}\tilde{G}_0(k)\mathrm{e}^{iE0^+} = \frac{1}{2\pi}\times(2\pi i)iN_{\boldsymbol{k}}, \qquad (43.54)$$

where $N_{\boldsymbol{k}}$ is the occupation number distribution which is unity for $|\boldsymbol{k}| < p_{\mathrm{F}}$ and 0 otherwise. We obtain expressions for our diagrams of

$$-i\tilde{\Sigma}(\text{Tadpole})_{\boldsymbol{p}} = (-1)\sum_{\boldsymbol{k}}(-i\tilde{V}_{\boldsymbol{q}=0})(-N_{\boldsymbol{k}}), \qquad (43.55)$$

and

$$-i\tilde{\Sigma}(\text{Oyster})_{\boldsymbol{p}} = \sum_{\boldsymbol{k}}(-i\tilde{V}_{\boldsymbol{p}-\boldsymbol{k}})(-N_{\boldsymbol{k}}). \qquad (43.56)$$

For the case of jellium metal, with its positive background charge, the tadpole contribution is cancelled and we only need consider the oyster contribution. We obtain for the oyster diagram

$$\tilde{\Sigma}(\text{Oyster})_{\boldsymbol{p}} = -\sum_{|\boldsymbol{k}|<p_{\mathrm{F}}}\frac{e^2}{\mathcal{V}\epsilon_0}\frac{1}{|\boldsymbol{p}-\boldsymbol{k}|^2} = -\frac{e^2}{\epsilon_0}\int_{|\boldsymbol{k}|<p_{\mathrm{F}}}\frac{\mathrm{d}^3k}{(2\pi)^3}\frac{1}{|\boldsymbol{p}-\boldsymbol{k}|^2},$$

$$= -\frac{p_{\mathrm{F}}}{\pi}\frac{e^2}{4\pi\epsilon_0}F\left(\frac{|\boldsymbol{p}|}{p_{\mathrm{F}}}\right), \qquad (43.57)$$

which is, of course, the same result as discussed earlier in this chapter.

Finally, the renormalization of the electron line from the oyster contribution, achieved with a sum to infinity, is shown in Fig. 43.10, showing that the quasiparticle energy is shifted to $\boldsymbol{p}^2/2m + \tilde{\Sigma}(\text{Oyster})_{\boldsymbol{p}}$.

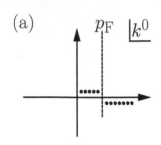

(a)

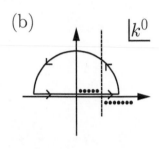

(b)

Fig. 43.9 The integration in eqn 43.53 can be carried out in the complex plane.

[16] Close to the Fermi level, we may expand the energy of an excitation as

$$E_{\boldsymbol{p}} = E_{\mathrm{F}} + \left(\frac{\partial E_{\boldsymbol{p}}}{\partial\boldsymbol{p}}\right)\Bigg|_{|\boldsymbol{p}|=p_{\mathrm{F}}}(|\boldsymbol{p}|-p_{\mathrm{F}})+\dots$$

which, for a non-interacting system, is given by

$$E_{\boldsymbol{p}} = \frac{p_{\mathrm{F}}^2}{2m} + \frac{p_{\mathrm{F}}}{m}(|\boldsymbol{p}|-p_{\mathrm{F}})+\dots$$

from which we identify $m^* = \frac{p_{\mathrm{F}}}{\left(\frac{\partial E_{\boldsymbol{p}}}{\partial|\boldsymbol{p}|}\right)}$.

[17] We ignore the heavy fermion materials where $m^* \approx 10^3 m$. See P. Coleman, Chapter 16, for a discussion of these systems.

So far we have only re-created the same results as seen earlier. To see the motivation for going further one can calculate the effective mass for Hartree–Fock theory.[16]

The result of computing the mass m^* corresponding to the dispersion $E_{\boldsymbol{p}}$ following from our expression for the Hartree–Fock energy (eqn 43.57) is that $m^* = 0$ close to the Fermi energy, which is clearly wrong: electrons should have effective masses of order[17] m. The problem here is that the Hartree–Fock theory is completely static. It treats an electron as if it's propagating in the static field of all of the others. In reality electrons will alter their configuration dynamically resulting in *time-dependent correlations*. These may be included in the calculation

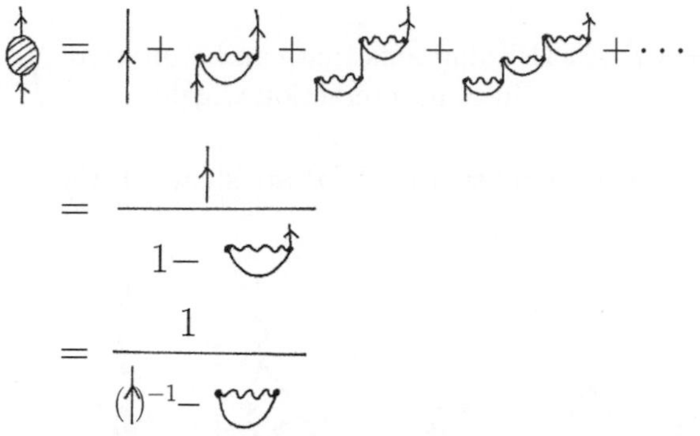

Fig. 43.10 The summation of the Fock terms to infinity.

through the inclusion of more processes in the electron self-energy as shown in Fig. 43.8.

The next logical step would be to evaluate contributions to the self-energy that include two interaction wiggles, including the pair-bubble diagram in Fig. 43.11. However examining the amplitude for this diagram we find that it is divergent for small q. This looks like a problem! However, this divergence is avoided if we take a side-step and, rather than concentrating on corrections to the electron self-energy, we consider a correction analogous to the photon self-energy of Chapter 41. This will have the effect of summing up a class of diagrams that are similar to that shown in Fig. 43.11 which will remove the divergence in the electron self-energy. This summation is the subject of the next section.

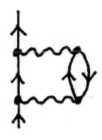

Fig. 43.11 The pair-bubble contribution to the self-energy.

[18]The randomness of a phase is unimportant in our description of the RPA. The reason for the name comes from an alternative treatment where a term $\sum_l e^{i\boldsymbol{q}\cdot\boldsymbol{x}_l}$ is shown to be neglectable, where $\boldsymbol{x}_l$ labels an electron's position. The physics of this approximation is that if $\boldsymbol{x}_l$ is distributed over a large range then the random phases that result will tend to cancel.

[19]David Pines (1924–)

43.7   The random phase approximation

For an electron gas with a high density of particles the most important correction to the Hartree–Fock approximation is provided by the lowest order correction to the wiggle in the interaction vertex. For historical reasons[18] this is known as the **random phase approximation** or RPA and was first formulated by David Bohm and David Pines.[19] This correction to the interaction vertex has much in common with the self-energy correction to the photon propagator in QED and represents a change in the potential felt by the electrons due to the effect of shielding. The idea is that the interaction wiggle creates an electron–hole pair, which annihilates shortly afterwards.

This state of affairs is known here, as in QED, as a **polarization process** and is shown in Fig. 43.12. It is one example of a class of 1PI self-energy processes that can be inserted into an interaction wiggle whose sum is a Green's function denoted $i\tilde{\Pi}(q)$, which we define similarly

Fig. 43.12 A first-order polarization process showing a bubble in the interaction line.

to its cousin in QED as

$$i\tilde{\Pi}(q) = \sum \left(\begin{array}{c} \text{All 1PI diagrams that can be inserted} \\ \text{into an interaction wiggle} \end{array} \right). \qquad (43.58)$$

Some of the terms contributing to $i\tilde{\Pi}(q)$ are shown in Fig. 43.13.

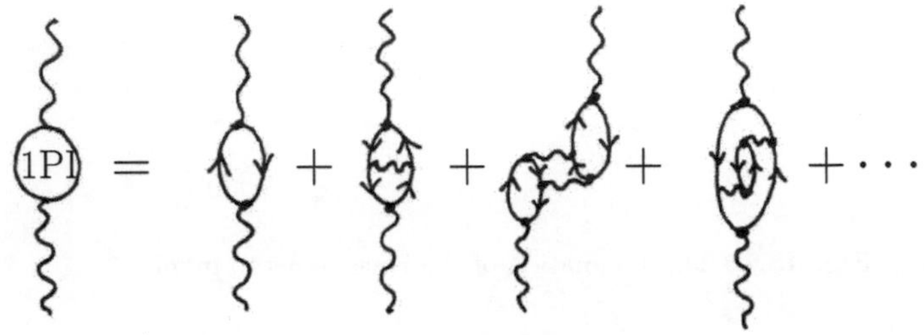

Fig. 43.13 Some of the diagrams contributing to $i\tilde{\Pi}(q)$.

In order to fully take account of $i\tilde{\Pi}(q)$ we need to sum the contribution to infinity. Since we don't have a photon propagator in this theory, these renormalize the potential, changing the interaction strength from $-i\tilde{V}_q$ to the effective potential $-i\tilde{V}_{\text{eff}}(q)$. The sum to infinity is simply carried out in the usual manner:

$$\begin{aligned} -i\tilde{V}_{\text{eff}}(q) &= -i\tilde{V}_q + \left[-i\tilde{V}_q\right]\left[i\tilde{\Pi}(q)\right]\left[-i\tilde{V}_q\right] \\ &\quad + \left[-i\tilde{V}_q\right]\left[i\tilde{\Pi}(q)\right]\left[-i\tilde{V}_q\right]\left[i\tilde{\Pi}(q)\right]\left[-i\tilde{V}_q\right] + \dots \\ &= \frac{-i\tilde{V}_q}{1 - \tilde{V}_q\tilde{\Pi}(q)}. \end{aligned} \qquad (43.59)$$

The interaction potential $\tilde{V}_q$ therefore becomes an effective potential $\tilde{V}_{\text{eff}}(q) = \tilde{V}_q/[1 - \tilde{V}_q\tilde{\Pi}(q)]$, which is often written $\tilde{V}_{\text{eff}} = \tilde{V}_q/\tilde{\epsilon}(q)$, where the permittivity $\tilde{\epsilon}(q) = 1 - \tilde{V}_q\tilde{\Pi}(q)$ is a function of energy q^0 and wave vector q. So, just as in QED, the vacuum state of the metal takes on dielectric properties caused by electron–hole pairs buzzing in and out of existence. The physics of these dielectric properties lies in the screening of the electrons from each other's charge by the electron–hole pairs. Finally we note that since $\tilde{\epsilon}(q)$ is a function of energy, the inclusion of polarization diagrams will provide the all-important time-dependent correlations whose neglect resulted in a nonsensical effective mass in the previous section.

Our treatment of $\tilde{\Pi}(q)$ has so far been quite general. In the RPA we evaluate the effect of only the lowest order contribution to $\tilde{\Pi}(q)$, that is, the electron–hole bubble shown in Fig. 43.12, whose amplitude we will call $\tilde{\pi}(q)$. (The sum to infinity of this interaction that renormalizes the photon line in the RPA is shown in Fig 43.14.)

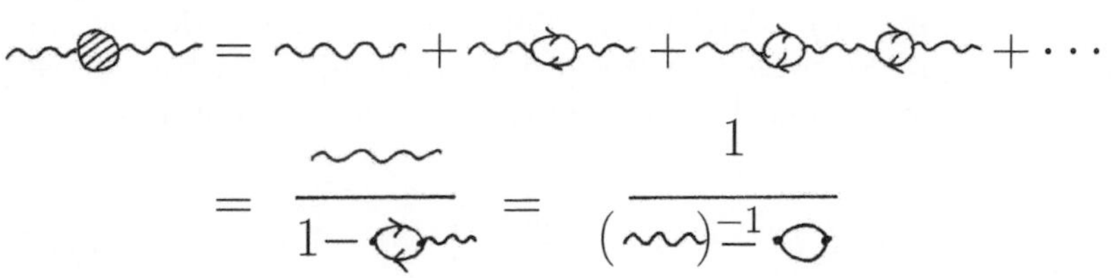

Fig. 43.14 The sum of bubbles in the interaction line.

Example 43.10

Including a minus sign for the fermion loop, the amplitude of the bubble in Fig. 43.12 is given by

$$
\begin{aligned}
\mathrm{i}\tilde{\pi}(q) &= -\mathcal{V}\int\frac{\mathrm{d}^4 p}{(2\pi)^4}\,\tilde{G}_0(p+q)\tilde{G}_0(p)\\
&= -\mathcal{V}\int\frac{\mathrm{d}^3 p}{(2\pi)^3}\int_{-\infty}^{\infty}\frac{\mathrm{d}p^0}{2\pi}\,\frac{\mathrm{i}}{p^0+q^0-E_{\boldsymbol{p}+\boldsymbol{q}}+\mathrm{i}\delta_{\boldsymbol{p}+\boldsymbol{q}}}\,\frac{\mathrm{i}}{p^0-E_{\boldsymbol{p}}+\mathrm{i}\delta_{\boldsymbol{p}}}\\
&= \mathcal{V}\int\frac{\mathrm{d}^3 p}{(2\pi)^3}\,\frac{1}{(E_{\boldsymbol{p}+\boldsymbol{q}}-\mathrm{i}\delta_{\boldsymbol{p}+\boldsymbol{q}})-(E_{\boldsymbol{p}}-\mathrm{i}\delta_{\boldsymbol{p}})-q^0}\\
&\quad\times\int_{-\infty}^{\infty}\frac{\mathrm{d}p^0}{2\pi}\left(\frac{1}{p^0+q^0-E_{\boldsymbol{p}+\boldsymbol{q}}+\mathrm{i}\delta_{\boldsymbol{p}+\boldsymbol{q}}}-\frac{1}{p^0-E_{\boldsymbol{p}}+\mathrm{i}\delta_{\boldsymbol{p}}}\right),\quad (43.60)
\end{aligned}
$$

where, in the final line, we have rewritten the amplitude in a form which allows us to do a contour integral (see Appendix B). The integrals over p^0 have poles at $p^0=-q^0+E_{\boldsymbol{p}+\boldsymbol{q}}-\mathrm{i}\delta_{\boldsymbol{p}+\boldsymbol{q}}$ and $p^0=E_{\boldsymbol{p}}-\mathrm{i}\delta_{\boldsymbol{p}}$ respectively. The residues for these are Fermi functions and so we pick up factors $2\pi\mathrm{i}N_{\boldsymbol{p}+\boldsymbol{q}}$ and $2\pi\mathrm{i}N_{\boldsymbol{p}}$ respectively and obtain the result[20]

$$
\tilde{\pi}(q)=\mathcal{V}\int\frac{\mathrm{d}^3 p}{(2\pi)^3}\,\frac{N_{\boldsymbol{p}+\boldsymbol{q}}-N_{\boldsymbol{p}}}{E_{\boldsymbol{p}+\boldsymbol{q}}-E_{\boldsymbol{p}}-q^0}. \qquad (43.61)
$$

[20]Dropping the infinitesimal in the denominator, since we won't use it further.

The integrand in eqn 43.61 is known as the **Lindhard function**. The evaluation of the integral involving the Lindhard function is carried out by P. Coleman, Chapter 8. The result is a rather complicated expression for $\tilde{\epsilon}(q)$. Analytically continuing to imaginary energies and doing the integral it is found that

$$
\tilde{\pi}(\mathrm{i}q^0,\boldsymbol{q})=-\frac{1}{2}g(E_{\mathrm{F}})\mathcal{F}\left(\frac{\mathrm{i}q^0}{4E_{\mathrm{F}}},\frac{|\boldsymbol{q}|}{2p_{\mathrm{F}}}\right), \qquad (43.62)
$$

where $g(E_{\mathrm{F}})=\left.\frac{\mathrm{d}N_{\boldsymbol{p}}}{\mathrm{d}E_{\boldsymbol{p}}}\right|_{E_{\mathrm{F}}}$ is the density of states at the Fermi level and

$$
\mathcal{F}(y,x)=1+\frac{1}{4x}\left[1-\left(x-\frac{y}{x}\right)^2\right]\ln\left|\frac{x-\frac{y}{x}+1}{x-\frac{y}{x}-1}\right|+\frac{1}{4x}\left[1-\left(x+\frac{y}{x}\right)^2\right]\ln\left|\frac{x+\frac{y}{x}+1}{x+\frac{y}{x}-1}\right|,
$$
$$(43.63)$$

is a dynamicized version of the function $F(x)$ that featured in the evaluation of the Fock self-energy. This latter expression must be analytically continued back to the result we want for real q^0 (see Mahan, Chapter 5 for the details). However, for our purposes we will limit our attention to the important limit $q^0\to 0$, where $\mathcal{F}(\mathrm{i}q^0/4E_{\mathrm{F}},|\boldsymbol{q}|/2p_{\mathrm{F}})$ reduces to $F(|\boldsymbol{q}|/2p_{\mathrm{F}})$.

We will extract three great results from the RPA by examining three special cases of $\tilde{\epsilon}(q^0,\boldsymbol{q})$.

Case I: In the static ($q^0 \to 0$) limit we have $\tilde{\pi}(0, \boldsymbol{q}) = -\frac{1}{2} g(E_F) F(|\boldsymbol{q}|/2p_F)$. Particularly illuminating is the static and small $\boldsymbol{q}$ limit, where $F(|\boldsymbol{q}|/2q_F) \to 2$ and $\lim_{\boldsymbol{q}\to0} \tilde{\pi}(0, \boldsymbol{q}) = -g(E_F) = -\frac{3N}{2E_F}$ (where we have included a factor of 2 to account for spin degeneracy) and so

$$\lim_{\boldsymbol{q}\to0} \epsilon(0, \boldsymbol{q}) = 1 + \left(\frac{3ne^2}{2\epsilon_0 E_F} \right) \frac{1}{|\boldsymbol{q}|^2} = 1 + \frac{q_{\mathrm{TF}}^2}{|\boldsymbol{q}|^2}, \qquad (43.64)$$

[21] The order of the limits is crucial here. Here we have started with the static limit and examined what happens as $q \to 0$. The other order, corresponding to considering a uniform field (i.e. $\boldsymbol{q} = 0$) then taking the limit $q^0 \to 0$ tells us about the response of a metal to ac fields and hence its transport properties.

where q_{TF} is known as the Thomas–Fermi wave vector. Notice that the permittivity diverges as $|\boldsymbol{q}| \to 0$. Perversely this is good news! It says that a uniform electric field can't penetrate a metal, which is something we certainly expect.[21] Plugging $\tilde{\epsilon}(0, \boldsymbol{q})$ into eqn 43.59 we see that the static interaction potential is changed to an effective potential

$$\lim_{\boldsymbol{q}\to0} \tilde{V}_{\mathrm{eff}}(0, |\boldsymbol{q}|) = \frac{e^2}{\epsilon_0 \mathcal{V}} \frac{1}{|\boldsymbol{q}|^2 + q_{\mathrm{TF}}^2}, \qquad (43.65)$$

which is shown in Fig. 43.15. Notice that this is well behaved as $|\boldsymbol{q}| \to 0$. We started this section by arguing that the RPA was going to tell us about screening. The connection with screening arises because our result for $\tilde{V}_{\mathrm{eff}}$ is the Fourier transform of the real-space interaction potential

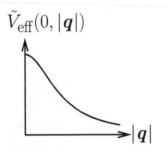

$\tilde{V}_{\mathrm{eff}}(0, |\boldsymbol{q}|)$

Fig. 43.15 The static effective potential $\lim_{|\boldsymbol{q}|\to0} \tilde{V}_{\mathrm{eff}}(0, \boldsymbol{q})$.

$$\lim_{|\boldsymbol{x}|\to\infty} V_{\mathrm{eff}}(\boldsymbol{x}) \propto \frac{e^2}{|\boldsymbol{x}|} e^{-q_{\mathrm{TF}}|\boldsymbol{x}|}, \qquad (43.66)$$

which is, of course, also known as the Yukawa potential, perhaps the simplest screened Coulomb potential. We conclude that the creation and annihilation of electron–hole pairs screens the electronic charge and in the RPA approximation this screening results in the Coulomb potential being replaced by an effective Yukawa potential.

Case II: In the static limit for nonzero $\boldsymbol{q}$ the infinite slope of $F(x)$ at $x = 1$ means that $\tilde{\epsilon}(0, \boldsymbol{q})$ has a weak singularity at $|\boldsymbol{q}|/2p_F = 1$. Near this point $\tilde{\pi}(0, \boldsymbol{q}) \propto (|\boldsymbol{q}| - 2p_F) \ln(|\boldsymbol{q}| - 2p_F)$. When inserted into eqn 43.59 and Fourier transformed this leads to an effective potential

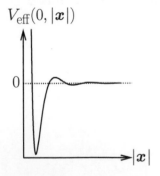

$V_{\mathrm{eff}}(0, |\boldsymbol{x}|)$

Fig. 43.16 The static effective potential $V_{\mathrm{eff}}(0, |\boldsymbol{x}|)$ given by the wave vector behaviour close to p_F.

$$V_{\mathrm{eff}}(0, |\boldsymbol{x}|) \propto \frac{\cos(2p_F|\boldsymbol{x}|)}{|\boldsymbol{x}|^3}, \qquad (43.67)$$

shown in Fig. 43.16. This long-range, oscillatory potential reflects the fact that the metal as a whole reacts to the presence of each charge. The sharpness of the Fermi surface causes this reaction to 'ring' in space, seen as the oscillations in the effective potential of an electron.[22]

Case III: Finally we examine a case with nonzero q^0 in the limit $|\boldsymbol{q}| \to 0$.

[22] These are known as **Friedel oscillations** after Jacques Friedel (1921–).

[23] Strictly one should consider the real part of $\tilde{\pi}(q)$ and so this derivation should be regarded with caution!

Example 43.11

In this case we may expand[23] the Lindhard function for small $\boldsymbol{q}$:

$$\frac{N_{\boldsymbol{p}+\boldsymbol{q}} - N_{\boldsymbol{p}}}{E_{\boldsymbol{p}+\boldsymbol{q}} - E_{\boldsymbol{p}} - q^0} \approx \frac{\boldsymbol{q} \cdot \boldsymbol{v}_{\boldsymbol{p}}}{\boldsymbol{q} \cdot \boldsymbol{v}_{\boldsymbol{p}} - q^0} \left(\frac{\mathrm{d}N_{\boldsymbol{p}}}{\mathrm{d}E_{\boldsymbol{p}}} \right), \qquad (43.68)$$

where $\boldsymbol{v_p} = \boldsymbol{\nabla}_{\boldsymbol{p}} E_{\boldsymbol{p}}$. Expanding in powers of momentum, and including a factor 2 for spin degeneracy, we have

$$\lim_{|\boldsymbol{q}| \to 0} \tilde{\pi}(q^0, \boldsymbol{q}) \approx -2\mathcal{V} \int \frac{\mathrm{d}^3 p}{(2\pi)^3} \left[\frac{\boldsymbol{q} \cdot \boldsymbol{v_p}}{q^0} + \frac{(\boldsymbol{q} \cdot \boldsymbol{v_p})^2}{(q^0)^2} \right] \left(\frac{\mathrm{d}N_{\boldsymbol{p}}}{\mathrm{d}E_{\boldsymbol{p}}} \right), \quad (43.69)$$

$$= 2\mathcal{V} \int \frac{\mathrm{d}|\boldsymbol{p}|\mathrm{d}(\cos\theta)d\phi}{(2\pi)^3} |\boldsymbol{p}|^2 \left[\frac{|\boldsymbol{q}||\boldsymbol{v_p}|\cos\theta}{q^0} + \left(\frac{|\boldsymbol{q}||\boldsymbol{v_p}|\cos\theta}{q^0} \right)^2 \right] \frac{m}{|\boldsymbol{p}|} \delta(|\boldsymbol{p}| - p_{\mathrm{F}}),$$

where we've used the fact that $\left(\frac{\mathrm{d}N_{\boldsymbol{p}}}{\mathrm{d}E_{\boldsymbol{p}}} \right) = -\frac{m}{|\boldsymbol{p}|} \delta(|\boldsymbol{p}| - p_{\mathrm{F}})$. The integration over angle kills the first term and we obtain from the second that

$$\lim_{|\boldsymbol{q}| \to 0} \tilde{\pi}(q^0, \boldsymbol{q}) = g(E_{\mathrm{F}}) \frac{|\boldsymbol{v}_{\mathrm{F}}|^2}{3} \frac{|\boldsymbol{q}|^2}{(q^0)^2}. \quad (43.70)$$

Inserting into our expression for the permittivity yields

$$\lim_{|\boldsymbol{q}| \to 0} \tilde{\epsilon}(q^0, \boldsymbol{q}) = 1 - \frac{\omega_{\mathrm{p}}^2}{(q^0)^2}, \quad (43.71)$$

where $\omega_{\mathrm{p}}^2 = \frac{ne^2}{m\epsilon_0}$ is known as the **plasma frequency**. The fact that $\lim_{|\boldsymbol{q}| \to 0} \tilde{\epsilon}(q^0, \boldsymbol{q})$ vanishes at $q^0 = \omega_{\mathrm{p}}$ is telling us about the presence of a new, long-wavelength, mode of excitation in the system. This is the **plasma oscillation** and corresponds to the entire system of electrons oscillating with respect to the positive background.

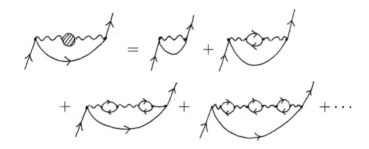

Fig. 43.17 The oyster diagram within the RPA.

Finally we mention the reason we took the diversion into the RPA in the first place: the correction to the excitation energy $E_{\boldsymbol{p}}$. This is obtained from the RPA by inserting the renormalized interaction shown by the shaded blob in Fig. 43.14 into our Feynman diagrams. For example, the corrected oyster diagram is shown in Fig. 43.17. The evaluation of the numbers is, as might be imagined, quite involved. It may be shown that the quasiparticle energy is predicted to be

$$E_{\boldsymbol{p}} = \frac{\boldsymbol{p}^2}{2m} - 0.166 r_{\mathrm{s}} (\ln r_{\mathrm{s}} + 0.203) \frac{|\boldsymbol{p}| p_{\mathrm{F}}}{2m} + \text{const.}, \quad (43.72)$$

which leads to a nonzero effective mass.

The insights provided by the RPA allow us to draw a diagram of the dispersion relation of excitations of the metal, shown in Fig. 43.18. The

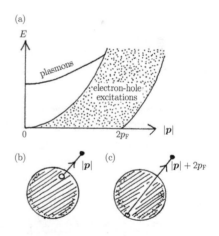

Fig. 43.18 (a) The excitation spectrum of a metal showing low-energy particle–hole quasiparticle excitations and the higher energy (collective) plasmon mode. A quasiparticle excitation with energy $E_{\boldsymbol{p}}$ may be created with: (b) a momentum transfer $|\boldsymbol{p}|$, from one side of the Fermi surface, and (c) $\approx$ $|\boldsymbol{p}| + 2p_{\mathrm{F}}$ from the other (and all momenta in between) leading to the continuum of states shown in the shaded region.

low-energy excitations are quasielectron–quasihole pairs, while at higher energies we see the plasmon mode. Immediately noticeable is the fact that the low-energy excitations form a continuum of width $2p_F$. This arises because excitations involve promoting an electron from within a Fermi sphere which has a diameter $2p_F$. For a given excitation energy we therefore have a range $2p_F$ of possible momenta, with the limits corresponding to forming the excitation from filled states on the nearest and furthest points on the Fermi sphere, as shown in Figs. 43.18(b) and (c).

Chapter summary

- A treatment of interactions in metals has been introduced at the level of the Hartree–Fock approximation. The Hartree term gives the energy due to the constant part of the potential and is represented by a tadpole diagram. The Fock term describes exchange and is represented by an oyster diagram.

- The next correction to the Hartree–Fock approximation is the RPA, introducing a bubble into the interaction line due to virtual electron–hole pairs and this allows us to treat screening.

Exercises

(43.1) Verify that $-\frac{3}{4\pi}\frac{e^2}{4\pi\epsilon_0}p_F = -\frac{0.916}{r_s}$, where the right-hand side is in units of Rydbergs per electron.

(43.2) (a) Using the Hartree–Fock expression for the quasiparticle excitation energy in eqn 43.37, show that the Hartree–Fock effective mass is given by

$$\frac{m}{m^*} = \frac{e^2 m}{2\pi p_F}\frac{1}{x^2}\left(\frac{1+x^2}{x}\ln\left|\frac{1+x}{1-x}\right| - 2\right), \quad (43.73)$$

with $x = |p|/p_F$ and show that this diverges at $x = 1$.

(b) Using the RPA expression for the quasiparticle excitation energy in eqn 43.72 show that the RPA effective mass is given by

$$\frac{m}{m^*} = 1 - 0.083 r_s(\ln r_s + 0.203). \quad (43.74)$$

(43.3) Including spin leads to a many-body Hamiltonian for electrons of

$$\hat{H} = \sum_{p\sigma}\frac{p^2}{2m}\hat{a}^\dagger_{p\sigma}\hat{a}_{p\sigma} \quad (43.75)$$

$$+ \frac{1}{2}\sum_{pkq\sigma\sigma'}\tilde{V}_q\hat{a}^\dagger_{p-q\sigma}\hat{a}^\dagger_{k+q\sigma'}\hat{a}_{k\sigma'}\hat{a}_{p\sigma}.$$

(a) Why does the kinetic energy term favour equal numbers of spin-up and spin-down electrons?
(b) Show that the Hartree contribution to the total energy is independent of the relative populations of spin-up and spin-down electrons.
(c) Show that the Fock term gives a negative contribution to the energy for electrons with like spins and no contribution from the interaction of electrons with unlike spins.
(d) Under what circumstances will this system be a ferromagnet?

(43.4) From the definition of the free electron–hole propagator

$$G_0(x,y) = \langle 0|T\hat{\psi}(x)\hat{\psi}^\dagger(y)|0\rangle, \qquad (43.76)$$

along with the electron–hole mode expansion, show that the electron–hole propagator is given by eqn 43.46.

(43.5) Use the fact (which was proved in Chapter 22) that $\langle\Omega(\infty)|\Omega(-\infty)\rangle = \langle 0|\hat{S}|0\rangle = \exp\left[\sum\left(\begin{array}{c}\text{Connected} \\ \text{vacuum diagrams}\end{array}\right)\right]$ to show that

$$-\mathrm{i}\frac{E}{V} = \frac{\sum\left(\begin{array}{c}\text{Connected} \\ \text{vacuum diagrams}\end{array}\right)}{VT}. \qquad (43.77)$$

Notice that dividing by the volume and time VT removes the factor of $\int \mathrm{d}^4x$ that we showed in Exercise 19.4 accompanies all vacuum diagrams.

(43.6) (a) Write down the amplitude for the pair-bubble diagram in Fig. 43.11 and show that it varies as $\int \frac{\mathrm{d}q}{|q|^4}$, which is divergent at small q.
(b) By considering the other diagrams containing two interaction wiggles, show that the pair-bubble is the most divergent diagram at second order.
(c) What is the most divergent diagram at third order?
(d) Show that the RPA corresponds to a summation of the most divergent diagrams in the electron self-energy.
See Mattuck for help.

(43.7) Consider an electron gas in (1+1)-dimensional spacetime.
(a) Show that the equation of motion for electrons near the Fermi energy may be written

$$\left(\frac{\partial}{\partial t} \pm v_F\frac{\partial}{\partial x}\right)\psi = 0, \qquad (43.78)$$

where the + sign is applied for electrons moving to the right and the − for electrons moving to the left.
(b) What is the Lagrangian for this system of left- and right-moving electrons?
(c) Using the (1 + 1)-dimensional representation of the γ matrices from Exercise 36.5, $\gamma^0 = \sigma^2$ and $\gamma^1 = \mathrm{i}\sigma^1$, show that the Lagrangian may be written

$$\mathcal{L} = \mathrm{i}\psi^\dagger\left(\frac{\partial}{\partial t} - v_F\sigma_3\frac{\partial}{\partial x}\right)\psi, \qquad (43.79)$$

where $\psi = \begin{pmatrix}\psi_L \\ \psi_R\end{pmatrix}$.
(d) Employing units where $v_F - 1$, show that this may be written the form of a massless Dirac Lagrangian.
This problem is discussed in more depth in Zee, Chapter V.5.

44 Superconductors

Fritz London (1900–1954)

[1] By 'breaking global symmetry' we mean that the ground state picks out a unique direction for the phases to point for the entire system.

[2] The electron–phonon interaction is quite similar to the electron–photon interaction in QED, with the interaction consisting of an electron emitting a phonon, rather than a photon as shown in Fig. 44.1.

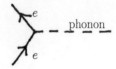

Fig. 44.1 The electron–phonon interaction vertex.

Herbert Frölich (1905–1991)

I'll use **super-mathematics**! *One bean weighs 1/20 of an ounce! The jar weighs 12 pounds! Allowing two pounds for the jar, that makes 20 times 16 times 10 ... or* **32,000 beans**! *Next request, please!*
Superman (1938–1992, 1993–)

Upon cooling, it is found that many metals undergo a transition to a superconducting state where all magnetic flux is expelled from their interior and their electrical resistance is zero. Experiments have probed the excitation spectra of these materials, which are found to have an energy gap Δ per particle that separates the ground state from excited, quasiparticle states. In this chapter we will use quantum field theory to describe this phase of matter.

44.1 A model of a superconductor

Like superfluidity, superconductivity involves a dissipationless flow of particles. In the late 1930s, Fritz London recognized that both effects involve the condensation of particles into a macroscopically occupied state. We saw that in the case of superfluidity this went hand in hand with the spontaneous breaking of a *global* $U(1)$ symmetry. This will also turn out to be the case for superconductors. However, since the particles in a superconductor are charged (they must be since they carry electrical current), we are motivated to look for a theory which includes electromagnetism. Electrodynamics is a gauge theory which is invariant with respect to *local* $U(1)$ transformations; it is the global breaking of this symmetry[1] that leads to superconductivity and the expulsion of flux and measured dispersion follow as consequences:

> A superconductor is a state resulting from breaking global phase symmetry in a system with *local* $U(1)$ invariance.

Which particles should we be considering as the constituents of the superconducting state? Although originally it was assumed that these were the electrons of a metal, this is incorrect. The superconductor is built, not from electrons, but from pairs of electrons. In 1949 Herbert Fröhlich formulated the theory of the electron–phonon interaction[2] and this allowed the possibility of an attractive interaction between electrons mediated by phonons. This interaction between two electrons looks like a t-channel electron–electron scattering process as shown in Fig. 44.2. What makes an attractive interaction a realistic proposition is the vast difference in time scales between the moving electrons and the movement of ion cores. Electrons move at the Fermi velocity and so the time scale over which they move through the crystal is E_F^{-1}. The time scale of the deformation of the lattice is set by the phonon Debye frequency[3] as ω_D^{-1}. Since $E_F^{-1} \ll \omega_D^{-1}$ then the original electron has moved well clear by the time the phonon finds the second electron. In the field theory picture, this difference in time scales corresponds to an effective electron–electron interaction that is strongly 'retarded', which is to say, strongly dependent on the phonon frequency.

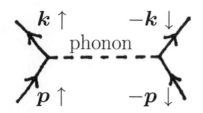

Fig. 44.2 The t-channel process in a superconductor. The effective interaction is mediated by a virtual phonon.

Example 44.1

We can piece together the form of the effective electron–electron interaction. By analogy with QED, the electron–phonon interaction vertex will be that shown in Fig. 44.1 with a coupling constant κ. The effective electron–electron interaction will be given by the t-channel process, mediated by the phonon, which has a propagator resembling the free propagator for the scalar field $D_0(q) \propto (\omega^2 - \omega_q^2)^{-1}$, where ω_q is the dispersion of the phonons. This results in an effective interaction between electrons given by

$$V_{\text{eff}}(\omega, \boldsymbol{q}) \propto \frac{\kappa^2}{\omega^2 - \omega_q^2}. \tag{44.1}$$

Taking the characteristic frequency of the phonons as the Debye frequency $\omega_q \approx \omega_D$, we deduce that the interaction is attractive for $\omega < \omega_D$. Leon Cooper showed more rigorously that an arbitrarily small electron–electron attraction would lead to pairing of electrons into bound states and that this pairing would occur between electrons across the whole of the metallic Fermi surface, making the entire metal unstable to the formation of pairs! Specifically, Cooper showed that there is an attractive interaction between two electrons with equal and opposite momenta and opposite spins. This is mediated by phonons and occurs chiefly between electrons within an energy ω_D of the Fermi energy E_F.

We will explore the consequences of the existence of these so-called **Cooper pairs** of electrons further by using an effective Hamiltonian given by[4]

$$\hat{H} = \sum_{\boldsymbol{p}\sigma} \varepsilon_{\boldsymbol{p}} \hat{c}_{\boldsymbol{p}\sigma}^\dagger \hat{c}_{\boldsymbol{p}\sigma} - \kappa^2 \sum_{\boldsymbol{p}\boldsymbol{k}} \hat{c}_{\boldsymbol{k}\uparrow}^\dagger \hat{c}_{-\boldsymbol{k}\downarrow}^\dagger \hat{c}_{-\boldsymbol{p}\downarrow} \hat{c}_{\boldsymbol{p}\uparrow}, \tag{44.2}$$

where $\sigma(=\uparrow,\downarrow)$ labels the electron spin. This Hamiltonian describes the attractive potential energy between pairs of electrons, with opposite spins and momenta, whose energies will be taken in a shell of width ω_D at the Fermi surface.[5]

This Hamiltonian in eqn 44.2 is the model that was used to solve the problem of superconductivity by John Bardeen, Leon Cooper and Robert Schrieffer, known colloquially as the **BCS model**. Perhaps the most significant feature of the BCS model is not the Hamiltonian itself, but the method used to solve it. Schrieffer invented a trial many-particle state $|\Psi_{\text{BCS}}\rangle$, known as the BCS wave function, which ingeniously captures the crucial physics. The BCS wave function is a coherent state whose $\boldsymbol{p} = 0$ mode is macroscopically occupied when the pairs condense to form

Peter Debye (1884–1966)

[3]The Debye frequency ω_D is the maximum energy that a phonon is allowed to take. Any larger and the phonon has a wavelength so small that it begins to explore length scales smaller than the distance between atoms and the notion of a phonon as a long-wavelength, collective excitation of a lattice breaks down. Phonons move at around the speed of sound while electron velocities are typically a factor of 10^2 larger.

Leon Cooper (1930–)

[4]Comparing our notation from Chapter 42, one can identify κ^2 and $\frac{g}{2V}$.

[5]Note that this form of the potential was motivated by the mean-field approach of Chapter 43. There we saw that the mean-field Hamiltonian for electrons in metals throws up terms such as a contribution to the energy of $C_0 = \langle \hat{c}_{\boldsymbol{p}-\boldsymbol{q}}^\dagger \hat{c}_{\boldsymbol{k}+\boldsymbol{q}}^\dagger \rangle \langle \hat{c}_{\boldsymbol{k}} \hat{c}_{\boldsymbol{p}} \rangle$, which we discarded for an ordinary metal since ordinary single-electron states give zero for each expectation value. This is not the case for Cooper pairs of spin-$\frac{1}{2}$ electrons.

John Bardeen (1908–1991) is the only person to have been awarded the Nobel Prize in physics twice.

J. Robert Schrieffer (1931–)

the superconducting ground state. It is used as a variational state with the model Hamiltonian to find the ground state energy of the system. We therefore start by building the BCS coherent state function.

44.2　The ground state is made of Cooper pairs

The building blocks of the superconductor are Cooper pairs. These are pairs of electrons with opposite spin and momenta. We can create a Cooper pair from the vacuum $|0\rangle$ using the operator

$$\hat{P}_{\boldsymbol{p}}^{\dagger} = \hat{c}_{\boldsymbol{p}\uparrow}^{\dagger}\hat{c}_{-\boldsymbol{p}\downarrow}^{\dagger}. \tag{44.3}$$

The BCS coherent state is built from the vacuum using this operator as follows:

$$|\Psi_{\mathrm{BCS}}\rangle = \prod_{\boldsymbol{p}} C_{\boldsymbol{p}}\, \mathrm{e}^{\alpha_{\boldsymbol{p}}\hat{P}_{\boldsymbol{p}}^{\dagger}}|0\rangle. \tag{44.4}$$

In eqn 44.4, the $C_{\boldsymbol{p}}$'s are normalization constants and $\alpha_{\boldsymbol{p}}$ is a complex number which depends on $\boldsymbol{p}$.

> Commutation relations for $\hat{P}_{\boldsymbol{p}}$
> $$[\hat{P}_{\boldsymbol{p}}, \hat{P}_{\boldsymbol{q}}] = 0$$
> $$[\hat{P}_{\boldsymbol{p}}^{\dagger}, \hat{P}_{\boldsymbol{q}}^{\dagger}] = 0$$
> $$\left[\hat{P}_{\boldsymbol{p}}, \hat{P}_{\boldsymbol{q}}^{\dagger}\right] = \delta_{\boldsymbol{pq}}(1 - \hat{N}_{\boldsymbol{p}\uparrow} - \hat{N}_{-\boldsymbol{q}\downarrow}).$$

The superconducting state is often described in terms of Cooper pairs of electrons in some sense 'becoming' bosons which subsequently undergo Bose condensation. This can't be quite right, since the commutation relations for the pair operator aren't those of a boson (see box). The pair operators do, however, commute with each other. Remember that the $\hat{c}_{\boldsymbol{p}}^{\dagger}$ operators create fermions which obey the exclusion principle and, in particular, we have

$$\hat{P}_{\boldsymbol{p}}^{\dagger}\hat{P}_{\boldsymbol{p}}^{\dagger} = \hat{c}_{\boldsymbol{p}\uparrow}^{\dagger}\hat{c}_{-\boldsymbol{p}\downarrow}^{\dagger}\hat{c}_{\boldsymbol{p}\uparrow}^{\dagger}\hat{c}_{-\boldsymbol{p}\downarrow}^{\dagger} = 0. \tag{44.5}$$

Example 44.2

The fact that $(\hat{P}_{\boldsymbol{p}}^{\dagger})^2 = 0$ gives us enough information to simplify the BCS coherent state. We notice that only the first two terms of the expansion of the exponential can be nonzero (since they contain no powers and one power of $\hat{P}_{\boldsymbol{p}}^{\dagger}$ respectively) and we have

$$|\Psi_{\mathrm{BCS}}\rangle = \prod_{\boldsymbol{p}} C_{\boldsymbol{p}}\mathrm{e}^{\alpha_{\boldsymbol{p}}\hat{P}_{\boldsymbol{p}}^{\dagger}}|0\rangle = \prod_{\boldsymbol{p}} C_{\boldsymbol{p}}\left(1 + \alpha_{\boldsymbol{p}}\hat{P}_{\boldsymbol{p}}^{\dagger}\right)|0\rangle. \tag{44.6}$$

We can also normalize the wave function:

$$1 = |C_{\boldsymbol{p}}|^2\langle 0|\left(1 + \alpha_{\boldsymbol{p}}^{*}\hat{P}_{\boldsymbol{p}}\right)\left(1 + \alpha_{\boldsymbol{p}}\hat{P}_{\boldsymbol{p}}^{\dagger}\right)|0\rangle = |C_{\boldsymbol{p}}|^2(1 + |\alpha_{\boldsymbol{p}}|^2), \tag{44.7}$$

giving $C_{\boldsymbol{p}} = (1 + |\alpha_{\boldsymbol{p}}|^2)^{-\frac{1}{2}}$. Defining

$$u_{\boldsymbol{p}} = \frac{1}{(1+|\alpha_{\boldsymbol{p}}|^2)^{\frac{1}{2}}}, \quad v_{\boldsymbol{p}} = \frac{\alpha_{\boldsymbol{p}}}{(1+|\alpha_{\boldsymbol{p}}|^2)^{\frac{1}{2}}}, \tag{44.8}$$

we have that $|u_{\boldsymbol{p}}|^2 + |v_{\boldsymbol{p}}|^2 = 1$, we can write[6]

$$|\Psi_{\mathrm{BCS}}\rangle = \prod_{\boldsymbol{p}}(u_{\boldsymbol{p}} + v_{\boldsymbol{p}}\hat{P}_{\boldsymbol{p}}^{\dagger})|0\rangle. \tag{44.9}$$

[6]It's important to notice that $|\Psi_{\mathrm{BCS}}\rangle$ is very different from the usual situation in a metal in which the wave function is a single specific configuration of empty and filled states. The product of terms $(u_{\boldsymbol{p}} + v_{\boldsymbol{p}}\hat{P}_{\boldsymbol{p}}^{\dagger})$ ensures that $|\Psi_{\mathrm{BCS}}\rangle$ contains a coherent sum containing all possible configurations in which every pair state is filled or empty. The BCS state combines all possibilities in a single wave function.

[7]See Exercise 44.3.

[8]Similarly $\langle\hat{N}_{\boldsymbol{p}\downarrow}\rangle = |v_{\boldsymbol{p}}|^2$. Of course, $|u_{\boldsymbol{p}}|^2 = 1 - |v_{\boldsymbol{p}}|^2$ tells us the average non-occupation of the state (since a fermion state can either be unoccupied or singly occupied).

Finally we can give meaning to $v_{\boldsymbol{p}}$. If we evaluate the ground state expectation value of the number operator for spin-up electrons, we obtain[7]

$$\langle\hat{N}_{\boldsymbol{p}\uparrow}\rangle = \langle\Psi_{\mathrm{BCS}}|\hat{c}_{\boldsymbol{p}\uparrow}^{\dagger}\hat{c}_{\boldsymbol{p}\uparrow}|\Psi_{\mathrm{BCS}}\rangle = |v_{\boldsymbol{p}}|^2, \tag{44.10}$$

which tells us that $|v_{\boldsymbol{p}}|^2$ gives the average pair occupation of a state labelled with momentum $\boldsymbol{p}$ with spin up.[8]

Written in terms of our single-fermion operators we have a simplified expression for the coherent state given by

$$|\Psi_{\mathrm{BCS}}\rangle = \prod_{\boldsymbol{p}} (u_{\boldsymbol{p}} + v_{\boldsymbol{p}} \hat{c}_{\boldsymbol{p}\uparrow}^{\dagger} \hat{c}_{-\boldsymbol{p}\downarrow}^{\dagger})|0\rangle. \qquad (44.11)$$

The number operator $\hat{N}$ is given by

$$\hat{N} = \sum_{\boldsymbol{p}\sigma} \hat{c}_{\boldsymbol{p}\sigma}^{\dagger} \hat{c}_{\boldsymbol{p}\sigma} = \sum_{\boldsymbol{p}} \left(\hat{N}_{\boldsymbol{p}\uparrow} + \hat{N}_{\boldsymbol{p}\downarrow} \right). \qquad (44.12)$$

Thus the total number of electrons in the superconducting state is found by summing over spin-up and spin-down occupancies and so

$$\langle \Psi_{\mathrm{BCS}} | \hat{N} | \Psi_{\mathrm{BCS}} \rangle = \sum_{\boldsymbol{p}} \left(\langle \hat{N}_{\boldsymbol{p}\uparrow} \rangle + \langle \hat{N}_{\boldsymbol{p}\downarrow} \rangle \right) = 2 \sum_{\boldsymbol{p}} |v_{\boldsymbol{p}}|^2. \qquad (44.13)$$

We now have enough ammunition to attack the variational problem of finding the ground state energy.

44.3   Ground state energy

We will now use the BCS ground state as a variational wave function for the model Hamiltonian (eqn 44.2) describing the superconductor. By minimizing the energy with respect to the parameters $u_{\boldsymbol{p}}$ and $v_{\boldsymbol{p}}$ we will extract the properties of the ground state of the superconductor. The expectation value of the energy is

$$E = \langle \Psi_{\mathrm{BCS}} | \hat{H} | \Psi_{\mathrm{BCS}} \rangle, \qquad (44.14)$$

which, upon substitution,[9] gives us

$$E = \sum_{\boldsymbol{p}} 2\varepsilon_{\boldsymbol{p}} |v_{\boldsymbol{p}}|^2 - \kappa^2 \sum_{\boldsymbol{p}\boldsymbol{k}} v_{\boldsymbol{p}}^* v_{\boldsymbol{k}} u_{\boldsymbol{k}}^* u_{\boldsymbol{p}}. \qquad (44.15)$$

We will minimize this subject to two constraints. The first fixes the total particle number via

$$N = 2 \sum_{\boldsymbol{p}} |v_{\boldsymbol{p}}|^2, \qquad (44.16)$$

which we impose as a constraint using a Lagrange multiplier[10] μ. We also introduce a Lagrange multiplier $E_{\boldsymbol{p}}$ to enforce the constraint that $|u_{\boldsymbol{p}}|^2 + |v_{\boldsymbol{p}}|^2 = 1$.

[9]The first term is easy since it is the kinetic energy term in the BCS Hamiltonian (eqn 44.2). It can be written

$$\sum_{\boldsymbol{p}\sigma} \varepsilon_{\boldsymbol{p}} \hat{N}_{\boldsymbol{p}\sigma},$$

and so we can use eqn 44.13 to reduce it to

$$\sum_{\boldsymbol{p}} 2\varepsilon_{\boldsymbol{p}} |v_{\boldsymbol{p}}|^2.$$

The second term is evaluated in the exercises.

[10]As will be obvious to those with expertise in statistical mechanics, the Lagrange multiplier μ is the chemical potential. The Lagrange multiplier $E_{\boldsymbol{p}}$ will turn out to give the quasiparticle energy.

Example 44.3

Using the method of Lagrange multipliers, we write a function f using

$$f = E - \mu N + \sum_{\boldsymbol{p}} E_{\boldsymbol{p}} (|u_{\boldsymbol{p}}|^2 + |v_{\boldsymbol{p}}|^2 - 1), \qquad (44.17)$$

and look for solutions to $\partial f/\partial u_{\boldsymbol{p}} = 0$ and $\partial f/\partial v_{\boldsymbol{p}} = 0$. This yields

$$\frac{\partial E}{\partial u_{\boldsymbol{p}}} - \mu \frac{\partial N}{\partial u_{\boldsymbol{p}}} + E_{\boldsymbol{p}} u_{\boldsymbol{p}}^* = 0,$$
$$\frac{\partial E}{\partial v_{\boldsymbol{p}}} - \mu \frac{\partial N}{\partial v_{\boldsymbol{p}}} + E_{\boldsymbol{p}} v_{\boldsymbol{p}}^* = 0. \qquad (44.18)$$

Doing the derivatives we find, after a little algebra (examined in Exercise 44.5), the matrix equation

$$\begin{pmatrix} (\varepsilon_{\boldsymbol{p}} - \mu) & \Delta \\ \Delta^* & -(\varepsilon_{\boldsymbol{p}} - \mu) \end{pmatrix} \begin{pmatrix} u_{\boldsymbol{p}}^* \\ v_{\boldsymbol{p}}^* \end{pmatrix} = E_{\boldsymbol{p}} \begin{pmatrix} u_{\boldsymbol{p}}^* \\ v_{\boldsymbol{p}}^* \end{pmatrix}, \qquad (44.19)$$

where we've written $\Delta = \kappa^2 \sum_{\boldsymbol{p}} u_{\boldsymbol{p}}^* v_{\boldsymbol{p}}$. This matrix has eigenvalues

$$E_{\boldsymbol{p}} = \pm \left[(\varepsilon_{\boldsymbol{p}} - \mu)^2 + |\Delta|^2 \right]^{\frac{1}{2}}. \qquad (44.20)$$

Finding the eigenvectors is made easier if we define

$$\cos 2\theta_{\boldsymbol{p}} = \frac{\varepsilon_{\boldsymbol{p}} - \mu}{E_{\boldsymbol{p}}}, \quad \sin 2\theta_{\boldsymbol{p}} = \frac{\Delta}{E_{\boldsymbol{p}}}. \qquad (44.21)$$

Then we find eigenvectors

$$u_{\boldsymbol{p}}^* = \cos \theta_{\boldsymbol{p}}, \quad v_{\boldsymbol{p}}^* = \sin \theta_{\boldsymbol{p}}, \qquad (44.22)$$

or

$$|u_{\boldsymbol{p}}|^2 = \frac{1}{2}\left(1 + \frac{\varepsilon_{\boldsymbol{p}} - \mu}{E_{\boldsymbol{p}}}\right),$$
$$|v_{\boldsymbol{p}}|^2 = \frac{1}{2}\left(1 - \frac{\varepsilon_{\boldsymbol{p}} - \mu}{E_{\boldsymbol{p}}}\right). \qquad (44.23)$$

The behaviour of $|u_{\boldsymbol{p}}|^2$ and $|v_{\boldsymbol{p}}|^2$ is shown in Fig. 44.3. We also find that $u_{\boldsymbol{p}}^* v_{\boldsymbol{p}} = \cos \theta_{\boldsymbol{p}} \sin \theta_{\boldsymbol{p}} = \frac{1}{2}\sin 2\theta_{\boldsymbol{p}} = \Delta/2E_{\boldsymbol{p}}$. We conclude that the BCS ground state is $|\Omega_{\mathrm{BCS}}\rangle = \prod_{\boldsymbol{p}}(\cos \theta_{\boldsymbol{p}} + \sin \theta_{\boldsymbol{p}} \hat{c}_{\boldsymbol{p}\uparrow}^{\dagger} \hat{c}_{-\boldsymbol{p}\downarrow}^{\dagger})|0\rangle.$

We will see in the next section that Δ is the energy gap between the ground state and excited quasiparticle states. We may now eliminate $u_{\boldsymbol{p}}$ and $v_{\boldsymbol{p}}$ to reveal the physics of the BCS state. Using $u_{\boldsymbol{p}}^* v_{\boldsymbol{p}} = \Delta/2E_{\boldsymbol{p}}$ with the definition of the gap Δ, we find our first result which is that

$$\Delta = \kappa^2 \sum_{\boldsymbol{p}} \frac{\Delta}{2E_{\boldsymbol{p}}}. \qquad (44.24)$$

Since the electrons that form the pairs only lie within ω_{D} of the Fermi energy, we can restrict our sum to $|(\varepsilon_{\boldsymbol{p}} - \mu)| < \omega_{\mathrm{D}}$. We then make the replacement $\sum_{\boldsymbol{p}} \to g(\varepsilon_{\mathrm{F}}) \int \mathrm{d}\varepsilon$, where $g(\varepsilon_{\mathrm{F}})$ is the density of states at the Fermi energy, and writing $\varepsilon \equiv \varepsilon_{\boldsymbol{p}} - \mu$, eqn 44.24 becomes

$$\Delta = \underbrace{\kappa^2 g(\varepsilon_{\mathrm{F}})}_{\Lambda} \int_{-\omega_{\mathrm{D}}}^{\omega_{\mathrm{D}}} \mathrm{d}\varepsilon \frac{\Delta}{2(\varepsilon^2 + \Delta^2)^{\frac{1}{2}}}, \qquad (44.25)$$

where we have defined Λ (the **effective coupling constant**) using $\Lambda = \kappa^2 g(\varepsilon_{\mathrm{F}})$. This integral gives

$$\frac{1}{\Lambda} = \sinh^{-1}\left(\frac{\omega_{\mathrm{D}}}{\Delta}\right), \qquad (44.26)$$

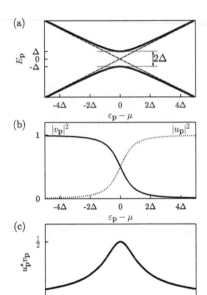

Fig. 44.3 (a) The function $E_{\boldsymbol{p}}$ (which we will later call the bogolon dispersion). (b) The functions $|u_{\boldsymbol{p}}|^2$ and $|v_{\boldsymbol{p}}|^2$ which show that well below μ the state is predominantly electron-like and well above it is predominantly hole-like. Near to μ it has a mixed electron-like and hole-like character. (c) $u_{\boldsymbol{p}}^* v_{\boldsymbol{p}} = \Delta/2E_{\boldsymbol{p}}$, which has a maximum when there are plenty of electron states and available states to scatter into.

but since $\Delta \ll \omega_D$ we obtain

$$|\Delta| \approx 2\omega_D\, e^{-1/\Lambda}. \tag{44.27}$$

This tells us how the energy gap that promotes superconductivity depends on the effective coupling and the Debye energy. Superconductivity is most robust if there is a large energy gap and so, if we want to find the most robust superconductor with the highest possible[11] critical temperature T_c, we need a material with a large Debye frequency and a large coupling constant Λ. The latter is achieved by having a large κ and a large density of states at the Fermi energy.

[11]This assumes BCS-like electron–phonon pairing. A higher T_c may be found using other mechanisms.

44.4 The quasiparticles are bogolons

We now make good on our claim of the last section: we want to confirm that Δ is indeed the energy gap in the particle spectrum of the superconducting ground state. To find the excited states we employ a similar approach to that of the metal. We consider the model Hamiltonian with averaged pairs of operators in the interaction part:

$$\hat{H} = \sum_{p\sigma}(\varepsilon_p - \mu)\hat{c}^\dagger_{p\sigma}\hat{c}_{p\sigma} - \kappa^2 \sum_p \left(\langle \hat{c}^\dagger_{p\uparrow}\hat{c}^\dagger_{-p\downarrow}\rangle \hat{c}_{-p\downarrow}\hat{c}_{p\uparrow} + \langle \hat{c}_{-p\downarrow}\hat{c}_{p\uparrow}\rangle \hat{c}^\dagger_{p\uparrow}\hat{c}^\dagger_{-p\downarrow} \right), \tag{44.28}$$

where the expectation values are taken with respect to the BCS ground state $|\Omega_{BCS}\rangle$. Substitution of $|\Omega_{BCS}\rangle$ allows us to identify the quantity $\kappa^2 \sum_p \langle \Omega_{BCS}|\hat{c}_{-p\downarrow}\hat{c}_{p\uparrow}|\Omega_{BCS}\rangle$ as the energy gap[12] $\Delta = \kappa^2 \sum_p u^*_p v_p$, and we have

$$\hat{H} = \sum_{p\sigma}(\varepsilon_p - \mu)\hat{c}^\dagger_{p\sigma}\hat{c}_{p\sigma} - \sum_p \left(\Delta^* \hat{c}_{-p\downarrow}\hat{c}_{p\uparrow} + \Delta \hat{c}^\dagger_{p\uparrow}\hat{c}^\dagger_{-p\downarrow} \right). \tag{44.29}$$

We need to diagonalize this Hamiltonian, which contains $\hat{c}\hat{c}$ and $\hat{c}^\dagger\hat{c}^\dagger$ bilinears. As in the last chapter, the Bogoliubov transformation gives us a method to do this.

[12]In manipulating these expressions involving the energy gap, note that if Δ is not real then, writing $\Delta = |\Delta|e^{i\phi}$, we may always make it real with the global transformation $\hat{c}_i \rightarrow e^{i\phi/2}\hat{c}_i$, $\hat{c}^\dagger_i \rightarrow e^{-i\phi/2}\hat{c}^\dagger_i$. In the next section we will confirm that choosing a particular value for Δ corresponds to breaking symmetry.

Example 44.4

Written in matrix form we have the Hamiltonian[13]

$$\hat{H} = \sum_p \begin{pmatrix} \hat{c}^\dagger_{p\uparrow} & \hat{c}_{-p\downarrow} \end{pmatrix} \begin{pmatrix} \varepsilon_p - \mu & -\Delta \\ -\Delta^* & -(\varepsilon_p - \mu) \end{pmatrix} \begin{pmatrix} \hat{c}_{p\uparrow} \\ \hat{c}^\dagger_{-p\downarrow} \end{pmatrix}. \tag{44.30}$$

This may be diagonalized using the Bogoliubov procedure, whose details are examined in Exercise 44.7, to give

$$\hat{H} = \sum_p \begin{pmatrix} \hat{b}^\dagger_{p\uparrow} & \hat{b}_{-p\downarrow} \end{pmatrix} \begin{pmatrix} E_p & 0 \\ 0 & -E_p \end{pmatrix} \begin{pmatrix} \hat{b}_{p\uparrow} \\ \hat{b}^\dagger_{-p\downarrow} \end{pmatrix}, \tag{44.31}$$

where the operators $\hat{b}^\dagger_{p\sigma}$ and $\hat{b}_{p\sigma}$ create and destroy bogolon quasiparticle excitations. The bogolon operators obey anticommutation laws:

$$\left\{ \hat{b}_{p_1\sigma_1}, \hat{b}^\dagger_{p_2\sigma_2} \right\} = \delta_{p_1 p_2}\delta_{\sigma_1\sigma_2}, \tag{44.32}$$

$$\left\{ \hat{b}^\dagger_{p_1\sigma_1}, \hat{b}^\dagger_{p_2\sigma_2} \right\} = 0, \quad \left\{ \hat{b}_{p_1\sigma_1}, \hat{b}_{p_2\sigma_2} \right\} = 0.$$

[13]This looks a lot like eqn 44.19 but with minus signs on the off-diagonal elements.

The diagonalized Hamiltonian for the excitations in the superconductor is given by

$$\hat{H} = \sum_{\boldsymbol{p}} E_{\boldsymbol{p}}(\hat{b}^{\dagger}_{\boldsymbol{p}\uparrow}\hat{b}_{\boldsymbol{p}\uparrow} + \hat{b}^{\dagger}_{-\boldsymbol{p}\downarrow}\hat{b}_{-\boldsymbol{p}\downarrow}), \tag{44.33}$$

where $E_{\boldsymbol{p}} = \sqrt{(\varepsilon_{\boldsymbol{p}} - \mu)^2 + |\Delta|^2}$. This is shown in Fig. 44.3. Notice that the bogolons have a dispersion relation that starts at an energy Δ above the ground state energy. This is the energy gap:[14] there are no allowed quasiparticle excitations between $E_{\boldsymbol{p}} = -\Delta$ and $E_{\boldsymbol{p}} = \Delta$.

[14]This is key to superconductivity. Excitations will dissipate momentum and prevent the superflow of current. The presence of a gap prevents their creation at low energies.

44.5   Broken symmetry

As in the superfluid state examined earlier, superconductivity involves the macroscopic occupation of a coherent ground state $|\Omega\rangle$. Recall that the order parameter for the superfluid was given by

$$\langle\Omega|\hat{\Phi}(\boldsymbol{x})|\Omega\rangle = \frac{1}{\sqrt{\mathcal{V}}}\sum_{\boldsymbol{p}}\langle\Omega|\hat{a}_{\boldsymbol{p}}|\Omega\rangle\mathrm{e}^{\mathrm{i}\boldsymbol{p}\cdot\boldsymbol{x}}$$

$$\approx \frac{1}{\sqrt{\mathcal{V}}}\langle\Omega|\hat{a}_{\boldsymbol{p}=0}|\Omega\rangle = \sqrt{n}\mathrm{e}^{\mathrm{i}\theta_0}. \tag{44.34}$$

For the superconductor we have (by the same token)

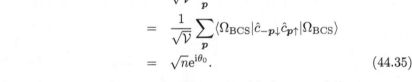

$$\langle\Omega_{\mathrm{BCS}}|\hat{\Psi}(\boldsymbol{x})|\Omega_{\mathrm{BCS}}\rangle = \frac{1}{\sqrt{\mathcal{V}}}\sum_{\boldsymbol{p}}\langle\Omega_{\mathrm{BCS}}|\hat{c}_{-\boldsymbol{p}\downarrow}\hat{c}_{\boldsymbol{p}\uparrow}|\Omega_{\mathrm{BCS}}\rangle\mathrm{e}^{\mathrm{i}(\boldsymbol{p}-\boldsymbol{p})\cdot\boldsymbol{x}}$$

$$= \frac{1}{\sqrt{\mathcal{V}}}\sum_{\boldsymbol{p}}\langle\Omega_{\mathrm{BCS}}|\hat{c}_{-\boldsymbol{p}\downarrow}\hat{c}_{\boldsymbol{p}\uparrow}|\Omega_{\mathrm{BCS}}\rangle$$

$$= \sqrt{n}\mathrm{e}^{\mathrm{i}\theta_0}. \tag{44.35}$$

From these expressions we see that we could also regard the superconducting energy gap as the order parameter for the system since we have

$$\Delta = \kappa^2 \sum_{\boldsymbol{p}}\langle\Omega_{\mathrm{BCS}}|\hat{c}_{-\boldsymbol{p}\downarrow}\hat{c}_{\boldsymbol{p}\uparrow}|\Omega_{\mathrm{BCS}}\rangle. \tag{44.36}$$

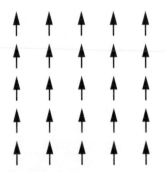

Fig. 44.4 Order in the superconductor involves the lining up of phase angles, just as in the case of an uncharged superfluid.

Returning to the real space expectation value $\langle\psi(\boldsymbol{x})|\hat{\Psi}(\boldsymbol{x})|\psi(\boldsymbol{x})\rangle = \sqrt{\rho(\boldsymbol{x})}\mathrm{e}^{\mathrm{i}\theta(\boldsymbol{x})}$ for a coherent state $|\psi(\boldsymbol{x})\rangle$, we see that for the superconducting ground state $|\Omega_{\mathrm{BCS}}\rangle$ has the phase field $\theta(\boldsymbol{x})$ breaking symmetry by becoming a constant θ_0. (As in the case of the superfluid we can visualize this with an XY-model, as shown in Fig. 44.4.) As alluded to before, the difference between superfluidity and superconductivity comes from the fact that the phase symmetry that's broken is a local one. Prior to breaking symmetry the superconductor enjoys *local* $U(1)$ symmetry: we could alter the phase by $\theta(x) \to \theta(x) + \alpha(x)$, where $\alpha(x)$ is different at all points in spacetime and the Lagrangian describing the system would remain invariant. A gauge field is needed in the Lagrangian to guarantee this. We have seen in Chapter 26 that the gauge field will consume the Goldstone boson and become massive. This is the Higgs mechanism and occurs in superconductors! We will examine the Higgs phenomenon in superconductivity in the next section.

44.6   Field theory of a charged superfluid

The superfluid Lagrangian is invariant with respect to the global transformation $\Psi \to e^{i\alpha}\Psi$. We now promote this global symmetry to a local one. As we have seen, this requires the introduction of a gauge field A_μ through the replacement of the derivative ∂_μ by the covariant derivative $D_\mu = \partial_\mu + iqA_\mu$. The coupling constant q is called the charge of the Ψ field, so by making a local phase transformation we are studying the physics of a charged superfluid. Starting with the non-relativistic superfluid Lagrangian (and denoting the coupling by $g = 2\mathcal{V}\kappa^2$):

$$\mathcal{L} = i\Psi^\dagger \partial_0 \Psi - \frac{1}{2m}\boldsymbol{\nabla}\Psi^\dagger \cdot \boldsymbol{\nabla}\Psi - \frac{g}{2}(n - \Psi^\dagger\Psi)^2, \qquad (44.37)$$

we gauge the theory by introducing the covariant derivative D_μ. This will be easier in polar coordinates, so we introduce $\Psi(x) = \sqrt{\rho(x)}e^{i\theta(x)}$ and $D_\mu\Psi = \left[\frac{1}{2\sqrt{\rho}}\partial_\mu\rho + i\sqrt{\rho}\left(\partial_\mu\theta + qA_\mu\right)\right]e^{i\theta}$ to obtain the gauged Lagrangian

$$\begin{aligned}\mathcal{L} &= -\rho(\partial_0\theta + qA_0) - \frac{1}{2m}\left[\frac{1}{4\rho}(\boldsymbol{\nabla}\rho)^2 + \rho\left(\boldsymbol{\nabla}\theta - q\boldsymbol{A}\right)^2\right] \\ &\quad -\frac{g}{2}\left(n - \rho\right)^2 - \frac{1}{4}F_{\mu\nu}F^{\mu\nu},\end{aligned} \qquad (44.38)$$

where, as before, we drop the total time derivative term $i\partial_0\rho/2$. Note that we have included the contribution of the gauge field via $F_{\mu\nu} = \partial_\mu A_\nu - \partial_\nu A_\mu$. The Lagrangian is now invariant with respect to the change $\theta(x) \to \theta(x) + \alpha(x)$ provided $A_\mu(x) \to A_\mu(x) - \frac{1}{q}\partial_\mu\alpha(x)$.

As we've seen before, when we break the symmetry we'll find that the massless photon field $A_\mu(x)$ gobbles up the Goldstone mode $\theta(x)$ and acquires a mass. This is the Higgs mechanism, at play in a charged superfluid.

Example 44.5

Here's how it unfolds.[15] We notice that we can cast our Lagrangian in terms of a gauge invariant contribution $C_\mu(x) = \frac{1}{q}\partial_\mu\theta(x) + A_\mu(x)$, and the contribution of the gauge field is $F_{\mu\nu} = \partial_\mu A_\nu - \partial_\nu A_\mu = \partial_\mu C_\nu - \partial_\nu C_\mu$. We have

$$\mathcal{L} = -\rho q C_0 - \frac{1}{2m}\left[\frac{1}{4\rho}(\boldsymbol{\nabla}\rho)^2 + \rho q^2 \boldsymbol{C}^2\right] - \frac{g}{2}(n-\rho)^2 - \frac{1}{4}F_{\mu\nu}F^{\mu\nu}. \qquad (44.39)$$

Now break the symmetry of the ground state.[16] We choose a ground state $\Psi_0(x) = \sqrt{n}e^{i\theta_0}$, where we'll make our usual choice of $\theta_0 = 0$. We expand about the ground state by examining small displacements $\sqrt{\rho} = \sqrt{n} + h$, which yields

$$\mathcal{L} = -\frac{nq^2}{2m}\boldsymbol{C}^2 - \frac{1}{2m}(\boldsymbol{\nabla}h)^2 - 2gnh^2 - 2q\sqrt{n}C_0 h - \frac{1}{4}F_{\mu\nu}F^{\mu\nu}, \qquad (44.40)$$

dropping the term $-qnC_0$ as we did for superfluids. Already notice that the $\boldsymbol{C}$-field appears to have acquired a mass $(nq^2/m)^{\frac{1}{2}}$. To complete our calculation we should integrate out the energetic field h to obtain a low-energy field theory. Just as in Chapter 42, we recognize that the h dependent part can be written in the form

$$-h\left(-\frac{1}{2m}\boldsymbol{\nabla}^2 + 2gn\right)h - (2q\sqrt{n}C_0)h. \qquad (44.41)$$

[15]See Example 42.4 for the simpler version of this argument appropriate for the superfluid.

[16]Remember, you don't break the symmetry of a Lagrangian; that always retains the full symmetry. We simply expand the Lagrangian around the broken symmetry ground state.

Again, just as before, we will integrate out the field h with our path integral via $-\frac{1}{2}\phi K\phi + J\phi \to \frac{1}{2}JK^{-1}J$. Setting $\phi = -\sqrt{2}h$ and $J = \sqrt{2nq}C_0$ we have

$$
\begin{aligned}
\mathcal{L} &= -\frac{nq^2}{2m}\boldsymbol{C}^2 + q\sqrt{n}C_0 \frac{1}{(-1/2m)\boldsymbol{\nabla}^2 + 2gn} q\sqrt{n}C_0 - \frac{1}{4}F_{\mu\nu}F^{\mu\nu} \\
&= \frac{q^2}{2g}C_0^2 - \frac{nq^2}{2m}\boldsymbol{C}^2 - \frac{1}{4}F_{\mu\nu}F^{\mu\nu} \\
&= \frac{q^2}{2g}\left(C_0^2 - v^2\boldsymbol{C}^2\right) - \frac{1}{4}F_{\mu\nu}F^{\mu\nu},
\end{aligned}
\tag{44.42}
$$

where $v^2 = ng/m$ and we have neglected the term $(1/2m)\boldsymbol{\nabla}^2$ as small compared to $2gn$. This is a non-relativistic equation telling us that the fields in a charged superfluid give rise to excitations which are spin-1 vector particles.[17]

[17]See Chapter 13 if in doubt.

The consequence of the Higgs mechanism in a superconductor is the expulsion of magnetic flux from the interior of the material. This is known as the **Meissner effect**. Starting with the low-energy Lagrangian from eqn 44.42 we recognize that we can eliminate the component C_0 (since a massive vector field has only three degrees of freedom, see Chapter 13). We may therefore integrate out C_0 (in the same manner as above) with the result that

$$
\mathcal{L} = -\frac{nq^2}{2m}\boldsymbol{C}^2 + \left(\begin{array}{c}\text{Terms involving} \\ \text{derivatives of } \boldsymbol{C}\end{array}\right).
\tag{44.43}
$$

Notice that the three massive fields described by this equation have mass $M^2 = nq^2/m$.

See Exercise 44.8.

Our low-energy Lagrangian in eqn 44.43 may be used to find the current in the superconductor, which is given by

$$
\boldsymbol{J} = \frac{nq}{m}\left(\boldsymbol{\nabla}\theta - q\boldsymbol{A}\right).
\tag{44.44}
$$

We conclude that the current in the superconductor responds to gradients in the phase field (just like the superfluid) and also to the applied field $\boldsymbol{A}$.

Example 44.6

Upon taking the curl of the current, we obtain an equation describing the magnetic induction field $\boldsymbol{B}$ in the superconductor:[18]

$$
\boldsymbol{\nabla}\times\boldsymbol{J} = -\frac{nq^2}{m}\boldsymbol{B},
\tag{44.45}
$$

[18]This is one of the London equations; the other is

$$
\frac{\partial\boldsymbol{J}}{\partial t} = \frac{nq^2}{m}\boldsymbol{E},
$$

which also follows from eqn 44.44 assuming $\boldsymbol{\nabla}\theta$ has no time-dependence.

which, combined with the static Maxwell equation $\boldsymbol{\nabla}\times\boldsymbol{B} = \boldsymbol{J}$, yields

$$
\boldsymbol{\nabla}^2\boldsymbol{B} = \frac{nq^2}{m}\boldsymbol{B}.
\tag{44.46}
$$

In our units, the quantity nq^2/m must have units (length)$^{-2}$, so we define $\lambda = (m/nq^2)^{1/2}$.

In one dimension eqn 44.46 now reads

$$
\frac{\mathrm{d}^2\boldsymbol{B}}{\mathrm{d}x^2} = \frac{1}{\lambda^2}\boldsymbol{B},
\tag{44.47}
$$

whose solutions have the form $\boldsymbol{B}(x) = \boldsymbol{B}(0)\mathrm{e}^{-x/\lambda}$. In other words, the field in a superconductor falls away as we enter the interior of the superconductor over a length scale determined by λ. Since λ tells us how far magnetic flux can penetrate the superconductor it is known as the **penetration depth**. Our treatment has all been aimed at describing a charged superfluid. A superconductor is quite similar. In a superconductor we have Cooper pairs, so we conventionally set $q = -2|e|$ and write $n_s = 2n$ and $m = 2m^*$ for the superfluid density and effective mass respectively. This gives a penetration depth[19] for a superconductor of

[19]In SI units, $\frac{1}{\lambda^2} = \frac{\mu_0 n_s e^2}{m^*}$.

$$\frac{1}{\lambda^2} = \frac{n_s e^2}{m^*}. \tag{44.48}$$

Chapter summary

- The BCS wave function is given by $|\Psi_{\mathrm{BCS}}\rangle = \prod_{\boldsymbol{p}}(u_{\boldsymbol{p}} + v_{\boldsymbol{p}}\hat{c}_{\boldsymbol{p}\uparrow}^\dagger\hat{c}_{-\boldsymbol{p}\downarrow}^\dagger)|0\rangle$ and is a coherent state of Cooper pairs.
- The quasiparticle excitations are bogolons, with dispersion given by $E_{\boldsymbol{p}} = \sqrt{(\varepsilon_{\boldsymbol{p}} - \mu)^2 + |\Delta|^2}$.
- The superconducting energy gap Δ can be regarded as the order parameter of a superconductor, and is given by $\Delta = \kappa^2 \sum_{\boldsymbol{p}} \langle \Omega_{\mathrm{BCS}}|\hat{c}_{-\boldsymbol{p}\downarrow}\hat{c}_{\boldsymbol{p}\uparrow}|\Omega_{\mathrm{BCS}}\rangle$.
- The Higgs mechanism in a superconductor leads to the London equation, which can be written as $\boldsymbol{J} = \frac{nq}{m}\left(\boldsymbol{\nabla}\theta - q\boldsymbol{A}\right)$.

Exercises

(44.1) Show that the amplitude

$$\langle\Psi_{\mathrm{BCS}}|\hat{c}_{\boldsymbol{p}\uparrow}^\dagger\hat{c}_{-\boldsymbol{p}\downarrow}^\dagger|\Psi_{\mathrm{BCS}}\rangle\langle\Psi_{\mathrm{BCS}}|\hat{c}_{-\boldsymbol{p}\downarrow}\hat{c}_{\boldsymbol{p}\uparrow}|\Psi_{\mathrm{BCS}}\rangle \tag{44.49}$$

is nonzero.

(44.2) Verify the commutator $\left[\hat{P}_{\boldsymbol{p}}, \hat{P}_{\boldsymbol{q}}^\dagger\right] = \delta_{\boldsymbol{p}\boldsymbol{q}}(1 - \hat{N}_{\boldsymbol{p}\uparrow} - \hat{N}_{-\boldsymbol{q}\downarrow})$.

(44.3) Verify $\langle\Psi_{\mathrm{BCS}}|\hat{N}_{\boldsymbol{p}\uparrow}|\Psi_{\mathrm{BCS}}\rangle = |v_{\boldsymbol{p}}|^2$.

(44.4) Verify eqn 44.15:

$$\langle E\rangle = \sum_{\boldsymbol{p}} 2\varepsilon_{\boldsymbol{p}}|v_{\boldsymbol{p}}|^2 - \kappa^2 \sum_{\boldsymbol{p}\boldsymbol{k}} v_{\boldsymbol{p}}^* v_{\boldsymbol{k}} u_{\boldsymbol{k}}^* u_{\boldsymbol{p}}. \tag{44.50}$$

(44.5) (a) Starting with the total energy and total particle number written in the form

$$E = \sum_{\boldsymbol{p}} \varepsilon_{\boldsymbol{p}}\left(|v_{\boldsymbol{p}}|^2 - |u_{\boldsymbol{p}}|^2 + 1\right) - \kappa^2 \sum_{\boldsymbol{p}\boldsymbol{k}} v_{\boldsymbol{p}}^* v_{\boldsymbol{k}} u_{\boldsymbol{k}}^* u_{\boldsymbol{p}},$$

$$N = \sum_{\boldsymbol{p}} \left(|v_{\boldsymbol{p}}|^2 - |u_{\boldsymbol{p}}|^2 + 1\right), \tag{44.51}$$

derive eqn 44.19:

$$\begin{pmatrix} (\varepsilon_{\boldsymbol{p}} - \mu) & \Delta \\ \Delta^* & -(\varepsilon_{\boldsymbol{p}} - \mu) \end{pmatrix} \begin{pmatrix} u_{\boldsymbol{p}}^* \\ v_{\boldsymbol{p}}^* \end{pmatrix} = E_{\boldsymbol{p}} \begin{pmatrix} u_{\boldsymbol{p}}^* \\ v_{\boldsymbol{p}}^* \end{pmatrix}. \tag{44.52}$$

(b) Diagonalize the Hamiltonian and find $|u_{\boldsymbol{p}}|^2$ and $|v_{\boldsymbol{p}}|^2$ using the method suggested in the text.

(44.6) (a) Define the **Nambu spinor** as $\hat{\psi}_{\boldsymbol{p}} = \begin{pmatrix} \hat{c}_{\boldsymbol{p}\uparrow} \\ \hat{c}_{-\boldsymbol{p}\downarrow}^\dagger \end{pmatrix}$ and the energy gap as $\Delta = \Delta_1 - \mathrm{i}\Delta_2$. Show that the BCS Hamiltonian can be written

$$\hat{H} = \sum_p \hat{\psi}_p^\dagger h_p \cdot \boldsymbol{\tau} \hat{\psi}_p, \qquad (44.53)$$

where $h_p = (\Delta_1, \Delta_2, \varepsilon_p - \mu)$ and $\boldsymbol{\tau}$ is a vector made up of the Pauli matrices.

(b) Show that eqn 44.19 is identical to the Dirac equation upon making the replacements:

$$\begin{aligned} \psi_L &\rightarrow v_p^*, \\ \psi_R &\rightarrow u_p^*, \\ \boldsymbol{\sigma} \cdot \boldsymbol{p} &\rightarrow \varepsilon_p - \mu, \\ m &\rightarrow \Delta. \end{aligned} \qquad (44.54)$$

This is the basis of **Nambu's analogy**.

(c) When local $U(1)$ symmetry is globally broken in the superconductor Δ takes on a nonzero value. What is the equivalent of this symmetry breaking in the Dirac problem?

Nambu's analogy is discussed further in Aitchison and Hey, Chapter 18.

(44.7) We will find the excitations out of the BCS ground state by diagonalizing eqn 44.30 using the Bogoli-

ubov procedure. The transformation to use in this case is

$$\begin{pmatrix} \hat{c}_{p\uparrow} \\ \hat{c}_{-p\downarrow}^\dagger \end{pmatrix} = \begin{pmatrix} u_p^* & v_p \\ -v_p^* & u_p \end{pmatrix} \begin{pmatrix} \hat{b}_{p\uparrow} \\ \hat{b}_{-p\downarrow}^\dagger \end{pmatrix}, \quad (44.55)$$

and the algebra is simplified by using the substitutions

$$u_p = \cos\theta_p, \qquad (44.56)$$
$$v_p = \sin\theta_p. \qquad (44.57)$$

(a) Check that the transformation maintains the anticommutation relations.

(b) Verify that this does the job of diagonalizing the Hamiltonian.

(c) Confirm that the bogolons are really the excitations by checking $\hat{b}_{p\uparrow}|\Omega_{BCS}\rangle = 0$ and $\hat{b}_{-p\downarrow}|\Omega_{BCS}\rangle = 0$, that is, that there are no bogolons in the BCS ground state.

(44.8) Show that the current in a superconductor is given by eqn 44.44.

The fractional quantum Hall fluid

45

The more success the quantum theory has, the sillier it looks.
Albert Einstein (1879–1955)

The application of a magnetic field to electrons in condensed matter has surprising consequences, none more so than the integer and fractional quantum Hall effects. The latter effect is a product of interactions, but there are some interesting features even in the non-interacting case which we review first.

Throughout this chapter, it will be helpful to keep in mind that the relevant length scale in the problem is the **magnetic length** $\ell_{\rm B}$, given by[1]

$$\ell_{\rm B} = \left(\frac{\hbar}{eB} \right)^{\frac{1}{2}}. \tag{45.1}$$

Note that[2] the flux Φ through a circle with radius $\sqrt{2}\ell_{\rm B}$ is equal to $\Phi = B(2\pi\ell_{\rm B}^2) = h/e$. This is double the value of the magnetic flux quantum $\Phi_0 = h/2e$ which is important in superconductivity (where the Cooper pairs have charge $-2e$) and plays the role of a flux quantum for problems involved with unpaired electrons. It is often called the **Dirac flux quantum**.

45.1 Magnetic translations

Problems involving a lattice in solid state physics make use of translational symmetry. Recall from Chapter 9 that the translation operator for a translation by $\boldsymbol{a}$ is $\mathrm{e}^{-\mathrm{i}\hat{\boldsymbol{p}}\cdot\boldsymbol{a}}$, where $\hat{\boldsymbol{p}}$ is the momentum operator. For a periodic system we would expect $\mathrm{e}^{-\mathrm{i}\hat{\boldsymbol{p}}\cdot\boldsymbol{a}}$ to commute with the Hamiltonian $\hat{H}$. Writing $\hat{U}(\boldsymbol{a}) = \mathrm{e}^{-\mathrm{i}\hat{\boldsymbol{p}}\cdot\boldsymbol{a}}$, we would then expect $\left[\hat{U}(\boldsymbol{a}), \hat{H}\right] = 0$ or $\hat{U}^{-1}(\boldsymbol{a})\hat{H}\hat{U}(\boldsymbol{a}) = \hat{H}$. This expectation is not realized in the presence of a magnetic field where we find that $\hat{U}(\boldsymbol{a})$ does not commute with $\hat{H}$. This is because $\hat{H}$ has become a function of the magnetic vector potential $\boldsymbol{A}$ and this changes as you translate. We should therefore expect[3]

$$\hat{U}^{-1}(\boldsymbol{a})\hat{H}(\boldsymbol{A})\hat{U}(\boldsymbol{a}) = \hat{H}\left(\hat{U}^{-1}(\boldsymbol{a})\boldsymbol{A}\right). \tag{45.2}$$

However, if the energy of the system is independent of global translations then this change in $\boldsymbol{A}$ can't affect the magnetic field we would measure with a magnetometer. We have, of course, met such shifts of the vector potential before: they simply represent gauge transformations.

[1] For ease of notation we follow convention in condensed matter physics and write $-e$ for the charge on the electron in this chapter.

[2] The fact that we have to make the circle have radius $\sqrt{2}\ell_{\rm B}$ and not $\ell_{\rm B}$ is annoying, but we will have to live with it. As we shall see, sometimes the relevant length scale is $\ell_{\rm B}$ and sometimes $\sqrt{2}\ell_{\rm B}$.

[3] Treating the magnetic vector potential as a function, rather than a second quantized field, we have $\hat{U}^{-1}(\boldsymbol{a})\boldsymbol{A}(\boldsymbol{r}) = \mathrm{e}^{\mathrm{i}\hat{\boldsymbol{p}}\cdot\boldsymbol{a}}\boldsymbol{A}(\boldsymbol{r}) = \boldsymbol{A}(\boldsymbol{r} - \boldsymbol{a})$.

Recall from Chapter 14 that a gauge transformation involves changes $\psi \rightarrow \psi e^{i\alpha}$ and $A_\mu \rightarrow A_\mu - \frac{1}{q}\partial_\mu \alpha$ and we sometimes write $\chi = \alpha/q$. Note also that the covariant derivative $D_\mu = \partial_\mu + iqA_\mu$ implies $\hat{P} = \hat{p} - q\boldsymbol{A}$. In this chapter we will often deal with electrons, where $q = -e$.

Example 45.1

Consider[4] a uniform field $\boldsymbol{B}$. In the gauge $\boldsymbol{A}(\boldsymbol{r}) = \frac{1}{2}\boldsymbol{B} \times \boldsymbol{r}$ we have

$$\hat{U}^{-1}(\boldsymbol{a})\boldsymbol{A}(\boldsymbol{r}) = \boldsymbol{A}(\boldsymbol{r} - \boldsymbol{a}) = \boldsymbol{A}(\boldsymbol{r}) - \frac{1}{2}\boldsymbol{B} \times \boldsymbol{a}. \qquad (45.3)$$

This is a gauge transformation $A_\mu \rightarrow A_\mu - \frac{1}{q}\partial_\mu \alpha(x)$ with $\frac{1}{q}\boldsymbol{\nabla}\alpha(x) = -\frac{1}{2}\boldsymbol{B} \times \boldsymbol{a}$ or

$$\frac{\alpha(\boldsymbol{r})}{q} = \chi(\boldsymbol{r}) = -\frac{1}{2}(\boldsymbol{B} \times \boldsymbol{a}) \cdot \boldsymbol{r}. \qquad (45.4)$$

Along with a shift in A_μ, a gauge transformation also involves a shift of the matter field $\psi \rightarrow \psi e^{i\alpha(x)}$ which here is $\psi \rightarrow \psi \exp\left[\frac{ie}{2}(\boldsymbol{B} \times \boldsymbol{a}) \cdot \boldsymbol{r}\right]$.

To summarize: in the presence of a magnetic field, we have a slightly unusual form of translational symmetry in that it is invariant under a combination of a translation with a gauge transformation.

We saw in Chapter 14 that the momentum operator $\hat{p}_\mu = i\partial_\mu$ doesn't transform properly under gauge transformations, but the covariant version $\hat{P}_\mu = iD_\mu = i(\partial_\mu + iqA_\mu)$ does. It therefore comes as little surprise to find that an upgraded translation operator that does commute with the gauged Hamiltonian is given by

$$\hat{U}(\boldsymbol{a}) = e^{-i\hat{\boldsymbol{P}} \cdot \boldsymbol{a}}, \qquad (45.5)$$

where $\hat{\boldsymbol{P}} = -i\boldsymbol{\nabla} - q\boldsymbol{A}$.

To see some consequences of this formalism, we now return to condensed matter physics and consider a two-dimensional square lattice (with lattice constant a) of non-interacting electrons in the presence of a magnetic field. The Hamiltonian is[5]

$$\hat{H} = \frac{1}{2m}\left(\hat{p}_x + eA_x\right)^2 + \frac{1}{2m}\left(\hat{p}_y + eA_y\right)^2 + V(x, y), \qquad (45.6)$$

where $V(x, y) = V(x + a, y) = V(x, y + a)$ is a periodic potential. With a uniform magnetic field parallel to z we can write the magnetic vector potential $\boldsymbol{A} = \frac{B}{2}(-y, x, 0)$. In this case the magnetic translation operators that translate us through one lattice spacing in the x- and y-directions are, respectively,

Since we are discussing two-dimensional space here we will revert to the ordinary notation of vector components where momentum along the x-direction is written p_x and so on. We will return to four-vector notation in Section 45.4.

This shows that the magnetic translation operators form a non-abelian group. See Chapter 46.

The proof of eqn 45.8 follows from $\left[-\frac{ia}{\hbar}\left(\hat{p}_x - \frac{eB\hat{y}}{2}\right), -\frac{ia}{\hbar}\left(\hat{p}_y + \frac{eB\hat{x}}{2}\right)\right] = 2\pi i\phi$ and the use of the identity

$$e^{\hat{A}}e^{\hat{B}} = e^{\hat{B}}e^{\hat{A}}e^{[\hat{A},\hat{B}]},$$

which holds if $\left[\hat{A}, \hat{B}\right]$ commutes with $\hat{A}$ and $\hat{B}$. See Exercise 45.1.

$$\begin{aligned}
\hat{U}_x &= \exp\left[-\frac{ia}{\hbar}\left(\hat{p}_x - \frac{eB\hat{y}}{2}\right)\right], \\
\hat{U}_y &= \exp\left[-\frac{ia}{\hbar}\left(\hat{p}_y + \frac{eB\hat{x}}{2}\right)\right],
\end{aligned} \qquad (45.7)$$

where we have restored factors of $\hbar$. As advertised, these operators commute with $\hat{H}$. However, they do not commute with each other and, in fact,[6]

$$\hat{U}_x\hat{U}_y = e^{2\pi i\phi}\hat{U}_y\hat{U}_x, \qquad (45.8)$$

where $\phi = ea^2B/h$ is the dimensionless magnetic flux.[7]

The quantity ϕ represents the number of flux quanta (h/e) passing through a square on the lattice (area a^2, usually called a **plaquette**). Thus

$$\phi = \frac{\Phi}{h/e}, \tag{45.9}$$

where $\Phi = Ba^2$ is the magnetic flux through one plaquette. If $\phi = p/q$ where p and q are integers then

$$\left[\hat{U}_x^q, \hat{U}_y\right] = 0, \tag{45.10}$$

because $(\mathrm{e}^{2\pi i\phi})^q = 1$. This means that a rectangle consisting of q adjacent plaquettes contains an integer number of flux quanta. Thus, the system has an effective periodicity equal to q times what would be the case for $B = 0$.

This idea is beautifully illustrated by the **Hofstadter butterfly**[8] shown in Fig. 45.1, which shows the energy level spectrum for a two-dimensional tight binding model (recall Example 4.8) with a magnetic field perpendicular to it. When $B = 0$ (and consequently $\phi = 0$), the allowed energies satisfy $-4t \leq E \leq 4t$ as you would expect from eqn 4.50. For $\phi = p/q$, the allowed energies form q sub-bands (note that sometimes the sub-bands touch, so the two sub-bands at $\phi = 1/2$ meet at $E = 0$). The resulting band structure is fractal (and was published only a year after Benoit Mandelbrot coined the term). It has the extraordinary feature that the spectrum depends on whether ϕ is rational or irrational (and can only be plotted for rational ϕ).

45.2 Landau Levels

In order to examine the quantum mechanics of charges in applied magnetic fields we return to the problem of two-dimensional free electrons. The Hamiltonian for this system may be written

$$\hat{H} = \frac{\hat{P}_x^2 + \hat{P}_y^2}{2m}, \tag{45.11}$$

where

$$\hat{P}_x = \hat{p}_x + eA_x, \quad \hat{P}_y = \hat{p}_y + eA_y. \tag{45.12}$$

Classically, electrons with momentum p will orbit in circles of radius p/eB in a magnetic field. Consequently, we write the position operators of an electron as

$$\hat{x} = \hat{X} + \tfrac{1}{eB}\hat{P}_y, \quad \hat{y} = \hat{Y} - \tfrac{1}{eB}\hat{P}_x, \tag{45.13}$$

where (X, Y) are the coordinates of the **guiding centre** of the orbit. We find that (X, Y) and (P_x, P_y) are independent coordinates because

$$\left[\hat{X}, \hat{P}_x\right] = \left[\hat{X}, \hat{P}_y\right] = \left[\hat{Y}, \hat{P}_x\right] = \left[\hat{Y}, \hat{P}_y\right] = 0. \tag{45.14}$$

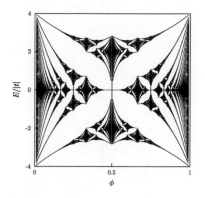

Fig. 45.1 The Hofstadter butterfly.

[8]For more discussion of the Hofstadter butterfly see Douglas Hofstadter's paper [Phys. Rev. B **14**, 2239 (1976)] and his well-known book *Gödel, Escher, Bach*, Chapter 5, which also features his take on renormalization.

Benoit B. Mandelbrot (1924–2010) chose his own middle initial, which doesn't stand for anything. There's an old joke that says that the 'B' in 'Benoit B. Mandelbrot' stands for 'Benoit B. Mandelbrot', thereby making the name of the inventor of fractals a recursively-defined fractal.

However, we also find that

$$\left[\hat{X}, \hat{Y}\right] = i\ell_{\mathrm{B}}^2, \quad \left[\hat{P}_x, \hat{P}_y\right] = -\frac{i\hbar^2}{\ell_{\mathrm{B}}^2}. \tag{45.15}$$

Thus the guiding centre cannot be determined more accurately than an area $\approx \ell_{\mathrm{B}}^2$, i.e. the area occupied by approximately one quantum of flux.

Example 45.2

Operators $\hat{X}$ and $\hat{Y}$ commute with $\hat{H}$ and hence $\frac{\mathrm{d}\hat{X}}{\mathrm{d}t} = \frac{\mathrm{d}\hat{Y}}{\mathrm{d}t} = 0$. However,

$$
\begin{aligned}
\frac{\mathrm{d}\hat{P}_x}{\mathrm{d}t} &= \frac{1}{i\hbar}\left[\hat{P}_x, \hat{H}\right] = -\omega_{\mathrm{c}}\hat{P}_y, \\
\frac{\mathrm{d}\hat{P}_y}{\mathrm{d}t} &= \frac{1}{i\hbar}\left[\hat{P}_y, \hat{H}\right] = \omega_{\mathrm{c}}\hat{P}_x,
\end{aligned} \tag{45.16}
$$

where $\omega_{\mathrm{c}} = eB/m$ is the cyclotron frequency. Thus (ignoring hats)

$$
\begin{aligned}
\ddot{P}_x + \omega_{\mathrm{c}}^2 P_x &= 0, \\
\ddot{P}_y + \omega_{\mathrm{c}}^2 P_y &= 0,
\end{aligned} \tag{45.17}
$$

and one can easily show that electrons perform circular orbits around (X, Y).

As with many subjects in quantum field theory, the physics of this problem may be described using the machinery of the simple harmonic oscillator. To map this problem onto the simple harmonic oscillator we introduce the operators

$$
\begin{aligned}
\hat{a} &= \frac{\ell_{\mathrm{B}}}{\sqrt{2}\hbar}\left(\hat{P}_x - i\hat{P}_y\right), & \hat{a}^\dagger &= \frac{\ell_{\mathrm{B}}}{\sqrt{2}\hbar}\left(\hat{P}_x + i\hat{P}_y\right), \\
\hat{b} &= \frac{1}{\sqrt{2}\ell_{\mathrm{B}}}\left(\hat{X} + i\hat{Y}\right), & \hat{b}^\dagger &= \frac{1}{\sqrt{2}\ell_{\mathrm{B}}}\left(\hat{X} - i\hat{Y}\right),
\end{aligned} \tag{45.18}
$$

satisfying $\left[\hat{a}, \hat{a}^\dagger\right] = \left[\hat{b}, \hat{b}^\dagger\right] = 1$ and $\left[\hat{a}, \hat{b}\right] = \left[\hat{a}^\dagger, \hat{b}\right] = 0$. The Hamiltonian then assumes the expected form: $\hat{H} = \left(\hat{a}^\dagger\hat{a} + \frac{1}{2}\right)\hbar\omega_{\mathrm{c}}$ and so there are eigenstates $|N\rangle$, where N is the number of a-quanta in the oscillator. The energy levels of this system are called **Landau levels**.

However, note that we can write the states

$$|N, n\rangle = \frac{1}{\sqrt{N!n!}}(\hat{a}^\dagger)^N(\hat{b}^\dagger)^n|0\rangle, \tag{45.19}$$

and so, even though the energy depends only on N, we shouldn't forget about n. This contributes to the degeneracy and, in fact, from this we will now show that each electron occupies a real space area of $\pi\ell_{\mathrm{B}}^2$.

Example 45.3

The density of states in momentum space is given by

$$g(k)\mathrm{d}k = \frac{\mathrm{d}k}{(2\pi/L)^2} \times 2, \tag{45.20}$$

where we write $k \equiv |\boldsymbol{k}| = (k_x^2 + k_y^2)^{\frac{1}{2}}$, the factor of 2 is for spin degeneracy and L is the linear size of the system. The radius k_N of the Nth Landau level in k-space is found by writing

$$\frac{\hbar^2 k_N^2}{2m} = \left(N + \frac{1}{2}\right)\hbar\omega_c, \tag{45.21}$$

and hence

$$k_N^2 = \frac{1 + 2N}{\ell_B^2}. \tag{45.22}$$

The area between two successive circles in k-space tells us how much momentum space is taken up by each Landau level. It is given by

$$\Delta A_k = \pi(k_{N+1}^2 - k_N^2) = \frac{2\pi}{\ell_B^2}, \tag{45.23}$$

where the factor of 2 again accounts for spin. We conclude that, in each Landau level, the number of states[9] is given by

$$\frac{2\pi}{\ell_B^2}\frac{1}{(2\pi/L)^2} \times 2 = \frac{L^2}{\pi\ell_B^2} = L^2\left(\frac{2eB}{h}\right), \tag{45.24}$$

which has the form:

$$\text{Number of states in one Landau level} = \frac{(\text{Area taken up by all of them})}{(\text{Area taken up by one of them})}. \tag{45.25}$$

So the effective real-space area occupied by one Landau state within a Landau level is $\pi\ell_B^2$.

However, we also need to discuss the effect of **spin splitting**. Each Landau level contains electrons in both spin states and a magnetic field will cause these to have different energies. The energy can be written

$$E = (N + \frac{1}{2})\hbar\omega_c \pm \frac{1}{2}g^*\mu_B B, \tag{45.26}$$

where g^* is the effective g-factor. This effect will now double the number of levels (these are now spin-split half Landau levels) and each of these levels now has half the number of states in it.[10] The effective area occupied by one state in a spin-split half Landau level is then[11] $2\pi\ell_B^2$.

[9]Thus the degeneracy of a Landau level is $2eB/h = 1/(\pi\ell_B^2)$ per unit area of the sample.

[10]Thus the degeneracy of a spin-split half Landau level is $eB/h = 1/(2\pi\ell_B^2)$ per unit area of the sample.

[11]Note that $2\pi\ell_B^2$ also equals the area occupied by one Dirac flux quantum.

45.3   The integer quantum Hall effect

The Hall effect depends on the number density of carriers and so our treatment of the integer quantum Hall effect will begin by considering how many states are available for occupation by electrons in a two-dimensional metal[12] when subject to a large magnetic field.

[12]Recall from Chapter 34 that we expect the conductance to vary as $G(L) \propto L^{(d-2)}$, where d is the dimensionality. For $d = 2$ we therefore expect behaviour that is independent of the size of the system, making the quantum Hall phenomena we will describe completely universal for two-dimensional systems.

Example 45.4

A reminder of the physics of the conventional Hall effect may be found in Exercise 45.4. In brief, for an isotropic material Ohm's law is written

$$\begin{pmatrix} J_x \\ J_y \end{pmatrix} = \begin{pmatrix} \sigma_{xx} & \sigma_{xy} \\ -\sigma_{xy} & \sigma_{xx} \end{pmatrix} \begin{pmatrix} E_x \\ E_y \end{pmatrix}, \tag{45.27}$$

and the Hall coefficient is defined as $R_H = \sigma_{xy}/B$. In experiments we measure the resistivities ρ_{xx} and ρ_{xy}, which are given by

$$\rho_{xx} \equiv \frac{E_x}{J_x} = \frac{\sigma_{xx}}{\sigma_{xx}^2 + \sigma_{xy}^2},$$

$$\rho_{xy} \equiv \frac{E_y}{J_x} = \frac{\sigma_{xy}}{\sigma_{xx}^2 + \sigma_{xy}^2}, \tag{45.28}$$

and in the high magnetic field limit we find $|\sigma_{xy}| \gg |\sigma_{xx}|$ giving

$$\rho_{xx} \approx \frac{\sigma_{xx}}{\sigma_{xy}^2}, \qquad \text{and} \qquad \rho_{xy} \approx \frac{1}{\sigma_{xy}} = R_{\mathrm{H}} B. \qquad (45.29)$$

For a conventional system we expect ρ_{xx} to be a constant and ρ_{xy} to increase linearly with field. As shown in Fig. 45.2, systems approximating a two-dimensional electron gas show something rather different with increasing field:

- Plateaux in ρ_{xy} occurring at values $\rho_{xy} = \frac{1}{\nu}\frac{h}{e^2}$, with ν an integer [Fig 45.2(b)].

- Dramatic drops in ρ_{xx} which takes very small values when we have a plateau in ρ_{xy} [Fig 45.2(a)].

- Maxima in ρ_{xx} in the transition region between the plateaux.

This is known as the integer quantum Hall effect.[13]

[13]See Singleton, *Band Theory and Electronic Properties of Solids*, for more detail on Hall effect physics.

Fig. 45.2 The quantum Hall effect in a GaAs-(Ga,Al)As heterojunction, shown in resistivity components (a) ρ_{xx} and (b) ρ_{xy}, measured at temperatures between 0.03 and 1.5 K. (Figure from J. Singleton, *Band Theory and Electronic Properties of Solids*, reprinted with permission.) (c) The fractional quantum Hall effect in a semiconductor heterojunction, shown in resistivity components ρ_{xy} (dotted line) and ρ_{xx} (solid line). Figure from P. Gee, D.Phil. Thesis, Clarendon Laboratory, University of Oxford (1997).

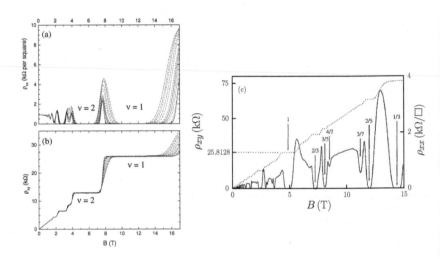

[14]Experimentally, this is accomplished by ensuring that electrons occupy the lowest sub-band in a quantum well in a semiconductor heterostructure.

The first step in studying the quantum Hall effect is to confine an electron gas into a two-dimensional region.[14] A magnetic field is then applied perpendicular to this two-dimensional electron gas. As shown in Example 45.3 each spin-split Landau level has a number of states per unit area of sample equal to $1/(2\pi\ell_{\mathrm{B}}^2)$. If the number of electrons per unit area is $n_{2\mathrm{d}}$, then we define the **filling factor** ν by

[15]The filling factor ν counts the number of spin-split half Landau levels. However, from now on we will simply refer to the spin-split levels as Landau levels. Whether the electrons interact with the magnetic field via their orbital (cyclotron) motion or via their spin angular momentum, the end result is the same: orbits with discrete energy levels; hence we will call these energy levels Landau levels, and ν of them are filled.

$$\nu = \frac{\text{(Areal density of electrons)}}{\text{(Areal density of states in a (spin-split half) Landau level)}}$$
$$= n_{2\mathrm{d}} 2\pi\ell_{\mathrm{B}}^2 = \frac{n_{2\mathrm{d}}h}{Be}. \qquad (45.30)$$

As a result, whenever the magnetic field takes a value

$$B = \frac{n_{2\mathrm{d}}h}{e} \times \frac{1}{p}, \qquad (45.31)$$

where p is an integer, then the filling factor ν will be an integer and hence ν (spin-split half) Landau levels[15] will be completely filled.

This then has the interesting consequence that the system realizes a macroscopic quantum state in which there is an energy gap for the creation of all charged excitations. The system is an **incompressible liquid**, essentially because the energy of the system becomes independent of area when the Landau levels are full (i.e. integer ν). In this case, we can use the fact that the Hall coefficient $R_\mathrm{H} = -1/(n_\mathrm{2d}e)$ to find that the Hall resistivity ρ_{xy} is[16]

$$\rho_{xy} = R_\mathrm{H}B = \frac{B}{n_\mathrm{2d}e} = \frac{1}{\nu}\frac{h}{e^2}. \qquad (45.32)$$

Thus we find plateaux in ρ_{xy} whenever ν Landau levels are full. Remarkably, the resistivity ρ_{xy} which is measured with a battery and some lengths of wire depends only on two fundamental constants of Nature. The quantity $h/e^2 = 25812.8\ \Omega$ has been measured with incredible precision and is often called the resistance quantum.[17] Equivalently, one can write eqn 45.32 in terms of the Hall conductivity σ_{xy} as

$$\sigma_{xy} = \nu\frac{e^2}{h}. \qquad (45.33)$$

Before feeling too pleased with ourselves we should remember that this analysis only tells us that ρ_{xy} is quantized at those values of magnetic field at which an integer number of Landau levels are completely filled, but does not explain the occurrence of the plateaux nor the vanishing of ρ_{xx}. In fact, these two features rely crucially on the fact that some of the states in a Landau level are *localized* via Anderson localization (described[18] in Chapter 34). The idea is that we imagine starting at a field where a single Landau level is exactly filled and then further increase the magnetic field. This reduces the degeneracy of the Landau level, forcing some electrons into the next Landau level at a higher energy. These electrons will end up in the localized states of that upper level and hence cannot contribute to the conductivity. As a result $\rho_{xx} = 0$, while ρ_{xy} remains at the quantized value corresponding to the complete filling of the lower level, resulting in a plateau.

45.4 The fractional quantum Hall effect

When $\nu = 1$ only a single Landau level is occupied and this maximizes the effect of electron–electron interactions. Although our treatment of the integer quantum Hall effect ignored these and treated the electrons as non-interacting, this will no longer be sufficient. The intriguing physics that is revealed when $\nu < 1$ is accessed by using magnetic fields with $B > n_\mathrm{2d}h/e$. In addition to the plateaux observed at integer values of ν, experiments in high-quality samples revealed further plateaux at fractional values of ν [see Fig. 45.2(c)]. These extra plateaux are due to the formation of a state of matter which can support fractional excitations and which is descibed by a topological field theory of the sort we met in Chapter 30.[19]

[16]To prevent a minus sign here, we use the convention that $B < 0$, so that $(-e)B > 0$. This convention is also followed in the discussion in Chapter 7 of Wen.

[17]It is also known as the von Klitzing contant, after Klaus von Klitzing (1943–) the discoverer of the quantum Hall effect.

[18]See Singleton for a simple explanation of how this works in this context or, for more detail, the article by J.T. Chalker in *Ecole des Houches: Topological Aspects of Low Dimensional Systems*, Eds. A. Comtet, T. Jolicoeur and S. Ouvry. (1998).

[19]Although we haven't stressed topology in our treatment of the integer quantum Hall effect, it may also be understood in these terms. See Altland and Simons, Section 9.3 for the details of Pruisken's field theory of the integer quantum Hall effect and its link to topology.

[20]The original explanation of the FQH effect was made by Robert Laughlin based on a trial wave function approach. We won't pursue this method here. The interested reader should consult Wen, Chapter 7 for a more detailed description of the other possible approaches to describing the physics.

[21]This is a form of **emergence**, where the properties and excitations of a phase of matter arise upon the condensation of that phase.

[22]A slightly more general, but basically similar, theory is needed to explain the others. Again, see Wen, Chapter 7 for the full story.

[23]We stress again that these quasiparticles can't be expected to be electrons with a shielded charge and increased mass as we had for quasiparticles in QED and in metals. In fact, the FQH quasiparticles will turn out to have quite different properties from electrons and holes, including fractional statistics and even fractional electric charge!

[24]Recall from Chapter 30 that in (2+1) dimensions we may write the magnetic field as $B = \partial_1 A^2 - \partial_2 A^1$ and, following convention, we take $B < 0$ so that $(-e)B > 0$.

We will assemble a quantum field theory that describes this **fractional quantum Hall** (FQH) effect.[20] This will be an **effective theory** in that we won't start with details of the electrons and holes found in a metal, rather with a set of fields that will model the ground state and the low energy or **elementary excitations**.

This is an important point. Nature doesn't allow us the luxury of being able to derive the properties of complex phases of matter from first principles. In the same way that the rigidity of solids is not derivable from the physics of those atoms in a gas, so the FQH effect is a different phase of matter from the two-dimensional electron gas from which it is created.[21] The properties of the phase of matter that give rise to a fractional quantum Hall effect cannot, therefore, be derived from the electron gas in zero magnetic field. So while most phases are separated by a breaking of symmetry, as discussed in Chapter 26, this one is very different. The FQH phase is *topologically distinct* from the electron gas. The phase transition between electron gas and FQH state is therefore topological and cannot be described by a broken symmetry theory as we had for the magnet, for example.

In the remainder of this chapter we will construct a quantum field theory of the FQH fluid, which is a phase of matter showing a FQH effect in its ground state. Specifically, we will explain the existence of a subset of the possible FQH states: those with $1/\nu$ equal to an odd integer.[22] We will also examine the elementary excitations of the FHQ fluid, which are the quasiparticles of this phase.[23]

We begin by asking what we want from our theory. Obviously we want the theory to predict the fundamental facts of the FQH effect. These are:

- In units where $\hbar = 1$, we want to predict a conductivity $\sigma_{xy} = J_x/E_y = \frac{e^2}{2\pi}\nu$ with $1/\nu$ taking odd integer values.
- The FQH fluid is incompressible with a charge density given by

$$\rho = -eJ^0_{\text{FQH}} = -e \times \left(\begin{array}{c} \text{Number of states per} \\ \text{Landau level} \end{array} \right) \times \frac{\nu}{L^2}$$
$$= -\frac{e^2 B}{2\pi}\nu, \tag{45.34}$$

where we've used the fact that the number of states per unit area L^2 in a Landau level is $1/(2\pi\ell_B^2)$.

These two facts are embodied in an equation describing the electromagnetic FQH current[24]

$$-eJ^\mu_{\text{FQH}} = -\frac{\nu e^2}{2\pi}\varepsilon^{\mu\nu\lambda}\partial_\nu A_\lambda, \tag{45.35}$$

where A^μ is the usual $U(1)$ gauge field of electromagnetism. We will cook up a Lagrangian whose equation of motion, at the very least, predicts this current.

The Lagrangian which will do this is of the Chern–Simons form and is thus a topological theory. It will describe another $U(1)$ gauge field a_μ, which couples to the electromagnetic field A_μ. We know from Chapter 30 that in a Chern–Simons theory the field a_μ gives rise to a current $J_{\text{CS}}^\mu \propto \varepsilon^{\mu\nu\lambda}\partial_\nu a_\lambda$. We choose to adjust our normalization slightly[25] here and will define the Chern–Simons current as

[25]We simply replace κ in Chapter 30 with $1/2\pi$ here.

$$J_{\text{CS}}^\mu = \frac{1}{2\pi}\varepsilon^{\mu\nu\lambda}\partial_\nu a_\lambda. \qquad (45.36)$$

The key to constructing the theory is to say that the FQH current in eqn 45.35 will contribute to the total Chern–Simons current. Our task is now to find a Lagrangian whose equation of motion tells us that the Chern–Simons current J_{CS} in eqn 45.36 is given by a contribution from the FQH current in eqn 45.35 added to a contribution from any source j^μ of quasiparticles that we put in the system. This is achieved with the Lagrangian

$$\mathcal{L} = -\frac{1}{2}\frac{s}{2\pi}\varepsilon^{\mu\nu\lambda}a_\mu\partial_\nu a_\lambda + \frac{e}{2\pi}\varepsilon^{\mu\nu\lambda}A_\mu\partial_\nu a_\lambda + j^\mu a_\mu, \qquad (45.37)$$

where s is a number. This Lagrangian is formed from three terms: **I**: a Chern–Simons term; **II**: a term where the electromagnetic gauge field A_μ couples to the Chern–Simons current; **III**: a term where a source of quasiparticles j^μ couples to the topological gauge field a_μ.

Example 45.5

We can check that this works by feeding the Lagrangian through the Euler–Lagrange equation to extract the equation of motion. We find that

$$\partial_\nu \frac{\partial\mathcal{L}}{\partial(\partial_\nu a_\mu)} = \frac{1}{2}\frac{s}{2\pi}\varepsilon^{\mu\nu\lambda}\partial_\nu a_\lambda - \frac{e}{2\pi}\varepsilon^{\mu\nu\lambda}\partial_\nu A_\lambda, \qquad \frac{\partial\mathcal{L}}{\partial a_\mu} = -\frac{1}{2}\frac{s}{2\pi}\varepsilon^{\mu\nu\lambda}\partial_\nu a_\lambda + j^\mu. \qquad (45.38)$$

We obtain

$$\frac{s}{2\pi}\varepsilon^{\mu\nu\lambda}\partial_\nu a_\lambda = \frac{e}{2\pi}\varepsilon^{\mu\nu\lambda}\partial_\nu A_\lambda + j^\mu, \qquad (45.39)$$

or

$$J_{\text{CS}}^\mu = \frac{e}{2\pi s}\varepsilon^{\mu\nu\lambda}\partial_\nu A_\lambda + \frac{1}{s}j^\mu. \qquad (45.40)$$

If we identify the dimensionless parameter s from the Lagrangian as being $1/\nu$, and temporarily set the quasiparticle source $j^\mu = 0$ then this is the fractional quantum Hall result we set out to predict.

Our low-energy Lagrangian would be of little use if it didn't provide some further insight into the workings of the fractional quantum Hall fluid. We will obtain these by asking about the quasiparticle excitations in the FQH ground state which are created by the conserved current j^μ. Recall from Chapter 30 that coupling a current like j^μ to the topological gauge field a_μ constrains the properties of these particles, attaching the flux of the field a_μ to each particle created by j^μ and imbuing them with fractional statistics. Specifically we will use the theory to show that:

- the FQH plateaux appear for half odd integer s;
- the FQH quasiparticles carry fractional electromagnetic charge;
- the FQH quasiparticles have fractional statistics.

[26] Potential for confusion arises here as there are two gauge fields in this problem: (i) the $U(1)$ gauge field of electromagnetism A_μ, which gives rise to the electric field $\boldsymbol{E}$ and magnetic flux density B, whose excitations couple to the charge $-e$; and (ii) the topological $U(1)$ gauge field a_μ, giving rise to an electric field $\boldsymbol{e}$ and magnetic flux density b, whose excitations couple to the charge l. The FQH quasiparticles carry integral a_μ-charge l and also A_μ-charge q, whose value we will determine. As the current j^μ of FQH quasiparticles is coupled to the topological field a_μ we expect, from Chapter 30, that they will also carry flux of the b-field (which we refer to as a_μ-flux).

[27] In Chapter 30 we saw that the Chern–Simons charge density is related to the flux density via $\rho (= J^0) = -\kappa b$ where, for our quasiparticles $\rho = l/s$ per unit area. Note also that $\kappa = \frac{1}{2\pi}$ and we take $b < 0$.

Our source of FQH quasiparticles j^μ will create excitations with a_μ-charge[26] l, where l is constrained to be an integer. We will construct the source term $j^\mu a_\mu = l a_0 \delta^{(2)}(\boldsymbol{x} - \boldsymbol{x}_0)$, where we have chosen to localize the quasiparticle source at position $\boldsymbol{x}_0$. The zeroth component of the equations of motion predicts a total electric charge density of

$$-e J_{\mathrm{CS}}^0 = \frac{e^2}{2\pi s}B - \frac{el}{s}\delta^{(2)}(\boldsymbol{x} - \boldsymbol{x}_0), \qquad (45.41)$$

where the first term on the right-hand side is the electric charge density expected from the FQH ground state and the second term is the extra electric charge of the quasiparticle. This equation implies two things: firstly that the FQH quasiparticles have electric charge $-el/s$; secondly, returning to the properties of Chern–Simons theories from Chapter 30, we see that this excitation carries a_μ-flux of $2\pi l/s$. This is progress, but we have yet to establish the values of s.

Next we find the statistics of the FQH quasiparticles. We picture two such excitations with charges l_1 and l_2. Move one in a circle around the other and we obtain an Aharonov–Bohm phase:

$$\begin{aligned}
\Delta\Phi &= (a_\mu \text{ flux}) \times (\text{quasiparticle charge}) \\
&= 2\pi\frac{l_1}{s} \times l_2. \qquad (45.42)
\end{aligned}$$

Now we set $l_1 = l_2 = l$ to obtain identical quasiparticles. Using the method of Chapter 30,[27] we note that exchanging these particles means rotating one through an angle π around the other so that we expect a phase change $\pi\eta$, where η will determine the statistics of the quasiparticles. The corresponding phase change $\Delta\Phi$ is half that in eqn 45.42, implying that we have statistics given by

$$\eta = \frac{l^2}{s}. \qquad (45.43)$$

Our conclusions so far are that the excitations of the FQH ground state are quasiparticles with a_μ-charge of l, which carry an electric charge $-el/s$, carry a_μ-flux of $2\pi/s$, and cause a phase change on exchange governed by $\eta = l^2/s$. Since we know that $\nu = 1/s$, which experiment teaches us takes values such as $1/3$, this suggests that the quasiparticles carry a fraction of an electron charge and possess fractional statistics! However we need to complete the puzzle by establishing the necessity for s to be an odd integer.

In order to complete the description it turns out that we must appeal to the presence of electrons and holes in the metal and link these to our FQH quasiparticles. Following convention, we will call our FQH quasiparticle excitations quasielectrons and quasiholes since, despite the fact that we cannot adiabatically tune between the FQH quasielectrons and the quasielectrons of a metal, we hope that their properties can at least be *related*. To establish the relation, we note that a single electron has electric charge $-el/s = -e$, it therefore has $l/s = 1$ and carries $l = s$ units of a_μ-charge. Its exchange property follows as $\eta =$

s. However, we know that electrons are fermions and must therefore have an exchange angle η of an odd integer. We therefore deduce that $s =$ odd integer, which is exactly the property we wanted to explain. We may now conclude that the FQH quasiparticle has fractional electric charge $-el/s$ and fractional statistics governed by $\eta = l^2/s$.

Example 45.6

Let's examine the case of $s = 3$. Electrons with $l = s = 3$ units of a_μ-charge form the FQH fluid with filling factor $\nu = 1/3$. The excitations of this liquid are FQH quasielectrons with a_μ-charge $l = 1$ and quasiholes with $l = -1$. Quasielectrons have fractional electric charge $-e/3$, they carry a_μ-flux of $2\pi/3$ and have statistics governed by $\eta = 1/3$.

Chapter summary

- In a magnetic field the Hamiltonian is invariant under a combination of a translation and a gauge transformation. This leads to the physics of the Hofstadter butterfly
- The integer quantum Hall effect follows from considering the occupancy of Landau levels as a function of applied magnetic field in a two-dimensional electron gas.
- The fractional quantum Hall effect is caused by interactions and may be explained using an effective, topological field theory.

Exercises

(45.1) Verify eqn 45.8 using the method suggested in the text.

(45.2) Verify eqns 45.14 and 45.15.

(45.3) (a) Use the creation and annihilation operators defined in eqns 45.18 to diagonalize the Hamiltonian in eqn 45.11.
(b) Evaluate $\hat{X}^2 + \hat{Y}^2$ to show that each electron occupies an area of $2\pi\ell_B^2$.

(45.4) *A reminder of the Hall effect.*
Consider driving a current of density J_x in the x-direction along a finite sized conductor, in the presence of a magnetic field B directed along z. Consider the equations of motion of a charge in the relaxation time approximation

$$m\ddot{v} = -\frac{mv}{\tau} + qE + qv \times B, \qquad (45.44)$$

where τ is the relaxation time.
(a) In the steady state, show that $E_y/E_x = -\omega_c\tau$.
(b) Verify that the Hall coefficient is given by

$$R_H = \frac{E_y}{J_x B} = -\frac{1}{ne}, \qquad (45.45)$$

where n is the electron density.

Part XI

Some applications from the world of particle physics

Quantum field theory is foundational for particle physics, since particles are simply excitations in quantum fields. In this part, we review several important particle physics theories which use the ideas developed in this book.

- Our treatment of gauge theory in Chapter 14 assumed that the underlying symmetry group is abelian (i.e. its elements commute). This is the case for $U(1)$, the group that describes electromagnetism. In Chapter 46 we introduce non-abelian gauge theory and show that the non-commutativity of group elements leads to a non-linear field tensor.

- A non-abelian gauge theory is at the heart of the electroweak theory due to Weinberg and Salam and this is presented in Chapter 47. Symmetry breaking of the Higgs field results in the emergence of the photon, the charged W^+ and W^- particles, and the neutral Z^0 particle. The photon is massless, but the other three acquire mass and so the weak force becomes short-ranged.

- Majorana fermions, particles that are their own antiparticles, are described in Chapter 48.

- Chapter 49 introduces two varieties of magnetic monopole: one due to Dirac which provides a link between the quantization of electric charge and the quantization of the magnetic charge of a monopole; the other due to 't Hooft and Polyakov which arises from a non-abelian gauge theory.

- The final chapter, Chapter 50, introduces the concept of instantons, which allows us to treat problems involving tunnelling. We ask what would happen if the Universe were to tunnel from a false vacuum into the true vacuum.

Non-abelian gauge theory

Physics is very muddled again at the moment; it is much too hard for me anyway, and I wish I were a movie comedian or something like that and had never heard anything about physics.
Wolfgang Pauli (1900–1958)

QED is a very successful field theory built on the gauge principle which accurately describes electromagnetism. Applying this principle involved taking a theory with a global (internal) $U(1)$ symmetry and promoting this to a local symmetry, necessitating the introduction of a gauge field $A_\mu(x)$. Turning our attention now to particle physics, we would also like to explain the weak and strong forces using fields. The weak and strong interactions can also be described by gauge theories, but unfortunately neither are based on the simple case of local $U(1)$ symmetry. In fact, both are based on local symmetry under more complicated groups of transformations: for the weak interaction it is the group $U(1) \otimes SU(2)$ and for the strong interaction it is $SU(3)$. Unlike $U(1)$, these groups are **non-abelian**.

A non-abelian group contains elements that don't commute: a transformation by element a followed by one by element b can have a different effect to the transformation by b followed by a. Non-abelian groups are quite familiar from everyday life, the rotation group $SO(3)$ being a simple example. Figure 46.1 shows that a rotation R_1 of a book by $\pi/2$ about x followed by rotation R_2 by $\pi/2$ about z leads to a different final orientation of the book from the rotations R_2 then R_1. In this chapter we examine the features of a gauge theory with a local non-abelian symmetry. We will concentrate on the simplest possible non-abelian cases $SU(2)$ and $SO(3)$. We start by reviewing the important properties of an abelian gauge theory.

Niels Hendrik Abel (1802–1829), after whom abelian groups are named, died aged 26 from tuberculosis. Non-abelian group theory was introduced by Evariste Galois (1811–1832) who was killed in a duel aged 20, having spent the entire night before writing down a revolutionary theory involving the use of groups to investigate the solubility of algebraic equations.

46.1 Abelian gauge theory revisited

In Chapter 14 we looked at the simplest possible gauge theory. Fundamentally it is a theory with a local symmetry. It is symmetric with respect to $U(1)$ transformations which are different at every point in space time. $U(1)$ is abelian since two transformations, applied one after another, always lead to the same result, no matter the order in which they're applied. As a reminder, here is the complex scalar field La-

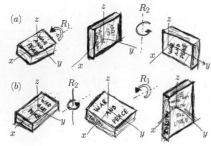

Fig. 46.1 A demonstration that the elements of the non-abelain group $SO(3)$ don't commute.

grangian:

$$\mathcal{L} = (D^\mu \psi)^\dagger (D_\mu \psi) - m^2 \psi^\dagger \psi - \frac{1}{4} F_{\mu\nu} F^{\mu\nu}, \qquad (46.1)$$

where $D_\mu \psi = (\partial_\mu \psi + iq A_\mu \psi)$. This is a locally gauge invariant quantity under the simultaneous transformations

$$\psi \to \psi e^{i\alpha(x)}, \quad A_\mu \to A_\mu - \tfrac{1}{q} \partial_\mu \alpha(x). \qquad (46.2)$$

In order to make the jump to more complicated symmetries we will deal with these in the form of infinitesimal transformations.

Example 46.1

We'll replay the proofs that the Lagrangian is invariant with respect to local gauge transformations using infinitesimal transformations.

$$
\begin{aligned}
\psi &\to (1 + i\alpha)\psi, \\
\partial_\mu \psi &\to \partial_\mu \psi + i(\partial_\mu \alpha)\psi + i\alpha(\partial_\mu \psi), \\
A_\mu &\to A_\mu - \frac{1}{q} \partial_\mu \alpha(x).
\end{aligned}
\qquad (46.3)
$$

This works up to order $O(\alpha)$ because we have

$$\psi^\dagger \psi \to (\psi^\dagger - i\alpha \psi^\dagger)(\psi + i\alpha \psi) = \psi^\dagger \psi + O(\alpha^2), \qquad (46.4)$$

and

$$
\begin{aligned}
D_\mu \psi &\to \left[\partial_\mu \psi + i(\partial_\mu \alpha)\psi + i\alpha(\partial_\mu \psi) + iq \left(A_\mu - \frac{1}{q} \partial_\mu \alpha \right) (\psi + i\alpha \psi) \right] \\
&= [\partial_\mu \psi + iq A_\mu \psi] + i\alpha [\partial_\mu \psi + iq A_\mu \psi] + O(\alpha^2) \\
&= (1 + i\alpha) D_\mu \psi + O(\alpha^2).
\end{aligned}
\qquad (46.5)
$$

Equation 46.5 shows that $D_\mu \psi$ transforms exactly the same way as ψ. This, in turn, means that $(D^\mu \psi)^\dagger (D_\mu \psi) \to (D^\mu \psi)^\dagger (D_\mu \psi) + O(\alpha^2)$. So all of the terms involving ψ are invariant with respect to the transformations.

46.2 Yang–Mills theory

In 1954 Chen-Ning Yang and Robert Mills asked if the theory of local symmetry could be extended to more complicated groups of transformations than $U(1)$, such as $SU(2)$ and $SO(3)$. These groups are non-abelian. Although an $SU(2)$ gauge theory doesn't describe the weak or strong interactions, it does display the properties of a more complicated non-abelian theory and so we will consider it in this chapter.

We ask what needs to be done to make a theory with a global internal $SU(2)$ symmetry also symmetric under local internal $SU(2)$ transformations. We'll use exactly the same strategy we used to upgrade the global internal $U(1)$ symmetry to a local one. We will first work out the consequences of imposing a local symmetry and then we will add a field which cancels out any extra terms that arise.

Chen-Ning Yang (1922–)
Robert Mills (1927–1999).
A gauge theory based on the $SU(N)$ group is called a **Yang–Mills theory** in honour of Yang and Mills who extended gauge theory to non-abelian groups in 1954.

[1]We use $\boldsymbol{\tau}$ for the Pauli matrices here, rather than $\boldsymbol{\sigma}$, to emphasize that they act on our states of two sorts of fermions, $\bar{\Psi} = (\ \bar{f}\quad \bar{g}\ )$, rather than on the spin of a single fermion, i.e. they are Pauli 'isospin' matrices, rather than Pauli 'spin' matrices.

Noether's theorem: $SU(2)$ internal symmetry

$$D^a\Psi = \tfrac{i}{2}\tau^a\Psi \qquad (a:\text{internal})$$
$$\Pi^\mu_\Psi = \bar{\Psi}i\gamma^\mu \qquad (\mu:\text{vector})$$
$$D\mathcal{L} = 0 \qquad W^\mu = 0$$
$$J^{\mu a}_{\text{Nc}} = \tfrac{1}{2}\bar{\Psi}\gamma^\mu\tau^a\Psi$$
$$Q^a_{\text{Nc}} = \int d^3x\,\tfrac{1}{2}\Psi^\dagger\tau^a\Psi$$

As examined in the exercises, $\hat{I}_z = \frac{1}{2}\int d^3x(\hat{f}^\dagger\hat{f} - \hat{g}^\dagger\hat{g})$. We might say that the Ψ-field carries $I = 1/2$ units of isospin, with f-on particles corresponding to $I_3 = 1/2$ and g-on particles to $I_3 = -1/2$. Isospin will be important in the next chapter.

[2]Remember that the $U(1)$ version of this is $\partial_\mu\psi \to \partial_\mu\psi + i\alpha(\partial_\mu\psi) + i(\partial_\mu\alpha)\psi$.

Here's the Dirac Lagrangian for two sorts of fermions, each with mass m:

$$\mathcal{L} = \bar{f}(i\gamma^\mu\partial_\mu - m)f + \bar{g}(i\gamma^\mu\partial_\mu - m)g, \qquad (46.6)$$

which we write more compactly as

$$\mathcal{L} = \bar{\Psi}(i\gamma^\mu\partial_\mu - m)\Psi, \qquad (46.7)$$

where

$$\Psi = \begin{pmatrix} f \\ g \end{pmatrix}, \quad \bar{\Psi} = (\ \bar{f}\quad \bar{g}\ ). \qquad (46.8)$$

This Lagrangian is invariant under a *global $SU(2)$* transformation

$$\Psi \to e^{\frac{i}{2}\boldsymbol{\tau}\cdot\boldsymbol{\alpha}}\Psi, \quad \bar{\Psi} \to \bar{\Psi}e^{-\frac{i}{2}\boldsymbol{\tau}\cdot\boldsymbol{\alpha}}, \qquad (46.9)$$

where $\boldsymbol{\tau} = (\tau_x, \tau_y, \tau_z)$ are the Pauli matrices.[1] The Noether currents arising from this symmetry are given in the box. The conserved charge-like quantity for this theory is the isospin $\hat{\boldsymbol{I}} = \int d^3x\,\hat{\Psi}^\dagger\frac{\boldsymbol{\tau}}{2}\hat{\Psi}$.

To investigate further we'll be working with infinitesimal transformations, so in infinitesimal form we can write

$$\Psi \to \left(1 + \frac{i}{2}\boldsymbol{\tau}\cdot\boldsymbol{\alpha}\right)\Psi. \qquad (46.10)$$

Example 46.2

We can show that this is a global symmetry of our Lagrangian. As before we write

$$\bar{\Psi}\Psi \to \bar{\Psi}\left(1 - \frac{i}{2}\boldsymbol{\tau}\cdot\boldsymbol{\alpha}\right)\left(1 + \frac{i}{2}\boldsymbol{\tau}\cdot\boldsymbol{\alpha}\right)\Psi$$
$$= \bar{\Psi}\Psi\left[1 + O(\alpha^2)\right]. \qquad (46.11)$$

Since we're working with infinitesimals, we agree to jettison any term above first order. Since, for a global symmetry the derivative $\partial_\mu\Psi$ transforms exactly as Ψ does, (since $\boldsymbol{\alpha}$ has no dependence on spacetime) the entire Lagrangian is invariant.

Now for the pivotal point of this chapter. We upgrade this argument to a local transformation by letting $\boldsymbol{\alpha}$ be a function of spacetime x. This is no problem for the mass term in the Lagrangian $\bar{\Psi}m\Psi$, which is still invariant. The trouble comes from the derivative, since

$$\partial_\mu\Psi \to \partial_\mu\Psi + \frac{i}{2}\left[\boldsymbol{\tau}\cdot\boldsymbol{\alpha}(x)\right]\partial_\mu\Psi + \frac{i}{2}\left[\boldsymbol{\tau}\cdot\partial_\mu\boldsymbol{\alpha}(x)\right]\Psi, \qquad (46.12)$$

and the final term in eqn 46.12 prevents $\partial_\mu\Psi$ transforming like Ψ and therefore prevents the derivative term from being invariant.[2]

Just as we introduced the gauge field $A_\mu(x)$ for the $U(1)$ case, we need to introduce a new gauge field which will mop up the term involving $\partial_\mu\alpha(x)$. This gauge field is called $\boldsymbol{W}_\mu(x)$ and has three internal components [that is $(W^1_\mu(x), W^2_\mu(x), W^3_\mu(x))$], each of which have four components in Minkowski spacetime (labelled with the index μ). Just as before, the new field is combined with the derivative operator to make a covariant derivative. We therefore define a covariant derivative[3]

$$D_\mu = \partial_\mu - \frac{i}{2}g\boldsymbol{\tau} \cdot \boldsymbol{W}_\mu(x), \tag{46.13}$$

where g is the charge of the theory (it's the quantity which will eventually tell us how strongly the gauge field $\boldsymbol{W}_\mu$ interacts with Ψ). If this procedure is going to work, then $D_\mu \Psi$ must transform in exactly the same way as Ψ does. That is, the manner in which the gauge field transforms must cancel out the extra term in the transformation of the ordinary derivative.[4] This will indeed be the case, but the fact that the transformation is non-abelian leads to a new complication.

[3]Recall that for $U(1)$ we have $D_\mu = \partial_\mu + iqA_\mu$. The equation here is formulated analogously with, by convention, a sign change reflecting the fact that our chosen $SU(2)$ rotation $e^{\frac{i}{2}\boldsymbol{\tau}\cdot\boldsymbol{\alpha}}$ rotates in the opposite (internal) direction to the $U(1)$ rotation $e^{i\alpha}$.

[4]Note that, written out explicitly, the covariant derivative reads $D_\mu \Psi = \begin{pmatrix} \partial_\mu f \\ \partial_\mu g \end{pmatrix} - \frac{i}{2}g \begin{pmatrix} W_\mu^3 & W_\mu^1 - iW_\mu^2 \\ W_\mu^1 + iW_\mu^2 & -W_\mu^3 \end{pmatrix}\begin{pmatrix} f \\ g \end{pmatrix}$.

Example 46.3

We want the derivative to transform in the same way as Ψ, which is to say that it should transform as

$$D_\mu \Psi \to \left(1 + \frac{i}{2}\boldsymbol{\tau} \cdot \boldsymbol{\alpha}(x)\right) D_\mu \Psi. \tag{46.14}$$

This implies

$$
\begin{aligned}
D_\mu \Psi \quad \to \quad & \left(1 + \frac{i}{2}\boldsymbol{\tau} \cdot \boldsymbol{\alpha}(x)\right)\left(\partial_\mu \Psi - \frac{i}{2}g\boldsymbol{\tau} \cdot \boldsymbol{W}_\mu(x)\Psi\right) \\
= \quad & \partial_\mu \Psi - \frac{i}{2}g\boldsymbol{\tau} \cdot \boldsymbol{W}_\mu(x)\Psi + \frac{i}{2}\boldsymbol{\tau} \cdot \boldsymbol{\alpha}(x)\partial_\mu \Psi \\
& - \left(\frac{i}{2}\right)^2 g\left[\boldsymbol{\tau} \cdot \boldsymbol{\alpha}(x)\right]\left[\boldsymbol{\tau} \cdot \boldsymbol{W}_\mu(x)\Psi\right]. \tag{46.15}
\end{aligned}
$$

So how do we get it to work? We suppose that the covariant derivative transforms as

$$D_\mu \to \partial_\mu - \frac{i}{2}g\boldsymbol{\tau} \cdot \boldsymbol{W}_\mu(x) - \frac{i}{2}g\boldsymbol{\tau} \cdot \delta\boldsymbol{W}_\mu(x), \tag{46.16}$$

where $\delta\boldsymbol{W}_\mu(x)$ is the change in the gauge field $\boldsymbol{W}_\mu(x)$ with the transformation. We also know that the field transforms as $\Psi \to \left[1 + \frac{1}{2}\boldsymbol{\tau} \cdot \boldsymbol{\alpha}(x)\right]\Psi$. Combining these transformations shows that

$$
\begin{aligned}
D_\mu \Psi \quad \to \quad & \left(\partial_\mu - \frac{i}{2}g\boldsymbol{\tau} \cdot \boldsymbol{W}_\mu(x) - \frac{i}{2}g\boldsymbol{\tau} \cdot \delta\boldsymbol{W}_\mu(x)\right)\left(1 + \frac{i}{2}\boldsymbol{\tau} \cdot \boldsymbol{\alpha}(x)\right)\Psi \\
= \quad & \partial_\mu \Psi + \frac{i}{2}\boldsymbol{\tau} \cdot \left[\partial_\mu\boldsymbol{\alpha}(x)\right]\Psi + \frac{i}{2}\boldsymbol{\tau} \cdot \boldsymbol{\alpha}(x)\partial_\mu \Psi \\
& - \frac{i}{2}g\boldsymbol{\tau} \cdot \boldsymbol{W}_\mu(x)\Psi - \left(\frac{i}{2}\right)^2 g\left[\boldsymbol{\tau} \cdot \boldsymbol{W}_\mu(x)\right]\left[\boldsymbol{\tau} \cdot \boldsymbol{\alpha}(x)\right]\Psi \\
& - \frac{i}{2}g\boldsymbol{\tau} \cdot \delta\boldsymbol{W}_\mu(x)\Psi, \tag{46.17}
\end{aligned}
$$

ignoring the $\alpha\delta\boldsymbol{W}_\mu$ term as it's second order.

By comparing terms in eqns 46.15 and 46.17 we find that for our gauge field to work we require

$$\boldsymbol{\tau} \cdot \delta\boldsymbol{W}_\mu(x) = \frac{1}{g}\boldsymbol{\tau} \cdot \left[\partial_\mu\boldsymbol{\alpha}(x)\right] + \frac{i}{2}\left\{\left[\boldsymbol{\tau} \cdot \boldsymbol{\alpha}(x)\right]\left[\boldsymbol{\tau} \cdot \boldsymbol{W}_\mu(x)\right] - \left[\boldsymbol{\tau} \cdot \boldsymbol{W}_\mu(x)\right]\left[\boldsymbol{\tau} \cdot \boldsymbol{\alpha}(x)\right]\right\}. \tag{46.18}$$

Notice that this last term survives only because the transformation is non-abelian and the two transformations involve $\boldsymbol{\tau}$ entering in different orders. Finally, we can use the identity

$$(\boldsymbol{\tau} \cdot \boldsymbol{a})(\boldsymbol{\tau} \cdot \boldsymbol{b}) = (\boldsymbol{a} \cdot \boldsymbol{b}) + i\boldsymbol{\tau} \cdot \boldsymbol{a} \times \boldsymbol{b}, \tag{46.19}$$

and conclude that

$$\boldsymbol{\tau} \cdot \delta\boldsymbol{W}_\mu(x) = \frac{1}{g}\boldsymbol{\tau} \cdot \left[\partial_\mu\boldsymbol{\alpha}(x)\right] - \boldsymbol{\tau} \cdot \left[\boldsymbol{\alpha}(x) \times \boldsymbol{W}_\mu(x)\right]. \tag{46.20}$$

[5]For comparison, in the $U(1)$ case this transformation is $A_\mu \to A_\mu - \frac{1}{q}\partial_\mu\alpha$.

The result of the preceding example is that the gauge field must transform[5] according to

$$\boldsymbol{\tau} \cdot \boldsymbol{W}_\mu \to \boldsymbol{\tau} \cdot \boldsymbol{W}_\mu + \frac{1}{g}\boldsymbol{\tau} \cdot (\partial_\mu\boldsymbol{\alpha}) - \boldsymbol{\tau} \cdot (\boldsymbol{\alpha} \times \boldsymbol{W}_\mu). \quad (46.21)$$

46.3   Interactions and dynamics of $\boldsymbol{W}_\mu$

Next we ask how the gauge field couples to the fermion particle field Ψ. Minimal coupling, which we first met in Chapter 14, tells us that this is simply a matter of multiplying out the term in the Dirac Lagrangian involving the covariant derivative D_μ. We have

$$\bar{\Psi}\mathrm{i}\gamma^\mu D_\mu\Psi = \bar{\Psi}\mathrm{i}\gamma^\mu\partial_\mu\Psi + \frac{g}{2}\bar{\Psi}\gamma^\mu\boldsymbol{\tau} \cdot \boldsymbol{W}_\mu\Psi. \quad (46.22)$$

Compare this to the $U(1)$ version

$$\bar{\psi}\mathrm{i}\gamma^\mu D_\mu\psi = \bar{\psi}\mathrm{i}\gamma^\mu\partial_\mu\psi - q\bar{\psi}\gamma^\mu A_\mu\psi, \quad (46.23)$$

which told us that the excitations in the $U(1)$ gauge field $A_\mu(x)$ mediated interactions between electrons by coupling to the charge q. Furthermore, the Feynman rule for the interaction was that a $\bar{\Psi}\Psi A_\mu$ interaction vertex contributed a factor $-\mathrm{i}q\gamma^\mu$. By analogy, the $SU(2)$ version suggests that the Feynman rule for a $\bar{\Psi}\Psi\boldsymbol{W}_\mu$ vertex is the contribution of a factor proportional to $\mathrm{i}g\boldsymbol{\tau}\gamma^\mu$. After quantization, there will be excitations in the $\boldsymbol{W}_\mu$ gauge field which mediate the interactions between the fermions of the theory. Explicitly, we expect there to be massless W^1, W^2 and W^3 photon-like particles, each with two possible transverse polarizations. The particles couple to the isospin of the Ψ field with a strength set by the charge g.

When we looked at the $U(1)$ theory, an additional, but very important contribution to the Lagrangian was provided by the gauge field $A_\mu(x)$ itself, whose dynamics gave us electromagnetism through the term $-\frac{1}{4}F_{\mu\nu}F^{\mu\nu}$. We now ask what contribution the gauge field $\boldsymbol{W}_\mu(x)$ makes to the Lagrangian. An obvious guess would be that we should add a term of the form $-\frac{1}{4}\boldsymbol{G}_{\mu\nu} \cdot \boldsymbol{G}^{\mu\nu}$ where $\boldsymbol{G}_{\mu\nu}$ is a tensor analogous to $F_{\mu\nu}$ in the electromagnetic case. The obvious guess for the form of this tensor is $\boldsymbol{G}_{\mu\nu} = \partial_\mu\boldsymbol{W}_\nu - \partial_\nu\boldsymbol{W}_\mu$, but since we will want $\boldsymbol{G}_{\mu\nu}$ to transform like Ψ and $\partial_\mu\Psi$, it turns out that the tensor we need is[6]

$$\boldsymbol{G}_{\mu\nu} = \partial_\mu\boldsymbol{W}_\nu - \partial_\nu\boldsymbol{W}_\mu + g(\boldsymbol{W}_\mu \times \boldsymbol{W}_\nu). \quad (46.24)$$

[6]This is a surprising result: the field strength $\boldsymbol{G}_{\mu\nu}$ is a *nonlinear* function of the gauge field $\boldsymbol{W}_\mu$. We thus have found ourselves a nonlinear interacting theory, even in the absence of matter!

Our final expression for the Lagrangian of a locally invariant $SU(2)$ theory is then given by

$$\mathcal{L} = \bar{\Psi}(\mathrm{i}\gamma^\mu D_\mu - m)\Psi - \frac{1}{4}\boldsymbol{G}_{\mu\nu} \cdot \boldsymbol{G}^{\mu\nu}. \quad (46.25)$$

The incredible thing about this Lagrangian is that the nonlinearity of $\boldsymbol{G}_{\mu\nu}$ allows a new sort of process to take place, namely, the direct interaction of particle excitations in one of the $\boldsymbol{W}_\mu$ fields with other $\boldsymbol{W}_\mu$

particles, as shown in the interaction vertex in Fig. 46.2, where we have three $\boldsymbol{W}_\mu$ particles interacting. (The Lagrangian also allows the interaction of four $\boldsymbol{W}_\mu$ particles.[7]) Note that photon–photon interactions are not possible in abelian electromagnetism as the photon carried no electromagnetic charge of its own. In contrast, the isospin field $\boldsymbol{W}_\mu$ carries an isospin charge of one unit ($I = 1$) so remarkably it may act as a source of itself.[8]

This self-interaction of the gauge field provides an explanation for a feature that makes Yang–Mills theories so interesting: the asymptotic freedom described in Chapter 34. As we probe further away from an electromagnetic charge (described by abelian $U(1)$ electromagnetism) the charge is shielded by dipoles formed from virtual electron–positron pairs and the electric charge appears to get smaller. However, the non-abelian gauge field produces a marked difference as we probe further away from a quark (where the strong interaction of colour charge is described by a non-abelian $SU(3)$ gauge theory). The screening effect would be the same for the quark as for the electron, except that we have the gauge field self-interaction which allows the gauge field quantum (a *gluon*) to decay into a pair of gluons. As the gluons themselves carry colour charge they lead to a proliferation of like colour charges around the quark leading to an increased colour charge at large distances. This is asymptotic freedom: the interaction gets stronger as we probe further from the quark.

Example 46.4

A free, globally $SO(3)$ invariant scalar field theory was introduced in Chapter 13. Its Lagrangian is

$$\mathcal{L} = \frac{1}{2}\partial^\mu \boldsymbol{\Phi} \cdot \partial_\mu \boldsymbol{\Phi} - \frac{m^2}{2} \boldsymbol{\Phi} \cdot \boldsymbol{\Phi}, \qquad (46.26)$$

where $\boldsymbol{\Phi} = (\phi_1, \phi_2, \phi_3)$. This theory is invariant with respect to the global (internal) rotation given by $\boldsymbol{\Phi}(x) \rightarrow \boldsymbol{\Phi}(x) - \boldsymbol{\theta} \times \boldsymbol{\Phi}(x)$. Again, we upgrade this to a local transformation by letting $\boldsymbol{\theta}$, the internal angle of rotation, be a function of spacetime x. That is, the transformation is now $\boldsymbol{\Phi}(x) \rightarrow \boldsymbol{\Phi}(x) - \boldsymbol{\theta}(x) \times \boldsymbol{\Phi}(x)$. To make a locally invariant theory we need a gauge field, leading to the covariant derivative:

$$D_\mu \boldsymbol{\Phi} = \partial_\mu \boldsymbol{\Phi} - g \boldsymbol{\Phi} \times \boldsymbol{W}_\mu. \qquad (46.27)$$

The correct $\boldsymbol{W}_\mu$ to choose is one that transforms as

$$\boldsymbol{W}_\mu \rightarrow \boldsymbol{W}_\mu + \frac{1}{g}\partial_\mu \boldsymbol{\theta} - (\boldsymbol{\theta} \times \boldsymbol{W}_\mu), \qquad (46.28)$$

where g is the charge of the theory, and as for $SU(2)$ we add a term $-\frac{1}{4}\boldsymbol{G}_{\mu\nu} \cdot \boldsymbol{G}^{\mu\nu}$ to the Lagrangian in eqn 46.26, where for this theory $\boldsymbol{G}_{\mu\nu} = \partial_\mu \boldsymbol{W}_\nu - \partial_\nu \boldsymbol{W}_\mu + g(\boldsymbol{W}_\mu \times \boldsymbol{W}_\nu)$. This theory describes three massive scalar fields (the three components of $\boldsymbol{\Phi}$) and three massless gauge fields (the three internal components of $\boldsymbol{W}_\mu$).

This completes our discussion of the properties of a non-abelian gauge theory.[9] As a summary, the gauge fields that we have met so far are listed in the margin on page 430.

[7]The terms leading to the interactions of three and four $\boldsymbol{W}_\mu$ particles can be seen by expanding $\boldsymbol{G}_{\mu\nu} \cdot \boldsymbol{G}^{\mu\nu}$.

[8]Since the algebra of isospin mirrors that of ordinary spin, we say that the excitations of the $I = 1$ field $\boldsymbol{W}_\mu$ are particles with $I_z = 0$ created (and destroyed) by the field $\hat{W}^3_\mu$; particles with $I_z = 1$ created by the combination $\frac{1}{\sqrt{2}}(\hat{W}^1_\mu + i\hat{W}^2_\mu)$ [and destroyed by $\frac{1}{\sqrt{2}}(\hat{W}^1_\mu - i\hat{W}^2_\mu)$]; and particles with $I_z = -1$ created by $\frac{1}{\sqrt{2}}(\hat{W}^1_\mu - i\hat{W}^2_\mu)$ [and destroyed by $\frac{1}{\sqrt{2}}(\hat{W}^1_\mu + i\hat{W}^2_\mu)$].

Fig. 46.2 The interaction of three W particles is possible in a gauge theory with local $SU(2)$ symmetry.

[9]The next logical step would be to quantize the Yang–Mills Lagrangian. This can be done using the path integral but requires some tricks that we won't pursue here. See Peskin and Schroeder, Chapter 16 for the details.

46.4 Breaking symmetry with a non-abelian gauge theory

We saw in Chapter 26 that spontaneous symmetry breaking in a gauge theory led to the famous Higgs mechanism. Here we treat exactly the same problem with the more complicated example of a non-abelian field. This is useful, not only because it provides yet another example of the wonderful physics of symmetry breaking, but also because the electroweak theory examined in the next chapter relies on exactly this physics.

We will examine the case of $SO(3)$ gauge theory with an ϕ^4-type interaction and, crucially, a positive, symmetry-breaking mass term, with a Lagrangian given by

$$\mathcal{L} = \frac{1}{2}D^\mu\boldsymbol{\Phi} \cdot D_\mu\boldsymbol{\Phi} + \frac{m^2}{2}\boldsymbol{\Phi}\cdot\boldsymbol{\Phi} - \lambda(\boldsymbol{\Phi}\cdot\boldsymbol{\Phi})^2 - \frac{1}{4}\boldsymbol{G}_{\mu\nu}\cdot\boldsymbol{G}^{\mu\nu}. \quad (46.29)$$

The potential possesses a spherical shell of minima at a radius (in internal $\boldsymbol{\Phi}$ space) of $|\boldsymbol{\Phi}_0| = \left(\frac{m^2}{4\lambda}\right)^{\frac{1}{2}}$. Symmetry breaking involves picking one of the infinite number of equivalent vacua, which breaks global (and local) symmetry. For simplicity, we chose the one that points along the 3-direction in isospin space, that is $\boldsymbol{\Phi}_0 = \left(\frac{m^2}{4\lambda}\right)^{\frac{1}{2}}\hat{e}_3$. For the excitations above this ground state, we write

$$\boldsymbol{\Phi} = \begin{pmatrix} \Phi_1(x) \\ \Phi_2(x) \\ |\boldsymbol{\Phi}_0| + \chi(x) \end{pmatrix}, \quad (46.30)$$

where $\chi(x)$ describes deviations in the 3-direction. Then, after a bout of tedious algebra, we find

$$\begin{aligned} \mathcal{L} = & \frac{1}{2}\left[(\partial_\mu\Phi_1)^2 + (\partial_\mu\Phi_2)^2 + (\partial_\mu\chi)^2\right] - 4|\boldsymbol{\Phi}_0|^2\lambda\chi^2 \\ & + g|\boldsymbol{\Phi}_0|\left[(\partial^\mu\Phi_1)W_\mu^2 - (\partial^\mu\Phi_2)W_\mu^1\right] \\ & + \frac{g^2|\boldsymbol{\Phi}_0|^2}{2}\left[(W_\mu^1)^2 + (W_\mu^2)^2\right] - \frac{1}{4}\boldsymbol{G}_{\mu\nu}\cdot\boldsymbol{G}^{\mu\nu} + \dots, (46.31) \end{aligned}$$

where we ignore higher order terms. The first line describes the dynamics of the $\chi(x)$ scalar field, the final line the dynamics of the $\boldsymbol{W}_\mu(x)$ vector field, but the terms that are mixed in components of $\boldsymbol{\Phi}$ and $\boldsymbol{W}_\mu$ aren't easy to interpret. However, there is a useful trick we can use: we select a gauge which ensures that the excited state field $\boldsymbol{\Phi}$ points along the 3-direction in isospace at every point in spacetime. (This is known as **unitary gauge**.) This removes all mention of the components Φ_1 and Φ_2 and means that the excited state field becomes simply

$$\boldsymbol{\Phi}(x) = [|\boldsymbol{\Phi}_0| + \chi(x)]\,\hat{e}_3. \quad (46.32)$$

Example 46.5

Since the term $D_\mu \boldsymbol{\Phi} = \partial_\mu \boldsymbol{\Phi} - g\boldsymbol{\Phi} \times \boldsymbol{W}_\mu$ mixes up the components of $\boldsymbol{\Phi}$, our choice of unitary gauge yields

$$
\begin{aligned}
D_\mu \Phi_1 &= g W_\mu^2 (|\boldsymbol{\Phi}_0| + \chi), \\
D_\mu \Phi_2 &= -g W_\mu^1 (|\boldsymbol{\Phi}_0| + \chi), \\
D_\mu \Phi_3 &= \partial_\mu \chi,
\end{aligned} \tag{46.33}
$$

giving,

$$
(D_\mu \boldsymbol{\Phi})^2 = (\partial_\mu \chi)^2 + g^2 |\boldsymbol{\Phi}_0|^2 \left[(W_\mu^1)^2 + (W_\mu^2)^2 \right] + \cdots \tag{46.34}
$$

This solves the problem of the mixing of components of $\boldsymbol{\Phi}$ and $\boldsymbol{W}_\mu$ fields.

The broken symmetry Lagrangian is now given by

$$
\begin{aligned}
\mathcal{L} = \ & \frac{1}{2} (\partial_\mu \chi)^2 - 4 |\boldsymbol{\Phi}_0| \lambda \chi^2 \\
& + \frac{g^2 |\boldsymbol{\Phi}_0|^2}{2} \left[(W_\mu^1)^2 + (W_\mu^2)^2 \right] - \frac{1}{4} \boldsymbol{G}_{\mu\nu} \cdot \boldsymbol{G}^{\mu\nu} + \cdots, (46.35)
\end{aligned}
$$

which is the sum of Lagrangians for a massive, single-component scalar field χ and a massive vector field with two components W_μ^1 and W_μ^2. Symmetry breaking has caused two massive scalar fields (Φ_1 and Φ_2) to disappear from our original Lagrangian and we are left with one, called χ. The vector fields W_μ^1 and W_μ^2 have grown massive, taking on a mass $g|\boldsymbol{\Phi}_0|$ (see Fig. 46.3). The field W_μ^3 remains massless.

In summary, symmetry breaking of the $SO(3)$ gauge theory causes:

$$
\begin{pmatrix} \text{3 massive scalar fields } \boldsymbol{\Phi} \\ \text{3 massless photon-like fields } \boldsymbol{W}_\mu \end{pmatrix} \rightarrow \begin{pmatrix} \text{1 massive scalar field } \chi \\ \text{2 massive vector fields } W_\mu^1, W_\mu^2 \\ \text{1 massless photon-like field } W_\mu^3 \end{pmatrix}.
$$

We may check that we haven't lost any degrees of freedom. Massive scalar fields only have a single degree of freedom, while massless photon-like fields have two.[10] This makes nine on the left. Noting that massive vector fields have three degrees of freedom we see that we also have nine on the right and all is well. Notice, finally, that we could describe electromagnetism by taking $W_\mu^3(x) = A_\mu(x)$. This immediately leads to the question of whether electromagnetism in our Universe results from breaking symmetry in an $SO(3)$ or $SU(2)$ non-abelian theory.

Is all this non-abelian gauge theory simply a mathematical exercise to amuse the theoretically inclined? On the contrary, it seems that the Universe has some non-abelian symmetries etched into it. In the next chapter we will apply what we have learnt about symmetry breakdown in a non-abelian gauge theory to formulating the electroweak theory of Steven Weinberg and Abdus Salam, and we will discover that the symmetry group we break is a little more complicated than $SO(3)$.

Fig. 46.3 The gauge fields W_μ^1 and W_μ^2 eat the Φ Goldstone modes and acquire mass when global symmetry is broken in a non-abelian gauge theory with local $SU(2)$ or $SO(3)$ symmetry.

[10]Remember the case of electromagnetism!

Chapter summary

- A non-abelian group contains elements which do not commute. The introduction of a non-abelian local symmetry leads to a non-abelian gauge field. The field tensor has an extra term which makes it nonlinear.
- The consequences of symmetry breaking were demonstrated for $SO(3)$ theory.

Exercises

(46.1) Show that $\hat{I}_z = \frac{1}{2} \int \mathrm{d}^3 x (\hat{f}^\dagger \hat{f} - \hat{g}^\dagger \hat{g})$.

(46.2) (a) Write an $SO(3)$ rotation about the z-axis in infinitesimal form.
(b) Show that the Lagrangian in eqn 46.26 is invariant with respect to global $SO(3)$ transformations.
(c) We want to arrange matters so that the theory is invariant with respect to local $SO(3)$ transformation. As in the abelian case, the trouble comes from the derivative term. Show that

$$\partial_\mu \boldsymbol{\Phi}(x) \to \partial_\mu \boldsymbol{\Phi}(x) - \partial_\mu \boldsymbol{\theta}(x) \times \boldsymbol{\Phi}(x) - \boldsymbol{\theta}(x) \times \partial_\mu \boldsymbol{\Phi}(x). \tag{46.36}$$

(d) Verify that $D_\mu \boldsymbol{\Phi}$ transforms in exactly the same way as $\boldsymbol{\Phi}$.

(46.3) (a) Show that, for $U(1)$ gauge theory, the commutator of the covariant derivatives gives

$$[D^\mu, D^\nu] = \mathrm{i}q(\partial^\mu A^\nu - \partial^\nu A^\mu) = \mathrm{i}q F^{\mu\nu}. \tag{46.37}$$

This suggests that we can extract the analogue of $F^{\mu\nu}$ for non-abelian gauge theory by evaluating the commutator. This turns out to be the case!
(b) Show that the analogous commutator for $SU(2)$ gauge theory is

$$[D^\mu, D^\nu] = -\frac{\mathrm{i}}{2} g\boldsymbol{\tau} \cdot (\partial^\mu \boldsymbol{W}^\nu - \partial^\nu \boldsymbol{W}^\mu + g\boldsymbol{W}^\mu \times \boldsymbol{W}^\nu)$$
$$= -\frac{\mathrm{i}}{2} g\boldsymbol{\tau} \cdot \boldsymbol{G}^{\mu\nu}.$$

(46.4) Verify eqn 46.31 and the simplified form eqn 46.35.

The Weinberg–Salam model

47

As theorists sometimes do, I fell in love with this idea. But as often happens with love affairs, at first I was rather confused about its implications.
Steven Weinberg (1933–) on symmetry breaking

I remember travelling back to London on an American Air Force (MATS) transport flight. Although I had been granted, for the night, the status of a Brigadier or a Field Marshal – I forget which – the plane was very uncomfortable; full of crying servicemen's children – that is, the children were crying, not the servicemen. I could not sleep. I kept reflecting on why Nature should violate left-right symmetry in the weak interactions.
Abdus Salam (1926–1996)

The Universe as we know it is the result of a symmetry breaking phase transition. In this chapter we will explain the features of a Universe in which:

- Electrons have mass, but neutrinos do not.[1]

- Electrons may occur with left- or right-handed chirality, neutrinos are only observed with left-handed chirality.

- The photon is massless.

The explanation of these fundamental properties emerges from a model formulated by Abdus Salam and Steven Weinberg which unites the electromagnetic and weak interactions. At the heart of the Weinberg–Salam **electroweak theory** lies a proposition: we live in a broken symmetry Universe and the particles that we observe in Nature are a result of the symmetry breaking.[2] In order to make progress with this idea we will write down a Lagrangian for the electroweak Universe before this symmetry breaking phase transition took place. We will propose a set of *local* symmetries for the Lagrangian. As usual, this will require that we introduce gauge fields to force the systems to be locally symmetric. We will then examine the consequences of the spontaneous breaking of symmetry on the particle spectrum of the system. Let's make a start!

[1] Of course neutrinos are not really massless, as shown by neutrino flavour oscillation experiments. Their mass is, however, undoubtedly small compared to the electron mass and in this chapter the Universe we describe is an approximation to ours.

Abdus Salam (1926–1996) is notable, not only for his many achievements in quantum field theory, but for a tireless advocacy for the development of science in the third world.

Steven Weinberg's (1933–) influence on quantum field theory is difficult to overstate. See his three volume masterwork on the subject for a profound and different take on the material.

[2] Although we won't consider muons and tauons, these leptons may also be included in addition to the electron.

47.1   The symmetries of Nature before symmetry breaking

[3]Strictly speaking this should be called a $U(2)$ symmetry. See Penrose for details.

Imagine the Universe a moment after the Big Bang. It is more symmetrical than the Universe in which we live. Specifically, the lepton fields enjoy an internal $SU(2) \otimes U(1)$ symmetry.[3] In around 10^{-12} seconds the Universe will cool below a temperature 10^{16} K and undergo a symmetry breaking phase transition. Before this occurs we will assess the state of the lepton fields of the Universe as described by the Weinberg–Salam model.

[4]That this model doesn't provide an explanation for this feature of Nature has led to much work over the last few decades regarding more general 'supersymmetries' that may be required to describe the Universe more completely.

Before symmetry breaking neutrinos and electrons are both massless particles. Electrons are possible in both left- and right-handed forms, but neutrinos only exist in left-handed form.[4] Neutrinos and electrons are fermions, and are therefore excitations in Fermi fields. We know what to expect from their Lagrangian: it's a Dirac Lagrangian for each field, without any mass terms:

$$\mathcal{L} = \bar{\nu}_e(x)\mathrm{i}\partial\!\!\!/\nu_e(x) + \bar{e}_{\mathrm{L}}(x)\mathrm{i}\partial\!\!\!/e_{\mathrm{L}}(x) + \bar{e}_{\mathrm{R}}(x)\mathrm{i}\partial\!\!\!/e_{\mathrm{R}}(x), \qquad (47.1)$$

where we've written the right-handed electron field as $e_{\mathrm{R}}(x)$ and so forth.

The interesting thing about having several fields in a theory is the possibility of internal symmetries. As we have seen, symmetries lead to conserved charges and one of our tasks in formulating the theory is to decide the values of the conserved charges that excitations in each of our fields will carry. Below we will assign two mysterious charges to our fields, Y and I, which we apparently pluck from the air on the grounds that they lead to a successful theory that agrees with all experimental observations. It should be noted that these charges are related to the more familiar electromagnetic charge Q via the Gell-Mann–Nishijima relation:

$$Q = I_3 + \frac{Y}{2}. \qquad (47.2)$$

Kazuhiko Nishijima (1926–2009). The relation was found by Nishijima and Tadao Nakano, and independently by Murray Gell-Mann. Thus it is often also known as the NNG rule.

The known values of Q for the particles in our Universe will constrain the values of I and Y we assign.

We begin by writing the fields as components of a large column vector:

$$\Psi(x) = \begin{pmatrix} \nu_e(x) \\ e_{\mathrm{L}}(x) \\ e_{\mathrm{R}}(x) \end{pmatrix}. \qquad (47.3)$$

Next we ask what symmetries this theory has. The first is a local $U(1)$ symmetry. The $U(1)$ transformation causes the field to pick up a phase factor $\mathrm{e}^{\mathrm{i}\beta(x)}$. To ensure invariance we introduce a gauge field $B_\mu(x)$ which transforms as $B_\mu(x) \to B_\mu(x) + \frac{1}{Yg'}\partial_\mu\beta(x)$, where Y is the charge of the theory and the parameter g' tells us how strongly particles will couple to the hypercharge (just as we write $q = Q|e|$ and $|e|$ tells us how strongly a photon couples to a single electronic charge). Rearranging the position of charge Y slightly in the equations and rescaling for future convenience, we'll rewrite the transformation as $\mathrm{e}^{\frac{\mathrm{i}}{2}Y\beta(x)}$ and

$B_\mu \to B_\mu + \frac{1}{g'}\partial_\mu\beta$. The former expression tells us that the strength of the transformation at a point in spacetime depends on how much charge a field carries.[5] Note carefully that the $U(1)$ charge Y isn't the electric charge; instead, Y is known as the **weak hypercharge**. We assign the fields hypercharges as shown in the table in the margin. Thus under $U(1)$ transformations we have

$$\begin{pmatrix} \nu_e \\ e_L \\ e_R \end{pmatrix} \to \begin{pmatrix} e^{-\frac{i\beta(x)}{2}} & 0 & 0 \\ 0 & e^{-\frac{i\beta(x)}{2}} & 0 \\ 0 & 0 & e^{-i\beta(x)} \end{pmatrix} \begin{pmatrix} \nu_e \\ e_L \\ e_R \end{pmatrix}. \qquad (47.4)$$

To summarize so far: we may turn the internal dial labelled '$U(1)$ weak hypercharge' without changing the properties of the state of the early Universe. The dial serves to multiply the fields by position-dependent phase factors, whose effects are cancelled by the gauge field $B_\mu(x)$.

The internal $U(1)$ symmetry is not the only local symmetry possessed by the early Universe. There is also a local $SU(2)$ symmetry. A $SU(2)$ transformation results in the field components acquiring phases whose values depend on the details of the conserved charge[6] I, known as **weak isospin,** carried by each field component. The weak isospin I obeys the usual rules of spin angular momentum so, for example, the third component of isospin for a field with $I = 1/2$ has eigenvalues $I_3 = \pm 1/2$. We assign isospin quantum numbers to particles as shown in the table in the margin, with the result that the fields undergo the transformation:

$$\begin{pmatrix} \nu_e \\ e_L \end{pmatrix} \to e^{\frac{i}{2}\boldsymbol{\tau}\cdot\boldsymbol{\alpha}(x)} \begin{pmatrix} \nu_e \\ e_L \end{pmatrix} \qquad e_R \to e_R. \qquad (47.5)$$

For $SU(2)$ transformations the gauge field needed to guarantee local invariance is $\boldsymbol{W}_\mu(x)$, which transforms according to $\boldsymbol{\tau}\cdot\boldsymbol{W}_\mu \to \boldsymbol{\tau}\cdot\boldsymbol{W}_\mu + \frac{1}{g}\boldsymbol{\tau}\cdot(\partial_\mu\boldsymbol{\alpha}) - \boldsymbol{\tau}\cdot(\boldsymbol{\alpha}\times\boldsymbol{W}_\mu)$, where g is the coupling to the weak isospin.[7]

Summarizing again: the dial labelled '$SU(2)$ weak isospin' may be turned at will in the early Universe without any physical consequence. This dial mixes up the massless fields ν_e and e_L and multiplies them by position dependent phases. Its effects are cancelled by the gauge field $\boldsymbol{W}_\mu(x)$. The transformation has no effect on e_R.

We have now described a set of local $U(1)$ and $SU(2)$ transformations for our set of field components. Together they make the theory locally invariant with respect to the enlarged group of transformations $SU(2) \otimes U(1)$. In order to make these symmetries of the Lagrangian we now need to replace derivatives ∂_μ in the Lagrangian with covariant derivatives. We write

$$\begin{array}{llll} U(1) & : & D_\mu\Psi & = & \partial_\mu\Psi - \frac{i}{2}g'Y B_\mu(x)\Psi, \\ SU(2) & : & D_\mu\Psi & = & \partial_\mu\Psi - igI\boldsymbol{\tau}\cdot\boldsymbol{W}_\mu(x)\Psi. \end{array} \qquad (47.6)$$

Since the original column vector (ν_e, e_L, e_R) is distinguished by the first two (left-handed) components having the same Y charge and I charge while the right-handed components have different ones, we will rewrite

[5]The reason for doing this here is that the different components of the field are going to carry different amounts of charge. We will call Y a '$U(1)$ charge' because it controls the coupling between the fermions and the gauge fields we need to introduce to guarantee local $U(1)$ symmetry.

Field	ν_e	e_L	e_R
Y	-1	-1	-2

Table of weak hypercharge Y

[6]See the previous chapter for further details of how this arises.

Field	ν_e	e_L	e_R
I	$\frac{1}{2}$	$\frac{1}{2}$	0
I_3	$\frac{1}{2}$	$-\frac{1}{2}$	0

Table of isospin quantum numbers I_3 and I

[7]As in the previous chapter, the $\boldsymbol{W}_\mu$ field itself has $I = 1$. Note also that this field has $Y = 0$ units of weak hypercharge.

[8]The logic of this notation is that it puts the field with $I_3 = 1$ in the upper slot of the L doublet and the one with $I_3 = -1$ in the lower slot.

[9]Note that the addition of mass terms to the Lagrangian in eqn 47.10 would violate the local symmetry we have set up. As shown in eqn 47.26, a mass term links the left- and right-handed parts of the electron field via a term $-m(\bar{e}_L e_R + \bar{e}_R e_L)$ but, since the $SU(2)$ transformation affects the left- and right-handed parts of the field differently, this term cannot be admitted.

the fields as

$$\Psi = \begin{pmatrix} L \\ R \end{pmatrix}, \quad \text{where} \quad L = \begin{pmatrix} \nu_e \\ e_L \end{pmatrix}, \quad R = e_R. \tag{47.7}$$

This prevents our having to keep writing Is and Ys in the Lagrangian.[8] In terms of these new fields, the covariant derivatives are written:

$$D_\mu L = \partial_\mu L - \frac{i}{2} g \boldsymbol{\tau} \cdot \boldsymbol{W}_\mu L + \frac{i}{2} g' B_\mu L, \tag{47.8}$$

$$D_\mu R = \partial_\mu R + i g' B_\mu R. \tag{47.9}$$

Our resulting, locally symmetric, Lagrangian is given by

$$\mathcal{L} = \bar{R} i \gamma^\mu \left(\partial_\mu + i g' B_\mu \right) R + \bar{L} i \gamma^\mu \left(\partial_\mu - \frac{i}{2} g \boldsymbol{\tau} \cdot \boldsymbol{W}_\mu + \frac{i}{2} g' B_\mu \right) L$$

$$- \frac{1}{4} \boldsymbol{G}_{\mu\nu}^{(W)} \cdot \boldsymbol{G}^{(W)\mu\nu} - \frac{1}{4} F_{\mu\nu}^{(B)} F^{(B)\mu\nu}, \tag{47.10}$$

where $\boldsymbol{G}_{\mu\nu}^{(W)} = \partial_\mu \boldsymbol{W}_\nu - \partial_\nu \boldsymbol{W}_\mu + g \boldsymbol{W}_\mu \times \boldsymbol{W}_\nu$ and $F_{\mu\nu}^{(B)} = \partial_\mu B_\nu - \partial_\nu B_\mu$, which are the correct contributions from the respective gauge fields for $SU(2)$ and $U(1)$ theories.[9]

Example 47.1

Although apparently plucked from the air, the assignment of charges to the field components may be physically motivated as follows. We want our theory of electroweak interactions to have both the weak interaction and electromagnetism correctly embedded within it. In terms of left- and right-handed components, we may write the weak isospin currents as

$$\boldsymbol{J}^\mu = \frac{1}{2} \bar{L} \gamma^\mu \boldsymbol{\tau} L, \tag{47.11}$$

and the electromagnetic current as

$$J_{\text{em}}^\mu = Q \left(\bar{e}_L \gamma^\mu e_L + \bar{e}_R \gamma^\mu e_R \right). \tag{47.12}$$

Note that, in terms of net charge transfer, J_{em}^μ and J_3^μ are *neutral currents*, whereas $J_{1,2}^\mu$ are *charged currents*. Since J_3^μ does not involve e_R, but J_{em}^μ does, we need to add an extra current to the problem. This should ensure that we have a consistent gauge theory that has electromagnetism correctly embedded within it. The simplest solution is to write

$$J_{\text{em}}^\mu = J_3^\mu + \frac{1}{2} J_Y^\mu, \tag{47.13}$$

where J_Y^μ is the weak hypercurrent and the factor $1/2$ is included by convention. Equation 47.13 then immediately implies the Gell-Mann-Nishijima relation $Q = I^3 + \frac{Y}{2}$. If we now expand out the expressions for J_3^μ and J_Y^μ we obtain

$$J_3^\mu = \frac{1}{2} \left(\bar{\nu}_e \gamma^\mu \nu_e - \bar{e}_L \gamma^\mu e_L \right)$$

$$J_Y^\mu = -\bar{\nu}_e \gamma^\mu \nu_e - \bar{e}_L \gamma^\mu e_L - 2 \bar{e}_R \gamma^\mu e_R, \tag{47.14}$$

where we have taken $Q = -1$ for the electron field in the final line. The values of Q, I_3 and Y may then be read off from the coefficients of the $\bar{\nu}_e \gamma^\mu \nu_e$, $\bar{e}_L \gamma^\mu e_L$ and $\bar{e}_R \gamma^\mu e_R$ terms in the expressions for J_{em}^μ, J_3^μ and J_Y^μ, in agreement with those listed in the tables above.

So far this seems merely an elaborate example of writing down a Lagrangian with some potentially interesting local symmetries. The amazing things start happening when we add a further field into the mix. It is this field which, from behind the scenes, will determine the properties of our electroweak Universe.

47.2   Introducing the Higgs field

We now introduce a massive, complex scalar field called the **Higgs field** into the Universe. It is the interaction of this field with e_L and e_R which will make electrons massive particles. The Higgs field has four components which are conveniently arranged into a two-component vector[10] as follows:

$$\phi = \begin{pmatrix} \phi^+ \\ \phi^0 \end{pmatrix} = \frac{1}{\sqrt{2}} \begin{pmatrix} \phi_3 + \mathrm{i}\phi_4 \\ \phi_1 + \mathrm{i}\phi_2 \end{pmatrix}. \tag{47.15}$$

This arrangement works since we have

$$\phi^\dagger \phi = (\phi^+)^* \phi^+ + (\phi^0)^* \phi^0 = \frac{1}{2}(\phi_1^2 + \phi_2^2 + \phi_3^2 + \phi_4^2), \tag{47.16}$$

giving something that looks like a magnitude.

   We need to know how the Higgs field transforms under $U(1)$ and $SU(2)$ transformations. The Higgs field is defined to have weak hypercharge $Y = +1$ and weak isospin $I = 1/2$. It therefore transforms according to

$$\begin{pmatrix} \phi^+ \\ \phi^0 \end{pmatrix} \rightarrow \begin{pmatrix} \mathrm{e}^{\mathrm{i}\frac{\beta}{2}} & 0 \\ 0 & \mathrm{e}^{\mathrm{i}\frac{\beta}{2}} \end{pmatrix} \begin{pmatrix} \phi^+ \\ \phi^0 \end{pmatrix}, \quad \begin{pmatrix} \phi^+ \\ \phi^0 \end{pmatrix} \rightarrow \mathrm{e}^{\frac{\mathrm{i}}{2}\boldsymbol{\tau}\cdot\boldsymbol{\alpha}} \begin{pmatrix} \phi^+ \\ \phi^0 \end{pmatrix}. \tag{47.17}$$

This choice of charges requires a covariant derivative for the Higgs field of the form

$$D_\mu \phi = \partial_\mu \phi - \frac{\mathrm{i}}{2} g \boldsymbol{\tau} \cdot \boldsymbol{W}_\mu \phi - \frac{\mathrm{i}}{2} g' B_\mu \phi. \tag{47.18}$$

The Higgs field will give a contribution $\mathcal{L}_\phi$ to the Lagrangian of

$$\mathcal{L}_\phi = (D^\mu \phi)^\dagger (D_\mu \phi) + \frac{m_\mathrm{h}^2}{2} \phi^\dagger \phi - \frac{\lambda}{4}(\phi^\dagger \phi)^2. \tag{47.19}$$

The positive mass term gives rise to the potential shown in Fig. 47.1, from which we see that we have set the Higgs field up for a fall: it's unstable to symmetry breaking. Recall that an ordinary scalar field, with negative mass term, has a minimum in potential at $\phi(x) = 0$ and so, classically at least, it takes the value zero in the vacuum. However, with its Mexican hat potential (Fig. 47.1), the ground state of the Higgs field is at $(\phi)_0 \neq 0$, implying that the ground state of a Universe containing the broken symmetry Higgs field, will be permeated by the uniform field $(\phi)_0$. As we shall see, it is the fact that electrons are moving through (and interacting with) this ether that gives them mass.

   Of course, all things related to the Higgs field would be irrelevant to our discussion of electrons and neutrinos if we didn't have an interaction between the Higgs field and the electron/neutrino field. This is given by

$$\mathcal{L}_\mathrm{I} = -G_\mathrm{e}(\bar{L}\phi R + \bar{R}\phi^\dagger L), \tag{47.20}$$

[10] Again the upper slot contains the fields with $I_3 = 1$, while the $I_3 = -1$ fields sit in the lower one.

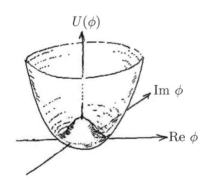

Fig. 47.1 The potential of the Higgs field.

where G_e is the coupling strength. Putting everything together, the Lagrangian for the Weinberg–Salam model is given by

$$
\begin{aligned}
\mathcal{L} \;=\;& \bar{L}i\gamma^\mu D_\mu L + \bar{R}i\gamma^\mu D_\mu R + (D^\mu\phi)^\dagger(D_\mu\phi) \\
&+\frac{m_h^2}{2}\phi^\dagger\phi - \frac{\lambda}{4}(\phi^\dagger\phi)^2 - G_e(\bar{L}\phi R + \bar{R}\phi^\dagger L) \\
&-\frac{1}{4}\boldsymbol{G}_{\mu\nu}^{(W)}\cdot\boldsymbol{G}^{(W)\mu\nu} - \frac{1}{4}F_{\mu\nu}^{(B)}F^{(B)\mu\nu},
\end{aligned}
\tag{47.21}
$$

where the covariant derivatives are those appropriate for each field, as discussed above. To be clear, the massless fields described by this Lagrangian are the left- and right-handed electron spinor fields, the left-handed neutrino spinor field, the massless $\boldsymbol{W}_\mu$ gauge field and the massless B_μ gauge field.

The Higgs field is unstable to symmetry breaking. What will happen? After $10^{-12}\,$s have elapsed the ground state of the Universe will break symmetry, with the Higgs field choosing a unique state from the infinity of possible ones in the gutter of the potential in Fig. 47.1. If the Weinberg–Salam theory describes reality then, as this happens, the electron should take on a mass, the neutrino should remain massless and the massless photon of electromagnetism should emerge. This will turn out to be how things fall and, as a bonus, the existence of some new massive particles, the $W^\pm$ and Z^0 bosons, will be predicted.

47.3 Symmetry breaking the Higgs field

It is now a time $t > 10^{-12}\,$s after the Big Bang, the Universe has cooled below $10^{16}\,$K and its ground state breaks the $SU(2)\otimes U(1)$ symmetry we have been discussing. Just as in Chapters 26 and 46 we're going to examine the consequences of this broken symmetry.[11] The minimum in potential of the Higgs field described in eqn 47.19 is not at $(\phi)_0 = 0$, but at $(\phi^\dagger\phi)_0 = v^2 = \left(\frac{m_h^2}{\lambda}\right)$. We'll choose to break the symmetry with a new ground state at

$$
(\phi_1)_0^2 = \tfrac{2m_h^2}{\lambda}, \quad (\phi_2)_0 = 0, \quad (\phi_3)_0 = 0, \quad (\phi_4)_0 = 0. \tag{47.22}
$$

Writing $(\phi_1)_0 = \left(\frac{2m_h^2}{\lambda}\right)^{\frac{1}{2}} = \sqrt{2}v$ we then have a ground state field

$$
(\phi)_0 = \begin{pmatrix} \phi^+ \\ \phi^0 \end{pmatrix}_0 = \begin{pmatrix} 0 \\ v \end{pmatrix}. \tag{47.23}
$$

An important point here is that when we break a non-abelian symmetry, our choice of vacuum may be invariant under a subset of the original symmetry transformations, meaning that those symmetries aren't broken at all. The vacuum chosen here has the property that it is still invariant with respect to the local transformation $\hat{U} = e^{i(\frac{Y}{2}+I_3\tau^3)\alpha(x)}$ which, using the Gell-Mann–Nishijima relation of eqn 47.2, we see is equivalent to the $U(1)$ transformation $e^{iQ\alpha(x)}$. Invariance with respect

[11]Since we have the non-abelian symmetry $SU(2)$ then this will bear a resemblance to the case examined in the last chapter.

to this transformation gives our broken symmetry Universe the gift of Maxwell's electromagnetism.

We could search for excitations from this ground state field in a very general way by allowing every component to vary.[12] However, just as in the last chapter, we are at liberty to perform a different gauge transformation at each point in spacetime, and thus reduce the excited field to the much simpler unitary gauge form

$$\phi = \begin{pmatrix} 0 \\ v + \frac{h(x)}{\sqrt{2}} \end{pmatrix}. \qquad (47.24)$$

We are now ready to examine the consequences of spontaneous symmetry breaking for this theory. We will demonstrate that it predicts[13] the existence of (i) massive electrons and massless neutrinos; (ii) massless photons; and (iii) massive $W^{\pm}$ and Z^0 particles.

47.4 The origin of electron mass

The physics of the Weinberg–Salam model is revealed by picking out the choice parts of the Lagrangian. Firstly we examine where we expect to find a fermion mass in a Lagrangian theory written in terms of left- and right-handed fields. If we start with the Dirac Lagrangian $\mathcal{L} = \bar{\psi}(\not{p}-m)\psi$ and write $\psi - \psi_{\mathrm{L}} + \psi_{\mathrm{R}}$, we have

$$\begin{aligned} \mathcal{L} = \ & \bar{\psi}_{\mathrm{L}}\not{p}\psi_{\mathrm{L}} + \bar{\psi}_{\mathrm{L}}\not{p}\psi_{\mathrm{R}} + \bar{\psi}_{\mathrm{R}}\not{p}\psi_{\mathrm{L}} + \bar{\psi}_{\mathrm{R}}\not{p}\psi_{\mathrm{R}} \\ & -\bar{\psi}_{\mathrm{L}}m\psi_{\mathrm{L}} - \bar{\psi}_{\mathrm{L}}m\psi_{\mathrm{R}} - \bar{\psi}_{\mathrm{R}}m\psi_{\mathrm{L}} - \bar{\psi}_{\mathrm{R}}m\psi_{\mathrm{R}}. \qquad (47.25) \end{aligned}$$

In fact a number of these terms cancel.[14] We are left with

$$\mathcal{L} = \bar{\psi}_{\mathrm{L}}\not{p}\psi_{\mathrm{L}} + \bar{\psi}_{\mathrm{R}}\not{p}\psi_{\mathrm{R}} - m(\bar{\psi}_{\mathrm{L}}\psi_{\mathrm{R}} + \bar{\psi}_{\mathrm{R}}\psi_{\mathrm{L}}). \qquad (47.26)$$

This provides us with a clue for where to search for the masses of the electrons in the Weinberg–Salam Lagrangian: we are looking for a term containing the combination $(\bar{\psi}_{\mathrm{L}}\psi_{\mathrm{R}} + \bar{\psi}_{\mathrm{R}}\psi_{\mathrm{L}})$, which we expect to be multiplied by a scalar quantity, which will be the electron mass for which we're searching. Just such a term is to be found in the interaction between the Higgs field and the lepton field, given by

$$\mathcal{L}_{\mathrm{I}} = -G_{\mathrm{e}}\left(\bar{L}\,\phi R + \bar{R}\,\phi^{\dagger}L\right). \qquad (47.27)$$

We will insert our broken symmetry excited state (eqn 47.24) into this equation and see what happens. The first term yields

$$\bar{L}\,\phi R = \bar{e}_{\mathrm{L}}\left(v + \frac{h(x)}{\sqrt{2}}\right)e_{\mathrm{R}}. \qquad (47.28)$$

Notice that the neutrino field has dropped out. This is good news since it implies that it will remain massless, just as we hope for a realistic theory. The second term gives

$$\bar{R}\,\phi^{\dagger}L = \bar{e}_{\mathrm{R}}\left(v + \frac{h(x)}{\sqrt{2}}\right)e_{\mathrm{L}}. \qquad (47.29)$$

[12] So, for example, we could write our excited state field as

$$\phi(x) = \begin{pmatrix} \frac{1}{\sqrt{2}}\,[\phi_3(x) + \mathrm{i}\phi_4(x)] \\ v + \frac{1}{\sqrt{2}}\,[\phi_1(x) + \mathrm{i}\phi_2(x)] \end{pmatrix}.$$

[13] Note that the masses of both the proton and the neutron (which contribute most to the masses of everyday objects) are dominated by the confinement energy of quarks inside them. Thus the popular notion of the Higgs field as some kind of unique giver of mass to everything in the Universe is rather wide of the mark.

[14] To show this we use the fact that the left- and right-handed parts are projections of the full field ψ, obtained via

$$\psi_{\mathrm{L}} = \left(\frac{1 - \gamma^5}{2}\right)\psi$$

and

$$\psi_{\mathrm{R}} = \left(\frac{1 + \gamma^5}{2}\right)\psi.$$

We commute the operators so that the projection parts sit together. We find that terms containing the momentum and mixed right- and left-handed fields always contain the combination

$$(1 - \gamma^5)(1 + \gamma^5) = 1 - (\gamma^5)^2$$

and since $(\gamma^5)^2 = 1$ we conclude that all mixed terms involving momentum cancel. Totally left-handed or right-handed terms involving the mass terms also contain this combination, so we lose those too.

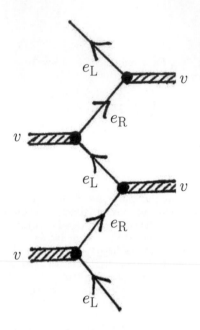

Fig. 47.2 The interaction of the left- and right-handed electron fields and the Higgs field from eqn 47.27.

The neutrino field doesn't feature here either and is destined to be massless. Putting the two halves together we obtain

$$\mathcal{L}_{\text{int}} = -G_{\text{e}}(\bar{e}_{\text{L}} v e_{\text{R}} + \bar{e}_{\text{R}} v e_{\text{L}}) - G_{\text{e}} \left(\bar{e}_{\text{L}} \frac{h(x)}{\sqrt{2}} e_{\text{R}} + \bar{e}_{\text{R}} \frac{h(x)}{\sqrt{2}} e_{\text{L}} \right). \quad (47.30)$$

The first term is in the predicted form of a scalar mass multiplied by $\bar{\psi}_{\text{L}} \psi_{\text{R}} + \bar{\psi}_{\text{R}} \psi_{\text{L}}$. It predicts the electron mass to be $m_{\text{e}} = G_{\text{e}} v$.

We conclude that, as claimed, the broken symmetry Higgs field, which takes the value $\sqrt{2} v$ throughout the vacuum, has provided a source of quantum mechanical treacle that causes the electrons to acquire mass. The nature of the interaction between the Higgs field and the left- and right-handed electrons is represented in Fig. 47.2, which depicts the interaction in eqn 47.27.

47.5 The photon and the gauge bosons

Next we need to show that the theory predicts a massless photon. We'll do this by looking at the place in the Lagrangian where the gauge fields interact with the Higgs field. Minimal coupling tells us that this is the term involving the covariant derivative of the Higgs field: $(D_\mu \phi)^\dagger (D^\mu \phi)$.

Example 47.2

Inserting the broken symmetry version of the Higgs field ϕ, the covariant derivative $D_\mu \phi$ becomes

$$\begin{pmatrix} 0 \\ \frac{1}{\sqrt{2}} \partial_\mu h(x) \end{pmatrix} - \left[\frac{ig}{2} \begin{pmatrix} W_\mu^3 & W_\mu^1 - iW_\mu^2 \\ W_\mu^1 + iW_\mu^2 & -W_\mu^3 \end{pmatrix} + \frac{ig'}{2} B_\mu \right] \begin{pmatrix} 0 \\ v + \frac{h(x)}{\sqrt{2}} \end{pmatrix}. \quad (47.31)$$

Multiplying out, we obtain

$$D_\mu \phi = -\frac{i}{2} \begin{pmatrix} gv(W_\mu^1 - iW_\mu^2) + \frac{gh(x)}{\sqrt{2}}(W_\mu^1 - iW_\mu^2) \\ i\sqrt{2} \partial_\mu h(x) + v(-gW_\mu^3 + g'B_\mu) + \frac{h(x)}{\sqrt{2}}(-gW_\mu^3 + g'B_\mu) \end{pmatrix}. \quad (47.32)$$

Finally we multiply this by its adjoint to form the term in the Lagrangian

$$\begin{aligned} (D_\mu \phi)^\dagger (D^\mu \phi) &= \frac{1}{2}[\partial_\mu h(x)]^2 + \frac{g^2 v^2}{4}(W_\mu^1)^2 + \frac{g^2 v^2}{4}(W_\mu^2)^2 \\ &\quad + \frac{v^2}{4}\left[(gW_\mu^3 - g'B_\mu)^2\right] + (\text{higher order terms}). \quad (47.33) \end{aligned}$$

Noting, as usual, that a boson mass always enters the Lagrangian in the form $\frac{1}{2}(\text{mass})^2 \times (\text{field})^2$, we see that in breaking the symmetry of the Higgs field we have developed massive W_μ^1 and W_μ^2 vector particles, with masses $M_{\text{W}}^2 = g^2 v^2 / 2$. The linear combination[15] of gauge fields $(gW_\mu^3 - g'B_\mu)$ has also grown massive, but this component has a mass that depends on the relative sizes of the coupling constants g and g'.

Notice that three Goldstone modes of the Higgs field have been consumed with the result that three gauge fields have grown massive. The excitations of these three fields will become the W^+, the W^- and the Z^0 vector bosons in the next step. We therefore have three massive vector particles. Where's the massless photon? To find it, we note the *absence* of the orthogonal linear combination $(g'W_\mu^3 + gB_\mu)$ and propose that this combination of fields is (proportional to) $A_\mu(x)$, the photon field of electromagnetism.

Since our particles depend, in a non-trivial manner, on the ratio of g and g', we may simplify things by drawing the right-angled triangle shown in Fig. 47.3, with the coupling constants g and g' as the lengths of the orthogonal sides and define the **Weinberg angle** $\tan\theta_\mathrm{W} = g'/g$. We then define two new fields Z_μ and A_μ by[16]

$$\begin{pmatrix} Z_\mu \\ A_\mu \end{pmatrix} = \begin{pmatrix} \cos\theta_\mathrm{W} & -\sin\theta_\mathrm{W} \\ \sin\theta_\mathrm{W} & \cos\theta_\mathrm{W}, \end{pmatrix} \begin{pmatrix} W_\mu^3 \\ B_\mu \end{pmatrix}, \qquad (47.35)$$

where, as suggested by the notation, A_μ is to be interpreted as our old friend, the electromagnetic field. In terms of the new fields, our original equation becomes

$$(D_\mu\phi)^\dagger(D^\mu\phi) = \frac{1}{2}[\partial_\mu h(x)]^2 + \frac{g^2v^2}{4}(W_\mu^1)^2 + \frac{g^2v^2}{4}(W_\mu^2)^2$$

$$+ \frac{g^2v^2}{4\cos^2\theta_\mathrm{W}}Z_\mu^2 + \text{(higher order terms)}, (47.36)$$

and from this equation we can read off[17] that the mass of the Z^0 particle is $M_\mathrm{W}/\cos\theta_\mathrm{W}$.

The fact that the W_μ and Z_μ fields have taken on a mass means that their propagators will have the form $\frac{G_{\mu\nu}}{p^2 - M^2 + \mathrm{i}\epsilon}$. Recall from our discussion of the Yukawa force in Chapter 17 that this implies a potential varying as $U(\boldsymbol{r}) \propto \mathrm{e}^{-M|\boldsymbol{r}|}$. We conclude that, compared to the electromagnetic interaction which is mediated by the massless photon, the weak interaction is *short-ranged*, falling off over a distance $1/M$.

We now have the ingredients for the non-interacting Lagrangian of the broken symmetry Universe (see margin on page 443). However, the only way to observe the predictions of the theory is via the interactions that the particles enter into with each other. Our final task is therefore to substitute our results back to see what the theory predicts for these interactions.

Example 47.3

As in the case of QED, minimal coupling allows us to see the interactions by considering the covariant derivatives. Starting with

$$\mathrm{i}\bar{R}\gamma^\mu(\partial_\mu + \mathrm{i}g'B_\mu)R + \mathrm{i}\bar{L}\gamma^\mu\left(\partial_\mu - \frac{\mathrm{i}}{2}g\boldsymbol{\tau}\cdot\boldsymbol{W}_\mu + \frac{\mathrm{i}}{2}g'B_\mu\right)L, \qquad (47.37)$$

and expanding out and using our expressions for A_μ, Z_μ and Weinberg's trigonometry, we may eliminate B_μ, W_μ^3 and g' and we obtain

$$\mathrm{i}\bar{e}\gamma^\mu\partial_\mu e + \mathrm{i}\bar{\nu}_e\gamma^\mu\partial_\mu\nu_e - g\sin\theta_\mathrm{W}\bar{e}\gamma^\mu eA_\mu$$

$$+ \frac{g}{\cos\theta_\mathrm{W}}\left(\sin^2\theta_\mathrm{W}\bar{e}_\mathrm{R}\gamma^\mu e_\mathrm{R} - \frac{1}{2}\cos 2\theta_\mathrm{W}\bar{e}_\mathrm{L}\gamma^\mu e_\mathrm{L} + \frac{1}{2}\bar{\nu}_e\gamma^\mu\nu_e\right)Z_\mu$$

$$+ \frac{g}{\sqrt{2}}\left[\bar{\nu}_e\gamma^\mu e_\mathrm{L}W_\mu^\dagger + \bar{e}_\mathrm{L}\gamma^\mu\nu_e W_\mu\right], \qquad (47.38)$$

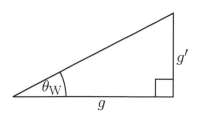

Fig. 47.3 Definition of the Weinberg angle θ_W in terms of g' and g.

[16]The fields may be written equivalently as

$$Z_\mu = \frac{(gW_\mu^3 - g'B_\mu)}{(g^2 + g'^2)^{\frac{1}{2}}},$$

$$A_\mu = \frac{(g'W_\mu^3 + gB_\mu)}{(g^2 + g'^2)^{\frac{1}{2}}}. \quad (47.34)$$

[17]The mass of the W boson is given by $M_\mathrm{W}^2 = g^2v^2/2$. The term premultiplying Z_μ^2 in eqn 47.36 is $\frac{g^2v^2}{4\cos^2\theta_\mathrm{W}} = \frac{1}{2}M_\mathrm{Z}^2$ and hence $M_\mathrm{Z} = M_\mathrm{W}/\cos\theta_\mathrm{W}$.

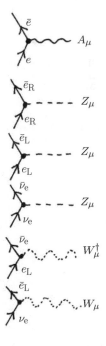

Fig. 47.4 The electroweak interaction vertices predicted by the Weinberg–Salam theory.

where $e = e_L + e_R$, likewise for ν and we define the field $W_\mu = (W_\mu^1 + iW_\mu^2)/\sqrt{2}$. We expect the quanta of the W_μ field to be two species of oppositely charged, massive vector particles: the W^+ (the quanta created by the $\hat{W}_\mu(x)$-field, with $I_3 = 1$) and the W^- (annihilated by $\hat{W}_\mu(x)$, with $I_3 = -1$). The Gell-Mann–Nishijima formula (eqn 47.2) tells us that W^+ has $Q = 1$ units of electric charge, while the W^- has $Q = -1$ units. In contrast the $Z_\mu(x)$ field is uncharged and we expect a single species of massive quantum Z^0.

[18]For a more complete description of the 18 interactions described by the theory see Mandl and Shaw, Chapter 19.

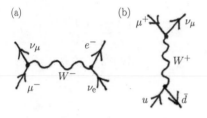

Fig. 47.5 Some example electroweak processes. (a) $\mu^- + \nu_e \rightarrow \nu_\mu + e^-$; (b) positive pion decay involving quarks u and $\bar{d}$ decaying into $\mu^+ + \nu_\mu$.

[19]Historically, a value for the Weinberg angle had been estimated from earlier measurements of lepton scattering cross-sections and provided a prediction for the masses of the $W^\pm$ and Z^0 particles which were subsequently discovered at the predicted energies. See Gary Taubes' highly recommended *Nobel Dreams* for the frightening history.

[20]Often incorrectly pronounced 'Boatswain' by the UK media, possibly as some kind of nautical throwback.

In eqn 47.38 we have an expression for some of the interactions[18] in the electroweak theory in terms of the coupling g and the Weinberg angle θ_W. Some important points to note from our result is that A_μ only couples to electrons and not to neutrinos, just as experiment shows the photon does. The vector field W_μ only couples left-handed electrons to neutrinos and doesn't touch right-handed ones. The vector field Z_μ couples to left- and right-handed electrons with different strengths. Notice also that that parity operation $\mathsf{P}^{-1}\mathcal{H}_I\mathsf{P}$, which exchanges left- and right-handed fields, is not a symmetry of those parts of the interaction involving the $W_\mu(x)$ and $Z_\mu(x)$ fields. This is the sense in which, famously, the weak interaction is said to 'violate parity'.

The vertices for these interactions are shown in Fig. 47.4. The most startling of these involve the W_μ field and show that the emission or absorption of a $W^\pm$ particle can result in the transformation of an electron into a neutrino (and vice versa). Some examples of particle processes allowed by the weak interaction are shown in Fig. 47.5.

Note what the theory predicts for the electric charge. In QED the charge is the coupling constant in front of the fermion–photon interaction term: $-q\bar{\psi}\gamma^\mu\psi A_\mu$. Taking $q = |e|$ for a single electronic charge we read off from eqn 47.38 that the electric charge is given by

$$|e| = g\sin\theta_W. \tag{47.39}$$

The Weinberg–Salam electroweak theory and its verification are two of the greatest achievements of twentieth century science. The Z^0 and $W^\pm$ particles were detected with masses $M_Z = 91.19$ GeV and $M_W = 80.40$ GeV and correspond to a Weinberg angle (via $\cos\theta_W = M_W/M_Z$) of $\theta_W \sim 28°$.[19] The final prediction of the theory is the existence of a scalar field excitation in the Higgs field, known as the Higgs boson.[20] At the time of writing, experiments at the Large Hadron Collider have found a particle, thought to be the Higgs, with a mass ≈ 125 GeV.

Thus in summary our current picture is of a symmetry-broken Higgs field that permeates the Universe. Other fields are set to zero in the vacuum and fluctuate around that, giving rise to particle–antiparticle excitations that wink in and out of existence; the Higgs field is nonzero by default. With no Higgs field, the electron and electron neutrino would be identical particles, and the W and Z particles, and in fact all standard model fermions, would be massless. That indeed *was* the situation in the very early Universe before the Higgs field became symmetry broken. But immediately after the electroweak phase transition (that occurred perhaps 10^{-12} s after the Big Bang) the present broken-symmetry state emerged and the masses of these particles became simply a function of how strongly they couple to the Higgs field. And couple they will, because the Higgs field is both nonzero and *everywhere*.

Chapter summary

- In the period $t \lesssim 10^{-12}$s after the Big Bang the leptons of the early Universe enjoy an internal $SU(2) \otimes U(1)$ local symmetry.
- The Weinberg–Salam electroweak theory describes the breaking of the symmetry of the Higgs field with the result that a nonzero Higgs field permeates the vacuum, providing electrons with mass.
- This also results in the emergence of the photon and three-vector particles W^+, W^- and Z^0. The excitations of the electromagnetic field A_μ do not acquire a mass, whereas the $W^\pm$ and Z^0 particles become massive.

Non-interacting electroweak Lagrangian

$$
\begin{aligned}
\mathcal{L}_0 \;=\;\; & \bar{e}(i\slashed{\partial} - m_e)e + \bar{\nu}_e(i\slashed{\partial})\nu_e \\
& -\frac{1}{4}F_{\mu\nu}F^{\mu\nu} \\
& -\frac{1}{2}W^\dagger_{\mu\nu}W^{\mu\nu} + M_W^2 W^\dagger_\mu W^\mu \\
& -\frac{1}{4}Z_{\mu\nu}Z^{\mu\nu} + \frac{1}{2}M_Z^2 Z_\mu Z^\mu \\
& +\frac{1}{2}(\partial_\mu h)^2 - \frac{1}{2}m_h^2 h^2
\end{aligned}
$$

where $W_{\mu\nu} = \partial_\mu W_\nu - \partial_\nu W_\mu$ and similarly for $Z_{\mu\nu}$.

Exercises

(47.1) Show that the broken symmetry ground state vacuum expectation value $(\phi)_0 = \begin{pmatrix} 0 \\ v \end{pmatrix}$ is invariant with respect to the transformation $\hat{U} = e^{i(\frac{Y}{2}+I_3\tau^3)\alpha(x)}$.

(47.2) Verify eqn 47.26.

(47.3) By considering the covariant derivative of the Higgs field, verify eqn 47.33.

(47.4) Verify eqn 47.38.

(47.5) *Fermi's theory of the weak interaction.*
(a) Read off the interaction vertices for weak interactions between electrons and neutrinos from eqn 47.38.
(b) Draw a Feynman diagram for negative muon decay $\mu^- \to e^- + \bar{\nu}_e + \nu_\mu$ and show that, at low momentum, the process is described by an interaction Hamiltonian

$$
\hat{H}_I = \frac{g^2}{2}\bar{\nu}_\mu\gamma^\alpha \mu_L \frac{g_{\alpha\beta}}{M_W^2}\bar{e}_L\gamma^\beta\nu_e. \tag{47.40}
$$

(c) Defining the current for the lepton species l as $J_l^\alpha = \bar{l}\gamma^\alpha(1-\gamma_5)\nu_l$, show that the interaction for the muon decay can be recast as

$$
\hat{H}_I = \frac{G}{\sqrt{2}}J_\mu^{\alpha\dagger}J_{e\alpha}, \tag{47.41}
$$

where $G = \frac{g^2}{4\sqrt{2}M_W^2}$. This is known as the Fermi current–current interaction.
(d) For muon decay $G \approx 10^{-5}m_p^{-2}$, where m_p is the proton mass. Comment on the 'weakness' of the weak interaction.

48

Majorana fermions

Tragically I was an only twin
Peter Cook (1937–1995), title of an unrealized project.

Can a fermion particle be its own antiparticle? That is to say, if we take the charge conjugate of the creation operator of a Fermi particle $c_{\boldsymbol{p}}^{\dagger}$, can it return the same operator $\mathsf{C}^{-1}\hat{c}_{\boldsymbol{p}}^{\dagger}\mathsf{C} = \hat{c}_{\boldsymbol{p}}^{\dagger}$ without creating an inconsistency in the theory? We know that particle excitations of the real scalar field are identical to antiparticle excitations of the scalar field. We also know that the photon excitations in the electromagnetic field are identical to antiphotons. Scalar particles (spin-0), photons (spin-1) and gravitons (spin-2) can be described by real fields: $\hat{\phi} = \hat{\phi}^{\dagger}$. Since $\hat{\phi}^{\dagger}$ creates a particle and $\hat{\phi}$ an antiparticle, we expect these particles to be identical to their antiparticles. The Dirac theory of the electron (Chapter 36) seems to require *complex* fields (though see below), and so it might seem at first glance as if fermion fields necessarily involve distinct particle and antiparticle parts. This would explain why electrons and positrons (anti-electrons) are distinct particles, and also why even the electrically-neutral neutron is not identical to its antiparticle, the antineutron. However, this is not the whole story, and in this chapter we examine the steps required to formulate a theory containing fermions whose particles and antiparticles are identical or, if you prefer, symmetric.

Fig. 48.1 A Majorana fermion is its own antiparticle.

Etorre Majorana (1906–[1938+x]) had a short, but brilliant, career in theoretical physics which was cut short when he disappeared whilst travelling by passenger ship from Palermo to Naples. It has been variously suggested that he committed suicide, was murdered, or escaped to Argentina.

48.1   The Majorana solution

Etorre Majorana came up with a theory in which a fermion particle was its own antiparticle (see Fig. 48.1). Such particles are called **Majorana fermions** and are unchanged by the act of charge conjugation.

Example 48.1

For charged particles we know that the charge conjugation operator swaps particles for antiparticles, so for a complex scalar field we have, for example: $\mathsf{C}^{-1}\hat{a}_{\boldsymbol{p}}^{\dagger}\mathsf{C} = \hat{b}_{\boldsymbol{p}}^{\dagger}$.

Let's start by taking a naive view and examining a mode expansion of a classical scalar field (that is, the expansion of a function; not an operator):

$$\phi(x) = \int \frac{\mathrm{d}^3 p}{(2\pi)^{\frac{3}{2}}} \frac{1}{(2E_{\boldsymbol{p}})^{\frac{1}{2}}} \left(a_{\boldsymbol{p}} \mathrm{e}^{-\mathrm{i}p\cdot x} + a_{\boldsymbol{p}}^* \mathrm{e}^{\mathrm{i}p\cdot x} \right). \tag{48.1}$$

Notice that the second term is the complex conjugate of the first and that $\phi(x)$ is real. Compare this with the case of a (classical) complex scalar field:

$$\psi(x) = \int \frac{\mathrm{d}^3p}{(2\pi)^{\frac{3}{2}}} \frac{1}{(2E_{\boldsymbol{p}})^{\frac{1}{2}}} \left(a_{\boldsymbol{p}}\mathrm{e}^{-\mathrm{i}p\cdot x} + b_{\boldsymbol{p}}^*\mathrm{e}^{\mathrm{i}p\cdot x} \right). \tag{48.2}$$

Here the complex conjugate of the first term yields something which is not identical to the second term, i.e. the particle is different from the antiparticle.

Speaking very roughly, charge conjugation involves swapping each field for its complex conjugate and so a real field like $\phi(x)$ will describe particles that are their own antiparticles. This will not be the case for the complex field $\psi(x)$.

Motivated by the previous example, we will search for a real (rather than complex) solution to the Dirac equation. Such a solution follows if the Dirac equation itself is real. Whether the Dirac equation $(\mathrm{i}\not\partial - m)\Psi = 0$ is real or complex depends on the choice of the γ matrices. Recall that there are many possible γs out there, constrained only by the requirement that $\{\gamma^\mu, \gamma^\nu\} = 2g^{\mu\nu}$. If we can find a set of γs which are purely imaginary, then the Dirac equation itself will be real and consequently so will its solutions.

Majorana found such a set of purely imaginary γ matrices, given in eqn 48.3 in the margin. Using the Dirac equation given by these matrices we will find solutions $\tilde{\nu}(x)$ which will have the property that $\tilde{\nu}(x) = \tilde{\nu}^*(x)$, which is known as the **Majorana condition** and reflects the fact that the solutions are identical to their complex, or more precisely, to their charge conjugates.

Having demonstrated the possibility of Majorana fermions existing as viable quantum mechanical entities, we would now like to make contact with our previous approach and describe Majorana's solution in the chiral representation, which involves stacking up pairs of two-component Weyl spinors to make four-component Dirac, or in this case Majorana, spinors. We therefore need to consider how charge conjugation works for Weyl spinors. (Note that we're treating our spinors as c-number wave functions for the moment; fields will follow shortly.)[1]

Majorana's γ matrices:

$$\tilde{\gamma}^0 = \begin{pmatrix} 0 & \sigma^2 \\ \sigma^2 & 0 \end{pmatrix}$$
$$\tilde{\gamma}^1 = \begin{pmatrix} \mathrm{i}\sigma^1 & 0 \\ 0 & \mathrm{i}\sigma^1 \end{pmatrix}$$
$$\tilde{\gamma}^2 = \begin{pmatrix} 0 & \sigma^2 \\ -\sigma_2 & 0 \end{pmatrix}$$
$$\tilde{\gamma}^3 = \begin{pmatrix} \mathrm{i}\sigma^3 & 0 \\ 0 & \mathrm{i}\sigma^3 \end{pmatrix}. \tag{48.3}$$

[1]Note also that, throughout the rest of this chapter, we reserve the symbol ψ for two-component Weyl spinors only, which may be left-handed (ψ_L) or right-handed (ψ_R), while the symbols Ψ and ν denote four-component Dirac and Majorana spinors respectively.

Example 48.2

There is a recipe for obtaining the charge conjugate $\Psi_\mathrm{C}(x)$ of a Dirac spinor $\Psi(x)$ in chiral representation. It is

$$\Psi_\mathrm{C}(x) = \mathsf{C}^{-1}\Psi(x)\mathsf{C} = C_0\Psi^*(x), \tag{48.4}$$

with $C_0 = -\mathrm{i}\gamma^2$. As an example, we could start with a Dirac spinor $\Psi = \begin{pmatrix} \psi_\mathrm{L} \\ 0 \end{pmatrix}$ containing only a left-handed Weyl spinor ψ_L and take the charge conjugate as follows:

$$\Psi_\mathrm{C} = -\mathrm{i}\gamma^2 \begin{pmatrix} \psi_\mathrm{L} \\ 0 \end{pmatrix}^* = \begin{pmatrix} 0 & -\mathrm{i}\sigma^2 \\ \mathrm{i}\sigma^2 & 0 \end{pmatrix} \begin{pmatrix} \psi_\mathrm{L}^* \\ 0 \end{pmatrix} = \begin{pmatrix} 0 \\ \mathrm{i}\sigma^2\psi_\mathrm{L}^* \end{pmatrix}. \tag{48.5}$$

This gives a component in the lower slot involving the original left-handed Weyl spinor.[2] You may verify that conjugating the charge twice returns the original spinor.

[2]Note that charge conjugation of the right-handed spinor gives $\begin{pmatrix} 0 \\ \psi_\mathrm{R} \end{pmatrix} \rightarrow \begin{pmatrix} -\mathrm{i}\sigma^2\psi_\mathrm{R}^* \\ 0 \end{pmatrix}$.

Here's the solution: in the chiral basis a Majorana spinor may be built out of a left-handed Weyl spinor and its charge conjugate as follows:

$$\nu(x) = \begin{pmatrix} \psi_{\rm L}(x) \\ 0 \end{pmatrix} + \begin{pmatrix} 0 \\ i\sigma^2 \psi_{\rm L}^*(x) \end{pmatrix} = \begin{pmatrix} \psi_{\rm L}(x) \\ i\sigma^2 \psi_{\rm L}^*(x) \end{pmatrix}, \quad (48.6)$$

that is $\nu(x) = \begin{pmatrix} \psi_{\rm L}(x) \\ \psi_{{\rm L},{\rm C}}(x) \end{pmatrix}$, where we write the charge conjugate of a Weyl spinor as $\psi_{{\rm L},{\rm C}} = i\sigma^2 \psi_{\rm L}^*$. This solution $\nu(x)$ obeys the all-important property that $\nu_{\rm C}(x) = \nu(x)$, which is a more general expression of the Majorana condition above. Notice that a Majorana particle may be built starting with only a left-handed Weyl spinor, putting its conjugate part into the slot where the right-handed part usually lives.[3] This is different to the ordinary Dirac fermion we met in Chapter 36, which has independent left- and right-handed parts.

48.2 Field operators

So far we've been working with wave functions. To make the theory respectable we should write things in terms of field operators. Like the wave functions, these fields must obey the Majorana condition, which expressed in terms of charge conjugation operators is $\hat{\nu}(x) = \hat{\nu}_{\rm C}(x) = C_0 \hat{\nu}^*(x)$. A slightly jarring abuse of notation here is the complex conjugate of the field operator, since the complex conjugate of a creation or annihilation operator is not, strictly speaking, defined. In taking the complex conjugate we are seeking to avoid the matrix transpose that goes along with Hermitian conjugation. Therefore a fussier, but more correct, way of writing the conjugation operation would be $\hat{\nu}^{\dagger {\rm T}}$. However, we'll stick with the complex conjugate notation where one should note that $u^{s*}(p) = \begin{pmatrix} u_{\rm L}^{s*}(p) \\ u_{\rm R}^{s*}(p) \end{pmatrix}$ and we freely abuse the notation with the understanding that here $\hat{a}_{\boldsymbol{p}}^* \equiv \hat{a}_{\boldsymbol{p}}^\dagger$.

[3]We could also have started with a right-handed spinor and then built a Majorana fermion $\nu' = \begin{pmatrix} -i\sigma^2 \psi_{\rm R}^* \\ \psi_{\rm R} \end{pmatrix}$. The point is that we only need to start with either a left-handed spinor or right-handed spinor.

Example 48.3

For a Dirac particle, recall that the field annihilation operator is given by

$$\hat{\Psi}(x) = \int \frac{{\rm d}^3 p}{(2\pi)^{\frac{3}{2}}} \frac{1}{(2E_{\boldsymbol{p}})^{\frac{1}{2}}} \sum_s \left(u^s(p) \hat{a}_{s\boldsymbol{p}} e^{-ip\cdot x} + v^s(p) \hat{b}_{s\boldsymbol{p}}^\dagger e^{ip\cdot x} \right), \quad (48.7)$$

whose charge conjugate is $C^{-1} \hat{\Psi}(x) C = C_0 \hat{\Psi}^*(x)$

$$= \int \frac{{\rm d}^3 p}{(2\pi)^{\frac{3}{2}}} \frac{1}{(2E_{\boldsymbol{p}})^{\frac{1}{2}}} \sum_s \left(-i\gamma^2 u^{s*}(p) \hat{a}_{s\boldsymbol{p}}^\dagger e^{ip\cdot x} - i\gamma^2 v^{s*}(p) \hat{b}_{s\boldsymbol{p}} e^{-ip\cdot x} \right). \quad (48.8)$$

Moreover, using the explicit forms of $u^s(p)$ and $v^s(p)$ it may be shown[4] that $-i\gamma^2 u^{s*}(p) = v^s(p)$ and $-i\gamma^2 v^{s*}(p) = u^s(p)$ leading to a charge conjugate field

$$C^{-1} \hat{\Psi}(x) C = \int \frac{{\rm d}^3 p}{(2\pi)^{\frac{3}{2}}} \frac{1}{(2E_{\boldsymbol{p}})^{\frac{1}{2}}} \sum_s \left(v^s(p) \hat{a}_{s\boldsymbol{p}}^\dagger e^{ip\cdot x} + u^s(p) \hat{b}_{s\boldsymbol{p}} e^{-ip\cdot x} \right). \quad (48.9)$$

For what follows, note that this can be made the same as the original Dirac field if we were to make the replacements $\hat{a}_{s\boldsymbol{p}}^\dagger \to \hat{b}_{s\boldsymbol{p}}^\dagger$ and $\hat{b}_{s\boldsymbol{p}} \to \hat{a}_{s\boldsymbol{p}}$, from which we conclude that the prescription indeed enacts charge conjugation on the Dirac field in that it returns a Dirac field with particle operators exchanged for antiparticle operators.

[4]Consider, for example, the charge conjugation of the positive helicity antiparticle spinor from Chapter 36: $-i\gamma^2 v^{s*}(p) =$

$$\begin{pmatrix} 0 & -i\sigma^2 \\ i\sigma^2 & 0 \end{pmatrix} \begin{pmatrix} 0 \\ \sqrt{E+|\boldsymbol{p}|} \\ 0 \\ -\sqrt{E-|\boldsymbol{p}|} \end{pmatrix}$$

$$= \begin{pmatrix} \sqrt{E-|\boldsymbol{p}|} \\ 0 \\ \sqrt{E+|\boldsymbol{p}|} \\ 0 \end{pmatrix} = u^s(p),$$

i.e. a positive helicity particle spinor, as claimed.

Finally we will write down a Majorana field. We start by noting that the general form of a quantum field is

$$
\begin{aligned}
\text{(Quantum Field)} \;&=\; \text{(Particle part)} + \text{(Antiparticle part)} \\
&=\; \text{(Particle part)} + \mathsf{C}^{-1}\text{(Particle part)}\mathsf{C}.
\end{aligned}
$$

If particles and antiparticles are identical (which is what we're after) then $\mathsf{C}^{-1}\hat{a}^{\dagger}_{sp}\mathsf{C} = \hat{a}^{\dagger}_{sp}$ and so $\mathsf{C}^{-1}u^s(p)\hat{a}^{\dagger}_{sp}\mathsf{C} = -i\gamma^2 u^{s*}(p)\hat{a}^{\dagger}_{sp}$. To construct the Majorana field we therefore replace the antiparticle part of the Dirac field $v^s(p)\hat{b}^{\dagger}_{sp}$ with $-i\gamma^2 u^{s*}(p)\hat{a}^{\dagger}_{sp}$. The result is the Majorana field

$$
\hat{\nu}(x) = \int \frac{\mathrm{d}^3 p}{(2\pi)^{\frac{3}{2}}} \frac{1}{(2E_{\boldsymbol{p}})^{\frac{1}{2}}} \sum_s \left(u^s(p)\hat{a}_{s\boldsymbol{p}}\mathrm{e}^{-ip\cdot x} - i\gamma^2 u^{s*}(p)\hat{a}^{\dagger}_{s\boldsymbol{p}}\mathrm{e}^{ip\cdot x} \right).
$$

(48.10)

As may be checked, the Majorana field does indeed enjoy the Majorana condition:

$$
\begin{aligned}
\mathsf{C}^{-1}\hat{\nu}(x)\mathsf{C} &= \int \frac{\mathrm{d}^3 p}{(2\pi)^{\frac{3}{2}}} \frac{1}{(2E_{\boldsymbol{p}})^{\frac{1}{2}}} \sum_s \left(-i\gamma^2 u^{s*}(p)\hat{a}^{\dagger}_{s\boldsymbol{p}}\mathrm{e}^{ip\cdot x} + u^s(p)\hat{a}_{s\boldsymbol{p}}\mathrm{e}^{-ip\cdot x} \right) \\
&= \hat{\nu}(x).
\end{aligned}
$$

(48.11)

Example 48.4

If we start with a left-handed (momentum space) Weyl spinor $u^s_{\mathrm{L}}(p)$ the Majorana field takes the form

$$
\hat{\nu}(x) = \int \frac{\mathrm{d}^3 p}{(2\pi)^{\frac{3}{2}}} \frac{1}{(2E_{\boldsymbol{p}})^{\frac{1}{2}}} \sum_s \left[\begin{pmatrix} u^s_{\mathrm{L}}(p) \\ 0 \end{pmatrix} \hat{a}_{s\boldsymbol{p}}\mathrm{e}^{-ip\cdot x} + \begin{pmatrix} 0 \\ i\sigma^2 u^{s*}_{\mathrm{L}}(p) \end{pmatrix} \hat{a}^{\dagger}_{s\boldsymbol{p}}\mathrm{e}^{ip\cdot x} \right].
$$

(48.12)

As in the case of wave functions, we see that Majorana fields may be built out of Weyl spinors of only one handedness.

Although the argument of this chapter has been very formal, the point is a simple one: it is indeed quite possible to build a Majorana field whose excitations are Majorana fermions. We will now discuss the physics of these particles.

48.3 Majorana mass and charge

First we consider the mass of the particle excitations of the Majorana field. Recall that Weyl spinors are necessarily massless. Massive Dirac spinors must contain independent left- and right-handed parts and the particles may be thought of as oscillating between the two. The mass term in the Dirac Lagrangian may be written in terms of Weyl spinors as[5]

$$
m_{\mathrm{D}}(\bar{\Psi}_{\mathrm{L}}\Psi_{\mathrm{R}} + \bar{\Psi}_{\mathrm{R}}\Psi_{\mathrm{L}}),
$$

(48.13)

[5] See the previous chapter for an explanation of this.

[6]Notice that we don't have terms $\bar{\Psi}_L\Psi_L$ and $\bar{\Psi}_R\Psi_R$ which are not Lorentz invariant and therefore are forbidden.

[7]Note that the N objects in this equation are built from four-component Majorana spinors ν, so themselves have eight components.

where we write $\Psi_L = \begin{pmatrix} \psi_L \\ 0 \end{pmatrix}$ and $\Psi_R = \begin{pmatrix} 0 \\ \psi_R \end{pmatrix}$. Vitally, eqn 48.13 is a Lorentz invariant quantity.[6]

As Majorana solutions may be written in terms of Weyl spinors of a single chirality we might question whether it is possible to identify a massive Majorana field. This worry looks to be valid since, as shown in the exercises, a mass term in the Lagrangian of the form $\bar{\nu}(x)\nu(x)$ vanishes if we treat the fields in the Lagrangian as c-numbers, as we do for the Dirac case. However, it turns out that we may define massive Majorana fields, as long as they anticommute. In that case we may have massive Majorana fields built out of left-handed Weyl spinors only, with mass m_L, or Majorana fields built from right-handed Weyl fields only, with mass m_R. Turning to the Lagrangian, we may write the most general, Lorentz invariant mass term for Majorana fields as[7]

$$-\frac{1}{2}\left(\bar{N}_{L,C}MN_L + \bar{N}_L M N_{L,C}\right),\qquad(48.14)$$

where

$$N_L = \begin{pmatrix} \nu_L \\ \nu_{R,C} \end{pmatrix}, \quad N_{L,C} = \begin{pmatrix} \nu_{L,C} \\ \nu_R \end{pmatrix}, \quad M = \begin{pmatrix} m_L & m_D \\ m_D & m_R \end{pmatrix},\qquad(48.15)$$

and where $\nu_L = \begin{pmatrix} \psi_L \\ 0 \end{pmatrix}$, $\nu_{L,C} = \begin{pmatrix} 0 \\ i\sigma^2\psi_L^* \end{pmatrix}$, $\nu_R = \begin{pmatrix} 0 \\ \psi_R \end{pmatrix}$ and $\nu_{R,C} = \begin{pmatrix} -i\sigma^2\psi_R^* \\ 0 \end{pmatrix}$. Here we have three types of mass: the Dirac mass m_D along with the right- and left-handed Majorana masses m_R and m_L.

Example 48.5

Working with Majorana fields $\nu(x)$ built from left-handed Weyl spinors, we set $m_D = m_R = 0$ and $m_L \neq 0$. We may then write a Majorana Lagrangian as

$$\mathcal{L} = \frac{1}{2}\bar{\nu}i\slashed{\partial}\nu - \frac{1}{2}(\bar{\nu}_{L,C}m_L\nu_L + \bar{\nu}_L m_L\nu_{L,C}),\qquad(48.16)$$

where $\nu = \nu_L + \nu_{L,C}$. Varying this Lagrangian with respect to $\psi_L^\dagger$, leads to the **Majorana equation** for the field ψ_L, given by

$$i\bar{\sigma}^\mu\partial_\mu\psi_L - m_L i\sigma^2\psi_L^* = 0,\qquad(48.17)$$

which may be recast in the form of a Dirac equation

$$(i\slashed{\partial} - m_L)\nu(x) = 0.\qquad(48.18)$$

To complete this cavalcade of formalism, we briefly turn to the electromagnetic charge of the Majorana field: it has none. This is, of course, necessary for the particle excitations of this field to be identical to the antiparticles. It can also be seen from the field equations by noting that if $\nu = \nu_L + \nu_{L,C}$, then if we make the transformation $\nu_L \to \nu_L e^{i\alpha}$, we must have $\nu_{L,C} \to \nu_{L,C}e^{-i\alpha}$ (because ν_L and $\nu_{L,C}$ are related by complex conjugation). It is therefore impossible to define a $U(1)$ transformation for

Majorana spinors which simultaneously provides both upper and lower slots with the same phase factor $e^{i\alpha}$. The Majorana equation cannot, therefore, be made invariant under local $U(1)$ transformations. Put the other way: a particle carrying the conserved $U(1)$ charge of electromagnetism cannot be a Majorana particle.

Having formulated this theory we now ask whether it's of any use. Indeed, for many years it seemed to be an interesting solution in need of a problem. However, in recent years, several such problems have arisen.

- The vast number of emergent quasiparticles in condensed matter physics present us with an ideal playground for searching for exotic excitations such as Majorana fermions. In a semiconductor or a metal, electrons and holes look different because they are oppositely charged, and so it does not seem possible that the particles (electrons) and antiparticles (holes) can be symmetrically related. A superconductor, on the other hand, blurs the distinction between electrons and holes and so seems like a possible environment for realizing Majorana fermions. One of the great things about superconductors is that they screen electric and magnetic fields, and so charge is not a good observable. The bogolon excitations of a superconductor superpose electrons and holes and, under rather special circumstances, it is possible that the quasiparticles of a superconducting system are Majorana fermions. An example of such a circumstance involves a superconductor in the presence of vortices, which changes the equations of motion of the electrons and can lead to the trapping of electron–hole pairs which can be described as Majorana fermions.

- At first sight neutrinos seem to be well described as solutions to Weyl's equation. All neutrinos are left-handed massless particles with negative helicity whereas all antineutrinos are left-handed massless particles with positive helicity. However, the discovery that neutrinos emitted from the Sun with one flavour may be detected with a different flavour suggests that these particles actually possess a nonzero, but small, mass. Whether this is a Dirac mass is an open question and there exists the possibility that neutrinos are actually Majorana particles with Majorana mass.

- Consider the possibility that for every species of boson in the Universe there exists a corresponding species of fermion (and vice versa) with the same mass. This would require the Lagrangian describing our Universe to exhibit a symmetry known as **supersymmetry**. In a supersymmetric Universe we should expect the existence of the 'selectron': a spin-0 particle with the mass of an electron; and the 'photino': a spin-1/2 massless particle. If the photino mirrors the properties of the photon then it must be its own antiparticle. This implies that the photino is a Majorana fermion, as will be the 'Higgsino' and various types of 'gaugino'. (Supersymmetry is a subject replete with wonderful terms, and happily the supersymmetric partner of the W boson is called a 'wino'.) Despite much effort, any direct evidence for supersymmetric particles has so far proved to be elusive.

Chapter summary

- A Majorana fermion is its own antiparticle and is a solution to a version of the Dirac equation in which the γ-matrices are purely imaginary.
- We have shown how to build Majorana fields from Weyl spinors of a single handedness and we have written down a Lagrangian that leads to their equation of motion.

Exercises

(48.1) Show that two operations of the charge conjugation operator used in eqn 48.5 return the original particle.

(48.2) Show that $-i\gamma^2 u^{s*}(p) = v^s(p)$ for a negative helicity particle spinor.

(48.3) (a) Show that, for a Majorana field ν built from left-handed Weyl spinors:
$$\bar{\nu}\nu = -i\psi_{\mathrm{L}}^{\mathrm{T}}\sigma^2\psi_{\mathrm{L}} + \text{h.c.}, \qquad (48.19)$$

where h.c. denotes the Hermitian conjugate. Explain why this must vanish if the fields are represented by c-numbers.

(b) Show that $\bar{\nu}\nu$ won't vanish if the fields are represented by anticommuting Grassmann numbers.

(48.4) Verify eqns 48.16 and 48.17.

(48.5) Verify that eqn 48.17 may be recast in the form of a Dirac equation as claimed.

Magnetic monopoles

49

One would be surprised if nature had made no use of it.
Paul Dirac (1902–1984), on magnetic monopoles.

Do **magnetic monopoles** exist? We are taught at our Mother's knee that Maxwell's equation $\nabla \cdot B = 0$ insists that they do not. However it is more the case that they have never been observed rather than that their existence is a physical impossibility. In fact there seems to be nothing preventing the existence of the magnetic monopole and in this chapter we will investigate the properties of these objects. We will present descriptions of two rather different sorts of monopole: those of Paul Dirac and those of Gerard 't Hooft and Alexander Polyakov. The existence of Dirac's monopoles are not *mandated* by any theory. However, we'll see that if we really do live in a Universe that results from the sort of non-abelian gauge theories with symmetry breaking discussed in previous chapters, then the existence of 't Hooft–Polyakov magnetic monopoles is almost inevitable.

49.1 Dirac's monopole and the Dirac string

The discovery of a certain kind of magnetic monopole with a magnetic charge g wouldn't be a disaster for Maxwell's electromagnetism. We could include them in his equations, which would become rather more symmetrical:

Remember we are using Heaviside units.

$$\nabla \cdot E = \rho_{\mathrm{e}}, \qquad \nabla \cdot B = \rho_{\mathrm{m}},$$
$$\nabla \times E = -J_{\mathrm{m}} - \frac{\partial B}{\partial t}, \quad \nabla \times B = J_{\mathrm{e}} + \frac{\partial E}{\partial t}.$$

The extra terms involve ρ_{m}, the magnetic charge density, and J_{m}, the magnetic current.[1] We would also need to adjust the Lorentz force law which would become

[1] In conventional electromagnetism $\rho_{\mathrm{m}} = 0$ and $J_{\mathrm{m}} = 0$.

$$F = q(E + v \times B) + g\,(B - v \times E)\,. \tag{49.1}$$

This variety of magnetic monopole was first considered by Paul Dirac and is therefore known as a **Dirac monopole**.

As a first step in our investigation of the properties of these objects, let's recap the physics of *electric* monopoles, which certainly do exist!

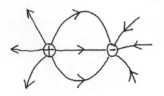

Fig. 49.1 Monopoles (electric or magnetic) are sources or sinks of field, because $\nabla \cdot \boldsymbol{E} = q\delta^{(3)}(\boldsymbol{r})$ and $\nabla \cdot \boldsymbol{B} = g\delta^{(3)}(\boldsymbol{r})$.

Fig. 49.2 You can think of a monopole as originating from an extremely long bar magnet.

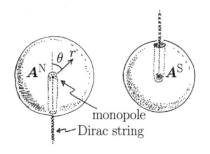

monopole
Dirac string

Fig. 49.3 The vector field from a monopole can be described everywhere except along a line of singularities (the Dirac string). $\boldsymbol{A}^{\mathrm{N}}$ is defined for $\theta < \pi$. $\boldsymbol{A}^{\mathrm{S}}$ is defined for $\theta > 0$.

[2] The expression for the curl of a vector function only containing an azimuthal ($\hat{\boldsymbol{e}}_\phi$) component in spherical polars is

$$\nabla \times \boldsymbol{A} = \frac{1}{r\sin\theta}\frac{\partial}{\partial\theta}\left[A_\phi \sin\theta\right]\hat{\boldsymbol{e}}_r$$
$$-\frac{1}{r}\frac{\partial}{\partial r}\left[rA_\phi\right]\hat{\boldsymbol{e}}_\theta.$$

They are the charges of classical electromagnetism (see Fig. 49.1). For an electrical charge q at the origin, Maxwell's equation yields

$$\nabla \cdot \boldsymbol{E} = q\,\delta^{(3)}(\boldsymbol{r}), \qquad (49.2)$$

whose solution, due to Monsieur Coulomb, takes the form

$$\boldsymbol{E} = \frac{q\boldsymbol{r}}{4\pi|\boldsymbol{r}|^3} = -\nabla\left(\frac{q}{4\pi|\boldsymbol{r}|}\right). \qquad (49.3)$$

The electric flux $\Phi_{\mathrm{E}} = \int \boldsymbol{E}\cdot\mathrm{d}\boldsymbol{S}$ through a spherical surface surrounding the electric monopole is then just q.

If analogous *magnetic* monopoles exist then their behaviour should be governed by similar equations. Instead of the Maxwell equation $\nabla \cdot \boldsymbol{B} = 0$, we should have for a magnetic monopole of charge g at the origin

$$\nabla \cdot \boldsymbol{B} = g\,\delta^{(3)}(\boldsymbol{r}), \qquad (49.4)$$

and a radial field equation (that is, the analogue of Coulomb's law):

$$\boldsymbol{B} = \frac{g\boldsymbol{r}}{4\pi|\boldsymbol{r}|^3}. \qquad (49.5)$$

The magnetic flux through a sphere surrounding the magnetic monopole is given by $\Phi_{\mathrm{M}} = g$.

Example 49.1

The problem with the Dirac monopole comes when you try and write down the magnetic vector potential $\boldsymbol{A}$. There is no single function, without singularities, that does the job everywhere; actually, this shouldn't be a surprise. If you could find such a function, then the flux Φ_{M} out of a closed region S (of volume V) would be

$$\Phi_{\mathrm{M}} = \int_S \underbrace{\nabla \times \boldsymbol{A}}_{\boldsymbol{B}} \cdot \mathrm{d}\boldsymbol{S} = \int \nabla\cdot(\nabla \times \boldsymbol{A})\,\mathrm{d}V \equiv 0, \qquad (49.6)$$

whereas this should come out as $\Phi_{\mathrm{M}} = g$. One way of understanding this physically is to think of the monopole originating from an infinitely long and very thin solenoid, or permanent magnet, which has one end at the origin and the other end very far away (somewhere out at infinity). This looks like a monopole at the origin (see Fig. 49.2), and the physical Universe can be modelled everywhere, except at a line of singularities where the solenoid or permanent magnet lies. To illustrate this problem (see Fig. 49.3), consider the function $\boldsymbol{A}^{\mathrm{N}}$ given by

$$\boldsymbol{A}^{\mathrm{N}} = \frac{g}{4\pi r}\frac{(1-\cos\theta)}{\sin\theta}\hat{\boldsymbol{e}}_\phi = \frac{g}{4\pi r}\tan\frac{\theta}{2}\hat{\boldsymbol{e}}_\phi \quad (\theta < \pi). \qquad (49.7)$$

This is well defined everywhere except at $\theta = \pi$ and taking the curl of it quickly[2] brings back eqn 49.5. The function blows up at $\theta = \pi$ for all radii r, and so this magnetic vector potential is singular along the whole of the negative z-axis. It's as if the monopole carries around a tail of singularities. This tail is known as a **Dirac string**. Similarly, we can define another function

$$\boldsymbol{A}^{\mathrm{S}} = -\frac{g}{4\pi r}\frac{(1+\cos\theta)}{\sin\theta}\hat{\boldsymbol{e}}_\phi = -\frac{g}{4\pi r}\cot\frac{\theta}{2}\hat{\boldsymbol{e}}_\phi \quad (\theta > 0), \qquad (49.8)$$

which does the same job but has a singularity at $\theta = 0$, so although we've removed the tail of singularities along the negative z-axis, we now have them along the positive z-axis: we've only managed to move the problem, not remove it. This illustrates the point that no single function will work everywhere.

One interesting consequence of these results is that the two magnetic vector potential functions $\boldsymbol{A}^{\mathrm{N}}$ and $\boldsymbol{A}^{\mathrm{S}}$ only differ by[3]

$$\boldsymbol{A}^{\mathrm{N}} - \boldsymbol{A}^{\mathrm{S}} = \frac{g}{2\pi r}\frac{1}{\sin\theta}\hat{\boldsymbol{e}}_\phi = \boldsymbol{\nabla}\left(\frac{g\phi}{2\pi}\right), \tag{49.9}$$

which looks right since $\boldsymbol{A}^{\mathrm{N}}$ and $\boldsymbol{A}^{\mathrm{S}}$ correspond to the same field $\boldsymbol{B}$ and so they should differ only by a gradient of a scalar function.

The two functions $\boldsymbol{A}^{\mathrm{N}}$ and $\boldsymbol{A}^{\mathrm{S}}$ represent different choices of electromagnetic gauge. The wave function ψ of an electric charge q surrounding the monopole can be written in either gauge, but the two expressions will be related by a gauge transformation:

$$\psi^{\mathrm{S}}(\boldsymbol{r}) = \psi^{\mathrm{N}}(\boldsymbol{r})\exp\left[-\mathrm{i}\frac{q}{\hbar}\left(\frac{g\phi}{2\pi}\right)\right]. \tag{49.10}$$

Thus as we take ϕ from 0 to 2π, the wave function will only be single-valued if

$$g = n\frac{h}{q}, \tag{49.11}$$

where n is an integer.[4] This equation tells us that if the electric charge q is quantized, then so is the magnetic charge g. The argument also works in the other direction: if we were to discover quantized magnetic charges then this would explain the existence of quantized electric charge. This is Dirac's formulation of monopoles. There is no necessity for them to exist, but if they are found, they provide a rationale for the quantization of electric charge.

49.2   The 't Hooft–Polyakov monopole

So far we have been considering single particles: the electrically charged particle and, by analogy, the magnetically charged particle.[5] Now we ask whether there is a quantum field theory for which a magnetic monopole represents a stable solution? There is: magnetic monopoles are topologically stable solutions of a non-abelian gauge theory with spontaneous symmetry breaking.[6] As we will show, these field theory monopoles, proposed in 1974 by 't Hooft and Polyakov, are rather different entities to the Dirac monopoles of the last section.

Example 49.2

The monopole solution is most easily seen if we review the topological objects of Chapter 29. There we started with a Lagrangian which was unstable to spontaneous symmetry breaking. Then the topological solutions arose by considering time-independent solutions which sat in a different vacuum at different points in space:

- The simplest was the kink in (1+1) dimensions, which was a field existing in different vacuum states at $\pm\infty$.

[3]Use has been made of the expression for grad in spherical polars:
$$\boldsymbol{\nabla}f = \frac{\partial f}{\partial r}\hat{\boldsymbol{e}}_r + \frac{1}{r}\frac{\partial f}{\partial \theta}\hat{\boldsymbol{e}}_\theta + \frac{1}{r\sin\theta}\frac{\partial f}{\partial \phi}\hat{\boldsymbol{e}}_\phi.$$

[4]This is called **Dirac's quantization condition**.

[5]In fact, taking the Dirac monopole to be a point particle is something of a necessity since, like the electric monopole it represents a singular solution to the Maxwell equations.

[6]Since, in Chapter 47, we saw that such a theory seems to describe the electroweak part of Nature, it seems likely that magnetic monopoles might be realized in our Universe! There should also, we hope, be a good reason why we haven't observed them yet.

Gerard 't Hooft (1946–)
Alexander Polyakov, (1945–)

- The second simplest was the vortex in (2+1) dimensions, where a field with broken $U(1)$ symmetry enjoyed different vacua at all points on a circle at spatial infinity.

Obviously the next step is to examine (3+1) dimensions and a configuration with broken $SO(3)$ symmetry which enjoys a different vacuum at all points at a spherical shell at infinity.

The solution with this topological property in (3+1)-dimensional space-time is the monopole, which has the form as $|\boldsymbol{r}| \to \infty$ that

$$\boldsymbol{\Phi}(\boldsymbol{r}) = A\frac{\boldsymbol{r}}{|\boldsymbol{r}|} \quad (|\boldsymbol{r}| \to \infty), \tag{49.12}$$

with A a constant. The thing to notice here (as with the vortex) is the distinction between internal and real space components. The vector $\boldsymbol{\Phi}$ lives in internal isospace but the radial vector $\boldsymbol{r}$ lives in real space. However, eqn 49.12 links these two together. In components we have, for example $\phi^1(|\boldsymbol{r}| \to \infty) = A\frac{x}{|\boldsymbol{r}|}$, that is, along the spatial 1-direction (also known as the x-direction) the field points along the internal 1-direction. Similarly along the y-direction the field points along the internal 2-direction and so on. In this sense, the field points radially outwards at infinity, which is why Polyakov called this the **hedgehog solution**. This is represented in Fig. 49.4.

Our task is now to find a theory for which the hedgehog is a solution and then to show that it has the electromagnetic properties of a monopole. Let's consider a gauge theory with local $SO(3)$ symmetry, defined by a Lagrangian

$$\mathcal{L} = \frac{1}{2}(D^\mu\boldsymbol{\Phi}) \cdot (D_\mu\boldsymbol{\Phi}) + \frac{m^2}{2}\boldsymbol{\Phi}\cdot\boldsymbol{\Phi} - \lambda(\boldsymbol{\Phi}\cdot\boldsymbol{\Phi})^2 - \frac{1}{4}\boldsymbol{G}_{\mu\nu}\cdot\boldsymbol{G}^{\mu\nu}, \tag{49.13}$$

where $\boldsymbol{G}_{\mu\nu}$ is the gauge field tensor and $D_\mu\boldsymbol{\Phi}$ is the covariant derivative given, respectively, by

$$\begin{aligned} \boldsymbol{G}_{\mu\nu} &= \partial_\mu\boldsymbol{W}_\nu - \partial_\nu\boldsymbol{W}_\mu + q\boldsymbol{W}_\mu \times \boldsymbol{W}_\nu, \\ D_\mu\boldsymbol{\Phi} &= \partial_\mu\boldsymbol{\Phi} - q\boldsymbol{\Phi} \times \boldsymbol{W}_\mu, \end{aligned} \tag{49.14}$$

with q the $SO(3)$ charge of the theory (which will be shown to be equivalent to electric charge a little later). Notice that the Lagrangian we have written is unstable to spontaneous symmetry breaking so $\boldsymbol{\Phi}$ is a Higgs field.[7] As stated above, we seek a topological field configuration for this theory with a different ground state at each point on the spatial boundary of the theory, which is the spherical shell S^2 at $|\boldsymbol{r}| \to \infty$. The resulting hedgehog solution of the Lagrangian in eqn 49.14 is given by

$$\boldsymbol{\Phi} = \left(\frac{m^2}{4\lambda}\right)^{\frac{1}{2}} \frac{\boldsymbol{r}}{|\boldsymbol{r}|} \quad (|\boldsymbol{r}| \to \infty). \tag{49.15}$$

As with all time-independent soliton-like solutions in more than one spatial dimension,[8] Derrick's theorem[9] tells us that this object, taken alone, gives rise to an infinite energy. We therefore need to appeal to the gauge field to stabilize the monopole and guarantee that its energy

Fig. 49.4 The hedgehog solution.

[7]Recall what happens in such a theory. Owing to the positive mass term, the symmetry is spontaneously broken and the Higgs field ϕ takes on a constant value in the ground state $|\boldsymbol{\Phi}_0| = \left(\frac{m^2}{4\lambda}\right)^{\frac{1}{2}}$. The broken symmetry ground states form a spherical shell S^2 in $\boldsymbol{\Phi}$-space of radius $|\boldsymbol{\Phi}_0|$. Previously we have chosen to direct the $\boldsymbol{\Phi}$-field along the $\boldsymbol{\Phi}_3$-direction. In that case W_μ^3 is equivalent to the usual electromagnetic field A^μ and we recover Maxwell's electromagnetism. We therefore take the charge of the theory to be q: the electromagnetic charge. (Although of course we know that our electromagnetism resulted from symmetry breaking in a $SU(2) \times U(1)$ gauge theory, the $SO(3)$ theory will be sufficient for examining monopoles solutions.)

[8]Recall that the vortex is one such object.

[9]See Section 29.3.

remains finite. For this to be true we require the gauge field $\boldsymbol{W}_\mu$ for $|\boldsymbol{r}| \to \infty$ to be (see Exercise 49.1)

$$W_i^a = -\varepsilon^{iab}\frac{r^b}{qr^2},$$
$$W_0^a = 0, \tag{49.16}$$

where here a is the internal index and i is the real-space vector index.

Example 49.3

We have identified a topological object that can exist in (3+1)-dimensional spacetime, but we've yet to find its electromagnetic properties. This turns out to be a rather non-trivial matter since we have three internal components of the field strength tensor $G_{\mu\nu}$ and, therefore, three different choices of magnetic field[10] $B^{ai} = -\varepsilon^{ijk}G_{jk}^a$ (with no sum implied over repeated indices). For the more usual uniform broken symmetry ground state, where $\boldsymbol{\Phi}$ points along the 3-direction, the choice is simple, the $a = 3$ component of $G_{\mu\nu}^a$ reduces to the usual form $F_{\mu\nu}$. More generally for an arbitrarily oriented ground state (i.e. one not necessarily pointing along the 3-direction) we have

$$F_{\mu\nu} = \frac{\boldsymbol{\Phi}}{|\boldsymbol{\Phi}|}\cdot\boldsymbol{G}_{\mu\nu}. \tag{49.17}$$

For any situation other than the ground state of $\boldsymbol{\Phi}$ we need a more complicated expression containing terms in $D_\mu\boldsymbol{\Phi}$ which vanish in the ground state.[11] Fortunately, these complexities may be ignored since we only require the magnetic flux emerging from the hedgehog solution, for which we find

$$F_{ij} = F^{ij} = -\varepsilon^{ijk}\frac{r^k}{q|\boldsymbol{r}|^3}. \tag{49.18}$$

Finally we're ready to see what the hedgehog field predicts for the magnetic field of a monopole: it is

$$\boldsymbol{B} = \frac{1}{q}\frac{\boldsymbol{r}}{|\boldsymbol{r}|^3}, \tag{49.19}$$

which is exactly the radial field expected for a monopole. It results in a magnetic flux

$$\Phi_{\mathrm{M}} = \frac{4\pi}{q}. \tag{49.20}$$

The hedgehog solution is therefore confirmed as a magnetic monopole. Comparing the flux from the hedgehog with the flux expected from a magnetic monopole $\Phi = g$, we must have that $\frac{4\pi}{q} = g$ or, on restoring factors of $\hbar$,

$$g = \frac{2h}{q}. \tag{49.21}$$

Comparing this with the result for a single Dirac monopole, for which $g = h/q$, we see that the magnetic charge of the 't Hooft–Polyakov monopole is twice the magnitude of the Dirac monopole.

---

[10]Here, as elsewhere in this chapter, $a = 1, 2, 3$ is the internal index and i, j, k are vector indices. Compare these three choices with the single choice from conventional electromagnetism: $B^i = -\varepsilon^{ijk}F_{jk}$ (no sum implied over repeated indices).

[11]See Exercise 49.3.

[12]The fact that these monopoles are solutions to the sort of broken symmetry theory that describes our Universe leads us to question why we have not yet observed any of them. 't Hooft estimated the mass of these beasts as $M_{\mathrm{W}}/\alpha \approx 137M_{\mathrm{W}} \approx 11\,\mathrm{TeV}$, making them very heavy indeed and outside the range of our experiments. Their existence therefore remains an open question.

Notice how the 't Hooft–Polyakov monopole and Dirac monopole are very different beasts. Dirac monopoles are singular mathematical solutions of electrodynamics which necessitate the introduction of point particles as the sources of the magnetic flux. These particles have arbitrary spin and mass. In contrast 't Hooft–Polyakov monopoles are non-singular solutions arising from the interaction of a non-abelian gauge theory and a scalar field. All of their properties, such as their mass, are determined by the original theory.[12]

Chapter summary

- Dirac's magnetic monopole leads to a singularity in the magnetic vector potential which cannot be removed (the so-called Dirac string). The quantization of magnetic charge of this monopole is inextricably linked to the quantization of electric charge.
- The 't Hooft–Polyakov monopole is a topological object in a non-abelian gauge theory.

Exercises

(49.1) Show that in order to kill the divergent derivative of the monopole field $\boldsymbol{\Phi} = \left(\frac{m^2}{4\lambda}\right)^{\frac{1}{2}} \frac{\boldsymbol{r}}{|\boldsymbol{r}|}$ at $r \to \infty$ we must introduce a covariant derivative with a gauge field with components $\boldsymbol{W}_\mu$

$$
\begin{aligned}
W_i^a &= -\varepsilon^{iab} \frac{r^b}{qr^2}, \\
W_0^a &= 0.
\end{aligned} \tag{49.22}
$$

(49.2) Verify eqn 49.19.

(49.3) We can generalize the definition of the Maxwell electromagnetic field $F_{\mu\nu}$ so that it describes electromagnetism in the presence of the topological monopole and reduces to ordinary electromagnetism under the normal circumstance of a broken symmetry Universe with components $\Phi^3 \neq 0$ and $\Phi^{(1,2)} = 0$. We define the modified electromagnetic field tensor as

$$
\mathcal{F}_{\mu\nu} = \frac{\boldsymbol{\Phi}}{|\boldsymbol{\Phi}|} \cdot \boldsymbol{G}_{\mu\nu} - \frac{\boldsymbol{\Phi}}{q|\boldsymbol{\Phi}|^3} \cdot (D_\mu \boldsymbol{\Phi} \times D_\nu \boldsymbol{\Phi}). \tag{49.23}
$$

(a) Check that this reduces to the ordinary form of $F^{\mu\nu}$ as claimed.

We may also define $A_\mu = \frac{1}{|\boldsymbol{\Phi}|} \boldsymbol{\Phi} \cdot \boldsymbol{W}_\mu$, which, in a broken symmetry Universe without monopoles, picks out the part of the potential that gives ordinary electromagnetism.

(b) Show that this leads to a cleaned up field equation

$$
\mathcal{F}_{\mu\nu} = \partial_\mu A_\nu - \partial_\nu A_\mu - \frac{\boldsymbol{\Phi}}{q|\boldsymbol{\Phi}|^3} \cdot (\partial_\mu \boldsymbol{\Phi} \times \partial_\nu \boldsymbol{\Phi}). \tag{49.24}
$$

(c) Starting with eqn 49.24 verify the algebra that leads to the value of the B-field in eqn 49.19.

Instantons, tunnelling and the end of the world

There is a theory which states that if ever anybody discovers exactly what the Universe is for and why it is here, it will instantly disappear and be replaced by something even more bizarre and inexplicable. There is another theory which states that this has already happened.
Douglas Adams (1952–2001)

In the true vacuum, the constants of nature, the masses and couplings of the elementary particles, are all different from what they were in the false vacuum, and thus the observer is no longer capable of functioning biologically, or even chemically.
Sidney Coleman (1937–2007)

One aspect of conventional quantum mechanics that we have not yet addressed with quantum field theory is tunnelling. In this chapter we will discuss a class of objects known as **instantons** which are found in the path integral version of quantum mechanics. Their field theory analogues allow us to address the problem of tunnelling in quantum field theory. In particular we will address the apocalyptic question of the end of the world! We know that when a system breaks a symmetry the result is that the vacuum of a system is one of (potentially) a number of equivalent vacua. This is presumably the case in our own Universe. But what would happen if the vacuum that the system adopts is not actually the lowest energy state? What if there is one with a slightly lower energy? If this is the case with our own Universe then we might worry that the Universe will undergo a transition, via a tunnelling event, into the true vacuum with catastrophic consequences for those of us who have grown dependent on the physics of the current (false) vacuum.[1]

We start by examining instantons. These are very similar to the kinks we examined in Chapter 29 which exist as stable entities localized in *space*. Instantons are constant energy solutions to quantum mechanical equations of motion with a characteristic, stable structure localized in *time*. The name 'instanton', coined by 't Hooft, reflects this particle-like existence in time, rather than space. Instantons are associated with critical points (local maxima, minima or saddle points) of the action.

[1]This problem was called the 'The fate of the false vacuum' by Sidney Coleman and our approach to this problem follows his treatment closely. See the superb lecture in his collection *Aspects of Symmetry* for the full story.

50.1   Instantons in quantum particle mechanics

To reveal the physics of the instanton we will temporarily leave behind fields and deal with single-particle quantum mechanics, albeit described by the path integral and Green's functions. We will examine a single particle in one spatial dimension described by a Lagrangian $L = p^2/2m - V(x)$. This gives rise to an action $S = \int_{-T/2}^{T/2} \mathrm{d}t\, L$ and a path integral $G = \int \mathcal{D}[x(t)]\, \mathrm{e}^{\mathrm{i}S/\hbar}$ which gives us the Green's function, or amplitude, for any given scenario. (For simplicity we will henceforth assume the particle described here has unit mass.) We now make a switch to Euclidean space,[2] where we have a Euclidean action

$$S_\mathrm{E} = \int_{-T_\mathrm{E}/2}^{T_\mathrm{E}/2} \mathrm{d}\tau \left[\frac{1}{2}\left(\frac{\mathrm{d}x}{\mathrm{d}\tau}\right)^2 + V(x)\right], \tag{50.1}$$

and a path integral

$$G(x_\mathrm{f}, T_\mathrm{E}/2, x_\mathrm{i}, -T_\mathrm{E}/2) = \int \mathcal{D}[x(\tau)]\, \mathrm{e}^{-S_\mathrm{E}/\hbar}, \tag{50.2}$$

where G is the Green's function describing the amplitude for starting at $(-T_\mathrm{E}/2, x_\mathrm{i})$ and ending up at $(T_\mathrm{E}/2, x_\mathrm{f})$, and the integral is carried out over all trajectories that have this property.

Let's suppose a stationary trajectory exists called $\bar{x}(\tau)$. Applying the Euler–Lagrange equations to the Euclidean action leads to

$$\frac{\mathrm{d}^2\bar{x}}{\mathrm{d}\tau^2} = \frac{\mathrm{d}V(\bar{x})}{\mathrm{d}x}, \tag{50.3}$$

which differs from the usual equation of motion for a particle in a potential by a minus sign. Doing our analysis in Euclidean space has resulted in equations of motion that seem to describe the motion of a particle in a potential $-V(x)$. As a result, the constant of the motion corresponding to energy is given by

$$E = \frac{1}{2}\left(\frac{\mathrm{d}\bar{x}}{\mathrm{d}\tau}\right)^2 - V(\bar{x}). \tag{50.4}$$

The solutions to the equations of motion that minimize the action have constant energy and so this equation for E enables us to identify them very conveniently.

[2]In Chapter 25 we used the Wick rotation, which turns Minkowski spacetime into Euclidean spacetime via the rotation taking $x^0 \to -\mathrm{i}\tau$, purely as a means of relating quantum field theory to statistical physics. In this chapter we will see that working in Euclidean space has the advantage of leading to new and useful solutions to the equations of motion of quantum particles. Note that here we do not impose the periodic boundary conditions of statistical physics and so imaginary time stretches from $-\infty \le \tau \le \infty$. We will also call the time interval over which the calculations are carried out T in Minkowski space and T_E in Euclidean space.

[3]See Chapter 23.

Example 50.1

In order to find the Green's functions which describe amplitudes in this potential we will use what is known as the **stationary phase approximation** whose guts are described here. Since we know that the path integral will be dominated by those paths closest to the stationary trajectory[3], we split the integral into

$$G = \mathrm{e}^{-(\text{Stationary action})/\hbar} \times (\text{quantum corrections}). \tag{50.5}$$

The key to solving problems is then to find the stationary action for the situation we are trying to describe and expanding around this point. The implementation of this approximation becomes clear if we look at a simple integral $\mathcal{I} = \int \mathrm{d}x \, \mathrm{e}^{-f(x)}$. Expanding $f(x)$ about a stationary point at $\bar{x}$ we have

$$f(x) = f(\bar{x}) + \frac{1}{2} f''(\bar{x})(x - \bar{x})^2 + \dots \qquad (50.6)$$

The second term will present us with a Gaussian integral and so we obtain

$$\mathcal{I} \approx \mathrm{e}^{-f(\bar{x})} \left(\frac{2\pi}{f''(\bar{x})} \right)^{\frac{1}{2}}. \qquad (50.7)$$

This result carries over to the case of the functional integral and we obtain an expression for the Green's function, for our Euclidean theory of a unit mass particle, given by the stationary phase approximation as

$$G(x_{\mathrm{f}}, T_{\mathrm{E}}/2, x_{\mathrm{i}}, -T_{\mathrm{E}}/2) = N \mathrm{e}^{-S[\bar{x}]/\hbar} \left[\det \left(-\frac{\partial^2}{\partial \tau^2} + V''(0) \right) \right]^{-\frac{1}{2}}, \qquad (50.8)$$

which is accurate to order $O(\hbar)$, which will be sufficient for our purposes, and where N supplies the normalization.

The conclusion from all of this scene-setting is simply that using the Euclidean action has the effect of flipping the sign of the potential. The point of this is that there are stationary solutions which exist in the Euclidean world of upside-down potentials which may be analytically continued back to Minkowski space. We therefore potentially gain access to lots of new possibilities for the motion of particles to which we were previously ignorant. One of these is tunnelling, which would never be predicted by calculating the quantum corrections to a stationary path found in Minkowski space.

In the next sections we will build some confidence in this approach by examining the stationary solutions that follow from some simple upside-down potentials.

50.2 A particle in a potential well

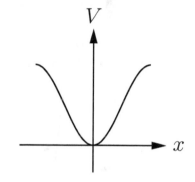

Fig. 50.1 A potential well.

Example 50.2

As a warm up exercise in getting used to the formalism, we will examine the case of a particle in a potential well such as that shown in Fig. 50.1, which has $V(x = 0) = 0$.

In the Euclidean version of the story the well is turned upside down into the potential hill shown in Fig. 50.2. If we impose boundary conditions that we want $x_{\mathrm{i}} = x_{\mathrm{f}} = 0$ then clearly the only solution is for a particle to balance at the top of the hill. This involves the particle stopped at $x = 0$ for the entire interval T_{E} and so has action $S(\bar{x}) = 0$. Using our equation for the Green's function we find the quantum amplitude that the particle starts at $x = 0$ at time $\tau = -T_{\mathrm{E}}/2$ and is found there again at $\tau = T_{\mathrm{E}}/2$ is given by

$$G(0, T_{\mathrm{E}}/2, 0, -T_{\mathrm{E}}/2) = N \left[\det \left(-\frac{\partial^2}{\partial \tau^2} + \omega^2 \right) \right]^{-\frac{1}{2}}, \qquad (50.9)$$

where $\omega^2 = V''(0)$. Recall that we've previously solved a related (exactly solvable) path integral problem: that of the simple harmonic oscillator potential. In that case we concluded that (in Minkowski spacetime) we have

$$G(0, T/2, 0, -T/2) = \langle 0 | \mathrm{e}^{-\mathrm{i}\hat{H}T/\hbar} | 0 \rangle = N \left[\det \left(-\frac{\partial^2}{\partial t^2} - \omega_0^2 \right) \right]^{-\frac{1}{2}}, \qquad (50.10)$$

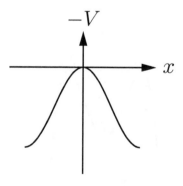

Fig. 50.2 An upside-down potential well, inverted by the Wick rotation to Euclidean space.

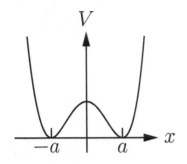

Fig. 50.3 The double potential well.

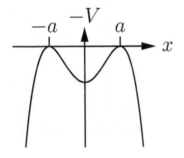

Fig. 50.4 The double potential well, turned upside-down by the Euclidean rotation.

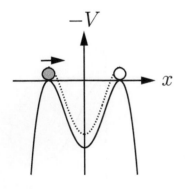

Fig. 50.5 The instanton solution.

where $\omega_0 = V''(0)$ and $V(x) = \frac{1}{2}\omega_0^2 x^2$. The stationary phase version above is simply an approximate generalization of this result with a sign change due to our working in Euclidean space.

Returning to the more general potential and to Euclidean space we see that working out G is made difficult by the determinant. However, we can argue the general form of the answer. We know that in Euclidean space the Green's function will look like $G = \langle 0|e^{-\hat{H}T_E/\hbar}|0\rangle = e^{-E_0 T_E/\hbar}$, and we expect that the ground state energy should be something close to $E_0 \approx \frac{1}{2}\hbar\omega$ for small oscillations in a potential well, suggesting that $G \propto e^{-\omega T_E/2}$. In fact, for large T_E, it may be shown that

$$G(0, T_E/2, 0, -T_E/2) = N\left[\det\left(-\frac{\partial^2}{\partial\tau^2} + \omega^2\right)\right]^{-\frac{1}{2}} = \left(\frac{\omega}{\pi\hbar}\right)^{\frac{1}{2}} e^{-\frac{\omega T_E}{2}}, \quad (50.11)$$

and we read off the ground state energy as $E_0 = \frac{1}{2}\hbar\omega\left[1 + O(\hbar)\right]$, as we expect.

We'll follow the same procedure again for more interesting potentials. The procedure is to write down the potential and shift it to Euclidean space, find the stationary configuration, work out its action and then find the approximate Green's function using eqn 50.8.

50.3   A particle in a double well

The **instanton** solution is found when we repeat the procedure outlined above for the double potential well. We define the well with minima at $\pm a$ and take $V(-x) = V(x)$. We also define $V''(\pm a) = \omega^2$. The potential for the double well is that shown in Fig. 50.3 and shifting to Euclidean space turns it upside down, as shown in Fig. 50.4. We look for solutions to the equations of motion with constant energy with boundary conditions that the particle itself is stationary at $\tau = \pm T_E/2$. These are useful since we will eventually make T_E very large and so they will represent well defined solutions of finite energy. There are two obvious sets of possible solutions: one involves a particle fixed at a or at $-a$, just as we had in the previous section. However, there is a more interesting possibility: the particle starts at $-a$ at $-T_E/2$ and ends up at a at $T_E/2$ as shown in Fig. 50.5. This is the instanton solution. Viewed with the potential flipped the right way up, the end points of this solution correspond to a particle tunnelling though the barrier.

Example 50.3

We may calculate the properties of the instanton. With the potential as defined in Fig. 50.5 this solution has $E = 0$, so we have

$$\frac{dx}{d\tau} = (2V)^{\frac{1}{2}}. \quad (50.12)$$

The action of an instanton is given by

$$S_0 = \int dt\left[\frac{1}{2}\left(\frac{dx}{d\tau}\right)^2 + V\right] = \int d\tau\left(\frac{dx}{d\tau}\right)^2 = \int_{-a}^{a} dx\,(2V)^{\frac{1}{2}}. \quad (50.13)$$

At large T_E we have that $\dot{x} = \omega(a - x)$ leading to

$$(a - x) \propto e^{-\omega\tau}, \quad (50.14)$$

which tells us that the (temporal) size of an instanton is $1/\omega$.

The previous example demonstrates that the instanton has a well defined structure in time of size $\tau \sim 1/\omega$. It is shown in Fig. 50.6, where its temporal extent is given by the width of the region where it crosses the axis. Notice the similarity between this and the kink in Chapter 29. (The kink represented a well defined structure in space, with a spatial extent $l \sim 1/m$.)

In order to use the instanton to predict the properties of the double well we must allow the possibility of the existence of more than one tunnelling event. This corresponds to examining more than one instanton. This is the subject of the next example.

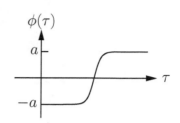

Fig. 50.6 The instanton as a tightly bound structure in (imaginary) time.

Example 50.4

We will now examine the case of a dilute gas of instanton objects. Using the same double well, the 'gas' corresponds to a particle on top of one hill that occasionally rolls down the hill and up to its neighbour, and then some time later rolls back down the hill and up to where it started. Some time later it undergoes another rolling event, and so on. The gas then comprises n rolls, constrained so that an instanton $a \to -a$ must be followed by an anti-instanton $-a \to a$. (Recall that we had the same constraints on kinks and antikinks.) The gas is dilute in that each instanton has a size (in time) far smaller than the gaps between instantons and anti-instantons. The gas is shown in Fig. 50.7. This may be the strangest picture of a gas it's possible to draw!

The action of n dilute instantons is given by nS_0, where S_0 is the action of one instanton that we calculated above. For a single well we had $G(0, T_E/2, 0, -T_E/2) = (\omega/\pi\hbar)^{\frac{1}{2}} e^{-\omega T_E/2}$. For the gas in the double well we need to consider this amplitude multiplied by the contribution K from each of the n instantons:

$$\left(\frac{\omega}{\pi\hbar}\right)^{\frac{1}{2}} e^{-\omega T_E/2} K^n, \qquad (50.15)$$

where K is the amplitude for the occurrence of a single instanton. Since we don't know where each instanton is centred in time, we must integrate this quantity over the centres of all of the instantons. This is conveniently carried out by noting that

$$\int_{-T_E/2}^{T_E/2} d\tau_1 \int_{-T_E/2}^{\tau_1} d\tau_2 ... \int_{-T_E/2}^{\tau_{n-1}} d\tau_n = \frac{T_E^n}{n!}. \qquad (50.16)$$

Putting all of this together we have, for example, to order $O(\hbar)$, an amplitude for the particle to start and finish at $x = -a$ of

$$G(-a, T_E/2, -a, -T_E/2) = \left(\frac{\omega}{\pi\hbar}\right)^{\frac{1}{2}} e^{-\omega T_E/2} \sum_{n_e} \frac{(Ke^{-S_0/\hbar}T_E)^{n_e}}{n_e!}, \qquad (50.17)$$

where n_e means that n only includes even numbers, ensuring that, however large we make n, the particle ends up where it started.[4] These sorts of sums may be done and the general result is that

$$G\left(\pm a, \frac{T_E}{2}, -a, -\frac{T_E}{2}\right) = \frac{1}{2}\left(\frac{\omega}{\pi\hbar}\right)^{\frac{1}{2}} e^{-\frac{\omega T_E}{2}} \left[\exp(Ke^{-\frac{S_0}{\hbar}}T_E) \mp \exp(-Ke^{-\frac{S_0}{\hbar}}T_E)\right]. \qquad (50.18)$$

We may read off the physics by comparing with $e^{-ET_E/\hbar}$ as we did for the single well. In this case we find that we have two low-lying energy states with energies corresponding to whether we finished at $-a$ (summing over even n) or $+a$ (summing over odd n). The energies of the two states are given by

$$E_\pm = \frac{1}{2}\hbar\omega \pm \hbar K e^{-S_0/\hbar}. \qquad (50.19)$$

It's worth pausing at this stage to take stock of what we've done by returning to Minkowski space and flipping the potential back the right way up.

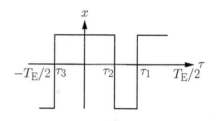

Fig. 50.7 A gas of instantons for the double well problem.

Note that $\hbar$ has been reintroduced for clarity.

[4]The same argument says that amnesia may be avoided if one is hit over the head an even number of times.

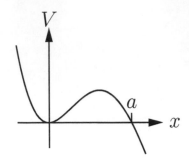

Fig. 50.8 A potential with a barrier.

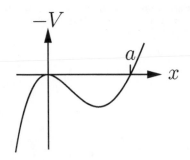

Fig. 50.9 The barrier potential turned upside-down.

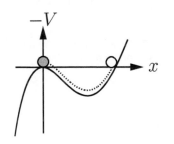

Fig. 50.10 The bounce solution: another example of an instanton.

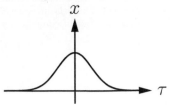

Fig. 50.11 The trajectory of the particle during the bounce has a maximum, signalling a node in the wave function.

The conventional quantum mechanics of the double well is instructive here. We have solutions that are localized in the left-hand well $|-a\rangle$ and in the right-hand well $|a\rangle$. We'll call the energy for sitting in a potential minimum U_0, but we allow the Hamiltonian to permit barrier penetrating transitions between minima with matrix element $\langle -a|\hat{H}|a\rangle = \langle a|\hat{H}|-a\rangle = -V$. The solutions are

$$
\begin{array}{ll}
\frac{1}{\sqrt{2}}\Big(|-a\rangle + |a\rangle\Big) & E = U_0 - V, \\
\frac{1}{\sqrt{2}}\Big(|-a\rangle - |a\rangle\Big) & E = U_0 + V.
\end{array} \tag{50.20}
$$

The system saves some energy through the wave function spreading out over the two wells. The symmetric combination lies lowest, separated by an energy $2V$ from the antisymmetric combination.

In the present case, we can identify $U_0 = \frac{1}{2}\hbar\omega$ as the energy for sitting in the bottom of a well and $V = \hbar K e^{-S_0/\hbar}$. This makes sense as the difference in energies is proportional to the factor accounting for barrier penetration $e^{-S_0/\hbar}$.

Example 50.5

We now examine tunnelling to freedom through a barrier as another example of the use of instantons. We start with the potential shown in Fig. 50.8 and up-end it by working in Euclidean space to obtain Fig. 50.9. The stationary solution of interest is a different instanton to those considered thus far. This instanton, shown in Fig. 50.10, involves the particle starting at $x_i = 0$, travelling to $x = a$, where it bounces, reverses its direction and returns to its starting point at $x_f = 0$.

This is rather similar to the previous problem of the instanton gas, except that the Green's function is given by summing over all n, since any number of instantons result in the particle back where it started at $x = 0$. We may carry out the sum to give

$$
G(0, T_E/2, 0, -T_E/2) = \left(\frac{\omega}{\pi\hbar}\right)^{\frac{1}{2}} e^{-\omega T_E/2} \exp\left[K e^{-S_0/\hbar} T_E\right], \tag{50.21}
$$

which we might think tells us of an energy eigenvalue of $E_0 = \frac{1}{2}\hbar\omega + \hbar K e^{-S_0/\hbar}$. However, the interpretation of this solution is slightly more subtle.

Let's return to the right-way-up potential of Fig. 50.8. We've seen that the factor $\hbar K e^{-S_0/\hbar}$ tells us about barrier penetration, but penetrating through the barrier in this problem results in the particle rolling away to infinity. This suggests that the bound state in the potential is unstable. Additional evidence arises when we plot the trajectory of the bouncing particle, which is shown in Fig. 50.11. The fact that the trajectory has a maximum tells us that the wave function must have a node. This means it's not the lowest energy wave function and there must be a lower one with a negative energy. Since K is a function of the square-root of energy then this implies that K is imaginary. This fits with the idea of barrier penetration: we know that unstable states have energies with imaginary parts which, when plugged into e^{-iEt} give $e^{-iE_0 t - \Gamma t}$ where 2Γ is the decay rate of the state. What our instanton tells us is not a correction to the energy of a bound state, but rather the imaginary part of the energy, which gives is the decay rate. We conclude therefore that $\mathrm{Im}E_0 \approx \hbar|K|e^{-S_0/\hbar}$ and so the decay rate of the state is roughly given by

$$
\Gamma \approx \hbar|K|e^{-S_0/\hbar}. \tag{50.22}
$$

50.4   The fate of the false vacuum

Finally we turn to fields. As stated in the introduction, our goal is to describe what happens to a broken symmetry system in the case that the vacuum that the system has chosen (known as the false vacuum) lies slightly higher in energy than another minimum in the potential, which we call the true vacuum. A potential describing this state of affairs is shown in Fig. 50.12. To be more colourful, we imagine that the vacuum of the Universe in which we live is a false vacuum (at $\phi = \phi_+$) and ask whether the Universe is about to tunnel through a potential maximum into the real vacuum (at $\phi = \phi_-$).

By analogy with the case of the magnet in Chapter 26, the catastrophic quantum collapse of the false vacuum would seem rather unlikely. In the magnet example the process of tunnelling to the true vacuum would rely on each electronic spin in the system simultaneously tunnelling. We might have expected this to have a vanishingly small probability since there are so many particles. However, the possibility of instanton solutions in the field allow a scenario where a cluster of spins finds the true vacuum and then an instanton causes the spins forming the boundary of this 'bubble' to flip, causing the Universe to pass a tipping-point beyond which there is sufficient volume of true vacuum to make this the preferred phase. Whether a change in phase of the Universe actually occurs depends on the relative balance of the energy saving in realizing the true vacuum versus the cost of the domain wall between true and false vacua.

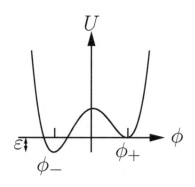

Fig. 50.12 False vacuum potential.

Example 50.6

Supercooled liquid presents us with a similar problem. In that case we cool a liquid which reaches a metastable state of liquidity, separated by a small energy from the true vacuum of solidity. A similar situation occurs for the superheated liquid, which remains in a metastable state of liquidity rather than evaporating.[5]

For the superheated liquid, thermal fluctuations often result in a small bubble of gas appearing. The inside of the bubble is the true ground state and so results in an energy saving ϵ, per unit volume. On the other hand, the surface energy of the bubble costs energy σ per unit surface area. We might write the energy

$$E = -\frac{4}{3}\pi r^3 \epsilon + 4\pi r^2 \sigma, \qquad (50.23)$$

shown in Fig. 50.13. Small bubbles, with a small volume energy saving compared to surface area cost will disappear. At some point a bubble will be large enough that it is energetically favourable for it to increase in size. By finding the stationary point in E we can work out how large a bubble must be to become ever-expanding. Extremizing, we find a maximum in E at

$$r = \frac{2\sigma}{\epsilon}. \qquad (50.24)$$

Bubbles larger than this find it energetically favourable to expand. As this occurs, more and more space is taken up by the true vacuum, which eventually spreads throughout all space.

[5] Although in this case, you should note that there's no difference in symmetry between a liquid and a gas, so the analogy with symmetry breaking no longer carries.

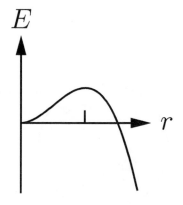

Fig. 50.13 The energy of a bubble of gas of radius r in a superheated liquid.

[6] again setting $\hbar = 1$.

We examine the decay of the false vacuum by analogy with the particle problems looked at so far in this chapter. The decay process is described by a bouncing instanton, which gives a decay rate. Since in quantum field theory we deal with a path integral[6] $\int \mathcal{D}\phi\, \mathrm{e}^{-\int \mathrm{d}^4 x\, \mathcal{L}_{\mathrm{E}}}$, then the quantity we will find is $\Gamma/\mathcal{V}$, a decay rate per unit volume, which we expect to be given by

$$\Gamma/\mathcal{V} = |K|\mathrm{e}^{-S_0}, \qquad (50.25)$$

where S_0 is the action of the instanton bounce. Calculating the approximate action of the bounce will be possible: since the minima of the false vacuum potential are only separated by a small energy difference, they closely resemble the exact double well problem for which we know the action.

Let's now be specific. We consider a quantum field theory in four-dimensional Euclidean space. This has an action

$$S = \int \mathrm{d}^4 x \left[\frac{1}{2}(\partial_\mu \phi)^2 + U(\phi) \right]. \qquad (50.26)$$

We cook up a symmetric, double-well potential U_{sym}, which we then modify slightly to obtain the false vacuum potential U:

$$U = U_{\mathrm{sym}} + \varepsilon(\phi - a)/2a. \qquad (50.27)$$

The potential has minima at $\phi_\pm = \pm a$, although that at $\phi_+ = a$ lies at a slightly higher energy, by an amount ε (Fig. 50.12).

We will follow the procedure we developed for identifying instantons in particle mechanics. We may employ the results from the particle, as long as we remember that the imaginary time τ in the Euclidean particle problem has become a spacetime point r in four-dimensional Euclidean space. The Euclidean potential is upside-down as shown in Fig. 50.14. The bouncing instanton clearly starts on the hill of the false vacuum and bounces from the point ϕ_{c}, where $U(\phi)$ cuts the ϕ-axis.

We can also imagine this running backwards, with the particle starting at a point ϕ_{c} near the true vacuum at ϕ_-. We choose to look at it this way around because the bubble that will result in 4-space then resembles the bubble in the superheated liquid with true vacuum within and false vacuum without (shown in Fig. 50.15). The particle will roll down the hill and come to a rest exactly on top of the hill at ϕ_+ (the false vacuum). We'll imagine this takes place *very rapidly* around some point in time r that we'll call $r = R$. Back in field language we ask what the hill-roll looks like. In Euclidean space we have a large spherical bubble of radius $r = R$. This has a thin wall representing the rolling process. (The wall is thin because the roll takes place very quickly.) The bubble wall separates the false vacuum on the outside of the bubble from the true vacuum on the inside. The wall of the bubble contains the instanton. If R, the spacetime point where the instanton is centred, is greater than the critical radius R_{c} above which the bubble is self-sustaining, then the instanton creates a bubble that will drag all of spacetime into the true vacuum.

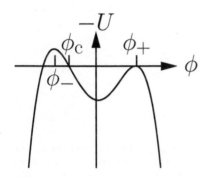

Fig. 50.14 False vacuum potential in Euclidean space.

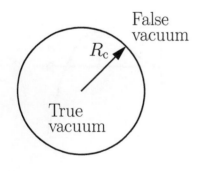

Fig. 50.15 The bubble in Euclidean space.

We will now construct an action S for the rolling process. As r increases we start inside the true vacuum bubble, move through the wall and end up outside. The action is given for this spherically symmetric problem by

$$S_0 = 2\pi^2 \int_0^\infty \mathrm{d}r\, r^3 \left[\frac{1}{2} \left(\frac{\mathrm{d}\bar\phi}{\mathrm{d}r} \right)^2 + U \right]. \tag{50.28}$$

Unlike the previous examples, where only the instanton part contributed to the action, we need to be more careful here. Inside the bubble we are in the true vacuum where we have ϕ_- and $U = -\epsilon$, so only the second term in the integral contributes, yielding an action

$$S_{\text{true vacuum}} = -\frac{1}{2}\pi^2 R^4 \epsilon. \tag{50.29}$$

Over the thin surface of the bubble we have a contribution from the instanton part:

$$S_{\text{instanton}} = 2\pi^2 R^3 \int \mathrm{d}r \left[\frac{1}{2} \left(\frac{\mathrm{d}\bar\phi}{\mathrm{d}r} \right)^2 + U_{\text{sym}} \right], \tag{50.30}$$

where, as advertised, we choose to neglect the difference ϵ in energy between the minima and so only need consider the symmetric double well potential U_{sym}. We know that the instanton solution takes us from $-a$ to a as r increases through R. By analogy with the particle problem, such an instanton has an action $S_1 = \int_{-a}^{a} \mathrm{d}\phi \, (2U_{\text{sym}})^{\frac{1}{2}}$. This gives us a contribution to the action from the instanton of

$$S_{\text{instanton}} = 2\pi^2 R^3 S_1. \tag{50.31}$$

Outside the bubble we are in the false vacuum where $\phi = \phi_+$ and $U = 0$, which makes no contribution to the action. The total action is therefore given by

$$S_{\text{tot}} = -\frac{1}{2}\pi^2 R^4 \epsilon + 2\pi^2 R^3 S_1. \tag{50.32}$$

We may now extremize the action to find the critical radius R_c at which the tunnelling destroys the false vacuum. We vary the total action with respect to R and find a radius $R_c = 3S_1/\epsilon$. Substituting this back into the expression for the total action of the bubble we predict a rate of tunnelling to annihilation of

$$\Gamma/\mathcal{V} \approx |K| e^{-S_{\text{tot}}}, \tag{50.33}$$

where $S_{\text{tot}} = 27\pi^2 S_1^4/2\varepsilon^3$.

Now we swap back into the Minkowski space of our Universe, and ask how the Universe will end. An analytical continuation of these results back to Minkowski space shows that the shape of the bubble in Euclidean 4-space is the same as that of the bubble in (3+1) dimensions. As R increases, the bubble expands, sweeping out a region in spacetime given by the hyperboloid

$$|\boldsymbol{x}|^2 - (ct)^2 = R_c^2. \tag{50.34}$$

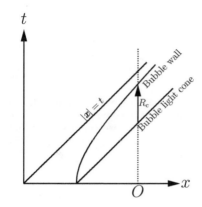

Fig. 50.16 Spacetime diagram showing the growth of the bubble. An observer at O only knows of the bubble when its forward light-cone intersects the observer's world-line. A time R_c/c seconds later the bubble wall intersects the observer's world-line.

We might expect that since the initial bubble is created by quantum mechanical fluctuations then R_c should be a very small number, perhaps of the order of 1 fm. This would mean, from eqn 50.34, that $|x|^2 \approx (ct)^2$ and the bubble expands at almost the speed of light. This, rather inconveniently, gives an observer effectively no warning of the coming of the true vacuum. Looked at on the spacetime diagram in Fig. 50.16 we see that the observer O doesn't know about the existence of the bubble until the light from the forward light-cone of the bubble's outer wall meets her world-line (see Fig. 50.17). We see that the true vacuum itself collides with her world-line a time R/c seconds later, and she ceases to exist. This time (by assumption, ≈ 1 fm/c seconds) is orders of magnitude shorter than the time it takes a neuron to fire. In Sidney Coleman's words she 'has literally nothing to worry about'.

Fig. 50.17 We have essentially no warning of the coming of the true vacuum. However, we can give some advance notice of the imminent end of this book.

Chapter summary

- Instantons are kink-like solutions of the equation of motion associated with critical points in the action and allow tunnelling problems to be treated.
- Instantons can be used to study the fate of the false vacuum in a symmetry broken system, such as the Universe, where the true vacuum state is only a tunnelling event away.

Exercises

(50.1) Verify the expression for the stationary phase approximation in eqn 50.8.

(50.2) Verify the solution of the double well problem in eqns 50.20.

Further reading

Many of the ideas in the preceding essay have been taken from another...
John Berger (1926–)

There are many other excellent books which introduce and describe quantum field theory.[1] Whether you are after the friendly approachability of Mattuck or Zee, or perhaps the elegant but unforgiving grandeur of Weinberg, or something in between, there is something for everyone. We particularly like Peskin and Schroeder, and also Ryder, and either (or both) could serve as a useful complement to this book. Many of the arguments we have used are adapted from these books (Zee and Peskin and Schroeder deserve special mention here) and also from the legendary lecture course given by Sidney Coleman.[2] Finally, for those eager to practise, a useful set of solved problems may be found in Radovanovic.

Further reading by chapter:

Chapter 1: An accessible introduction to Lagrangian mechanics is given in Feynman, Leighton and Sands. See Landau and Lifshitz (vol. I) for the full story. *Chapters 2 and 3*: The simple harmonic oscillator and occupation numbers are explained in most books on quantum field theory. The books by Aitchison and Hey,[3] P. Coleman and Feynman are especially clear. *Chapter 4*: Several examples of non-relativistic applications of second quantization may be found in P. Coleman. *Chapter 5*: Classical field theory is covered in Maggiore, Ryder and Itzykson and Zuber amongst many others. *Chapter 6*: The Klein–Gordon equation and the problems with its interpretation are explored in Aitchison and Hey. *Chapter 7*: The Lagrangians we describe are all examined in most books on quantum field theory. Ryder, Huang, and Zee are good sources of information. *Chapter 8*: Time evolution operators and related issues are discussed in most books on advanced quantum mechanics (Ziman, for example) and in the lectures by S. Coleman. *Chapter 9*: Transformations are described very clearly in Ryder. For a more sophisticated treatment see Weinberg. *Chapter 10*: An in-depth discussion of Noether's theorem may be found in Neuenschwander. Both Ryder and Huang provide accessible introductions. *Chapter 11*: Our treatment of canonical quantization follows S. Coleman's lectures. Aitchison and Hey, Bjorken and Drell[4] (RQF) and Schiff are other sources of clear information. *Chapter 12*: The properties of the complex scalar field are examined in nearly all books on quan-

[1] See the detailed list below, beginning on page 470.

[2] These mid-1970s lectures are, at least at the time of writing, viewable on the Harvard Physics website.

[3] Assume volume I of multivolume texts unless stated otherwise.

[4] We abbreviate Bjorken and Drell's *Relativistic Quantum Fields* as here RQF to avoid confusion with their *Relativistic Quantum Mechanics* (RQM).

tum field theory [see Aitchison and Hey or Bjorken and Drell (RQF)]. Our discussion of its non-relativistic properties follows Zee. *Chapter 13*: Internal symmetries and vector fields are discussed in Ryder, Zee and in Aitchison and Hey. *Chapter 14*: Gauge theory and its consequences are very clearly described in Aitchison and Hey who also give a good exposition of the difficulties inherent in the canonical quantization of the electromagnetic field. The solution to these difficulties is discussed in Peskin and Schroeder. *Chapter 15*: Charge conjugation, parity and time inversion are described in Aitchison and Hey and in Bjorken and Drell (RQF). A more sophisticated account is Weinberg. Relevant background on the mathematics may be found in the books by Georgi and Nakahara. *Chapter 16*: Green's functions are introduced in Mattuck in a similar way to that described here. An approach to perturbation theory based on propagators may be found in Schiff. Classical Green's functions are described in Barton. *Chapter 17*: The Feynman propagator is discussed in Aitchison and Hey. The non-relativistic version is discussed in Mattuck. *Chapter 18*: Useful information on the S-matrix may be found in Aitchison and Hey, and also Peskin and Schroeder. Weinberg provides the philosophy. *Chapter 19*: The expansion of the S-matrix is carried out for a simple toy theory in Aitchison and Hey, while the ϕ^4 theory is discussed in Peskin and Schroeder and in Ryder. *Chapter 20*: Scattering is covered by most books on quantum field theory and on particle physics. An introduction to scattering for particle physics applications is given in Griffiths. Clear treatments from the field theory perspective appear in Peskin and Schroeder (who discuss $\psi^\dagger \psi \phi$ theory) and also in Weinberg. *Chapter 21*: Statistical physics is described in Blundell and Blundell and at a more advanced level in Chaikin and Lubensky. Linear response theory is described in Binney, Dowrick, Fisher and Newman. The analogy between field theory and magnetism is described in more detail in Peskin and Schroeder. See also Zinn-Justin. *Chapter 22*: Generating functionals are described in Ryder, Peskin and Schroeder, and in Binney, Dowrick, Fisher and Newman. *Chapter 23*: The path integral is described in Zee and Ryder. A different approach is described by its inventor in Feynman and Hibbs. *Chapter 24*: Functional integrals are clearly described in Zee, Peskin and Schroeder and in S. Coleman. *Chapter 25*: The Wick rotation is discussed in Peskin and Schroeder and Altland and Simons. Kapusta and Gale is a good book on the applications of statistical field theory and Mahan describes how these techniques are used in solid state physics. *Chapter 26*: Broken symmetry is discussed in Blundell and in more detail in Anderson. A lively, advanced, field-centred treatment may be found in S. Coleman. *Chapter 27*: Coherent states are introduced in Annett and in Loudon. The history of the phase operator problem is described in the paper by Nieto, arXiv:hep-th/9304036v1. Altland and Simons and P. Coleman describe coherent states for field applications. *Chapter 28*: Grassmann variables are treated in Chapter 44 of Srednicki, and in great detail by Negele and Orland. *Chapter 29*: Zee is a very clear source on topological effects and we have followed his treatment in our presentation. Topological objects are described in Zee and Ryder, with more advanced treatments in S. Coleman, Weinberg, Nakahara, and in Altland and Simons. *Chapter 30*: Our description of topological field theory follows Zee and the notes by Dunne (arXiv:hep-th/9902115v1). It is examined in more advanced form in Wen. *Chapter 31*: Quasiparticles are described in Mattuck and P. Coleman. The Landau Fermi liquid is introduced in P. Coleman. *Chapter 32*: The philosophy of renormalization is made clear in Peskin and Schroeder and in Weinberg. *Chapter 33*: The use of renormalization in perturbation theory is stressed in Peskin and Schroeder. *Chapter 34*: Our presentation of the philosophy of the renormalization group is similar to that of Zee and of Peskin and Schroeder. Good sources on the use of the method may be found in Altland and Simons, Peskin and Schroeder, Binney, Dowrick, Fisher and Newman and McComb. *Chapter 35*: More detail on ferromagnetism and the renormalization group may be found in Altland and Simons. *Chapter 36*: There are many different approaches to introducing the Dirac equation. Aitchison and Hey and Ryder are good places to start. A more modern approach is found in Maggiore. We have followed the approach in Penrose. *Chapter 37*: The transformation of spinors is discussed in Ryder and Maggiore and in more detail in Weinberg. *Chapter 38*: The quantum mechanics of fermion fields is introduced very clearly in Peskin and Schroeder and in Bjorken and Drell (RQM and RQF). *Chapter 39*: Simple examples from QED feature in Aitchison and Hey, Mandl and Shaw and Peskin and Schroeder. The history of the subject is explored in Schweber. *Chapter 40*: The examples of scattering in QED covered in this chapter are more fully explained in Peskin and Schroeder. See also Berestetskii, Lifshitz and Pitaevskii for lots more examples. *Chapter 41*: The renormalization of QED is covered clearly and in detail in Peskin and Schroeder. *Chapter 42*: Bogoliubov's treatment is described in Ziman. The Lagrangian treatment of a superfluid may be found in Zee and in more detail in Wen. See also Fulde. *Chapter 43*: An introduction to the field theory of metals may be found in P. Coleman. See the detailed tome by Mahan for the full story and also Giuliani and Vignale for more background. *Chapter 44*: Our treatment of superconductivity in terms of second quantized operators follows Annett. The Lagrangian formulation is described by Wen. See also Weinberg (vol. II) and Aitchison and Hey (vol. II). *Chapter 45*: The field theory of the fractional quantum Hall effect is described straightforwardly in Zee and in more detail in Ezawa and Wen. *Chapters 46 and 47*: Non-abelian gauge theory and the Weinberg–Salam model are described in Ryder, in Peskin and Schroeder and, of course, in Weinberg (vol. II). An approach motivated by particle physics phenomenology is presented in Aitchison and Hey (vol. II). *Chapter 48*: Majorana fermions are described in Aitchison and Hey (vol. II) and in a review by Pal (arXiv:1006.1718v2). *Chapter 49*: Magnetic monopoles are described in Ryder, Zee and by S. Coleman. *Chapter 50*: Our treatment of instantons is based very closely on the lecture in the collection by S. Coleman. *Appendix B*: Useful background may be found in Penrose. Our treatment follows that of Boas.

Bibliography

- A. A. Abrikosov, L. P. Gorkov and I. E. Dzyaloshinski, *Methods of quantum field theory in statistical physics*, Dover, New York (1963).
- I. J. R. Aitchison and A. J. G. Hey, *Gauge theory in particle physics*, vol. I, 3rd edition, IOP, Bristol (2003).
- I. J. R. Aitchison and A. J. G. Hey, *Gauge theory in particle physics*, vol. II, 3rd edition, Taylor and Francis, New York (2004).
- A. Altland and B. D. Simons, *Condensed matter field theory*, CUP, Cambridge (2006).
- P. W. Anderson, *Basic notions of condensed matter physics*, Benjamin-Cummings, Menlo Park (1984).
- J. F. Annett, *Superconductivity, superfluids and condensates*, OUP, Oxford (2004).
- A. Auerbach, *Interacting electrons and quantum magnetism*, Springer-Verlag, New York (1994).
- T. Banks, *Modern quantum field theory*, CUP, Cambridge (2008).
- G. Barton, *Elements of Green's function and propagation*, OUP, Oxford (1989).
- V. B. Berestetskii, E. M. Lifshitz and L. P. Pitaevskii, *Quantum electrodynamics*, 2nd edition, (volume IV of Landau and Lifshitz), Butterworth-Heinemann, Oxford (1982).
- J. J. Binney, N. J. Dowrick, A. J. Fisher and M. E. J. Newman, *The theory of critical phenomena*, OUP, Oxford (1992).
- J. D. Bjorken and S. Drell, *Relativistic quantum mechanics*, Dover, New York (2012).
- J. D. Bjorken and S. Drell, *Relativistic quantum fields*, Dover, New York (2012).
- S. J. Blundell, *Magnetism in condensed matter*, OUP, Oxford (2001).
- S. J. Blundell and K. M. Blundell, *Concepts in thermal physics*, 2nd edition, OUP, Oxford (2010).
- M. L. Boas, *Mathematical methods in the physical sciences*, 2nd edition, Wiley, New York (1983).
- P. M. Chaikin and T. C. Lubensky, *Principles of condensed matter physics*, CUP, Cambridge (1995).
- P. Coleman, *Introduction to many body physics*, CUP, Cambridge (2015)
- S. Coleman, *Aspects of symmetry*, CUP, Cambridge (1985).
- S. Coleman, *Physics 253: Quantum field theory: lectures by Sidney Coleman*. Lectures from 1975-1976 available to stream from http://www.physics.harvard.edu/events/videos/Phys253.
- S. Doniach and E. H. Sondheimer, *Green's functions for solid state physicists*, Imperial College Press, London (1998).
- Z. F. Ezawa, *Quantum Hall effects*, World Scientific, Singapore (2008).
- R. P. Feynman, *Statistical mechanics*, Westview, Boulder (1998).
- R. P. Feynman and A. R. Hibbs, *Quantum mechanics and path integrals*, Emended edition, Dover, New York (2005).
- R. P. Feynman, R. B. Leighton and M. L. Sands, *Lectures in physics*, vol. II, Addison-Wesley, Reading (1963).
- P. Fulde, *Correlated electrons in quantum matter*, World Scientific, Singapore (2012).
- H. Georgi, *Lie algebras in particle physics*, 2nd edition, Westview, Boulder (1999).

471

- G. F. Giuliani and G. Vignale, *Quantum theory of the electron liquid*, CUP, Cambridge (2005).

- D. J. Griffiths, *Introduction to elementary particles*, 2nd edition, Wiley VHC, Weinheim (2008).

- F. Halzen and A. D. Martin, *Quarks and leptons*, John Wiley and Sons, Hoboken (1984).

- K. Huang, *Quantum field theory*, Wiley VCH, Weinheim (2010).

- C. Itzykson and J.-B. Zuber, *Quantum field theory*, Dover, New York (1980).

- J. I. Kapusta and C. Gale, *Finite-temperature field theory*, CUP, Cambridge (2006).

- M. Kaku, *Quantum field theory*, OUP, Oxford (1993).

- L. D. Landau and E. M. Lifshitz, *Mechanics* (volume I of Landau and Lifshitz), Pergamon, Oxford (1976).

- E. M. Lifshitz and L. P. Pitaevskii, *Statistical Physics, part 2* (volume IX of Landau and Lifshitz), Pergamon, Oxford (1980).

- R. Loudon, *The quantum theory of light*, OUP, Oxford (2000).

- M. Maggiore, *A modern introduction to quantum field theory*, OUP, Oxford (2005).

- G. D. Mahan, *Many-particle physics*, Plenum, New York (1990).

- F. Mandl and G. Shaw, *Quantum field theory*, 2nd edition, Wiley, Chichester (2010).

- R. D. Mattuck, *A guide to Feynman diagrams in the many-body problem*, Dover, New York (1967).

- W. D. McComb, *Renormalization methods*, OUP, Oxford (2004).

- V. F. Mukhanov and S. Winitzki, *Introduction to quantum effects in gravity*, CUP, Cambridge (2007).

- M. Nakahara, *Geometry, topology and physics*, Adam Hilger, Bristol (1990).

- D. E. Neuenschwander, *Emmy Noether's wonderful theorem*, Johns Hopkins, Baltimore (2011).

- J. W. Negele and H. Orland, *Quantum many-particle systems*, Addison Wesley, Reading (1988).

- R. Penrose, *The road to reality*, Vintage, London (2004).

- M. E. Peskin and D. V. Schroeder, *An introduction to quantum field theory*, Westview Press, Boulder (1995).

- V. Radovanovic, *Problem book in quantum field theory*, 2nd edition, Springer, Berlin (2008).

- L. H. Ryder, *Quantum field theory*, CUP, Cambridge (1985).

- J. J. Sakurai, *Modern quantum mechanics*, revised edition, Addison-Wesley, Reading (1994).

- L. I. Schiff, *Quantum mechanics*, 3rd edition, McGraw-Hill, New York (1968).

- S. S. Schweber, *QED and the men who made it*, Princeton University Press, New Jersey (1994).

- M. Srednicki, *Quantum field theory*, CUP, Cambridge (2007).

- R. Ticciati, *Quantum field theory for mathematicians*, CUP, Cambridge (1999).

- A. M. Tsvelik, *Quantum field theory in condensed matter physics*, CUP, Cambridge (1995).

- S. Weinberg, *The quantum theory of fields*, vol. I, CUP, Cambridge (1995).

- S. Weinberg, *The quantum theory of fields*, vol. II, CUP, Cambridge (1996).

- X.-G. Wen, *Quantum field theory of many-body systems*, OUP, Oxford (2004).
- A. Zee, *Quantum field theory in a nutshell*, Princeton University Press, Princeton (2003).
- J. M. Ziman, *Elements of advanced quantum mechanics*, CUP, Cambridge (1969).
- J. Zinn-Justin, *Quantum field theory and critical phenomena*, OUP, Oxford (1989).

Useful complex analysis

For every complex problem there is an answer that is clear, simple, and wrong.
H. L. Mencken (1880–1956)

Throughout the book we have tried to keep the amount of complex analysis to a minimum. This appendix provides a simple guide to some of the complex analysis commonly employed in quantum field theory. The guide is illustrated by examples drawn from the subject, including the most important function of a complex variable in quantum field theory: the propagator.

B.1 What is an analytic function?

If a function is analytic in a region close to a point z, then it has a derivative at every point in that region. We define the derivative of a complex number as

$$f'(z) = \frac{\mathrm{d}f}{\mathrm{d}z} = \lim_{\Delta z \to 0} \frac{f(z + \Delta z) - f(z)}{\Delta z}. \tag{B.1}$$

Importantly, for the function to be analytic, the derivative shouldn't depend on the way the interval in the complex plane Δz is selected.

Example B.1

The function $f(z) = z^2$ is analytic; $g(z) = |z|^2$ is not. To see the first write

$$
\begin{aligned}
f'(z) &= \lim_{\Delta z \to 0} \frac{(z + \Delta z)^2 - z^2}{\Delta z} \\
&= \lim_{\Delta z \to 0} \frac{z^2 + 2z\Delta z + (\Delta z)^2 - z^2}{\Delta z} \\
&= 2z, \tag{B.2}
\end{aligned}
$$

just as for a normal derivative. Note that we didn't have a choice of Δz, this procedure works whatever we choose.

On the other hand for $g(z) = |z|^2$ we have

$$g'(z) = \lim_{\Delta z \to 0} \frac{|z + \Delta z|^2 - |z|^2}{\Delta z}. \tag{B.3}$$

If $\Delta z = i\Delta y$ (with y real) then you will get a different derivative to the case of $\Delta z = \Delta x$, with x real. See Boas for more details.

B.2 What is a pole?

A pole is a type of singularity[1] that behaves like the singularity of $1/z^n$ at $z = 0$. Let $f(z)$ be analytic between two circles C_1 and C_2. In the region between them we can write $f(z)$ as a so-called Laurent series expanded about a point z_0:

$$f(z) = a_0 + a_1(z - z_0) + a_2(z - z_0)^2 + \ldots + \frac{b_1}{(z - z_0)} + \frac{b_2}{(z - z_0)^2} + \ldots \quad \text{(B.4)}$$

The part with b coefficients is known as the principal part of the series. Don't confuse this with the principal value of an integral, which is different, and discussed below.

A result and some definitions:

- If all b's are zero then $f(z_0)$ is analytic at $z = z_0$.
- If all b's after b_n are zero then we say that *we have a pole of order n at $z = z_0$*. If $n = 1$ we have a *simple pole*.
- The coefficient b_1 is called the residue of $f(z)$ at $z = z_0$.

An important example is the function

$$f(z) = \frac{\alpha}{z - \beta}, \quad \text{(B.5)}$$

which has residue $b_1 = \alpha$ and all other a_i and b_i zero. This function has a simple pole at $z = \beta$.

Example B.2

Let's examine the pole structure of two of our propagators. The non-relativistic, retarded, free electron propagator is given by

$$\tilde{G}_0^+(E) = \frac{i}{E - E_{\boldsymbol{p}} + i\epsilon}. \quad \text{(B.6)}$$

This has a first-order pole at $E_{\boldsymbol{p}} - i\epsilon$. This is shown in Fig. B.1(a).

The Feynman propagator for the free scalar field is usually written

$$\tilde{\Delta}(p) = \frac{i}{p^2 - m^2 + i\epsilon}; \quad \text{(B.7)}$$

this is helpfully rewritten (see Chapter 17) as a function of the complex variable p^0:

$$\tilde{\Delta}(p) = \frac{1}{2E_{\boldsymbol{p}}} \left[\frac{i}{(p^0) - E_{\boldsymbol{p}} + i\epsilon} - \frac{i}{(p^0) + E_{\boldsymbol{p}} - i\epsilon} \right]. \quad \text{(B.8)}$$

The first (particle) part has a simple pole at $p^0 = E_{\boldsymbol{p}} - i\epsilon$. The second (antiparticle) part has a simple pole at $-E_{\boldsymbol{p}} + i\epsilon$. This is shown in Fig. B.1(b).

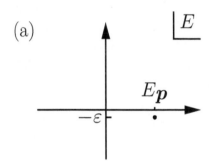

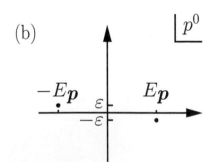

Fig. B.1 (a) Position of the pole in the complex E plane for $\tilde{G}_0^+(E)$. (b) Positions of the poles in the complex p_0 plane for the Feynman propagator for free scalar fields.

B.3 How to find a residue

You can find a residue $R(z_0)$ at the pole z_0 by writing a Laurent series. There are more direct methods too. For example, when we have a simple pole, we can write

$$R(z_0) = \lim_{z \to z_0} (z - z_0) f(z). \quad \text{(B.9)}$$

Example B.3

The residue of $\tilde{G}(E) = \frac{iZ}{E - E_{\boldsymbol{p}} + i\epsilon}$ at the simple pole $E = E_{\boldsymbol{p}} - i\epsilon$ is given by

$$R(E_{\boldsymbol{p}} - i\epsilon) = \lim_{E \to E_{\boldsymbol{p}} - i\epsilon} (E - E_{\boldsymbol{p}} + i\epsilon)\frac{iZ}{E - E_{\boldsymbol{p}} + i\epsilon} = iZ. \qquad \text{(B.10)}$$

B.4   Three rules of contour integrals

A contour C is a closed path in the complex plane with a finite number of corners which doesn't cross itself. Integrals around such contours have a number of useful properties.

Example B.4

We can get some practice with a contour integral by calculating

$$\oint_C dz \, z^2, \qquad \text{(B.11)}$$

where the contour is shown in Fig. B.2. We split the contour into two parts and start with the straight line along the real axis. Take $z = re^{i\theta}$ and this part becomes

$$(\text{straight line}) = \int_{r=-1}^{1} dr \, r^2 = \left[\frac{r^3}{3}\right]_{-1}^{1} = \frac{2}{3}. \qquad \text{(B.12)}$$

Now for the semicircle, described by $z = r_0 e^{i\theta}$, where $r_0 = 1$. We have $dz = ir_0 e^{i\theta} d\theta$ giving

$$(\text{semicircle}) = \int_{\theta=0}^{\pi} d\theta \, ir_0^3 e^{3i\theta} = \left[\frac{e^{3i\theta}}{3}\right]_0^{\pi} = -\frac{2}{3}. \qquad \text{(B.13)}$$

Adding the contributions we have

$$(\text{straight line}) + (\text{semicircle}) = 0. \qquad \text{(B.14)}$$

There are three useful theorems for evaluating integrals taken around contours. The first is **Cauchy's theorem**:

> If $f(z)$ is analytic on and inside C then
>
> $$\oint_C dz \, f(z) = 0. \qquad \text{(B.15)}$$

This is good news, since it says that if the region in a contour contains no poles then the integral gives zero. It also explains why the previous example gives zero: z^2 is analytic on and inside the contour.

The second theorem is known as **Cauchy's integral formula**:

Augustin-Louis Cauchy (1789–1857)

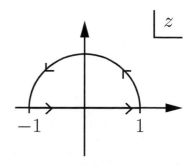

Fig. B.2 An example of a contour in the complex z plane.

If $f(z)$ is analytic on and inside a simple closed curve C, and the point a is inside C, then the value of $f(a)$ is given by

$$f(a) = \frac{1}{2\pi i} \oint_C dz\, \frac{f(z)}{z-a}. \tag{B.16}$$

The third is the **residue theorem**:

If $f(z)$ has singularities at points z_i, then, for a closed curve enclosing these points we have

$$\oint_C dz\, f(z) = 2\pi i \sum_i \left(\begin{array}{c} \text{Residue at } f(z_i) \\ \text{inside } C \end{array} \right), \tag{B.17}$$

where the integral around C is performed in the *anticlockwise* direction. (You merely change the sign of the answer if you perform the integral in the clockwise direction.)

Often we want to do difficult integrals over real variables. These may be turned into easier integrals if we form a contour in the complex plane which includes the original domain of integration and use the rules given above. The art is in choosing the best contour to do the integral.

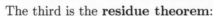

(a)

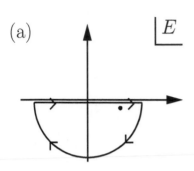

(b)

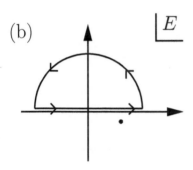

Fig. B.3 (a) The contour completed in the lower half-plane enclosing the pole. Note that the direction here is clockwise, earning us an extra minus sign. (b) The contour completed in the upper half-plane. No poles are enclosed so the answer is given by Cauchy's theorem.

---

[2]This disappearance of semicircular contours in the limit of infinite radius is a result of **Jordan's lemma**. This says that the integral $\int_{-\infty}^{\infty} dz\, f(z) e^{iaz}$ along the infinite upper semicircle is zero, provided (i) $a > 0$, and (ii) $f(z)$ is a well-behaved function satisfying $\lim_{R\to\infty} |f(Re^{i\theta})| = 0$.

Example B.5

We can use Cauchy's theorem along with the residue theorem to justify some of the more seemingly cavalier tricks employed in the discussion of propagators in Chapters 16 and 17. Let's find the inverse Fourier transform of the retarded propagator $\tilde{G}_0^+(E)$, given by

$$G_0^+(t-t') = \int_{-\infty}^{\infty} \frac{dE}{2\pi} \frac{i e^{-iE(t-t')}}{E - E_p + i\epsilon}, \tag{B.18}$$

for which the integration path is along the real axis. To use our contour integral rules we must complete the contour by joining up this path with a further section of path which will either be in the upper half of the complex E plane or in the lower half.

Suppose we take it in the lower half-plane. Then, as we take the limits along the real axis to $\pm\infty$ the semicircular path gets larger and larger. This will make a large, negative imaginary contribution to E. Let's call it $-i|\eta|$. The exponential will then involve a contribution $e^{-|\eta|(t-t')}$. If $(t-t')$ is positive then this contribution gets smaller, eventually vanishing as the contour becomes infinitely large.[2] We conclude that, for the case $t - t' > 0$, the integral above is equivalent to the contour shown in Fig. B.3(a).

Let's do that integral. The contour contains the pole at $E = E_p - i\epsilon$ so we use the residue theorem to say

$$\oint_C \frac{dE}{2\pi} \frac{i e^{-iE(t-t')}}{E - E_p + i\epsilon} = -i\left(\begin{array}{c} \text{Residue at} \\ E = E_p - i\epsilon \end{array} \right), \tag{B.19}$$

where the minus sign follows from our attempt to take the integral in the clockwise direction. The residue at the pole is $i e^{-iE_p(t-t')} e^{-\epsilon(t-t')}$ and the answer is

$$G^+(t-t') = e^{-iE_p(t-t')} e^{-\epsilon(t-t')}, \tag{B.20}$$

which we stress applies for $(t - t') > 0$.

What if we tried to complete the contour in the upper half-plane? Then we would have obtained a large, positive, imaginary contribution to the exponential resulting in a contribution $e^{|\eta|(t-t')}$ which blows up for $t - t' > 0$. Such a badly behaved integral is certainly not suitable for evaluating the Fourier transform. However, for $t - t' < 0$ the semicircular contour has a vanishing contribution at infinity and we again have the equivalence of the Fourier transform and the contour C' shown in Fig. B.3(b). Notice that C' contains no poles, so Cauchy's theorem says that the integral is zero.

We conclude that $G^+(t-t') = 0$ for $t-t' < 0$ and noting that ϵ is an infinitesimal quantity, we may replace both (i) zero for $t - t' < 0$ and (ii) $e^{\epsilon(t-t')}$ for $t - t' > 0$ with $\theta(t - t')$ and conclude

$$G^+(t - t') = \theta(t - t')e^{-iE_{\boldsymbol{p}}(t-t')}, \tag{B.21}$$

just as we had in Chapter 16 without the need for adding damping factors by hand!

The meaning of the $i\epsilon$ factors now becomes clear. These infinitesimals position the poles of the propagators in such a way as to ensure the correct causality relationships. Returning to the scalar field propagator of Fig. B.1 we see that closing the contour in the lower half-plane picks up the positive energy pole, leading to a factor $\theta(x^0 - y^0)$, while closing the contour in the upper half-plane picks up the negative energy pole and leads to the factor $\theta(y^0 - x^0)$. This motivates the definition of the Feynman propagator in Chapter 17.

B.5 What is a branch cut?

Some functions such as $\sqrt{z}$ and $\ln z$ are multivalued. For example, $\sqrt{4}$ is either 2 or -2. The function $\ln z$ is multivalued because its inverse has the property $e^{i\theta} = e^{i(\theta + 2\pi n)}$ where n is an integer. The fact that these functions are multivalued means that care must be taken when taking roots and logarithms.

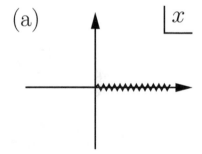

(a)

x

Example B.6

For the case of the logarithm we have, taking $z = re^{i\theta}$:

$$\ln z = \ln r + i\theta. \tag{B.22}$$

Clearly, for fixed r, $\ln z$ takes a different value if in the exponent θ is replaced by $\theta + 2\pi n$, even though both choices correspond to the same z.

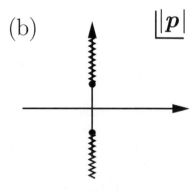

(b)

$|\boldsymbol{p}|$

We therefore agree that we should only consider angles in some interval in θ of size 2π, known as a *branch* of the function. In order to make this clear in the complex plane, we can define lines that we will agree not to cross with any of our operations. These are known as **branch cuts**. The points from which these emerge are known as branch points.

For $\ln z$ the branch point is the origin and the branch cut may be taken along the positive real axis as shown in Fig. B.4(a). It may also be taken along the negative real axis, or indeed any convenient line. Crossing the branch cut makes the function jump by $2\pi i$.

Fig. B.4 (a) The branch cut along the positive, real axis. (b) The complex plane for eqn. B.23.

Recall from Chapter 8 the integral

$$\frac{-i}{(2\pi)^2|\boldsymbol{x}|} \int_{-\infty}^{\infty} d|\boldsymbol{p}|\, |\boldsymbol{p}|e^{i|\boldsymbol{p}||\boldsymbol{x}|}e^{-it\sqrt{|\boldsymbol{p}|^2+m^2}}, \qquad (B.23)$$

which we consider in the complex $|\boldsymbol{p}|$ plane. The square root in this equation $\sqrt{\boldsymbol{p}^2 + m^2}$ must be restricted to a single branch. The square root vanishes for $|\boldsymbol{p}| = \pm im$. which are therefore the branch points. For convenience, we take the branch cuts to extend along the imaginary axis as shown in Fig. B.4(b). Notice that when we do the integral in Chapter 8 we can't cross the cuts with our contour, so we must direct the contour around the cuts.

Example B.7

Another occasion where we must consider a function with a branch cut is the full propagator discussed in Chapter 31, given by

$$\tilde{G}(p) = \frac{iZ}{p^2 - m^2 + i\epsilon} + \int_{\approx 4m^2}^{\infty} \frac{dM^2}{2\pi}\rho(M^2)\frac{i}{p^2 - M^2 + i\epsilon}, \qquad (B.24)$$

which has a pole structure shown in Fig. B.5. The second term in the expression tells us to expect multiparticle contributions. These lead to a line of poles with infinitesimal separation between them. Such a line of poles is another way of describing a branch cut and so we draw a cut extending along the real axis from the branch point given by the two-particle production threshold $(p^0)^2 \approx 4m^2$.

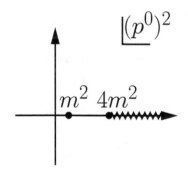

Fig. B.5 The pole structure of the full propagator in eqn. B.24.

[3]Specifically we require

$$\int_a^b dx\, f(x) = \pm\infty,$$

for $a < b$ and

$$\int_b^c dx\, f(x) = \mp\infty,$$

for $c > b$. (That is, one sign in front of the ∞ is plus and one minus.)

B.6   The principal value of an integral

The **Cauchy principal value** is a method of giving improper integrals a value. This is not complex analysis, but is often used in integrals involving the propagator. Suppose we want to evaluate the integral $\int_a^c dx\, f(x)$ but we have the problem that the integrand $f(x)$ diverges at $x = b$ (where $a < b < c$) and so both $\int_a^b dx\, f(x)$ and $\int_b^c dx\, f(x)$ will also blow up.[3] In that case we may take the Cauchy principal value of the integral, denoted by $\mathcal{P}$ and defined by

$$\mathcal{P}\int_a^c dx\, f(x) = \lim_{\epsilon \to 0^+}\left[\int_a^{b-\epsilon} dx\, f(x) + \int_{b+\epsilon}^c dx\, f(x)\right]. \qquad (B.25)$$

This gives the integral an unambiguous value, as demonstrated in the example below.

Example B.8

The integral

$$I = \int_0^{10} \frac{dx}{x-2}, \qquad (B.26)$$

is not well defined as the integrand diverges at $x = 2$. You can see that the two integrals $\int_0^2 \frac{dx}{x-2}$ and $\int_2^{10} \frac{dx}{x-2}$ both diverge, giving $-\infty$ and $+\infty$ respectively.

To get around this we integrate from 0 up to $2 - \epsilon$ and then from $2 + \epsilon$ up to 10, thereby cutting out the troublesome part of the problem. We obtain

$$I_1 = \int_0^{2-\epsilon} \frac{dx}{x-2} = \ln \epsilon - \ln 2, \quad I_2 = \int_{2+\epsilon}^{10} \frac{dx}{x-2} = \ln 8 - \ln \epsilon. \tag{B.27}$$

We find that $I = I_1 + I_2 = \ln 4$, independent of ϵ. We may therefore take the limit $\epsilon \to 0$ and obtain an unambiguous result. We conclude that

$$\mathcal{P} \int_0^{10} \frac{dx}{x-2} = \ln 4. \tag{B.28}$$

The principal value arises in quantum field theory when we want to do integrals of the form $\int_a^b dx \, \frac{f(x)}{x+i\epsilon}$, where $f(x)$ is a complex-valued function and a and b are real, obeying $a < 0 < b$. In this case we use the following theorem which says that

$$\lim_{\epsilon \to 0^+} \int_a^b dx \, \frac{f(x)}{x \pm i\epsilon} = \mathcal{P} \int_a^b dx \, \frac{f(x)}{x} \mp i\pi f(0). \tag{B.29}$$

Often we take $f(x - x_0) = \delta(x - x_0)$ and obtain the identity,[4]

$$\frac{1}{x_0 \pm i\epsilon} = \frac{\mathcal{P}}{x_0} \mp i\pi \delta(x_0), \tag{B.30}$$

as used in Exercises 22.1 and 31.3.

[4] Sometimes called the Dirac relation in the physics literature.

Index